Linear Programming Computation

Ping-Qi PAN

Linear Programming Computation

Springer

Ping-Qi PAN
Department of Mathematics
Southeast University
Nanjing
China

ISBN 978-3-662-51430-6 ISBN 978-3-642-40754-3 (eBook)
DOI 10.1007/978-3-642-40754-3
Springer Heidelberg New York Dordrecht London

Softcover reprint of the hardcover 1st edition 2014

Printed on acid-free paper

Springer is part of Springer Science+Business Media (www.springer.com)

Dedicated to
my parents, Chao Pan and Yiyun Yan
and to
my wife, Minghua Jiang, and my son,
Yunpeng Pan

Preface

Linear programming (LP) (Dantzig 1948, 1951a,b,c) might be one of the most well-known and widely used mathematical tools in the world. As a branch of optimization, it serves as the most important cornerstone of operations research, decision science, and management science.

This branch of study emerged when the American mathematician George B. Dantzig proposed the LP model and the simplex method in 1947. The computer, emerging around the same period, propelled the development of LP and the simplex method toward practical application. As a basic branch of study, LP orchestrated the birth of a number of new fields, such as nonlinear programming, network flow and combinatorial optimization, stochastic programming, integer programming, and complementary theory, and invigorated the whole field of operations research.

An outstanding feature of LP is its broad applications. Closely related to LP, a number of people have made pioneering contributions in their respective areas. In the area of economics, in particular, in 1973, Russian-American economist Wassily Leontief took the Nobel Economic Prize for his epoch-making contribution on quantitative analysis of economic activities. The academician L. V. Kantorovich of the former Soviet Academy of Science and American economist Professor T. C. Koopmans won the 1975 Nobel Prize for their optimal allocation theory of resources using LP. The same prize was also given to Professors K. Arrow, P. Samuelson, H. Simon, and L. Herwricz, several decades later when they paid close attention to LP at the starting days of their professional careers. In practice, on the other hand, the simplex method has achieved great success. Applications of the simplex method to the areas such as economy, commerce, production, science and technology, and defense and military affairs have brought about astonishing economic and social benefits. It is recognized as one of The Ten Algorithms in the Twenty Century (IEEE2002; see Cipra 2000).

Since W. Orchard-Hays worked out the first simplex-method-based commercial software, his implementation techniques were used and developed by many scholars, such as M. A. Saunders and R. E. Bixby. As a result, the simplex method became a powerful practical tool. However, the method is shown to be a non-polynomial-time one. In 1979, a former Soviet mathematician – L. G. Khachiyan proposed the first polynomial-time method, called "ellipsoid method." But unfortunately, it performed badly in computations and is not competitive with the simplex method. In 1984, an Indian mathematician – N. Karmarkar proposed another polynomial-time method. It is an interior-point method of lower polynomial order; it performed remarkably well in computations. The ensuing surge of research in interior-point methods has led to some very efficient interior-point algorithms for solving large sparse LP problems. During the same period, due to contributions on pivot rules made by P. M. J. Harris, J. J. H. Forest, and D. Goldfarb, among others, the efficiency of the simplex method made great progress as well, leading to an intense head-to-head competition between the two types of methods.

After more than 60 years since its birth, LP is now a relatively mature but rapidly developing discipline. Nevertheless, it is still facing great challenges. The importance of large-scale sparse LP models is nowadays enhanced further by the globalization. Everyday practice calls upon the research community to provide more powerful solution tools just to keep up with the ever-increasing problem size. This book attempts to respond to this reality and reflect the state of the art of LP by presenting the most valuable knowledge and results. It has been my long-lasting belief that research results, in operations research/management science in particular, should be of practical value, potentially at least. I therefore focus on theories, methods, and implementation techniques that are closely related to LP computation and hence applications.

This book consists of two parts. Part I mainly presents fundamental and conventional materials, such as geometric of feasible region, the simplex method, duality principle and dual simplex method, implementation of the simplex method, sensitivity analysis and parametric LP, variants of the simplex method, decomposition method, and interior-point method. In addition, integer linear programming (ILP), differing from LP in nature, is also considered in this chapter, not only because ILP models can be handled by solving a sequence of LP models but because they are so rich in practice as form a major application area of LP computation. Part II mainly covers the author's recent published and unpublished results, such as pivot rule, dual pivot rule, simplex phase-1 method, dual simplex phase-I method, reduced simplex method, D-reduced simplex method, criss-cross simplex method, generalized reduced simplex method, deficient-basis method, dual deficient-basis method, face method, dual face method, and pivotal interior-point method. The last chapter contains special topics, such as special forms of the LP problem, approaches to intercepting for primal and dual optimal sets, practical pricing schemes, relaxation principle, local duality, "decomposition principle," and ILP method based on the generalized reduced simplex framework.

To make materials easier to follow and understand, algorithms in this book are formulated and accompanied with illustrative examples wherever possible. If the book is used as a textbook for upper-level undergraduate or graduate course, Chaps. 1 and 3–6 may be used as basic course material.

Acknowledgements I deeply cherish the memory of Professor Xuchu He who, as my mentor at Nanjing University, sparked my interest in optimization. I would like to honor the father of linear programming and the simplex method Professor George. B. Dantzig for his encouragement given during the 16th International Symposium on Mathematical Programming, Lausanne EPFL, in August 1997. In particular, I am very grateful to Professor Michael Saunders of Stanford University for his thoughtful comments and suggestions given in the past. He selflessly offered the MINOS 5.51 package (the latest version of MINOS at that time) and related materials, from which my work and this book benefited greatly. Actually, MINOS has been the main benchmark and platform of our computational experiments. I would like to thank Professor R. Tyrrell Rockafellar of University of Washington, Professor Thomas F. Coleman of Cornell University, and Professor Weide Zheng of Nanjing University for their support and assistance. I would also like to thank the following colleagues for their support and assistance: Professor James V. Burke of University of Washington, Professor Liqun Qi of Hong Kong Polytechnic University, Professor Lizhi Liao of Hong Kong Baptist University, Professor Leena Suhl and Dr. Achim Koberstein of Paderborn Universität and Professor Uwe Suhl of Freie Universität Berlin, and Dr. Pablo Guerrero-Garcia of University of Malaga. This book was supported partially by projects 10871043 and 70971136 of National Natural Science Foundation of China.

Nanjing, China
11/12/2012

Ping-Qi Pan

Contents

1 **Introduction** 1
  1.1 Brief History of LP 2
  1.2 From Practical Issue to LP Model 5
  1.3 Illustrative Applications 8
  1.4 Errors of Floating Point Arithmetic 12
  1.5 The Standard LP Problem 15
  1.6 Canonical Form of Ax = b 18

Part I Fundamentals

2 **Geometry of the Feasible Region** 27
  2.1 Polyhedral Convex Set and the Feasible Region 28
  2.2 Geometric Structure of the Feasible Region 36
  2.3 Optimal Face and Vertex 49
    2.3.1 Graphic Approach 51
  2.4 Feasible Direction and Active Constraint 52
  2.5 Heuristic Characteristic of Optimal Solution 56

3 **Simplex Method** 61
  3.1 Simplex Tableau 61
  3.2 Simplex Method: Tableau Form 63
  3.3 Start-Up of the Simplex Method 71
  3.4 Revised Simplex Tableau 78
  3.5 Simplex Method 80
  3.6 Degeneracy and Cycling 85
  3.7 Finite Pivot Rule 88
  3.8 Computational Complexity 95
  3.9 On Features of the Simplex Method 98

4 **Duality Principle and Dual Simplex Method** 101
  4.1 Dual LP Problem 101
  4.2 Duality Theorems 104

4.3 Optimality Condition 107
4.4 Dual Simplex Method: Tableau Form 110
4.5 Dual Simplex Method 114
4.6 Economic Interpretation of Duality: Shadow Price 118
4.7 Notes 119

5 Implementation of the Simplex Method 123
5.1 Miscellaneous 123
5.2 Preprocessing: Scaling 125
5.3 LU Factorization of Basis 127
5.3.1 Use of LU Factorization 128
5.3.2 Sparse LU Factorization 131
5.4 Updating LU Factors 134
5.5 Crashing for Initial Basis 137
5.6 Harris Rule and Tolerance Expending 140

6 Sensitivity Analysis and Parametric LP 143
6.1 Change in Costs 144
6.2 Change in the Right-Hand Side 146
6.3 Change in Coefficient Matrix 148
6.3.1 Dropping a Variable 148
6.3.2 Adding a Variable 150
6.3.3 Dropping a Constraint 152
6.3.4 Adding a Constraint 154
6.3.5 Replacing a Row or Column 159
6.4 Parametric LP 162
6.4.1 Parameterized Cost 162
6.4.2 Parameterized Right-Hand Side 166

7 Variants of the Simplex Method 169
7.1 Primal-Dual Simplex Method 169
7.2 Self-Dual Parametric Simplex Method 175
7.3 General LP Problems 179
7.4 Generalized Simplex Method 180
7.4.1 Generalized Phase-I 187
7.5 Generalized Dual Simplex Method: Tableau Form 188
7.5.1 Generalized Dual Phase-I 192
7.6 Generalized Dual Simplex Method 193
7.7 Bound Flipping 201

8 Decomposition Method 205
8.1 D-W Decomposition 206
8.1.1 Generalization of D-W Decomposition 211
8.2 Start-Up of D-W Decomposition 211
8.3 Application of D-W Decomposition 212
8.4 Economic Interpretation of D-W Decomposition 218
8.5 Benders Decomposition 221

8.6 Application of Benders Decomposition ... 228
8.7 Primal Benders Decomposition ... 234

9 Interior-Point Method ... 239
9.1 Karmarkar Algorithm ... 240
9.1.1 Projective Transformation ... 241
9.1.2 Karmarkar Algorithm ... 242
9.1.3 Convergence ... 244
9.2 Affine Interior-Point Method ... 249
9.2.1 Formulation of the Algorithm ... 249
9.2.2 Convergence and Start-Up ... 252
9.3 Dual Affine Interior-Point Method ... 253
9.4 Path-Following Interior-Point Method ... 256
9.4.1 Primal-Dual Method ... 258
9.4.2 Infeasible Primal-Dual Method ... 262
9.4.3 Predictor-Corrector Primal-Dual Method ... 264
9.4.4 Homogeneous and Self-Dual Method ... 266
9.5 Notes ... 269

10 Integer Linear Programming (ILP) ... 275
10.1 Introduction ... 276
10.1.1 Graphic Approach ... 277
10.2 Branch-and-Bound Method ... 277
10.3 Cutting-Plane Method ... 285
10.4 Controlled-Branch Method ... 288
10.5 Controlled-Cut Method ... 293

Part II Advanced Topics

11 Pivot Rule ... 297
11.1 Partial Pricing ... 298
11.2 Steepest-Edge Rule ... 300
11.3 Approximate Steepest-Edge Rule ... 302
11.4 Largest-Distance Rule ... 304
11.5 Nested Rule ... 306
11.6 Nested Largest-Distance Rule ... 308

12 Dual Pivot Rule ... 311
12.1 Dual Steepest-Edge Rule ... 312
12.2 Approximate Dual Steepest-Edge Rule ... 315
12.3 Dual Largest-Distance Rule ... 317
12.4 Dual Nested Rule ... 318

13 Simplex Phase-I Method ... 321
13.1 Infeasibility-Sum Method ... 321
13.2 Single-Artificial-Variable Method ... 327

13.3 The Most-Obtuse-Angle Column Rule 332
13.4 Perturbation of Reduced Costs 336

14 Dual Simplex Phase-l Method 341
14.1 Dual Infeasibility-Sum Method 341
14.2 Dual Single-Artificial-Variable Method 345
14.3 The Most-Obtuse-Angle Row Rule 351
14.4 Perturbation of the Right-Hand Side 354

15 Reduced Simplex Method 357
15.1 Derivation 357
15.2 Reduced Simplex Method 363
15.3 Reduced Phase-I: Single-Artificial-Variable 367
15.4 Dual Reduced Simplex Method 370
15.5 Dual Reduced Phase-I: The Most-Obtuse-Angle 378
15.6 Notes 384

16 Improved Reduced Simplex Method 387
16.1 Variation of the Reduce Simplex Method 387
16.2 Phase-I for Improved Reduced Simplex Method 391
16.3 Dual Improved Reduced Simplex Method 394
16.4 Phase-I for the Dual Improved Reduced Simplex Method 398
16.5 Bisection Reduced Simplex Method 402

17 D-Reduced Simplex Method 409
17.1 D-Reduced Simplex Tableau 409
17.2 Dual D-Reduced Simplex Method 412
17.3 Dual D-Reduced Phase-I 418
17.4 Dual D-Reduced Phase-I: Single-Artificial-Variable 421
17.5 D-Reduced Simplex Method 424
17.6 D-Reduced Phase-I: The Most-Obtuse-Angle 431
17.7 Dual Bisection D-Reduced Simplex Method 435

18 Criss-Cross Simplex Method 441
18.1 Most-Obtuse-Angle Criss-Cross Method 442
18.2 Perturbation Simplex Method 445
18.3 Criss-Cross Reduced Simplex Method 449
18.4 Perturbation Reduced Simplex Method 453
18.5 Mixed Two-Phase Method 457

19 Generalizing Reduced Simplex Method 461
19.1 Generalized Reduced Simplex Method 461
19.2 Generalized Reduced Phase-I 469
19.3 Generalized Reduced Phase-I: Single-Artificial-Variable 476
19.4 Generalized Dual Reduced Simplex Method 482
19.5 Generalized Dual Reduced Phase-I 486
19.6 Generalized Dual D-Reduced Simplex Method 490

20 Deficient-Basis Method 503
20.1 Deficient-Basis and Tableau 503
20.2 Deficient-Basis Method: Tableau Form 505
20.3 Deficient-Basis Method 509
20.4 Phase-I: Single-Artificial-Variable 514
20.5 On Implementation 519
20.5.1 Initial Basis 519
20.5.2 LU Updating in Rank-Increasing Iteration 520
20.6 Deficient-Basis Reduced Method 521
20.6.1 Starting-Up 528
20.7 Phase-I: The Most-Obtuse-Angle Column Rule 529

21 Dual Deficient-Basis Method 535
21.1 Dual Deficient-Basis Method: Tableau Form 535
21.2 Dual Deficient-Basis Method 540
21.3 Dual Deficient-Basis D-Reduced Method: Tableau Form 544
21.4 Dual Deficient-Basis D-Reduced Method 553
21.5 Deficient-Basis D-Reduced Gradient Method: Tableau Form 556
21.6 Deficient-Basis D-Reduced Gradient Method 567
21.7 Notes 568

22 Face Method 571
22.1 Face Method Using Cholesky Factorization 572
22.1.1 The Steepest Downhill 572
22.1.2 Updating Solution 574
22.1.3 Face Contraction 575
22.1.4 Optimality Test 578
22.1.5 Face Expansion 579
22.1.6 Face Algorithm 580
22.2 Face Phase-I: Single-Artificial-Variable 583
22.3 Generalizing the Face Method 586
22.4 Affine Face Method 588
22.5 Notes 591

23 Dual Face Method 595
23.1 Dual Face Method Using Cholesky Factorization 595
23.1.1 The Steepest Uphill 596
23.1.2 Updating Dual Solution 598
23.1.3 Dual Face Contraction 599
23.1.4 Optimality Test 600
23.1.5 Dual Face Expansion 601
23.1.6 Dual Face Algorithm 602
23.2 Dual Face Phase-I 604
23.3 Dual Face Method Using Gauss-Jordan Elimination 608
23.4 Variant of the Dual Face Algorithm 618

24 Pivotal Interior-Point Method 623
24.1 Pivotal Affine Interior-Point Method 623
24.2 Pivotal Affine Face Interior-Point Method 628
24.3 Pivotal D-Reduced Gradient Interior-Point Method 631
24.4 Feasible-Point Simplex Method 639
24.4.1 Column Pivot Rule and Optimality Condition 640
24.4.2 Search Direction 641
24.4.3 Stepsize and Row Pivot Rule 642
24.4.4 Formulation of the Algorithm 643
24.4.5 Phase-1 and Purification 645
24.4.6 Computational Results 645

25 Special Topics 647
25.1 Special Forms of the LP Problem 647
25.1.1 Reduced Problem 648
25.1.2 D-Reduced Problem 649
25.1.3 Dual Elimination 649
25.1.4 Reduced Dual Elimination 655
25.2 Intercepting of Optimal Set and Bilevel LP 656
25.3 Pricing Scheme 660
25.4 Relaxation Principle 664
25.4.1 Illustrative Application: Partial Pricing 665
25.5 Local Duality 668
25.6 Decomposition Principle 670
25.6.1 Subprograms and Algorithms 670
25.6.2 The Basic Idea: "arena contest" 673
25.6.3 Practical Remarks 675
25.6.4 Illustration on the Standard LP Problem 676
25.6.5 Illustration on the Bounded-Variable LP Problem 679
25.6.6 Illustration on the ILP Problem 680
25.7 ILP: Based on the Reduced Simplex Framework 682

Appendix A MPS File 697

Appendix B Test LP Problems 703

Appendix C Empirical Evaluation of Nested Pricing 709

Appendix D Empirical Evaluation of Face Methods 727

Appendix E Empirical Evaluation of the Feasible-Point Simplex Method 733

References 737

Notation

In this book, in general, uppercase English letters are used to denote matrices, lowercase English letters are used to denote vectors, and lowercase Greek letters to denote reals. Set is designated by uppercase English or Greek letters. Unless indicated otherwise, all vectors are column ones.

LP :	Linear programming.
ILP :	Integer linear programming.
$\mathcal{R}^n$:	Euclidean space of dimension n.
$\mathcal{R}$:	Euclidean space of dimension 1.
0 :	Origin of $\mathcal{R}^n$ (or the zero vector).
e_i :	Unit vector with the ith component 1 (dimension will be clear from the context; the same below).
e :	Vector of all ones.
I :	Unit matrix.
A :	Coefficient matrix of the linear program; also stands for index set of A's columns, i.e., $A = \{1, \ldots, n\}$.
m :	Number of rows of A.
n :	Number of columns of A.
b :	Right-hand side of the equality constraint.
c :	Cost vector, i.e., coefficient vector of the objective function.
B :	Basis (matrix). Also the set of basic indices.
N :	Nonbasic columns of A. Also the set of nonbasic indices.
a_j:	The jth column of A.
a_{ij} :	The entry of the ith row and jth column of A.
v_j :	The jth component of vector v.
$\|v\|$:	Euclidean norm of vector v.
$\max(v)$:	The largest component of vector v.
A_J :	Submatrix consisting of columns corresponding to index set J.
v_I :	Subvector consisting of components corresponding to row index set I.
A^T :	Transpose of A.
X :	Diagonal matrix whose diagonal entries are components of vector x.

$\nabla f(x)$:	Gradient of function $f(x)$.
$\Pi \subset \Gamma$:	Π is a subset of Γ.
$\Pi \cup \Gamma$:	Union set, i.e., $\{\tau \mid \tau \in \Pi \text{ or } \tau \in \Gamma\}$.
$\Pi \cap \Gamma$:	Intersection set, i.e., $\{\tau \mid \tau \in \Pi, \tau \in \Gamma\}$.
$\Pi \backslash \Gamma$:	Complementary set, i.e., $\{\tau \mid \tau \in \Pi, \tau \notin \Gamma\}$.
$\emptyset$:	Empty set.
$\mid$:	Such that. For example, $\{x \mid Ax = b\}$ means the set of all x such that $Ax = b$ holds.
$\ll (\gg)$:	Far less(bigger) than.
$O(\alpha)$:	Implies a number $\leq k\alpha$, where k, a fixed constant independent of the value of α, is meant to convey the notion that k is some small integer value.
$\lvert\tau\rvert$:	Absolute value of τ if τ is a real, or cardinality of τ if τ is a set.
range H :	Column space of matrix H.
null H :	Null space of matrix H.
dim Γ :	Dimension of set Γ.
C_n^m :	Number of combinations of taking m from n elements.
int P :	Interior of set P.
$\lfloor\alpha\rfloor$:	The largest integer no more than α.
$\lceil\alpha\rceil$:	The smallest integer no less than α.

This book involves the following two simplex FORTRAN codes:

RSA: The author's code of Revised Simplex Algorithm 3.5.1, supported by Algorithm 3.3.1 as Phase-I. Without exploiting sparsity, this code uses Harris two-pass row Rule 5.6.1 in the place of Rule 3.2.1. *tolpiv* $= 10^{-8}$ for pivot tolerance, and 10^{-6} for both primal and dual feasibility tolerance.

MINOS: The sophisticated smooth optimization package developed by Bruce A. Murtagh and Michael A. Saunders (1998) at Department of Management Science and Engineering of Stanford University. Based on the simplex method, the LP options of this sparsity-exploiting code are used as the benchmark and platform for empirical evaluation of new LP methods.

Chapter 1
Introduction

As intelligent creatures, human beings plan and carry out activities with pre-set objectives. Early human ancestors relied on their experience, whereas their modern-day descendants use scientific methods to aid their decision making in order to be more efficient and effective.

The word "optimization" means to achieve the best result with minimum effort. For example, how would one to allocate limited resources (labor, money, materials, etc.) to maximize return, or to minimize cost while satisfying certain needs? People gather information and data, and represent practical problems in appropriate forms using mathematical language such as numbers, equations, and functions, as is so-called "mathematical modeling" process. Then, mathematical methods of optimization are applied to solve the resulting models, thereby providing a quantitative basis for sound decision making. This process of formalizing a problem and then solving it using mathematical methods has been further revolutionized by the advent of the electronic computer.

The LP model is of very simple mathematical structure, which involves only linear functions, equalities, and inequalities (Dantzig 1963). However, the scale of the model can be tremendous, commonly comprising hundreds of thousands of variables and equations, which renders the problem difficult to solve. Fortunately, with the aid of computer, computational techniques for LP are capable solving problems of very large scale; for example, R.E. Bixby et al. (1992) solved an LP problem with 12,753,313 variables and found an optimal scheme for an airline crew scheduling application. As the globalization intensifies, there is growing demand for more powerful, more efficient algorithms and software that can solve increasingly larger LP problems aiming at achieving optimal benefits over large systems as a whole.

The aim of this book is to present from a practical point of view the theory, methods, and implementation of LP, including not only the fundamentals and conventional contents, but also reflections of newest research results and progress in this area.

P.-Q. PAN, *Linear Programming Computation*, DOI 10.1007/978-3-642-40754-3_1,

1.1 Brief History of LP

It would be interesting, inspiring, and beneficial to briefly recall the origin and development of LP.

LP can be traced back to as early as 1820s. French mathematician, J.B.J. Fourier (well-known due to the mathematical series named after him) in 1823 and Belgian mathematician V. Poussin in 1911 each wrote a paper related to LP. Unfortunately, these isolated efforts went unnoticed. In 1939, the academician L.V. Kantorovich (1960) of former Soviet Union academy of sciences published a book entitled *Mathematical Method of Production Management and Planning*, concerning LP models and their solution, but it was unfortunately neglected by the authority, and had not been known by the academic communities. In 1941, F.L. Hitchcock's original paper on the transportation problem once again did not draw any attention, until in the late 1940s and early 1950s when it was finally rediscovered after the simplex method came around.

Indeed, human activities is the only source of all scientific theories and methodologies. Although the World War II led to devastating losses and casualties, the war efforts thrust science and technology forward and hastened emergence of many new branches of learning. In particular, Operations Research and Optimization are one area that has greatly benefited from the war-time demand.

After receiving his Ph.D., young George B. Dantzig became a mathematical advisor to the U.S. (Cottle et al. 2007) Air Force Controller in 1946. He was challenged to mechanize the planning process such that a time-staged deployment, training and logistical supply program can be more rapidly computed using desk calculators. Using an active analysis approach, he formulated a model without an objective function, and tried to solve it. Dantzig recalled the situation at that time at the XI-th International Symposium on Mathematical Programming (held at the University of Bonn, W. Germany, from August 23 to 27 in 1982), by giving the following example:

> How to assign 70 men to 70 jobs ?

It is required that each job must be filled and each man must be assigned. An "activity" is defined as a 1-to-1 correspondence between the men and the jobs. There are $70! > 10^{100}$ activities. It is impracticable to determine the best among them by comparing all, because the number is too large even though being finite. Dantzig said:

> Suppose we had an IBM 370-168 available at the time of the Big Bang 15 billion years ago. Would it have been able to look at all the 70! combinations by the year 1981? No! Suppose instead it could examine 1 billion assignments per second? The answer is still no. Even if the Earth were filled with such computers all working in parallel, the answer would still be no. If, however, the were 10^{44} suns all filled with nano-second speed computers all programmed in parallel from the time of the big bang until sum grows cold, then perhaps the answer is yes.

This example illustrates the ordeals faced by decision makers before 1947. Therefore, "experience", "mature judgement", and "rules-of-thumb" had to be relied on to come up with an "acceptable" scheme, which was often far from the optimal.

Dantzig proposed the simplex method in the summer of 1947, as marked the birth of LP. On October 3rd, he visited one of the greatest scientists of the twentieth century, J. von Neumann, and talked about his results. The latter immediately got the basic idea behind the method, and indicated potential relationship between the method and game theory which he was working on. In 1948, Dantzig attended a conference of the Econometric Society in Wisconsin, meeting well-known statisticians, mathematicians and economists such as H. Hotelling, T. Koopmans, J. von Newmann and many others who became famous later, but were just starting their careers back then. After young Dantzig presented his simplex method and LP, the chairman called for discussion. There was silence for a moment. Then, Hotelling said, "But we all know the world is nonlinear." When Dantzig was frantically trying to properly reply, Von Neumann came to his rescue, "The speaker titled his talk 'Linear Programming' and he carefully stated his axioms. if you have an application that satisfied the axioms, use it. If it does not, then don't." Of course, what Hotelling said is right, the world is highly nonlinear; but, fortunately, the vast majority of nonlinear relations encountered in practical planning can be approximated by linear ones. As well-known now, subsequent development turned out to be entirely beyond people's imagination, even not expected by Dantzig himself.

The electronic computer, born in the late 1940s, has changed the world fundamentally. Its milestone and revolutionary impact can never be overstated, as it extended the capability of Man's brain unprecedentedly. It has deeply changed, and is changing visage of almost all branches of learning (including LP and the simplex method) and is driving the development at an unprecedented speed.

Implementation of the simplex method began at National Bureau of Standards in USA. Around 1952, a team led by A. Hoffman of National Bureau of Standards tested the simplex method with some problems, and compared with T. Motzkin's method prevalent at that time. The former defeated the latter unambiguously. In 1953–1954, W. Orchard-Hays (1954) started his pioneer work. He worked out the first simplex-method-based commercial software, performed on a primitive computer by today's standard. Subsequently, his implementation techniques were used and enhanced by many scholars, such as M.A. Saunders and R.E. Bixby, et al. As a result, the simplex method became a powerful practical tool. If the simplex method is used, the assignment problem mentioned above can be solved in 1 s on IBM370-168, which is even not the fastest computer at that time when Dantzig told his story.

A number of Nobel Economic Prizes awarded in the past are related to LP. Russian-American economist Wassily Leontief took the Nobel Economic Prize in 1973 for developing the input-output model, representing the interdependencies between different branches of a national economy or different regional economies. Kantorovich and Koopmans shared Nobel Economic prize in 1975 for their contribution of resources allocation theory based on LP. The award was also given to Professors K. Arrow, P. Samuelson, H. Simon, and L. Herwricz et al., several decades later when they paid close attention to LP during the early days of their professional career. On the practice side, applications of LP and the simplex method have brought about enormous economic and social benefits in the past. The simplex method was selected as one of the "Top Ten Algorithms of the Century." Highlighted in the January/February 2000 issue of Computing in Science & Engineering, a joint publication of the American Institute of Physics and the IEEE Computer Society, the ten algorithms were regarded as algorithms with the greatest influence on the development and practice of science and engineering in the twentieth century. Again and again, history confirms that it is practice that spawns theories and methods that are deep-rooted and exuberant.

Nevertheless, the simplex method has exponential time computational complexity (Klee and Minty 1972), while a "good" method is thought to be a polynomial time one. Furthermore, even finiteness of the simplex method is not guaranteed; in other words, the process may not terminate in finitely many iterations (Beale 1955; Hoffman 1953). Along another line, a break-though was made by a former Soviet Union mathematician L.G. Khachiyan (1979) who proposed a first polynomial time LP solver, the ellipsoid algorithm. Unfortunately, it performed poorly, and is not competitive with the simplex method. This is because the so-called "polynomial-time complexity" is only the worst case complexity, and such a case may hardly occur in practice.

Soon Indian mathematician N. Karmarkar (1984) published an interior-point algorithm, which is not only of lower order of polynomial time complexity than the ellipsoid algorithm, but also appeared promisingly fast in practice. The algorithm initiated an upsurge of interior-point algorithms. Some implementations of such type of algorithms are so efficient that many scholars thought the interior-point method to be superior to the simplex method in solving large-scale sparse LP problems. Unlike simplex methods, however, interior-point methods cannot be "warmly" started, and hence not applicable for solving integer linear programming (ILP) problems. Since interior-point methods cannot produce an optimal basic solution, moreover, there is a need for an additional purification process if a vertex solution is required (see, e.g., Mehrotra 1991; Tapia and Zhang 1991).

Meanwhile, the art of the simplex method did not stand still, but rather, was developing and improving. For example, P.M.J. Harris (1973) proposed, and tested approximate steepest-edge pivot rules successfully; J.J.H. Forrest and D. Goldfarb (1992) described several variants of the steepest-edge pivot rule and reported very favorable numerical results with their recurrence formulas. Consequently, simplex methods and interior-point methods were neck-and-neck in competition.

As for computational complexity, the worst case complexity does not reflect real experience well. The average time complexity might be more relevant with this respect. K.-H. Borgwardt (1982a,b) was able to show that on the average the simplex method has a polynomial time complexity in a ceratin probabilistic model, and using a certain pivot rule. S. Smale (1983a,b) offered a similar result. These results coincide with the remarkable performance of the simplex method.

Essentially, the evaluation of a method is a practical issue, and rigidly adhering to theory is not wise. The merit of a method, its efficiency, precision, and reliability and/or stability is ultimately determined by its performance in practice. In this respect, method's finiteness or complexity is even misleading contingently. After all, finite or polynomial time methods usually perform far worse than infinite or non-polynomial time ones overall. In fact, the latter rather than the former, plays a major role in applications. In view of this, the book draws its material from a practical point of view, focusing on practically effective methods, theories and implementations closely related to LP computations and hence applications.

1.2 From Practical Issue to LP Model

In practice, there are usually many schemes to be chosen for decision makers, and different schemes may lead to widely divergent results, as is of vital importance in keen competitions. This situation provides a wide stage for decision makers' intelligence and wisdom to play. Making decisions just based on experience can not be mentioned in the same breath with through mathematical tools.

Formation of a mathematical model is the first step toward the application of LP. Using full and reliable data gathered, a good model comes from one's knowledge, skill, experience and wisdom. It is beyond the scope of this book to handle this topic in detail. This section only gives the reader a taste by simple examples.

Example 1.2.1. A manufacturer who has 1,100 ton of log is committed by contract to supply 470 ton of lath to a company. There is a 6 % loss in log when it is processed into lath. In addition, the production costs have risen since the signing of the contract, but the selling price of lath remains the same. In fact, he can make more profit by selling the log as raw material to another firm. What can the manufacturer do to have the maximum profit while still honoring the contract?

This problem can be solved algebraically. Let x be the number of ton of log the manufacturer can sell as raw material and let y be the number of log available to produce lath. It is clear that

$$x + y = 1{,}100.$$

where x should be maximized to gain the largest profit. On the other hand, since 6 % out of y is lost during the manufacturing process, the actual amount of lath produced is

$$y - 0.06y,$$

which, associated to the contract, must be equal to 470 ton, i.e.,

$$y - 0.06y = 470.$$

Consequently, we have the following system of linear equations:

$$\begin{cases} y - 0.06y = 470 \\ 1{,}100 - y = \quad x \end{cases} \tag{1.1}$$

which has a unique solution

$$x = 600, \qquad y = 500.$$

Therefore, the manufacturer should sell 600 ton log as raw material, and use the rest 500 ton for production of lath.

In the preceding, the manufacturer has a unique scheme. For most problems, however, there are usually multiple or even infinitely many choices facing the decision maker, as is the following example:

Example 1.2.2. A manufacturer produces lath and sheet piles. Profit per lath is 2 dollars and that per sheet pile is 5 dollars. The equipment capability of the manufacturer can at most produce 6,000 lath or 4,000 sheet piles per day. According to the sales contract, he can sell 3,000 sheep piles at most, and the sales volume of laths and sheet piles can not excess 5,000 per day. How should he arrange to produce lath and sheep piles to gain a maximum profit?

Let x be the amount of lath and let y be that of sheep piles to be produced per day (in 1,000 as unit). For values of each pair variables (x, y) corresponding to a decision, the associated profit is $f(x, y) = 2x + 5y$ 1,000 dollars. So, what the manufacturer want to do is to maximize the function value of $f(x, y)$.

It is clear that the yielded profit would be infinitely large if there was no any restriction on the production. It is not the case, of course. According to the equipment capability, the average rate of producing lath and sheet piles is 6,000 and 4,000 per day, respectively. Therefore, variables x and y should satisfy inequality $x/6 + y/4 \leq 1$, or equivalently,

$$2x + 3y \leq 12.$$

According to the selling limitation, in addition, there is another inequality

$$x + y \leq 5,$$

and finally,

$$y \leq 3.$$

In addition, the amount of output should be certainly nonnegative integers. As the amount is large, we use restriction $x, \; y \geq 0$ instead for simplicity, and finally obtain integer results by rounding.

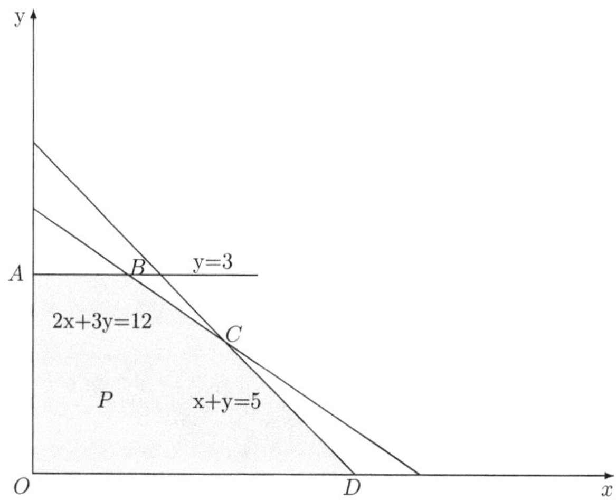

Fig. 1.1 The feasible region of Example 1.2.2

Preceding analysis can be summarized into the following problem:

$$
\begin{array}{ll}
\max & f(x, y) = 2x + 5y, \\
\text{s.t.} & -2x - 3y \geq -12, \\
& -x - \ \ y \geq -5, \\
& \quad\ \ - \ \ y \geq -3, \\
& x, \ y \geq 0,
\end{array}
\tag{1.2}
$$

which is a mathematical model for Example 1.2.2. x and y are so-called "decision variables", and $f(x, y)$ is so-called "objective function". The inequalities are "constraints". (x, y) satisfying constraints is "feasible solution", and the set of all feasible solutions is "feasible region". Solving a model is to find its "optimal solution", a feasible solution maximizing the objective function over the feasible region.

In contrast with Example 1.2.1 that has a single feasible solution, and hence no any optimization method is needed, Example 1.2.2 involves infinitely many feasible solutions. In fact, the problem has the following feasible region:

$$
P = \{(x, y) \mid 2x + 3y \leq 12, \ x + y \leq 5, \ y \leq 3, \ x \geq 0, y \geq 0\}.
$$

Since there is a 1-to-1 correspondence between pairs of real numbers and points in a rectangular coordinate system, the feasible region P can be represented geometrically by the shaded area of Fig. 1.1, where x (thousand) represents the amount of laths and y (thousand) that of sheep piles, and where each boundary is labeled by an equality corresponding to a closed half plane. The feasible region P is just the intersection of these half planes, each point within which corresponds to a feasible solution.

Therefore, the problem now is how to find a point in P, corresponding to the largest possible value of function $2x + 5y$. Since there are infinitely many points in P, it is impossible to pick up and compare them one by one, not to mention much more challenging large-scale problems with hundreds of thousands of variables and constraints. Beyond doubt, powerful mathematical solvers, like the well-known simplex method, are essential and indispensable.

1.3 Illustrative Applications

LP has a very wide range of applications, as touch upon almost all areas related to decision making and management. This section only brings up a few illustrative instances. Strictly speaking, some of these instances involve variables of nonnegative integer value, and hence belong to so-called ILP or mixed ILP models, though the "integer" requirement is ignored here for simplicity.

Example 1.3.1 (Production planning). A factory produces furniture A, B, C. Each furniture production goes through three procedures: component processing, electroplating and assembling. The production capacity of each procedure per day is converted into effective working hours. Below are the required effective working hours and getatable profit for each piece of the furniture.

Procedure	Hours per piece			Available hours per day
	Product A	Product B	Product C	
Component processing	0.025	0.05	0.3	400
Electrofacing	0.20	0.05	0.1	900
Assembling	0.04	0.02	0.20	600
Profit (dollars/piece)	1.25	1.5	2.25	

How can the factory achieve the highest profit?

Answer Let x_1, x_2 and x_3 be respectively the amount of furniture A,B,C, and let f be the total profit. To determine values of these decision variables to gain the highest profit, construct the following mathematical model:

$$\begin{aligned}
\max \quad & f = 1.25x_1 + 1.5x_2 + 2.25x_3, \\
\text{s.t.} \quad & 0.025x_1 + 0.05x_2 + 0.3x_3 \leq 400, \\
& 0.20x_1 + 0.05x_2 + 0.1x_3 \leq 900, \\
& 0.04x_1 + 0.02x_2 + 0.20x_3 \leq 600, \\
& x_1, x_2, x_3 \geq 0.
\end{aligned}$$

An optimal solution obtained by the simplex method is

$$x_1 = 2{,}860, \; x_2 = 6{,}570, \; x_3 = 0,$$

and the associated objective function value is $f = 13{,}430$. That is to say, the factory should produce 2,860 pieces of product A, 6,570 pieces of product B and no piece of product C, with the highest profit 13,430 dollars.

Example 1.3.2 (Transportation). A company has 8,800, 7,200 and 5,700 containers at ports A, B, C, respectively. These containers should be transported to plants 1, 2, 3, 4, 5, whose working ability is respectively 3,200, 5,300, 4,100, 6,200 and 2,900 containers. The following table lists the freight rate (dollars/container) of transport service from the ports to the plants:

Port	1	2	3	4	5
A	55	150	78	38	90
B	25	60	25	43	35
C	102	80	52	74	60

How should the company arrange to achieve the least total freight charges.

Answer This is a "balanced" transportation problem, as the total number of containers at all the three ports is equal to the total number of containers the five plants can handle, i.e.,

$$8{,}800 + 7{,}200 + 5{,}700 = 3{,}200 + 5{,}300 + 4{,}100 + 6{,}200 + 2{,}900 = 21{,}700$$

Let x_{ij}, $i = 1, 2, 3$; $j = 1, \ldots, 5$ be the number of containers which are transported from port A, B, C to plant 1–5 and let f be the total freight charges to be minimized. The model is as follows:

$$\begin{aligned}
\min \; f = \; & 55x_{11} + 150x_{12} + 78x_{13} + 38x_{14} + 90x_{15} \\
& + 25x_{21} + 60x_{22} + 25x_{23} + 43x_{24} + 35x_{25} \\
& + 102x_{31} + 80x_{32} + 52x_{33} + 74x_{34} + 60x_{35}, \\
\text{s.t.} \quad & x_{11} + x_{12} + x_{13} + x_{14} + x_{15} = 8{,}800, \\
& x_{21} + x_{22} + x_{23} + x_{24} + x_{25} = 7{,}200, \\
& x_{31} + x_{32} + x_{33} + x_{34} + x_{35} = 5{,}700, \\
& x_{11} + x_{21} + x_{31} = 3{,}200, \\
& x_{12} + x_{22} + x_{32} = 5{,}300, \\
& x_{13} + x_{23} + x_{33} = 4{,}100, \\
& x_{14} + x_{24} + x_{34} = 6{,}200, \\
& x_{15} + x_{25} + x_{35} = 2{,}900, \\
& x_{ij} \geq 0, \; i = 1, 2, 3; \; j = 1, \ldots, 5.
\end{aligned}$$

An optimal solution obtained by the simplex method is

$$x_{11} = 2{,}600,\ x_{14} = 6{,}200,\ x_{21} = 600,\ x_{23} = 4{,}100,\ x_{25} = 2{,}500,\ x_{32} = 5{,}300,\ x_{35} = 400,$$

and all the other variables take value 0. Then the company pays the minimum total freight 1,031,600 dollars.

Example 1.3.3 (Burdening). Someone feeds animals, as need 30 g of mineral, 700 g of protein and 100 mg of vitamin per day, at least. There are five types of feed to purchase, the nutrition and price per kilogram for each are as follows:

Fodder	Mineral(g)	Protein(g)	Vitamin(mg)	Price(dollars/kg)
A	0.5	18.0	0.8	0.8
B	2.0	6.0	2.0	0.3
C	0.2	1.0	0.2	0.4
D	0.5	2.0	1.0	0.7
E	1.0	3.0	0.5	0.2

Find a purchasing scheme to achieve a minimum cost to serve animals' needs.

Answer Let $x_j\,(kg)\ j = 1,\dots,5$ be the purchase quantities of the five type of feed, respectively, and let f be the total cost. The goal is to determine values of these decision variables corresponding to the minimum cost. The model is as follows:

$$\begin{array}{ll} \min & f = 0.8x_1 + 0.3x_2 + 0.4x_3 + 0.7x_4 + 0.2x_5, \\ \text{s.t.} & 0.5x_1 + 2.0x_2 + 0.2x_3 + 0.5x_4 + \quad x_5 \geq \ 30, \\ & 18.0x_1 + 6.0x_2 + \quad x_3 + 2.0x_4 + 0.3x_5 \geq 700, \\ & 0.8x_1 + 2.0x_2 + 0.2x_3 + \quad x_4 + 0.5x_5 \geq 100, \\ & x_j \geq 0,\ j = 1,\dots,5. \end{array}$$

An optimal solution is obtained by the simplex method, i.e.,

$$x_1 = 25.6,\ x_2 = 39.7,\ x_3 = x_4 = x_5 = 0,$$

and the associated objective function value is $f = 32.28$. So, he should purchase 25.6 kg of feed A, 39.7 kg of feed B, and none of the other three types of feed, with the lowest cost 32.28 dollars.

Example 1.3.4 (Human resource arrangement). A round-the-clock supermarket arranges salespersons. Every day falls into six time intervals, 4 h each. Every salesperson starts work at the beginning of the intervals, and then continuously works for 8 h. The required numbers of salesperson for each of the intervals are as follows:

Interval	1	2	3	4	5	6
	2–6 o'clock	6–10 o'clock	10–14 o'clock	14–18 o'clock	18–22 o'clock	22–2 o'clock
Number	7	15	25	20	30	7

How to arrange to achieve the least number of salespersons per day.

Answer Let $x_j,\ j = 1,\ldots,6$ be the number of assistants who start work at the beginning of interval j and let f be the total number of salespersons. The goal is to determine variable values to achieve the minimum number of salespersons working per day. The model is as follows:

$$\begin{array}{lll}
\min & f = x_1 + x_2 + x_3 + x_4 + x_5 + x_6, & \\
\text{s.t.} & x_1 + x_6 \ge 7, & \\
& x_1 + x_2 \ge 15, & \\
& x_2 + x_3 \ge 25, & \\
& x_3 + x_4 \ge 20, & \\
& x_4 + x_5 \ge 30, & \\
& x_5 + x_6 \ge 7, & \\
& x_j \ge 0,\ j = 1,\ldots,6. &
\end{array}$$

An optimal solution obtained by the simplex method is

$$x_2 = 25,\ x_4 = 30,\ x_6 = 7,\ x_1 = x_3 = x_5 = 0,$$

and the associated objective function value is $f = 62$. So, the business should arrange 25 salespersons working from 6 o'clock, 30 ones working from 14 o'clock, and 7 from 22 o'clock. This scheme requires only 62 salespersons.

Example 1.3.5 (Laying-off). A shop is asked to make 450 steel frames. Each frame needs an 1.5, an 2.1 and an 2.9 m long angle steel. Each raw material is 7.4 m long. Find the optimum laying-off way.

Answer A direct way is to make a frame using a raw material to get the three types of angle steel. If so, there would be a need for 450 raw materials, each of which would leave an 0.9 m long offcut. To obtain the optimum way, consider the following schemes, each of which would leave an offcut of no more than 0.9 m long.

Length (m)	Scheme					
	A	B	C	D	E	F
2.9	1	2	0	1	0	1
2.1	0	0	2	2	1	1
1.5	3	1	2	0	3	1
Total	7.4	7.3	7.2	7.1	6.6	6.5
Biscuit	0.0	0.1	0.2	0.3	0.8	0.9

These schemes should be mixed and matched to gain the minimum total length of offcuts. Let $x_j,\ j = 1,\ldots,6$ be numbers of raw materials laying-off according to schemes A, B, C, D, E, F, respectively, and let f be the total length of offcuts. Construct the following model:

$$\begin{array}{ll} \min & f = 0.1x_2 + 0.2x_3 + 0.3x_4 + 0.8x_5 + 0.9x_6, \\ \text{s.t.} & x_1 + 2x_2 \qquad + x_4 \qquad + x_6 = 450, \\ & \qquad 2x_3 + 2x_4 + x_5 + x_6 = 450, \\ & 3x_1 + x_2 + 2x_3 \qquad + 3x_5 + x_6 = 450, \\ & \qquad x_j \geq 0,\ j = 1,\ldots,6. \end{array}$$

The optimal solution obtained by the simplex method is

$$x_1 = 135,\ x_2 = 45,\ x_4 = 225,\ x_3 = x_5 = x_6 = 0,$$

with objective function value $f = 72$. This is to say that the shop should lay-off 135 raw materials of scheme A, 45 of scheme B and 225 of scheme D, with length 72 m of all offcuts. Consequently, the minimum total number of used raw materials is 405.

The size of the preceding problems are very small,[1] by which the reader should have a taste about LP applications. Nowadays, real world problems are often much larger. Hereafter, however, small problems will still be utilized to convey ideas and show computational steps illustratively.

1.4 Errors of Floating Point Arithmetic

As a basic tool for modern numerical computation, a computer is indispensable. LP without it is not imaginable. In using a computer, however, numerical errors are unavoidable, because it can only represent a proper subset F of real numbers, and

[1] In this book, $m + n$ is taken as an index for problem size.

as a result data can only stored and calculated in finite digits. On one hand, primary data may be out of F, and expressed approximately; on the other hand, numerical results yielded from subsequent operations of arithmetic may have to be rounded. Errors arising therefrom are called *rounding errors*.

The system F of floating point represented by a computer is characterized by four positive integers: the *base* β, *precision* (mantissa digits) t, *exponent range* $[L, U]$ (Forsythe et al. 1977, pp. 10–29). More precisely, a computer can represent all nonzero numbers of the following form:

$$fl(x) = \pm .\alpha_1 \cdots \alpha_t \, \beta^e, \qquad 0 \leq \alpha_i < \beta, \; i = 1, \ldots, t, \quad \alpha_1 \neq 0; \qquad L \leq e \leq U, \tag{1.3}$$

which together with 0 constitute F, and where $\alpha_1 \cdots \alpha_t$ is called *mantissa part*, β^e is exponent part. Nowadays, a computer usually uses the binary system, i.e., $\beta = 2$.

It is clear that taking notation

$$m = \beta^{L-1} \qquad\qquad M = \beta^U(1 - \beta^{-t}),$$

then for any floating point number $fl(x) \in F$, it holds that $m \leq |x| \leq M$. Thus, F is a set of non-uniformly distributed rational numbers. Define set

$$S = \{x \in \mathcal{R} \mid m \leq |x| \leq M\}.$$

Then, for any $0 \neq x \in S$, the approximate number $fl(x)$, yielded from the computer, only holds mantissa of t digits, via the following two ways:

(i) Rounding method: is based on the $(t + 1)$th figure of the mantissa of x. The figures at this position and after are all taken off if the figure is less than a half of β; in the other case, these figures are taken off after a unit is added to the previous figure, and the associated absolute error bound is therefore

$$|fl(x) - x| \leq \frac{1}{2}\beta^{e-t}. \tag{1.4}$$

(ii) Truncation method: The $(t + 1)$th figure and after are all taken off. Thereby, $fl(x)$ is the floating point number nearest to x in those of F, whose the absolute values are no more than $|x|$, and hence the associated absolute error bound is

$$|fl(x) - x| \leq \beta^{e-t}. \tag{1.5}$$

The t is a key quantity to numerical computations. To see this, define

$$\epsilon = \begin{cases} \frac{1}{2}\beta^{1-t} & \text{runding} \\ \beta^{1-t} & \text{Truncating} \end{cases} \tag{1.6}$$

Then, from either (1.4) or (1.5) it follows that

$$|fl(x) - x| \le \epsilon\beta^{e-1}.$$

Since

$$|x| \ge .\alpha_1\beta^e > 0,$$

it holds that

$$\left|\frac{fl(x) - x}{x}\right| \le \frac{\epsilon\beta^{e-1}}{.\alpha_1\beta^e} = \epsilon/\alpha_1 \le \epsilon,$$

that is

$$fl(x) = x(1 + \delta), \qquad |\delta| \le \epsilon.$$

Therefore, ϵ is just a relative error bound of the approximate number $fl(x)$. Called *machine precision*, ϵ is the smallest positive value in F, satisfying

$$fl(1 + \epsilon) \ne 1,$$

based on which one may get to know the precision of a computer.

Assume $a, b \in F$, and denote any of the four arithmetic operation $+, -, *, /$ by $\Box$. It is clear that if $a\Box b \notin S$, then the computer can not handle, and display *overflow* (when $|a\Box b| > M$) or *underflow* (when $0 < |a\Box b| < m$), and terminate the computation (some computer and compiler place value 0 in the case of underflow). In the normal cases, for $a\Box b$'s computed value $fl(a\Box b)$, it is clear that

$$fl(a\Box b) = (a\Box b)(1 + \delta), \qquad |\delta| < \epsilon,$$

Therefore, the relative error bound for a single arithmetic operation is ϵ, a small number. Unfortunately, the situation may be entirely contrary when executing an algorithm or a series of operations.

Usually, algorithm's performance may be affected by rounding or truncation errors greatly. Due to error accumulations, final results obtained could be entirely meaningless, as they could be far from its theoretical value. It is therefore important to take care of effects of numerical errors in computations. In this aspect, the following simple rules should be followed:

1. Avoid using a number close to 0 as denominator or divisor.
2. Avoid subtracting a number from another that is close to it.
3. Avoid operating numbers with very different order of magnitude so that "the large swallows the small".

4. Avoid results close to m or M.
5. Reduce the number of operations, especially multiplications or divisions.

When $\delta \ll 1$, for instance, calculating according to the right-hand side of one of the following equalities result in a more accurate result than according to its left-hand side.

$$\sqrt{x+\delta} - \sqrt{x} = \delta/(\sqrt{x+\delta} + \sqrt{x}),$$
$$\cos(x+\delta) - \cos x = -2\sin(x+\delta/2)\sin(\delta/2).$$

Due to different errors introduced in computations, algorithms which are equivalent theoretically are often not equivalent numerically. An algorithm is said to be of good *(numerical) stability* if resulting numerical errors are under control, and hence affecting results not much. An algorithm of bad stability is unreliable.

Usually, it is not a easy task to accurately evaluate error bound of computational results, obtained from executing an algorithm. Instead, the so-called *backward error analysis* method is applied for such purpose. Interested readers are referred to Wilkinson (1971).

1.5 The Standard LP Problem

Some of LP problems in Sect. 1.3 seeks for maximizing the objective function values, while others for minimizing. Besides, some of their constraints are equalities, while others are inequalities with "≥" or "≤". For convenience of handling, we introduce the following problem of form:

$$\begin{array}{ll} \min & f = c_1x_1 + c_2x_2 + \cdots + c_nx_n, \\ \text{s.t.} & a_{11}x_1 + a_{12}x_2 + \cdots + a_{1n}x_n = b_1, \\ & a_{21}x_1 + a_{22}x_2 + \cdots + a_{2n}x_n = b_2, \\ & \qquad \vdots \\ & a_{m1}x_1 + a_{m2}x_1 + \cdots + a_{mn}x_n = b_m, \\ & \qquad x_j \geq 0, \; j = 1, \ldots, n, \end{array} \tag{1.7}$$

which is termed "standard (LP) problem".

As shown as follows, all kinds of LP problems can be transformed to standard forms:

1. Since maximizing a quantity is equivalent to minimizing its negative, any maximization problem can be transformed to a minimization problem.

   Using $f = -f'$, e.g.,

$$\max\ f' = c_1x_1 + c_2x_2 + \cdots + c_nx_n$$

can be replaced by

$$\min\ f = -c_1x_1 - c_2x_2 - \cdots - c_nx_n.$$

Clearly, this does not matter to the optimal solution but changes the sign of the optimal value only.

2. Any inequality can be transformed to an equality by introducing an extra nonnegative variable.

   Introducing a so-called *slack variable* $x_{k+1} \geq 0$, e.g., the "≤" type of inequality

$$\alpha_1x_1 + \alpha_2x_2 + \cdots + \alpha_kx_k \leq \beta$$

   can be transformed to equality

$$\alpha_1x_1 + \alpha_2x_2 + \cdots + \alpha_kx_k + x_{k+1} = \beta,$$

   and "≥" type of inequality

$$\alpha_1x_1 + \alpha_2x_2 + \cdots + \alpha_kx_k \geq \beta$$

   can be transformed to equality

$$\alpha_1x_1 + \alpha_2x_2 + \cdots + \alpha_kx_k - x_{k+1} = \beta.$$

3. A standard LP problem involves the nonnegative constraints, as is usually the case in practice. In exceptional cases, the problem can be handled by introducing auxiliary variables.

Using variable transformation $x'_k = -x_k$, e.g., non-positive constraint $x'_k \leq 0$ is turned to $x_k \geq 0$. "free variable" (without any sign restriction) can be denoted as the difference of two nonnegative variables; e.g., the free variable x'_k can be eliminated by the following variable substitution:

$$x'_k = x_k - x_{k+1}, \qquad x_k, x_{k+1} \geq 0$$

In convention, theories and methods are usually developed only for the standard problem, though they are essentially applicable to all LP problems.

Example 1.5.1. Transform the following LP problem to standard form:

$$\begin{array}{ll} \max & f' = -2x_1 + x_2 - 5x'_3, \\ \text{s.t.} & -3x_1 + x_2 + x'_3 \leq 2, \\ & x_1 - 7x_2 + 4x'_3 \geq -3, \\ & x_1 - 3x_2 - 2x'_3 \leq 1, \\ & x_1, x_2 \geq 0. \end{array}$$

Answer Maximizing f' is transformed to minimizing $f = -f'$. The free variable x_3' is substituted by $x_3' = x_3 - x_4,\ x_3, x_4 \geq 0$. In addition, the first three inequalities are transformed to equalities by introducing slack variables x_5, x_6, x_7 into them, respectively. Consequently, we obtain the following standard problem:

$$
\begin{array}{ll}
\min & f = 2x_1 - x_2 + 5x_3 - 5x_4, \\
\text{s.t.} & -3x_1 + x_2 + x_3 - x_4 + x_5 = 2, \\
& x_1 - 7x_2 + 4x_3 - 4x_4 - x_6 = -3, \\
& x_1 - 3x_2 - 2x_3 + 2x_4 + x_7 = 1, \\
& x_j \geq 0,\ j = 1, \ldots, 7.
\end{array}
$$

Using vector and matrix notation $A = (a_{ij})$

$$
x = (x_1, \ldots, x_n)^T, \qquad c = (c_1, \ldots, c_n)^T, \qquad b = (b_1, \ldots, b_m)^T,
$$

the standard problem (1.7) can be put into the compact form below:

$$
\begin{array}{ll}
\min & f = c^T x, \\
\text{s.t.} & Ax = b, \qquad x \geq 0,
\end{array}
\tag{1.8}
$$

where $A \in \mathcal{R}^{m \times n}$, $c \in \mathcal{R}^n$, $b \in \mathcal{R}^m$, $m < n$. Speaking for itself, each row or column of the coefficient matrix A does not vanish. Without confusion, thereafter symbol A will also be used to denote the set of column indices of the coefficient matrix, i.e.,

$$
A = \{1, \cdots, n\}.
$$

$f = c^T x$ is called *objective function*, the first row of the problem indicates minimization of it. c is called *price (cost)* vector, components of which are called *prices (costs)*. The other rows called *constraints (conditions)*: $Ax = b$ called *constraint system*; "$x \geq 0$" *nonnegative constraints*. The left-hand side functions of the constraints are *constraint functions*.

A solution fulfilling all constraints is a *feasible solution*, and the set of all feasible solutions, i.e.,

$$
P = \{x \in \mathcal{R}^n \mid Ax = b,\ x \geq 0\} \tag{1.9}
$$

is *feasible region*. Objective function value corresponding to a feasible solution is *feasible value*. The minimum value of the objective function over the feasible region is *(global) optimal value*, and the associated feasible solution is *optimal solution*. The set of all optimal solutions is *optimal (solution) set*. The problem is said (lower) unbounded if the objective value can be arbitrarily small over the feasible region, hence there exists no optimal solution to the problem.

1.6 Canonical Form of Ax = b

The standard LP problem, introduced in the preceding section, involves a system $Ax = b$ of equations as a part of its constraints. Neglecting the nonnegative constraints, we will handle the system $Ax = b$ alone at the moment. Note that $Ax = b$ is a so-called "underdetermined" system, as the number of equalities is less that of involved unknowns. We will see that this system has infinitely many solutions if any.

The set of all solutions is called system's solution set. Systems are equivalent if they have the same solution set. There are two types of basic equivalent transformations to linear systems. With this respect, the validity of the following propositions is evident.

Proposition 1.6.1. *A system, resulting from multiplying any equality by a nonzero, is equivalent to the original.*

Proposition 1.6.2. *A system, resulting from adding a multiple of any equality to another, is equivalent to the original.*

Any of the preceding operations is called an *elementary (row) transformation*. The second type of elementary transformation is especially important, as it can eliminate a nonzero entry of the coefficient matrix. Via a series of such transformations, e.g., the *Gauss-Jordan elimination* converts a linear system to a so-called *canonical form* that is readily solvable.

Let us bring up an example of 3 × 5 standard LP problem:

$$
\begin{array}{llrrrrrcr}
\min & f = x_1 + 2x_2 - x_4 + x_5, \\
\text{s.t.} & 2x_1 &+ x_2 &+ 3x_3 &+ 2x_4 & &= & 5, \\
& x_1 &- x_2 &+ 2x_3 &- x_4 &+ 3x_5 &= & 1, \\
& x_1 & &- 2x_3 & &- 2x_5 &= & -1, \\
& & & x_j \geq 0, \ j = 1, \ldots, 5.
\end{array}
\tag{1.10}
$$

For simplicity of expression, the coefficients are peeled off from the system, and filled in the following tableau:

x_1	x_2	x_3	x_4	x_5	RHS
2	1	3	2		5
1	−1	2	−1	3	1
1		−2		−2	−1

where blank spaces stand for value 0 (the same below). From comparing between the preceding tableau and the related constraint system, it is seen that the coefficients of the system form the first five columns of the tableau, that is, the 3 × 5 *coefficient matrix*. The right-hand side of the system forms the sixth column. All together, the six columns constitute a 3 × 6 *augmented* matrix of the system. The head of the tableau only indicates the positions of entries of the matrix; e.g., x_1 indicates that the related column contains the coefficients of variable x_1, and RHS marks the right-hand side of the system.

Carrying out elementary transformations with the tableau amounts to doing the same with the system itself, and hereafter, therefore, the system and the tableau will be regarded as the same.

A *pivot* is a nonzero entry of the coefficient matrix, which determines a certain series of elementary transformations. As an example, it might be well to take the first entry 2 of the first column as a pivot. Multiply by 1/2 the first (pivot) row of the augmented matrix to turn the entry to 1, obtaining the following tableau:

x_1	x_2	x_3	x_4	x_5	RHS
1	1/2	3/2	1		5/2
1	−1	2	−1	3	1
1		−2		−2	−1

Then add the −1 times of the first row to the second and third (non-pivot) rows, so that all the components of the first column becomes 0 except for the first, obtaining the following tableau:

x_1	x_2	x_3	x_4	x_5	RHS
1	1/2	3/2	1		5/2
	−3/2	1/2	−2	3	−3/2
	−1/2	−7/2	−1	−2	−7/2

Thereby, the first column becomes a unit vector with the first component 1. Now take the second entry of the second row as the next pivot. Multiply by −2/3 the second row of the augmented matrix to turn the entry to 1, leading to the following tableau:

x_1	x_2	x_3	x_4	x_5	RHS
1	1/2	3/2	1		5/2
	1	−1/3	4/3	−2	1
	−1/2	−7/2	−1	−2	−7/2

Now respectively add the −1/2 and 1/2 times of the second row to the first and the third row, so that all the components of the second column becomes 0, except for the second, giving the following tableau:

x_1	x_2	x_3	x_4	x_5	RHS
1		5/3	1/3	1	2
	1	−1/3	4/3	−2	1
		−11/3	−1/3	−3	−3

So, the second column becomes a unit vector with the second component 1. Then, take the third entry of the third row as the next pivot. Multiply by $-3/11$ the third row of the augmented matrix to turn the entry to 1, leading to the following tableau:

x_1	x_2	x_3	x_4	x_5	RHS
1		5/3	1/3	1	2
	1	−1/3	4/3	−2	1
		1	1/11	9/11	9/11

Finally, respectively add the $-5/3$ and $1/3$ times of the third row to the first and the second row, so that all the components of the third column becomes 0, except for the third, giving the following tableau:

x_1	x_2	x_3	x_4	x_5	RHS
1			2/11	−4/11	7/11
	1		15/11	−19/11	14/11
		1	1/11	9/11	9/11

Consequently, the tableau includes a 3×3 unit matrix, and Gauss-Jordan elimination is completed. The corresponding so-called "canonical" system can be written

$$\begin{cases} x_1 = 7/11 - (2/11)x_4 + (4/11)x_5 \\ x_2 = 14/11 - (15/11)x_4 + (19/11)x_5 \\ x_3 = 9/11 - (1/11)x_4 - (9/11)x_5 \end{cases} \tag{1.11}$$

which is equivalent to the original (Propositions 1.6.1 and 1.6.2). Nevertheless, the preceding is so simple that itself can be regarded as a explicit expression of the general solutions of the system. In fact, x_4 and x_5 there can be taken as parametric variables; by giving any values to them, one can calculate the corresponding values of x_1 and x_2 easily, obtaining a special solution of the system. Specifically, giving zero to both x_4 and x_5 leads to

$$x_1 = 7/11, \quad x_2 = 14/11, \quad x_3 = 9/11, \quad x_4 = x_5 = 0,$$

which is called a *basic solution* (see below). Note that this solution can be directly read from the RHS column of the last tableau.

It is noted that Gauss-Jordan elimination, described above, does not always go smoothly. Taken as a denominator, each pivot in the process must be nonzero; otherwise, the elimination procedure breaks off. In fact, even when all the pivots are nonzero and the procedure can perform throughout, the final result may still be useless, as is distorted by outgrown numerical errors whenever the modulus of some pivot is too small (Sect. 1.4). Therefore, practical Gauss-Jordan elimination needs some kind of *pivoting* operations.

Let us turn to the so-called *Gauss-Jordan elimination with complete pivoting* in general. Put the augmented coefficient matrix of the system $Ax = b$ in the following tableau:

x_1	x_2	$\cdots$	x_n	RHS
a_{11}	a_{12}	$\cdots$	a_{1n}	b_1
a_{21}	a_{22}	$\cdots$	a_{2n}	b_2
$\vdots$	$\vdots$	$\vdots$	$\vdots$	$\vdots$
a_{m1}	a_{m2}	$\cdots$	a_{mn}	b_m

Every step of the procedure comprises two parts: selecting a nonzero pivot from the coefficient matrix and performing eliminations to the whole augmented matrix. The row and column, where a pivot located at, are called *pivot row* and *pivot column*, respectively. The rows and columns other than pivot rows and columns are called *non-pivot* (excluding RHS column). Once a pivot is determined, certain elementary transformations are performed to turn the associated pivot column to a unit vector with the component 1 at the pivotal position. To do so, the pivot row is multiplied by the reciprocal of the pivot, and then relevant times of the pivot row are added to the other rows to eliminate the other nonzeros in the pivot column.

Suppose that after $r < m$ steps, the first r pivot columns become unit vectors, as is shown below:

x_1	x_2	$\cdots$	x_r	x_{r+1}	$\cdots$	x_n	RHS
1				$\bar{a}_{1\,r+1}$	$\cdots$	$\bar{a}_{1\,n}$	$\bar{b}_1$
	1			$\bar{a}_{2\,r+1}$	$\cdots$	$\bar{a}_{2\,n}$	$\bar{b}_2$
		$\ddots$		$\vdots$	$\vdots$	$\vdots$	$\vdots$
			1	$\bar{a}_{r\,r+1}$	$\cdots$	$\bar{a}_{r\,n}$	$\bar{b}_r$
				$\bar{a}_{r+1\,r+1}$	$\cdots$	$\bar{a}_{r+1\,n}$	$\bar{b}_{r+1}$
				$\vdots$	$\vdots$	$\vdots$	$\vdots$
				$\bar{a}_{m\,r+1}$	$\cdots$	$\bar{a}_{m\,n}$	$\bar{b}_m$

At step $r + 1$, a pivot $\bar{a}_{pq}$ is determined such that

$$|\bar{a}_{p\,q}| = \max\{|\bar{a}_{i\,j}| \mid i = r + 1, \ldots, m;\ j = r + 1, \ldots, n\}. \tag{1.12}$$

If $\bar{a}_{pq} \neq 0$, it is taken as a pivot. It is noted that the pivot is selected from entries within the current non-pivot rows and columns. Then, row p is multiplied by $1/a_{p\,q}$, and then $-a_{j\,q}$ times of the new row p is added to rows $j = 1, \ldots, m; j \neq p$,

Table 1.1 Canonical tableau

x_{j_1}	x_{j_2}	$\cdots$	x_{j_r}	$x_{j_{r+1}}$	$\cdots$	x_{j_n}	RHS
1				$\bar{a}_{i_1\, j_{r+1}}$	$\cdots$	$\bar{a}_{i_1\, j_n}$	$\bar{b}_{i_1}$
	1			$\bar{a}_{i_2\, j_{r+1}}$	$\cdots$	$\bar{a}_{i_2\, j_n}$	$\bar{b}_{i_2}$
		$\ddots$		$\vdots$	$\vdots$	$\vdots$	$\vdots$
			1	$\bar{a}_{i_r\, j_{r+1}}$	$\cdots$	$\bar{a}_{i_r\, j_n}$	$\bar{b}_{i_r}$
							$\bar{b}_{i_{r+1}}$
							$\vdots$
							$\bar{b}_{i_m}$

respectively. Thereby, column q becomes a unit vector with the pth component 1, while the first r pivot columns remain unchanged. Finally, step $r + 1$ is completed by moving row p to the position of row $r + 1$, and column q to that of column $r + 1$ via row and column exchanges. Note that the row and column exchanges amount to changing the order of equalities and variables, and therefore do not change the solution set of the system.[2]

The elimination terminates when $r = m$, or, otherwise, $r < m$ but $\bar{a}_{pq} = 0$, i.e., all entries of the non-pivot rows and columns are zero.

The final tableau, resulting by Gauss-Jordan elimination, is called *canonical.* Assume that the procedure terminates at step r. The tableau includes an $r \times r$ permutation matrix, one that becomes a unit matrix by relevant row and column exchanges. Such a permuted canonical tableau is of the following (standard) form: where the north-west corner is the unit matrix and all the entries of the last $m - r$ (non-zero) rows are zero, except perhaps for the last $m - r$ entries, $\bar{b}_{i_t}, t = r + 1, \ldots, m$, of the right-hand side RHS (when $\bar{b}_{i_t}, t = r + 1, \ldots, m$ vanish, the identities, represented by the non-pivot rows, can be dropped from the system actually). It is clear that any constraint system $Ax = b$ has a canonical tableau.

The (ordered) index sets $B = \{j_1, \ldots, j_r\}$ of pivot columns and $N = A \backslash B = \{j_{r+1}, \ldots, j_n\}$ of non-pivot columns are called *basic* and *nonbasic*, respectively. Accordingly, variables corresponding to B and N are called *basic* and nonbasic.

From Canonical Tableau 1.1, it is easy to determine cases of solutions to system $Ax = b$.

Theorem 1.6.1. *System $Ax = b$ $(m < n)$ has solutions if and only if $r = m$, or $r < m$ but all the entries in the non-pivot rows of RHS column vanish. If so, it has infinitely many solutions.*

According to linear algebra, such a key positive integer r is equal to the rank of A. Elementary transformations change neither the rank of A or $(A|b)$ nor the solution set to $Ax = b$. The preceding theorem equivalently says that $Ax = b$ has infinitely many solutions if and only if the rank of A and that of $(A|b)$ are equal. In

[2]The row and column exchanges is not really necessary in practice if one uses two integer arrays to record labels of pivot rows and columns.

convention, the following assumption is made to rule out the case of nonexistence of solution to $Ax = b$.

Assumption. The rank of the coefficient matrix is equal to the number of its rows, i.e., rank $A = m$.

Unless specified otherwise, we will always discuss under this assumption, especially in Part I of this book. Including an $m \times m$ permutation matrix, therefore, a canonical tableau corresponds to a general expression of solutions to $Ax = b$, i.e.,

$$\begin{cases} x_B &= \bar{b} - \bar{N} x_N, \\ x_N &= x_N, \end{cases}$$

where the nonbasic variables x_N in the right-hand side is viewed as parametric.

Definition 1.6.1. The solution, resulting from the canonical form by setting the non-basic variables to zero, is a basic solution. If its basic components are all nonnegative, it is a basic feasible solution.

So, the canonical tableau corresponds to basic solution

$$\bar{x}_B = \bar{b}, \qquad \bar{x}_N = 0,$$

which is a basic feasible solution if $\bar{b} \geq 0$.

Theorem 1.6.2. *A feasible solution is a basic feasible solution if and only if columns of A corresponding to its positive components are linearly independent.*

Proof. Necessity. It is clear that all positive components of a basic feasible solution are basic ones, and hence correspond to linearly independent columns.

Sufficiency. Assume that a feasible solution has s positive components corresponding to linearly independent columns. Since rank $A = m$, it holds that $s \leq m$. If $s = m$, then the feasible solution is clearly a basic solution. If, otherwise, $s < m$, there exist additional $m - s$ columns of A, which and the s columns together constitute a linearly independent set of columns. Clearly, the solution is just the basic one associated with the canonical form, corresponding to the set taken as basic columns. □

Definition 1.6.2. A nonsingular $m \times m$ submatrix of A is a basis (matrix).

By Gauss-Jordan elimination with column and row exchanges, any given basis can be converted to the $m \times m$ unit matrix, so that $(A|b)$ converted to a canonical tableau. It is to say, a basis corresponds to a canonical tableau, and hence an unique basic solution. But it is not an 1-to-1 correspondence: a basic solution could correspond multiple bases, as is closely related to so-called "degeneracy" in the simplex method (Sect. 3.6).

A basis is called *feasible basis* if the corresponding basic solution is feasible. Such type of basis (or solution) plays a central role in the simplex method.

For sparse computations, (1.12) used to determine the pivot should be loosened by requiring $|\bar{a}_{pq}|$ appropriately large to widen the range of selection. Such doing enables one to take into account sparsity of basis matrix in pivoting (see Sect. 5.5). It is noted however that different choices of pivots lead to different canonical forms and basic solutions, despite solution sets of these canonical forms are the same.

Carrying out an elementary transformation on a matrix amounts to premultiplying it by a Gauss transformation (or elementary) matrix. So, the Gauss-Jordan elimination turns $Ax = b$ to a canonical form, involving an unit submatrix in its coefficient matrix, by premultiplying a series of Gauss transformations.

Alternatively, the *Gauss elimination* turns $Ax = b$ to a *triangular form*, involving an upper triangular submatrix in the coefficient matrix, which is easy to solve to gain a basic solution. Also, the triangularization of a system can be realized via premultiplying by a series of elementary orthogonal matrices, such as Householder matrices and Givens rotations or reflections (for more details, see Golub and Van Loan 1989). As conventional, we will use the canonical form for the development of the simplex method to solve LP problems, though the goal can be achieved otherwise via the triangular form.

Part I
Fundamentals

Chapter 2
Geometry of the Feasible Region

Feasible region is of a basic significance to optimization. Due do the linear structure of the standard LP problem (1.8), the feasible region P, defined by (1.9), has special features. In this chapter, we explore this theme from a geometric point of view.

Denote the n-dimensional Euclidean space by $\mathcal{R}^n$, as usual. Points or column vectors are the basic geometrical elements. Denote a point, or the vector from the origin to this point by

$$x = (x_1, \ldots, x_n)^T \in \mathcal{R}^n,$$

whose components (coordinates) are $x_1, \ldots, x_n$. Thereafter, points and vectors will not be distinguished. Denote by e_j, $j = 1, \ldots, n$ the jth coordinate vector, i.e., the unit vector with its jth component 1, and denote by e the vector of all ones. The reader is referred to related literature for basic concepts and operations in Euclidean space, such as linear dependency and independency, set of points and its boundedness and unboundedness, the inner product $x^T y$ of vectors $x, y \in \mathcal{R}^n$, $\cos < x, y >= x^T y/(\|x\| \|y\|)$ of the angle between them, their orthogonality $x \perp y$ or $x^T y = 0$, i.e., $< x, y >= \pi/2$, the Euclidean module or norm $\|x\| = \sqrt{x^T x}$ of vector x, and so on.

Involved in the standard LP problem, both c and x may be viewed as vectors in $\mathcal{R}^n$, columns a_j, $j = 1, \ldots, n$ of A as vectors in $\mathcal{R}^m$, b as a vector in $\mathcal{R}^m$, and the feasible region P as a closed polyhedral in $\mathcal{R}^n$ in general (as will be clear a little later), though P could be degenerate, or even empty. The following lemma, proposed by Farkas (1902), renders a sufficient and necessary condition for nonempty P (the proof is delayed to at the end of Sect. 4.2).

Lemma 2.1 (Farkas). *Assume $A \in \mathcal{R}^{m\times n}$ and $b \in \mathcal{R}^m$. The feasible region P is nonempty if and only if*

$$b^T y \geq 0, \qquad \forall\, y \in \{y \in \mathcal{R}^m \mid A^T y \geq 0\}.$$

P.-Q. PAN, *Linear Programming Computation*, DOI 10.1007/978-3-642-40754-3_2,

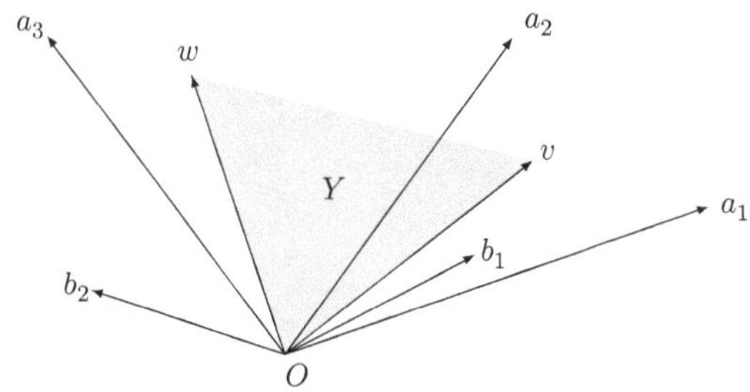

Fig. 2.1 A geometrical explanation for Farkas lemma

Fig. 2.1 serves as a geometric explanation for the preceding Lemma, where a_1, a_2, a_3 are columns of $A \in \mathcal{R}^{2\times3}$. Y is the set of all vectors which forms with every column of A an angle no more than $\pi/2$ (corresponding to the shaded area between vectors v and w in the figure). b_1 forms with each $y \in Y$ an angle no more than $\pi/2$ but b_2 does not. Thereby, P is nonempty when $b = b_1$, whereas empty when $b = b_2$.

In addition to rank $A = m$, discussions in this chapter will be based on the following.

Assumption. The feasible region P is nonempty and infinite.

2.1 Polyhedral Convex Set and the Feasible Region

For any given two points $x, y \in \mathcal{R}^n$, set

$$S = \{\alpha x + (1 - \alpha)y \mid \alpha \in \mathcal{R}\}$$

is a *straight line*; if $0 < \alpha < 1$, it is an *open segment* with end points x and y, denoted by (x, y); if $0 \leq \alpha \leq 1$, it is a *closed segment*, denoted by $[x, y]$. Hereafter so-called "segment" will be all closed.

Definition 2.1.1. Π is an affine set if, whenever it includes any two points, it includes the whole straight line passing through them. The smallest affine set including a set is the affine hull of the latter.

Straight lines in $\mathcal{R}^2$ and planes in $\mathcal{R}^3$ are instances of affine sets. The whole space $\mathcal{R}^n$ is an affine set. An empty set and a single point set are viewed as affine sets. It is clear that the intersection of affine sets is an affine set.

For any given $\alpha_i,\ i = 1, \ldots, k$ satisfying $\sum_{i=1}^k \alpha_i = 1$, point

$$x = \sum_{i=1}^{k} \alpha_i x^i$$

is called an *affine combination* of $x^1, \ldots, x^k$. It is easy to show that the set of all such affine combinations, i.e.,

$$\left\{\sum_{i=1}^{k} \alpha_k x^i \mid \sum_{i=1}^{k} \alpha_i = 1,\ \alpha_i \in \mathcal{R},\ i = 1, \ldots, k\right\}$$

is an affine set, called *affine hull* of these points. The two points in Definition 2.1.1 can be generalized to multiple points: it is easy to show that Π is an affine set if and only if the affine hull of any finitely many points within Π belongs to Π.

Set L is a *subspace* of $\mathcal{R}^n$ if it is closed for all linear operations, that is, for any $x, y \in L$ and $\alpha, \beta \in \mathcal{R}$ it holds that $\alpha x + \beta y \in L$. An affine set is an affine subspace if it is a subspace.

Theorem 2.1.1. *An affine set is an affine subspace if and only if it includes the origin.*

Proof. The necessity is clear. Sufficiency. Let Π be an affine set including the origin. Then for any $x \in \Pi$ and $\alpha \in \mathcal{R}$, it holds that

$$\alpha x = \alpha x + (1 - \alpha)0 \in \Pi.$$

On the other hand, it holds for any $x, y \in \Pi$ that

$$\frac{x + y}{2} = \frac{1}{2}x + (1 - \frac{1}{2})y \in \Pi.$$

Therefore, Π is closed for linear operations, and is thus an affine subspace. □

Theorem 2.1.2. *For any nonempty affine set Π, there exists vector p so that*

$$L = \{x + p \mid x \in \Pi\}$$

is an affine subspace, and such subspace is unique.

Proof. According to Theorem 2.1.1, Π is an affine subspace if $0 \in \Pi$. Note that $L = \Pi$ corresponds to $p = 0$. Assume that $0 \notin \Pi$. Since $\Pi \neq \emptyset$, there exists $0 \neq y \in \Pi$. Letting $p = -y$, it is clear that

$$L = \{x + p \mid x \in \Pi\}$$

is an affine set including the origin, and is hence an affine subspace.

Now let us show the uniqueness. Assume that L_1, L_2 are affine subspaces such that

$$L_1 = \{y + p_1 \mid y \in \Pi\}, \qquad L_2 = \{y + p_2 \mid y \in \Pi\}.$$

It is clear that

$$\Pi = \{x - p_1 \mid x \in L_1\}.$$

If $p = -p_1 + p_2$, therefore, it holds that

$$L_2 = \{x + p \mid x \in L_1\},$$

from which it follows that $x + p \in L_2$ for any $x \in L_1$. Since $0 \in L_1$, $p \in L_2$ holds. Further, since L_2 is a subspace, $x = (x + p) - p \in L_2$ holds. Therefore, $L_1 \subset L_2$. $L_2 \subset L_1$ can be similarly derived. So it can be asserted that $L_1 = L_2$. □

Geometrically, the affine subspace L may be viewed as a parallelism of affine set Π along vector p. It is therefore called *parallel subspace* of Π. The dimension of L is said to be that of Π, which is equal to the number of independent components (coordinates) of elements in Π. It is clear that an affine set with one or more than one dimension is unbounded.

Let a be a nonzero vector and let η be a real number. Set

$$H = \{x \in \mathcal{R}^n \mid a^T x = \eta\}$$

is called *superplane*, whose normal vector is a ($a \perp H$); in fact, for any two points $x, y \in H$, it holds that

$$a^T(x - y) = a^T x - a^T y = \eta - \eta = 0.$$

It is easy to show that any superplane is an affine set.

Any straight line in $\mathcal{R}^2$ and any plane in $\mathcal{R}^3$ are instances of superplane.

The "signed" distance from any point $\bar{x}$ to superplane H is defined by $r/\|a\|$, where r is the residual

$$r = a^T \bar{x} - \eta.$$

If $r = 0$, point $\bar{x}$ is within H. It might be well to assume $\eta > 0$. Then, if $r < 0$, the origin and $\bar{x}$ are in the same side of H; if $r > 0$, the two points are in different sides of H.

Superplanes associated with the objective function are of significance to LP. Regarding objective value f as a parameter, the sets

$$H(f) = \{x \in \mathcal{R}^n \mid c^T x = f\}$$

are a family of contour surfaces of the objective function. The gradient $\nabla f = c$ of the objective function is the common normal vector of all the contour surfaces, pointing to the increasing side of objective value f.

The following gives a mathematical expression of an affine set.

Theorem 2.1.3. *Set Π is an affine set if and only if there exists $W \in \mathcal{R}^{k\times n}$ and $h \in \mathcal{R}^k$ such that*

$$\Pi = \{x \mid Wx = h\}. \tag{2.1}$$

Proof. It might be well to rule out the trivial case when Π is empty or the whole space.

Sufficiency. Let Π is defined by (2.1). For any $x, y \in \Pi$ and $\alpha \in R^1$, it holds that

$$W(\alpha x + (1-\alpha)y) = \alpha Wx + (1-\alpha)Wy = \alpha h + (1-\alpha)h = h,$$

leading to $\alpha x + (1-\alpha)y \in \Pi$. Therefore, Π is an affine set.

Necessity. Let Π be an affine set. Assume that L is a parallel affine subspace, and $w^1, \ldots, w^k$ are basis of the orthogonal complementary space of it. Then, it follows that

$$L = \{y \mid (w^i)^T y = 0,\ i = 1, \ldots, k\} \triangleq \{y \mid Wy = 0\},$$

where rows of $W \in \mathcal{R}^{k\times n}$ are $(w^1)^T, \ldots, (w^k)^T$. Introduce notation $h = Wp$. Since L is a parallel subspace of Π, there exists a vector p such that

$$L = \{x - p \mid x \in \Pi\}.$$

Thus, for any $x \in \Pi$, it holds that $x - p \in L$; and $W(x - p) = 0$ means that $x \in \{x \mid Wx = h\}$. If $x \in \{x \mid Wx = h\}$, conversely, then $Wx - Wp = W(x - p) = 0$, hence from $x - p \in L$ it follows that $x \in \Pi$. So, Π has expression (2.1). □

The preceding Theorem says that a set is affine set if and only if it is the intersection of finitely many superplanes. It is easy to show that the dimension of the affine set is equal to $n - \text{rank}(W)$. In particular, the solution set

$$\Delta = \{x \mid Ax = b\}$$

of the constraint system of the standard LP problem is an affine set. Since rank $A = m$, $\Delta \neq \emptyset$ and dim $\Delta = n - m$.

Definition 2.1.2. C is a convex set if it includes the whole segment whenever it includes its two end points. The smallest convex set including a set is the convex hull of the latter.

Any disks and the first quadrant in $\mathcal{R}^2$ and spheres in $\mathcal{R}^3$ are instances of convex sets. Clearly, segments and the whole space are convex sets too. Empty sets and single point sets are regarded as convex sets. Intersections of convex sets are convex. Any affine set is convex, but a convex set is not an affine set in general; e.g., disks and spheres are not affine sets. It is clear that any convex set has an affine hull. The dimension of the latter is said to be the dimension of the former. Hereafter, so-called "convex set" is a closed convex set.

For any $\alpha_i \geq 0,\ i = 1, \ldots, k$ satisfying $\sum_{i=1} \alpha_i = 1$, the point

$$x = \sum_{i=1}^{k} \alpha_k x^i$$

is called *convex combination* of points $x^1, \ldots, x^k$. It is easy to show that the set of all such convex combinations, i.e.,

$$\left\{ \sum_{i=1}^{k} \alpha_k x^i \mid \sum_{i=1}^{k} \alpha_i = 1; \alpha_i \geq 0,\ i = 1, \ldots, k \right\}$$

is the convex hull of these points. It is easy to show that C is a convex set if and only if the convex hull of any finitely many points within C belongs C.

Any superplane H divides the whole space to two *closed half spaces*, i.e.,

$$H_L = \{x \mid a^T x \leq \eta\}, \quad \text{and} \quad H_R = \{x \mid a^T x \geq \eta\}. \tag{2.2}$$

The intersection of infinitely many closed half spaces is called *polyhedral*, and bounded polyhedral called *polyhedron*. It could degenerate to a segment or point, or even empty set.

It is clear that a half space is convex. Therefore, a polyhedral or polyhedron is convex, as termed *polyhedral convex set*.

A convex set C is called *polyhedral convex cone* if $\alpha x \in C$ holds for any $x \in C$ and $\alpha \geq 0$. It is easy to show that a set is a polyhedral convex cone if and if it is the intersection of finitely many closed half spaces passing through the origin, as expressed $\{x \in \mathcal{R}^n \mid Ax \geq 0\}$.

The nonnegative constraints $x \geq 0$ in the standard problem correspond to the positive octant, which is the intersection of the n closed half spaces with the coordinate planes as its boundary. Therefore, the feasible region P is the intersection of the affine set Δ and the positive octant. As any superplane $a^T x = \eta$ may be viewed as the intersection of two closed half spaces (2.2), P may be also viewed as the intersection of finitely many closed half spaces. Therefore, we make the following statement.

Proposition 2.1.1. *The feasible region P is a polyhedral convex set.*

Such a set could be degenerate, however. The following result concerns the dimension of P.

Proposition 2.1.2. *Define index set*

$$J' = \{j \in A \mid x_j = \mu_j,\ x \in P\}, \tag{2.3}$$

where $\mu_j,\ j \in A$ *are nonnegative constants and denote by* $I_{J'}$ *the coefficient matrix of system* $x_j = \mu_j,\ j \in J'$*. Then it holds that*

$$n - \min\{m + |J'|, n\} \le \dim P = n - r \le n - \max\{m, |J'|\}, \tag{2.4}$$

where

$$r = \text{rank}\begin{pmatrix} A \\ I_{J'} \end{pmatrix}.$$

Proof. Note that rank $A = m$ and P is a nonempty infinite set, according to the basic assumption.

It is clear that $|J'| \le r$ and

$$P = \{x \in \mathcal{R}^n \mid Ax = b,\ x \ge 0;\ x_j = \mu_j,\ j \in J'\}.$$

IF $|J'| = r$, then $x_j = \mu_j,\ j \in J'$ is a canonical form of system

$$Ax = b; \qquad x_j = \mu_j, \quad j \in J'.$$

If $|J'| < r$, besides $x_j\ j \in J'$, there are additional $r - |J'|$ basic variables, and hence a canonical form of the preceding system. It is to say that there exists a canonical form, whose $n - r$ nonbasic variables do not belong to J'. Therefore, it holds that $\dim P = n - r$. Hence (2.4) follows from $\min\{m + |J'|, n\} \ge r \ge \max\{m, |J'|\}$. □

In particular, $\dim P = n - m$ if $J' = \emptyset$.

The special case of $\mu_j = 0$ is of significance to the standard LP problem. Introduce sets

$$J = \{j \in A \mid x_j = 0,\ x \in P\}, \qquad \bar{J} = A \backslash J, \tag{2.5}$$

where J is said to be *index set of zero components.*

According to the preceding definition, it is clear that $P \subset P'$, and in addition

$$x_j = 0, \qquad \forall\, j \in J,\ x \in P. \tag{2.6}$$

If $x \in P$ and $x_j > 0$, then $j \in \bar{J}$.

If $J \ne \emptyset$, the feasible region P has no interior point in the normal sense, as is often the case for real problems. For convenience of applications, the following concept is introduced instead.

Definition 2.1.3. Assume $\bar{x} \in P$. If there exists $\delta > 0$ such that a neighborhood of $\bar{x}$ is included in P, i.e.,

$$\Omega(\delta) = \{x \in \Delta \mid x_j = 0,\ j \in J;\ \|x - \bar{x}\| < \delta\} \subset P, \tag{2.7}$$

then $\bar{x}$ is an *interior point* of P; otherwise, it is a *boundary point.*

The set of all interior points of P is called its *interior*, denoted by int P. P and its interior have the same dimension.

For sake of distinguishing, the point is said to be *relative interior point* when $J \neq \emptyset$. The set of relative interior points is *relative interior*, while *(strict) interior point* or *interior* stands for the case of $J = \emptyset$.

Theorem 2.1.4. *Assume $\bar{x} \in P$. Then $\bar{x} \in \text{int } P$ if and only if*

$$\bar{x}_j > 0, \qquad j \in \bar{J}. \tag{2.8}$$

Proof. Sufficiency. Let point $\bar{x} \in P$ satisfy (2.8). Using

$$\delta = \min_{j \in \bar{J}} \bar{x}_j > 0, \tag{2.9}$$

and

$$\Omega(\delta) = \{x \in \Delta \mid x_j = 0, \ j \in J; \ \|x - \bar{x}\| < \delta\},$$

then for any $x \in \Omega(\delta)$, it holds that

$$x \in \Delta \qquad \text{and} \qquad x_j = 0, \ j \in J.$$

Moreover, sine

$$\|x - \bar{x}\| = \sqrt{\sum_{j=1}^{n} (x_j - \bar{x}_j)^2} < \delta,$$

it holds that

$$|\bar{x}_j - x_j| \leq \|x - \bar{x}\| < \delta, \quad j \in \bar{J},$$

which and (2.9) together give

$$x_j > \bar{x}_j - \delta \geq 0, \quad j \in \bar{J}.$$

Thus $x \in P$. Therefore $\Omega(\delta) \subset P$, and $\bar{x}$ is an interior point of P.

Necessity. Assuming $\bar{x} \in \text{int } P$, then there is $\delta > 0$ such that $\Omega(\delta) \subset P$; also there is $p \in \bar{J}$ such that $\bar{x}_p = 0$. According to the definition of $\bar{J}$, there is $x' \in P$ such that $x'_p > 0$. It is clear that for any $\alpha > 0$ there exists

$$x = -\alpha x' + (1 + \alpha)\bar{x} = \bar{x} + \alpha(\bar{x} - x') \in \Delta. \tag{2.10}$$

Hence, when α is sufficiently small, it holds that

$$\|x - \bar{x}\| = \alpha\|\bar{x} - x'\| < \delta.$$

In addition, it is clear that $x'_j,\ \bar{x}_j = 0,\ j \in J$, and hence $x_j = 0,\ j \in J$. Therefore $x \in \Omega(\delta)$. On the other hand, from (2.10), $\bar{x}_p = 0$ and $\alpha > 0,\ x'_p > 0$ it follows that

$$x_p = -\alpha x'_p + (1+\alpha)\bar{x}_p = -\alpha x'_p < 0,$$

which contradicts $\Omega(\delta) \subset P$. Therefore, (2.8) holds if $\bar{x} \in \text{int } P$. □

Proposition 2.1.3. *If* $\dim P \geq 1$, *then* P *has a relative interior point.*

Proof. The assumption of the proposition implies $\bar{J} \neq \emptyset$, because otherwise it holds that $J = A$ and hence $\dim P = 0$, leading to contradiction. On the other hand, for any $j \in \bar{J}$ there exists $x \in P$ such that $x_j > 0$; hence from the convexity of P, it follows that there is $x \in P$ satisfying $x_j > 0,\ j \in \bar{J}$. According to Theorem 2.1.4, it is known that $x \in \text{int } P \neq \emptyset$. □

Example 2.1.1. Investigate the interior of the feasible region of the following problem:

$$\begin{aligned} \min\ & x_1 + x_2 + x_3 + x_4 + x_5, \\ \text{s.t.}\ & x_1 \qquad\quad - x_3 + x_4 + x_5 = 6, \\ & \qquad x_2 - x_3 - x_4 - x_5 = 0, \\ & \quad -x_2 + 3x_3 + x_4 + x_5 = 0, \\ & \quad x_j \geq 0,\ j = 1, \ldots, 5. \end{aligned}$$

Answer Adding the second constraint equality to the third gives

$$2x_3 = 0.$$

It is clear that the feasible region is nonempty, and the x_3 component of all feasible points equals 0. Eliminating x_3 from the problem leads to

$$\begin{aligned} \min\ & x_1 + x_2 + x_4 + x_5, \\ \text{s.t.}\ & x_1 \qquad + x_4 + x_5 = 6, \\ & \qquad x_2 - x_4 - x_5 = 0, \\ & x_j \geq 0,\ j = 1, 2, 4, 5, \end{aligned}$$

the interior of the feasible region of which is clearly nonempty, corresponding to the relative interior of the feasible region of the original problem.

2.2 Geometric Structure of the Feasible Region

This section will discuss the geometric structure of the feasible region P. Actually, most results in this section are also valid for general polyhedral convex sets.

Definition 2.2.1. Let P' be a nonempty convex subset of P. It is a face of P if for any $x \in P'$, satisfying $x \in (y, z) \subset P$, it holds that $y, z \subset P'$.

The preceding means that a face includes the whole segment if it includes a interior point of any segment of P.

The following renders a mathematical expression of P.

Theorem 2.2.1. *A nonempty convex subset $P(Q)$ of P is its face if and only if there exists index set $Q \subset A$ such that*

$$P(Q) = \{x \in \mathcal{R}^n \mid Ax = b,\ x \geq 0;\ x_j = 0,\ j \in Q\}. \tag{2.11}$$

Proof. Sufficiency. Let $\emptyset \neq P(Q) \subset P$ be defined by (2.11). If $v \in P(Q)$ is an interior point of segment (y, z) and $y, z \in P$, then there exists $0 < \alpha < 1$ such that

$$v = \alpha y + (1 - \alpha) z.$$

Hence from $\alpha > 0,\ 1 - \alpha > 0,\ y, z \geq 0$ and

$$v_j = \alpha y_j + (1 - \alpha) z_j = 0, \qquad j \in Q,$$

it follows that

$$y_j, z_j = 0, \qquad j \in Q,$$

Therefore, $y, z \in P(Q)$, and hence $P(Q)$ is a face of P.

Necessity. Let $P(Q) \neq \emptyset$ be a face of P. Introduce notation

$$Q = \{j \in A \mid x_j = 0,\ x \in P(Q)\}. \tag{2.12}$$

Now we show that $P(Q)$ is equivalent to

$$P(Q)' = \{x \in \mathcal{R}^n \mid Ax = b,\ x \geq 0;\ x_j = 0,\ j \in Q\}. \tag{2.13}$$

It is clear that $P(Q) \subset P(Q)' \subset P$.

If $P(Q)$ includes the origin 0 only, it follows that $b = 0$ and $Q = A$. And $P(Q)'$ clearly includes 0, hence $P(Q) = P(Q)'$. Now assuming that $P(Q)$ does not include 0, we will show $P(Q)' \subset P(Q)$.

Assume $x \in P(Q)'$. If $x = 0$ $(b = 0)$, then for any $0 \neq v \in P(Q)$ and

$$y = 2v$$

it holds that

$$v = \frac{y}{2} + \frac{0}{2},$$

which means that $v \in (y, 0)$. Since $P(Q)$ is a face and $y, 0 \in P$, we have $x = 0 \in P(Q)$. On the other hand, if $x \neq 0$, i.e.,

$$S = \{j \in A \mid x_j > 0\}$$

is nonempty, then from $x \in P(Q)'$ and (2.13), it is known that $j \notin Q$ for any $j \in S$. Therefore there exists $u \in P(Q)$ such that $u_j > 0$. If S includes two or more indices, then $u, w \in P(Q)$ satisfy $u_i, w_j > 0$ for any $i, j \in S$, hence $z = u/2 + w/2 \in P(Q)$ satisfies $z_i, z_j > 0$. This means that there exists $v \in P(Q)$ such that $v_j > 0$ for all $j \in S$. As for the relation between x and v, there are the following two cases only:

(i) $\{j \in S \mid x_j > v_j\} = \emptyset$. It is clear that

$$z = 2v - x = v + (v - x) \in P.$$

Since $P(Q)$ is a face of P, and $v = z/2 + x/2 \in (x, z)$, it holds that $x \in P(Q)$.

(ii) $\{j \in S \mid x_j > z_j\} \neq \emptyset$. Define

$$z = x + \beta(v - x), \qquad \beta = \min\{x_j/(x_j - v_j) \mid x_j - v_j > 0, j \in S\} > 1.$$

It is easy to verify that $z \in P$, and

$$v = \alpha z + (1 - \alpha)x \in (x, z), \qquad 0 < \alpha = 1/\beta < 1.$$

In addition, $P(Q)$ is a face of P, hence $x \in P(Q)$. Therefore $P(Q)' \subset P(Q)$. □

Clearly, P itself is a face ($Q = \emptyset$). If a face $P(Q) \neq P$, it is called a *proper face*. It is easy to show that face $P(Q)$ is a proper face if and only if dim $P(Q) <$ dim P. From the proof of Proposition 2.1.2, it is know that dim $P(Q) \leq n - \max\{m, |Q|\}$. Face of face is a face.

If dim $P(Q) =$ dim $P - 1$, face $P(Q)$ is called *facet* of P. An 1-dimensional face is also called *edge*; an 0-dimensional face is called *vertex* or *extreme point*.

It is clear that the feasible region P has infinitely many faces. In fact, it is known that the number of $(n - m - k)$-dimensional faces of P is no more than C_n^k ($k = 1, \ldots, n - m$); in particular, there exist, at most, an $(n - m)$-dimensional face (that is P itself), C_n^{m+1} edges and $C_n^{n-m} = C_n^m$ vertices.

"Vertex" can also be defined by the following alternatively.

Definition 2.2.2. x is a vertex of P if $x \in [y, z]$ leads to $x = y$ or $x = z$ for any $y, z \in P$.

The preceding implies that a vertex is not an interior point of any segment of P. It is clear that a vertex of face is a vertex of P, and that the origin is a vertex if it belongs to P.

Vertex has its distinctive algebraical attribute.

Lemma 2.2.1. *A point $x \in P$ is a vertex if and only if columns of A corresponding to its positive components are linearly independent.*

Proof. It might be well to let the first s components of x be great than 0 and let the rest be 0. Assume that $\bar{x}$ is the subvector consisting of the first s components of x and $\bar{A}$ is the submatrix consisting of the first s columns of A. Then $\bar{A}\bar{x} = b$.

Necessity. Let x be a vertex of P. If columns of $\bar{A}$ are linearly dependent, there is a nonzero vector $\bar{v}$ such that $\bar{A}\bar{v} = 0$. Introduce notation

$$\bar{y} = \bar{x} + \alpha\bar{v}, \qquad \bar{z} = \bar{x} - \alpha\bar{v}.$$

It is clear that for any real α it holds that

$$\bar{A}\bar{y} = \bar{A}\bar{z} = b.$$

Take sufficiently small $\alpha > 0$, such that $\bar{y}, \bar{z} \geq 0$. Construct vectors y and z, so that the first s components of them respectively constitute $\bar{y}$ and $\bar{z}$, and the others are 0. Then it is clear that $y, z \in P$ and $x = y/2 + z/2$. Thus x is not a vertex of P, as leads to a contradiction. Therefore, columns corresponding to all positive components of x are linear independent if x is a vertex of P.

Sufficiency. Assume that columns of $\bar{A}$ are linearly independent. If $x \in P$ is not a vertex, there are two points $y, z \in P$ and a real $\alpha \in (0, 1)$ such that

$$x = \alpha y + (1 - \alpha)z,$$

from which it is known that the last $n - s$ components of y and z are both 0. Therefore, $v = x - y \neq 0$ and

$$\bar{A}\bar{v} = Av = Ax - Ay = b - b = 0,$$

which means that columns of $\bar{A}$ are linearly dependent, as a contradiction. Therefore x is a vertex if columns of $\bar{A}$ are linearly independent. □

In view of rank$(A) = m$, the preceding theorem implies that the number of positive components of a vertex of P is no more than m.

Lemma 2.2.2. *$\bar{x}$ is a vertex of the feasible region if and only if it is a basic feasible solution.*

Proof. It is from Theorem 1.6.2 and Lemma 2.2.1. □

The preceding Lemma says that vertex and basic feasible solution of P are the same; thus, the two may be regarded as geometrical and algebraical names of the same mathematical item. It is usually difficult to determine if a point is a vertex based on the definition itself, while a basic feasible solution can be conveniently determined algebraically. Recall that every canonical form of the constraint system $Ax = b$ corresponds to a basic solution, which is a basic feasible solution if it is nonnegative (Sect. 1.6).

Lemma 2.2.3. *Nonempty feasible region has a vertex.*

Proof. It is clearly the case when P is a single point set. Let P be infinite set, and $\bar{x} \in P$. If $\bar{x}$ is not a vertex, there are two distinct points $y, z \in P$ and a real $\bar{\alpha} \in (0, 1)$ such that

$$\bar{x} = \bar{\alpha} y + (1 - \bar{\alpha})z = z + \bar{\alpha}(y - z).$$

Thus, a component of $\bar{x}$ is 0 if and only if the corresponding components of both y, z are 0. Introduce

$$T = \{j \in A \mid \bar{x}_j > 0\}.$$

Then $T \neq \emptyset$, since $T = \emptyset$ implies that $\bar{x} = 0$ is a vertex. It might be well to assume that for some $i \in \{1, \ldots, n\}$ it holds that $z_i > y_i$, and hence $i \in T$. This means that

$$\{j \in T \mid z_j - y_j > 0\} \neq \emptyset.$$

It is easy to show that redefined

$$\bar{x} = \alpha_1 y + (1 - \alpha_1)z, \quad \alpha_1 = z_q/(z_q - y_q) = \min\{z_j/(z_j - y_j) \mid z_j - y_j > 0,\ j \in T\}$$

satisfies $\bar{x} \in P$ and $\bar{x}_q = 0$. Thus $|T|$ for the new $\bar{x}$ is less than that for the old by 1, at least. Repeating no more than n times, therefore, the preceding process terminates at a vertex. □

The above proof produces a series of feasible points, corresponding to faces, each of which is a proper face of its predecessor, until reaching a 0-dimensional face (vertex). Such a technique for shifting to faces of lower dimensions will often be used.

For any given point x and vector $d \neq 0$, set $\{x + \alpha d \mid \alpha \geq 0\}$ is said to be *ray (or half-line)*, emanating from x along the direction of d. It is clear that a ray is an infinite set.

Definition 2.2.3. A nonzero vector d is an unbounded direction if P includes rays, emanating from all $x \in P$ along d.

Two unbounded directions having the same direction are regarded as the same.

Theorem 2.2.2. *Vector d is an unbounded direction of the nonempty feasible region if and only if*

$$Ad = 0, \qquad d \neq 0, \qquad d \geq 0. \tag{2.14}$$

Proof. Sufficiency. With the assumptions, it is easy to verify that for any given $x \in P$ and $\alpha \geq 0$ it holds that $x + \alpha d \in P$, therefore $d \neq 0$ is an unbounded direction.

Necessity. Let $d \neq 0$ be an unbounded direction. Thus, there is $x \in P$, satisfying $x + \alpha d \in P$ for any $\alpha \geq 0$. Hence, from $Ax = b$ and $A(x + \alpha d) = b$, it follows that $Ad = 0$. In addition, $d \geq 0$ holds, because, otherwise, d has a negative component, and hence the corresponding component of $x + \alpha d$ is negative whenever $\alpha > 0$ becomes sufficiently large, as contradicts $x + \alpha d \in P$. □

Corollary 2.2.1. *A nonzero vector d is a unbounded direction if P includes the ray, emanating from some $x \in P$ along d.*

It is clear that any nonnegative linear combination of finitely many unbounded directions is an unbounded direction if the combination coefficients are not all zero. Note that "unbounded direction" is meaningless to an empty P.

Theorem 2.2.3. *The feasible region is unbounded if and only if it has an unbounded direction.*

Proof. Sufficiency is clear, it is only needed to show necessity.

If $v \in P$, then the translation of P, i.e.,

$$C = \{x - v \mid x \in P\}$$

clearly includes the origin, and P is unbounded if and only if C is unbounded. It might be well to assume that $0 \in P$.

Let $S' = \{x^k\} \in P$ be an unbounded sequence of points. Without loss of generality, assume that

$$\|x^k\| \to \infty \quad \text{as} \quad k \to \infty.$$

Then the sequence

$$S'' = \{x^k / \|x^k\|\}$$

on the unit sphere is bounded, hence has a cluster point. Letting x be its cluster point, then S'' includes a subsequence converging to x. It might be well to assume

$$x^k / \|x^k\| \to x.$$

Now it should be shown that x is an unbounded direction of P. Let M be a given positive number. Since $\|x^k\| \to \infty$, there is a positive integer K such that $\|x^k\| > M$ or $M / \|x^k\| < 1$ when $k \geq K$. Introduce

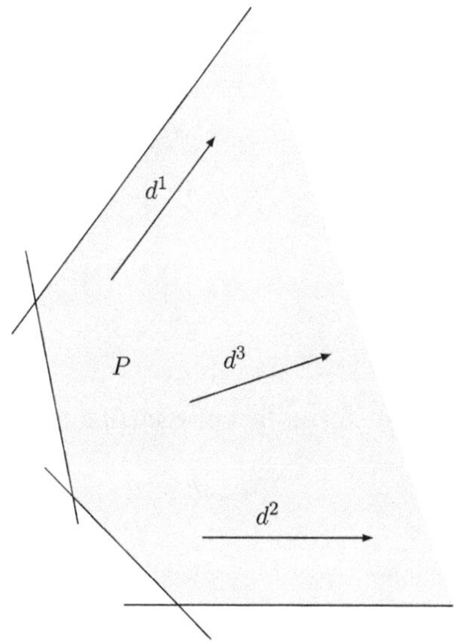

Fig. 2.2 d^1, d^2, d^3 are unbounded directions. d^1, d^2 are extreme directions, but d^3 not

$$y^k = (M/\|x^k\|)x^k, \qquad k = K, K+1, \ldots.$$

Since P is convex, we have $0, x^k \in P$ and $y^k \in (0, x^k)$, and hence $y^k \in P$ and $y^k \to Mx$ when $k \geq K$. As P is closed, it can be asserted that $Mx \in P$. □

Definition 2.2.4. An unbounded direction of P is extreme direction if it can not be expressed by a positive linear combination of two distinct unbounded directions.

According to the preceding, that d is an extreme direction means that if there are unbounded directions d', d'' and positive number $\sigma_1, \sigma_2 > 0$, satisfying $d = \sigma_1 d' + \sigma_2 d''$, then there must be $d' = \sigma d'$, where $\sigma > 0$. Two extreme directions having the same direction are regarded as the same.

In Fig. 2.2, d^1, d^2, d^3 are unbounded directions of P. d^1 and d^2 are extreme directions, but d^3 is not.

Theorem 2.2.4. *An unbounded direction is extreme direction if and only if the rank of columns, corresponding to its positive components, is less than the number of columns by* 1.

Proof. It might be well to assume that k positive components of unbounded direction d correspond to the set of columns $a_1, \ldots, a_k$. The satisfaction of $Ad = 0$ implies that the columns are linearly dependent. Denote the rank of the set by is r, then it is clear that $r < k$. Without loss of generality, assume that the first r columns are linear independent. Introduce $B = (a_1, \ldots, a_r)$. It is clear that

$$\text{rank } B = r \leq \text{rank } A = m.$$

Note that $k \geq 2$, because if $k = 1$, otherwise, then from $Ad = 0$ it follows that $a_1 = 0$, as leads to a contradiction.

Necessity. Assume that d is an extreme direction, but $r \neq k-1$, that is, $a_1, \ldots, a_{k-1}$ are linearly dependent. Thus, there exists point

$$y = (y_1, \ldots, y_{k-1}, 0, \ldots, 0)^T \neq 0$$

such that

$$Ay = \sum_{j=1}^{k-1} y_j a_j = 0.$$

Clearly, for sufficiently small $\delta > 0$, it holds that

$$0 \neq d' = d + \delta y \geq 0, \qquad 0 \neq d'' = d - \delta y \geq 0,$$

and $Ad' = Ad'' = 0$. Therefore, d', d'' are unbounded directions of P, and hence not of the same direction. But $d = (d' + d'')/2$, as contradicts that d is an extreme direction. It therefore holds that $r = k - 1$.

Sufficiency. Assume $r = k - 1$. If there exist unbounded directions d', d'' and $\sigma_1, \sigma_2 > 0$ such that

$$d = \sigma_1 d' + \sigma_2 d'',$$

then zero components of d clearly correspond to zero components of d' and d''. So, the last $n - k$ components of d' and d'' are all zero. In addition, since d', d'' are unbounded directions, it holds that $Ad' = Ad'' = 0$ (Theorem 2.2.2), and hence that

$$Bd'_B + d'_k a_k = 0, \qquad Bd''_B + d''_k a_k = 0.$$

Note that $d'_k, d''_k > 0$; because if $d'_k = 0$, otherwise, then $d'_B = 0$, and hence $d' = 0$, as is a contradiction. Premultiplying the two sides of the preceding two equalities by B^T gives

$$B^T B d'_B + d'_k B^T a_k = 0, \qquad B^T B d''_B + d''_k B^T a_k = 0,$$

from which it follows that

$$d'_B = -d'_k (B^T B)^{-1} B^T a_k, \qquad d''_B = -d''_k (B^T B)^{-1} B^T a_k.$$

Therefore, it holds that

$$d'' = (d''_k / d'_k) d',$$

which means that unbounded directions d' and d'' have the same direction, and hence d is an extreme direction. □

The extreme direction and 1-dimensional face (edge) have close relationship.

Theorem 2.2.5. *A vector is an extreme direction if and only if it is the unbounded direction of a edge.*

Proof. Necessity. Assume that d is an extreme direction. Based on Theorem 2.2.4, assume that columns corresponding to its positive components are

$$a_1, \ldots, a_r, a_{m+1},$$

where the first $r \leq m$ columns are linearly independent. Thus

$$d_1, \ldots, d_r, d_{m+1} > 0, \qquad d_{r+1}, \ldots, d_m, d_{m+2}, \ldots, d_n = 0. \tag{2.15}$$

Since the rank of A is m, there are $m - r$ columns which together with $a_1, \ldots, a_r$ form a basis when $r < m$. Without loss of generality, assume that the first m columns of A constitute a basis, i.e.,

$$B = (a_1, \ldots, a_r, a_{r+1}, \ldots, a_m), \qquad N = \{a_{m+1}, \ldots, a_n\}. \tag{2.16}$$

From (2.15) and $Ad = 0$, it is follows that

$$\sum_{i=1}^{r} d_i a_i + d_{m+1} a_{m+1} = 0,$$

hence

$$a_{m+1} = -\sum_{i=1}^{r} (d_i / d_{m+1}) a_i.$$

Assume that basis B corresponds to the canonical form below:

$$x_B = \bar{b} - \bar{N} x_N. \tag{2.17}$$

Since its augmented matrix $(I \;\; \bar{N} \mid \bar{b})$ comes from $(A \mid b) = (B \;\; N \mid b)$ by elementary transformations, and a_{m+1} is a linear combination of $a_1, \ldots, a_r$, the last $m - r$ components of column $\bar{a}_{m+1}$ in the canonical form are all zero.

Let $\bar{x}$ belong to P. If $\bar{x}_N = 0$, then $\bar{x}$ is the basic feasible solution corresponding to basis B. Now assume, otherwise, that $\bar{x}_N \neq 0$. We will create a new basis B associated with a basic feasible solution by a series of elementary transformations and some solution updating.

Assume that $\bar{x}_j > 0$ holds for some $j \in N,\ j \neq m+1$. Reducing $\bar{x}_j$ and keep the other nonbasic components unchanged, we determine the corresponding value of $\bar{x}_B$ such that $\bar{x}$ satisfies (2.17), resulting in a new solution $\bar{x}$. For the column corresponding to $\bar{x}_j$, of the canonical form (2.17), there are only two cases arising:

(i) $\bar{a}_j \geq 0$.
It is clear that $\bar{x}_j$ may decrease to 0 and associated $\bar{x}_B$ remains nonnegative. Thus, setting $\bar{x}_j = 0$ gives a new feasible solution $\bar{x}$.

(ii) $\bar{a}_j \not\geq 0$.
If the first r components of $\bar{a}_j$ are nonnegative, one of the last $m-r$ components of $\bar{x}_B$ decreases to 0 first (so-called "blocking") as $\bar{x}_j$ decreases. Assume that the blocking is component i $(r+1 \leq i \leq m)$. Set $\bar{x}_j$ to the according value, and interchange a_j and a_i to update B and N. By exchanging their indices at the same time, the nonbasic component $\bar{x}_j$ of the new solution $\bar{x}$ becomes 0.
If some of the first r components of $\bar{a}_j$ are negative, then we can determine a $\sigma > 0$ such that the first r basic components of

$$\bar{x} := \bar{x} + \sigma d$$

are sufficiently large and all the nonbasic components remain unchanged, except for $\bar{x}_{m+1}$, such that no broking happens to the first r components of $\bar{x}_B$ as $\bar{x}_j$ decreases. If no broking happens to the last $m-r$ components either, we set $\bar{x}_j = 0$. If broking happens to component $r+1 \leq i \leq m$, we set $\bar{x}_j$ to the according value, and interchange a_j and a_i to updated B and N. Consequently, by exchanging their indices, the nonbasic component $\bar{x}_j$ of the new solution $\bar{x}$ is now equal to 0.

As such, we can transform all $\bar{x}_j,\ j \in N,\ j \neq m+1$ to 0, without affecting the first r indices of B. Then, if $\bar{x}_{m+1} = 0$, we are done.

If $\bar{x}_{m+1} > 0$, we reduce it and keep the other nonbasic components unchanged. Since the last $m-r$ components of $\bar{a}_{m+1}$ are zero, the corresponding components of $\bar{x}_B$ remain unchanged, and there will be only two cases arising:

(i) The first r components of $\bar{a}_{m+1}$ are all nonnegative. Then it is clear that $\bar{x}_{m+1}$ can decrease to 0 and according $\bar{x}_B$ remain nonnegative, thus we set $\bar{x}_{m+1} = 0$.
(ii) Some of the first r components of $\bar{a}_{m+1}$ are negative. If the according $\bar{x}_B$ remains nonnegative as $\bar{x}_{m+1}$ decreases to 0, we set $\bar{x}_{m+1} = 0$; otherwise, if broking happens for component $1 \leq i \leq r$ of $\bar{x}_B$, we set $\bar{x}_{m+1}$ to the associated value, and exchange a_{m+1} and a_i to update B and N. Then, by exchanging their indices, the nonbasic component $\bar{x}_{m+1}$ of the new $\bar{x}$ is equal to zero.

Therefore, it might be well to assert that the basis B, defined by (2.16), corresponds to basic feasible solution $\bar{x}$.

Consider the following face

$$\begin{aligned}P' &= \{x \in \mathcal{R}^n \mid Ax = b,\ x \geq 0,\ x_j = 0,\ j = m+2, \ldots, n\} \\ &= \{x \in \mathcal{R}^n \mid Bx_B + x_{m+1}a_{m+1} = b,\ x_B, x_{m+1} \geq 0; \\ &\quad x_j = 0,\ j = m+2, \ldots, n\} \\ &= \{x \in \mathcal{R}^n \mid x_B + x_{m+1}\bar{a}_{m+1} = \bar{b},\ x_B, x_{m+1} \geq 0; \\ &\quad x_j = 0,\ j = m+2, \ldots, n\}.\end{aligned} \tag{2.18}$$

It is clear that $\bar{x} \in P'$. Hence, from (2.15) and

$$Bd_B + d_{m+1}a_{m+1} = 0, \tag{2.19}$$

it is known that

$$\bar{x} + \alpha d \in P', \qquad \forall\, \alpha \geq 0.$$

Therefore, d is a unbounded direction of P'. It is now only needed to show dim $P' = 1$.

For any $x' \in P'$ and $x' \neq \bar{x}$, introduce $d' = x' - \bar{x}$. It is clear that

$$d'_1, \ldots, d'_r, d'_{m+1} > 0, \quad d'_{r+1}, \ldots, d'_m, d'_{m+2}, \ldots, d'_n = 0 \tag{2.20}$$

and

$$Bd'_B + d'_{m+1}a_{m+1} = 0. \tag{2.21}$$

From (2.19) and (2.21) it follows respectively that

$$d_B = -d_{m+1}B^{-1}a_{m+1}, \qquad d'_B = -d'_{m+1}B^{-1}a_{m+1},$$

Therefore, $d' = (d'_{m+1}/d_{m+1})d$, where $d'_{m+1}/d_{m+1} > 0$. This implies that dim $P' = 1$. Therefore, P' is an 1-dimensional face or edge, and d is an unbounded direction of it.

Sufficiency. Assume that d is an unbounded direction of edge P' (2.19), and hence satisfies (2.19). If d is a positive linear combination of unbounded directions d', d'' of P, then there exists a correspondence between zero components of d and of d', d'', and hence d', d'' are also unbounded directions of P'. Since dim $P' = 1$, in addition, d' and d'' have the same direction. Therefore, d is an extreme direction. □

Lemma 2.2.4. *An unbounded direction that is not an extreme one is a positive linear combination of two unparallel unbounded directions.*

Proof. Without loss of generality, assume that columns $a_1, \ldots, a_k$ correspond to positive components of an unbounded direction d, and the first r columns are linearly independent. As d is not an extreme direction, it holds by Theorem 2.2.4 that $r < k - 1$, or

$$k - r \geq 2. \tag{2.22}$$

Introduce matrix $B_1 = (a_1, \ldots, a_r)$. Within the set of $a_{k+1}, \ldots, a_n$, determine $m-r$ columns, which might be assumed to correspond to $B_1 = (a_{k+1}, \ldots, a_{k+m-r})$, to construct basis $B = (B_1, B_1)$ (it is clear that $m - r \leq n - k$). Then the nonbasic matrix is $N = (N_1, N_2)$, where $N_1 = (a_{r+1}, \ldots, a_k)$, $N_2 = (a_{k+m-r+1}, \ldots, a_n)$. Assume that B corresponds to the canonical form $x_B = \bar{b} - \bar{N}x_N$. Then d satisfies

$$B_1 d_{B_1} = -\bar{N}_1 d_{N_1}, \qquad d_{N_2} = 0.$$

or equivalently,

$$d_{B_1} = -(B_1^T B_1)^{-1} B_1^T \bar{N}_1 d_{N_1}, \qquad d_{N_2} = 0.$$

Introduce $e \in \mathcal{R}^{k-r}$ and

$$d'_{B_1} = -(B_1^T B_1)^{-1} B_1^T \bar{N}_1 (d_{N_1} + \epsilon e), \qquad d''_{B_1} = -(B_1^T B_1)^{-1} B_1^T \bar{N}_1 (d_{N_1} - \epsilon e).$$

Letting $\epsilon = (\epsilon_1, \ldots, \epsilon_{k-r})^T > 0$, then vectors

$$d' = \begin{pmatrix} d'_{B_1} \\ 0 \\ d_{N_1} + \epsilon \\ 0 \end{pmatrix}, \qquad d'' = \begin{pmatrix} d''_{B_1} \\ 0 \\ d_{N_1} - \epsilon \\ 0 \end{pmatrix}$$

satisfy $d = d'/2 + d''/2$. It is clear that $d', d'' \neq 0$ and $Ad' = 0, Ad'' = 0$. As (2.22) holds, it is known that there exists a sufficiently small ϵ such that $d', d'' \geq 0$ are unparallel unbounded directions. □

Theorem 2.2.6. *If P has an unbounded direction, it has a extreme direction.*

Proof. Let d be an unbounded direction. Assume that it has k positive components and the rank of the corresponding columns is $r \leq k - 1$. If $r = k - 1$, then d is an extreme direction (Theorem 2.2.4). Otherwise, by Lemma 2.2.4, d can be expressed as

$$d = \sigma_1 d' + \sigma_2 d'',$$

where d', d'' are unparallel unbounded directions, and $\sigma_1, \sigma_2 > 0$. Without loss of generality, assume that d'' has at least one component greater than the corresponding component of d'. Thus, the following vector

$$\hat{d} = d'' - \alpha(d'' - d'), \quad \alpha = \min\{d''_j/(d''_j - d'_j) \mid d''_j - d'_j > 0, \ j = 1, \ldots, n\} > 0.$$

is well-defined. Consider

$$A\tilde{d} = 0, \qquad \tilde{d} \geq 0, \qquad \tilde{d} \neq 0,$$

where the first two equalities clearly hold. If the third does not hold, then $d'' - \alpha(d'' - d') = 0$, as leads to

$$d' = \frac{\alpha - 1}{\alpha} d'',$$

implies that d', d'' are parallel, as a contradiction. Therefore, the third equality also holds, and hence $\tilde{d}$ is an unbounded direction. In addition, it is known that the number of zero components of $\tilde{d}$ is less than that of d by 1, at least. Then set $d = \tilde{d}$ and repeat the preceding steps. It is clear that such a process can only repeat finitely many times, and terminates at some extreme direction. □

According to Theorem 2.2.5, extreme directions and unbounded edges of P are 1-to-1 correspondent (extreme directions having the same direction are viewed as the same). As there are finitely many edges, the number of extreme directions are finite.

It is now time to lay a theoretical basis to Dantzig-Wolfe decomposition method for solving large-scale LP problems (Chap. 8).

Theorem 2.2.7 (Representation Theorem of the Feasible Region). *Let P be nonempty. Assume that $\{u^1, \ldots, u^s\}$ is the vertex set and $\{v^1, \ldots, v^t\}$ the extreme direction set. Then $x \in P$ if and only if*

$$\begin{aligned} x &= \sum_{i=1}^{s} \alpha_i u^i + \sum_{j=1}^{t} \beta_j v^j, \\ \sum_{i=1}^{s} \alpha_i &= 1, \qquad \alpha_i \geq 0, \quad i = 1, \ldots, s, \\ \beta_j &\geq 0, \qquad j = 1, \ldots, t. \end{aligned} \tag{2.23}$$

Proof. When (2.23) holds, it is clear that $x \in P$. So it is only needed to show necessity. We use inductive method to dimensions.

If dim $P = 0$, P is a single point set, including a vertex. The conclusion holds clearly. Assume that it holds for dim $P < k$. We will show that it holds for dim $P = k \geq 1$.

From Proposition 2.1.3, it follows that int $P \neq \emptyset$. Assume $x \in$ int P, and consider

$$x' = x - \lambda(u^1 - x). \tag{2.24}$$

Note that $u^1 \neq x$. There are the following two cases arising:

(i) $u^1 - x \not\leq 0$.

Determine λ such that

$$\lambda = x_q/(u_q^1 - x_q) = \min\{x_j/(u_j^1 - x_j) \mid u_j^1 - x_j > 0,\ j = 1, \ldots, n\} > 0.$$

Then, it is easy to verify that x', defined by (2.24), satisfies $x' \in P$ and $x'_q = 0$. Therefore, x' belongs to some proper face with its dimension less than k. According to the assumption of induction, x' can be expressed as the sum of a convex combination of vertices and a nonnegative combination of extreme directions of the proper face. Since vertices and extreme directions of a face are also that of P, therefore, x' can be expressed as the sum of a convex combination of vertices and a nonnegative combination of extreme directions of P, i.e.,

$$x' = \sum_{i=1}^{s_1} \alpha'_i u^i + \sum_{j=1}^{t_1} \beta'_j v^j,$$

$$\sum_{i=1}^{s_1} \alpha'_i = 1, \qquad \alpha'_i \geq 0, \quad i = 1, \ldots, s_1,$$

$$\beta'_j \geq 0, \qquad j = 1, \ldots, t_1,$$

where u^i and v^j are vertices and extreme directions of P, respectively. Substituting the preceding into

$$x = \frac{1}{1+\lambda} x' + (1 - \frac{1}{1+\lambda}) u^1,$$

which is equivalent to (2.24), leads to the expression of the form of (2.23), i.e.,

$$x = \sum_{i=1}^{s_1} \frac{1}{1+\lambda} \alpha'_i u^i + (1 - \frac{1}{1+\lambda}) u^1 + \sum_{j=1}^{t_1} \frac{1}{1+\lambda} \beta'_j v^j.$$

(ii) $u^1 - x \leq 0$.

Then x', defined by (2.24), are all feasible points for any $\lambda \geq 0$, hence $-(u^1 - x)$ is an unbounded direction. According to Lemma 2.2.6, there is an extreme direction, say v^1. Now take a sufficiently large μ such that

$$\tilde{x} = u^1 + \mu v^1$$

has at least one component greater than the corresponding component of x. Consequently, the point defined by

$$x' = x - \lambda(\tilde{x} - x),$$

$$\lambda = x_q/(\tilde{x}_q - x_q) = \min\{x_j/(\tilde{x}_j - x_j) \mid \tilde{x}_j - x_j > 0,\ j = 1, \ldots, n\} > 0.$$

is a feasible point, satisfying $x'_q = 0$. Therefore, this point belongs to some proper face with dimension less than k, and hence can be expressed as the sum of a convex combination of vertices and a nonnegative combination of extreme directions of P. As a result, an expression of the form (2.23) can be obtained in an analogous manner as case (i). □

The preceding Theorem indicates that a feasible point is the sum of a convex combination of vertices and a nonnegative combination of extreme directions, and vice versa. In particular, the following is a direct corollary.

Corollary 2.2.2. *Let the feasible region be bounded. A point is feasible if and only if it is a convex combination of vertices.*

2.3 Optimal Face and Vertex

We describe a basic result without proof (see, e.g., Rockafellar 1997).

Theorem 2.3.1 (Partition Theorem). *Let $\bar{x}$ be a boundary point of convex set S. Then there exists a superplane including $\bar{x}$ and partitioning the total space to two half spaces, one of which includes S.*

The superplane involved in the preceding Theorem is said to be *supporting superplane* of S. That is to say, there is a supporting superplane through every boundary point of a convex set. Although the result is applicable to any convex set, we are only concerned with the feasible region P, in particular.

Supporting superplane of P is closely related to its face, as the following reveals.

Lemma 2.3.1. *The intersection of P and a supporting superplane is a face.*

Proof. Assume that $H = \{x \in \mathcal{R}^n \mid a^T x = \eta\}$ is the superplane of P, and $P' = P \cap H$. Without loss of generality, let $a^T x \leq \eta$ hold for all $x \in P$.

Assume that $v \in P'$ is an interior point of segment (y, z), and $y, z \in P$, i.e.,

$$v = \alpha y + (1 - \alpha)z, \tag{2.25}$$

where $0 < \alpha < 1$. Note that, P' is a nonempty convex set.

It is only needed to show $y, z \in H$, as leads to $y, z \in P'$, hence P' is a face of P.

Assume $y, z \notin H$. Then it holds that

$$a^T y < \eta, \qquad a^T z < \eta.$$

Multiplying the preceding two formulas respectively by $\alpha > 0$ and $1 - \alpha > 0$, and then adding the results gives

$$a^T(\alpha y + (1 - \alpha)z) < \eta\alpha + \eta(1 - \alpha) = \eta,$$

combining which and (2.25) leads to $v \notin H$, and hence $v \notin P'$. This contradicts the assumption $v \in P'$. Therefore, at least one of y and z belongs to H.

Without loss of generality, assume $z \in H$. Then, it follows from (2.25) that

$$y = (1/\alpha)v + (1 - 1/\alpha)z,$$

implying that y is in the straight line through v and z. Moreover, $z, v \in H$ and H is a superplane, therefore $y \in H$. □

Lemma 2.3.2. *Let $\bar{f}$ be the optimal value of the standard LP problem. Set F is the set of optimal solutions if and only if it is the intersection of P and the objective contour plane*

$$\bar{H} = \{x \in \mathcal{R}^n \mid c^T x = \bar{f}\}. \tag{2.26}$$

Proof. Assume $F = P \cap \bar{H}$. It is clear that any optimal solution $\bar{x} \in P$ satisfies $c^T\bar{x} = \bar{f}$, implying $\bar{x} \in \bar{H}$, hence $\bar{x} \in F$. Therefore, F is the set of optimal solutions. If F is the set of optimal solutions, conversely, then $c^T\bar{x} = \bar{f}$ holds for any $\bar{x} \in F \subset P$. Therefore $\bar{x} \in \bar{H}$, and hence $\bar{x} \subset P \cap \bar{H}$. □

It is clear that $\bar{H}$ is a supporting superplane of P, as is referred to as *objective supporting superplane*.

A face is optimal if its elements are all optimal solutions. A vertex is optimal if it is an optimal solution.

Lemma 2.3.3. *If there exists an optimal solution, then there exists an optimal face.*

Proof. According to Lemma 2.3.2, a nonempty set of optimal solutions is the intersection of feasible region P and objective contour plane $\bar{H}$. Therefore it is an optimal face, according to Lemma 2.3.1 and the definition of an optimal face. □

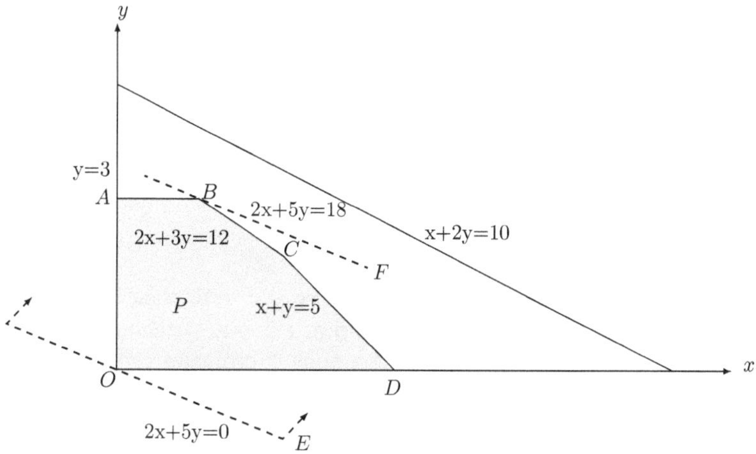

Fig. 2.3 Graphic solution to Example 1.2.2

Theorem 2.3.2. *If there exists a feasible solution, there exists a basic feasible solution. If there is an optimal solution, there is a basic optimal solution.*

Proof. By Lemmas 2.2.2 and 2.2.3, it is known that nonempty feasible region has a basic feasible solution. By Lemma 2.3.3, if there is an optimal solution, there is an optimal face, which is a nonempty polyhedral convex set, and hence having an optimal convex or basic optimal solution. □

In presence of optimal solution, there exists an optimal 0-dimensional face (or vertex). In general, there could exist optimal faces of higher dimensions. It is clear that the optimal face of the highest dimension is the set of all optimal solutions, as is referred to as *optimal set*. After a LP problem is solved by the simplex method, the optimal set can be obtained easily (Sect. 25.2).

2.3.1 Graphic Approach

A LP problem of 2-dimension can be solved via a graphic approach. To do so, let us return to Example 1.2.2. The shaded area enclosed by polygon OABCD in Fig. 1.1 is the feasible region (ignore the straight line $x + 2y = 10$, at the moment). It is required to determine a point over the area such that the objective function reaches the highest value at the point (Fig. 2.3).

In the figure, the equation $2x + 5y = 0$ of the contour line of the objection function corresponds to the dashed line OE, going through the origin, all points on which correspond to the same objective value 0. The line's slope, i.e., the tangent of the angle between it and x axis, is $-2/5 = -0.4$. Therefore, the corresponding contour line shifts parallel to the upper-right side as the objective value increases from 0. Points in the intersection of the line and the area of OABCD are all feasible points. The parallel shifting should be carried out as far as the intersection remains nonempty to attain the biggest possible objective value. It is seen from the figure that the contour line shifting the farthest is the dashed line BF though vertex B, that is, the "objective supporting plane". The optimal set, i.e., the intersection of the line and the feasible region includes a single vertex B, corresponding to the basic optimal solution. Consequently, problem (1.2) is solved after measuring coordinates of point B, and calculating the associated objective value.

If the figure or measurement is not relatively accurate, however, the end solution would involve unacceptable errors. For a graphic approach, therefore, it would be better to use coordinate paper, with the help of algebraic calculating. Once B is known to be the optimal vertex, for instance, its coordinates can be obtained by solving the following system of equations:

$$\begin{cases} 2x + 3y = 12 \\ \quad\quad\; y = \;\;3 \end{cases} \tag{2.27}$$

from which the optimal basic solution $\bar{x} = 1.5,\ \bar{y} = 3$ to problem (1.2) follows, with the optimal value $f = 18$. That is to say, the manufacturer should arrange production of 1,500 laths and 3,000 sheep piles daily, gaining 18,000 dollars profit.

The graphic approach is not suitable for cases of $n \geq 3$, though it is simple. Even for case of $n = 2$, in fact, its application is rare seen. However, it still offers some inspiration, as is the topic of Sect. 2.5.

2.4 Feasible Direction and Active Constraint

Methods for solving mathematical problems fall into two categories: direct and iterative methods. The latter produces a sequence of points by iterations, offering an exact or approximate solution. Methods presented in this book belong to the iterative category.

Line search is the mostly used iterative approach in optimization. At a current point $\bar{x}$, in each iteration, a new point $\hat{x}$ is determined along a ray starting from $\bar{x}$ along a nonzero vector d, i.e.,

$$\hat{x} = \bar{x} + \alpha d, \tag{2.28}$$

where d is referred to as *search direction*, $\alpha > 0$ as *stepsize*. Once the two are available, $\hat{x}$ can be calculated and then one iteration is complete. Repeating this process yields a sequence of points, until a solution is reached. Formula (2.28) is referred to as *line search* or *iterative scheme*.

The determination of a search direction is crucial. In presence of constraints, d should be such that the intersection of the ray (2.28) and the feasible region is nonempty. More precisely, we introduce the following concept:

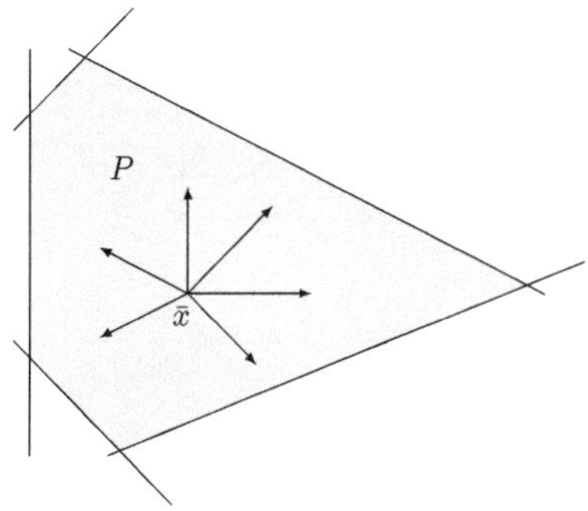

Fig. 2.4 Any direction is feasible at $\bar{x}$

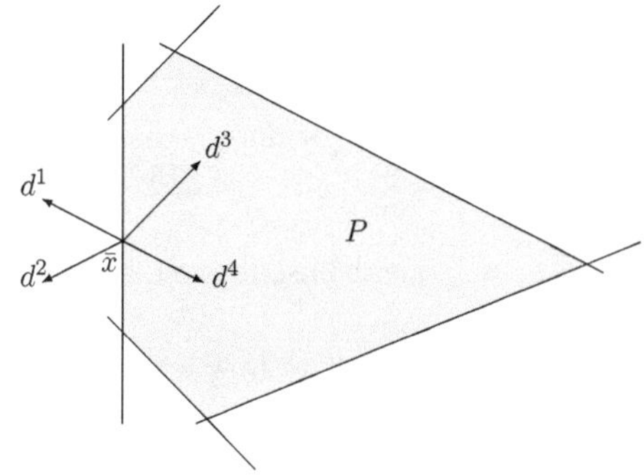

Fig. 2.5 d^3, d^4 are feasible at $\bar{x}$, but d^1, d^2 not

Definition 2.4.1. Let P be the feasible region. Assume that $\bar{x} \in P,\ d \neq 0$. If there exists $\bar{\alpha} > 0$ such that

$$\bar{x} + \alpha d \in P, \qquad \forall\, \alpha \in [0, \bar{\alpha}],$$

d is a feasible direction at $\bar{x}$.

The preceding is relevant to general constrained optimization problems, including the LP problem. The following are some instances, in conjunction with feasible region P.

Example 2.4.1. In Fig. 2.4, $\bar{x}$ is an interior point of P, and hence any direction is feasible at it.

Example 2.4.2. In Fig. 2.5, $\bar{x}$ is a boundary point of P. It is seen that d^3, d^4 are feasible directions at $\bar{x}$, but d^1, d^2 are not.

Example 2.4.3. In Fig. 2.6, $\bar{x}$ is a vertex of P. Any direction, e.g., d^4 or d^5, within the angle area between d^6 and d^7 (which are respectively along two sides of P) is feasible at $\bar{x}$. Vectors d^1, d^2, d^3 are not feasible.

Let d be a feasible search direction. It is possible to maintain feasibility of some new iterate $\hat{x}$. In order for $\hat{x}$ to be close to an optimal solution, it is needed to take

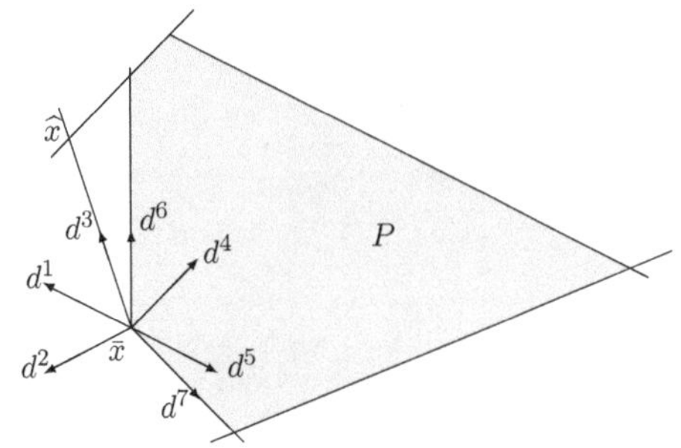

Fig. 2.6 d^4, d^5, d^6, d^7 are feasible at $\bar{x}$, but d^1, d^2, d^3 not

into account the objective function. To this end, it might be well to consider the following problem:

$$\begin{array}{lll} \min & f = c^T x, & \\ \text{s.t.} & a_i^T x \geq b_i, & i = 1, \ldots, m, \end{array} \tag{2.29}$$

where $m > n$, and whose feasible region is

$$P = \{x \in \mathcal{R}^n \mid a_i^T x \geq b_i,\ i = 1, \ldots, m\}.$$

Definition 2.4.2. Vector d satisfying condition $c^T d < 0$ is a descent direction. If d is also a feasible direction at $\bar{x} \in P$, it is a feasible descent direction at $\bar{x}$.

It is clear that d is a feasible descent direction at $\bar{x}$ if and only if it is a feasible direction and forms an obtuse angle with the objective gradient c. Once such a direction d is available, a stepsize $\alpha > 0$ can be determined, and hence a new iterate $\hat{x} \in P$, obtained by (2.28), corresponding to a smaller objective value. Then, one iteration is complete.

It is noticeable that not all constraints affect the determination of a feasible descent direction at a current point.

Definition 2.4.3. A constraint which is violated, or satisfied as an equality, by the current point is an active constraint.

Aimed at a *feasible* point, the "active" constraint is usually defined as one satisfied as equality, or binding at the point. But it seems to be useful to include infeasible point by regarding a constraint violated as active.

Let us bring up (2.29) as an example. If $a_i^T \bar{x} = b$, then $a_i^T x \geq b$ is active at $\bar{x}$; if, otherwise, $a_i^T \bar{x} > b$, then $a_i^T x \geq b$ is not active at that point. So, the current point is on a boundary of an active constraint. From the simple instances given in the preceding figures, it is seen that a feasible descent direction can be determined by taking into account active constraints only.

In practice, the preceding definition for active constraint does not come up to expectations, as a current point close to boundary could lead to too small stepsize, and hence insignificant progress. In view of this, Powell (1989) proposes the so-called "ϵ-active" constraint, where ϵ is a small positive number. For example, if $a_i^T\bar{x} - b \le \epsilon$, then $a_i^T x \ge b$ is an ϵ-active constraint at $\bar{x}$; whereas it is not if $a_i^T\bar{x} - b > \epsilon$.

The LP problem can be solved by the so-called "active set method", which is usually used for solving nonlinear programming problems though some scholars prefer it to be a LP problem solver (e.g., Fletcher 1981; Hager 2002). We outline the method in conjunction with problem (2.29) in the remainder of this section.

Assume that the current vertex $\bar{x}$ is an unique solution to the linear system below:

$$a_i^T x = b_i, \qquad i \in \mathcal{A},$$

where $\mathcal{A}$ is called *the active set (of constraints)*, consisting of n indices of (total or part) active constraints, with linearly independent gradients a_i. If $\bar{x}$ is judged to be optimal under some criterion, we are done; otherwise, some index $p \in \mathcal{A}$ is selected such that the $n-1$ equalities

$$a_i^T x = b_i, \qquad i \in \mathcal{A}\backslash\{p\}, \tag{2.30}$$

determines a descent edge (1-dimensional face).

In fact, since the rank of the coefficient matrix of (2.30) is $n-1$, the associated homogeneous system

$$a_i^T d = 0, \qquad i \in \mathcal{A}\backslash\{p\},$$

has a solution d such that

$$d \ne 0, \qquad c^T d < 0.$$

It is easily verified that for $\alpha \ge 0$, all points on the ray

$$\hat{x} = \bar{x} + \alpha d,$$

satisfy (2.30). Under the condition that the objective value is lower bounded over the feasible region, the following stepsize is well defined:

$$\alpha = (b_q - a_q^T\bar{x})/a_q^T d = \min\{(b_i - a_i^T\bar{x})/a_i^T d \mid a_i^T d < 0, \ i \notin \mathcal{A}\} \ge 0.$$

In fact, such an α is the largest stepsize possible to maintain feasibility of $\hat{x}$. Since $a_q^T\hat{x} = b_q$, the $a_q^T x \ge b_q$ is an active constraint at $\hat{x}$.

Consequently, setting $\bar{x} = \hat{x}$ and redefining $\mathcal{A} = \mathcal{A}\backslash\{p\} \cup \{q\}$ completes an iteration of the active set method. If $\alpha > 0$, the new vertex corresponds to a lower objective value. Note however that if there are multiple active constraints at the current vertex $\bar{x}$, then α defined by (2.30) would vanish, so that the descent edge degenerates to a vertex. That is to say, the "new vertex" actually coincides with the old, though set $\mathcal{A}$ changes. Such a vertex is called *degenerate*. In Fig. 2.6, e.g., $\bar{x}$ is a degenerate vertex, at which the three edges along respective d^3, d^6, d^7 intersect.

The simplex method can be viewed as a special scheme of the active set method, as will be discussed in Sect. 3.9. In history, however, the former emerged before the latter from another path by fully taking advantage of the linear structure.

2.5 Heuristic Characteristic of Optimal Solution

From the graphic approach demonstrated in Fig. 2.3, it is seen that the optimal solution to the LP problem is attained at a vertex of the feasible region. It is imaginable that if the feasible region has a side going through the vertex, which is parallel to objective contour lines, then the whole side corresponds the optimal set, associated with the same optimal value.

So, the solution key lies on how to determine lines intersecting at an optimal vertex. In other words, it is only needed to know which inequalities are active at an optimal solution. Once these active inequalities are known, what left to do is just to solve an linear system; for the instance in Fig. 2.3, the optimal solution was quickly calculated through solving the system (2.27).

Thereby, we make the observation that normal directions (pointing to the interior of the feasible region) of lines AB and BC, intersecting at the optimal vertex B, form *the largest angles* with the parallel shifting direction of the contour line BF, among all the lines.

Now turn to the more general minimization problem

$$\begin{array}{ll} \min & f = c^T x, \\ \text{s.t.} & Ax \geq b, \end{array} \tag{2.31}$$

where $A \in \mathcal{R}^{m\times n}$, $c \in \mathcal{R}^n$, $b \in \mathcal{R}^m$, $m > 1$. Note that constraints here are all inequalities of "$\geq$" type.

We could imagine analogously in the space of multiple dimensions. Now the constraint inequalities correspond to half spaces, and vertices correspond to intersection points formed by half spaces. What we should do is to examine angles between normal directions and parallel shifting direction of the contour plane. For the minimization problem, the shifting direction is the negative gradient direction of the objective function. This leads to the following plausible statement (Pan 1990).

Proposition 2.5.1 (Heuristic characteristic of optimal solution). *Gradients of active constraints at an optimal solution of a minimization problem tend to form the largest angles with the negative objective gradient.*

It is now needed to quantize magnitude of the angles. If we denote the ith row vector of A by $\bar{a}_i^T$, the cosine of the angle between the ith constraint gradient and negative objective gradient is then

$$\cos < \bar{a}_i, c >= -\bar{a}_i^T c/(\|\bar{a}_i\| \|c\|).$$

For simplicity, we ignore the constant factor $1/\|c\|$ and introduce the following

Definition 2.5.1. The pivoting-index of the ith constraint is defined as

$$\alpha_i = -\bar{a}_i^T c/\|\bar{a}_i\|. \tag{2.32}$$

Then, we are able to compare the angles by pivoting-indices, and hence Proposition 2.5.1 may be reformulated as

Gradients of active constraints at an optimal solution tend to have the smallest pivoting-indices.

Example 2.5.1. Investigate problem (1.2) via pivoting-indices:

$$\begin{array}{ll} \min & f = -2x - 5y, \\ \text{s.t.} & -2x - 3y \geq -12, \\ & -x - y \geq -5, \\ & -y \geq -3, \\ & x, y \geq 0. \end{array} \tag{2.33}$$

Answer Calculate indices of the constraints, and put them in the following table in the order of increasing pivoting-indices:

Constraints	α_i
$-2x - 3y \geq -12$	-5.26
$-y \geq -3$	-5.00
$-x - y \geq -5$	-4.95
$x \geq 0$	2.00
$y \geq 0$	5.00

From the preceding table, it is seen that $-2x - 3y \geq -12$ and $-y \geq -3$ are the two constraints with the smallest pivoting-indices. Thus, the two are active constraints at an optimal solution. This immediately leads to solving system (2.27), as coincides with the outcome from the graphic approach.

If Proposition 2.5.1 were true in general, solving the LP problem amounts to solving a system of linear equations by $O(m^3)$ basic arithmetics, in contrast to existing iterative methods (see Sect. 3.8)! Unfortunately, the characteristic of

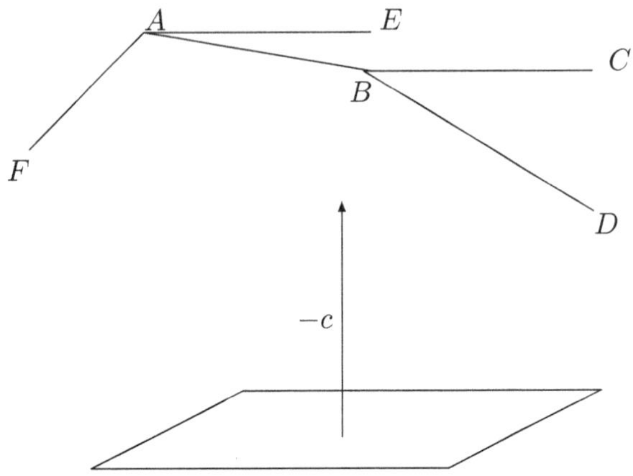

Fig. 2.7 The normal vectors (pointing to the interior of the polyhedron) of planes ABC, ABD and CBD form the most obtuse three angles with the negative objective gradient $-c$. But the optimal vertex is not their intersection B, but A, the intersection of planes ABC, ABD and (*vertical*) EAF

optimal solution is only heuristic. Counterexamples are easily constructed. If adding a constraint $x + 2y \leq 10$ to (1.2) (see Fig. 2.3), e.g., it is then clear that the added constraint is not active at the optimal solution, though its gradient forms the largest angle with the objective gradient (with pivoting-index −5.37); in fact, added is a redundant constraint, not affecting the feasible region at all. It is also not difficult to construct counterexamples without redundant constraint. Someday in the Fall of 1986 when the author was visiting the Mathematics Department of University of Washington, after he talked with Professor Rockafellar about his idea on the characteristic of an optimal solution, the latter quickly he showed a counterexample by sketching on a piece of paper, as is seen in Fig. 2.7.

For all that, Proposition 2.5.1 might still offer some clue toward optimal solution, shedding a light on LP research. Such a trick well be referred to as "the most-obtuse-angle heuristics" in this book.

In some cases, in fact, it is possible to detect unboundedness of a problem simply from signs of pivoting-indices.

Theorem 2.5.1. *Assume that the feasible region is nonempty. If pivoting-indices of constraints are all nonnegative, then problem (2.31) is unbounded.*

Proof. By contradiction. Assume that the problem is bounded under the assumptions. It follows that

$$c^T v \leq 0, \qquad \forall\, v \in \{v \in \mathcal{R}^n \mid Av \geq 0\}. \tag{2.34}$$

In fact, there is a vector v such that

$$Av \geq 0, \qquad c^T v > 0. \tag{2.35}$$

Thus, for any $\alpha \geq 0$ and feasible solution $\bar{x}$, vector $x = \bar{x} + \alpha v$ satisfies $Ax \geq b$, that is, v is an unbounded direction. Further, it is known from (2.35) that

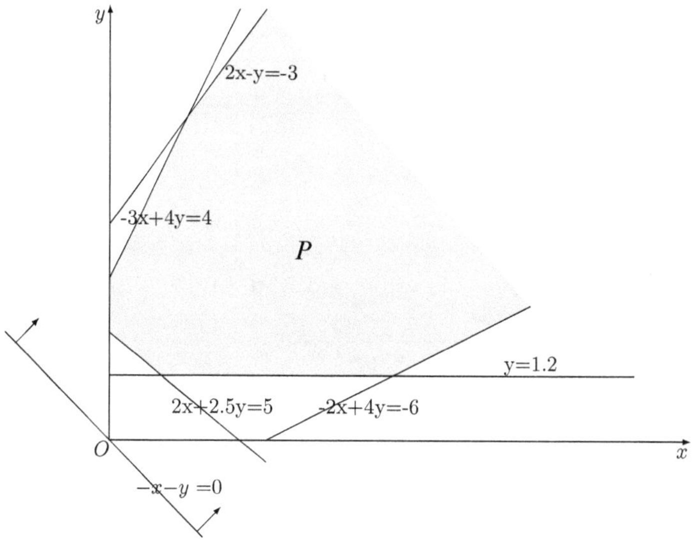

Fig. 2.8 Unbounded problem

$$c^T x = c^T \bar{x} + \alpha c^T v \rightarrow \infty, \quad (\text{as } \alpha \rightarrow \infty),$$

which contradicts that (2.31) is upper bounded. Therefore, (2.34) holds.

Thus, according to Farkas' Lemma 2.1, there is $y \geq 0$ such that

$$-c = A^T y,$$

premultiplying which by $-c^T$ gives

$$0 < c^T c = -y^T Ac.$$

Hence $Ac \geq 0$ follows from nonnegativeness of pivoting-indices. This implies that the right-hand side of the preceding is less than or equal to 0, as is a contradiction. Therefore, the problem is unbounded. □

An alternative proof of the preceding Theorem is via showing $-c$ to be an unbounded direction of the feasible region. The following Corollary gives a necessary condition for the existence of an optimal solution.

Corollary 2.5.1. *If there is an optimal solution, then there is at least one constraint bearing negative pivoting-index.*

Example 2.5.2. Investigate the following LP problem by pivoting-indices:

		α_i
min	$f = -x - y$	
s.t.	$2x - y \geq -3$	0.71
	$-3x + 4y \geq 4$	0.20
	$2x + 2.5y \geq 5$	1.41
	$-2x + 4y \geq -8$	0.45
	$y \geq 1.2$	1.00
	$x \geq 0$	1.00
	$y \geq 0$	1.00

(2.36)

Answer Calculate indices of the constraints, and fill in the right-hand side of the preceding table. According to Theorem 2.5.1, the problem is unbounded since all indices are nonnegative (see Fig. 2.8).

Chapter 3
Simplex Method

The simplex method is an efficient and widely used LP problem solver. Since proposed by George B. Dantzig in 1947, it has been dominating this area for more than 60 years.

The basic idea behind the simplex method is quite simple. In geometric words, it moves from a vertex to an adjacent vertex, while improving the objective value, until reaching an optimal vertex. Such doing is based on Theorem 2.3.2, guaranteeing the existence of a basic optimal solution if an optimal solution exists. It seems to be natural to hunt for an optimal solution among vertices in the feasible region, as it usually involves infinitely many point but only finitely many vertices (no more than C_n^m). So, such a strategy shrinks the hunting scope from the whole feasible region to a finite subset.

The idea may be traced back to as early as Fourier (1823). It was materialized algebraically by Dantzig (1951a). In this chapter, the simplex method will be presented in a tableau form first, then it is revised to a more applicable version. Discussed will also be related topics, such as how to get the method started, finiteness problem and finite pivot rules, and computational complexity. The last section will comment on features of the method.

3.1 Simplex Tableau

We begin with introduction of the so-called "simplex tableau" for problem (1.10). In Sect. 1.6, we already obtained the canonical form (1.11) of its constraint system, without touching the objective function at all. Now we put the objective function in the equation form

$$x_1 + 2x_2 - x_4 + x_5 - f = 0,$$

where f is called *objective variable*, at the bottom of the constraint system. The according tableau is

P.-Q. PAN, *Linear Programming Computation*, DOI 10.1007/978-3-642-40754-3_3,

x_1	x_2	x_3	x_4	x_5	f	RHS
1			2/11	−4/11		7/11
	1		15/11	−19/11		14/11
		1	1/11	9/11		9/11
1	2		−1	1	−1	

Then we eliminate the nonzero entries, corresponding to pivot columns (associated with variables x_1, x_2, x_3), in its bottom (objective) row. To do so, add −1 times of the first row to the bottom row first:

x_1	x_2	x_3	x_4	x_5	f	RHS
1			2/11	−4/11		7/11
	1		15/11	−19/11		14/11
		1	1/11	9/11		9/11
	2		−13/11	15/11	−1	−7/11

then add −2 times of row 2 to the bottom row:

x_1	x_2	x_3	x_4	x_5	f	RHS
1			2/11	−4/11		7/11
	1		15/11	−19/11		14/11
		1	1/11	9/11		9/11
			−43/11	53/11	−1	−35/11

where the north-west corner is the unit matrix, corresponding to zero entries in the bottom row. Thereby, the tableau offers not only a basic solution

$$x_1 = 7/11, \quad x_2 = 14/11, \quad x_3 = 9/11, \quad x_4 = x_5 = 0, \tag{3.1}$$

but also a reduced form of the objective function over the feasible region. Note that the solution is a basic feasible solution, associated with the objective value 35/11, which is equal to the opposite number of the south-east corner entry.

The same tableau may be obtained otherwise by putting coefficients of the constraint system and of the objective function together to form an initial tableau, then applying the relevant Gauss-Jordan elimination.

Such a tableau is called *simplex tableau*, whose general form is as shown by Table 3.1[1].

The associated terms coincide with those the same named for the canonical form of the system $Ax = b$ (Sect. 1.6):

[1]It is always possible to arrange the unit matrix at the north-west corner of the simplex tableau by column exchanges. Practically, however, this is not needed, and the matrix corresponding to basic variables is usually a permutation matrix.

Table 3.1 Simplex tableau

x_{j_1}	x_{j_2}	$\cdots$	x_{j_m}	$x_{j_{m+1}}$	$\cdots$	x_{j_n}	f	RHS
1				$\bar{a}_{1\ j_{m+1}}$	$\cdots$	$\bar{a}_{1\ j_n}$		$\bar{b}_1$
	1			$\bar{a}_{2\ j_{m+1}}$	$\cdots$	$\bar{a}_{2\ j_n}$		$\bar{b}_2$
		$\ddots$		$\vdots$	$\vdots$	$\vdots$		$\vdots$
			1	$\bar{a}_{m\ j_{m+1}}$	$\cdots$	$\bar{a}_{m\ j_n}$		$\bar{b}_m$
				$\bar{z}_{j_{m+1}}$	$\cdots$	$\bar{z}_{j_n}$	-1	$-\bar{f}$

Variables (components) corresponding to the unit matrix are *basic variables (components)*, and the rest are *nonbasic variables (components)*. The basic and nonbasic index sets

$$B = \{j_1, \ldots, j_m\}, \quad \text{and} \quad N = A \backslash B.$$

are *basis* and *nonbasis*, respectively. The sets of basic and nonbasic variables are also called basis and nonbasis. The importance of the simplex tableau lies in that it gives a *basic solution* $\bar{x}_B = \bar{b}$; $\bar{x}_N = 0$. If $\bar{x}_B \geq 0$, the solution and tableaus are *basic feasible solution* and *feasible (simplex) tableau*, respectively. If the objective function attains the minimum value over the feasible region, the solution and tableau are said to be *basic optimal solution and optimal (simplex) tableau.*

In addition, $\bar{z}_N$ in the simplex tableau is termed *reduced costs (coefficients).* The *opposite* number of the south-east corner entry gives the according objective value $\bar{f}$.

Throughout this book, it is stipulated that the bottom row of a simplex tableau always corresponds to the objective function. It will be seen that the f column does not change in solution process by the simplex method, and hence can be omitted. However, it is indispensable in the context of the"reduced simplex method", presented in Chap. 15.

3.2 Simplex Method: Tableau Form

In the previous section, a simplex tableau of the LP problem (1.10) together with the associated basic feasible solution (3.1) were obtained. But it can no not be asserted that the solution is optimal, since the reduced cost of variable x_4 is negative. As the value of x_4 increases from 0 while the value of x_5 fixed at 0, in fact, the objective function would decrease further, reaching a lower value than the current.

The new value of x_4 should be as large as possible, so that the associated objective value becomes as low as possible, subject to maintaining nonnegativity of corresponding values of x_1, x_2, x_3, satisfying

$$\begin{cases} x_1 = 7/11 - (2/11)x_4 \geq 0, \\ x_2 = 14/11 - (15/11)x_4 \geq 0, \\ x_3 = 9/11 - (1/11)x_4 \geq 0, \end{cases} \tag{3.2}$$

The preceding set of inequalities are equivalent to

$$\begin{cases} x_4 \leq 7/2 \\ x_4 \leq 14/15 \\ x_4 \leq 9 \end{cases}$$

whose solution set is

$$x_4 \leq \min\{7/2, 14/15, 9\} = 14/15.$$

Thereby, $\bar{x}_4 = 14/15$ is the largest possible value taken by x_4. Substituting it to (3.2) gives the new feasible solution

$$\bar{x} = (7/15, 0, 11/15, 14/15, 0)^T, \tag{3.3}$$

corresponding to objective value $\bar{f} = -7/15$ lower than 35/11.

The according new simplex tableau is obtained by taking entry 15/11 at row 2 and column 4 as the pivot. To this end, firstly multiply row 2 by 11/15 to turn the pivot to 1, leading to

x_1	x_2	x_3	x_4	x_5	f	RHS
1			2/11	−4/11		7/11
	11/15		1	−19/15		14/15
		1	1/11	9/11		9/11
			−43/11	53/11	−1	−35/11

Then add −2/11, −1/11 and 43/11 times of row 2 to rows 1, 3 and 4, respectively, giving the new simplex tableau

x_1	x_2	x_3	x_4	x_5	f	RHS
1	−2/15			−2/15		7/15
	11/15		1	−19/15		14/15
	−1/15	1		14/15		11/15
	43/15			−2/15	−1	7/15

which clearly corresponds to the new basic feasible solution (3.3).

As the reduced cost, associated with variable x_5, in the objective line is negative, still it cannot be asserted that the new solution is optimal. Similarly as in the previous step, we consider the following set of inequalities to determine the new value of x_5 that can be increased to and an associated pivot:

$$\begin{cases} x_1 = 7/15 + (2/15)x_5 \geq 0 \\ x_4 = 14/15 + (19/15)x_5 \geq 0 \\ x_3 = 11/15 - (14/15)x_5 \geq 0 \end{cases}$$

Since coefficients of x_5 in the first two inequalities are positive, the according values of x_1 and x_4 remain nonnegative as x_5 increases from 0, while x_2 is fixed on zero. It is therefore only needed to consider the third inequality, associated with the negative coefficient of x_5. Setting $x_3 = 0$ in the third equation gives $x_5 = 11/14$, leading to the basic feasible solution

$$\bar{x} = (4/7, 0, 0, 27/14, 11/14)^T, \tag{3.4}$$

associated with objective value $\bar{f} = -4/7$ lower than $-7/15$.

To obtain the associated simplex tableau, it is only needed to enter x_5 to and drop x_3 from the basis by taking the entry 14/15 at row 3 and column 5 as the pivot. Multiply row 3 by 15/14 gives

x_1	x_2	x_3	x_4	x_5	f	RHS
1	−2/15			−2/15		7/15
	11/15		1	−19/15		14/15
	−1/14	15/14		1		11/14
	43/15			−2/15	−1	7/15

Then add 2/15, 19/15 and 2/15 times of row 3 to rows 1, 2 and 4, respectively, leading to

x_1	x_2	x_3	x_4	x_5	f	RHS
1	−1/7	1/7				4/7
	9/14	19/14	1			27/14
	−1/14	15/14		1		11/14
	20/7	1/7			−1	4/7

where reduced costs in the bottom row are all nonnegative. As will be proved a little later, it is now can be asserted that the corresponding basic feasible solution is optimal, which is just (3.4), and we are done.

Now turn to the general standard LP problem (1.7). Following the preceding example, we describe an iteration by determining a pivot, and then updating the tableau by relevant elementary transformations.

Assume at the current iteration that we are faced with the feasible Tableau 3.1, the right-hand side of which gives the basic feasible solution

$$\bar{x}_B = \bar{b} \geq 0, \qquad \bar{x}_N = 0, \tag{3.5}$$

associated with the objective value $\bar{f}$ equal to the opposite number of the south-east corner entry of the tableau.

Lemma 3.2.1. *If reduced costs are all nonnegative, the feasible simplex tableau is optimal, giving a basic optimal solution.*

Proof. The simplex tableau results from a series of elementary transformations, and hence equivalent to the original problem. Its bottom row represents equality

$$f = \bar{f} + \bar{z}_N^T x_N. \tag{3.6}$$

Assume that $\tilde{x}$ is any feasible solution, associated with objective value $\tilde{f}$. Substituting it to (3.6) leads to

$$\tilde{f} = \bar{f} + \bar{z}_N \tilde{x}_N \geq \bar{f},$$

where the inequality is from $\bar{z}_N \geq 0$ and $\tilde{x} \geq 0$. Therefore, $\bar{x}$ is a basic optimal solution. □

The reduced costs are often called *check numbers*, as their sign can be used to jude the optimality of a simplex tableau. Usually, there are negative check numbers in the tableau.

Lemma 3.2.2. *Assuming that $\bar{z}_q < 0$ holds for some $q \in N$, and that*

$$\bar{a}_{i,q} \leq 0, \qquad i = 1, \ldots, m, \tag{3.7}$$

then the LP problem is (lower) unbounded.

Proof. The simplex tableau is associated with the constraint system

$$x_{j_i} = \bar{b}_i - \sum_{j \in N} \bar{a}_{i\,j} x_j, \qquad i = 1, \ldots, m.$$

Setting $x_j = 0,\ j \in N,\ j \neq q$ in the preceding and combining the result with the nonnegative constrains gives

$$x_{j_i} = \bar{b}_i - \bar{a}_{i\,q} x_q \geq 0, \qquad i = 1, \ldots, m. \tag{3.8}$$

It is known from (3.7) that the set (3.8) of inequalities hold for all $x_q = \alpha \geq 0$, associated with feasible value

$$\hat{f} = \bar{f} + \alpha \bar{z}_q, \tag{3.9}$$

which, since $\bar{z}_q < 0$, can be arbitrarily low as α increases. Therefore, the problem is lower unbounded. □

If (3.7) does not hold, then the value that the nonbasic variable x_q takes on will be restricted by the set (3.8) of inequalities. It is not difficulty to verify that the following rule gives the largest possible value α of x_q subject to (3.8).

Rule 3.2.1 (Row pivot rule) Determine a row index p and stepsize α such that

$$\alpha = \bar{b}_p / \bar{a}_{p\,q} = \min\{\bar{b}_i / \bar{a}_{i\,q} \mid \bar{a}_{i\,q} > 0,\ i = 1, \ldots, m\} \geq 0. \tag{3.10}$$

which is often called *minimum-ratio test.*

Setting $x_q = \alpha$ in the equation part of (3.8) gives a new basic feasible solution, i.e.,

$$\hat{x}_{j_i} = \bar{b}_i - \alpha \bar{a}_{i\,q}, \quad i = 1, \ldots, m; \qquad \hat{x}_j = 0, \quad j \in N, \; j \neq q; \qquad \hat{x}_q = \alpha. \tag{3.11}$$

Taking $\bar{a}_{p\,q}$ as the pivot, the according simplex tableau is obtained by multiplying row p by $1/\bar{a}_{p\,q}$ to convert the pivot to 1, adding $-\bar{a}_{i\,q}$ times of row p to rows $i = 1, \ldots, m, \; i \neq p$, and adding $-\bar{z}_q$ times of row p to the objective row. Finally, (B, N) is updated by exchanging j_p and q, and an iteration is complete.

It is seen from (3.9) that the associated objective value decreases strictly if $\alpha > 0$; no real decrement is made if $\alpha = 0$.

Definition 3.2.1. If some components of $\bar{b}$ is equal to zero, then the associated basic feasible solution (or tableau) is degenerate.

A LP problems is said to be nondegenerate if all basic solutions are nondegenerate.

In degeneracy case, the stepsize α defined by (3.10) could vanish, and hence the objective function remains unchanged (see (3.9)). That is to say, the associated "new solution" (3.11) is actually the same as the old although the basis is changed.

In general, there are multiple choices for q, as any q with negative $\bar{z}_q$ is eligible to be chosen. Dantzig's original *minimum reduced cost rule* is as follows.

Rule 3.2.2 (Column pivot rule) Select a column index q such that

$$q \in \arg\min_{j \in N} \bar{z}_j. \tag{3.12}$$

Thus, this rule selects the column with the most negative reduced cost as the pivot column.[2] For unit increment of the nonbasic variable x_q, this choice leads to the largest amount of decrease in the objective value.

The overall steps are summarized to the following algorithm (Dantzig 1947).

Algorithm 3.2.1 (Simplex algorithm: tableau form). Initial: a feasible simplex tableau of the form Table 3.1. This algorithm solves the standard LP problem (1.7).

1. Determine a pivot column index $q \in \arg\min_{j \in N} \bar{z}_j$.
2. Stop if $\bar{z}_q \geq 0$.
3. Stop if $I = \{i = 1, \ldots, m \mid \bar{a}_{i\,q} > 0\} = \emptyset$.
4. Determine a pivot row index $p \in \arg\min_{i \in I} \bar{b}_i / \bar{a}_{i q}$.
5. Convert $\bar{a}_{p\,q}$ to 1, and eliminate the other nonzeros in the column by elementary transformations.
6. Go to Step 1.

[2]Any choice is eligible if there is a tie when the number of the most negative reduced costs is more than one. Similarly below.

Theorem 3.2.1. *Under the nondegeneracy assumption, the simplex algorithm terminates either at*

(i) Step 2, generating a basic optimal solution; or at
(ii) Step 3, detecting lower unboundedness of the problem.

Proof. Note that there are infinitely many basic feasible solutions. If is clear that Algorithm 3.2.1 generates a sequence of basic feasible solutions, while the associated objective value decreases, due to the nonnegative stepssize α. Under the nondegeneracy assumption, the stepssize α is positive, and hence the objective value decreases strictly in each iteration. In the solution process, therefore, any basic solution can only appear once at most. So, infiniteness of the solution process implies that there are infinitely many basic feasible solutions, as is a contradiction. Therefore, Algorithm 3.2.1 terminates.

The meanings of the exits of the Algorithm 3.2.1 comes from Lemmas 3.2.1 and 3.2.2. □

It should be aware that the nondegeneracy assumption is beyond reality at all. As practical problems are almost always degenerate, termination of the simplex Algorithm 3.2.1 is actually not guaranteed. In other words, the possibility is not ruled out that indices enter and leave the basis infinitely many times. In fact, few instances that cannot be solved by the simplex algorithm had been constructed (we will handle this topic in Sects. 3.6 and 3.7). Even so, the possibility for not terminating is very rare, so as dose not matter to broad applications of the simplex algorithm.

A simplex tableau is nothing but a concise expression of a standard LP problem. As they represent problems equivalent to the original problem itself, all the tableaus created by the simplex algorithm are viewed as equivalent. Recursive formulas between a simplex tableau and its predecessor are listed below:

1. The objective row

$$\begin{aligned} \beta &= -\bar{z}_q/\bar{a}_{p\,q}, \\ \hat{f} &= \bar{f} - \beta\bar{b}_p, \\ \hat{z}_j &= \bar{z}_j + \beta\bar{a}_{p\,j}, \qquad j \in N, \\ \hat{z}_{j_i} &= \begin{cases} \beta & i = p, \\ 0 & i = 1,\ldots,m,\ i \neq p. \end{cases} \end{aligned} \tag{3.13}$$

2. The right-hand side

$$\begin{aligned} \alpha &= \bar{b}_p/\bar{a}_{p\,q}, \\ \hat{b}_i &= \begin{cases} \bar{b}_i - \alpha\bar{a}_{i\,q} & i = 1,\ldots,m,\ i \neq p, \\ \alpha & i = p. \end{cases} \end{aligned} \tag{3.14}$$

3. Entries of the constraint matrix

$$\hat{a}_{t,j} = \begin{cases} 0, & t = 1, \ldots, m,\ t \neq p;\ j = q. \\ 1, & t = p;\ j = q, \\ \bar{a}_{t\,j} - (\bar{a}_{p\,j}/\bar{a}_{p\,q})\bar{a}_{t\,q}, & t = 1, \ldots, m,\ t \neq p; j \in N,\ j \neq q, \\ \bar{a}_{p\,j}/\bar{a}_{p\,q}, & t = p;\ j \in N,\ j \neq q. \end{cases} \tag{3.15}$$

$$\hat{a}_{t\,j_i} = \begin{cases} 0, & t = 1, \ldots, m;\ i = 1, \ldots, m,\ i \neq p;\ i \neq t, \\ 1, & t = i = 1, \ldots, m;\ i \neq p, \\ -\bar{a}_{t\,q}/\bar{a}_{p\,q}, & t = 1, \ldots, m,\ t \neq p;\ i = p, \\ 1/\bar{a}_{p\,q}, & t = i = p. \end{cases}$$

Example 3.2.1. Solve the following problem by Algorithm 3.2.1:

$$\begin{array}{lllr} \min & f = -4x_1 - 3x_2 - 5x_3, & & \\ \text{s.t.} & 2x_1 + x_2 + 3x_3 \qquad + x_5 & & = 15, \\ & x_1 + x_2 + \ x_3 + x_4 & & = 12, \\ & -2x_1 + x_2 - 3x_3 & + x_7 & = 3, \\ & 2x_1 + x_2 & +x_6 & = 9, \\ & & x_j \geq 0, j = 1, \ldots, 7. & \end{array}$$

Answer Initial: the following feasible simplex tableau can be directly obtained from the problem:

x_1	x_2	x_3	x_4	x_5	x_6	x_7	RHS
2	1	3*		1			15
1	1	1	1				12
−2	1	−3				1	3
2	1				1		9
−4	−3	−5					

Iteration 1:

1. $\min\{-4, -3, -5\} = -5 < 0,\ q = 3.$
3. $I = \{1, 2\} \neq \emptyset.$
4. $\min\{15/3,\ 12/1\} = 15/3,\ p = 1.$
5. Take 3 in row 1 and column 3 as the pivot (marked by "*", the same below).

Multiply row 1 by $1/3$, then add $-1, 3, 5$ times of row 1 to rows 2,3,5, respectively:

x_1	x_2	x_3	x_4	x_5	x_6	x_7	RHS
2/3	1/3	1		1/3			5
1/3	2/3		1	−1/3			7
	2*			1		1	18
2	1				1		9
−2/3	−4/3			5/3			25

Iteration 2:

1. $\min\{-2/3, -4/3, 5/3\} = -4/3 < 0,\ q = 2$.
3. $I = \{1, 2, 3, 4\} \neq \emptyset$.
4. $\min\{5/(1/3), 7/(2/3), 18/2, 9/1\} = 9/1,\ p = 3$.
5. Multiply row 3 by $1/2$, then add $-1/3, -2/3, -1, 4/3$ times of row 3 to rows 1,2,4,5, respectively:

x_1	x_2	x_3	x_4	x_5	x_6	x_7	RHS
2/3		1		1/6		−1/6	2
1/3			1	−2/3		−1/3	1
	1			1/2		1/2	9
2*				−1/2	1	−1/2	0
−2/3				7/3		2/3	37

Iteration 3:

1. $\min\{-2/3, 7/3, 2/3\} = -2/3 < 0,\ q = 1$.
3. $I = \{1, 2, 4\} \neq \emptyset$.
4. $\min\{2/(2/3), 1/(1/3), 0/2\} = 0,\ p = 4$.
5. Multiply row 4 by $1/2$, then add $-2/3, -1/3, 2/3$ times of row 4 to rows 1,2,5, respectively:

x_1	x_2	x_3	x_4	x_5	x_6	x_7	RHS
		1		1/3	−1/3		2
			1	−7/12	−1/6	−1/4	1
	1			1/2		1/2	9
1				−1/4	1/2	−1/4	
				13/6	1/3	1/2	37

Now all reduced costs in the preceding tableau are nonnegative, and hence the basic optimal solution and associated objective value are, respectively,

$$\bar{x} = (0, 9, 2, 1, 0, 0, 0)^T, \qquad \bar{f} = -37.$$

It is seen from the preceding example that the tableau in the second iteration already attained the basic optimal solution, but an additional iteration was performed, due to the existence of a negative reduced cost. This occurred because of degeneracy leading to a zero stepsize. So, the condition that reduced costs are all nonnegative is sufficient but not necessary for optimality.

3.3 Start-Up of the Simplex Method

Algorithm 3.2.1 must start from a feasible simplex tableau. In Example 3.2.1, there is a feasible simplex tableau available, as is not the case in general. A so-called *Phase-I* procedure is usually carried out to provide an initial feasible simplex tableau (if any), then Algorithm 3.2.1 is used to achieve optimality or detect unboundedness of the problem. Thus, the simplex algorithm described in the previous section is actually a "Phase-II" procedure. A standard LP problem is usually solved by the two procedures in succession, referred to as *two-phase simplex method*. In this section, a classical Phase-I procedure using artificial variables will be presented first; described is then a closely related start-up method, the so-called "big M".

Assume that all components of the right-hand side are nonnegative, i.e.,

$$b_i \geq 0, \; i = 1, \ldots, m.$$

If not so, multiply each constraint equation with negative right-hand side by -1 before hand. Then construct an auxiliary problem as follows.

For each $i = 1, \ldots, m$, introduce a nonnegative *artificial variable* x_{n+i} to the ith equation, and take the sum of all artificial variables as the auxiliary objective function, i.e.,

$$f' = \sum_{i=1}^{m} x_{n+i}.$$

Using the constraint system, we eliminate all artificial variables from the auxiliary objective, resulting in

$$\begin{array}{lllll} \min & f' = \sum_{i=1}^{m} b_i - (\sum_{i=1}^{m} a_{i1})x_1 - \cdots - (\sum_{i=1}^{m} a_{in})x_n, & & & \\ \text{s.t.} & a_{11}x_1 + a_{12}x_2 + \cdots + a_{1n}x_n + x_{n+1} & & & = b_1, \\ & a_{21}x_1 + a_{22}x_2 + \cdots + a_{2n}x_n & + x_{n+2} & & = b_2, \\ & & & & \vdots \\ & a_{m1}x_1 + a_{m2}x_2 + \cdots + a_{mn}x_n & & + x_{n+m} & = b_m, \\ & & x_j \geq 0, \quad j = 1, \ldots, n+m. & & \end{array} \tag{3.16}$$

Clearly, there is an available feasible simplex tableau to the preceding auxiliary program, corresponding to the basic feasible solution

$$x' = (\underbrace{0, \ldots, 0}_{n}, \underbrace{b_1, \ldots, b_m}_{m})^T \geq 0.$$

Thereby, the program can be solved by Algorithm 3.2.1.

Regarding the outcome, we have the following.

Theorem 3.3.1. *The auxiliary program has an optimal solution, associated with a nonnegative optimal value:*

(i) If the optimal value is strictly greater than zero, the original problem is infeasible;

(ii) If the optimal value is equal to zero, the first n components of the optimal solution to the auxiliary program form a feasible solution to the original problem.

Proof. Clearly, there exists a feasible solution to problem (3.16). Since artificial components of all feasible solutions are nonnegative, all feasible values of the auxiliary program are nonnegative too. Therefore, there exists an optimal solution, associated with a nonnegative objective value:

(i) If the optimal value is strictly greater than zero, it can be asserted that the original problem is infeasible, because if it had a feasible solution $\bar{x}_j \geq 0,\ j = 1, \ldots, n$, then

$$\bar{x}_1, \ldots, \bar{x}_n, \bar{x}_{n+1} = \cdots = \bar{x}_{n+m} = 0$$

clearly satisfied constraints of (3.16), and hence was a feasible solution to (3.16), corresponding to auxiliary objective value zero, as is a contradiction.

(ii) If the optimal value is zero, then artificial components of the optimal solution are 0. From substituting it to the constraints of (3.16), it is therefore seen that its first n components just satisfy the constraints of the original problem, and hence constitute a feasible solution to the latter. □

Corollary 3.3.1. *The original problem is feasible if and only if the optimal value of the auxiliary program vanishes.*

Once a feasible solution to the original problem is obtained by the preceding approach, a feasible simplex tableau can be yielded from the optimal auxiliary tableau by the "following-up steps" below. These steps come from the fact that setting all the artificial variables to zero in the system, corresponding to the auxiliary optimal tableau, leads to a system equivalent to the original one.

Following-up steps:

(A) Delete columns, associated to all nonbasic artificial variables (which can be deleted once the corresponding artificial variable leaves the basis).
(B) Go to step D if there is no basic artificial variable.
(C) Delete the row, associated to a basic artificial variable, if all its nonbasic entries are zero (see the Note below); otherwise, take a nonzero entry of it as pivot to let the artificial variable become nonbasic, and then delete the associated column. This is repeated until no artificial variable is basic.
(D) Cover the auxiliary objective row by the original costs, and then eliminate all basic entries of this row, giving a feasible simplex tableau to the original problem.

Note: In step C, the row is deleted because substituting 0 to the associated artificial variable turns the corresponding equation to an identity, as reflects dependence of the original constraint equations. So, the method can get rid of such dependency.

The preceding can be put into the following algorithm.

Algorithm 3.3.1 (Phase-1: artificial variable). This algorithm finds a feasible tableau.

1. Introduce artificial variables, and construct auxiliary program of form (3.16).
2. Call the simplex Algorithm 3.2.1.
3. If the optimal value of the auxiliary program is zero, create a feasible tableau via "Following-up steps".
4. The original problem is infeasible if the optimal value of the auxiliary program is strictly greater than zero.

Note that if a constraint matrix includes some columns of the unit matrix, such columns should be employed to reduce the number of artificial variables. The preceding discussions are still valid, though the auxiliary objective function involves artificial variables only.

Example 3.3.1. Find a feasible simplex tableau to the following problem:

$$
\begin{array}{lrcl}
\min & f = -x_1 + x_2 - 2x_3, & & \\
\text{s.t.} & x_1 - 3x_2 - 2x_3 + x_4 & = & -4, \\
& x_1 - x_2 + 4x_3 \qquad - x_5 & = & 2, \\
& -3x_1 + x_2 + x_3 \qquad\qquad + x_6 & = & 8, \\
& x_j \geq 0, \; j = 1, \ldots, 6. & &
\end{array}
$$

Answer Construct auxiliary program: the first constraint equation is multiplied by -1 to turn its right-hand side to nonnegative; as the coefficients of x_6 give a unit vector $(0, 0, 1)^T$, only two artificial variables x_7, x_8 are introduced.

$$\begin{array}{lrcl}
\min & f' = x_7 + x_8, & & \\
\text{s.t.} & -x_1 + 3x_2 + 2x_3 - x_4 \qquad\qquad + x_7 & = & 4, \\
& x_1 - \ x_2 + 4x_3 \qquad - x_5 \qquad\qquad + x_8 & = & 2, \\
& -3x_1 + \ x_2 + \ x_3 \qquad\qquad + x_6 \qquad\qquad & = & 8, \\
& x_j \geq 0, \ j = 1, \ldots, 8. & &
\end{array}$$

Put the preceding auxiliary program into the following tableau:

x_1	x_2	x_3	x_4	x_5	x_6	x_7	x_8	RHS
−1	3	2	−1			1		4
1	−1	4		−1			1	2
−3	1	1			1			8
						1	1	

Turn the preceding to a simplex tableau: eliminate nonzeros in x_7 and x_8 columns at the bottom (objective) row by adding -1 times of row 1 and of row 2 to that row:

x_1	x_2	x_3	x_4	x_5	x_6	x_7	x_8	RHS
−1	3	2	−1			1		4
1	−1	4*		−1			1	2
−3	1	1			1			8
	−2	−6	1	1				−6

which is a feasible simplex tableau to the auxiliary program. Call Algorithm 3.2.1 to solve it:

Iteration 1:

1. $\min\{0, -2, -6, 1, 1\} = -6 < 0, \ q = 3.$
3. $I = \{1, 2, 3\} \neq \emptyset.$
4. $\min\{4/2, 2/4, 8/1\} = 1/2, \ p = 2.$
5. Multiply row 2 by $1/4$, and then add $-2, -1, 6$ times of row 2 to rows 1,3,4, respectively

(Erase x_8 column after artificial variable x_8 becomes nonbasic):

x_1	x_2	x_3	x_4	x_5	x_6	x_7	RHS
−3/2	7/2*		−1	1/2		1	3
1/4	−1/4	1		−1/4			1/2
−13/4	5/4			1/4	1		15/2
3/2	−7/2		1	−1/2			−3

Iteration 2:

1. $\min\{3/2, -7/2, 1, -1/2\} = -7/2 < 0,\ q = 2.$
3. $I = \{1, 3\} \neq \emptyset.$
4. $\min\{3/(7/2), (15/2)/(5/4)\} = 6/7,\ p = 1.$
5. Multiply row 1 by $2/7$, and then add $1/4, -5/4, 7/2$ times of row 1 to rows 2,3,4, respectively

(Erase x_7 column after artificial variable x_7 becomes nonbasic):

Now, all the artificial variables become nonbasic, hence reached is the optimal objective value 0 of the auxiliary program.

Covering the bottom row by original costs leads to

x_1	x_2	x_3	x_4	x_5	x_6	RHS
−3/7	1		−2/7	1/7		6/7
1/7		1	−1/14	−3/14		5/7
−19/7			5/14	1/14	1	45/7
−1	1	−2				

Adding −1 times of row 1 and 2 times of row 2 to the bottom row gives a feasible tableau of the original problem, i.e.,

x_1	x_2	x_3	x_4	x_5	x_6	RHS
−3/7	1		−2/7	1/7		6/7
1/7		1	−1/14	−3/14		5/7
−19/7			5/14	1/14	1	45/7
−2/7			1/7	−4/7		4/7

Thus, the preceding can be taken as an initial feasible tableau to get Algorithm 3.2.1 started to solve the original problem. Solving LP problems usually requires two phases, both of which are carried out using the simplex algorithm.

On the other hand, it seems to be attractive to solve LP problems in a single phase, as leads to the following so-called *big-M method*.

The according auxiliary program shares the same constraints as before with (3.16), while its objective function is the sum of the original objective function and M times of the sum of all the artificial variables, i.e.,

$$\begin{array}{lllr}
\min & f' = c_1x_1 + c_2x_2 + \cdots + c_nx_n + M(x_{n+1} + x_{n+2} \cdots + x_{n+m}), & & \\
\text{s.t.} & a_{11}x_1 + a_{12}x_2 + \cdots + a_{1n}x_n + x_{n+1} & & = b_1, \\
& a_{21}x_1 + a_{22}x_2 + \cdots + a_{2n}x_n \qquad + x_{n+2} & & = b_2, \\
& & & \vdots \\
& a_{m1}x_1 + a_{m2}x_2 + \cdots + a_{mn}x_n \qquad\qquad + x_{n+m} & & = b_m, \\
& \qquad\qquad x_j \geq 0, \quad j = 1, \ldots, n+m. & &
\end{array}$$

Artificial variables in the objective function are eliminated using the constraint system. As a result, there will be a feasible simplex tableau to the auxiliary program, which can be taken as an initial one to get the simplex algorithm started.

The reason for using such an auxiliary objective function is as follows. Its artificial variable part may be regarded as a "penalty function", where M serves as a "penalty factor", is a sufficiently large positive number (far larger than the absolute value of any number involved in the computations). The big M inflicts penalty on possible increase of values of artificial variables, consequently forcing them minimized prior to the original objective.

It is difficult however to determine a suitable M in advance. Too large M could lead to bad numerical stability, while too small M degrades method's effect. It depends not only on the problem to be solved, but also the computer used. A practicable way is to take M as a parameter in the solution process.

To demonstrate, we again bring up Example 3.3.1. Its auxiliary program is of the following form:

x_1	x_2	x_3	x_4	x_5	x_6	x_7	x_8	RHS
-1	3	2	-1			1		4
1	-1	4		-1			1	2
-3	1	1			1			8
-1	1	-2				M	M	

Add M times of row 1 and of row 2 to the objective row, giving

x_1	x_2	x_3	x_4	x_5	x_6	x_7	x_8	RHS
-1	3	2	-1			1		4
1	-1	4*		-1			1	2
-3	1	1			1			8
-1	$1-2M$	$-2-6M$	M	M				$-6M$

Thereby, we can get the simplex algorithm stated from the preceding tableau. In selection of a pivot column index q, however, it should be noted that M is so large that the sign of reduced costs depends upon coefficients of M only. In the preceding tableau, e.g., x_3 column is selected as the pivot column, as term M's coefficients are $0, -2, -6, 1, 1$ for costs of nonbasic variables x_1 through x_5 respectively, and

$$\min\{0, -2, -6, 1, 1\} = -6, \qquad q = 3.$$

Row 2 is selected as the pivot row by the minimum-ratio test below:

$$\min\{4/2, 2/4, 8/1\} = 1//2, \qquad p = 2.$$

Then elementary transformations are performed to make a corresponding basis change, completing the first iteration.

If the process continued, it can be found that sequences of iterates created by Big M method and the two-phase simplex method are actually the same. This is not surprising because, as mentioned previously, the big "penalty factor" M forces values of artificial variables vanishing first before pursuing optimality of the original problem–the two methods are essentially the same. Practically, however, the two-phase method is certainly preferable to the big M method, as it involves no any parameter, and easier to realize.

Nevertheless, the auxiliary programs, presented previously in this section, are usually found in textbooks only. If the number m of rows is large, the scale of the programs would become unacceptable large. A somehow practicable approach is to use an auxiliary program with a single artificial variable as follows.

Introducing artificial variable x_{n+1}, we consider the following auxiliary program instead:

$$\begin{array}{ll} \min & f' = x_{n+1}, \\ \text{s.t.} & a_{11}x_1 + a_{12}x_2 + \cdots + a_{1n}x_n + b_1x_{n+1} = b_1, \\ & a_{21}x_1 + a_{22}x_2 + \cdots + a_{2n}x_n + b_2x_{n+1} = b_2, \\ & \vdots \\ & a_{m1}x_1 + a_{m2}x_2 + \cdots + a_{mn}x_n + b_mx_{n+1} = b_m, \\ & x_j \geq 0, \quad j = 1, \ldots, n+1, \end{array} \tag{3.17}$$

to which there is a feasible solution

$$x' = (\underbrace{0, \ldots, 0}_{n}, 1)^T \geq 0.$$

Results similar to Theorem 3.3.1 and Corollary 3.3.1 hold to the preceding auxiliary program. On the other hand, using the following auxiliary objective function leads to an analogue to the big M method:

$$f' = c_1x_1 + c_2x_2 + \cdots + c_nx_n + Mx_{n+1}.$$

A drawback of such auxiliary programs seems to be lack of a explicit feasible simplex tableau. This will be seen not essential, however. In Sect. 13.2, we will present other Phase-I methods as well as a more practicable single artificial variable approach.

Now it is known that the answer to a LP problem must be one of the following three cases:

(i) Infeasible problem: there exists no feasible solution;
(ii) Unbounded problem: there exists a feasible solution but the feasible value is lower unbounded over the feasible region;
(iii) There exists an optimal basic solution.

In principle, a two-phase simplex method can be used to solve any LP problem, achieving an basic optimal solution, if any, or detecting infeasibility or unboundedness, otherwise.

3.4 Revised Simplex Tableau

Simplex tableau is not an unique tool to implement the simplex method. In fact, getting rig of the tableau can lead to a more compact variant of the simplex method. For this purpose, we will employ vectors or matrices more from now on.

The standard LP problem (1.8) may be represented by the following tableau:

x^T	f	RHS
A		b
c^T	-1	

Assume that through some elementary transformations, the preceding table becomes the simplex tableau Table 3.1, which may be succinctly put into

x_B^T	x_N^T	f	RHS
I	$\bar{N}$		$\bar{b}$
	$\bar{z}_N^T$	-1	$-\bar{f}$

(3.18)

Unless specified otherwise, thereafter the associated basic and nonbasis index sets are assumed to be

$$B = \{j_1, \cdots, j_m\}, \qquad N = A \backslash B = \{j_{m+1}, \cdots, j_n\}. \tag{3.19}$$

Columns corresponding to B are said to be *basic*, and those to N *nonbasic*. Without confusion, B and N will also be used to respectively denote submatrices consisting of corresponding columns. The two submatrices are respectively called *basis matrix* and *nonbasis matrix*, or basis and nonbasis for short. It is clear that B is an invertible square matrix. The simplex tableau corresponds to the basic solution

$$\bar{x}_B = \bar{b}, \qquad \bar{x}_N = 0.$$

If $\bar{b} \geq 0$, $\bar{z}_N \geq 0$, then the tableau is an optimal (simplex) tableau, giving an basic optimal solution, and the according B and N are *optimal basis* and *optimal nonbasis*, respectively.

On the other hand, if $Ax = b$ is premultiplied by B^{-1}, and some transposition of terms is made, it follows that

$$x_B + B^{-1}Nx_N = B^{-1}b.$$

Table 3.2 Equivalence between the associated quantities

Quantity	Simplex tableau	Relation	Revised simplex tableau
Objective row	$\bar{z}_N^T$	$=$	$c_N^T - c_B^T B^{-1} N$
Pivot column	$\bar{a}_q$	$=$	$B^{-1} a_q$
Right-hand side	$\bar{b}$	$=$	$B^{-1} b$
Pivot row	$e_p^T \bar{N}$	$=$	$e_p^T B^{-1} N$

Substituting the preceding to

$$c_B^T x_B + c_N^T x_N - f = 0$$

gives

$$(c_N^T - c_B^T B^{-1} N) x_N - f = -c_B^T B^{-1} b,$$

which can put in

x_B^T	x_N^T	f	RHS
I	$B^{-1}N$		$B^{-1}b$
	$c_N^T - c_B^T B^{-1} N$	-1	$-c_B^T B^{-1} b$

(3.20)

corresponding to basic solution $\bar{x}_B = B^{-1}b,\ \bar{x}_N = 0$ (hereafter $\bar{x}_B = B^{-1}b$ is often said basic solution for short). The preceding, representing a problem equivalent to the original, is called *revised simplex tableau*, compared to the simplex tableau (3.18).

For simplicity, x_B^T and f columns in the preceding two tableaus may be omitted, as they remain unchanged as basis changes.

Proposition 3.4.1. *Any simplex tableau and revised simplex tableau, corresponding to the same basis, are equivalent.*

Proof. Denote by (3.18) and (3.20) the two tableaus, having the same basis B. Since problems represented by them are equivalent, the corresponding entries of the two tableaus are equal. □

Based on the preceding Proposition, Table 3.2 gives equivalence correspondence between quantities, involved in simplex steps, of tableaus (3.18) and (3.20):

In conventional simplex context, each iteration corresponds to a basis B (or its inverse B^{-1}), with which any entry in a simplex tableau can be calculated from the original data (A, b, c). Thereby, Table 3.2 will be used as a tool to derive common simplex variants, such as the (revised) simplex algorithm in the next section and the dual (revised) simplex algorithm in Sect. 4.5.

Notations in this section will be employed throughout this book.

3.5 Simplex Method

A simplex tableau has to be calculated in each iteration by the tableau simplex Algorithm 3.2.1. But its $(m+1)\times(n+1)$ entries are not all useful in an iteration. In fact, only the objective row is needed for the selection of a pivot column, while the pivot column and right-hand side needed for the determination of a pivot row. Using B^{-1}, therefore, a variant without any simplex tableau can be derived by calculating the first three items in Table 3.2.

Let us consider updating B^{-1}. Assume that pivot column index q and row index p are already determined. Putting the nonbasic column a_q in place of B's pth column a_{j_p} gives the new basis below:

$$\hat{B} = (a_{j_1}, \ldots, a_{j_{p-1}}, a_q, a_{j_{p+1}}, \ldots, a_{j_m}). \tag{3.21}$$

It is now needed to compute $\hat{B}^{-1}$ to go on the next iteration.

Note that $\bar{a}_q = B^{-1}a_q$. Taking $\bar{a}_{pq}$ as the pivot, the according elementary transformations amount to premultiplying the first m rows of the tableau by $m \times m$ elementary matrix

$$E_p = \begin{pmatrix} 1 & & -\bar{a}_{1\,q}/\bar{a}_{p\,q} & & \\ & \ddots & \vdots & & \\ & & -\bar{a}_{p-1,\,q}/\bar{a}_{p\,q} & & \\ & & 1/\bar{a}_{p\,q} & & \\ & & -\bar{a}_{p+1,\,q}/\bar{a}_{p\,q} & & \\ & & \vdots & \ddots & \\ & & -\bar{a}_{m\,q}/\bar{a}_{p\,q} & & 1 \end{pmatrix} \begin{matrix} \\ \\ \\ p \\ \\ \\ \\ \end{matrix} \tag{3.22}$$

$$p$$

which may also be obtained by executing the same elementary transformations on the unit matrix. It is seen that such a matrix, which is the same as the unit matrix except for the pth column, is determined only by $\bar{a}_q$. Combining (3.21) and (3.22) gives

$$\begin{aligned} E_pB^{-1}\hat{B} &= E_p(B^{-1}a_{j_1}, \ldots, B^{-1}a_{j_{p-1}}, B^{-1}a_q, B^{-1}a_{j_{p+1}}, \ldots, B^{-1}a_{j_m}) \\ &= E_p(e_1, \ldots, e_{p-1}, \bar{a}_q, e_{p+1}, \ldots, e_m) \\ &= (E_pe_1, \ldots, E_pe_{p-1}, E_p\bar{a}_q, E_pe_{p+1}, \ldots, E_pe_m) = I, \end{aligned}$$

from which the update of the basis' inverse follows, i.e.,

$$\hat{B}^{-1} = E_p B^{-1}. \tag{3.23}$$

Based on the preceding discussions and the equivalence between the simplex tableau and revised simplex tableau, we are able to revise Algorithm 3.2.1 to the following version (Dantzig and Orchard-Hays 1953):

Algorithm 3.5.1 (Simplex algorithm 1). Initial: $(B, N), B^{-1}, \bar{x}_B = B^{-1}b \geq 0$ and $\bar{f} = c_B^T \bar{x}_B$. This algorithm solves the standard LP problem (1.8).

1. Compute $\bar{z}_N = c_N - N^T \bar{y}, \quad \bar{y} = B^{-T} c_B$.
2. Determine pivot column index $q \in \arg\min_{j \in N} \bar{z}_j$.
3. Stop if $\bar{z}_q \geq 0$ (optimality achieved).
4. Compute $\bar{a}_q = B^{-1} a_q$.
5. Stop if $\bar{a}_q \leq 0$ (unbounded problem).
6. Determine stepsize α and pivot row index p such that
   $\alpha = \bar{x}_{j_p}/\bar{a}_{p\,q} = \min\{\bar{x}_{j_i}/\bar{a}_{i\,q} \mid \bar{a}_{i\,q} > 0;\ i = 1, \ldots, m\}$.
7. Set $\bar{x}_q = \alpha$, and update $\bar{x}_B = \bar{x}_B - \alpha \bar{a}_q$, $\bar{f} = \bar{f} + \alpha \bar{z}_q$ if $\alpha \neq 0$.
8. Update B^{-1} by (3.23).
9. Update (B, N) by exchanging j_p and q.
10. Go to step 1.

The preceding, usually called *revised simplex algorithm*, will be referred to as *simplex algorithm 1*.

In step 1, vector $\bar{y}$ is calculated first, then it is used to compute reduced costs $\bar{z}_N$, as is referred to as *pricing*. $\bar{y}$ is called *simplex multipliers (vector)*, whose additional meanings will be clear later.

See Sect. 3.3 for how to provide an initial basic feasible solution (or basis). This topic will be handled further in Chap. 13.

Example 3.5.1. Solve the following problem by Algorithm 3.5.1:

$$\begin{array}{lllr}
\min & f = -4x_1 - 3x_2 - 5x_3, & & \\
\text{s.t.} & 2x_1 + x_2 + 3x_3 \qquad + x_5 & & = 15, \\
& x_1 + x_2 + \ x_3 + x_4 & & = 12, \\
& \qquad x_2 - 3x_3 & + x_7 & = 3, \\
& 2x_1 + x_2 & + x_6 & = 9, \\
& & x_j \geq 0, \quad j = 1, \ldots, 7. &
\end{array}$$

Answer Initial: $B = \{5, 4, 7, 6\}$, $N = \{1, 2, 3\}$, $B^{-1} = I$, $\bar{x}_B = (15, 12, 3, 9)^T$, $\bar{f} = 0$.

Iteration 1:

1. $\bar{y} = B^{-T} c_B = (0, 0, 0, 0)^T$, $\bar{z}_N = c_N - N^T \bar{y} = (-4, -3, -5)^T$.
2. $\min\{-4, -3, -5\} = -5 < 0$, $q = 3$.

4. $\bar{a}_3 \ = B^{-1}a_3 = (3, 1, -3, 0)^T$.
6. $\alpha \ = \min\{15/3,\ 12/1\} = 15/3 = 5,\ \ p = 1$.
7. $\bar{x}_B = (15, 12, 3, 9)^T - 5 \times (3, 1, -3, 0)^T = (0, 7, 18, 9)^T,\ x_3 = 5,$
   $f \ = 5 \times (-5) = -25$.

$$8.\ B^{-1} = \begin{pmatrix} 1/3 & & & \\ -1/3 & 1 & & \\ 1 & & 1 & \\ 0 & & & 1 \end{pmatrix}.$$

9. $B \ = \{3, 4, 7, 6\},\ N = \{1, 2, 5\},\ \bar{x}_B = (5, 7, 18, 9)^T$.

Iteration 2:

1. $\bar{y} \ = (-5/3, 0, 0, 0)^T,\ \bar{z}_N = (-4, -3, 0)^T - (-10/3, -5/3, -5/3)^T$
   $= (-2/3, -4/3, 5/3)^T$.
2. $\min\{-2/3, -4/3, 5/3\} = -4/3 < 0,\ q = 2$.
4. $\bar{a}_2 \ = (1/3, 2/3, 2, 1)^T$.
6. $\alpha \ = \min\{15,\ 21/2,\ 9, 9\} = 9,\ \ p = 3$.
7. $\bar{x}_B = (5, 7, 18, 9)^T - 9 \times (1/3, 2/3, 2, 1)^T = (2, 1, 0, 0)^T,\ x_2 = 9,$
   $f \ = -25 + 9 \times (-4/3) = -37$.

$$8.\ B^{-1} = \begin{pmatrix} 1 & & -1/6 & \\ 0 & 1 & -1/3 & \\ 0 & & 1/2 & \\ 0 & & -1/2 & 1 \end{pmatrix} \begin{pmatrix} 1/3 & & & \\ -1/3 & 1 & & \\ 1 & & 1 & \\ 0 & & & 1 \end{pmatrix} = \begin{pmatrix} 1/6 & & -1/6 & \\ -2/3 & 1 & -1/3 & \\ 1/2 & & 1/2 & \\ -1/2 & & -1/2 & 1 \end{pmatrix}.$$

9. $B \ = \{3, 4, 2, 6\},\ N = \{1, 7, 5\}, \bar{x}_B = (2, 1, 9, 0)^T$.

Iteration 3:

1. $\bar{y} \ = (-7/3, 0, -2/3, 0)^T,\ \bar{z}_N = (-4, 0, 0)^T - (-10/3, -2/3, -7/3)^T$
   $= (-2/3, 2/3, 7/3)^T$.
2. $\min\{-2/3, 2/3, 7/3\} = -2/3,\ q = 1$.
4. $\bar{a}_1 \ = (2/3, 1/3, 0, 2)^T$.
6. $\alpha \ = \min\{2/(2/3), 1/(1/3), 0/2\} = 0,\ \ p = 4$.
7. $\bar{x}_B = (2, 1, 9, 0)^T,\ x_6 = 0,\ f = -37$.

$$8.\ B^{-1} = \begin{pmatrix} 1 & & & -1/3 \\ 0 & 1 & & -1/6 \\ 0 & & 1 & 0 \\ 0 & & & 1/2 \end{pmatrix} \begin{pmatrix} 1/6 & & -1/6 & \\ -2/3 & 1 & -1/3 & \\ 1/2 & & 1/2 & \\ -1/2 & & -1/2 & 1 \end{pmatrix} = \begin{pmatrix} 1/3 & & 0 & -1/3 \\ -7/12 & 1 & -1/4 & -1/6 \\ 1/2 & & 1/2 & 0 \\ -1/4 & & -1/4 & 1/2 \end{pmatrix}.$$

9. $B \ = \{3, 4, 2, 1\},\ N = \{6, 7, 5\}, \bar{x}_B = (2, 1, 9, 0)^T$.

Iteration 4:

1. $\bar{y} \ = (-13/6, 0, -1/2, -1/3)^T,\ \bar{z}_N \ = \ (0, 0, 0)^T - (-1/3, -1/2, -13/6)^T \ =$ $(1/3, 1/2, 13/6)^T \geq 0$.

2. The optimal basic solution and optimal value:

$$\bar{x} = (0, 9, 2, 1, 0, 0, 0)^T, \qquad \bar{f} = -37.$$

If some practicable pivot rule (Chap. 11) or pricing scheme (Sect. 25.3) is used in the simplex method, there will be a need for computing row p. In order not to increase the number of systems to be solved, modern LP codes are often based on the following variant, where the objective row is computed in recurrence (see (3.13)).

Algorithm 3.5.2 (Simplex algorithm 2). Initial: (B, N), B^{-1}, $\bar{x}_B = B^{-1}b \geq 0$, $\bar{z}_N = c_N - N^T B^{-T} c_B$ and $\bar{f} = c_B^T \bar{x}_B$. This algorithm solves the standard LP problem (1.8).

1. Determine pivot column index $q \in \arg\min_{j \in N} \bar{z}_j$.
2. Stop if $\bar{z}_q \geq 0$ (optimality achieved).
3. Compute $\bar{a}_q = B^{-1} a_q$.
4. Stop if $\bar{a}_q \leq 0$ (unbounded).
5. Determine stepsize α and pivot row index p such that
   $\alpha = \bar{x}_{j_p} / \bar{a}_{p\,q} = \min\{\bar{x}_{j_i} / \bar{a}_{i\,q} \mid \bar{a}_{i\,q} > 0;\ i = 1, \ldots, m\}$.
6. Set $\bar{x}_q = \alpha$, and update $\bar{x}_B = \bar{x}_B - \alpha \bar{a}_q$, $\bar{f} = \bar{f} + \alpha \bar{z}_q$ if $\alpha \neq 0$.
7. Compute $\sigma_N = N^T v$, where $v = B^{-T} e_p$.
8. Update by: $\bar{z}_N = \bar{z}_N + \beta \sigma_N$, $\bar{z}_{j_p} = \beta$, where $\beta = -\bar{z}_q / \bar{a}_{p\,q}$.
9. Update B^{-1} by (3.23).
10. Update (B, N) by exchanging j_p and q.
11. Go to step 1.

Although they are equivalent in theory, the revised Algorithms differ from the tableau algorithm numerically. For solving large-scale LP problems, they are certainly superior to the latter (especially when $m \ll n$, see Sect. 3.8). In fact, it serves as a basis for designing practicable simplex variants, though the formulation of the latter is simpler, providing a suitable tool for illustration.

Algorithm 3.5.1 was previously derived based on the equivalence of the simplex tableau and revised simplex tableau. It may be derived alternatively by taking a downhill edge, emanating from a current vertex, as a search direction to form a line search scheme, as follows.

Without loss of generality, let $B = \{1, \ldots, m\}$ and $N = \{m + 1, \ldots, n\}$ be respectively the basis and nonbasis, associated with basic feasible solution $\bar{x}$. Assume that a pivot column index $q \in N$ has been determined such that

$$\bar{z}_q = c_q - a_q^T B^{-T} c_B < 0.$$

Introduce vector

$$\Delta x = \begin{pmatrix} B^{-1} a_q \\ -e_{q-m} \end{pmatrix}, \tag{3.24}$$

where e_{q-m} is the $(n-m)$-dimensional unit vector with the $(q-m)$th component 1. It is clear that

$$-c^T\Delta x = c_q - a_q^T B^{-T} c_B = \bar{z}_q < 0. \quad (3.25)$$

Therefore, $-\Delta x$ is a downhill with respect to $c^T x$. Taking it as search direction gives the following line search scheme:

$$\hat{x} = \bar{x} - \alpha\Delta x, \quad (3.26)$$

where $\alpha \geq 0$ is a stepsize to be determined.

Since $\bar{x}$ is feasible, it holds for any $\alpha \geq 0$ that

$$A\hat{x} = A\bar{x} - \alpha[B, N]\Delta x = A\bar{x} = b.$$

Therefore, what should do is to maximize α subject to $\hat{x}_B \geq 0$. When $B^{-1}a_q \not\leq 0$, such doing results in α and p such that

$$\alpha = \bar{x}_{j_p}/(B^{-1}a_{p\,q}) = \min\{\bar{x}_{j_i}/(B^{-1}a_{i\,q}) \mid B^{-1}a_{i\,q} > 0,\ i = 1,\ldots,m\}. \quad (3.27)$$

It is clear that the according new solution $\hat{x}$ is still feasible. In fact, it is verified that $\hat{x}$ is just the basic feasible solution, associated with the new basis resulting from the old by exchanging j_p and q.

The relation between the new and old basis matrices is

$$\hat{B} = B + (a_q - a_{j_p})e_p^T.$$

In view of that a_{j_p} is the pth column of B and that $B^{-1}a_{j_p} = e_p$ and $B^{-1}a_q = \bar{a}_q$ hold, it is not difficult to derive the following result from Sherman-Morrison formula (Golub and Van Loan 1989):

$$\hat{B}^{-1} = B^{-1} - \frac{B^{-1}(a_q - a_{j_p})e_p^T B^{-1}}{1 + e_p^T B^{-1}(a_q - a_{j_p})} = \left(I - \frac{(\bar{a}_q - e_p)e_p^T}{\bar{a}_{pq}}\right) B^{-1}, \quad (3.28)$$

which may serve as an update of B^{-1}. In fact, it is easily verified that the preceding and (3.23) are actually equivalent.

The search direction $-\Delta x$, defined by (3.24), can be further investigated geometrically. Regarding set

$$\begin{aligned} E &= \{x \in \mathcal{R}^n \mid Ax = b,\ x \geq 0,\ x_j = 0,\ q \neq j \in N\} \\ &= \{x \in \mathcal{R}^n \mid x_B = B^{-1}b - x_q(B^{-1}a_q) \geq 0,\ x_q \geq 0,\ x_j = 0,\ q \neq j \in N\}, \end{aligned} \quad (3.29)$$

we have the following result.

Proposition 3.5.1. *Set E is a downhill edge, emanating from the current vertex $\bar{x}$, and $-\Delta x$ is its direction. If $B^{-1}a_q \leq 0$, then $-\Delta x$ is an (unbounded) extreme direction.*

Proof. It is clear that E is a half-line or edge, emanating from $\bar{x}$. By (3.29), (3.27) and (3.24), for any $x \in E \subset P$ it holds that

$$\begin{aligned} E &= \{x \in \mathcal{R}^n \mid x_B = B^{-1}b - x_q(B^{-1}a_q),\ 0 \leq x_q \leq \alpha,\ x_j = 0,\ q \neq j \in N\} \\ &= \{x \in \mathcal{R}^n \mid x = \bar{x} - x_q \Delta x,\ x_q \in [0, \alpha]\}. \end{aligned}$$

By (3.25), it is known that the associated objective value satisfies

$$f = c^T x = c^T \bar{x} - x_q c^T \Delta x = c^T \bar{x} + x_q \bar{z}_q \leq c^T \bar{x}.$$

Note that (3.27) is well-defined when $B^{-1}a_q \not\leq 0$. If, in addition, $\bar{x}_{j_p} = 0$, then $\alpha = 0$, and hence E degenerates to vertex $\bar{x}$. If $\bar{x}_{j_p} > 0$, hence $\alpha > 0$, then the associated objective value strictly decreases with $x_q \in [0, \alpha]$. Therefore, $-\Delta x$ is a direction of the downhill edge E. When $B^{-1}a_q \leq 0$, it is clear that $\alpha = +\infty$ corresponds to the edge $E \in P$, and hence $-\Delta x$ is an extreme direction. □

Note that edge E, defined by (3.29), could degenerate to the current vertex $\bar{x}$ if some component of $B^{-1}b$ vanishes and that the objective value is lower unbounded over the feasible region if $-\Delta x$ is an extreme direction.

3.6 Degeneracy and Cycling

It was seen that a zero stepsize leads to the same basic feasible solution, and hence the unchanged objective value. Thus, finiteness of the simple method is questionable (see, e.g., Ryan and Osborne 1988; Wolfe 1963). Soon after its emerging, in fact, the simplex method is found not to terminate in few cases. E.M.L. Beale (1955) and A.J. Hoffman (1953) offered such instances independently. The following is due to Beale.

Example 3.6.1. Solve the following problem by Algorithm 3.2.1:

$$\begin{array}{llllllll} \min & f = & & & -3/4x_4 & +20x_5 & -1/2x_6 & +6x_7, \\ \text{s.t.} & x_1 & & & +1/4x_4 & -8x_5 & -x_6 & +9x_7 = 0, \\ & & x_2 & & +1/2x_4 & -12x_5 & -1/2x_6 & +3x_7 = 0, \\ & & & x_3 & & & +x_6 & = 1, \\ & & & & x_j \geq 0, & j = 1, \ldots, 7. & & \end{array}$$

Answer Initial: the following feasible tableau is available from the preceding:

x_1	x_2	x_3	x_4	x_5	x_6	x_7	RHS
1			1/4*	−8	−1	9	
	1		1/2	−12	−1/2	3	
		1			1		1
			−3/4	20	−1/2	6	

Iteration 1:

1. $\min\{-3/4, 20, 1, -1/2, 6\} = -3/4 < 0,\ q = 4.$
3. $I = \{1, 2\} \neq \emptyset.$
4. $\min\{0/(1/4), 0/(1/2)\} = 0,\ p = 1.$
5. Multiply row 1 by 4, and then add $-1/2, 3/4$ times of row 1 to rows 2,4, respectively:

x_1	x_2	x_3	x_4	x_5	x_6	x_7	RHS
4			1	−32	−4	36	
−2	1			4*	3/2	−15	
		1			1		1
3				−4	−7/2	33	

Iteration 2:

1. $\min\{3, -4, -7/2, 33\} = -4 < 0,\ q = 5.$
3. $I = \{2\} \neq \emptyset.$
4. $\min\{0/4\} = 0,\ p = 2.$
5. Multiply row 2 by 1/4, and then add $32, 4$ times of row 2 to rows 1,4, respectively:

x_1	x_2	x_3	x_4	x_5	x_6	x_7	RHS
−12	8		1		8*	−84	
−1/2	1/4			1	3/8	−15/4	
		1			1		1
1	1				−2	18	

Iteration 3:

1. $\min\{1, 1, -2, 18\} = -2 < 0,\ q = 6.$
3. $I = \{1, 2, 3\} \neq \emptyset.$
4. $\min\{0/8, 0/(3/8), 1/1\} = 0,\ p = 1.$
5. Multiply row 1 by 1/8, and then add $-3/8, -1, 2$ times of row 1 to rows 2,3,4, respectively:

x_1	x_2	x_3	x_4	x_5	x_6	x_7	RHS
−3/2	1		1/8		1	−21/2	
1/16	−1/8		−3/64	1		3/16*	
3/2	−1	1	−1/8			21/2	1
−2	3		1/4			−3	

Iteration 4:

1. $\min\{-2, 3, 1/4, -3\} = -3 < 0,\ q = 7$.
3. $I = \{2, 3\} \neq \emptyset$.
4. $\min\{0/(3/16), 1/(21/2)\} = 0,\ p = 2$.
5. Multiply row 2 by 16/3, and then add $21/2, -21/2, 3$ times of row 2 to rows 1,3,4, respectively:

x_1	x_2	x_3	x_4	x_5	x_6	x_7	RHS
2*	−6		−5/2	56	1		
1/3	−2/3		−1/4	16/3		1	
−2	6	1	5/2	−56			1
−1	1		−1/2	16			

Iteration 5:

1. $\min\{-1, 1, -1/2, 16\} = -1 < 0,\ q = 1$.
3. $I = \{1, 2\} \neq \emptyset$.
4. $\min\{0/2, 0/(1/3)\} = 0,\ p = 1$.
5. Multiply row 1 by 1/2, and then add $-1/3, 2, 1$ times of row 1 to rows 2,3,4, respectively:

x_1	x_2	x_3	x_4	x_5	x_6	x_7	RHS
1	−3		−5/4	28	1/2		
	1/3*		1/6	−4	−1/6	1	
		1			1		1
	−2		−7/4	44	1/2		

Iteration 6:

1. $\min\{-2, -7/4, 44, 1/2\} = -2 < 0,\ q = 2$.
3. $I = \{2\} \neq \emptyset$.
4. $\min\{0/(1/3)\} = 0,\ p = 2$.

5. Multiply row 2 by 3, and then add 3, 2 times of row 2 to rows 1,4, respectively:

x_1	x_2	x_3	x_4	x_5	x_6	x_7	RHS
1			1/4	−8	−1		
	1		1/2	−12	−1/2	3	
		1			1		1
			−3/4	20	−1/2	6	

It is seen that stepsizes are equally zero in all the six iterations, and the last tableau is the same as the first one, consequently. Therefore, continuing the process must generate the same sequence of tableaus, as is a phenomena called *cycling*. So, the simplex algorithm failed to solve Beale's problem. It is clear that such a hated infinite case to the simplex method could occur only when degeneracy presents.

At the early days of the simplex method, some scholars thought that degeneracy hardly happens in practice, and up to now the nondegeneracy is still frequently assumed in theory. However, it turns out that degeneracy almost always presents when the simplex method is applied to solving real-world LP problems. Even so, fortunately, cycling rarely occurs, except for few artificial instances, and the simplex method has achieved great success in practice.

The real problem caused by degeneracy seems to be *stalling*, as it degrades method's performance seriously when a large number of iterations stay at a vertex for too long a time before exiting it. It is especially a headache for highly degenerate problems, where vanished basic components occupy a large proportion, as leads to a huge number of iterations. But this hard problem is only with the simplex method using the conventional pivot rule, not with variants using rules, presented in Chap. 11.

3.7 Finite Pivot Rule

As was shown int the previous section, the finiteness of the simplex method is not guaranteed in general. An approach or pivot rule that turns the simplex method to a finite one is called *finite*.

Is there any finite approach or pivot rule?

The answer is positive. Charnes (1952) proposed a "perturbation approach" by adding a perturbation term to the right-hand side of the initial feasible simplex tableau, i.e.,

$$w = (\epsilon, \epsilon^2, \ldots, \epsilon^m)^T,$$

where $\epsilon > 0$ is a sufficiently small parameter (while still using Dantzig's original rule for pivot column selection).

Theorem 3.7.1. *The perturbation approach is finite.*

Proof. The perturbation term added to the right-hand side can be put in the form $w = Iw$. In any iteration, the right-hand side can be written

$$v \triangleq \bar{b} + Uw,$$

where U is a permutation, resulting from performing elementary transformations on I. Note that U and I have the same rank m, and every row of U is nonzero. Firstly, it holds that $\bar{b} \geq 0$, because if, otherwise, $\bar{b}_i < 0$ for some $i \in \{1, \ldots, m\}$, then it follows that $v_i < 0$, as contradicts to problem's feasibility. Further, it is clear that $v_i > 0$ holds for all row indices i, satisfying $\bar{b}_i > 0$; on the other hand, $v_i > 0$ also holds for all row index i, satisfying $\bar{b}_i = 0$, because the first nonzero of the ith row of U is positive (otherwise, it contradicts the feasibility). Therefore, $v > 0$ holds. Since each tableau corresponds to a nondegenerate basic feasible solution, there is no any possibility of cycling, hence the process terminates within finitely many iterations. Consequently, eliminating all parameter terms in the end tableau leads to the final tableau of the original problem. □

The order of two vectors, determined by their first different components, is called lexicographic order. Equal vectors are regarded as equal in the lexicographic order. $(\lambda_1, \ldots, \lambda_t) \prec (\mu_1, \ldots, \mu_t)$ means that the former is less than the latter in lexicographic order, that is, for the smallest subscript i, satisfying $\lambda_i \neq \mu_i$, it holds that $\lambda_i < \mu_i$. Similarly, "$\succ$" is used to denote "greater than" in lexicographic order.

Once a pivot column index q is determined, the perturbation approach amounts to determining a pivot row index p by

$$p \in \arg\min\{(\bar{b}_i + u_{i\,1}\epsilon + u_{i\,2}\epsilon^2 + \cdots + u_{i\,m}\epsilon^m)/\bar{a}_{i\,q} \mid \bar{a}_{i\,q} > 0;\ i = 1, \ldots, m\}. \tag{3.30}$$

As ϵ is sufficiently small, the preceding is equivalent to the following so-called *lexicographic rule* (Dantzig et al. 1955):

$$p \in \arg\mathit{min}\{(\bar{b}_i, u_{i\,1}, u_{i\,2}, \ldots, u_{i\,m})/\bar{a}_{i\,q} \mid \bar{a}_{i\,q} > 0;\ i = 1, \ldots, m\}, \tag{3.31}$$

where $u_{i\,j}$ is the entry at the ith row and the jth column of U, and "min" is minimization in the sense of lexicographic order.

Among existing finite rules, Bland (1977) rule draws great attention due to its simplicity (also see Avis and Chvatal 1978).

Rule 3.7.1 (Bland rule) Among nonbasic variables, corresponding to negative reduced costs, select the smallest-indexed one to enter the basis. When there are multiple rows, attaining the same minimum-ratio, select the basic variable with the smallest index to leave the basis.

Theorem 3.7.2. *Bland rule is finite.*

Proof. Assume that cycling occurs with the simplex algorithm using Bland rule. If some variable leaves the basis in a circle, it must enter the basis gain. Denote by T the index set of such shuttling variables, and define

$$t = \max\{j \in T\}.$$

Note that the stepsize is always equal to 0 in each iteration in the circle, and hence leads to the same basic feasible solution; besides, the h-indexed component of the basic feasible solution is 0 for any $h \in T$.

Assume that x_t is selected to enter the basis for simplex tableau

$$\begin{array}{cc|c} \bar{A} & 0 & \bar{b} \\ \hline \bar{z}^T & -1 & \bar{\beta} \end{array} \tag{3.32}$$

thus $\bar{z}_t < 0$, and $\bar{z}_j \geq 0$ for any reduced cost's index $j < t$.

Assume that at another simplex tableau

$$\begin{array}{cc|c} \hat{A} & 0 & \hat{b} \\ \hline \hat{c}^T & -1 & \hat{\beta} \end{array} \tag{3.33}$$

basic variable x_t in row p leaves and nonbasic variable x_s enters the basis. Let $x_{j_1}, \ldots, x_{j_m}$ be basic variables ($x_{j_p} \equiv x_t$). It follows that $\hat{c}_s < 0$, and $\hat{c}_j \geq 0$ for any reduced cost's index $j < s$. Note that pivot is positive, i.e., $\hat{a}_{ps} > 0$; since $s \in T$, it holds that $s < t$.

Define v_k, $k = 1, \ldots, n, n+1$ as follows:

$$v_k = \begin{cases} 1, & k = s, \\ -\hat{a}_{is}, & k = j_i, \quad i = 1, \ldots, m, \\ \hat{c}_s, & k = n+1, \\ 0, & \text{otherwise.} \end{cases} \tag{3.34}$$

Note that basic columns of $\hat{A}$ constitute a permutation. Nonbasic components of vector

$$v = (v_1, \ldots, v_n, v_{n+1})^T$$

are all 0, except for $v_s = 1$. For $i = 1, \ldots, m$, on the other hand, the basic entries in row i of $\hat{A}$, except for $\hat{a}_{ij_i} = 1$, are all zero; basic entries of $\hat{c}$ are all zero. Therefore it holds that

$$\begin{pmatrix} \hat{A} & 0 \\ \hat{c}^T & -1 \end{pmatrix} v = \begin{pmatrix} \hat{a}_{1,s} - \hat{a}_{1,s} \\ \vdots \\ \hat{a}_{m,s} - \hat{a}_{m,s} \\ \hat{c}_s - \hat{c}_s \end{pmatrix} = 0. \tag{3.35}$$

Since (3.32) can be obtained from (3.33) by premultiplying a series of elementary matrices, it follows that

$$\begin{pmatrix} \bar{A} & 0 \\ \bar{z}^T & -1 \end{pmatrix} v = 0, \tag{3.36}$$

where the last equality is

$$\sum_{k=1}^{n} \bar{z}_k v_k - v_{n+1} = 0,$$

hence

$$\sum_{k=1}^{n} \bar{z}_k v_k = v_{n+1} = \hat{c}_s < 0.$$

Therefore, there exists some index $h < n + 1$ such that

$$\bar{z}_h v_h < 0, \tag{3.37}$$

giving $\bar{z}_h \neq 0$ and $v_h \neq 0$.

On the other hand, it is known by $v_h \neq 0$ and the definition of v that $h \in \{j_1, \ldots, j_m, s\}$. Thus, there are only following three cases arising:

(i) $h = s$. $v_h = 1$ in this case. Since x_t is an entering variable for simplex tableau (3.32) and $h = s < t$, hence $\bar{z}_h > 0$, it follows that $\bar{z}_h v_h = \bar{z}_h > 0$, contradicting (3.37).

(ii) $h = j_p = t$. In this case, from $\bar{z}_h = \bar{z}_t < 0$ and $v_h = -\hat{a}_{ps} < 0$, it follows that $\bar{z}_h v_h > 0$, contradicting (3.37).

(iii) $h = j_i \neq j_p$ or $h \neq t$. Now x_h is a nonbasic variable of simplex tableau (3.32) (otherwise, $\bar{z}_h = 0$); it is also a basic index of simplex tableau (3.33), hence $h \in T$. It follows that

$$\hat{b}_i = 0, \qquad h < t, \tag{3.38}$$

and hence $\bar{z}_h > 0$. Further, it holds that

$$v_h = -\hat{a}_{i,s} > 0,$$

since, otherwise, $v_h \neq 0$ gives $\hat{a}_{i,s} > 0$, from which and (3.38) it follows that x_h, rather than x_t, were selected to leave the basis for simplex tableau (3.33), as a contradiction to (3.37). Therefore, Bland rule is finite. □

Chang (1979), Terlaky (1985) and Wang (1987) independently proposed a so-called "criss-cross" finite variant of Bland rule, which is embedded in a somehow different context, compared with the simplex method (see Chap. 18).

Unfortunately, it turns out that these finite rules are very slow in practice, not be mentioned in the same breath with the conventional rule. This is not surprising, however. For example, Rule 3.7.1 gives nonbasic variables with small index priority to enter the basis, while we all know that basic variables of an optimal solution are not necessarily small indexed.

Rule 3.7.1 actually uses a priority order, coinciding with decreasing indices, for selection of an entering variable. It is clear that the "ideal" order, if any, should enter the basic variables of an optimal solution to the basis. According to the heuristic Proposition 2.5.1, inequality constraints with small pivoting-indices should be satisfied as equations by an optimal solution, therefore the corresponding variables would be better to be nonbasic (zero-valued). In other words, variables with large pivoting-indices should have the priority to enter the basis (stipulation: among variables with equal pivoting-indices, select one with the largest index). Thus, we have the following variant of Bland rule (Pan 1990, 1992c).

Rule 3.7.2. Among nonbasic variables, corresponding to negative reduced costs, select the largest pivoting-indexed one to enter the basis. When there are multiple rows, attaining the same minimum-ratio, select the largest pivoting-indexed basic variable to leave the basis. When multiple variables correspond to the same largest pivoting-index, take the largest indexed one.

Theorem 3.7.3. *Rule 3.7.2 is finite.*

Proof. This rule is equivalent to Rule 3.7.1 if variables are re-given indices in accordance with their pivoting-indices. □

Preliminary computational experiments with small test problems showed that performance of Rule 3.7.2 is much better than Bland's Rule 3.7.1. It might be the best among known finite rules. However, it is still inferior to the conventional rule, as requiring more iterations than the latter in general (Pan 1990).

Bland's Rule can be easily generalized to the following finite rule.

Rule 3.7.3. Given any order for variables. Among nonbasic variables, corresponding to negative reduced costs, select one the smallest in this order to enter the basis. When there are multiple rows attaining the same minimum-ratio, select the basic variable smallest in the order to leave the basis.

In Example 3.6.1 (Beale problem), we have seen that cycling occurred with the simplex algorithm. The situation will be different if Rule 3.7.2 is used in the place of the conventional rule.

Example 3.7.1. Solve Beale problem by Algorithm 3.2.1 using Rule 3.7.2:

$$\begin{aligned}
\min\quad & f = -3/4x_4 + 20x_5 - 1/2x_6 + 6x_7,\\
\text{s.t.}\quad & x_1 \qquad + 1/4x_4 - 8x_5 - x_6 + 9x_7 = 0,\\
& \quad x_2 \quad + 1/2x_4 - 12x_5 - 1/2x_6 + 3x_7 = 0,\\
& \qquad x_3 \qquad\qquad + x_6 = 1,\\
& x_j \geq 0, \quad j = 1, \ldots, 7.
\end{aligned}$$

Answer As the coefficient matrix includes a unit matrix, it is easy to transform the constraints to "$\geq$" type of inequalities:

$$\begin{aligned}
\min\quad & f = -3/4x_4 + 20x_5 - 1/2x_6 + 6x_7,\\
\text{s.t.}\quad & -1/4x_4 + 8x_5 + x_6 - 9x_7 \geq 0,\\
& -1/2x_4 + 12x_5 + 1/2x_6 - 3x_7 \geq 0,\\
& - x_6 \geq -1,\\
& x_j \geq 0, \quad j = 4, \ldots, 7.
\end{aligned}$$

Note that the first three constraints correspond to the original variables x_1, x_2, x_3, respectively; pivoting-indices of constraints may be regarded as those for the associated variables.

The gradient of the objective function is $c = (-3/4, 20, -1/2, 6)^T$. The gradient of the first constraint is $a_1 = (-1/4, 8, 1, -9)^T$. The pivoting-index of this constraint (or corresponding variable x_1) is $\alpha_1 = -a_1^T c/\|a_1\| = -8.74$. Similarly, calculate all pivoting-indices and put them in the following table in decreasing order:

Variable	Constraint		α_i
x_4	x_4	≥ 0	0.75
x_6	x_6	≥ 0	0.50
x_3	$-x_6$	≥ -1	−0.50
x_7	x_7	≥ 0	−6.00
x_1	$-1/4x_4 + 8x_5 + x_6 - 9x_7$	≥ 0	−8.74
x_5	x_5	≥ 0	−20.00
x_2	$-1/2x_4 + 12x_5 + 1/2x_6 - 3x_7$	≥ 0	−114.78

Now call Algorithm 3.2.1 with Rule 3.7.2.

Initial: The following feasible simplex tableau is obtained directly from the preceding problem:

x_1	x_2	x_3	x_4	x_5	x_6	x_7	RHS
1			1/4*	−8	−1	9	
	1		1/2	−12	−1/2	3	
		1			1		1
			−3/4	20	−1/2	6	

Iteration 1:

1. Among nonbasic variables x_4 ($\alpha_4 = 0.75$) and x_6 ($\alpha_6 = 0.50$) with negative reduced costs, select the largest pivoting-indexed x_4 to enter the basis, $q = 4$.
3. $I = \{1, 2\} \neq \emptyset$.
4. $\min\{0/(1/4), 0/(1/2)\} = 0$. Among basic variables in rows 1 and 2, select the largest pivoting-indexed x_1 ($\alpha_1 = -8.74 > -114.78 = \alpha_2$) to leave the basis, $p = 1$.
5. Multiply row 1 by 4, and then add $-1/2, 3/4$ times of row 1 to rows 2,4, respectively:

x_1	x_2	x_3	x_4	x_5	x_6	x_7	RHS
4			1	−32	−4	36	
−2	1			4	3/2*	−15	
		1			1		1
3				−4	−7/2	33	

Iteration 2:

1. Among nonbasic variables x_5 ($\alpha_5 = -20.00$) and x_6 ($\alpha_6 = 0.50$) with negative reduced costs, select the largest-pivoting-indexed x_6 to enter the basis, $q = 6$.
3. $I = \{2, 3\} \neq \emptyset$.
4. $\min\{0/(3/2), 1/1\} = 0$, only x_2 is eligible for leaving the basis, $p = 2$.
5. Multiply row 2 by $2/3$, and then add $4, -1, 7/2$ times of row 2 to rows 1,3,4, respectively:

x_1	x_2	x_3	x_4	x_5	x_6	x_7	RHS
−4/3	8/3		1	−64/3		−4	
−4/3	2/3			8/3	1	−10	
4/3	−2/3	1		−8/3		10*	1
−5/3	7/3			16/3		−2	

Iteration 3:

1. Among nonbasic variables x_1 ($\alpha_1 = -8.74$) and x_7 ($\alpha_7 = -6.00$) with negative reduced costs, select the largest-pivoting-indexed x_7 to enter the basis, $q = 7$.
3. $I = \{3\} \neq \emptyset$.
4. Only x_3 is eligible for leaving the basis, $p = 3$.
5. Multiply row 3 by $1/10$, and then add $4, 10, 2$ times of row 3 to rows 1,2,4, respectively:

x_1	x_2	x_3	x_4	x_5	x_6	x_7	RHS
$-4/5$	$12/5$	$2/5$	1	$-112/5$			$2/5$
		1			1		1
$2/15^*$	$-1/15$	$1/10$		$-4/15$		1	$1/10$
$-7/5$	$11/5$	$1/5$		$24/5$			$1/5$

Iteration 4:

1. Only nonbasic variable x_1 is eligible for entering the basis, $q = 1$.
3. $I = \{3\} \neq \emptyset$.
4. Only x_7 is eligible for leaving the basis, $p = 3$.
5. Multiply row 3 by $15/2$, and then add $4/5, 7/5$ times of row 3 to rows 1,4, respectively:

x_1	x_2	x_3	x_4	x_5	x_6	x_7	RHS
	2	$2/5$	1	-24		6	1
		1			1		1
1	$-1/2$	$3/4$		-2		$15/2$	$3/4$
	$3/2$	$5/4$		2		$21/2$	$5/4$

All reduced costs are now nonnegative. The optimal solution and optimal value are

$$\bar{x} = (3/4, 0, 0, 1, 0, 1, 0)^T, \qquad \bar{f} = -5/4.$$

Thus, Beale problem is solved without cycling.

3.8 Computational Complexity

The evaluation of an algorithm is concerned with the amount of required arithmetics and storages, numerical stability and degree of difficulty for programming. In this section, we will discuss the simplex method's computational complexity, including time complexity (estimate of the number of required four basic arithmetics and comparisons), and storage complexity (estimate of the number of memory locations).

Either time or storage complexity is closely related to the scale of the problem handled: the larger the problem is, the higher the complexity. Therefore, analyzing complexity must be done with fixed problem's size. As for a standard LP problem, it is convenient to use m and n to determine its size roughly. Further, problem's size also depends on concrete values of (A, b, c), as can be characterized by total number L, called *input length* of binary digits of input data. For reaching a certain solution precision, the amount of arithmetics is a function of m, n, L. If the number of arithmetics required by solving some type of problems is bounded above by some function $\tau f(m, n, L)$, the algorithm is said to have order $O(f(m, n, L))$ of (time) complexity, where $\tau > 0$ is a constant and $f(m, n, L)$ is *complexity function*. If $f(m, n, L)$ is a polynomial in m, n and L, it is said to be of polynomial time complexity. Usually, such algorithms are regarded as "good" ones, and the lower the order of the polynomial is, the better the algorithm. On the other hand, if $f(m, n, L)$ is exponential in m, n or L, it is said to be of exponential time complexity. Such algorithms are regarded as "bad", as they can fail to solve larger problems by consuming unacceptable amount of time. Note, however, that such complexity is the worst case complexity, as the amount of arithmetics never exceeds $\tau f(m, n, L)$.

The following table lists numbers of arithmetics per iteration and storage locations, required by tableau simplex Algorithm 3.2.1 vs. simplex Algorithm 3.5.1:

	Algorithm 3.2.1	Algorithm 3.5.1
Multiplications	$(m+1)(n-m+1)$	$m(n+2m)+2m+1$
Additions	$m(n-m+1)$	$m(n+2m)-2m+1$
Storages	$(m+1)(n+1)$	$(m+1)(m+2)$

In the preceding table, the amount of storage locations required by Algorithm 3.2.1 excludes that for original data (A, b, c), though these data should be stored, in practice. In fact, both algorithms have to restart from scratch periodically after a certain number of iterations (see Sect. 5.1), let alone Algorithm 3.5.1 utilizes a part of them in each iteration. Therefore, storage requirement of the tableau simplex algorithm is significantly high, relative to that of the revised version, especially when $n \gg m$.

As iterative algorithms, their time complexity depends on the required number of iterations, as well as that of arithmetics per iteration. As they are equivalent theoretically, the two algorithms would require the same iterations in solving any standard LP problem, if rounding errors are neglected. Thus, we only compare the amount of arithmetics, mainly multiplications, in a single iteration. It is seen from the table that Algorithm 3.5.1 is much superior to the tableau version if $m \ll n$. In fact, the latter is not applied in practice, but only seen in textbooks. Note that all listed in the table is for dense computations. As for sparse computations, the former and its variants are even very much superior to the latter (see Chap. 5).

In addition, it is seen that the numbers of arithmetics per iteration are polynomial functions in m, n. Therefore, required iterations are a key to their time complexity. Note that each iteration corresponds to a basis, and the number of bases is no more than C_n^m. If $n \geq 2m$, then $C_n^m \geq (n/m)^m \geq 2^m$, as indicates that the required number of iterations could attain an exponent order. Indeed, Klee and Minty (1972) offered an example, indicating that the simplex method using the conventional pivot rule passes through all the 2^m vertices. Thus, the conventional rule is not polynomial, in the sense that it does not turn the simplex method to a polynomial-time one. Moreover, it turns out that Bland's Rule 3.7.1, the "most improvement rule", and many other subsequent rules, like the steepest edge rule, are all not polynomial (Chap. 11). Actually, it has not been clear whether there exists a polynomial rule though such possibility seems to be very low, if any.

Computational experiences indicate that the conventional simplex algorithm is slow for solving certain type of LP problems, such as some hard large-scale problems or problems with combinatorial constraints; e.g., with zero or one coefficients or those from "Krawchuk polynomials" (Schrijver 1986, p. 141; also see Klee 1965). Nevertheless, its average efficiency is quite high. For solving small or medium LP problems, in particular, it usually requires iterations no more than $4m$ to $6m$ (including Phase-1).

The fact that the non-polynomial-time simplex algorithm and its variants perform very well in practice reveals that the worst case complexity is of limited reference value, even could be misguiding. In fact, the worst case hardly happens in practice, and complexity under some probability sense would be closer to reality. In this aspect, Borgwardt (1982a,b) showed that an average complexity of the simplex algorithm is polynomial. Specifically, for LP problem

$$\begin{array}{ll} \min & c^T x, \\ \text{s.t.} & Ax \leq b, \end{array}$$

where $A \in \mathcal{R}^{m\times n}$, $b > 0$, and components of c are random under certain assumptions, he proved that the mathematical expectation of iterations, required by the simplex algorithm using a special pivot rule, is

$$O(n^3 m^{1/(n-1)}).$$

Using a different probability model and pivot rule, Smale (1983a,b) proved that average complexity of the simplex algorithm when solving

$$\begin{array}{lll} \min & c^T x, & \\ \text{s.t.} & Ax \leq b, & x \geq 0, \end{array}$$

is bounded above by

$$O((log\, n)^{m^2+m}),$$

which is not polynomial, but still better than Borgwardt's result when $m \ll n$. Combining Borgwardt's pivot rule and a generalized Smale's probability model, Haimovich (1996) proved that the average complexity of iterations required is linear polynomial. These theoretical results coincide with real situation.

Finally, we stress that algorithms' evaluation is basically a practical issue. In a word, practice is the unique touchstone: the value and vitality of an algorithm lie on its performance only.

3.9 On Features of the Simplex Method

In this final section of the chapter, we focuss on some features of the simplex method.

It is interesting that the method's prefix "simplex" came from a chat between G.B. Dantzig and T. Motzkin (Dantzig 1991) in the early days of LP. The latter indicated that the m columns of the basis matrix and the entering column just form a "simplex" in the m-dimensional space. Thus, each iteration in Dantzig's method may be viewed as a movement from a simplex to an adjacent simplex. Dantzig accepted his suggestion by consorting with the "simplex".

Accordingly, the simplex method is pivotal and basis-based, as is closely related to the linear structure of the LP model. Each iteration of it is characterized by a basis: once a basis is determined, so is done the corresponding basic feasible solution. If optimality cannot be asserted, a pivot is selected to make a basis change to improve the solution, or unboundedness of the problem is detected. Consequently, computational work per iteration, involved in the simplex method, is much less than that required by the interior-point method (Chap. 9).

If an optimal basis is available, an LP problem can be handled by just solving a single system of linear equations. Even if this is not the case, a basis close to an optimal one is useful: less iterations are usually required starting from a basis, yielded from a previously interrupted solution process. Such a so-called "warm start" features a source of main bonus of the simplex method. For instance, it is applied to sensitivity analysis and parametric programs (Chap. 6), the restarting tactic used in implementation (Chap. 5), as well as the decomposition principle (Sect. 25.6). In addition, the warm start is of great importance to the methodology for solving ILP problems.

It is noted that each iteration of the simplex method consists of a pivot selection and a basis change. Since it emerged, in fact, research on the method has not been beyond the scope of the two aspects. On one side, the pivot rule used in it is, no doubt, crucial to method's efficiency. As a result, new pivot rules were suggested from time to time, though Dantzig's original rule, because of its simplicity, had gained broad applications for a long time, as is a situation that has changed only about 20 years ago. More efficient rules will be presented in Chap. 11. On the other side, the computation related to pivot and basis change has been improved continually. Related results will be presented in later chapters, especially in Part II of this book.

As for concern whether an index enters and leaves the basis too many times, the following property seems to be favorable.

Proposition 3.9.1. *A leaving column in a simplex iteration does not enter the basis in the next iteration.*

Proof. Since an entering column corresponds to a negative reduced cost and the pivot determined is positive (see (3.10)), a leaving column corresponds to a positive reduced cost, after the associated elementary transformations carried out, and hence never enters the basis in the next iteration. □

Nevertheless, it is not difficult to construct an instance, in which a column that just entered the basis leaves it immediately.

As was known, nonnegativity of nonbasic reduced costs is not a necessary condition for optimality. The following indicates that it is necessary if the nondegeneracy is ensured.

Proposition 3.9.2. *If a basic optimal solution is nondegenerate, reduced costs in the associated simplex tableau are all nonnegative.*

Proof. Assume that there are negative reduced costs in the simplex Tableau 3.1. Without loss of generality, assume $\bar{z}_q < 0$. If (3.7) holds, then unboundedness of the problem follows from Theorem 3.2.2, as contradicts the existence of an optimal solution; if, otherwise, (3.7) does not hold, it is known from the nondegeneracy assumption that $\alpha > 0$, and hence there is a feasible value strictly less than the optimal value, as is a contradiction. Therefore, reduced costs are all nonnegative. □

The preceding and Lemma 3.2.1 together imply that the condition of nonnegativity of nonbasic reduced costs is not only sufficient but also necessary to optimality under the nondegeneracy assumption. The following result concerns presence of multiple optimal solutions.

Proposition 3.9.3. *If reduced costs are all positive, there is a unique optimal solution to the LP problem. If a basic optimal solution is nondegenerate and there is a zero-valued reduced cost, then there are infinitely many optimal solutions; in the case when the feasible region is bounded, there are multiple basic optimal solutions.*

Proof. We prove the first half first. Assume that reduced costs in an optimal tableau are all positive, corresponding to the basic optimal solution $\bar{x}$. For any feasible solution $\hat{x} \geq 0$ different from $\bar{x}$, there is an index $s \in N$ such that $\hat{x}_s > 0$ (otherwise, the two are the same). Therefore, substituting $\hat{x}$ to (3.6) leads to

$$\hat{f} = \bar{f} + \sum_{j \in N} \bar{z}_j \hat{x}_j > \bar{f},$$

which implies that $\hat{x}$ is not optimal, as is a contradiction. Therefore, there is an unique optimal solution.

To prove the seconde half, assume that a tableau, say Table 3.1, gives a nondegenerate basic optimal solution and has zero reduced costs. Without loss of generality, assume $\bar{z}_q = 0$. If (3.7) holds, then inequalities of the right-hand side of (3.8) hold for any $x_q = \alpha > 0$, that is, there are infinitely many feasible solutions, corresponding to the same optimal value $-\bar{f}$ (see (3.9)); if the feasible region is bounded, then (3.7) does not hold, hence it is known from $\bar{b}_p > 0$ that the stepsize α, defined by (3.10), is positive. Thus, for any value of x_q in $[0, \alpha]$, a feasible solution can be determined by the equalities of (3.8), corresponding to optimal value $-\bar{f}$. Therefore, there are infinitely many optimal solutions. It is clear that entering x_q to and dropping x_{j_p} from the basis give a different basic optimal solution. □

The last half of the proof actually describes an approach to obtain multiple basic optimal solutions by entering the nonbasic indices, corresponding to zero-valued reduced costs, to the basis. With this respect, an approach to intercepting for the optimal set will be described in Sect. 25.2.

As was mentioned in Sect. 2.4, the simplex method can be explained in terms of the active set method. In each iteration, in fact, a vertex is determined by $Ax = b$ and $x_j = 0,\ j \in N$, corresponding to n active constraints. Since it has zero basic components, a degenerate vertex is the intersection of superplanes, the number of which is greater than n. At first glance, this case seems rarely to occur in practice. Surprisingly, however, the situation is just the opposite: problems stemming from practice are almost all degenerate.

The simplex tableau is essentially the canonical form of $Ax = b$ (together with reduced costs), which may be initially created by the Gauss-Jordan elimination. Such a tableau was used to develop the simplex method previously, although the same can be done alternatively via the triangular form, involving an upper triangular submatrix rather than unit matrix. As it is associated with the Gauss elimination, in fact, the latter should be more relevant to implementation (see also the last paragraph of Sect. 1.6).

Finally, there are two issues that are not guaranteed by the simplex method.

As was well-known, the method is not a polynomial time one; even finiteness of it is, in presence of degeneracy, not guaranteed in theory. Practically, however, this might not be a serious problem, as the method performs well overall if implemented properly although some authors do not agree with this point (see, e.g., Kotiah and Steinberg 1978).

More seriously, the method in its very form is numerically unstable, because the selected pivot may be arbitrarily small in module (see, e.g., Chan 1985; Maros 2003b; Ogryczak 1988). Refer to Rule 3.2.1 used in steps 4 of Algorithm 3.2.1. The pivot $\bar{a}_{pq}$, selected by the minimum-ratio test, could be too small to carry out subsequent computations. Indeed, the simplex method in its very form can only solve few (even very small) LP problems.

Instead of Rule 3.2.1, the following rule may serve as a remedy for solving highly degenerate LP problems.

Rule 3.9.1 (Row rule) Define $I = \{i \mid \bar{a}_{iq} > 0,\ i = 1, \cdots, m\}$, $I_1 = \{i \mid \bar{b}_i = 0,\ i \in I\}$. Determine pivot row index p and stepsize α by

$$p \in \begin{cases} \arg\max\{\bar{a}_{iq} \mid i \in I_1\}, & \alpha = 0, \quad \text{if } I_1 \neq \emptyset, \\ \arg\min\{\bar{b}_i/\bar{a}_{iq} \mid i \in I\}, & \alpha = \bar{b}_p/\bar{a}_{pq}, \quad \text{otherwise.} \end{cases} \tag{3.39}$$

A more favorable and applicable remedy is Harris two-pass Rule 5.6.1 though somehow cumbersome (see also Greenberg 1978). Even so, the stability problem is still not overcome yet entirely, as is the source of many troubles encountered in practice. With this aspect, alternative methods presented in Chaps. 15 and 16 might be "terminators".

There are analogues to the preceding issues and remedies for various simplex variants, including the dual simplex method presented in the next chapter.

Chapter 4
Duality Principle and Dual Simplex Method

The duality features a special relationship between a LP problem and another, both of which involve the same original data (A, b, c), located differently (except for the self-duality, see below). The former is referred to as *primal problem* while the latter as *dual problem*. It is important that there exists a close relationship between their feasible regions, optimal solutions and optimal values. The duality together with optimality conditions, yielding from it, constitute a basis for the LP theory. On the other hand, an economic interpretation of duality features its applications to practice. This chapter is devoted to these topics.

On the other hand, once any of primal and dual problems is solved, the problems are both solved due to duality. Thereby, a so-called *dual simplex method* will be derived by handling the dual problem in this chapter. Its tableau version will still proceed with the same simplex tableau.

From now on, "primal" will be added as a prefix, if necessary, to the simplex method and associated items to distinguish with their dual counterparts, introduced in this chapter.

4.1 Dual LP Problem

If the standard LP problem (1.8), i.e.,

$$\begin{array}{lll} (P) & \min \; f = c^T x, & \\ & \text{s.t.} \;\; Ax = b, & x \geq 0, \end{array} \tag{4.1}$$

is referred to as "primal", the following problem

$$\begin{array}{lll} (D) & \max \; g = b^T y, & \\ & \text{s.t.} \;\; A^T y + z = c, & z \geq 0, \end{array} \tag{4.2}$$

P.-Q. PAN, *Linear Programming Computation*, DOI 10.1007/978-3-642-40754-3_4,

constructed with the same data (A, b, c), is the "dual problem" of (4.1). There is 1–1 correspondence between variables of one of them and constraints of another.

Values of y, z, satisfying $A^T y + z = c$, are called *dual* solution. Set

$$D = \{(y, z) \in \mathcal{R}^m \times \mathcal{R}^n \mid A^T y + z = c,\ z \geq 0\}$$

is called *dual feasible region*, elements in which are *dual feasible solutions*.

It is clear that $\bar{y} = 0,\ \bar{z} = c$ is a dual solution; if, in addition, $c \geq 0$, it is a dual feasible solution. Given basis B, setting $z_B = 0$ in $B^T y + z_B = c_B$ gives

$$\bar{y} = B^{-T} c_B, \qquad \bar{z}_B = 0, \qquad \bar{z}_N = c_N - N^T \bar{y}, \tag{4.3}$$

called *dual basic solution*. $\bar{z}$ is just reduced costs; and $\bar{y}$ the simplex multiplier (see Note on Algorithm 3.5.1). If $\bar{z}_N \geq 0$, $(\bar{y}, \bar{z})$ is a dual basic feasible solution, corresponding to a vertex in D. For simplicity, thereafter $\bar{z}_N$ alone is often said to be dual basic solution.

The following alternative form of the dual problem (4.2):

$$\begin{array}{llll} (D) & \max & g = b^T y, & \\ & \text{s.t.} & A^T y \leq c, & \end{array} \tag{4.4}$$

is useful in some cases. Problems (4.2) and (4.4) will be regarded as the same.

As it can be converted to a standard one, any LP problem corresponds to a dual problem. By introducing slack variables $u \geq 0$, e.g., the problem

$$\begin{array}{ll} \max & c^T x, \\ \text{s.t.} & Ax \leq b, \quad x \geq 0, \end{array} \tag{4.5}$$

can be turned to the standard problem

$$\begin{array}{ll} \min & -c^T x, \\ \text{s.t.} & Ax + u = b, \quad x, u \geq 0, \end{array}$$

the dual problem of which is

$$\begin{array}{ll} \max & b^T y', \\ \text{s.t.} & \begin{pmatrix} A^T \\ I \end{pmatrix} y' \leq \begin{pmatrix} -c \\ 0 \end{pmatrix}. \end{array}$$

Setting $y = -y'$ turns the preceding to (4.5)'s dual problem below:

$$\begin{array}{ll} \min & b^T y, \\ \text{s.t.} & A^T y \geq c, \quad y \geq 0. \end{array}$$

Correspondence between primal and dual problems are summarized to the following table:

Primal problem		Dual problem	
Objective function	min	Objective function	max
Variables	Nonnegative nonpositive free	Constraints	$\leq$ $\geq$ $=$
Constraints	$\geq$ $\leq$ $=$	Variables	Nonnegative nonpositive free

Note: In applications of the preceding table, sign restriction is not handled as a constraint, but attributed to the associated variable

For example, the so-called "bounded-variable" LP problem

$$\begin{array}{ll} \min & c^T x, \\ \text{s.t.} & Ax = b, \quad l \leq x \leq u, \end{array} \tag{4.6}$$

can be transformed to

$$\begin{array}{llrcl} \min & c^T x, & & & \\ \text{s.t.} & Ax & & = & b, \\ & x + s & & = & u, \\ & -x & + t & = & -l, \\ & & s, t & \geq & 0. \end{array} \tag{4.7}$$

According the preceding table, the dual problem of (4.6) is

$$\begin{array}{ll} \max & b^T y - u^T v + l^T w, \\ \text{s.t.} & A^T y - v + w = c, \quad v, w \geq 0. \end{array} \tag{4.8}$$

The so-called *self-duality* referees to a special case when the dual problem of an LP problem is just itself. Combining the primal problem (4.1) and the dual problem (4.4), we construct the following problem:

$$\begin{array}{lll} \min & c^T x - b^T y, & \\ \text{s.t.} & Ax \qquad = b, & x \geq 0, \\ & \qquad A^T y \ \leq c. & \end{array} \tag{4.9}$$

According the preceding table, the dual problem of it is

$$\begin{array}{lllll} \max & b^T y + c^T v, & & & \\ \text{s.t.} & A^T y & & \le c, & \\ & & Av & = -b, & v \le 0. \end{array}$$

By setting $v = -x$ and handling the objective function properly, the preceding can be transformed to the original problem (4.9). Therefore, (4.9) is a self-dual problem.

4.2 Duality Theorems

This section only focuses on the duality of (P) and (D), as obtained results are valid for more general cases.

Theorem 4.2.1 (Symmetry). *The dual problem of a dual problem is the primal problem.*

Proof. Introduce slack variable $u \ge 0$ to dual problem (D), and make a variable transformation $y = y_1 - y_2$ to convert it to

$$\begin{array}{lll} \max & b^T(y_1 - y_2), & \\ \text{s.t.} & A^T(y_1 - y_2) + u = c, & y_1, y_2, u \ge 0, \end{array}$$

or equivalently,

$$\begin{array}{lll} \min & (-b^T, b^T, 0)(y_1^T, y_2^T, u^T)^T, & \\ \text{s.t.} & (A^T \vdots - A^T \vdots\, I)(y_1^T, y_2^T, u^T)^T = c, & y_1, y_2, u \ge 0. \end{array}$$

The dual problem of the preceding is

$$\begin{array}{ll} \max & c^T x', \\ \text{s.t.} & \begin{pmatrix} A \\ -A \\ I \end{pmatrix} x' \le \begin{pmatrix} -b \\ b \\ 0 \end{pmatrix}, \end{array}$$

that is,

$$\begin{array}{lll} \max & c^T x', & \\ \text{s.t.} & Ax' = -b, & x' \le 0, \end{array}$$

which becomes (P) by setting $x' = -x$. □

The preceding says that any of the primal and dual problems is the dual problem of the other. So, the two are symmetric in position. This is why any fact, holding for one of the primal and dual problems, has its counterpart for the other. It is of great importance that there is a close relationship between feasible or optimal solutions of the pair.

Theorem 4.2.2 (Weak duality). *If x and y are feasible solutions to primal and dual problems, respectively, then $c^T x \geq b^T y$.*

Proof. Premultiplying $c \geq A^T y$ by $x \geq 0$ gives $c^T x \geq y^T Ax$, substituting $b = Ax$ to which leads to $c^T x \geq b^T y$. □

According to the preceding, if there are feasible solutions to both primal and dual problems, any feasible value of the former is an upper bound of any feasible value of the latter, whereas any feasible value of the latter is a lower bound of any feasible value of the former.

Corollary 4.2.1. *If any of the primal and dual problems is unbounded, there exists no feasible solution to the other.*

Proof. By contradiction. Assume that there is a feasible solution to the dual problem. Then it follows from Theorem 4.2.2 that feasible values of the primal problem is bounded below. Analogously, if the primal problem is feasible, the dual problem is bounded above. □

Corollary 4.2.2. *Let $\bar{x}$ and $\bar{y}$ be feasible solutions to the primal and dual problems, respectively. If $c^T \bar{x} = b^T \bar{y}$, they are optimal solutions to the pair, respectively.*

Proof. According to Theorem 4.2.2, for any feasible solution x to the primal problem, it holds that $c^T x \geq b^T \bar{y} = c^T \bar{x}$, therefore $\bar{x}$ is an optimal solution to the primal problem. Analogously, $\bar{y}$ is an optimal solution to the dual problem. □

Theorem 4.2.3 (Strong duality). *If there exists an optimal solution to any of the primal and dual problems, then there exists an optimal one to the other, and the associated optimal values are equal.*

Proof. Assume that there is an optimal solution to the primal problem. According to Theorem 2.3.2, there is a basic optimal solution. Let B and N are optimal basis and nonbasis, respectively. Then

$$c_N^T - c_B^T B^{-1} N \geq 0, \qquad B^{-1} b \geq 0.$$

Thus, setting

$$\bar{y} = B^{-T} c_B, \tag{4.10}$$

leads to

$$A^T \bar{y} - c = \begin{pmatrix} B^T \\ N^T \end{pmatrix} \bar{y} - \begin{pmatrix} c_B \\ c_N \end{pmatrix} \leq \begin{pmatrix} 0 \\ 0 \end{pmatrix}.$$

Therefore $\bar{y}$ is a feasible solution to the dual problem. By (4.10), on the other hand, the basic feasible solution ($\bar{x}_B = B^{-1}b,\ \bar{x}_N = 0$) satisfies

$$c^T\bar{x} = c_B^T\bar{x}_B = c_B^T B^{-1}b = b^T\bar{y}.$$

By Corollary 4.2.2, therefore, $\bar{x}$ and $\bar{y}$ are respective optimal solutions to the primal and dual problems with the same optimal value. Moreover, it is known by Theorem 4.2.1 that if there is an optimal solution to the dual problem, so is to the primal problem, with the same optimal value. □

Based on the strong duality, thereafter primal and dual optimal values will not be distinguished.

It is clear that if there is an optimal solution to one of the pair of (4.1) and (4.4), so is the self-dual problem (4.9). Moreover, the optimal value of the latter is equal to zero, and the optimal solution of the latter gives the primal and dual optimal solutions to the pair. A variation of it will be used to derive the so-called "homogeneous and self-dual interior-point method" (Sect. 9.4.4).

In case when any of the primal and dual problems is infeasible, it can be asserted that the other problem is infeasible or unbounded. The computation would be finished in this case. In some applications, however, it would be needed to determine which case the problem is. This can be resolved via the duality as follows.

Assume now that the primal problem (4.1) is infeasible. To determine whether the dual problem (4.2) is infeasible or unbounded, consider

$$\begin{array}{ll} \min & c^T x, \\ \text{s.t.} & Ax = 0, \qquad x \geq 0, \end{array} \tag{4.11}$$

which has a feasible solution $x = 0$. Solve the preceding program by the simplex algorithm. If (4.11) is unbounded, then the program

$$\begin{array}{ll} \min & 0, \\ \text{s.t.} & A^T y + z = c, \qquad z \geq 0, \end{array}$$

is infeasible (Corollary 4.2.1). Therefore, (4.2) is infeasible either. If an optimal solution to (4.11) is reached, then there exists an optimal solution to the preceding program (Theorem 4.2.3), as indicates that (4.2) is feasible; thereby, it can be further asserted that (4.2) is unbounded.

Using the duality, now we are able to prove Farkas Lemma 2.1 concisely.

Lemma 4.2.1 (Farkas). *Assume $A \in \mathcal{R}^{m\times n}$ and $b \in \mathcal{R}^m$. The feasible region P is nonempty if and only if*

$$b^T y \geq 0, \qquad \forall\, y \in \{y \in \mathcal{R}^m \mid A^T y \geq 0\}.$$

Proof. Consider the following LP problem

$$\begin{array}{ll} \min & 0, \\ \text{s.t.} & Ax = b, \qquad x \geq 0, \end{array} \tag{4.12}$$

the dual problem of which is

$$\begin{array}{ll} \max & b^T y', \\ \text{s.t.} & A^T y' \leq 0. \end{array} \tag{4.13}$$

Note that $y' = 0$ is a feasible solution to it, with feasible value 0.

Necessity. Assume that the feasible region P of (4.12) is nonempty, hence all feasible solutions correspond to the same objective value 0. According to Theorem 4.2.2, for any $y' \in \{y' \in \mathcal{R}^m \mid A^T y' \leq 0\}$ it holds that $b^T y' \leq 0$. By setting $y = -y'$, it is known that $b^T y \geq 0$ holds for $y \in \{y \in \mathcal{R}^m \mid A^T y \geq 0\}$.

Sufficiency. Assume that for any $y \in \{y \in \mathcal{R}^m \mid A^T y \geq 0\}$ it holds that $b^T y \geq 0$. Then, for $y' \in \{y' \in \mathcal{R}^m \mid A^T y' \leq 0\}$ we have $b^T y' \leq 0$, hence there is an optimal solution to (4.13). According to Theorem 4.2.3, therefore, there is an optimal solution to (4.12), as implies nonempty of P. □

4.3 Optimality Condition

From duality theorems presented in the previous section, it is possible to derive a set of conditions for primal and dual solutions to be optimal, as stands as a theoretical basis for LP. We consider the standard LP problem first, and then more general LP problems.

Assume that x and (y, z) are primal and dual (not necessarily feasible) solutions.

Definition 4.3.1. Difference $c^T x - b^T y$ between the primal and dual objective values is *duality gap* between x and (y, z).

Definition 4.3.2. If $x^T z = 0$, x and (y, z) are *complementary*; if $x + z > 0$, in addition, the two are strictly complementary.

Quantity $x^T z$ is termed *complementarity residual.*

Lemma 4.3.1. *The duality gap and complementarity residual of x and (y, z) are equal; x and (y, z) are complementary if and only if their duality gap equals zero.*

Proof. Since x and (y, z) satisfy the equality constrains, it is easy to derive that

$$c^T x - b^T y = x^T c - (Ax)^T y = x^T (c - A^T y) = x^T z.$$

□

If it holds as an equality, an "$\geq$" or "$\leq$" type of inequality is said to be tightly satisfied, and if it does as a strict inequality, it is said to be slackly satisfied.

In case when the nonnegative constraints are satisfied, the complementarity of x and (y, z) is equivalent to satisfaction of

$$x_j z_j = 0, \qquad \forall \quad j = 1, \ldots, n. \tag{4.14}$$

It is clear that for $j = 1, \ldots, n$, it holds that $x_j = 0$ (or $z_j = 0$) if $z_j > 0$ (or $x_j > 0$). If a component of x (or z) slackly satisfies the associated nonnegativity constraint, therefore, the corresponding component of z (or x) must satisfy the associated nonnegativity constraint tightly.

Theorem 4.3.1 (Optimality conditions for the standard LP problem). *x is an optimal solution of the standard LP problem if and only if there exist y, z such that*

$$\begin{array}{llll} (i) & Ax = b, & x \geq 0, & \text{(primal feasibility)} \\ (ii) & A^T y + z = c, & z \geq 0, & \text{(dual feasibility)} \\ (iii) & x^T z = 0. & & \text{((slackness) complementarity)} \end{array} \tag{4.15}$$

Proof. Note that for x and (y, z), zero duality gap is equivalent to complementarity (Lemma 4.3.1).

Sufficiency. By Corollary 4.2.2 and the equivalence of zero duality gap and complementarity, it follows from (4.15) that x and (y, z) are primal and dual optimal solutions, respectively.

Necessity. If x is a primal optimal solution, then it satisfies condition (i). By Theorem 4.2.3, in addition, there is a dual optimal solution (y, z) such that the duality gap is zero, hence conditions (ii) and (iii) are satisfied. □

The following result is stated without proof (Goldman and Tucker 1956b).

Theorem 4.3.2 (Strict complementarity). *If there exists a pair of primal and dual optimal solutions, then there exists a strictly complementary pair of such solutions.*

The significance of the optimal conditions speaks for itself. As for algorithm research, any type of optimality criterions in various contexts must coincide with these conditions. From them, it is understandable that LP algorithms always solve the pair of problems at the same time. For instance, once the simplex algorithm reaches primal optimal solution

$$\bar{x}_B = B^{-1} b, \qquad \bar{x}_N = 0,$$

it also gives a dual optimal solution

$$\bar{y} = B^{-T} c_B, \qquad \bar{z}_B = 0, \quad \bar{z}_N = c_N - N^T \bar{y}.$$

In view of the symmetry between primal and dual problems, moreover, it is not surprising why LP algorithms often present in pair: if there is an algorithm for solving the primal problem, there is an according algorithm for solving the dual problem, and vise versa. As an example, the dual algorithm, presented in Sect. 4.5, matches the simplex algorithm.

Interior-point methods often judge the degree of approaching optimality through the complementarity residual: the smaller the residual is, the closer to optimality; when it vanishes, primal and dual optimal solutions are attained respectively. It is noticeable, moreover, that direct dealing with the optimal conditions as a system of equalities and inequalities can lead to some interior-point algorithms (Sect. 9.4). Such algorithms usually generate a strictly complementary pair of optimal solutions in the limit, as is of importance for asymptotic analysis (Güler and Ye 1993).

For the bounded-variable LP problem (4.6), we have the following result.

Theorem 4.3.3 (Optimal conditions for the bounded-variable problem). *x is an optimal solution of the bounded-variable LP problem if and only if there exist y, v, w such that*

$$\begin{array}{llll} (i) & Ax = b, & l \leq x \leq u, & \text{(primal feasibility)} \\ (ii) & A^T y - v + w = c, & v, w, \geq 0, & \text{(dual feasibility)} \\ (iii) & (x-l)^T w = 0, & (u-x)v = 0. & \text{((slackness) complementarity)} \end{array} \tag{4.16}$$

Proof. It is derived from (4.6) to (4.8) and Theorem 4.3.1. □

Finally, consider the general problem of form

$$\begin{array}{ll} \min & c^T x, \\ \text{s.t.} & x \in \Omega, \end{array} \tag{4.17}$$

where $\Omega \subset \mathcal{R}^n$ is a convex set.

Lemma 4.3.2. *It is an optimal solution to (4.17) if and only if x^* satisfies*

$$c^T(x - x^*) \geq 0, \qquad \forall \quad x \in \Omega \tag{4.18}$$

Proof. Assume that x^* is an optimal solution. If (4.18) does not hold, i.e., there is a point $\bar{x} \in \Omega$ such that

$$c^T(\bar{x} - x^*) < 0, \tag{4.19}$$

then it holds that

$$c^T \bar{x} < c^T x^*, \tag{4.20}$$

which contradicts optimality of x^*. Conversely, assume that (4.18) holds. If x^* is not optimal to (4.17), then there exists $\bar{x} \in \Omega$ satisfying (4.20), which implies (4.19), as contradicts satisfaction of (4.18). □

It is noted from the proof that the preceding Lemma is actually valid for an arbitrary set Ω. Geometrically, it says that a sufficient and necessary condition for x^* to be an optimal solution to (4.17) is that the angle between the vector from x^* to any point $x \in \Omega$ and the gradient of the objective function is no more than $\pi/2$.

Vector $x^* \in \Omega$ is termed *local optimal solution* if it is an optimal solution over some spherical neighborhood of it; or more precisely, there exists $\gamma > 0$ such that

$$c^T x^* = \min \{c^T x \mid x \in (\Omega \cap \Sigma)\}, \qquad \Sigma = \{x \in \mathcal{R}^n \mid \|x - x^*\| \leq \gamma\}. \tag{4.21}$$

Theorem 4.3.4. *A vector is an optimal solution to (4.17) if and only if it is a local optimal solution to it.*

Proof. The necessity is clear. The sufficiency. Assume that x^* is a local optimal solution. So, there is $\gamma > 0$ so that (4.21) holds. For any $\bar{x} \in \Omega$ with $\bar{x} \neq x^*$, define

$$x' = x^* + \gamma/\|\bar{x} - x^*\|(\bar{x} - x^*). \tag{4.22}$$

Then, $\|x' - x^*\| = \gamma$, hence $x' \in \Omega \cap \Sigma$. Thereby, it follows from (4.22) that $c^T x' - c^T x^* \geq 0$ together with

$$(\bar{x} - x^*) = \|\bar{x} - x^*\|/\gamma c^T (x' - x^*),$$

gives

$$c^T (\bar{x} - x^*) = \|\bar{x} - x^*\|/\gamma c^T (x' - x^*) \geq 0.$$

According to Lemma 4.3.2, x^* is an optimal to (4.17). □

4.4 Dual Simplex Method: Tableau Form

The dual simplex method is of great importance, as it can be even more efficient than the simplex method, and serve as a basic tool to solve integer or mixed LP problems. This section will derive its tableau version.

As was well-known, simplex tableaus created by a series of elementary transformations are equivalent in the sense that they represent problems equivalent to the original one. No matter how pivots are selected, a resulting simplex tableau offers a pair of complementary primal and dual solutions, which are optimal whenever entries in the right-hand side and the objective row are all nonnegative, or in other words, both primal and dual feasibility achieved. Starting from a feasible simplex tableau, e.g., the tableau simplex algorithm generates a sequence of feasible simplex tableaus, until dual feasibility achieved.

Using an alternative pivot rule, the so-called "dual simplex algorithm" presented in this section generates a sequence of dual feasible simplex tableaus, until primal feasibility achieved. To do so, of course, it has to start from a dual feasible simplex tableau.

Consider the standard LP problem (4.1). Let (3.18) be a current dual feasible simplex tableau, satisfying $\bar{z}_N \geq 0$ but $\bar{b} \not\geq 0$.

In contrast to the simplex method, we first determine pivot row rather than column.

Rule 4.4.1 (Dual row rule) Select row index by

$$p \in \arg\min\{\bar{b}_i \mid i = 1, \ldots, m\}.$$

It is clear that the preceding will drop the basic variable x_{j_p} from the basis, turning it to primal feasible.

Lemma 4.4.1. *Assume that* $\bar{z}_N \geq 0$ *and* $\bar{b}_p < 0$. *If column index set*

$$J = \{j \in N \mid \bar{a}_{pj} < 0\} \tag{4.23}$$

is empty, then the LP problem is infeasible.

Proof. $\bar{z}_N \geq 0$ indicates dual feasibility. Assume that the dual problem is bounded, hence there is an optimal dual solution. According to the strong duality Theorem, there is a optimal primal solution. Assume that $\hat{x} \geq 0$ is such an optimal primal solution, which satisfies the equality, corresponding to the pth row of the simplex tableau, i.e.,

$$\hat{x}_{j_p} + \sum_{j \in N} \bar{a}_{pj} \hat{x}_j = \bar{b}_p.$$

From (4.23) and $\hat{x} \geq 0$, it follows that the left-side of the preceding is nonnegative, as contradicts $\bar{b}_p < 0$. Therefore, the problem is dual unbounded, and hence infeasible. □

Assume that p has been determined and (4.23) does not hold. Then the following is well-defined.

Rule 4.4.2 (Dual column rule) Determine β and column index q such that

$$\beta = -\bar{z}_q / \bar{a}_{pq} = \min_{j \in J} -\bar{z}_j / \bar{a}_{pj} \geq 0. \tag{4.24}$$

β is referred to as *dual stepsize*.

Once a pivot, say $\bar{a}_{pq}$, is determined, perform elementary transformations to turn it to 1 and eliminate all other nonzeros in the q-indexed column. Then adding β times of the pth row to the objective row results in a new simplex tableau (see (3.13)). It is not difficult to show that the new tableau remains dual feasible, that is, its objective row remains nonnegative.

The objective value of the resulting tableau is then

$$-\hat{f} = -\bar{f} + \beta\bar{b}_p \leq -\bar{f}, \tag{4.25}$$

Therefore, $\hat{f} \geq \bar{f}$, indicating that the objective value does not decrease. When $\bar{z}_N > 0$, the objective value strictly increases, as is a case in which the simplex tableau (or the dual feasible solution) is said to be *dual nondegenerate*.

The overall steps can be put in the following algorithm (Beale 1954; Lemke 1954).

Algorithm 4.4.1 (Dual simplex algorithm: tableau form). Initial: a dual feasible simplex tableau of form (3.18). This algorithm solves the standard LP problem (1.7).

1. Select pivot row index $p \in \arg\min\{\bar{b}_i \mid i = 1, \ldots, m\}$.
2. Stop if $\bar{b}_p \geq 0$.
3. Stop if $J = \{j \in N \mid \bar{a}_{pj} < 0\} = \emptyset$.
4. Determined pivot column $q \in \arg\min_{j \in J} -\bar{z}_j / \bar{a}_{pj}$.
5. Convert $\bar{a}_{pq}$ to 1, and eliminate the other nonzeros in the column by elementary transformations.
6. Go to step 1.

Theorem 4.4.1. *Under the dual nondegenerate assumption, Algorithm 4.4.1 terminates either at*

(i) Step 2, achieving a pair of primal and dual optimal solutions; or at
(ii) Step 3, detecting infeasibility of the problem.

Proof. The proof on the termination is similar to the simplex algorithm. The meanings of its exits come from Lemmas 3.2.1 and 4.4.1 and discussions preceding the algorithm. □

If dual degeneracy presents, the dual simplex method would stall in solution process, even fail to solve a problem due to cycling (Beale 1955). Despite the dual degeneracy almost always occurs, the dual simplex method perform successfully in practice, as is just in the primal simplex context.

Algorithm 4.4.1 starts from a dual feasible simplex tableau. In general, there is a need for a dual Phase-I procedure to serve for this purpose. This topic will be delayed to Chap. 14.

Example 4.4.1. Solve the following LP problem by Algorithm 4.4.1:

$$\begin{array}{ll} \min & f = x_1 + 2x_2 + x_3, \\ \text{s.t.} & 2x_1 + \ \ x_2 + x_3 - x_4 = 1, \\ & -x_1 + 4x_2 + x_3 \qquad \geq 2, \\ & \ \ x_1 + 3x_2 \qquad\qquad \leq 4, \\ & x_j \geq 0, \ j = 1, \ldots, 4. \end{array}$$

Answer Initial: turn the problem to the standard form by introducing slack variables $x_5, x_6 \geq 0$ in constrains. Then, premultiply the first two constraints by -1 respectively:

$$\begin{array}{ll} \min & f = x_1 + 2x_2 + x_3, \\ \text{s.t.} & -2x_1 - \ \ x_2 - x_3 + x_4 \qquad\qquad = -1, \\ & \ \ x_1 - 4x_2 - x_3 \qquad + x_5 \qquad = -2, \\ & \ \ x_1 + 3x_2 \qquad\qquad\qquad + x_6 = \ \ 4, \\ & \qquad x_j \geq 0, \quad j = 1, \ldots, 6, \end{array}$$

which corresponds to an available dual feasible simplex tableau, i.e.,

x_1	x_2	x_3	x_4	x_5	x_6	RHS
−2	−1	−1	1			−1
1	−4*	−1		1		−2
1	3				1	4
1	2	1				

Iteration 1:

1. $\min\{-1, -2, 4\} = -2 < 0, \ p = 2.$
3. $J = \{2, 3\}.$
4. $\min\{-2/(-4), -1/(-1)\} = 1/2, \ q = 2.$
5. Multiply row 2 by $-1/4$, and then add $1, -3, -2$ times of row 2 to rows 1,3,4, respectively:

x_1	x_2	x_3	x_4	x_5	x_6	RHS
−9/4*		−3/4	1	−1/4		−1/2
−1/4	1	1/4		−1/4		1/2
7/4		−3/4		3/4	1	5/2
3/2		1/2		1/2		−1

Iteration 2:

1. $\min\{-1/2, 1/2, 5/2\} = -1/2 < 0,\ p = 1.$
3. $J = \{1, 3, 5\}.$
4. $\min\{-(3/2)/(-9/4), -(1/2)/(-3/4), -(1/2)/(-1/4)\} = 2/3,\ q = 1.$
5. Multiply row 1 by $-4/9$, and then add $1/4, -7/4, -3/2$ times of row 1 to rows 2,3,4, respectively:

x_1	x_2	x_3	x_4	x_5	x_6	RHS
1		1/3	−4/9	1/9		2/9
	1	1/3	−1/9	−2/9		5/9
		−4/3	7/9	5/9	1	19/9
			2/3	1/3		−4/3

The right-hand side of the preceding is now nonnegative, hence obtained is an optimal simplex tableau. The optimal solution and associated objective value are

$$\bar{x} = (2/9, 5/9, 0, 0)^T, \qquad \bar{f} = 4/3.$$

4.5 Dual Simplex Method

In the preceding section, the tableau dual simplex algorithm was formulated. In this section, we first derive its revised version based on equivalence between the simplex tableau and the revised simplex tableau, just as what we have done for deriving the simplex method from its tableau version. Then, we derive it alternatively to reveal the fact that it essentially solves the dual problem.

Like the simplex Algorithm 3.5.2, in each iteration the dual simplex algorithm computes the objective row, the right-hand side, pivot column and row. The objective row and/or the right-hand side can be computed in a recurrence manner (see (3.13) and (3.14)). The pivot column and row can be computed through B^{-1} and the original data, just as in the simplex method.

If nonbasic entries in the pivot row are all nonnegative, i.e.,

$$\sigma_N^T \triangleq e_p^T \bar{N} = e_p^T B^{-1} N \geq 0,$$

the dual problem is unbounded, hence the original problem is infeasible. B^{-1} will be updated in the same way as in the simplex method.

Based on Table 3.2, therefore, Algorithm 4.4.1 can be revised to the following algorithm.

Algorithm 4.5.1 (Dual simplex algorithm). Initial: (B, N), B^{-1}. $\bar{z}_N \geq 0$, $\bar{x}_B = B^{-1}b$ and $\bar{f} = c_B^T \bar{x}_B$. This algorithm solves the standard LP problem (1.8).

1. Select row index $p \in \arg\min\{\bar{x}_{j_i} \mid i = 1, \ldots, m\}$.
2. Stop if $\bar{x}_{j_p} \geq 0$ (optimality achieved).
3. Compute $\sigma_N = N^T B^{-T} e_p$.
4. Stop if $J = \{j \mid \sigma_j < 0,\ j \in N\} = \emptyset$ (infeasible problem).
5. Determine β and column index q such that
   $\beta = -\bar{z}_q/\sigma_q = \min_{j \in J} -\bar{z}_j/\sigma_j$.
6. Set $\bar{z}_{j_p} = \beta$, and update $\bar{z}_N = \bar{z}_N + \beta\sigma_N,\ \bar{f} = \bar{f} - \beta\bar{x}_{j_p}$ if $\beta \neq 0$.
7. Compute $\bar{a}_q = B^{-1}a_q$.
8. Update by $\bar{x}_B = \bar{x}_B - \alpha\bar{a}_q,\ \bar{x}_q = \alpha$, where $\alpha = \bar{x}_{j_p}/\sigma_q$.
9. Update B^{-1} by (3.23).
10. Update (B, N) by exchanging j_p and q.
11. Go to step 1.

Alternatively, Algorithm 4.5.1 can be derived by solving the dual problem itself as follows.

Consider the dual problem

$$(D) \qquad \begin{array}{ll} \max & g = b^T y, \\ \text{s.t.} & A^T y \leq c. \end{array}$$

Given (B, N), B^{-1}. It is easy to verify that $\bar{y} = B^{-T} c_B$ satisfies the dual constraints, i.e.,

$$\bar{z}_B = c_B - B^T \bar{y} = 0, \qquad \bar{z}_N = c_N - N^T \bar{y} \geq 0. \tag{4.26}$$

$\bar{y}$ is a dual basic feasible solution, or geometrically a vertex in the dual feasible region

$$D = \{y \mid A^T y \leq c\}.$$

In the primal simplex context, $\bar{y}$ is usually called "simplex multipliers".

Consider the associated primal basic solution

$$\bar{x}_B = B^{-1}b, \qquad \bar{x}_N = 0.$$

If $\bar{x}_B \geq 0$, then $\bar{x}$ and $(\bar{y}, \bar{z})$ satisfy the optimality condition, and are therefore a pair of primal and dual basic optimal solutions.

Now assume that $\bar{x}_B = B^{-1}b \not\geq 0$. Determine row index p such that

$$\bar{x}_{j_p} = \min\{\bar{x}_{j_i} \mid i = 1, \ldots, m\} < 0. \tag{4.27}$$

Introduce vector

$$h = B^{-T} e_p, \tag{4.28}$$

which with (4.27) gives

$$-b^T h = -b^T B^{-T} e_p = -e_p^T (B^{-1} b) = \bar{x}_{j_p} > 0, \tag{4.29}$$

implying that $-h$ is an uphill direction, with respect to the dual objective function g.

Now consider the line search scheme below:

$$\hat{y} = \bar{y} - \beta h, \tag{4.30}$$

where β is a dual stepsize to be determined. From (4.30), (4.28) and (4.26), it follows that

$$\hat{z}_B = c_B - B^{\mathrm{T}} \hat{y} = c_B - B^{\mathrm{T}} (\bar{y} - \beta h) = \beta e_p \geq 0, \tag{4.31}$$

$$\hat{z}_N = c_N - N^{\mathrm{T}} \hat{y} = c_N - N^{\mathrm{T}} (\bar{y} - \beta h) = \bar{z}_N + \beta N^{\mathrm{T}} h. \tag{4.32}$$

If $\sigma_N = N^T h \not\geq 0$, then it is seen from (4.32) that a too large $\beta > 0$ will lead to $\hat{z}_N \not\geq 0$, as violates the dual feasibility. It is easy to determine the largest possible β and according column index q, subject to $\hat{z}_N \geq 0$ (see step 5 of Algorithm 4.4.1). Then, drop j_p from and enter q to the basis. It is easy to verify that $\hat{y}$ is just the dual basic feasible solution, corresponding to the resulting basis.

The following is valid in the other case.

Proposition 4.5.1. *If $\sigma_N = N^T h \geq 0$, the dual problem is unbounded, and $-h$ is an uphill extreme direction in the dual feasible region D.*

Proof. If $\sigma_N \geq 0$, it is seen from (4.26) and (4.31), (4.32) that

$$\hat{z} = c - A^{\mathrm{T}} \hat{y} \geq 0, \qquad \forall \beta \geq 0,$$

implying feasibility of the new solution $\hat{y}$ given by (4.30). On the other hand, it is known from (4.30) and (4.29) that the associated new objective value is

$$b^{\mathrm{T}} \hat{y} = b^{\mathrm{T}} \bar{y} - \beta \bar{x}_{j_p}, \tag{4.33}$$

which goes to ∞, as β tends to ∞. Thus the dual problem is unbounded. This means that $-h$ is a uphill unbounded direction of D. In fact, it is seen that $-h$ is the direction of 1-dimensional face or edge

$$\{y \in \mathcal{R}^m \mid A^{\mathrm{T}} y \leq c;\ a_{j_i}^{\mathrm{T}} y = c_{j_i},\ i = 1, \cdots, m,\ i \neq p\},$$

and therefore a uphill extreme direction. □

The preceding analysis is actually valid for any given negative $\bar{x}_{j_p}$. Though far from the best, Rule (4.27) is simple and easy to use, and the corresponding objective increment is the largest possible for a unit stepsize. More efficient dual pivot rules will be presented in Chap. 12.

Since the beginning of 1990s of the last century, successful applications of the dual steepest-edge rule (Forrest and Goldfarb 1992), some of its approximations and bound-flipping (Kirillova et al. 1979) have injected fresh vigoure to the dual simplex method, so that it becomes one of the most powerful methods for solving LP problems (Bixby 2002; Koberstein 2008).

Example 4.5.1. Solve the following problem by Algorithm 4.5.1:

$$\begin{array}{llllllll}
\min & f = x_1 + 2x_2 + x_3, & & & & & & \\
\text{s.t.} & -2x_1 & - \; x_2 & - \; x_3 & + \; x_4 & & & = -1, \\
& x_1 & - \; 4x_2 & - \; x_3 & & + \; x_5 & & = -2, \\
& x_1 & + \; 3x_2 & & & & + \; x_6 & = 4, \\
& & & & x_j \geq 0, & j = 1, \ldots, 6. & &
\end{array}$$

Answer Initial: $B = \{4, 5, 6\}$, $N = \{1, 2, 3\}$, $B^{-1} = I$, $\bar{z}_N = (1, 2, 1)^T$, $\bar{x}_B = (-1, -2, 4)^T$, $f = 0$.

Iteration 1:

1. $\min\{-1, -2, 4\} = -2 < 0$, $p = 2$, x_5 leaves the basis.
3. $\sigma_N = (1, -4, -1)^T$.
5. $\beta = \min\{2/4, 1/1\} = 1/2$, $q = 2$, $x_2$enters the basis.
6. $\bar{z}_N = (3/2, 0, 1/2)^T$, $\bar{z}_{j_p} = 1/2$, $\bar{f} = 0 - (1/2)(-2) = 1$.
7. $\bar{a}_q = (-1, -4, 3)^T$.
8. $\alpha = -2/-4 = 1/2$, $\bar{x}_B = (-1, -2, 4)^T - (1/2)(-1, -4, 3)^T = (-1/2, 0, 5/2)^T$, $\bar{x}_2 = \alpha = 1/2$.
9. $B^{-1} = \begin{pmatrix} 1 & 1/-4 & \\ & 1/-4 & \\ & 3/4 & 1 \end{pmatrix}$.
10. $B = \{4, 2, 6\}$, $N = \{1, 5, 3\}$, $\bar{z}_N = (3/2, 1/2, 1/2)^T \geq 0$, $\bar{x}_B = (-1/2, 1/2, 5/2)^T$.

Iteration 2:

1. $\min\{-1/2, 1/2, 5/2\} = -1/2 < 0$, $p = 1, x_4$ leaves the basis.
3. $\sigma_N = (-9/4, -1/4, -3/4)^T$.
5. $\beta = \min\{(3/2)/(9/4), (1/2)/(1/4), (1/2)/(3/4)\} = 2/3$, $q = 1$, x_1 enters the basis.
6. $\bar{z}_N = (3/2, 1/2, 1/2)^T + (2/3)(-9/4, -1/4, -3/4)^T = (0, 1/3, 0)^T$. $\bar{z}_{j_p} = 2/3$, $\bar{f} = 1 - (2/3)(-1/2) = 4/3$.
7. $\bar{a}_q = (-9/4, -1/4, 7/4)^T$.

8. $\alpha = (-1/2)/(-9/4) = 2/9$, $\bar{x}_B = (-1/2, 1/2, 5/2)^T - (2/9)(-9/4, -1/4, 7/4)^T = (0, 5/9, 19/9)^T$, $\bar{x}_1 = 2/9$.

9. $B^{-1} = \begin{pmatrix} -4/9 & & \\ -1/9 & 1 & \\ 7/9 & & 1 \end{pmatrix} \begin{pmatrix} 1 & 1/-4 & \\ & 1/-4 & \\ & 3/4 & 1 \end{pmatrix} = \begin{pmatrix} -4/9 & 1/9 & \\ -1/9 & -2/9 & \\ 7/9 & 5/9 & 1 \end{pmatrix}$.

10. $B = \{1, 2, 6\}$, $N = \{4, 5, 3\}$, $\bar{z}_N = (2/3, 1/3, 0)^T \geq 0$, $\bar{x}_B = (2/9, 5/9, 19/9)^T \geq 0$.
The optimal solution and objective value:

$$\bar{x} = (2/9, 5/9, 0, 0, 0, 19/9)^T, \qquad \bar{f} = 4/3.$$

4.6 Economic Interpretation of Duality: Shadow Price

The dual problem is of an interesting economic interpretation. Assume that the primal problem

$$\begin{array}{ll} \max & f = c^T x, \\ \text{s.t.} & Ax \leq b, \quad x \geq 0, \end{array}$$

is a plan model for a manufacturer to produce n products using m resources. The available amount of resource i is b_i, $i = 1, \ldots, m$, units; producing an unit of product j consumes a_{ij} units of resource i. The profit of an unit of product j is c_j, $j = 1, \ldots, n$. The goal is to achieve the highest profit with the limited resources.

The dual problems is

$$\begin{array}{ll} \min & b^T y, \\ \text{s.t.} & A^T y \geq c, \quad y \geq 0. \end{array}$$

Let $\bar{x}$ and $\bar{y}$ be primal and dual optimal solutions, respectively. According to the strong duality Theorem, associated primal and dual optimal values are equal, i.e.,

$$v = c^T \bar{x} = b^T \bar{y}.$$

Optimal value's partial derivative with respect to b_i is

$$\frac{\partial v}{\partial b_i} = \bar{y}_i.$$

Therefore, $\bar{y}_i$ is equal to the increment of the highest profit, created by adding one unit of resource i, and can be taken as manufacturer's assessment for resource i, as is named *shadow price* by Paul Samuelson.[1] Shadow price $\bar{y}_i$ is the upper price limit that the manufacturer can afford to buy resource i. When market price of resource i is lower than shadow price $\bar{y}_i$, the manufacturer should consider to buy it to expand the production scale, whereas he should consider to sell it to reduce the production scale in the other case. The manufacturer will not buy resource i any more, no matter how low its price is, whenever the optimal solution $\bar{x}$ satisfies the ith primal inequality constraint slackly, as implies that resource i is not fully used. In fact, the shadow price $\bar{y}_i$ vanishes in this case.

Let x and y be any primal and dual feasible solutions respectively, hence $c^T x \leq b^T y$ holds according to the weak duality. Inequality

$$c^T x < b^T y,$$

implies that the total profit (output) of the plan is less than the available value (input) of the resources. In economic terms, the input-output system is said "instable (non-optimal)" in this case. It is a stable (optimal) system only when output is equal to input.

Consider economic implication of the dual constraints. The manufacturer negotiates with the supplier at price y_i for resource i, as is calculated to purchase resources $b_i, i = 1, \ldots, m$ by overall payment $b^T y$. For $j = 1, \ldots, n$, on the other hand, the supplier asks for resource prices to produce an unit product j being no less than the profit of an unit of product j, as satisfies the jth dual constraint

$$\sum_{i=1}^{m} a_{i,j} y_i \geq c_j.$$

If the suppler asks for too high prices, that is, the dual optimal solution $\bar{y}$ satisfies the jth dual constraint slackly, then $\bar{x}_j$ vanishes, as implies that the manufacturer should not arrange for producing product j at all, no matter how high the profit of an unit of the product is.

4.7 Notes

The concept and theorems of duality were first proposed by famous mathematician von Neumann. In October 1947, he made foundational discussions on the topic in a talk with George B. Dantzig and in a working paper, finished a few weeks later. In 1948, Dantzig provided a rigorous proof on the duality theorems in a report. Subsequently, Gale et al. (1951) formulated the duality theorems and proved them

[1]Paul Samuelson (1915–2009), American economist, the winner of The Nobel Economics Prize (1970), the first American winning this prize.

using Farkas Lemma, independently. Goldman and Tucker (1956b) and Balinski and Tucker (1969) discussed theoretical properties of the dual problem systematically.

As was stressed, the simplex tableau is just a concise expression of a LP problem itself, and all such tableaus created by the primal or dual simplex algorithm are equivalent in the sense of their representation of the LP problem. Then the following question arises:

Are dual problems corresponding to the simplex tableaus equivalent?

Consider the dual problem, corresponding to tableau (3.18), i.e.,

$$\begin{array}{ll} \max & \bar{f} + \bar{b}^{\mathrm{T}} y', \\ \text{s.t.} & \begin{pmatrix} I \\ \bar{N}^{\mathrm{T}} \end{pmatrix} y' + \begin{pmatrix} z'_B \\ z'_N \end{pmatrix} = \begin{pmatrix} 0 \\ \bar{z}_N \end{pmatrix}, \quad z'_B, z'_N \geq 0. \end{array}$$

As their $\bar{b}, \bar{N}, \bar{z}_N$ are not the same, the dual problems corresponding to different simplex tableaus are also different. However, such differences are not essential. In fact, Making variable transformations $y' = B^{\mathrm{T}} y - c_B$, $z' = z$ and noting

$$\bar{b} = B^{-1}b, \quad \bar{N} = B^{-1}N, \quad \bar{z}_N = c_N - N^{\mathrm{T}} B^{-\mathrm{T}} c_B, \quad \bar{f} = c_B^{\mathrm{T}} B^{-1} b,$$

the dual problem can be converted to

$$\begin{array}{ll} \max & b^{\mathrm{T}} y, \\ \text{s.t.} & \begin{pmatrix} B^{\mathrm{T}} \\ N^{\mathrm{T}} \end{pmatrix} y + \begin{pmatrix} z_B \\ z_N \end{pmatrix} = \begin{pmatrix} c_B \\ c_N \end{pmatrix}, \quad z_B, z_N \geq 0, \end{array}$$

which is the original dual problem. Therefore, all the generated simplex tableaus can be regarded as equivalent with respect to represented dual problems.

In summary, elementary transformations generate equivalent simplex tableaus. On the primal side, the right-hand sides give primal basic solutions and the bottom rows give primal reduced objective functions. On the dual side, the right-hand sides render dual reduced objective functions and the bottom rows dual basic solutions.

Based on duality, the following is also valid.

Proposition 4.7.1. *If it has a dual solution* $(\bar{y}, \bar{z})$*, the standard problem (4.1) is equivalent to*

$$\begin{array}{lll} \max & f = \bar{y}^T b + \bar{z}^T x, & \\ \text{s.t.} & Ax = b, & x \geq 0. \end{array} \tag{4.34}$$

Proof. $(\bar{y}, \bar{z})$ satisfies $A^{\mathrm{T}} \bar{y} + \bar{z} = c$ or equivalently,

$$c^T = \bar{y}^T A + \bar{z}^T.$$

Substituting the preceding to the objective of the standard problem and noting the constraint system gives the objective of (4.34), i.e.,

$$f = c^T x = \bar{y}^T Ax + \bar{z}^T x = \bar{y}^T b + \bar{z}^T x,$$

and vice versa. □

The preceding says that the cost vector c in the standard problem can be replaced by any dual solution $\bar{z}$, with only a constant difference in objective value.

Chapter 5
Implementation of the Simplex Method

All algorithms formulated in this book, such as the simplex algorithm and the dual simplex algorithm, are theoretical or conceptional, and can not be put to practical use via programming directly. Softwares, resulting by following the algorithms, step by step naively, would only solve textbook instances, involving only few variables and constraints, not real-world problems, especially large-scale sparse problems. Experiences indicate that implementation details are crucial to algorithms' success (Orchard-Hays 1954; Tomlin 1974; Todd 1982, 1983; Bixby 1994).

This chapter will highlight effective implementation techniques, without which the simplex algorithm will become useless. The issue is mainly two-fold: one is to improve numerical stability and the other to reduce involved computational efforts. Only limited sparse techniques will be touched, as a *full* treatment of sparsity is beyond the scope of this book.

In view of the complexity disadvantage of the tableau version (Sect. 3.8), this chapter will handle the topic based on the (revised) simplex algorithm. We stress that materials presented here are of general value, as also fit other simplex variants, perhaps with some relevant modifications.

Based on the simplex algorithm, implementation techniques may lead to different codes. In particular, MINOS is the main reference in writing this chapter (see Notation).

5.1 Miscellaneous

As rounding errors affect computational procedures and results (see Sect. 1.4), restricting such effects to improve the numerical stability is important to implementation of the simplex algorithm.

Firstly, theoretical 0 or ∞ involved in the algorithm should be treated properly by giving tolerances, in accordance with the computer precision used. Based on

P.-Q. PAN, *Linear Programming Computation*, DOI 10.1007/978-3-642-40754-3_5,

MINOS, for instance, main parameters are list below, where the computer precision is assumed to be

$$\epsilon = 2^{-52} \approx 2.22 \times 10^{-16}$$

Introduce notation

$$\begin{aligned} \epsilon_0 &= \epsilon^{0.8} \approx 3.00 \times 10^{-13}, \\ \epsilon_1 &= \epsilon^{0.67} \approx 3.25 \times 10^{-11}, \\ \text{featol} &= 10^{-6}, \\ \text{plinfy} &= 10^{20}, \end{aligned}$$

where

- Quantities in solution process, whose absolute values are no more than ϵ_0, are treated as machine 0.
- Quantities, whose absolute values are more than plinfy, are treated as ∞.
- tolpiv $= \epsilon_1$ is used to avoid a too small pivot; e.g., the minimum-ratio test (3.10) is actually replaced by

$$\alpha = \bar{b}_p/\bar{a}_{pq} = \min\{\bar{b}_i/\bar{a}_{iq} \mid \bar{a}_{iq} > \text{tolpiv}, i = 1, \cdots, m\} \geq 0. \tag{5.1}$$

Thereby, any pivot chosen satisfies $\bar{a}_{pq} > \text{tolpiv}$. The problem is regarded as unbounded if no such a pivot is found.

The magnitude of a pivot seriously affects the numerical stability of the algorithm. Determination of a numerical nonzero pivot should not only depend on the computer precision, but also the problem to be solved. It is difficult to give a "best" tolpiv. Some value between ϵ_1 and 10^{-7} seems to be feasible, though still far from satisfactory. This topic will be discussions further in Sect. 5.6.

- featol is a primal and dual feasibility tolerance. For $j = 1, \cdots, n$, $x_j \geq -\text{featol}$ is used to replace the nonnegativity condition $x_j \geq 0$ (for primal feasibility). For $j \in N$, in optimality test, $\bar{z}_j \geq -\text{featol}$ is used to replace $\bar{z}_j \geq 0$ (for dual feasibility). Thus, the column rule (3.12) is replaced by

$$q \in \arg\min\{\bar{z}_j \mid \bar{z}_j < -\text{featol}, j \in N\}. \tag{5.2}$$

If there is no such an q found, i.e., $\bar{z}_j \geq -\text{featol}$, $\forall j \in N$, then optimality is declared.

In solution process, however, it is not easy to ensure all iterates to keep within the feasibility tolerance, let alone degeneracy being intertwined with this. Solution of large-scale LP problems usually involves a large number of iterations. After many iterations, due to accumulated errors, intermediate results could be too inaccurate to be used. It is therefore necessary to perform a so-called "restarting strategy":

rebooting the process from scratch by recomputing (from original data) inverse B^{-1} or the LU factorization of the current basis. It might be well to restart every 100 iterations. In addition, residuals of equality constraints with respect to iterates should be monitored in solution process: the restarting strategy is carried out whenever residuals are found to exceed some tolerance. Practically, it is also necessary to predetermine an upper bound for the number of iterations to limit running time. Once iterations exceed such a bound, the solution process is terminated after information of the end basis is recorded for a possible later "warm start".

Real-world demand is a powerful pushing force for implementation techniques for solving large-scale LP problems. Nowadays, LP problems yielding from practice become larger and larger. Very large problems appear frequently. Solution of such type of problems is very challenging: required storage would be too large to meet; even the internal memory of a computer is enough, running time consumed by solution process would still be too high to afford.

Like many other models, fortunately, most of large-scale LP problems are sparse, i.e., data A, b, c involve large amount of zeros. Usually, the proportion of nonzero entries of A is no more than $1\,\%$, hardly $5\,\%$. This is so because variables, involved in large systems, usual belong to different groups (corresponding to different regions, departments or properties, and so on), and constraints mainly reflect relations within interior of a group, so that each variable would appear in few constraints. With the help of sparse matrix techniques, which are quite mature nowadays, it is possible to proceed without storing and manipulating zeros at all. Thereby, the key for solving large-scale sparse problems lies in taking advantage of sparsity to reduce storage and arithmetics requirements. For many professional optimization packages, original nonzero data are expressed in a so-called "MPS format" in an input data file (see Appendix A).

As for special methods for solving large-scale LP problems, some decomposition techniques have been proposed in the past (Chap. 8). The so-called "decomposition principle" will be presented in Sect. 25.6.

5.2 Preprocessing: Scaling

Large magnitude differences within original data affect computational results significantly, as is related with accumulation of rounding errors. To reduce such differences, original data should be preprocessed before hand by some so-called *scaling* technique (Fulkerson and Wolfe 1962; Hamming 1971; Curtis and Reid 1972; Tomlin 1975).

Adjusting measurement unit is a common scaling method. It amounts to performing variable transformation $x = D_1 x'$, where D, called *scaling factor*, is a diagonal square matrix with positive diagonals. For instance, simply taking

$$D_1 = \begin{pmatrix} 1/\|a_1\| & & & \\ & 1/\|a_2\| & & \\ & & \ddots & \\ & & & 1/\|a_n\| \end{pmatrix} \tag{5.3}$$

converts the standard LP problem to

$$\begin{array}{ll} \min & f = (c^{\mathrm{T}} D_1)x', \\ \text{s.t.} & (AD_1)x' = b, \quad x' \geq 0, \end{array} \tag{5.4}$$

where each column of AD_1 is of norm 1, so that magnitude differences between columns are balanced. Then once an optimal solution $\bar{x}'$ to (5.4) is obtained, the according optimal solution of the original is readily available, i.e., $\bar{x} = D_1\bar{x}'$.

The preceding scaled problem is usually amenable to be solved, compared to the original. We will see that such doing enhances the pivot column selection, consequently reducing the number of required iterations (Sect. 11.4).

An analogous scaling is to balance magnitude differences among rows. If d_i^{T} is used to denote row i of A and

$$D_2 = \begin{pmatrix} 1/\|d_1\| & & & \\ & 1/\|d_2\| & & \\ & & \ddots & \\ & & & 1/\|d_m\| \end{pmatrix}, \tag{5.5}$$

then the standard LP problem is equivalent to

$$\begin{array}{ll} \min & f = c^{\mathrm{T}} x, \\ \text{s.t.} & (D_2A)x = D_2b, \quad x \geq 0, \end{array} \tag{5.6}$$

where each row of the coefficient matrix D_2A is of norm 1.

Professional codes for handling large-scale sparse problems usually use more sophisticated scaling methods, based on nonzero data only. MINOS applies an iterative procedure based on Fourer (1979), which alternately balances nonzeros of rows and columns for obtaining scaling factors by geometric mean. After the scaled problem is solved, relevant transformations are carried out to recover an optimal solution of the original. More precisely, the main steps are put in the following algorithm.

Algorithm 5.2.1 (Scaling). Given a scaling tolerance scltol < 1. In the following, $a_{i,j}$ is nonzero entry of A.

1. Compute aratio $= \max_j \{\max_i |a_{i,j}| / \min_i |a_{i,j}|\}$.
2. Divide rows $i = 1, \cdots, m$ of $Ax = b$ by $\sqrt{\min_j |a_{i,j}| \times \max_j |a_{i,j}|}$.

3. Divide columns $i = 1, \cdots, n$ of $\begin{pmatrix} A \\ c^{\mathrm{T}} \end{pmatrix}$ by $\sqrt{\min_i |a_{i,j}| \cdot \max_i |a_{i,j}|}$.
4. Compute sratio by formula in step 1.
5. If sratio $\geq$ scltol $\cdot$ aratio, return; else, go to step 1.

As for more general LP problems, scaling factors for bounded variables (i.e., there are constraints: $l_j \leq x_j \leq u_j,\ j = 1, \ldots, n$), free or fixed variables should be determined differently, as is omitted here.

The simplex method is not invariant to different scaling methods. Effects of scaling are problem-dependent, and difficult to make a theoretical analysis, as is not clear what kind of scaling is the best. At any events, scaling is indispensable in real applications. Even simple scaling would improve algorithm's performance significantly. Without scaling, in fact, only few real-world problems can be handled.

5.3 LU Factorization of Basis

For large-scale sparse computations, it is important to maintain sparsity of matrices and vectors in solution process (see. e.g., Gay 1978; George and Liu 1981). To this end, in this section we firstly present a product form of the inverse of basis matrix, and then turn to the LU factorization of basis matrix, as well as its sparsity version in particular.

The simplex Algorithms 3.5.1 or 3.5.2 utilizes an explicit expression of the inverse of the basis matrix, and updates it iteration by iteration. Such doing is just for convenience of composition, not suitable for sparse computations, because the inverse would be very dense even though the basis matrix itself is sparse. The following instance was given by Orchard-Hays (1971).

Example 5.3.1. It is verified that the inverse of matrix

$$B = \begin{pmatrix} 1 & & & & 1 \\ 1 & 1 & & & \\ & 1 & 1 & & \\ & & 1 & 1 & \\ & & & 1 & 1 \end{pmatrix}$$

is

$$B^{-1} = \begin{pmatrix} 1 & 1 & -1 & 1 & -1 \\ -1 & 1 & 1 & -1 & 1 \\ 1 & -1 & 1 & 1 & -1 \\ -1 & 1 & -1 & 1 & 1 \\ 1 & -1 & 1 & -1 & 1 \end{pmatrix},$$

all entries of which are nonzero.

That an original zero entry becomes nonzero is said to be *fill-in*. In the preceding instance, a large proportion of fill-ins lead to the inverse losing sparsity completely. An approach to bypassing this difficulty is to express inverse in product form of elementary matrices (Dantzig and Orchard-Hays 1953).

As was known, the Gauss-Jordan elimination turns a basis B to the unite matrix. Such doing amounts to premultiplying B by a sequence of Gauss transformation (elementary) matrices $E_1, \cdots, E_m$ such that

$$E_m \cdots E_1 B = I.$$

Thereby, the product form of B^{-1} follows, i.e.,

$$B^{-1} = E_m \cdots E_1.$$

Then, all manipulations involving B^{-1} can be carried out through the factors smoothly. Instead of B^{-1} itself, consequently, there is only a need for storing elementary matrix factors, which is usually sparse, relative to B^{-1}. Note that only a single vector needs to be stored for each factor, as an elementary matrix differs from the unit matrix only by a column.

The preceding approach was considered in early days of the simplex algorithm. We will not go into details here, but turn to a desirable and commonly used alternative, the LU factorization.

5.3.1 Use of LU Factorization

As far as computations in the simplex algorithm are concerned, the inverse itself is not its goal (Bartels and Golub 1969). In fact, what really needed are only vectors

$$\bar{a}_q = B^{-1} a_q, \qquad \text{and} \qquad \bar{y} = B^{-\mathrm{T}} c_B,$$

which can be obtained by solving the following two systems, respectively:

$$B\bar{a}_q = a_q, \qquad \text{and} \qquad B^{\mathrm{T}} \bar{y} = c_B. \tag{5.7}$$

It is well known that a linear system can be solved through the LU factorization of the coefficient matrix efficiently. In particular, sparse matrix techniques are available for solving very large-scale sparse systems.

Assume that the LU factorization $B = LU$ is available, where L is an unit lower triangular matrix (with all diagonals 1), and U is an upper triangular matrix. Then, the first system of (5.7) is converted to the following two triangular systems:

$$Lu = a_q, \quad U\bar{a}_q = u. \tag{5.8}$$

and the second to the two triangular systems below:

$$U^{\mathrm{T}} v = c_B, \quad L^{\mathrm{T}} \bar{y} = v. \tag{5.9}$$

It is a easy task to solve such a system, as only $m^2/2$ multiplications are required (for dense computations).

Matrix factorization is closely related to elimination. The LU factorization of a matrix can be obtained by Gauss elimination. In each step, Gauss elimination zeros nonzeros below the diagonal, as amounts to premultiplying by a Gauss transformation.

Assume that at the end of the $(k-1)$th step $(k < m)$, there are Gauss transformations $G_1, \cdots, G_{k-1}$ such that the entries below the diagonal of the first $k-1$ columns are all zero, i.e.,

$$B^{(k-1)} = G_{k-1} \cdots G_1 B = \begin{pmatrix} B_{11}^{(k-1)} & B_{12}^{(k-1)} \\ & B_{22}^{(k-1)} \end{pmatrix}, \tag{5.10}$$

where $B_{11}^{(k-1)}$ is upper triangular of order $k-1$. Assume that

$$B_{22}^{(k-1)} = \begin{pmatrix} b_{kk}^{(k-1)} & \cdots & b_{km}^{(k-1)} \\ \vdots & & \vdots \\ b_{mk}^{(k-1)} & \cdots & b_{mm}^{(k-1)} \end{pmatrix}, \tag{5.11}$$

and that $b_{kk}^{(k-1)} \neq 0$. Thus, the following factors are well defined:

$$l_{ik} = b_{ik}^{(k-1)} / b_{kk}^{(k-1)}, \quad i = k+1, \cdots, m. \tag{5.12}$$

It is clear that Gauss transformation

$$G_k = I - h^{(k)} e_k^{\mathrm{T}}, \quad h^{(k)} = (\underbrace{0, \cdots, 0}_{k}, l_{k+1,k}, \cdots, l_{m,k})^{\mathrm{T}} \tag{5.13}$$

satisfies

$$B^{(k)} = G_k B^{(k-1)} = \begin{pmatrix} B_{11}^{(k)} & B_{12}^{(k)} \\ & B_{22}^{(k)} \end{pmatrix},$$

where $B_{11}^{(k)}$ is upper triangular of order k. Thus, the kth step of Gauss elimination is complete, provided the kth diagonal (pivot) is nonzero. In such a manner, a sequence $G_1, \cdots, G_{m-1}$ of Gauss transformations can be determined such that matrix

$$B^{(m-1)} = G_{m-1} \cdots G_1 B \triangleq U$$

is an upper triangular with nonzero diagonals.

It is easy to verify that $G_k^{-1} = I + h^{(k)}e_k^{\mathrm{T}}$. Thereby, the LU factorization follows:

$$B = LU,$$

where

$$L = G_1^{-1} \cdots G_{m-1}^{-1} = \prod_{k=1}^{m-1}(I + h^{(k)}e_k^{\mathrm{T}}) = I + \sum_{k=1}^{m-1} h^{(k)}e_k^{\mathrm{T}} = I + (h^{(1)} \cdots, h^{(m-1)}, 0)$$

is a unit lower triangular matrix. The preceding LU factorization may be of a practicable form, i.e.,

$$L^{-1}B = U, \quad L^{-1} = G_{m-1} \cdots G_1. \tag{5.14}$$

Consequently, (5.8) becomes

$$U\bar{a}_q = G_{m-1} \cdots G_1 a_q, \tag{5.15}$$

and (5.9) becomes

$$U^{\mathrm{T}}v = c_B, \quad \bar{y} = G_1^{\mathrm{T}} \cdots G_{m-1}^{\mathrm{T}} v. \tag{5.16}$$

Thereby, there is no need for an explicit expression of L^{-1}, but holding Gauss factors $G_k, k = 1, \cdots, m-1$ is enough.

It is often the case when L^{-1} is dense even though the Gauss factors are sparse. Expressing L^{-1} in product form is therefore suitable for sparse computations. The required storage of G_k is not high, as only nonzeros of vector h_k are stored. In dense computations, these nonzeros are usually stored in the lower triangular part of the matrix while U stored in the upper triangular part.

It is noticeable that the LU factorization is not always a clear road. The existence condition for the LU factorization of a matrix is that the principal submatrices are all nonsingular, as guarantees a nonzero pivot in each step (see Golub and Van Loan 1989); e.g., there is no LU factorization for the following simple nonsingular matrix:

$$\begin{pmatrix} & 1 \\ 1 & 1 \end{pmatrix}.$$

because its first diagonal is 0, not eligible to be a pivot.

Fortunately, there exists the LU factorization for a matrix, resulting from properly rearranging rows and columns, and such doing does not matter essentially, but amounts to exchanging the order of equations and variables.

5.3.2 *Sparse LU Factorization*

On the other hand, rearranging rows and columns does affect sparsity of the LU factors significantly. Let us bring up a simple example.

Example 5.3.2. Use Gauss elimination to convert matrix

$$B = \begin{pmatrix} 2 & 4 & 2 \\ 3 & & \\ -1 & & 1 \end{pmatrix}$$

to upper triangular.

Answer

$$G_1 = \begin{pmatrix} 1 & & \\ -3/2 & 1 & \\ 1/2 & & 1 \end{pmatrix} \longrightarrow G_1 B = \begin{pmatrix} 2 & 4 & 2 \\ & -6 & -3 \\ & 2 & 2 \end{pmatrix},$$

$$G_2 = \begin{pmatrix} 1 & & \\ & 1 & \\ & 1/3 & 1 \end{pmatrix} \longrightarrow G_2 G_1 B = \begin{pmatrix} 2 & 4 & 2 \\ & -6 & -2 \\ & & 1 \end{pmatrix} \triangleq U.$$

The right-hand side U is already upper triangular. Vectors h_1, h_2, related to the Gauss transformations, may be stored in the lower triangular part of U:

$$\begin{pmatrix} 2 & 4 & 2 \\ -3/2 & -6 & -2 \\ 1/2 & 1/3 & 1 \end{pmatrix}.$$

It is seen that the resulting matrix is full filled with nonzeros. Let us now rearrange rows and columns, according to the order of row indices $2, 3, 1$ and of column indices $1, 3, 2$ as follows first:

$$\bar{B} = \begin{pmatrix} 3 & & \\ -1 & 1 & \\ 2 & 2 & 4 \end{pmatrix}.$$

Then carry normal eliminations in order:

$$\bar{G}_1 = \begin{pmatrix} 1 & & \\ 1/3 & 1 & \\ -2/3 & & 1 \end{pmatrix} \longrightarrow \bar{G}_1 \bar{B} = \begin{pmatrix} 3 & & \\ & 1 & \\ & 2 & 4 \end{pmatrix},$$

$$\bar{G}_2 = \begin{pmatrix} 1 & & \\ & 1 & \\ & -2 & 4 \end{pmatrix} \quad \longrightarrow \quad \bar{G}_2\bar{G}_1\bar{B} = \begin{pmatrix} 3 & & \\ & 1 & \\ & & 4 \end{pmatrix} \triangleq \bar{U}.$$

Vectors $\bar{h}_1, \bar{h}_2$ related to the Gauss transformations is stored in the lower triangular part of $\bar{U}$, i.e.,

$$\begin{pmatrix} 3 & & \\ 1/3 & 1 & \\ -2/3 & -2 & 4 \end{pmatrix}.$$

It is seen that the resulting matrix involves three zeros, just as the original B; in other words, no fill-in is introduced. This is the case because the rearranged matrix is lower triangular, as is amenable to Gauss elimination.

For the instance below, the left side is the original matrix (where nonzeros are denoted by "×"). If classical Gauss elimination is carried out, its sparsity is completely lost. But if rows and columns are rearranged as the matrix on the right hand side, then no any fill-in results at all.

$$\begin{pmatrix} \times & \times & \times & \times & \times & \times \\ \times & \times & & & & \\ \times & & \times & & & \\ \times & & & \times & & \\ \times & & & & \times & \\ \times & & & & & \times \end{pmatrix} \longrightarrow \begin{pmatrix} \times & & & & & \times \\ & \times & & & & \times \\ & & \times & & & \times \\ & & & \times & & \times \\ & & & & \times & \times \\ \times & \times & \times & \times & \times & \times \end{pmatrix}.$$

Note that the right-hand side matrix is close to lower triangular, with the bratty bump in the right-most column.

So-called "preassigned pivot procedure (P^3)" and "partitioned preassigned pivot procedure (P^4)" describe how to rearrange rows and columns toward lower triangular (Hellerman and Rarick 1971, 1972). Subsequently, a stable revision, P^5, of P^4 was proposed (Duff et al. 1986). All such type of approaches provide the LU factorization of some permuted version of a matrix, i.e.,

$$L^{-1}PBQ = U, \quad L^{-1} = G_{m-1}\cdots G_1, \tag{5.17}$$

where P and Q are permutation matrices (resulting from the unit matrix by permutation). By (5.17), it is easy to convert the two systems in (5.7) to the equivalent forms

$$Uw = G_{m-1}\cdots G_1 Pa_q, \quad \bar{a}_q = Qw \tag{5.18}$$

and

$$U^{\mathrm{T}}v = Q^{\mathrm{T}}c_B, \quad \bar{y} = P^{\mathrm{T}}G_1^{\mathrm{T}} \cdots G_{m-1}^{\mathrm{T}}v. \tag{5.19}$$

Combination of such approaches and the classical Gauss elimination perform very well in practice. We will not go into details here though, interested readers are referred to related literature.

We prefer another approach selecting pivots in elimination process rather than preassigned. In this aspect, Markowitz rule (1957) is especially successful in practice.

Assume that at the end of step $k-1$, Gauss elimination generates (5.10) and (5.11). Currently, only the $(m-k+1)(m-k+1)$ submatrix at the south-east corner of the matrix is "active", the elimination process proceeds only within it without touching the other part. At step k, the standard Gauss elimination would take $b_{kk}^{(k-1)}$ as the pivot, provided it is nonzero.

Denote by r_i^k the number of nonzeros of row i of $B_{22}^{(k-1)}$, by c_j^k the number of nonzeros of column j. Markowitz selects a pivot, within nonzeros of $B_{22}^{(k-1)}$, such that $(r_i^k - 1)(c_j^k - 1)$ is minimized. The pivot chosen is then moved to the kth diagonal's position by row and column permutations, and step k is complete. It is noted that here $(r_p^k - 1)(c_q^k - 1)$ rather than $r_p^k \, c_q^k$ is used, thereby a row or column involving only a single nonzero is certainly selected, if any, so as no any fill-in occurs at step k.

Nevertheless, so-called "zero" or "nonzero" is only of theoretical sense, as a pivot too close to zero could lead to computational failure. Considering numerical stability, the pivot chosen should not be too small, compared with other nonzeros. There is a contradiction between sparsity and stability. In the following rule, the two aspects are balanced to some extent.

Rule 5.3.1 (Markowitz pivot rule) Given constant $0 < \sigma \le 1$, select the pivot $b_{pq}^{(k-1)}$ such that

$$(r_p^k - 1)(c_q^k - 1) = \min\{(r_i^k - 1)(c_j^k - 1) \mid |b_{ij}^{(k-1)}| \ge \sigma\theta,\ i, j = k, \cdots, m\}, \tag{5.20}$$

where

$$\theta = \max\{|b_{ij}^{(k-1)}| \mid i, j = k, \cdots, m\}. \tag{5.21}$$

It is clear that large σ value contributes to stability, while small one contributes sparsity. Value $\sigma = 0.5$ or so seems to be suitable. In order to avoid going through every nonzero of the active submatrix, on the other hand, some variants of the preceding rule is designed, in which only part of rows or/and columns are involved.

Such kind of approaches actually lead to the following LU factorization:

$$G_{m-1}P_{m-1}\cdots G_1P_1BQ_1\cdots Q_{m-1}=U, \tag{5.22}$$

where $P_i, Q_i, i=1,\ldots,m-1$ are permutations. Setting

$$P=P_{m-1}\cdots P_1, \quad Q=Q_1\cdots Q_{m-1},$$

then it is known from (5.22) that pivot determinations and row and column permutations, carried out alternately, are of the same effects as according prepermutations followed by the standard factorization, i.e.,

$$L^{-1}PBQ=U, \quad L^{-1}=G_{m-1}P_{m-1}\cdots G_1P_1P^{\mathrm{T}}.$$

Therefore, the two systems in (5.7) can still be respectively converted to

$$Uw=G_{m-1}P_{m-1}\cdots G_1P_1a_q, \quad \bar{a}_q=Qw,$$

and

$$U^{\mathrm{T}}v=Q^{\mathrm{T}}c_B, \quad \bar{y}=P_1^{\mathrm{T}}\cdots G_1^{\mathrm{T}}\cdots P_{m-1}^{\mathrm{T}}G_{m-1}^{\mathrm{T}}v.$$

For either row and column prepermutations, or permutations in the process, there is no need for actually moving around related entries in computations, as two integer arrays are enough to record the positions of pivot rows and columns, respectively.

The LU factorization of a basis requires large amount of operations, compared with those in a simplex iteration. Nevertheless, there is no need for such doing in each iteration. After an initial basis matrix is factorized, LU factors of subsequent bases can be obtained by recurrence or updating methods. A basis matrix are refactorized only when the number of iterations exceeds some predetermined upper bound so that LU factors are unacceptably inaccurate or dense, or the list of related Gauss transformation matrices are too long to store (see Sect. 5.1).

5.4 Updating LU Factors

As only one column of the basis is different from that of its predecessor, it is possible to obtain LU factors by updating old ones cheaply. In this section, we will present updating approaches commonly used in practice.

Assume that the LU factorization of the current basis is

$$L^{-1}B = U, \quad L^{-1} = G_s P_s \cdots G_1 P_1. \tag{5.23}$$

and that a pivot column index q and row index p have been determined.

The first updating approach, proposed by Bartels and Golub (1969), drops the pth column a_{j_p} from B, then moves the $p+1$ through mth columns forward by one column space, and adds a_q at the end as the mth column, as results in the new basis

$$\hat{B} = (a_{j_1}, \cdots, a_{j_{p-1}}, a_{j_{p+1}}, \cdots, a_{j_m}, a_q).$$

Consider matrix

$$H \triangleq L^{-1}\hat{B} = (L^{-1}a_{j_1}, \cdots, L^{-1}a_{j_{p-1}}, L^{-1}a_{j_{p+1}}, \cdots, L^{-1}a_{j_m}, L^{-1}a_q).$$

Note that the last column $\tilde{a}_q = L^{-1}a_q$ of the new basis is available, as it was computed as an intermediate result when solving $B\bar{a}_q = a_q$. It is also seen that H may result from dropping the pth column from the upper triangular matrix U, moving forward the $p+1$ through mth columns by one column space, and putting $\tilde{a}_q$ as the last column. Therefore, H is an upper-Hessenberg matrix with nonzero subdiagonals in the p through $(m-1)$th columns, the patten of which looks like:

$$H = \begin{pmatrix} \times & \times & \times & \times & \times & \times & \times & \times & \times & \times \\ & \times & \times & \times & \times & \times & \times & \times & \times & \times \\ & & \times & \times & \times & \times & \times & \times & \times & \times \\ & & & \times & \times & \times & \times & \times & \times & \times \\ & & & & * & * & * & * & * & * \\ & & & & \times & \times & \times & \times & \times & \times \\ & & & & & \times & \times & \times & \times & \times \\ & & & & & & \times & \times & \times & \times \\ & & & & & & & \times & \times & \times \\ & & & & & & & & \times & \times \end{pmatrix}.$$

Then eliminating H's nonzero sub-diagonals in the p through $m-1$th columns, in turn, by elementary transformations to lead to upper triangular. However, this process is impossible when some diagonals are zero. Bartels-Golub's approach overcomes this difficulty by exchanging rows: in the elimination process it takes a diagonal as pivot whenever its absolute value is large enough, compared with the according subdiagonal; otherwise, it exchanges the two rows where the diagonal and subdiagonal are located. Such doing leads to better numerical stability, though exchanging rows frequently could degrade efficiency.

As a variant of Bartels-Golub approach, Forrest-Tomlin (1972) moves the pth row of H (marked by $*$) to the bottom (the p through $(m-1)$th rows are moved up by one row space). The resulting matrix differs from upper triangular only in p through $(m-1)$-indexed possible nonzeros in row m:

$$\tilde{H} = \tilde{P}H = \begin{pmatrix} \times & \times & \times & \times & \times & \times & \times & \times & \times & \times \\ & \times & \times & \times & \times & \times & \times & \times & \times & \times \\ & & \times & \times & \times & \times & \times & \times & \times & \times \\ & & & \times & \times & \times & \times & \times & \times & \times \\ & & & & \times & \times & \times & \times & \times & \times \\ & & & & & \times & \times & \times & \times & \times \\ & & & & & & \times & \times & \times & \times \\ & & & & & & & \times & \times & \times \\ & & & & & & & & \times & \times \\ & & & & * & * & * & * & * & * \end{pmatrix}$$

Such doing amounts to premultiplying H by a permutation, say $\tilde{P}$. Assume that the resulting matrix is $\tilde{H} = \tilde{P}H$, in which the first nonzero in row m is p-indexed, i.e., $\tilde{h}_{mp} \neq 0$. Then, taking diagonals in the p through $(m-1)$th columns as pivots, we eliminate nonzeros in row m, in turn, by elementary transformations. This results in an upper triangular matrix $\tilde{U}$. As all diagonals are nonzero in this case, no any row exchange is needed theoretically. For sake of numerical stability, however, if the ratio of the nonzero in the mth row to the associated diagonal is greater than σ in magnitude, exchange the mth row and the row in which the diagonal is located before eliminating. For this purpose, $\sigma = 10$ or so seems to be suitable.

It might be well to assume that there is no need for such a row exchange. Take elimination of $\tilde{h}_{mp}$ as an example. Such doing amounts to premultiplying $\tilde{H}$ by elementary matrix

$$\tilde{G}_p = I - h^p e_p^{\mathrm{T}}, \quad h^p = (\underbrace{0, \cdots, 0}_{m-1}, \tilde{h}_{mp}/\tilde{h}^p)^{\mathrm{T}}, \tag{5.24}$$

Thus, the LU factorization of the new basis $\hat{B}$ follows, i.e.,

$$\tilde{L}^{-1}\hat{B} = \tilde{U}, \quad \tilde{L}^{-1} = \tilde{G}_{m-1} \cdots \tilde{G}_p \tilde{P} L^{-1}. \tag{5.25}$$

In view of that the lower end components of vector $\tilde{a}_q$ is often zero in sparse computations, Reid (1982) suggests not putting $\tilde{a}_q$ as the last column of the new basis, but inserting it to as the $(t-1)$th column instead, if its first nonzero component (counting from the end) is in row t ($\bar{a}_{t,q} \neq 0$):

$$
\begin{array}{c}
\begin{array}{cccccccccc}
 & & & & p & & & t & & \\
\end{array}\\
H = \left(\begin{array}{cccccccccc}
\times & \times & \times & \times & \times & \times & \times & \times & \times & \times \\
 & \times & \times & \times & \times & \times & \times & \times & \times & \times \\
 & & \times & \times & \times & \times & \times & \times & \times & \times \\
 & & & \times & \times & \times & \times & \times & \times & \times \\
 & & & & * & * & * & * & * & * \\
 & & & & \times & \times & \times & \times & \times & \times \\
 & & & & & \times & \times & \times & \times & \times \\
 & & & & & & \times & \times & \times & \times \\
 & & & & & & & & \times & \times \\
 & & & & & & & & & \times
\end{array}\right)
\begin{array}{c}
\\ \\ \\ \\ p \\ \\ \\ t \\ \\ \\
\end{array}
\end{array}
$$

After that, remove the pth row, then move the $p+1$ through tth row up a row space, and put the original pth row in as the new tth row, i.e.,

$$
\begin{array}{c}
\begin{array}{cccccccccc}
 & & & & p & & & t & & \\
\end{array}\\
H = \left(\begin{array}{cccccccccc}
\times & \times & \times & \times & \times & \times & \times & \times & \times & \times \\
 & \times & \times & \times & \times & \times & \times & \times & \times & \times \\
 & & \times & \times & \times & \times & \times & \times & \times & \times \\
 & & & \times & \times & \times & \times & \times & \times & \times \\
 & & & & \times & \times & \times & \times & \times & \times \\
 & & & & & \times & \times & \times & \times & \times \\
 & & & & & & \times & \times & \times & \times \\
 & & & & * & * & * & * & * & * \\
 & & & & & & & & \times & \times \\
 & & & & & & & & & \times
\end{array}\right)
\begin{array}{c}
\\ \\ \\ \\ p \\ \\ \\ t \\ \\ \\
\end{array}
\end{array}
$$

Finally, eliminating nonzeros in the p through $(t-1)$th columns of the tth row leads to the wanted upper triangular matrix.

At present, Forrest-Tomlin approach and its variants, such as Reid tactics, are commonly used for updating LU factors in practice. Suhl and Suhl (1993) give some detailed considerations on the implementation. Interested readers are referred to related literature (see also Tomlin 1972; Saunders 1973, 1976; Gill et al. 1987).

5.5 Crashing for Initial Basis

The simplex method and its variants require an initial basis, which is usually neither primal nor dual feasible, to get them started. It turns out that the quality of an initial basis greatly affects required storage, the number of iterations and running time.

There are a number of so-called "crash" procedures for determination of an initial basis, which are based on some criterions, mainly toward a sparser initial basis, which tentatively reduces subsequent iterations required (see, e.g., Bixby 1992).

Let us bring up the crash procedure of MINOS, designed for solving general linear and nonlinear programming problems. Once data were input, it introduces a so-called "logical variable" to each equality of the constraint system to convert the problem to a bounded-variable one with a zero right-hand side. As a result, the coefficient matrix of logical variables form an unit matrix, which is taken by a crash option of MINOS as an initial basis, though not very commonly used in practice.

Considering that an initial basis will be LU-factorized, other crash options of MINOS select basic columns to form a nearly-lower triangular matrix to limit the number of subsequent fill-ins. A standard option of crash includes five stages. At the first stage, all free logical variables are entered the basis, corresponding to unit vectors, the number of which is usually less than m. At the second stage, other free variables (or those with very wide bounds) are entered. At the third stage, columns having only a single nonzero in the active submatrix are entered. At the fourth stage, columns having two nonzeros are entered. Finally, yet selected logical variables are selected to form the basis relevantly, if the number of basic columns is still less than m after the previous four stages.

On the other hand, an optimal basis, if any, is clearly an ideal choice if considering the number of required iterations. If B is optimal, e.g., what is needed to do is only to solve system $Bx_B = b$, without any subsequent iterations at all. This is of course impossible in general. What could be expected is to create a matrix close to optimal in some sense. With this respect, it seems to be favorable to determine an initial basis by the most-obtuse-angle heuristics. According to Proposition 2.5.1, inequality constraints with lower pivoting-indices tend to be satisfied as equalities by an optimal solution, that is, the associated (slack) variables should be taken as nonbasic variables. In other words, variables with larger pivoting-indices be taken as basic ones. Let us illustrate by the instance below.

Example 5.5.1. Determine an initial basis by the most-obtuse-angle heuristics:

$$\begin{array}{lrcl}
\min & f = 3x_1 - 2x_2 + x_3, & & \\
\text{s.t.} & x_1 + 2x_2 - x_3 - x_4 & = & 1, \\
& x_1 + x_2 - x_3 \qquad + x_5 & = & 3, \\
& -x_1 + x_2 + x_3 \qquad\qquad + x_6 & = & 2, \\
& x_j \geq 0, \quad j = 1, \cdots, 6. & &
\end{array}$$

Answer The problem can be converted to

$$\begin{array}{ll}
\min & f = 3x_1 - 2x_2 + x_3, \\
\text{s.t.} & x_4 = -1 + x_1 + 2x_2 - x_3 \geq 0, \\
& x_5 = \ \ 3 - x_1 - x_2 + x_3 \geq 0, \\
& x_6 = \ \ 2 + x_1 - x_2 - x_3 \geq 0, \\
& \qquad x_1, x_2, x_3 \geq 0.
\end{array}$$

The following table lists pivoting-indices α_i for variables.

Variable	Constraints	α_j
x_1	$x_1 \geq 0$	−3.00
x_6	$x_1 - x_2 - x_3 \geq -2$	−2.31
x_3	$x_3 \geq 0$	−1.00
x_5	$-x_1 - x_2 + x_3 \geq -3$	0.00
x_4	$x_1 + 2x_2 - x_3 \geq 1$	1.63
x_2	$x_2 \geq 0$	2.00

The first three pivoting-indices are the smallest, corresponding to constraints whose gradients form the most-obtuse angles with the negative objective gradient. Accordingly, x_1, x_3, x_6 are therefore taken as nonbasic variables, thus

$$B = \{2,4,5\},\ N = \{1,3,6\}. B^{-1} = \begin{pmatrix} 2 & -1 & 0 \\ 1 & 0 & 1 \\ 1 & 0 & 0 \end{pmatrix}^{-1} = \begin{pmatrix} 0 & 0 & 1 \\ -1 & 0 & 2 \\ 0 & 1 & -1 \end{pmatrix},$$

$$\bar{y} = B^{-\mathrm{T}} c_B = \begin{pmatrix} 0 & 0 & 1 \\ -1 & 0 & 2 \\ 0 & 1 & -1 \end{pmatrix}^{\mathrm{T}} \begin{pmatrix} -2 \\ 0 \\ 0 \end{pmatrix} = \begin{pmatrix} 0 \\ 2 \\ 0 \end{pmatrix},$$

$$\bar{z}_N = c_N - N^{\mathrm{T}} \bar{y} = \begin{pmatrix} 3 \\ 1 \\ 0 \end{pmatrix} - \begin{pmatrix} 1 & -1 & 0 \\ 1 & -1 & 0 \\ -1 & 1 & 1 \end{pmatrix}^{\mathrm{T}} \begin{pmatrix} 0 \\ 2 \\ 0 \end{pmatrix} = \begin{pmatrix} 1 \\ 3 \\ 0 \end{pmatrix} \geq 0,$$

$$\bar{x}_B = B^{-1} b = \begin{pmatrix} 0 & 0 & 1 \\ -1 & 0 & 2 \\ 0 & 1 & -1 \end{pmatrix} \begin{pmatrix} 1 \\ 3 \\ 2 \end{pmatrix} = \begin{pmatrix} 2 \\ 3 \\ 1 \end{pmatrix} \geq 0.$$

Then, it is seen that the according primal and dual solutions are both feasible, and exhibit complementarity. Consequently, the primal basic optimal solution and associated objective value are

$$\bar{x} = (0,2,0,3,1,0)^{\mathrm{T}}, \quad \bar{f} = c_B^{\mathrm{T}} \bar{x}_B = (-2,0,0)(2,3,1)^{\mathrm{T}} = -4.$$

Therefore, determined is just an optimal basis, and there is hence no need for any iteration.

Of course, we can not expect that such a favorable case always happens. It is interesting however that an optimal basis can often be obtained by the approach for small textbook instances; even when it is not the case, only few iterations are required subsequently to achieve optimality. It is noted that the approach produces a little effect if differences between pivoting-indices are insignificant. For large-scale LP problems, there is a serious conflict between following the most-obtuse-angle heuristics and pursuing sparsity, as is not easy to balance. Interested readers are referred to related literatures (see, e.g., Gould and Reid 1989; Hu 2007; Hu and Pan 2006; Li 2004; Pan 2008a).

5.6 Harris Rule and Tolerance Expending

For success of the simplex algorithm, control of pivot's magnitude is not only important, but also cumbersome. A code based on the standard row rule (see step 6 of Algorithm 3.5.1) is not practicable. It can solve few very small textbook problems only, because such a pivot could be too close to zero to ensure numerical stability.

It is therefore necessary to impose restriction on the magnitude of the pivot. It should be no more than a parameter *tolpiv* such that

$$\epsilon_1 \approx 10^{-11} < tolpiv < 10^{-7} \tag{5.26}$$

(see Sect. 5.1). It is still not enough though. In fact, $tolpiv = 10^{-11}$ is too small for many problems while $tolpiv = 10^{-7}$ is often too large. It is difficult to find a *tolpiv* value that is suitable for all LP problems. In this aspect, Harris (1973) rule with *tolpiv* satisfying (5.26) seems to be feasible overall.

She designed a so-called *two pass procedure*, which passes m rows twice in each iteration. In the first pass, a prospective stepsize α_1 is determined by tolerating bounds being violated by δ ($0 < \delta \ll 1$), which is equal to current feasible tolerance (this will be clear a little later). In the second pass, a pivot row is chosen from rows, corresponding to real stepsizes no more than α_1. The rule may be formulated as follows.

Rule 5.6.1 (Harris two-pass row rule) Firstly, determine

$$\alpha_1 = \min\{(\bar{b}_i + \delta)/\bar{a}_{i\,q} \mid \bar{a}_{i\,q} > \text{tolpiv}; i = 1, \cdots, m\}, \tag{5.27}$$

then select pivot row index

$$p \in \arg\max\{\bar{a}_{i\,q} \mid \bar{b}_i/\bar{a}_{i\,q} \leq \alpha_1,\ \bar{a}_{i\,q} > \text{tolpiv}; i = 1, \cdots, m\}. \tag{5.28}$$

The preceding rule widens the scope of pivot row candidates, so that a pivot with larger magnitude is chosen to improve numerical stability. Moreover, geometrically the gradient of the nonnegative constraint $x_{j_p} \geq 0$ and the descent edge indexed by q form a larger angle, as is conformable with the heuristic characteristic of optimal solution (Sect. 2.5). Harris two-pass rule performs well in practice, and is at present an indispensable part of modern LP codes. However, the stability difficulty still presents occasionally. Methods presented in Chaps. 15 and 16 might be good remedies with this respect.

On the other hand, the theoretical stepsize $\alpha = \bar{b}_p/\bar{a}_{pq}$ seems to be another source of troubles. Even if pivot $\bar{a}_{pq}$ is of a normal magnitude, there is still a somewhat sticky problem when $\bar{b}_p$ is numerically less than zero, as leads to the objective value increasing. Essentially, this is a result of degeneracy or near degeneracy. To guarantee a positive stepsize, and hence a strict decrease in objective value, Gill et al. (1989) suggest a tactic termed *tolerance expanding*.

In each iteration, the current feasible tolerance δ is increased by a tiny amount, say τ, i.e.,

$$\delta = \delta + \tau.$$

Assuming that upper-bound *featol* of δ is set to 10^{-6} (see Sect. 5.1), and initial value is set to

$$\delta_0 = 0.5 \times 10^{-6},$$

then the increment

$$\tau = \frac{(0.99 - 0.5) \times 10^{-6}}{10{,}000} = 0.49 \times 10^{-10}$$

seems to be appropriate. A positive stepsize is then set by

$$\alpha = \max\{\bar{b}_p/\bar{a}_{pq}, \tau/\bar{a}_{pq}\}. \tag{5.29}$$

After every 10,000 times of iterations, $\delta = \delta_0$ is reset, the basis matrix is refactorized, and the associated solution is recalculated. Of course, Phase-1 procedure will be carried out again if the solution becomes infeasible. After optimality achieved, in addition, the entire process has to be restarted by setting $\delta = \delta_0$ to guarantee quality of the final optimal solution.

Chapter 6
Sensitivity Analysis and Parametric LP

In some applications, after a LP problem was solved by the simplex method, there is a need for solving a new problem, resulting from changing some part of the solved problem, e.g., adding or dropping variables or constraints. This chapter discusses how to deal with such problems efficiently. As concerned with solution changes caused by problem changes, this topic belongs to so-called *sensitivity analysis*.

Sensitivity analysis is of diverging application background. For solving widely used integer or mixed ILP problems (yielding from LP models by confining variables, or a part of variables, to integers), e.g., present common methods usually involve a large number of correlating LP problems. In this chapter, we consider changes associated with the standard LP problem (1.8), and then handle closely related parametric problems.

A desirable strategy to resolve such problems is the "warm start" of the simplex method, as will be employed through out this chapter. As was mentioned in Sect. 3.9, even for a new problem differing significantly from the old problem, it is often a shortcut to start from the final basis, yielded from solving the old. Indeed, it is imaginable that the optimal basis would even remain unchanged if the problem changes within a sufficiently small range, as is a case in which a basic optimal solution to the resulting problem is readily available.

Let (3.18) be an optimal simplex tableau, corresponding to the following pair of primal and dual optimal solutions

$$\bar{x}_B = \bar{b} = B^{-1}b \geq 0, \quad \bar{x}_N = 0;$$
$$\bar{z}_B = 0, \quad \bar{z}_N = c_N - N^{\mathrm{T}}\bar{y} \geq 0, \quad \bar{y} = B^{-\mathrm{T}}c_B.$$

P.-Q. PAN, *Linear Programming Computation*, DOI 10.1007/978-3-642-40754-3_6,

6.1 Change in Costs

Assume that c is changed to c'.

For the resulting problem, the primal and dual basic solutions corresponding to the basis B are, respectively,

$$x'_B = B^{-1}b = \bar{x}_B \geq 0, \quad z'_N = c'_N - N^{\mathrm{T}}y', \quad y' = B^{-\mathrm{T}}c'_B.$$

As the primal basic solution does not change at all, primal feasibility remains. If $z'_N \geq 0$, therefore, the pair is just primal and dual basic optimal solutions to the new problem.

Now assume that $z'_N \not\geq 0$. In case when changes in costs are slight, corresponding changes in the dual basic solution should be slight too; therefore, it could be expected that the number of requited iterations for solving the new problem is low if starting from B. In particular, changes in nonbasic costs only cause changes in associated reduced costs. Specifically, assume that

$$c'_j = c_j + \Delta c_j, \qquad j \in T \subset N.$$

In this case, it is clear that $y' = \bar{y}$, and hence only the following reduced costs need to be computed:

$$z'_j = c'_j - a_j^{\mathrm{T}}\bar{y} = \bar{z}_j + \Delta c_j, \quad j \in T.$$

If they are all nonnegative, then optimality is already achieved; otherwise, simplex steps are carried out from this point.

In some applications, it is needed to determine a change range to maintain the optimality of B with changes in costs. More precisely, assume that

$$c' = c + \tau \Delta c, \tag{6.1}$$

where $\Delta c \neq 0$ is a given vector and τ is a scalar to be determined.

It is clear that B is an optimal basis to the resulting problem if the new reduced costs are nonnegative, i.e.,

$$z'_N = c_N + \tau \Delta c_N - N^{\mathrm{T}}B^{-\mathrm{T}}(c_B + \tau \Delta c_B) = \bar{z}_N + \tau \Delta z_N \geq 0, \tag{6.2}$$

where

$$\Delta z_N = \Delta c_N - N^{\mathrm{T}}B^{-\mathrm{T}}\Delta c_B. \tag{6.3}$$

Introducing

$$\beta = \min\{-\bar{z}_j/\Delta z_j \mid \Delta z_j < 0, \ j \in N\},$$
$$\alpha = \min\{\bar{z}_j/\Delta z_j \mid \Delta z_j > 0, \ j \in N\},$$

then nonnegative values of τ, satisfying inequality (6.2), is determined by

$$\tau \le \beta,$$

while nonpositive values of τ satisfying it is determined by

$$|\tau| \le \alpha.$$

Consequently, it is known that the optimal basis or solution remains optimal when τ satisfies

$$-\alpha \le \tau \le \beta. \tag{6.4}$$

Example 6.1.1. Find the change range of the cost, corresponding to x_1, such that the optimal solution remains optimal:

$$\begin{array}{lrl} \min & f = & -4x_1 - 3x_2 - 5x_3, \\ \text{s.t.} & & 2x_1 + x_2 + 3x_3 + x_5 = 15, \\ & & x_1 + x_2 + x_3 + x_4 = 12, \\ & & -2x_1 + x_2 - 3x_3 + x_7 = 3, \\ & & 2x_1 + x_2 + x_6 = 9, \\ & & x_j \ge 0, \quad j = 1, \ldots, 7. \end{array}$$

Answer Example 3.2.1 already gives an optimal simplex tableau to this problem, i.e.,

x_1	x_2	x_3	x_4	x_5	x_6	x_7	RHS
		1		1/3	−1/3		2
			1	−7/12	−1/6	−1/4	1
	1			1/2		1/2	9
1				−1/4	1/2	−1/4	
				13/6	1/3	1/2	37

The optimal basis and nonbasis are $B = \{3, 4, 2, 1\}$, $N = \{5, 6, 7\}$, corresponding to the optimal solution and objective value $\bar{x} = (0, 9, 2, 1, 0, 0, 0)^{\mathrm{T}}$, $\bar{f} = -37$; $c = (-4, -3, -5, 0, 0, 0, 0)^{\mathrm{T}}$, $\bar{z}_N = (13/6, 1/3, 1/2)^{\mathrm{T}}$.

Setting $\Delta c = (1, 0, 0, 0, 0, 0, 0)^{\mathrm{T}}$, then $\Delta c_B = (0, 0, 0, 1)^{\mathrm{T}}$, $\Delta c_N = (0, 0, 0)^{\mathrm{T}}$. By (6.3), it holds that

$$\Delta z_N = \begin{pmatrix} 0 \\ 0 \\ 0 \end{pmatrix} - \begin{pmatrix} 1 & 0 & 0 \\ 0 & 0 & 0 \\ 0 & 0 & 1 \\ 0 & 1 & 0 \end{pmatrix}^{\mathrm{T}} \begin{pmatrix} 3 & 0 & 1 & 2 \\ 1 & 1 & 1 & 1 \\ -3 & 0 & 1 & -2 \\ 0 & 0 & 1 & 2 \end{pmatrix}^{-\mathrm{T}} \begin{pmatrix} 0 \\ 0 \\ 0 \\ 1 \end{pmatrix} = \begin{pmatrix} 1/4 \\ -1/2 \\ 1/4 \end{pmatrix}.$$

$$\alpha = \min\{(13/6)/(1/4), (1/2)/(1/4)\} = 2, \quad \beta = \min\{-(1/3)/(-1/2)\} = 2/3.$$

By (6.4), the optimal solution remains optimal for

$$-2 \le \tau \le 2/3.$$

Further, from (6.1) and $\Delta c = (1,0,0,0,0,0,0)^{\mathrm{T}}$, it is known that the optimal solution remains optimal while the cost c_1 of x_1 satisfies

$$-4-2=-6 \le c_1' \le -4+2/3=-10/3.$$

For instance, when the cost of x_1 is set to the lower bound -6, it holds by (6.2) that

$$z_N' = (13/6, 1/3, 1/2)^{\mathrm{T}} + (-2)(1/4, -1/2, 1/4)^{\mathrm{T}} = (5/3, 4/3, 0)^{\mathrm{T}} \ge 0,$$

and the optimal simplex tableau is then

x_1	x_2	x_3	x_4	x_5	x_6	x_7	RHS
		1		1/3	−1/3		2
			1	−7/12	−1/6	−1/4	1
	1			1/2		1/2	9
1				−1/4	1/2	−1/4	
				5/3	4/3		37

6.2 Change in the Right-Hand Side

Let the right-hand vector b become

$$b' = b + \eta \Delta b, \tag{6.5}$$

where η is a scalar, and Δb is a given vector.

Corresponding to basis B, the primal and dual basic solutions of the resulting problem are

$$x_B' = B^{-1}(b+\eta\Delta b) = \bar{x}_B + \eta B^{-1}\Delta b, \quad z_N' = c_N - N^{\mathrm{T}}\bar{y} = \bar{z}_N \ge 0, \quad \bar{y} = B^{-\mathrm{T}}c_B. \tag{6.6}$$

It is clear that the dual solution remains unchanged, so does the dual feasibility. If $x_B' \ge 0$, then the primal solution is also feasible, and hence the two are just a pair of primal and dual optimal solutions for the resulting problem.

Now assume that $x_B' \not\ge 0$. When $|\eta|$ is small, the change in x_B' is small too, it can be expected to solve the resulting problem quickly by the dual simplex method, starting from B. In fact, B is an optimal basis for the resulting problem if η satisfies the following condition

$$\bar{x}_B + \eta v \geq 0, \quad v = B^{-1}\Delta b, \tag{6.7}$$

The nonpositive values of η, satisfying the preceding inequality, are determined by

$$|\eta| \leq \alpha = \min\{\bar{x}_i/v_i \mid v_i > 0,\ i = 1,\ldots,m\},$$

while nonnegative values of η are determined by

$$\eta \leq \beta = \min\{-\bar{x}_i/v_i \mid v_i < 0,\ i = 1,\ldots,m\}.$$

Therefore, (6.6) gives primal and dual basic optimal solutions of the new problem when η satisfies

$$-\alpha \leq \eta \leq \beta.$$

Example 6.2.1. Find the change range of the first component of the right-hand side such that the optimal basis remains optimal:

$$\begin{array}{ll} \min & f = x_1 + 2x_2 + x_3, \\ \text{s.t.} & -2x_1 - x_2 - x_3 + x_4 = -1, \\ & x_1 - 4x_2 - x_3 + x_5 = -2, \\ & x_1 + 3x_2 + x_6 = 4, \\ & x_j \geq 0, \quad j = 1,\ldots,6. \end{array}$$

Answer Example 4.4.1 gives the following optimal simplex tableau to the preceding program:

x_1	x_2	x_3	x_4	x_5	x_6	RHS
1		1/3	−4/9	1/9		2/9
	1	1/3	−1/9	−2/9		5/9
		−4/3	7/9	5/9	1	19/9
			2/3	1/3		−4/3

which corresponds to the optimal solution and objective value $\bar{x} = (2/9, 5/9, 0, 0, 0, 19/9)^{\mathrm{T}},\ \bar{f} = 4/3$.

The optimal basis and nonbasis are $B = \{1,2,6\},\ N = \{3,4,5\}$.

Set $\Delta b = (1,0,0)^{\mathrm{T}}$. By the second expression of (6.7), it holds that

$$v = \begin{pmatrix} -2 & -1 & 0 \\ 1 & -4 & 0 \\ 1 & 3 & 1 \end{pmatrix}^{-1} \begin{pmatrix} 1 \\ 0 \\ 0 \end{pmatrix} = \begin{pmatrix} -4/9 \\ -1/9 \\ 7/9 \end{pmatrix},$$

$$\alpha = \min\{(19/9)/(7/9)\} = 19/7, \quad \beta = \min\{-(2/9)/(-4/9), -(5/9)/(-1/9)\} = 1/2.$$

Therefore, the optimal solution remains optimal whenever

$$-19/7 \leq \eta \leq 1/2.$$

Further, from (6.5) and $\Delta b = (1, 0, 0)^{\mathrm{T}}$, it is known that the optimal basis remains optimal whenever the first component b'_1 of the right-hand side satisfies

$$-1 - (19/7) = -26/7 \leq b'_1 \leq -1 + 1/2 = -1/2.$$

If setting $b'_1 = -1/2$, then it holds by (6.6) that

$$x'_B = (2/9, 5/9, 19/9)^{\mathrm{T}} + (1/2)(-4/9, -1/9, 7/9)^{\mathrm{T}} = (0, 1/2, 5/2)^{\mathrm{T}} \geq 0.$$

The optimal simplex tableau to the new program is accordingly

x_1	x_2	x_3	x_4	x_5	x_6	RHS
1		1/3	−4/9	1/9		0
	1	1/3	−1/9	−2/9		1/2
		−4/3	7/9	5/9	1	5/2
			2/3	1/3		−1

6.3 Change in Coefficient Matrix

There will be several cases to be discussed separately.

6.3.1 *Dropping a Variable*

Assume that the basic variable x_p, in row p, is to be dropped. It is clear that if the component $\bar{x}_p$ of the optimal solution is equal to zero, deleting it just gives an optimal solution to the new problem.

Now assume that $\bar{x}_p > 0$. Thus, it holds that

$$\bar{x}_p = \bar{b}_p = e_p^{\mathrm{T}} B^{-1} b > 0.$$

Consider the nonbasic entries in row p, i.e.,

$$\bar{a}_{pj} = e_p^{\mathrm{T}} B^{-1} a_j, \quad j \in N.$$

Proposition 6.3.1. *The new program is infeasible if*

$$J' = \{j \in N \mid \bar{a}_{pj} > 0\} = \emptyset.$$

Proof. The feasibility of the new program implies that the original program has a feasible solution with the pth component vanishing. Assume that the feasible solution is $\hat{x} \geq 0$, $\hat{x}_p = 0$. Then it satisfies the equality, corresponding to row p of the optimal tableau of the original program, i.e.,

$$\hat{x}_p + \sum_{j \in N} \bar{a}_{pj}\hat{x}_j = \bar{b}_p,$$

the left-side of which is less than or equal to 0, but the right-side of which is strictly greater than 0, as is a contradiction. Therefore, the new program is infeasible. □

If J' is nonempty, determine column index q such that

$$q \in \arg\min_{j \in J'} \bar{z}_j / \bar{a}_{pj},$$

where $\bar{z}_j$, $j \in J'$ are reduced costs of the optimal simplex tableau of the original program. So $\bar{a}_{pq}$ is the pivot. Carry out elementary transformations to drop p from and enter q to the basis. Then deleting the column, corresponding to the nonbasic variable x_p, results in a dual feasible simplex tableau to the new problem, so as can be solved by the dual simplex method.

Example 6.3.1. Consider problem

$$\begin{array}{llllllll} \min & f = x_1 + x_2 - 3x_3, & & & & & & \\ \text{s.t.} & -2x_1 - x_2 + 4x_3 & & + x_5 & & = -4, & & \\ & x_1 - 2x_2 + x_3 & & & + x_6 = & 5, & & (6.8)\\ & -x_1 \quad + 2x_3 + x_4 & & & & = -3, & & \\ & x_j \geq 0, \quad j = 1, \ldots, 6, & & & & & & \end{array}$$

to which there is the optimal simplex tableau below:

x_1	x_2	x_3	x_4	x_5	x_6	RHS
1	−4/3		−1/3		2/3	13/3
	−2/3	1	1/3		1/3	2/3
	−1		−2	1		2
	1/3		4/3		1/3	−7/3

Solve the problem, resulting from (6.8) by dropping variable x_1.

Answer From the optimal tableau, it is known that $\bar{x}_1 = 13/3 > 0,\ J' = \{6\} \neq \emptyset,\ \min\{(1/3)/(2/3)\} = 1/2,\ q = 6$.

The following dual feasible simplex tableau is obtained by multiply row 1 by 3/2, and then adding $-1/3, -1/3$ times of row 1 to rows 2,4, respectively:

x_2	x_3	x_4	x_5	x_6	RHS
−2		−1/2		1	13/2
0	1	1/2			−3/2
−1		−2	1		2
1		3/2			−9/2

Call dual simplex Algorithm 4.4.1.

Iteration 1:

1. $\min\{13/2, -3/2, 2\} = -3/2 < 0,\ p = 2$.
3. $J = \emptyset$, detecting dual unboundedness. Therefore, the new problem is infeasible.

6.3.2 *Adding a Variable*

Assume that variable x_{n+1} is to be added. Let c_{n+1} and a_{n+1} be the associated cost and column, respectively. The new program is then

$$\begin{array}{ll} \min & c^{\mathrm{T}}x + c_{n+1}x_{n+1}, \\ \text{s.t.} & Ax + a_{n+1}x_{n+1} = b, \quad x, x_{n+1} \geq 0. \end{array}$$

It is clear that B is a feasible basis to the new program, corresponding to the basic feasible solution

$$\begin{pmatrix} \bar{x} \\ \bar{x}_{n+1} \end{pmatrix} = \begin{pmatrix} \bar{x} \\ 0 \end{pmatrix}.$$

Since variable x_{n+1} is nonbasic, the preceding is actually a basic optimal solution if according reduced cost

$$\bar{z}_{n+1} = c_{n+1} - a_{n+1}^{\mathrm{T}}\bar{y}$$

is nonnegative. If this not the case, solve the new program by the simplex method, starting from B.

Example 6.3.2. It is known that to program

$$\begin{aligned} \min \quad & f = -x_1 - 2x_2 - 4x_4 + 3x_5, \\ \text{s.t.} \quad & 2x_1 + x_3 = 3, \\ & x_1 + x_4 = 2, \\ & -x_2 + x_5 = 0, \\ & x_j \geq 0, \quad j = 1, \ldots, 5, \end{aligned}$$

there is the optimal simplex tableau below:

x_1	x_2	x_3	x_4	x_5	RHS
2		1			3
1			1		2
	−1			1	
3	1				8

Solve the problem, resulting from the program by adding variable x_6, with $a_6 = (1, -2, 3)^{\mathrm{T}}$, $c_6 = -2$.

Answer The optimal basis for the original program is $B = \{3, 4, 5\}$. $\bar{a}_6 = B^{-1}a_6 = (1, -2, 3)^{\mathrm{T}}$, $\bar{y} = B^{-\mathrm{T}}c_B = (0, -4, 3)^{\mathrm{T}}$, $\bar{z}_6 = -2 - (1, -2, 3)(0, -4, 3)^{\mathrm{T}} = -19$.

We solve the new program by the tableau simplex algorithm. The initial simplex tableau is obtained by modifying the optimal tableau to the original program, i.e.,

x_1	x_2	x_3	x_4	x_5	x_6	RHS
2		1			1	3
1			1		−2	2
	−1			1	3*	
3	1				−19	8

Call Algorithm 3.2.1.

Iteration 1:

1. $\min\{3, 1, -19\} = -19 < 0$, $q = 6$.
3. $I = \{1, 3\} \neq \emptyset$.
4. $\min\{3/1, 0/3\} = 0$, $p = 3$.
5. Multiply row 3 by 1/3, then add $-1, 2, 19$ times of row 3 to rows 1,2,4, respectively:

x_1	x_2	x_3	x_4	x_5	x_6	RHS
2	$1/3^*$	1		$-1/3$		3
1	$-2/3$		1	$2/3$		2
	$-1/3$			$1/3$	1	
3	$-16/3$			$19/3$		8

Iteration 2:

1. $\min\{3, -16/3, 19/3\} = -16/3 < 0,\ q = 2.$
3. $I = \{1\} \neq \emptyset.$
4. $\min\{3/(1/3)\},\ p = 1.$
5. Multiply row 1 by 3, then add $2/3, 1/3, 16/3$ times of row 1 to rows 2,3,4, respectively:

x_1	x_2	x_3	x_4	x_5	x_6	RHS
6	1	3		-1		9
5		2	1			8
2		1			1	3
35		16		1		56

Iteration 3:

1. $\min\{35, 16, 1\} \geq 0$. The basic optimal solution and objective value:

$$\bar{x} = (0, 9, 0, 8, 0, 3)^{\mathrm{T}}, \quad \bar{f} = -56.$$

6.3.3 Dropping a Constraint

The problem, yielding from dropping some constraints of an LP problem is the so-called "relaxation problem" (Sect. 25.4). According to Proposition 25.4.1, in case when dropped constraints are inactive at an optimal solution to the original problem, it remains optimal to the relaxation problem, hence nothing is needed to do in this case. Now assume that dropped is an active constraint.

We solve the relaxation problem from the basis if the optimal basis to the original problem is still a basis to the relaxation problem. Even this is not the case, it is advantageous to form an initial basis based on the original optimal basis as much as possible.

The preceding idea can be taken as a basis to develop a class of variants of the simplex method. We will handle this topic further in Sect. 25.4.

Example 6.3.3. Drop the third equality constraint from problem

$$
\begin{aligned}
\min\ & f = -2x_1 - x_2, \\
\text{s.t.}\ & x_1 - x_2 + x_3 = 2, \\
& x_1 + 2x_2 + x_4 = 8, \\
& -x_1 - x_2 + x_5 = -3, \\
& -3x_1 - 4x_2 + x_4 + 3x_5 = -14, \\
& x_j \geq 0, \quad j = 1, \ldots, 5,
\end{aligned}
$$

and solve the resulting problem.

Answer The tableau below is from the relaxation problem.

x_1	x_2	x_3	x_4	x_5	RHS
1	−1	1			2
1	2*		1		8
−3	−4		1	3	−14
−2	−1				

It is known from Example 6.3.5 that the final basis for the original problem is $B = \{5, 2, 1, 3\}$. Taking $\{5, 2, 3\}$ as the initial basis for the relaxation problem, we convert the tableau to a simplex one by multiplying row 2 by $1/2$, and then adding $1, 4, 1$ times of row 2 to rows 1,3,4, respectively:

x_1	x_2	x_3	x_4	x_5	RHS
3/2		1	1/2		6
1/2	1		1/2		4
−1			3	3	2
−3/2			1/2		4

Multiplying row 3 by $1/3$ leads to

x_1	x_2	x_3	x_4	x_5	RHS
3/2*		1	1/2		6
1/2	1		1/2		4
−1/3			1	1	2/3
−3/2			1/2		4

which is a feasible simplex tableau to the relaxation problem. Call Algorithm 3.2.1.

Iteration 1:

1. $\min\{-3/2, 1/2\} = -3/2,\ q = 1.$
3. $I = \{1, 2\}.$
4. $\min\{6/(3/2), 4/(1/2)\} = 4,\ p = 1.$
5. Multiply row 1 by $2/3$, then add $-1/2, 1/3, 3/2$ times of row 1 to rows 2,3,4, respectively:

x_1	x_2	x_3	x_4	x_5	RHS
1		2/3	1/3		4
	1	−1/3	1/3		2
		2/9	10/9	1	2
		1	1		10

Optimality achieved. The basic optimal solution and associate objective value are

$$\bar{x} = (4, 2, 0, 0, 2)^{\mathrm{T}}, \qquad \bar{f} = -10.$$

6.3.4 Adding a Constraint

In this case, the old problem can be viewed as a relaxation of the new problem. If the optimal solution $\bar{x}$ to the old problem is feasible to the new problem, therefore, it is also optimal to the new problem. In other words, that $\bar{x}$ satisfies the added constraint is a sufficient optimal condition to the new problem.

Assume now that the optimal condition is not fulfilled. Then, there are two cases arising:

(i) Adding an inequality constraint $v^{\mathrm{T}}x \leq \rho$ such that $v^{\mathrm{T}}\bar{x} = v_B^{\mathrm{T}}\bar{x}_B > \rho$.

Introduce slack variable x_{n+1} to turn the added inequality to equality, i.e.,

$$v_B^{\mathrm{T}}x_B + v_N^{\mathrm{T}}x_N + x_{n+1} = \rho.$$

Inserting the according data to the optimal simplex tableau to the old problem as row $m + 1$ gives the following tableau for the new problem:

x_B^{T}	x_N^{T}	x_{n+1}	f	RHS
I	$\bar{N}$			$\bar{b}$
v_B^{T}	v_N^{T}	1		ρ
	$\bar{z}_N^{\mathrm{T}}$		-1	$-\bar{f}$

For $i = 1, \ldots, m$, add relevant times of row i to row $m+1$ to eliminate v_B^{T}. The resulting simplex tableau is of form

x_B^{T}	x_N^{T}	x_{n+1}	f	RHS
I	$\bar{N}$			$\bar{b}$
	$\bar{v}_N^{\mathrm{T}}$	1		$\bar{\rho}$
	$\bar{z}_N^{\mathrm{T}}$		-1	$-\bar{f}$

(6.9)

It is noted that only row $m+1$ changed. Thus, the preceding is a dual feasible tableau to the new problem, and can be taken to get Algorithm 4.4.1 started. Note that only component $\bar{\rho} = \bar{x}_{n+1}$ of the right-hand side is negative, since the infeasible solution,

$$(\bar{x},\ \bar{x}_{n+1} = \rho - v_B^{\mathrm{T}} \bar{x}_B < 0),$$

to the new problem satisfies the equality associated with row $m+1$.

Example 6.3.4. Solve the problem, yielding from adding constraint $-2x_1 + x_3 - 3x_4 + 2x_6 \leq -25$ to Example 6.3.1.

Answer Example 6.3.1 gives the basic optimal solution $\bar{x} = (13/3, 0, 2/3, 0, 2, 0)^{\mathrm{T}}$, which does not satisfy the added constraint, because

$$-2 \times (13/3) + 2/3 = -24/3 > -25.$$

Introduce slack variable x_7 to the added constraint, and insert a row in accordance with the resulting equality to the optimal simplex tableau of the old problem, i.e,

x_1	x_2	x_3	x_4	x_5	x_6	x_7	RHS
1	$-4/3$		$-1/3$		$2/3$		$13/3$
	$-2/3$	1	$1/3$		$1/3$		$2/3$
	-1		-2	1			2
-2		1	-3		2	1	-25
	$1/3$		$4/3$		$1/3$		$-7/3$

Add 2 times of row 1 to row 4:

x_1	x_2	x_3	x_4	x_5	x_6	x_7	RHS
1	$-4/3$		$-1/3$		$2/3$		$13/3$
	$-2/3$	1	$1/3$		$1/3$		$2/3$
	-1		-2	1			2
	$-8/3$	1	$-11/3$		$10/3$	1	$-49/3$
	$1/3$		$4/3$		$1/3$		$-7/3$

5. Add −1 times of row 2 to row 4:

x_1	x_2	x_3	x_4	x_5	x_6	x_7	RHS
1	−4/3		−1/3		2/3		13/3
	−2/3	1	1/3		1/3		2/3
	−1		−2	1			2
	−2*		−4		3	1	−17
	1/3		4/3		1/3		−7/3

Now the preceding is a dual feasible simplex tableau to the new problem. Call dual simplex Algorithm 4.4.1.

Iteration 1:

1. $\min\{13/3, 2/3, 2, -17\} = -17 < 0,\ p = 4$.
3. $J = \{2, 4\} \neq \emptyset$.
4. $\min\{-(1/3)/-2, -(4/3)/-4\} = 1/6,\ q = 2$.
5. Multiply row 4 by $-1/2$, then add $4/3, 2/3, 1, -1/3$ times of row 4 to rows 1,2,3,5, respectively:

x_1	x_2	x_3	x_4	x_5	x_6	x_7	RHS
1			7/3		−4/3	−2/3	47/3
		1	5/3		−2/3	−1/3	19/3
				1	−3/2	−1/2	21/2
	1		2		−3/2	−1/2	17/2
			2/3		5/6	1/6	−31/6

Iteration 2:

1. $\min\{47/3, 19/3, 21/2, 17/2\} \geq 0$. The basic optimal solution and associated objective value:

$$\bar{x} = (47/3, 17/2, 19/3, 0, 21/2, 0, 0)^{\mathrm{T}}, \quad \bar{f} = 31/6.$$

(ii) Adding an equality constraint $v^{\mathrm{T}}x = \rho$ such that $v^{\mathrm{T}}\bar{x} \neq \rho$. It might be well to assume

$$v^{\mathrm{T}}\bar{x} < \rho. \tag{6.10}$$

Consider the auxiliary program, yielding from adding constraint $v^{\mathrm{T}}x + x_{n+1} = \rho$ to the old problem. Just as in case (i), a dual feasible simplex tableau, say (6.9), to the auxiliary problem can be obtained. Since $\bar{x}, \bar{x}_{n+1} = \rho - v_B^{\mathrm{T}} \bar{x}_B > 0$ satisfies the equality, associated with added row $m+1$, it holds that

$$\bar{\rho} = \bar{x}_{n+1} > 0. \tag{6.11}$$

Therefore, the tableau (6.9) is optimal to the auxiliary program.

Proposition 6.3.2. *Assume that (6.10) holds and that $\bar{v}_N^T$ are nonbasics in row $m+1$ of tableau (6.9). If*

$$J' = \{j \in N \mid \bar{v}_j > 0\}$$

is empty, then the new problem is infeasible.

Proof. That there is a feasible solution to the new problem implies that there is a feasible solution to the auxiliary program, whose $(n+1)$-indexed component vanishes. Assume that $(\hat{x} \geq 0,\ \hat{x}_{n+1} = 0)$ is such a feasible solution, satisfying the equality, associated with row $m+1$ of (6.9), i.e.,

$$\hat{x}_{n+1} + \sum_{j \in N} \bar{v}_j \hat{x}_j = \bar{\rho}.$$

Since $J' = \emptyset$ and $\hat{x}_N,\ \hat{x}_{n+1} \geq 0$, the left-hand side of the preceding equality is less than or equal to zero. But it is known by (6.11) that the right-hand side of the preceding is positive, as a contradiction. Therefore, the new problem is infeasible. □

Assume now that J' is nonempty. Select column index q such that

$$q \in \arg\min_{j \in J'} \bar{z}_j / \bar{a}_{m+1\,j}.$$

Taking $\bar{a}_{m+1\,q}$ as the pivot, we manipulate tableau (6.9) by elementary transformations to drop $n+1$ from and enter q to the basis. It is clear that deleting the column, associated with x_{n+1}, in the resulting tableau leads to a feasible simplex tableau to the new problem. Consequently, the dual simplex method can get itself start from it.

Example 6.3.5. It is known that to problem

$$\begin{array}{llrcl}
\min & f = -2x_1 - x_2, & & & \\
\text{s.t.} & x_1 - \ \ x_2 + x_3 & & = & 2, \\
 & x_1 + 2x_2 \qquad + x_4 & & = & 8, \\
 & -x_1 - \ \ x_2 \qquad\qquad + x_5 & & = & -3, \\
 & x_j \geq 0, \quad j = 1, \ldots, 5, & & &
\end{array}$$

there is the following optimal tableau:

x_1	x_2	x_3	x_4	x_5	RHS
		1/3	2/3	1	3
	1	−1/3	1/3		2
1		2/3	1/3		4
		1	1		10

Solve the problem, resulting from the preceding problem by adding constraint $-3x_1 - 4x_2 + x_4 + 3x_5 = -14$.

Answer The optimal solution $\bar{x} = (4, 2, 0, 0, 3)^{\mathrm{T}}$ to the preceding problem does not satisfies the added constraint, because

$$-3 \times 4 - 4 \times 2 + 3 \times 3 = -11 > -14.$$

Inserting row 4, associated with $3x_1 + 4x_2 - x_4 - 3x_5 + x_6 = 14$, to the optimal tableau gives the following tableau to the auxiliary program:

x_1	x_2	x_3	x_4	x_5	x_6	RHS
		1/3	2/3	1		3
	1	−1/3	1/3			2
1		2/3	1/3			4
3	4		−1	−3	1	14
		1	1			10

Executing relevant elementary transformations gives the simplex tableau below, corresponding to basis $B = \{5, 2, 1, 6\}$:

x_1	x_2	x_3	x_4	x_5	x_6	RHS
		1/3	2/3	1		3
	1	−1/3	1/3			2
1		2/3	1/3			4
		1/3*	−4/3		1	3
		1	1			10

$J' = \{3\} \neq \emptyset$, $\min\{1/(1/3)\} = 3$, $q = 3$. Multiply row 4 by 3, then add $-1/3, 1/3, -2/3, -1$ times of row 4 to rows 1,2,3,5, respectively:

x_1	x_2	x_3	x_4	x_5	x_6	RHS
			2	1	−1	
	1		−1		1	5
1			3		−2	−2
		1	−4		3	9
			5		−3	1

Deleting the column associated with x_6 from the preceding leads to a dual feasible simplex tableau to the new problem, i.e.,

x_1	x_2	x_3	x_4	x_5	RHS
			2	1	
	1		−1		5
1			3		−2
		1	−4		9
			5		1

Call Algorithm 4.4.1.

Iteration 1:

1. $\min\{0, 5, -2, 9\} = -2 < 0, \ p = 3$.
3. $J = \emptyset$, detecting infeasibility of the new problem.

6.3.5 Replacing a Row or Column

In case when a row of the coefficient matrix of a problem is replaced by a new row, the problem can be handled in two phases.

The first phase is to solve the problem, obtained by dropping the associated equality constraint. The second phase is to solve the new problem, obtained by adding the wanted constraint to the previous problem. Techniques presented in the previous two sections are applicable to each phase.

A problem, resulting from replacing a column of the coefficient matrix, may be handled by dropping and adding the associated columns in two phases similarly. In the rest of this section, a direct approach to such a problem will be presented.

Assume that column a_k is replaced by a'_k.

If $k \in N$, then the basis B, and hence the related basic solution are not affected, but reduced costs become

$$z'_k = c_k - (a'_k)^{\mathrm{T}} \bar{y}, \quad \bar{y} = B^{-\mathrm{T}} c_B.$$

If $z'_k \geq 0$, it is clear that the optimal solution and basis remains optimal to the new problem.

If, otherwise, $k \in B$, we solve the following auxiliary program:

$$\begin{array}{ll} \min & x_{n+1}, \\ \text{s.t.} & a_1x_1 + \cdots, a_{k-1}x_{k-1} + a'_k x_k + a_{k+1}x_{k+1} + \cdots + a_n x_n + a_k x_{n+1} = b, \\ & x, x_{n+1} \geq 0. \end{array} \tag{6.12}$$

It is clear that the new problem has a feasible solution if and only if the optimal objective value to the auxiliary program is equal to 0. Moreover, the optimal basis B to the old problem is feasible to the auxiliary program, and hence can be taken as an initial basis to get the simplex method started.

Example 6.3.6. From Example 4.5.1, an optimal basis $B = \{1, 2, 6\}$ is known to problem

$$\begin{array}{lrcl} \min & f = x_1 + 2x_2 + x_3, & & \\ \text{s.t.} & -2x_1 - x_2 - x_3 + x_4 & = & -1, \\ & x_1 - 4x_2 - x_3 \qquad + x_5 & = & -2, \\ & x_1 + 3x_2 \qquad\qquad + x_6 & = & 4, \\ & x_j \geq 0, \quad j = 1, \ldots, 6. & & \end{array}$$

Solve the problem, resulting from replacing the first column of the coefficient matrix by $(1, 1, -2)^{\mathrm{T}}$.

Answer The according auxiliary program of form (6.12) has the following tableau (keep the original objective row, while put the auxiliary objective row at the bottom).

Firstly, find the simplex tableau, corresponding to basis $B = \{7, 2, 6\}$.

1. Multiply row 1 by $-1/2$, then add $-1, -1, -1$ times of row 1 to rows 2,3,5, respectively:

x_1	x_2	x_3	x_4	x_5	x_6	x_7	RHS
1	−1	−1	1			−2*	−1
1	−4	−1		1		1	−2
−2	3				1	1	4
1	2	1					
						1	

x_1	x_2	x_3	x_4	x_5	x_6	x_7	RHS
−1/2	1/2	1/2	−1/2			1	1/2
3/2	−9/2*	−3/2	1/2	1			−5/2
−3/2	5/2	−1/2	1/2		1		7/2
1	2	1					
1/2	−1/2	−1/2	1/2				−1/2

2. Multiply row 2 by $-2/9$, then add $-1/2, -5/2, -2, 1/2$ times of row 2 to rows 1,3,4,5, respectively:

x_1	x_2	x_3	x_4	x_5	x_6	x_7	RHS
−1/3		1/3*	−4/9	1/9		1	2/9
−1/3	1	1/3	−1/9	−2/9			5/9
−2/3		−4/3	7/9	5/9	1		19/9
5/3		1/3	2/9	4/9			−10/9
1/3		−1/3	4/9	−1/9			−2/9

Call Algorithm 3.2.1.

Iteration 1:

1. $\min\{1/3, -1/3, 4/9, -1/9\} = -1/3 < 0,\ q = 3$.
3. $I = \{1, 2\} \neq \emptyset$.
4. $\min\{(2/9)/(1/3), (5/9)/(1/3)\} = 2/3,\ p = 1$.
5. Multiply row 1 by 3, then add $-1/3, 4/3, -1/3, 1/3$ times of row 1 to rows 2,3,4,5, respectively:

x_1	x_2	x_3	x_4	x_5	x_6	x_7	RHS
−1		1	−4/3	1/3		3	2/3
	1		1/3	−1/3		−1	1/3
−2			−1	1	1	4	3
2			2/3	1/3		−1	−4/3
						1	

It is seen that the optimal value of the auxiliary program vanishes. Thus, deleting x_7 column and the bottom row gives a feasible simplex tableau to the new problem, i.e.,

x_1	x_2	x_3	x_4	x_5	x_6	RHS
−1		1	−4/3	1/3		2/3
	1		1/3	−1/3		1/3
−2			−1	1	1	3
2			2/3	1/3		−4/3

As the objective row is nonnegative, the preceding is optimal to the new problem. The basic optimal solutio and objective value are

$$\bar{x} = (0, 1/3, 2/3, 0, 0, 3)^{\mathrm{T}}, \quad \bar{f} = 4/3.$$

6.4 Parametric LP

If original data of an LP program include parameters, it is called *parametric program* (see, e.g., Gass and Saaty 1955). Only the case will be considered when a parameter, say θ, emerges in linear form, and varies in some interval. The parametric program is solved for some fixed value of θ first, usually for an end of the interval; then a range of θ is determined, within which the optimal basis remains optimal, and so on.

6.4.1 *Parameterized Cost*

Given a vector c'. Consider the following parametric variant of the standard LP problem (1.8):

$$\begin{array}{ll} \min & f = (c + \theta c')^T x, \\ \text{s.t.} & Ax = b, \qquad x \geq 0. \end{array} \tag{6.13}$$

Without loss of generality, we are assumed to solve the preceding program for $\theta \in [0, +\infty)$ because the case of $\theta \in [-\infty, 0)$ can be handled by setting $c' := -c'$ instead. It is noted that the parameter has nothing to do with feasibility of the program.

We begin with the following example:

Example 6.4.1. Solve the following program for parameter $\theta \in [0, +\infty)$:

$$\begin{array}{lllll} \min & -(3 + 2\theta)x_1 - (5 - \theta)x_2, & & & \\ \text{s.t.} & 3x_1 + 2x_2 & & + x_5 & = 18, \\ & \quad\;\; 2x_2 & + x_4 & & = 12, \\ & x_1 \quad\;\; + x_3 & & & = \;\; 4, \\ & \quad x_j \geq 0, \; j = 1, \ldots, 5. & & & \end{array}$$

Answer The parametric program with $t = 0$ is solved with Algorithm 3.2.1 first. The following is the optimal simplex tableau, with parameter terms added as the bottom row afterward:

x_1	x_2	x_3	x_4	x_5	RHS
1			$-1/3$	$1/3$	2
	1		$1/2$		6
		1	$1/3$	$-1/3$	2
			$3/2$	1	36
-2θ	θ				

Eliminate parameter entries, corresponding to the basis, by elementary transformations to convert the preceding to the simplex tableau below:

x_1	x_2	x_3	x_4	x_5	RHS
1			$-1/3$	$1/3$	2
	1		$1/2$		6
		1	$1/3^*$	$-1/3$	2
			$3/2$	1	36
			$-7/6\theta$	$2/3\theta$	-2θ

To be optimal to the parametric program, nonbasic reduced costs have to be nonnegative, i.e.,

$$\bar{z}_4 = 3/2 - 7/6\theta \geq 0, \qquad \bar{z}_5 = 1 + 2/3\theta \geq 0.$$

As coefficient $2/3$ of θ in the second expression is nonnegative, $\bar{z}_5$ maintains nonnegative as θ increases from 0. But coefficient $-7/6$ of θ in the first is negative, the increase of θ is restricted for $\bar{z}_4$ to be nonnegative. The largest θ value is $(3/2)/(7/6) = 9/7$. Therefore the simplex tableau remains optimal while

$$\theta \in [0, 9/7],$$

corresponding to the same basis $B = \{1, 2, 3\}$ and basic optimal solution $\bar{x} = (2, 6, 2, 0, 0)^T$.

That $\bar{z}_4 < 0$ when $\theta > 9/7$ indicates that x_4 should enter the basis ($q = 4$). To maintain primal feasibility, perform the minimum-ratio test:

$$\min\{2/(1/3), 6/(1/2)\} = 6, \qquad p = 3.$$

Thus x_3 is determined to leave the basis. Carrying out the according elementary transformations results in

x_1	x_2	x_3	x_4	x_5	RHS
1		1		0	4
	1	$-3/2$		$1/2^*$	3
		3	1	-1	6
		$-9/2$		$5/2$	27
		$7/2\theta$		$-1/2\theta$	5θ

Now consider inequalities $\bar{z}_3 = -9/2 + 7/2\theta \geq 0$, $\bar{z}_5 = 5/2 - 1/2\theta \geq 0$. The largest value of θ is only determined by the second inequality, that is 5. Therefore, the tableau maintains optimal for

$$\theta \in [9/7, 5],$$

corresponding to the same basis $B = \{1, 2, 4\}$ and basic optimal solution $\bar{x} = (4, 3, 0, 6, 0)^T$.

That $\bar{z}_5 < 0$ when $\theta > 5$ suggests entering x_5 to the basis ($q = 5$). From minimum-ratio test $\min\{3/(1/2)\}$, $p = 2$, x_2 is determined to leave the basis. Then associated elementary transformations leads to

x_1	x_2	x_3	x_4	x_5	RHS
1	0	1			4
	2	-3		1	6
	2		1		12
	-5	3			12
	θ	2θ			8θ

It is clear that $\bar{z}_2 = -5 + \theta \geq 0$, $\bar{z}_3 = 3 + 2\theta \geq 0$ holds for all $\theta > 5$, corresponding to optimal basis $B = \{1, 4, 5\}$ and basic optimal solution $\bar{x} = (4, 0, 0, 12, 6,)^T$. In summary, the results are put in the following table:

Range of θ	Basic optimal solution	Optimal value
$[0, 9/7]$	$(2, 6, 2, 0, 0)^T$	$-36 + 2\theta$
$(9/7, 5]$	$(4, 3, 0, 6, 0)^T$	$-27 - 5\theta$
$(5, +\infty)$	$(4, 0, 0, 12, 6)^T$	$-12 - 8\theta$

It is seen from the preceding table that there are three ranges of θ, each of which corresponds to a fixed basic optimal solution and a varying optimal value in linear function of θ. It can be shown that the optimal value of the program is a continuous piecewise linear convex function of θ, in general.

From the preceding illustration, it is not difficult to describe a formal (revised) algorithm for solving the program with parameterized costs. Starting with a basic optimal solution for $\theta = 0$, the resulting algorithm uses a different column rule to clarify program's optimality over the whole interval $(0, +\infty)$, section by section, from the left to the right.

Algorithm 6.4.1. Initial: $(B, N), B^{-1}$ are optimal for $\theta = \theta_1 = 0$. $\bar{b} = B^{-1}b \geq 0$. This algorithm solves parametric program (6.13).

1. Compute $\bar{z}_N = c_N - N^T\bar{y} \geq 0$, where $\bar{y} = B^{-T}c_B$ and $\bar{z}'_N = c'_N - N^T\bar{y}'$, where $\bar{y}' = B^{-T}c'_B$.
2. Stop if $\bar{z}'_N \geq 0$. (*the current basis is optimal for* $\theta \in (\theta_1, +\infty)$).
3. Determine q and θ_2 such that

$$\theta_2 = -\bar{z}_q/\bar{c}'_q = \min\{-\bar{z}_j/\bar{z}'_j \mid \bar{z}'_j < 0,\ j \in N\}.$$

(*the current basis is optimal for* $\theta \in (\theta_1, \theta_2]$)
4. Compute $\bar{a}_q = B^{-1}a_q$.
5. Stop if $\bar{a}_q \leq 0$ (*unbounded for* $\theta \in [\theta_1, +\infty)$).
6. Determine stepsize α and pivot row index p such that

$$\alpha = \bar{b}_p/\bar{a}_{p\,q} = \min\{\bar{b}_i/\bar{a}_{i\,q} \mid \bar{a}_{i\,q} > 0;\ i = 1, \ldots, m\}.$$

7. Update B^{-1} by (3.23).
8. Update (B, N) by exchanging j_p and q and set $\theta_1 = \theta_2$.
9. Compute $\bar{b} = B^{-1}b$.
10. Go to step 1.

Proofs for finiteness of the preceding algorithm under nondegeneracy and for meanings of its exits are omitted.

For more understandable, take a look at the only parameterized part, the objective function

$$f = (c + \theta c')^T x,$$

which is a positive combination of $c^T x$ and $c'^T x$. As weight θ increases, the program shifts to minimizing $c'^T x$ subject to the same constraints. For large enough θ, therefore, the optimal basis coincides with that to the latter, if any, or the program is unbounded otherwise. If $c' = c$, in particular, the optimal basis for $\theta = 0$ are also optimal for $\theta \in [0, +\infty)$, but the associated optimal values are not constant.

It is seen that three linear systems have to be solved in each iteration of the preceding Algorithm, compared with two systems solved by conventional simplex algorithms. As they share the same coefficient matrix, however, systems $B^T y = c_B$ and $B^T y' = c'_B$ can be solved more efficiently.

6.4.2 Parameterized Right-Hand Side

Consider the following program with parameterized right-hand side:

$$\begin{array}{ll} \min & f = c^T x, \\ \text{s.t.} & Ax = b + \theta b', \qquad x \geq 0, \end{array} \tag{6.14}$$

where $b' \in \mathcal{R}^m$, and parameter $\theta \in [0, +\infty)$. The case of $\theta \in [-\infty, 0)$ can be handled by setting $b' = -b'$.

Again, we begin with an example.

Example 6.4.2. Solve the following program for parameter $\theta \in [0, +\infty)$:

$$\begin{array}{lrcl} \min & -2x_1 - 3x_2, & & \\ \text{s.t.} & x_1 - x_2 \qquad + x_4 & = & 6 - 2\theta, \\ & -x_1 + 4x_2 + x_3 & = & 3, \\ & -3x_1 + 2x_2 \qquad\qquad x_5 & = & -2 + \theta, \\ & x_j \geq 0, \ j = 1, \ldots, 4. & & \end{array}$$

Answer Put the program into the following tableau, where a combination of the last two columns gives the right-hand side of the parametric program:

x_1	x_2	x_3	x_4	x_5	$\bar{b}$	θ
1	−1		1		6	−2
−1	4	1			3	
−3	2			1	−2	1
−2	−3					

The program for $\theta = 0$ is solved by Algorithm 3.2.1, resulting in the optimal simplex tableau below:

x_1	x_2	x_3	x_4	x_5	$\bar{b}$	θ
1		1/3	4/3		9	−8/3
	1	1/3	1/3		3	−2/3
		1/3	10/3	1	19	−17/3
		5/3	11/3		27	−22/3

The increase of θ should maintain nonnegativity of the right-hand side, i.e.,

$$\bar{b}_1 = 9 - 8/3\theta \geq 0, \qquad \bar{b}_2 = 3 - 2/3\theta \geq 0 \bar{b}_3 = 19 - 17/3\theta \geq 0.$$

The largest θ value is determined by

$$\min\{-9/(-8/3), -3/(-2/3), -19/(-17/3)\} = -19/(-17/3) = 57/17, \ p = 3.$$

Therefore, the current basis and solution is optimal for $\theta \in [0, 57/17]$.

That $\bar{b}_3 < 0$ when $\theta > 57/17$ suggests that x_5 should leave the basis ($p = 3$). But the nonbasic entries in row 3 are all nonnegative, detecting dual unboundedness or primal infeasibility of the program for $\theta \in (57/17, +\infty)$.

The results are put in the table below:

Range of θ	Basic optimal solution	Optimal value
$[0, 57/17]$	$(9 - 8/3\theta, 3 - 2/3\theta, 0, 0, 19 - 17/3\theta)^T$	$-27 + 22/3\theta$
$(57/17, +\infty)$	infeasible	–

We formulate the following algorithm for solving the parametric program. It can be viewed as a variant of the dual simplex Algorithm. But it starts from an optimal basis for $\theta = 0$, and uses a different row rule.

Algorithm 6.4.2. Initial: $(B, N), B^{-1}, \bar{z}_N \geq 0, \bar{b} = B^{-1}b \geq 0$ and $\bar{b}' = B^{-1}b'$. This algorithm solves parametric program (6.14).

1. Stop if $\bar{b}' \geq 0$. (*the current basis is optimal for* $\theta \in (\theta_1, +\infty)$)
2. Determine p and θ_2 such that

$$\theta_2 = -\bar{b}_p/\bar{b}'_p = \min\{-\bar{b}_i/\bar{b}'_i \mid \bar{b}'_i < 0,\ i = 1, \ldots, m\}.$$

   (*the current basis is optimal for* $\theta \in (\theta_1, \theta_2]$)
3. Compute $\sigma_N = N^T B^{-T} e_p$.
4. Stop if $J = \{j \in N \mid \sigma_j < 0\} = \emptyset$ (*infeasible for* $\theta \in [\theta_1, +\infty)$).
5. Determine β and column index q such that

$$\beta = -\bar{z}_q/\sigma_q = \min_{j \in J} -\bar{z}_j/\sigma_j.$$

6. Update : $\bar{z}_N = \bar{z}_N + \beta\sigma_N$, $\bar{z}_{j_p} = \beta$.
7. Compute $\bar{a}_q = B^{-1}a_q$.
8. Update B^{-1} by (3.23).
9. Update (B, N) by exchanging j_p and q.
10. Compute $\bar{b} = B^{-1}b$ and $\bar{b}' = B^{-1}b'$, and set $\theta_1 = \theta_2$.
11. Go to step 1.

Discussions can be made, analogously to those for Algorithm 6.4.1. For θ large enough, for instance, the program amounts to minimizing the same objective function with constraint system $Ax = b'$ (in terms of the same optimal basis, if any, or program's unboundedness otherwise).

Although the previous two Algorithms handle parametric programs by starting from the left-hand end, 0, of the interval of θ, it is of course possible, without any essential difficulty, to start from the other end if the interval is finite (see Sect. 7.2).

Chapter 7
Variants of the Simplex Method

Besides the simplex method and dual simplex method, a number of their variants have been proposed in the past. To take advantages of both types, attempts were made to combine them. At first, two important variants will be presented in the following two sections respectively, both of which prefixed by "primal-dual" because they execute primal as well as dual simplex steps, though they are based on different ideas. More recent variants of such type will be presented later in Chap. 18.

In the other sections, the primal and dual simplex methods are generalized to handle bounded-variable LP problems, which are commonly used in practice.

7.1 Primal-Dual Simplex Method

The primal-dual method (Dantzig et al. 1956) will be presented in this section, which is an extension of the same named method (Ford and Fulkerson 1956) for solving transportation problems.

Just like the dual simplex method, this method proceeds toward primal feasibility while maintaining dual feasibility and complementarity. However, they pursue primal feasibility in different ways. The former attempts to fulfil $x \geq 0$ while maintaining $Ax = b$, whereas the latter attempts to get rid of artificial variables in the auxiliary Phase-I program to fulfil $Ax = b$ while keeping $x \geq 0$.

We are concerned with the standard LP problem (1.8), whose dual problem is (4.2). Let $(\bar{y}, \bar{z})$ be the current dual feasible solution, satisfying $A^T\bar{y} + \bar{z} \leq c$.

To obtain a primal solution matching $(\bar{y}, \bar{z})$, consider the auxiliary program (3.16), written as

$$\begin{array}{ll} \min & \zeta = e^T x_a, \\ \text{s.t.} & Ax + x_a = b, \qquad x, x_a \geq 0, \end{array} \tag{7.1}$$

P.-Q. PAN, *Linear Programming Computation*, DOI 10.1007/978-3-642-40754-3_7,

where $x_a = (x_{n+1}, \ldots, x_{n+m})^T$ is an artificial variable vector. It would be well to assume $b \geq 0$. Introducing index set

$$Q = \{j \in A \mid \bar{z}_j = 0\}, \tag{7.2}$$

define the so-called "restricted program":

$$\begin{array}{lll} \min & \zeta = e^T x_a, & \\ \text{s.t.} & Ax + x_a = b, & x_a \geq 0, \\ & x_j \geq 0, & j \in Q, \\ & x_j = 0, & j \notin Q. \end{array} \tag{7.3}$$

Since $b \geq 0$, it is clear that the feasible region of the preceding program is nonempty, and hence there is an optimal solution to it. The restricted program may be viewed as one formed by all artificial columns and columns indexed by j belonging to Q.

Assume that $(\bar{x}, \bar{x}_a)$ is an optimal solution to (7.3) with optimal value $\bar{\zeta}$, and that $\bar{w}$ is the associated optimal simplex multiplier.

Theorem 7.1.1. *If the optimal value $\bar{\zeta}$ vanishes, $\bar{x}$ and $(\bar{y}, \bar{z})$ are a pair of primal and dual optimal solutions.*

Proof. $\bar{\zeta} = e^T \bar{x}_a = 0$ and $\bar{x}_a \geq 0$ together imply that $\bar{x}_a = 0$. Thus, $\bar{x}$ is a feasible solution to the original problem (4.1). By the definition of Q, moreover, it holds that $\bar{x}^T \bar{z} = 0$, as exhibits complementarity. Therefore, $\bar{x}$ and $(\bar{y}, \bar{z})$ are a pair of primal and dual optimal solutions. □

When $\bar{\zeta} > 0$, otherwise, $\bar{x}$ could be regarded as the closest one to feasibility among all those complementary with $(\bar{y}, \bar{z})$. Nevertheless, the $\bar{x}$ is not feasible to the original problem because it does not satisfy $Ax = b$, but $x \geq 0$ only. In other words, it should be possible to improve $(\bar{y}, \bar{z})$ by increasing the associated dual objective value. To do so, consider the dual program of (7.3) in the form

$$\begin{array}{llll} \min & b^T w, & & \\ \text{s.t.} & a_j^T w + s_j = 0, & s_j \geq 0, & j \in Q, \\ & w \leq e. & & \end{array} \tag{7.4}$$

Since the simplex multiplier vector $\bar{w}$ is just an optimal solution to the preceding program, it follows from the duality that

$$b^T \bar{w} = \bar{\zeta} > 0,$$

which implies that $\bar{w}$ is an uphill with respect to the objective $b^T y$ of the dual problem (4.2). This leads to the following line search scheme for updating $(\bar{y}, \bar{z})$:

$$\hat{y} = \bar{y} + \beta \bar{w}, \qquad \hat{z} = c - A^T \hat{y}. \tag{7.5}$$

For being an improved dual feasible solution, it must satisfy the dual constraints for some $\beta > 0$, i.e.,

$$\hat{z} = c - A^T(\bar{y} + \beta\bar{w}) = \bar{z} + \beta\bar{s} \geq 0, \qquad \bar{s} = -\beta A^T\bar{w}. \tag{7.6}$$

Since $\bar{z} \geq 0$, and $\bar{w}$ satisfies the constrains of (7.4), it is known that

$$\bar{s}_j = -a_j^T\bar{w} \geq 0, \qquad \forall j \in Q.$$

Therefore, if index set

$$J = \{j \in A\backslash Q \mid \bar{s}_j = -a_j^T\bar{w} < 0\} \tag{7.7}$$

is empty, then (7.6) holds for all $\beta \geq 0$, giving a class of dual feasible solutions. Since $\bar{\zeta} > 0$, the associated dual objective value

$$b^T\hat{y} = b^T\bar{y} + \beta b^T\bar{w} = b^T\bar{y} + \beta\bar{\zeta}$$

tends to $+\infty$ as β infinitely increases. This implies dual unboundedness or primal infeasibility.

If, otherwise, there is some $j \in A\backslash Q$ such that $\bar{s}_j = -a_j^T\bar{w} < 0$, then (7.6) holds for the largest possible stepsize β such that

$$\beta = -\frac{\bar{z}_q}{\bar{s}_q} = \min\left\{-\frac{\bar{z}_j}{\bar{s}_j} \mid \bar{s}_j < 0,\ j \in A\backslash Q\right\} > 0. \tag{7.8}$$

Thus, the resulting dual solution is feasible, corresponding to a strictly larger dual objective value. It is then used for the next iteration.

Let B be the optimal basis of the restricted program. If a column of B is not artificial, it must be indexed by some $j \in Q$ such that $\bar{z}_j = 0$. Since the associated reduced cost is zero, i.e., $\bar{s}_j = 0 - a_j^T\bar{w} = 0$, it holds that

$$\hat{z}_j = \bar{z}_j + \beta\bar{s}_j = 0,$$

implying that the j also belongs to the next Q. Therefore, the optimal basis of the restricted program can be used as a starting basis for the next iteration. In addition, it is seen from (7.8) that there is at least one index (e.g., q) in $A\backslash Q$ belongs to the next Q, and the associated reduced cost is negative, i.e., $\bar{s}_q < 0$. In other words, there exist new candidates to enter the basis in the next iteration. Therefore, the restricted program in each iteration can be solved by applying primal simplex method to the original auxiliary program (7.1) itself, except the choice of columns entering the basis is restricted to those indexed by $j \in Q \cap N$. Once an artificial variable leaves the basis, it is dropped from the auxiliary program immediately.

It is clear that optimality of the restricted program is achieved if $Q \cap N = \emptyset$. In case when the initial set Q is empty, for instance, all the artificial columns just form an optimal basis and the optimal multiplier is $\bar{w} = e$; so, no any simplex step is needed for the first iteration.

The steps can be summarized into the following algorithm, the meanings of whose exists are clear.

Algorithm 7.1.1 (Primal-dual simplex algorithm). Initial: a dual feasible solution $(\bar{y}, \bar{z})$, and associated Q defined by (7.2). $B = \{n+1, \ldots, n+m\}$, $N = \{1, \ldots, n\}$. This algorithm solves the standard LP problem (1.8).

1. Carry out simplex steps to solve the restricted auxiliary program (7.1).
2. Stop if the optimal value of the restricted program vanishes (optimality achieved).
3. Stop if J defined by 7.7 is empty. (infeasible problem)
4. Compute β by (7.8).
5. Update $(\bar{y}, \bar{z})$ by (7.5).
6. Update Q by (7.2)
7. Go to step 1.

Although the simplex method was used to solve the restricted program, any method for solving it will apply. The primal-dual simplex method seems to be amenable to certain network flow problems, in particular, since the labeling method solves the restricted program more efficiently and an initial dual feasible solution is easy to obtain (Papadimitriou and Steiglitz 1982).

It is noted that the objective value, corresponding to the dual feasible solution, increases monotonically iteration by iteration. Therefore, the primal-dual method will terminate if each restricted program encountered is solved in finitely many subiterations, It is however not the case as the simplex method is utilized.

Example 7.1.1. Solve the following problem by Algorithm 7.1.1:

$$\begin{array}{lrcrcrcrcrcr}
\min & 2x_1 &+& 5x_2 &+& x_3 &+& 4x_4 &+& 8x_5, & & \\
\text{s.t.} & -x_1 &+& 4x_2 &-& 2x_3 &+& 2x_4 &-& 6x_5 &=& -1, \\
& x_1 &+& 2x_2 &+& 2x_3 & & &-& 4x_5 &=& 8, \\
& -x_1 &+& x_2 & & &+& 2x_4 &+& 2x_5 &=& 2, \\
& & & & & \multicolumn{7}{l}{x_j \geq 0, \ j = 1, \ldots, 5.}
\end{array}$$

Answer Construct the auxiliary program below:

$$\begin{array}{ll}
\min & \zeta = x_6 + x_7 + x_8, \\
\text{s.t.} & x_1 - 4x_2 + 2x_3 - 2x_4 + 6x_5 + x_6 = 1, \\
& x_1 + 2x_2 + 2x_3 \qquad\quad - 4x_5 \qquad + x_7 = 8, \\
& -x_1 + x_2 + \qquad\quad 2x_4 + 2x_5 \qquad\qquad + x_8 = 2, \\
& \qquad\qquad\qquad\qquad x_j \geq 0, \ j = 1, \ldots, 8.
\end{array}$$

Initial: $B = \{6, 7, 8\}$, $N = \{1, \ldots, 5\}$. Since the costs of the original problem are positive, a feasible dual solution ($\bar{y} = (0, 0, 0)^T, \bar{z} = (2, 5, 1, 4, 8)^T$) is available, with $Q = \emptyset$.

Iteration 1:

1. Since $Q = \emptyset$, no simplex step is needed.
2. The optimal value of the restricted program is positive, and the optimal simplex multiplier is $\bar{w} = (1, 1, 1)^T$.
3. $\bar{s}_J = (-1, -4, -4)^T,\ J = \{1, 3, 5\} \neq \emptyset$.
4. $\bar{z}_J = (2, 1, 8)^T,\ \theta = \min\{2/1, 1/4, 8/4\} = 1/4,\ q = 3$.
5. $\bar{y} = (0, 0, 0)^T + 1/4(1, 1, 1)^T = (1/4, 1/4, 1/4)^T$,

$$\bar{z}_N = \begin{pmatrix} 2 \\ 5 \\ 1 \\ 4 \\ 8 \end{pmatrix} - \begin{pmatrix} 1 & -4 & 2 & -2 & 6 \\ 1 & 2 & 2 & & -4 \\ -1 & 1 & & 2 & 2 \end{pmatrix}^T \begin{pmatrix} 1/4 \\ 1/4 \\ 1/4 \end{pmatrix} = \begin{pmatrix} 7/4 \\ 21/4 \\ 0 \\ 4 \\ 7 \end{pmatrix}.$$

6. $Q = \{3\}$.

Iteration 2:

1. Carry out restricted simplex steps of Algorithm 3.5.1:
   Subiteration 1:

   (2) Column selection is restricted to $Q \cap N = \{3\}.x_3$ enters the basis.
   (4) $\bar{a}_3 = a_3 = (2, 2, 0)^T \not\leq 0$.
   (6) $\bar{x}_B = (1, 8, 2)^T,\ \alpha = \min\{1/2, 8/2\} = 1/2,\ p = 1,\ x_6$ leaves the basis, and is dropped.
   (7) $\bar{x}_B = (1, 8, 2)^T - 1/2(2, 2, 0)^T = (0, 7, 2)^T.\bar{x}_3 = 1/2$.
   (8) $B^{-1} = \begin{pmatrix} 1/2 & & \\ -1 & 1 & \\ & & 1 \end{pmatrix}$.
   (9) $B = \{3, 7, 8\},\ N = \{1, 2, 4, 5\}$.

   Subiteration 2:

   (1) $\bar{w} = \begin{pmatrix} 1/2 & & \\ -1 & 1 & \\ & & 1 \end{pmatrix}^T \begin{pmatrix} 0 \\ 1 \\ 1 \end{pmatrix} = \begin{pmatrix} -1 \\ 1 \\ 1 \end{pmatrix}$,

   $$\bar{s}_N = -\begin{pmatrix} 1 & -4 & -2 & 6 \\ 1 & 2 & & -4 \\ -1 & 1 & 2 & 2 \end{pmatrix}^T \begin{pmatrix} -1 \\ 1 \\ 1 \end{pmatrix} = \begin{pmatrix} 1 \\ -7 \\ -4 \\ 8 \end{pmatrix}.$$

   (2) $Q \cap N = \{3\} \cap \{1, 2, 4, 5\} = \emptyset$.

2. The optimal value of the restricted program is positive.
3. $\bar{s}_J = \{-7, -4\}^T,\ J = \{2, 4\} \neq \emptyset$.
4. $\bar{z}_J = (21/4, 4)^T,\ \beta = \min\{(21/4)/7, 4/4\} = 21/28,\ q = 2$.
5. $\bar{y} = (1/4, 1/4, 1/4)^T + 21/28(-1, 1, 1)^T = (-1/2, 1, 1)^T$.

$$\bar{z}_N = \begin{pmatrix} 2 \\ 5 \\ 4 \\ 8 \end{pmatrix} - \begin{pmatrix} 1 & -4 & -2 & 6 \\ 1 & 2 & & -4 \\ -1 & 1 & 2 & 2 \end{pmatrix}^T \begin{pmatrix} -1/2 \\ 1 \\ 1 \end{pmatrix} = \begin{pmatrix} 5/2 \\ 0 \\ 1 \\ 13 \end{pmatrix}.$$

6. $Q = \{3, 2\}$.

Iteration 3:

1. Carry out simplex steps of Algorithm 3.5.1 restricted:
   Subiteration 1:

   (2) Column selection is restricted to $Q \cap N = \{2\}$. x_2 enters the basis.

   (4) $$\bar{a}_2 = \begin{pmatrix} 1/2 & & \\ -1 & 1 & \\ & & 1 \end{pmatrix} \begin{pmatrix} -4 \\ 2 \\ 1 \end{pmatrix} = \begin{pmatrix} -2 \\ 6 \\ 1 \end{pmatrix} \not\leq 0.$$

   (6) $\bar{x}_B = (1/2, 7, 2)^T$, $\alpha = \min\{7/6, 2/1\} = 7/6$, $p = 2$, x_7 leaves the basis, and is dropped.

   (7) $\bar{x}_B = (1/2, 7, 2)^T - 7/6(-2, 6, 1)^T = (17/6, 0, 5/6)^T, \bar{x}_2 = 7/6$.

   (8) $$B^{-1} = \begin{pmatrix} 1 & 1/3 & \\ & 1/6 & \\ & -1/6 & 1 \end{pmatrix} \begin{pmatrix} 1/2 & & \\ -1 & 1 & \\ & & 1 \end{pmatrix} = \begin{pmatrix} 1/6 & 1/3 & \\ -1/6 & 1/6 & \\ 1/6 & -1/6 & 1 \end{pmatrix}.$$

   (9) $B = \{3, 2, 8\}$, $N = \{1, 4, 5\}$.

   Subiteration 2:

   (1) $$\bar{w} = \begin{pmatrix} 1/6 & 1/3 & \\ -1/6 & 1/6 & \\ 1/6 & -1/6 & 1 \end{pmatrix}^T \begin{pmatrix} 0 \\ 0 \\ 1 \end{pmatrix} = \begin{pmatrix} 1/6 \\ -1/6 \\ 1 \end{pmatrix},$$

   $$\bar{s}_N = - \begin{pmatrix} 1 & -2 & 6 \\ 1 & & -4 \\ -1 & 2 & 2 \end{pmatrix}^T \begin{pmatrix} 1/6 \\ -1/6 \\ 1 \end{pmatrix} = \begin{pmatrix} 1 \\ -5/3 \\ -11/3 \end{pmatrix}.$$

   (2) $Q \cap N = \{3, 2\} \cap \{1, 4, 5\} = \emptyset$.

2. The optimal value of the restricted program is positive.
3. $s_J = (-5/3, -11/3)^T$, $J = \{4, 5\} \neq \emptyset$.
4. $\bar{z}_J = (1, 13)^T$, $\beta = \min\{1/(5/3), 13/(11/3)\} = 3/5$, $q = 4$.
5. $\bar{y} = (-1/2, 1, 1)^T + 3/5(1/6, -1/6, 1)^T = (-2/5, 9/10, 8/5)^T$,

$$\bar{z}_N = \begin{pmatrix} 2 \\ 4 \\ 8 \end{pmatrix} - \begin{pmatrix} 1 & -2 & 6 \\ 1 & & -4 \\ -1 & 2 & 2 \end{pmatrix}^T \begin{pmatrix} -2/5 \\ 9/10 \\ 8/5 \end{pmatrix} = \begin{pmatrix} 31/10 \\ 0 \\ 54/5 \end{pmatrix}.$$

6. $Q = \{3, 2, 4\}$.

Iteration 4:

1. Carry out simplex steps of Algorithm 3.5.1 restricted:
   Subiteration 1:

   (2) Column selection is restricted to $Q \cap N = \{3, 2, 4\} \cap \{1, 4, 5\}$. x_4 enters the basis.

   (4) $\bar{a}_4 = \begin{pmatrix} 1/6 & 1/3 & \\ -1/6 & 1/6 & \\ 1/6 & -1/6 & 1 \end{pmatrix} \begin{pmatrix} -2 \\ 0 \\ 2 \end{pmatrix} = \begin{pmatrix} -1/3 \\ 1/3 \\ 5/3 \end{pmatrix} \not\leq 0.$

   (6) $\bar{x}_B = (17/6, 7/6, 5/6)^T$, $\alpha = \min\{(7/6)/(1/3), (5/6)/(5/3)\} = 1/2$,
   $p = 3$, x_8 leaves the basis, and dropped.

   (7) $\bar{x}_B = (17/6, 7/6, 5/6)^T - 1/2(-1/3, 1/3, 5/3)^T = (3, 1, 0)^T$, $\bar{x}_4 = 1/2$.

   (8) $B^{-1} = \begin{pmatrix} 1 & & 1/5 \\ & & -1/5 \\ & 1 & 3/5 \end{pmatrix} \begin{pmatrix} 1/6 & 1/3 & \\ -1/6 & 1/6 & \\ 1/6 & -1/6 & 1 \end{pmatrix} = \begin{pmatrix} 1/5 & 3/10 & 1/5 \\ -1/5 & 1/5 & -1/5 \\ 1/10 & -1/10 & 3/5 \end{pmatrix}.$

   (9) $B = \{3, 2, 4\}$, $N = \{1, 5\}$.

2. The optimal value of the restricted program is zero, optimality achieved.
   The optimal solution and objective value are

$$\bar{x} = (0, 1, 3, 1/2, 0)^T, \qquad \bar{f} = 10.$$

7.2 Self-Dual Parametric Simplex Method

Based on discussions made in Sect. 6.4, it is not difficult to go over to a method for solving problems with the costs and the right-hand side both parameterized, i.e.,

$$\begin{array}{ll} \min f = (c + \theta c')^T x, & \\ \text{s.t.} \ \ Ax = b + \theta b', & x \geq 0. \end{array} \tag{7.9}$$

In this section, we will solve the standard LP program via handling the preceding parametric program. This method is closely related to Orchard-Hays' work (1956), and has been used by Smale (1983b) for investigating the worst-case complexity of the simplex method.

The method belongs to a more general approach, so-called "homotopy", which generates a continuous deformation, converting a given problem to a related but trivially solved one, and then proceeds backwards from the latter to the original by solving all the problems in between. It is seen that the standard problem (1.8) is just the parametric program (7.9) with $\theta = 0$.

Assume availability of a simplex tableau to the standard LP problem, which is neither primal nor dual feasible. It is a simple matter to determine a value $\theta = \theta_2 > 0$ such that the objective row and the right-hand side both become nonnegative after adding it relevantly. Such doing amounts to adding some terms

$\theta c'$ and $\theta b'$ respectively to the costs and the right-hand side of the original problem, corresponding to $\theta = \theta_1 = 0$. Then, θ is decreased from θ_2 down to 0 while maintaining optimality. If primal feasibility is violated first in this process, a row index p and a new θ_2 are determined; then a column index q is determined by the dual simplex ratio test. If, otherwise, dual feasibility violated first, a column index q and a new θ_2 are determined; a row index p is determined by the primal simplex ratio test. Subsequent operations in the iteration are just for a normal basis change.

Assume that the current simplex tableau is optimal to $\theta = \theta_2$, i.e.,

x_B^T	x_N^T	RHS
I	$\bar{N}$	$\bar{b} + \bar{b}'\theta$
	$\bar{z}_N^T + (\bar{z}')_N^T\theta$	$-\bar{f}$

(7.10)

The procedure is put into the following algorithm, where the parametric program with $\theta = 0$ corresponds to the original problem.

Algorithm 7.2.1 (Self-dual parametric algorithm: tableau form). Given $\theta_2 > 0$. Initial: a simplex tableau of the form (7.10), which is optimal for $\theta = \theta_2$. This algorithm solves the standard LP problem.

1. If $\bar{z}'_N \leq 0$, set $\beta = 0$; else, determine q and β such that

$$\alpha = -\bar{z}_q/\bar{c}'_q = \max\{-\bar{z}_j/\bar{z}'_j \mid \bar{z}'_j > 0,\ j \in N\}.$$

2. If $\bar{b}' \leq 0$, set $\alpha = 0$; else, determine p and α such that

$$\beta = -\bar{b}_p/\bar{b}'_p = \max\{-\bar{b}_i/\bar{b}'_i \mid \bar{b}'_i > 0,\ i = 1, \ldots, m\}.$$

3. If $\alpha \geq \beta$, do the following

   (1) If $\alpha \leq 0$, set $\theta_2 = 0$ and stop (optimality achieved);
   (2) Stop if $\bar{a}_q \leq 0$ (unbounded);
   (3) Determine row index p such that

$$(\bar{b}_p + \bar{b}'_p\theta)/\bar{a}_{pq} = \min\{(\bar{b}_i + \bar{b}'_i\theta)/\bar{a}_{iq} \mid \bar{a}_{iq} > 0,\ i = 1, \ldots, m\},$$

   where θ is close to θ_2;

   else

   (4) If $\beta \leq 0$, set $\theta_2 = 0$, and stop (optimality achieved);
   (5) Stop if $J = \{j \in N \mid \bar{a}_{pj} < 0\}$ (infeasible);
   (6) Determine column index q such that

$$-(\bar{z}_q + \bar{z}'_q\theta)/\bar{a}_{pq} = \min_{j \in J} -(\bar{z}_j + \bar{z}'_j\theta)/\bar{a}_{pj},$$

   where θ is close to θ_2.

4. If $\alpha \geq \beta$, set $\theta_2 = \alpha$ else set $\theta_2 = \beta$.
5. Convert $\bar{a}_{pq}$ to 1, and eliminate the other nonzeros in the column by elementary transformations.
6. Go to step 1.

An advantage of the preceding Algorithm is that it can solve problems in a single phase by starting from any basis. It is sometimes describe as "criss-cross" because of its shuttling between primal and dual sides, depending on which of α and β is larger (see step 3). Therefore, it seems critical to scale the costs and the right-hand side for equilibrium of their magnitudes before hand. On the other hand, the algorithm requires more computational effort per iteration, compared with the simplex algorithm. As a homotopy algorithm, it seems to be more suitable for solving hard problems. At least, it stands good as a tool for handling the parametric program (7.9) itself.

Discussions concerning the preceding Algorithm can be made similarly to Algorithms 6.4.1 and 6.4.2. The revised version of it is omitted.

Example 7.2.1. Solve the following problem by Algorithm 7.2.1:

$$\begin{array}{llr} \min & -2x_1 - 3x_2, & \\ \text{s.t.} & x_1 + 2x_2 \qquad\qquad + x_4 & = 2, \\ & -2x_1 - x_2 + x_3 & = -1, \\ & -3x_1 + 4x_2 \qquad\qquad\qquad x_5 & = -3, \\ & \qquad x_j \geq 0,\ j = 1, \ldots, 4. & \end{array}$$

Answer Put the program into the following tableau with the costs and the right-hand side both parameterized

x_1	x_2	x_3	x_4	x_5	RHS
1	2		1		2
−2	−1	1			$-1+\theta$
−3	4*			1	$-3+\theta$
$-2+\theta$	$-3+\theta$				

Given $\theta_2 = 4 > 0$.

Iteration 1:

1. $\alpha = \max\{-(-2)/1, -(-3)/1\} = 3,\ q = 2.$
2. $\beta = \max\{-(-1)/1, -(-3)/1\} = 3,\ p = 3.$
3. $\alpha \geq \beta$.

(3) $\min\{(-3+\theta)/4\} = (-3+\theta)/4,\ p = 3$, where θ is close to 4.

4. Set $\theta_2 = 3$.

5. Taking $q = 2,\ p = 3$, according basis change leads to

x_1	x_2	x_3	x_4	x_5	RHS
$5/2$			1	$-1/2$	$7/2 - 1/2\theta$
$-11/4$		1		$1/4$	$-7/4 + 5/4\theta$
$-3/4^*$	1			$1/4$	$-3/4 + 1/4\theta$
$-17/4 + 7/4\theta$				$3/4 - 1/4\theta$	$-9/4 + 3/2\theta - 1/4\theta^2$

Iteration 2:

1. $\alpha = \max\{-(-17/4)/(7/4)\} = 17/7,\ q = 1.$
2. $\beta = \max\{-(-7/4)/(5/4), -(-3/4)/(1/4)\} = 3,\ p = 3.$
3. $\alpha \not\geq \beta.$

   (6) $\min\{-(-17/4 + 7/4\theta)/(-3/4))\} = -17/3 + 7/3\theta,\ q = 1$, where θ is close to 3.

4. Set $\theta_2 = 3$ (a degenerate step).
5. Taking $p = 3,\ q = 1$, according basis change leads to

x_1	x_2	x_3	x_4	x_5	RHS
	$10/3^*$		1	$1/3$	$1 + 1/3\theta$
	$-11/3$	1		$-2/3$	$1 + 1/3\theta$
1	$-4/3$			$-1/3$	$1 - 1/3\theta$
	$-17/3 + 7/3\theta$			$-2/3 + 1/3\theta$	$2 - 5/3\theta + 1/3\theta^2$

Iteration 3:

1. $\alpha = \max\{-(-17/3)/(7/3), -(-2/3)/(1/3)\} = 17/7,\ q = 2.$
2. $\beta = \max\{-1/(1/3), -1/(1/3)\} = -3,\ p = 1$.
3. $\alpha > \beta.$

   (3) $\min\{(1 + 1/3\theta)/(10/3))\},\ p = 1$, where θ is close to 3.

4. Set $\theta_2 = 17/7.$
5. Taking $q = 2,\ p = 1$, according basis change leads to

x_1	x_2	x_3	x_4	x_5	RHS
	1		$3/10$	$1/10^*$	$3/10 + 1/10\theta$
		1	$11/10$	$-3/10$	$21/10 + 7/10\theta$
1			$2/5$	$-1/5$	$7/5 - 1/5\theta$
			$17/10 - 7/10\theta$	$-1/10 + 1/10\theta$	$37/10 - 9/5\theta + 1/10\theta^2$

Iteration 4:

1. $\alpha = \max\{-(-1/10)/(1/10)\} = 1,\ q = 5.$
2. $\beta = \max\{-(3/10)/(1/10), -(21/10)/(7/10)\} = -3,\ p = 1.$
3. $\alpha > \beta.$

   (3) $\min\{(3/10 + 1/10\theta)/(10/3))\},\ p = 1$, where θ is close to $17/7$.

4. Set $\theta_2 = 1$.
5. Taking $q = 5,\ p = 1$ as pivot, according basis change leads to

x_1	x_2	x_3	x_4	x_5	RHS
	10		3	1	$3+\theta$
	3	1	2		$3+\theta$
1	2		1		2
	$1-\theta$		$2-\theta$		$4-2\theta$

Iteration 5:

1. $\alpha = 0.$
2. $\beta = \max\{-3/1, -3/1\} = -3,\ p = 1.$
3. $\alpha > \beta.$

   (1) $\theta_2 = 0$. The basic optimal solution and associated objective value:

$$\bar{x} = (2, 0, 3, 0, 3)^T, \qquad \bar{f} = -4.$$

7.3 General LP Problems

Sa far we have presented methods for solving standard LP problems. Nevertheless, models from practice are various, as can be put in a more general from below:

$$\begin{array}{ll} \min & f = c^{\mathrm{T}}x, \\ \text{s.t.} & a \le Ax \le b, \\ & l \le x \le u, \end{array} \tag{7.11}$$

where $A \in \mathcal{R}^{m\times n}$, $c, l, u \in \mathcal{R}^n$, $a, b \in \mathcal{R}^m$, $m < n$, rank $A = m$, and a, b, l, u are all given vectors. Such type of problems have not only upper and lower bounds on variables, but also ranges, i.e., variation range of Ax. This type of problems are usually referred to as *problems with ranges and bounds*.

Ranges involved in the problems can be eliminated by introducing new variables. Setting $w = Ax$, in fact, the preceding problem is converted to

$$\begin{array}{ll} \min & f = c^{\mathrm{T}}x, \\ \text{s.t.} & Ax - w = 0, \\ & l \leq x \leq u, \\ & a \leq w \leq b. \end{array} \tag{7.12}$$

Components of x are said to be *structural variables*, whereas those of w said to be *logical variables*.

We will focus on the following *bounded-variable problem*:

$$\begin{array}{ll} \min & f = c^{\mathrm{T}}x, \\ \text{s.t.} & Ax = b, \quad l \leq x \leq u, \end{array} \tag{7.13}$$

where $A \in \mathcal{R}^{m\times n}$, $c, l, u \in \mathcal{R}^n$, $b \in \mathcal{R}^m$, rank $A = m$, $m < n$. Unless indicated otherwise, it is assume that l, u are finite, and $l_j < u_j$. Infinite upper or lower bounds can be represented by sufficiently large or small reals. Thereby, the standard LP problems (1.8) can be regarded as a special case of the preceding problem.

Clearly, such a problem can be converted to the standard form by variable transformations though such doing increases problem's scale. In the following sections, we will generalize the simplex method and dual simplex method to solve the bounded-variable problem directly.

In the sequel, the following sign function will be useful:

$$\text{sign}(t) = \begin{cases} 1, & \text{if } t > 0, \\ -1, & \text{if } t < 0, \\ 0, & \text{if } t = 0. \end{cases} \tag{7.14}$$

Assume that the current basis and nonbasis are

$$B = \{j_1, \cdots, j_m\}, \quad N = A\backslash B. \tag{7.15}$$

7.4 Generalized Simplex Method

Almost all terms for the standard LP problem are applicable for the bounded-variable problem. A solution to $Ax = b$ is said to be basic if nonbasic components of it attain one of the associated upper and lower bounds. It is clear that basic solution, associated with a basis, is not necessarily unique, in contrast to basic solution in the standard LP problem context.

The following results are similar to those for the standard problem, as are stated in the sequel without proofs.

Lemma 7.4.1. *If there exists a feasible solution to the bounded-variable problem, so does a basic feasible solution; if there exists an optimal solution to it, so does a basic optimal solution.*

Therefore, it is possible to find a basic optimal solution among basic feasible solutions, as is a basis for the generalized simplex algorithm.

Let $\bar{x}$ be a basic feasible solution, associated with B:

$$\bar{x}_j = l_j \text{ or } u_j, \quad j \in N, \tag{7.16}$$

$$l_B \leq \bar{x}_B = B^{-1}b - B^{-1}N\bar{x}_N \leq u_B. \tag{7.17}$$

The according reduced costs and objective value are

$$\bar{z}_N = c_N - N^{\mathrm{T}}B^{-T}c_B, \quad \bar{f} = c_B^{\mathrm{T}}B^{-1}b + \bar{z}_N^{\mathrm{T}}\bar{x}_N. \tag{7.18}$$

Define index set

$$\Gamma = \{\, j \in N \mid \bar{x}_j = l_j\}, \quad \Pi = \{\, j \in N \mid \bar{x}_j = u_j\}. \tag{7.19}$$

So it holds that

$$\Gamma \cup \Pi = N, \quad \Gamma \cap \Pi = \emptyset.$$

Without confusion, thereafter Γ and Π are also used to respectively denote submatrices, consisting of columns indexed by their elements.

Lemma 7.4.2. *A feasible solution $\bar{x}$ is optimal if the following set is empty:*

$$J = \{j \in \Gamma \mid \bar{z}_j < 0\} \cup \{j \in \Pi \mid \bar{z}_j > 0\}. \tag{7.20}$$

Proof. Let x' be any feasible solution. Thus it holds that

$$l_j \leq x'_j \leq u_j, \quad j \in N.$$

It is known by the assumption that

$$\bar{z}_j \geq 0, \ j \in \Gamma; \quad \bar{z}_j \leq 0, \ j \in \Pi.$$

Hence for any $j \in N$, there two cases arising:

(i) $j \in \Gamma$. It follows from $x'_j \geq l_j = \bar{x}_j$ that

$$\bar{z}_j x'_j \geq \bar{z}_j \bar{x}_j; \tag{7.21}$$

(ii) $j \in \Pi$. From $x'_j \leq u_j = \bar{x}_j$ again (7.21) follows. Therefore

$$\sum_{j \in N} \bar{z}_j x'_j \geq \sum_{j \in N} \bar{z}_j \bar{x}_j,$$

which implies that

$$c_B^{\mathrm{T}}B^{-1}b + \bar{z}_N^{\mathrm{T}}x'_N \geq c_B^{\mathrm{T}}B^{-1}b + \bar{z}_N^{\mathrm{T}}\bar{x}_N.$$

The preceding indicates that the objective value at x' is no less than that at $\bar{x}$, therefore $\bar{x}$ is optimal. □

Assume now that J is nonempty. Thus, a column index q can be determined by

$$q \in \arg\max_{j \in J} |\bar{z}_j|.$$

Assuming $q = j_t$, define vector

$$\Delta x \triangleq \begin{pmatrix} \Delta x_B \\ \Delta x_N \end{pmatrix} = \mathrm{sign}(\bar{z}_q) \begin{pmatrix} -B^{-1}a_q \\ e_{t-m} \end{pmatrix}, \tag{7.22}$$

where e_{q-m} is the $(n-m)$-dimensional unit vector with the $(q-m)$th component 1.

Proposition 7.4.1. Δx *satisfies*

$$A\Delta x = 0, \quad c^{\mathrm{T}}\Delta x > 0.$$

Proof. It is known by (7.22) that

$$A\Delta x = B\Delta x_B + N\Delta x_N = \mathrm{sign}(\bar{z}_q)(-a_q + a_q) = 0.$$

From the first formula of (7.18) together with (7.4) and (7.22), it follows that

$$-c^{\mathrm{T}}\Delta x = \mathrm{sign}(\bar{z}_g)(a_q^{\mathrm{T}}B^{-1}c_B - c_q) = -\mathrm{sign}(\bar{z}_q)\bar{z}_q = \mu(q)\bar{z}_q = -|\bar{z}_q| < 0. \tag{7.23}$$

□

The preceding Proposition says that $-\Delta x$ is a descent direction with respect to the objective $c^{\mathrm{T}}x$.

Let $\alpha \geq 0$ be a stepsize from $\bar{x}$ along the direction. The new iterate is then

$$\hat{x} = \bar{x} - \alpha\Delta x. \tag{7.24}$$

Thus, since $\bar{x}$ is feasible, it holds for any $\alpha \geq 0$ that

$$A\hat{x} = A\bar{x} - \alpha(B, N)\Delta x = A\bar{x} = b.$$

The value of stepsize α should be such that $\hat{x}$ satisfies $l \leq \hat{x} \leq u$. Thereby the largest possible stepsize is

$$\alpha = \min\{u_q - l_q, \min\{\alpha_i \mid i = 1, \cdots, m\}\}, \tag{7.25}$$

where

$$\alpha_i = \begin{cases} (\bar{x}_{j_i} - u_{j_i})/\Delta x_{j_i}, & \text{if } \Delta x_{j_i} < 0, \\ (\bar{x}_{j_i} - l_{j_i})/\Delta x_{j_i}, & \text{if } \Delta x_{j_i} > 0, \\ \infty, & \text{if } \Delta x_{j_i} = 0, \end{cases} \qquad i = 1, \cdots, m. \tag{7.26}$$

There are the following two cases arising:

(i) $\alpha = u_q - l_q$. In this case, if $\bar{x}_q = l_q$, then $\hat{x}_q = u_q$; and if $\bar{x}_q = u_q$, then $\hat{x}_q = l_q$. The new solution $\hat{x}$ is basic feasible, corresponding to the same basis. Therefore, there is no need for any basis change.
(ii) $\alpha < u_q - l_q$. Determine row index $p \in \{1, \cdots, m\}$ such that

$$\alpha = \alpha_p. \tag{7.27}$$

Then $\hat{x}_{j_p}$ attains its lower bound l_{j_p} or upper bound u_{j_p}. In this case, the new basis and nonbasis follows from B and N by exchanging j_p and q. In addition, it is verified that the new solution $\hat{x}$ is a basic solution, corresponding to the new basis.

It is known from (7.23) and (7.24) that the new objective value is

$$\hat{f} = c^T \hat{x} = c^{\mathrm{T}} \bar{x} - \alpha c^{\mathrm{T}} \Delta x = c^{\mathrm{T}} \bar{x} - \alpha |\bar{z}_q| \leq c^{\mathrm{T}} \bar{x},$$

which strictly decreases if $\alpha > 0$. The preceding expression leads to the recurrence formula of the objective value, i.e.,

$$\hat{f} = \bar{f} - \alpha |\bar{z}_q|,$$

The preceding formula will not be used in each iteration in the following algorithm, however; instead, the objective value will be computed at the end from the final basis and original data.

Definition 7.4.1. A feasible solution is *degenerate* (with respective to a basis) if a basic component of it is on one of its bounds.

Concerning stepsize α, the following two points should be noted.

(i) In case when a basic solution is degenerate, α value would vanish, and the basic solution remains unchanged even if the basis changes.
(ii) In practice, the problem should be deemed unbounded if the value of α exceeds some sufficiently large number.

From the discussions made above, the following conclusions are attained.

Lemma 7.4.3. *Let $\bar{x}$ be a basic feasible solution. Then the new solution, determined by (7.22), (7.24), (7.25) and (7.26), is a basis feasible solution. The corresponding objective value does not increase, while strictly decreases if nondegeneracy is assumed.*

The overall steps are put in the following algorithm.

Algorithm 7.4.1 (Generalized simplex algorithm). Initial: $(B, N), B^{-1}$ and associated basic feasible solution $\bar{x}$. This algorithm solves bounded-variable problem (7.13).

1. Compute $\bar{z}_N = c_N - N^{\mathrm{T}}\bar{y}$, where $\bar{y} = B^{-T}c_B$.
2. Compute $\bar{f} = c^{\mathrm{T}}\bar{x}$, and stop if set J defined by (7.20) is empty.
3. Select column index q such that $q \in \max_{j\in J} |\bar{z}_j|$.
4. Compute $\Delta x_B = -\mathrm{sign}(\bar{z}_q)B^{-1}a_q$.
5. Determine stepsize α by (7.25) and (7.26).
6. Update $\bar{x}$ by (7.24) and (7.22).
7. Go to step 1 if $\alpha = u_q - l_q$; else determine row index $p \in \{1, \cdots, m\}$ such that $\alpha = \alpha_p$.
8. Update B^{-1} by (3.23).
9. Update (B, N) by exchanging j_p and q.
10. Go to step 1.

Theorem 7.4.1. *Algorithm 7.4.1 generates a sequence of basic feasible solutions. Assuming nondegeneracy throughout the solution process, it terminates at step 2, giving a basic optimal solution.*

Proof. Its validity comes from Lemmas 7.4.2 and 7.4.3.

Example 7.4.1. Solve the following problem by Algorithm 7.4.1:

$$\begin{array}{ll} \min & f = -x_1 + 3x_2, \\ \text{s.t.} & 2 \le 2x_1 - 3x_2 \le 10, \\ & 1 \le x_1 - x_2 \le 5, \\ & -x_1 + 2x_2 \le 0, \\ & 0 \le x_1 \le 6, -2 \le x_2. \end{array}$$

Answer Introduce x_3, x_4, x_5 to convert the preceding to

$$\begin{array}{ll} \min & f = x_1 - 3x_2, \\ \text{s.t.} & -2x_1 + 3x_2 + x_3 = 0, \\ & -x_1 + x_2 + x_4 = 0, \\ & x_1 - 2x_2 + x_5 = 0, \\ & 0 \le x_1 \le 6, \ -2 \le x_2 \le \infty, \ 2 \le x_3 \le 10, \ 1 \le x_4 \le 5, \ -\infty \le x_5 \le 0. \end{array}$$

In the following, the unbounded variables will be handled as $-\infty$ or ∞, upon which only the determination of stepsize touches.

Initial: $B = \{3, 4, 5\}$, $N = \{1, 2\}$, $B^{-1} = I$, $\bar{x}_N = (0_{(-)}, -2_{(-)})^{\mathrm{T}}$, $\bar{x}_B = (6, 2, -4)^{\mathrm{T}}$, $\bar{f} = 6$. The initial solution is basic feasible (with subscript "(−)" to denote on the lower bound, and superscript "(+)" on the upper bound. The same below).

Iteration 1:

1. $y \;= B^{-\mathrm{T}} c_B = (0,0,0)^{\mathrm{T}},\ \bar{z}_N = (1,-3)^{\mathrm{T}}$.
2. $J \;= \{2\}$.
3. $\max_J |\bar{z}_j| = 3, q = 2,\ x_2$ enters the basis.
4. $\bar{a}_2 \;= B^{-1} a_2 = (3,1,-2)^{\mathrm{T}}$.
5. $\alpha_1 \;= (6-2)/3 = 4/3, \alpha_2 = (2-1)/1 = 1,$
   $\alpha_3 \;= (-4-0)/-2 = 2;\ \alpha = \min\{\infty, 4/3, 1, 2\} = 1.$
6. $\bar{x}_B = (6,2,-4)^{\mathrm{T}} - 1 \times (3,1,-2)^{\mathrm{T}} = (3,1,-2)^{\mathrm{T}},$
   $\bar{x}_N = (0_{(-)}, -2)^{\mathrm{T}} - 1 \times (0,-1) = (0_{(-)}, -1)^{\mathrm{T}}.$
7. $p \;= 2,\ x_4$ leaves the basis.
8. $$B^{-1} = \begin{pmatrix} 1 & -3 & \\ & 1 & \\ & 2 & 1 \end{pmatrix}.$$
9. $B \;= \{3,2,5\},\ N = \{1,4\};\ \bar{x}_B = (3,-1,-2)^{\mathrm{T}},\ \bar{x}_N = (0_{(-)}, 1^{\mathrm{T}}_{(-)}).$

Iteration 2:

1. $y \;= B^{-\mathrm{T}} c_B = (0,-3,0)^{\mathrm{T}},\ \bar{z}_N = c_N - N^{\mathrm{T}} y = (1,0)^{\mathrm{T}} - (3,-3)^{\mathrm{T}} = (-2,3)^{\mathrm{T}}.$
2. $J \;= \{1\}$.
3. $\max_J |\bar{z}_j| = 2, q = 1,\ x_1$ enters the basis.
4. $\bar{a}_1 \;= B^{-1} a_1 = (1,-1,-1)^{\mathrm{T}}$.
5. $\alpha_1 \;= (3-2)/1 = 1,\ \alpha_2 = (-1-\infty)/-1 = \infty,$
   $\alpha_3 \;= (-2-0)/-1 = 2, \alpha = \min\{6-0, 1, \infty, 2\} = 1.$
6. $\bar{x}_B = (3,-1,-2)^{\mathrm{T}} - 1 \times (1,-1,-1)^{\mathrm{T}} = (2,0,-1)^{\mathrm{T}},$
   $\bar{x}_N = (0,1)^{\mathrm{T}} - 1 \times (-1,0) = (1,1)^{\mathrm{T}}.$
7. $p \;= 1,\ x_3$ leaves the basis.
8. $$B^{-1} = \begin{pmatrix} 1 & & \\ 1 & 1 & \\ 1 & & 1 \end{pmatrix} \begin{pmatrix} 1 & -3 & \\ & 1 & \\ & 2 & 1 \end{pmatrix} = \begin{pmatrix} 1 & -3 & \\ 1 & -2 & \\ 1 & -1 & 1 \end{pmatrix}.$$
9. $B \;= \{1,2,5\},\ N = \{3,4\};\ \bar{x}_B == (1,0,-1)^{\mathrm{T}},\ \bar{x}_N = (2_{(-)}, 1_{(-)})^{\mathrm{T}}.$

Iteration 3:

1. $y \;= B^{-\mathrm{T}} c_B = (-2,3,0)^{\mathrm{T}},\ \bar{z}_N = c_N - N^{\mathrm{T}} y = (0,0)^{\mathrm{T}} - (-2,3)^{\mathrm{T}} = (2,-3)^{\mathrm{T}}.$
2. $J \;= \{4\}$.
3. $\max_J |\bar{z}_j| = 3, q = 4,\ x_4$ enters the basis.
4. $\bar{a}_4 \;= B^{-1} a_4 = (-3,-2,-1)^{\mathrm{T}}$.
5. $\alpha_1 \;= (1-6)/-3 = 5/3,\ \alpha_2 = (0-\infty)/-2 = \infty,$
   $\alpha_3 \;= (-1-0)/-1 = 1;\ \alpha = \min\{5-1, 5/3, \infty, 1\} = 1.$
6. $\bar{x}_B = (1,0,-1)^{\mathrm{T}} - 1 \times (-3,-2,-1)^{\mathrm{T}} = (4,2,0)^{\mathrm{T}},$
   $\bar{x}_N = (2,1)^{\mathrm{T}} - 1 \times (0,-1) = (2,2)^{\mathrm{T}}.$
7. $p \;= 3,\ x_5$ leaves the basis.
8. $$B^{-1} = \begin{pmatrix} 1 & & -3 \\ & 1 & -2 \\ & & -1 \end{pmatrix} \begin{pmatrix} 1 & -3 & \\ 1 & -2 & \\ 1 & -1 & 1 \end{pmatrix} = \begin{pmatrix} -2 & & -3 \\ -1 & & -2 \\ -1 & 1 & -1 \end{pmatrix}.$$
9. $B \;= \{1,2,4\},\ N = \{3,5\};\ \bar{x}_B = (4,2,2)^{\mathrm{T}},\ \bar{x}_N = (2_{(-)}, 0^{(+)})^{\mathrm{T}}.$

Iteration 4:

1. $y = B^{-\mathrm{T}}c_B = (1,0,3)^{\mathrm{T}}$, $\bar{z}_N = c_N - N^{\mathrm{T}}y = (0,0)^{\mathrm{T}} - (1,3)^{\mathrm{T}} = (-1,-3)^{\mathrm{T}}$.
2. $J = \{3\}$.
3. $\max_J |\bar{z}_j| = 1, q = 3$, x_3 enters the basis.
4. $\bar{a}_3 = B^{-1}a_3 = (-2,-1,-1)^{\mathrm{T}}$.
5. $\alpha_1 = (4-6)/-2 = 1, \alpha_2 = (2-\infty)/-1 = \infty$,
   $\alpha_3 = (2-5)/-1 = 3; \alpha = \min\{10-2, 1, \infty, 3\} = 1$.
6. $\bar{x}_B = (4,2,2)^{\mathrm{T}} - 1 \times (-2,-1,-1)^{\mathrm{T}} = (6,3,3)^{\mathrm{T}}$,
   $\bar{x}_N = (2,0)^{\mathrm{T}} - 1 \times (-1,0) = (3,0)^{\mathrm{T}}$.
7. $p = 1$, x_1 leaves the basis.
8. $B^{-1} = \begin{pmatrix} -1/2 & & \\ -1/2 & 1 & \\ -1/2 & & 1 \end{pmatrix} \begin{pmatrix} -2 & & -3 \\ -1 & & -2 \\ -1 & 1 & -1 \end{pmatrix} = \begin{pmatrix} 1 & & 3/2 \\ & & -1/2 \\ & 1 & 1/2 \end{pmatrix}$.
9. $\bar{x}_B = (3,3,3)^{\mathrm{T}}$, $B = \{3,2,4\}$; $\bar{x}_N = (6^{(+)}, 0^{(+)})^{\mathrm{T}}$, $N = \{1,5\}$.

Iteration 5:

1. $y = B^{-T}c_B = (0,0,3/2)^{\mathrm{T}}$,
   $\bar{z}_N = c_N - N^{\mathrm{T}}y = (1,0)^{\mathrm{T}} - (3/2,3/2)^{\mathrm{T}} = (-1/2,-3/2)^{\mathrm{T}}$.
2. $J = \emptyset$. The basic optimal solution and associated objective value:

$$\bar{x} = (6,3,3,3,0)^{\mathrm{T}}, \quad \bar{f} = 6 - 3 \times 3 = -3.$$

As for the tableau version of Algorithm 7.4.1, the associated simplex tableau is the same as the conventional, except there is no need for RHS column to display the corresponding basic solution. We add three additional rows $(u, \bar{x}, l)$, respectively, listing upper bounds, variable values and lower bounds. The simplex tableau is of the form below:

	x_B^{T}	x_N^{T}
	I	$\bar{N}$
		$\bar{z}_N$
u	u_B^{T}	u_N^{T}
$\bar{x}$	$\bar{x}_B^{\mathrm{T}}$	$\bar{x}_N^{\mathrm{T}}$
l	l_B^{T}	l_N^{T}

Based on Table 3.1, Algorithm 7.4.1 can be revised to a tableau form. As $\bar{a}_q = B^{-1}a_q$, (7.26) should be replaced by

$$\alpha_i = \begin{cases} (u_{j_i} - \bar{x}_{j_i})/\mathrm{sign}(\bar{z}_q)\bar{a}_{i\,q}, & \text{if } \mathrm{sign}(\bar{z}_q)\bar{a}_{i\,q} > 0, \\ (l_{j_i} - \bar{x}_{j_i})/\mathrm{sign}(\bar{z}_q)\bar{a}_{i\,q}, & \text{if } \mathrm{sign}(\bar{z}_q)\bar{a}_{i\,q} < 0, \\ \infty, & \text{if } \bar{a}_{i\,q} = 0, \end{cases} \qquad i = 1, \cdots, m. \quad (7.28)$$

Algorithm 7.4.2 (Generalized simplex algorithm: tableau form). Initial: feasible tableau of form (7.4), associated with $\bar{x}$. This algorithm solves the bounded-variable problem (7.13).

1. Compute $\bar{f} = c^{\mathrm{T}}\bar{x}$, and stop (optimality achieved) if J defined by (7.20) is empty.
2. Select column index q such that $q \in \max_{j\in J} |\bar{z}_j|$.
3. Determine stepsize α by (7.25), where α_i defined by (7.28).
4. Set $\bar{x}_q = -\mathrm{sign}(\bar{z}_q)\alpha$, and update $\bar{x}_B = \bar{x}_B + \alpha\mathrm{sign}(\bar{z}_q)\bar{a}_q$ if $\alpha \neq 0$.
5. If $\alpha = u_q - l_q$, go to step 1; else, determine row index $p \in \{1, \cdots, m\}$ such that $\alpha = \alpha_p$.
6. Convert $\bar{a}_{pq}$ to 1, and eliminate the other nonzeros in the column by elementary transformations.
7. Go to step 1.

Note The last three rows in the tableau should be updated in each iteration.

7.4.1 Generalized Phase-I

The following is devoted to generate an initial feasible tableau to Algorithm 7.4.2,

Assume that B and N are respectively basis and nonbasis at the current iteration, associated with basic solution $\bar{x}$. Introduce index set

$$\begin{aligned} I_1 &= \{i = 1, \cdots, m \mid \bar{x}_{j_i} < l_{j_i}\}, \\ I_2 &= \{i = 1, \cdots, m \mid \bar{x}_{j_i} > u_{j_i}\}, \\ I &= \{1, \cdots, m\} \backslash (I_1 \cup I_2). \end{aligned}$$

If $I_1 \cup I_2 = \emptyset$, then $\bar{x}$ is feasible. In the other case, construct the following auxiliary program:

$$\begin{array}{ll} \min & w = -\sum_{i\in I_1} x_{j_i} + \sum_{i\in I_2} x_{j_i}, \\ \text{s.t.} & Bx_B = b - Nx_N, \\ & l_I \le x_I \le u_I, \quad l_N \le x_N \le u_N, \end{array}$$

where the objective function is termed "infeasible-sum".

The according tableau of the auxiliary program is manipulated by one iteration of Algorithm 7.4.1 (in which the row pivot rule should be modified slightly, see below). Then a new auxiliary program is formed, and so on, until $I_1 \cup I_2$ becomes empty, or infeasibility is detected.

Related discussions are similar to those with the infeasible-sum Phase-I method for the standard LP problem (for details, see Sect. 13.1).

7.5 Generalized Dual Simplex Method: Tableau Form

Let B and N are given by (7.15) and let (7.4) be the according simplex tableau. Assume that the associated basic solution $\bar{x}$ are valued by

$$\bar{x}_j = l_j \text{ or } u_j, \quad j \in N,$$

and

$$\bar{x}_B = \bar{b} - \bar{N}\bar{x}_N, \quad \bar{f} = c^{\mathrm{T}}\bar{x}.$$

Index sets Γ and Π are defined by (7.19). If the following conditions hold:

$$\bar{z}_\Gamma \geq 0, \quad \bar{z}_\Pi \leq 0, \tag{7.29}$$

the simplex tableau is said to be dual feasible. If, further, $l_B \leq x_B \leq u_B$ holds, $\bar{x}$ is clearly a basic optimal solution.

Whether a simplex tableau of a bounded-variable problem is dual feasible dependents on the values taken by nonbasic components of the solution. In principle, in the case when components of l and u are finite, it is always possible to have nonbasic components valued, such that the resulting solution be dual feasible, though $l_B \leq x_B \leq \mu_B$ does not hold in general.

Introduce "bound-violation" quantities

$$\rho_i = \begin{cases} l_{j_i} - \bar{x}_{j_i}, & \text{if } \bar{x}_{j_i} < l_{j_i}, \\ u_{j_i} - \bar{x}_{j_i}, & \text{if } \bar{x}_{j_i} > u_{j_i}, \\ 0, & \text{if } l_{j_i} \leq \bar{x}_{j_i} \leq u_{j_i}, \end{cases} \quad i = 1, \cdots, m, \tag{7.30}$$

and determine row index p by the following rule:

$$p \in \arg\max\{|\rho_i| \mid i = 1, \cdots, m\}. \tag{7.31}$$

If $\rho_p = 0$, optimality is achieved. Now assume that $\rho_p \neq 0$: $\rho_p > 0$ indicates that $\bar{x}_p$ violates the lower bound while $\rho_p < 0$ indicates that it violates the upper bound. Introduce index set

$$J = \{j \in \Gamma \mid \mathrm{sign}(\rho_p)\bar{a}_{pj} < 0\} \cup \{j \in \Pi \mid \mathrm{sign}(\rho_p)\bar{a}_{pj} > 0\}. \tag{7.32}$$

It is not difficult to show that the original problem is infeasible if $J = \emptyset$; else, a column index q and a step size β are determined such that

$$\beta = -\bar{z}_q/(\mathrm{sign}(\rho_p)\bar{a}_{pq}) = \min_{j \in J} -\bar{z}_j/(\mathrm{sign}(\rho_p)\bar{a}_{pj}) \geq 0. \tag{7.33}$$

Takin $\bar{a}_{pq}$ as the pivot, convert the simplex tableau by relevant elementary transformations. Then the resulting simplex tableau corresponds to the new basis and nonbasis below:

$$B = \{j_1, \cdots, j_{p-1}, q, j_{p+1}, \cdots, j_m\}, \quad N = N \backslash q \cup \{j_p\}.$$

It might be well to still use (7.4) to denote the new simplex tableau, $\hat{x}$ denote the associated basic solution. As new tableau is equivalent to the old, $\hat{x}$ and $\bar{x}$ satisfy

$$\bar{x}_B = -\bar{N}\bar{x}_N, \quad \hat{x}_B = -\bar{N}\hat{x}_N. \tag{7.34}$$

Now set the new nonbasic component $\hat{x}_p$ to the violated bound, i.e.,

$$\hat{x}_{j_p} = \bar{x}_{j_p} + \rho_p, \tag{7.35}$$

and maintain other nonbasic components unchanged, i.e.,

$$\hat{x}_j = \bar{x}_j, \quad j \in N, \ j \neq j_p.$$

Then from subtraction of the two equalities of (7.34), the updating formula of $\bar{x}_B$ follows:

$$\hat{x}_B = \bar{x}_B - \rho_p \bar{a}_{j_p}. \tag{7.36}$$

It is not difficulty to show that the new simplex tableau with such a $\hat{x}$ is still dual feasible. The β is actually the largest possible stepsize maintaining dual feasibility.

Noting that $\bar{z}_{j_p} = \text{sign}(\rho_p)\beta$ holds for the new tableau, the following recurrence formula of the objective value can be derived from $\hat{x}$ and $\bar{x}$ satisfying (7.35) and the equality associated with the bottom row of the tableau:

$$\hat{f} = \bar{f} + \bar{z}_{j_p}(\hat{x}_{j_p} - \bar{x}_{j_p}) = \bar{f} + \rho_p \bar{z}_{j_p} = \bar{f} + \beta \geq \bar{f},$$

which indicates that the objective value increases. If all components of $\bar{z}_N$ are nonzero, the simplex tableau is said to be dual nondegenerate, and hence $\beta > 0$, so that the objective value strictly increases.

The overall steps are put into the following algorithm, in which the objective value is calculated at the end.

Algorithm 7.5.1 (Generalized dual simplex algorithm: tableau form). Initial: a dual feasible tableau of form (7.4), corresponding to $\bar{x}$. This algorithm solves the bounded-variable problem (7.13).

1. Select a row index p by (7.31) together with (7.30).
2. If $\rho_p = 0$, compute $\bar{f} = c^{\mathrm{T}}\bar{x}$, and stop (optimality achieved).
3. Stop if J defined by (7.32) is empty (infeasible problem).

4. Determine a column index q by (7.33).
5. Convert $\bar{a}_{pq}$ to 1, and eliminate the other nonzeros in the column by elementary transformations.
6. Update $\bar{x}$ by (7.35) and (7.36).
7. Go to step 1.

Note The last three rows in the simplex tableau should be updated in each iteration.

The proof regrading meanings of the algorithm's exits are delayed to the derivation of its revised version.

Example 7.5.1. Solve the following problem by Algorithm 7.5.1:

$$\begin{array}{ll} \min & f = 2x_1 - x_2 + 3x_3 - 6x_4, \\ \text{s.t.} & -2x_1 + 3x_2 - 4x_3 + 2x_4 + x_5 = 14, \\ & -3x_1 + 4x_2 - 5x_3 + 6x_4 + x_6 = 16, \\ & x_1 - 2x_2 + 2x_3 - 7x_4 + x_7 = -15, \\ & -15 \le x_1 \le 30, \quad -12 \le x_2 \le 20, \quad -17 \le x_3 \le 10, \\ & -8 \le x_4 \le 15, \quad -10 \le x_5 \le 26, \quad -13 \le x_6 \le 34, \\ & 0 \le x_7 \le 19. \end{array}$$

Answer Initial tableau:

	x_1	x_2	x_3	x_4	x_5	x_6	x_7
	−2	3	−4	2	1		
	−3	4*	−5	6		1	
	1	−2	2	−7			1
	2	−1	3	−6			
u	30	20	10	15	26	34	19
$\bar{x}$	−15	20	−17	15	−174	−284	179
l	−15	−12	−17	−8	−10	−13	0

Take

$$\bar{x}_N = (-15_{(-)}, 20^{(+)}, -17_{(-)}, 15^{(+)})^{\mathrm{T}} (N = \{1,2,3,4\}),$$
$$\bar{x}_B = \bar{b} - \bar{N}\bar{x}_N = (-174, -284, 179)^{\mathrm{T}} (B = \{5,6,7\}), \ \bar{f} = -191.$$

Iteration 1:

1. $\rho_1 = -10 - (-174) = 164, \rho_2 = -13 - (-284) = 271,$
   $\rho_3 = 19 - 179 = -160.\ \max\{|164|, |271|, |-160|\} = 271 \neq 0, p = 2, j_2 = 6.$
3. $J = \{1,2,3,4\} \neq \emptyset.$
4. $\min\{-2/(-3), -(-1)/4, -3/(-5), -(-6)/6\} = 1/4, q = 2.$

5. Multiply row 2 by $1/4$, and then add $-3, 2, 1$ times of row 2 to rows 1,3,4, respectively.
6. $\bar{x}_6 = -284 + 271 = -13$.
   $\bar{x}_B = (-174, 20, 179)^{\mathrm{T}} - 271(-3/4, 1/4, 1/2, -1/4)^{\mathrm{T}}$.
   $= (117/4, -191/4, 87/2)^{\mathrm{T}}, \ B = \{5, 2, 7\}$.

	x_1	x_2	x_3	x_4	x_5	x_6	x_7
	1/4		−1/4	−5/2	1	−3/4*	
	−3/4	1	−5/4*	3/2		1/4	
	−1/2		−1/2	−4		1/2	1
	5/4		7/4	−9/2		1/4	
u	30	20	10	15	26	34	19
$\bar{x}$	−15	−191/4	−17	15	117/4	−13	87/2
l	−15	−12	−17	−8	−10	−13	0

Iteration 2:

1. $\rho_1 = 26 - 117/4 = -13/4, \rho_2 = -12 - (-191/4) = 143/4$,
   $\rho_3 = 19 - 87/2 = -49/2. \max\{|-13/4|, |143/4|, |-49/2|\} = 143/4 \neq 0$,
   $p = 2, j_2 = 2$.
3. $J = \{1, 3, 4\} \neq \emptyset$.
4. $\min\{-(5/4)/(-3/4), -(7/4)/(-5/4), -(-9/2)/(3/2)\} = 7/5, \ q = 3$.
5. Multiply row 2 by $-4/5$, and then add $1/4, 1/2, -7/4$ times of row 2 to rows 1,3,4, respectively.
6. $\bar{x}_2 = -191/4 + 143/4 = -12$,
   $\bar{x}_B = (117/4, -17, 87/2)^{\mathrm{T}} - (143/4)(-1/5, -4/5, -2/5)^{\mathrm{T}}$.
   $= (182/5, 58/5, 289/5)^{\mathrm{T}}, \ B = \{5, 3, 7\}$.

	x_1	x_2	x_3	x_4	x_5	x_6	x_7
	2/5	−1/5		−14/5	1	−4/5	
	3/5	−4/5	1	−6/5		−1/5	
	−1/5	−2/5		−23/5*		2/5	1
	1/5	7/5		−12/5		3/5	
u	30	20	10	15	26	34	19
$\bar{x}$	−15	−12	58/5	15	182/5	−13	289/5
l	−15	−12	−17	−8	−10	−13	0

Iteration 3:

1. $\rho_1 = 26 - 182/5 = -52/5, \rho_2 = 10 - 58/5 = -8/5$,
   $\rho_3 = 19 - 289/5 = -194/5. \max\{|-52/5|, |-8/5|, |-194/5|\}$
   $= 194/5 \neq 0, p = 3, j_3 = 7$.
3. $J = \{4, 6\} \neq \emptyset$.

4. $\min\{-(-12/5)/(-23/5), -(3/5)/(2/5)\} = 12/23,\ q = 4.$
5. Multiply row 3 by $-5/23$, and then add $14/5, 6/5, 12/5$ times of row 3 to rows 1,2,4, respectively.
6. $\bar{x}_7 = 289/5 - 194/5 = 19.$
   $\bar{x}_B = (182/5, 58/5, 15)^{\mathrm{T}} - (-194/5)(-14/23, -6/23, -5/23)^{\mathrm{T}}.$
   $= (294/23, 34/23, 151/23)^{\mathrm{T}},\ B = \{5, 3, 4\}.$

	x_1	x_2	x_3	x_4	x_5	x_6	x_7
	12/23	1/23			1	−24/23	−14/23
	15/23	−16/23	1			−7/23	−6/23
	1/23	2/23		1		−2/23	−5/23
	7/23	37/23				9/23	−12/23
u	30	20	10	15	26	34	19
$\bar{x}$	−15	−12	34/23	151/23	294/23	−13	19
l	−15	−12	−17	−8	−10	−13	0

Iteration 4:

1. $\rho_1 = \rho_2 = \rho_3 = 0$. The basic optimal solution and optimal value are

$$\bar{x} = (-15, -12, 34/23, 151/23, 294/23, -13, 19)^{\mathrm{T}},$$

$$\bar{f} = (2, -1, 3, -6)(-15, -12, 34/23, 151/23)^{\mathrm{T}} = -1{,}218/23.$$

7.5.1 Generalized Dual Phase-I

It is not difficult to generalize dual Phase-I methods (Chap. 14) for standard problems to initiate the generalized dual simplex algorithm.

Using a generalized version of the most-obtuse-angle row rule (14.3), Koberstein and Suhl (2007) designed a dual Phase-I procedure, named by PAN, for solving generale problems. Taking MOPS[1] as a platform, they tested several main dual Phase-1 methods on 46 typical large-scale sparse problems, the largest among which involves more than 500,000 constraints and 1,000,000 variables. The numerical results show that for most of the tested problem, PAN required a small number of iterations; only for few most difficult problems, the required iterations exceeded an acceptable amount. In the latter cases, they turned to a version of the dual infeasibility-sum Phase-I, named by SDI. It turned out that such a combination, PAN + SDI, is the best among four commonly used Phase-I methods. Therefore, PAN+SDI was taken as the default option for MOPS dual simplex algorithm.

[1]MOPS is a developed package by Suhl et al. of College of Production Information Economy and Operations Research of Berlin Free University see Suhl (1994).

In view of the preceding facts, the author suggests generalizing Rule 14.3.2 by replacing (7.25) with

$$\alpha = \min\{u_q - l_q,\ \min\{\alpha_i \mid |\bar{a}_{i\,q}| \geq \tau\theta,\ i = 1, \cdots, m\}\}, \qquad (7.37)$$
$$\theta = \max\{|\bar{a}_{i\,q}| \mid i = 1, \cdots, m\},$$

where $0 < \tau \leq 1, \alpha_i,\ i = 1, \cdots, m$, are defined by (7.28). The basic idea of such doing is to restrict stepsizes to some extent.

This consideration leads to the following algorithm, yielding from modifying Algorithm 7.4.2.

Algorithm 7.5.2 (Tableau generalized dual Phase-I: the most-obtuse-angle rule). Given $0 < \tau \leq 1$. Initial: a dual feasible simplex tableau of form (7.4), associated with $\bar{x}$. This algorithm solves the bounded-variable problem (7.13).

1. If J defined by (7.20) is empty, compute $\bar{f} = c^{\mathrm{T}}\bar{x}$, and stop (optimality achieved).
2. Select column index q such that $q \in \max_{j \in J} |\bar{z}_j|$.
3. Determine stepsize α by (7.37).
4. Set $\bar{x}_q = -\text{sign}(\bar{z}_q)\alpha$, and update $\bar{x}_B = \bar{x}_B + \alpha\text{sign}(\bar{z}_q)\bar{a}_q$ if $\alpha \neq 0$.
5. If $\alpha = u_q - l_q$, go to step 1; else, determine row index $p \in \{1, \cdots, m\}$ such that $\alpha = \alpha_p$.
6. Convert $\bar{a}_{p\,q}$ to 1, and eliminate the other nonzeros in the column by elementary transformations.
7. Go to step 1.

7.6 Generalized Dual Simplex Method

According to Table 3.1, which gives the correspondence between entries of the simplex tableau and the revised simplex tableau, it is easy to formulate the revised version of Algorithm 7.5.1. However, we will not do so, but derive it based on local duality (Sect. 25.5), revealing that such an algorithm actually solves the dual bounded-variable problem.

Let $B = \{j_1, \cdots, j_m\}$ and $N = A \backslash B$ be the current basis and nonbasis, respectively, associated with primal basic solution $\bar{x}$, i.e.,

$$\begin{aligned} \bar{x}_s &= l_s \text{ or } u_s, \quad s \in N, \\ \bar{x}_B &= B^{-1}b - B^{-1}N\bar{x}_N. \end{aligned} \qquad (7.38)$$

Notation Γ, Π, are again defined by (7.19), and ρ_i is defined by (7.30). Assume that row index p has already been determined by (7.31).

Consider the following local problem at $\bar{x}$ (25.5):

$$\begin{array}{lll} \min & f = c^{\mathrm{T}}x, & \\ \text{s.t.} & Ax = b, & \\ & l_{\Gamma} \leq x_{\Gamma}, & \\ & x_{\Pi} \leq u_{\Pi}, & \\ & l_{j_p} \leq \bar{x}_{j_p}, & \text{if } \rho_p > 0, \\ & \bar{x}_{j_p} \leq u_{j_p}, & \text{if } \rho_p < 0. \end{array} \tag{7.39}$$

Using notation

$$h_p = \begin{cases} l_{j_p}, & \text{If } \rho_p > 0, \\ u_{j_p}, & \text{If } \rho_p < 0, \end{cases} \tag{7.40}$$

the local dual problem can be written

$$\begin{array}{ll} \max & b^{\mathrm{T}}y - u_{\Pi}^{\mathrm{T}}v_{\Pi} + l_{\Gamma}^{\mathrm{T}}w_{\Gamma} + h_p z_{j_p}, \\ \text{s.t.} & B^{\mathrm{T}}y + z_{j_p}e_p = c_B, \\ & \Gamma^{\mathrm{T}}y + w_{\Gamma} = c_{\Gamma}, \\ & \Pi^{\mathrm{T}}y - v_{\Pi} = c_{\Pi}, \\ & \rho_p z_{j_p}, v_{\Pi}, \ w_{\Gamma} \geq 0. \end{array}$$

Based on the equality constraints, eliminate variable v_{Π}, w_{Γ}, and combine (7.19) and (7.20) to reduce the objective function to

$$\begin{aligned} (b - \Pi u_{\Pi} - \Gamma l_{\Gamma})^{\mathrm{T}}y + h_p z_p &= c_{\Pi}^{\mathrm{T}}\bar{x}_{\Pi} + c_{\Gamma}^{\mathrm{T}}\bar{x}_{\Gamma} + (b - \Pi\bar{x}_{\Pi} - \Gamma\bar{x}_{\Gamma})^{\mathrm{T}}y + h_p z_{j_p} \\ &= c_N^{\mathrm{T}}\bar{x}_N + (b - N\bar{x}_N)^{\mathrm{T}}y + h_p z_{j_p}. \end{aligned} \tag{7.41}$$

Then setting $z_{\Gamma} = w_{\Gamma}, z_{\Pi} = -v_{\Pi}$, transform the local dual problem to the following equivalent form:

$$\begin{array}{ll} \max & g(y,z) = c_N^{\mathrm{T}}\bar{x}_N + (b - N\bar{x}_N)^{\mathrm{T}}y + h_p z_{j_p}, \\ \text{s.t.} & B^{\mathrm{T}}y + z_{j_p}e_p = c_B, \\ & N^{\mathrm{T}}y + z_N = c_N, \\ & \rho_p z_{j_p} \geq 0, \quad z_{\Pi} \leq 0, \quad z_{\Gamma} \geq 0. \end{array} \tag{7.42}$$

Now, define

$$\bar{y} = B^{-\mathrm{T}}c_B, \tag{7.43}$$

$$\bar{z}_N = c_N - N^{\mathrm{T}}\bar{y}, \ \bar{z}_B = 0. \tag{7.44}$$

and assume that the following conditions hold:

$$\bar{z}_\Pi \leq 0, \quad \bar{z}_\Gamma \geq 0, \tag{7.45}$$

under which it is not difficult to verify that the primal objective value at $\bar{x}$ and the dual objective value at $(\bar{y}, \bar{z})$ are equal, i.e., $\bar{f} = \bar{g}$. Using the preceding notation, moreover, the following is valid.

Lemma 7.6.1. *$(\bar{y}, \bar{z})$ is a basic feasible solution to the local dual problem, which exhibits complementarity with $\bar{x}$.*

Proof. It is clear that $(\bar{y}, \bar{z})$ is the basic solutio to (7.42), satisfying the sign constraints at the bottom. So, it is only needed to show

$$B^{\mathrm{T}}\bar{y} + \bar{z}_{j_p}e_p = c_B, \; N^{\mathrm{T}}\bar{y} + \bar{z}_N = c_N, \tag{7.46}$$

$$(\bar{x}_\Gamma - l_\Gamma)^{\mathrm{T}}\bar{z}_\Gamma = 0, \quad (u_\Pi - \bar{x}_\Pi)^{\mathrm{T}}\bar{z}_\Pi = 0, \tag{7.47}$$

$$(\bar{x}_{j_p} - l_{j_p})\bar{z}_{j_p} = 0, \quad \text{if } \rho_p > 0, \tag{7.48}$$

$$(\bar{x}_{j_p} - u_{j_p})\bar{z}_{j_p} = 0, \quad \text{if } \rho_p < 0. \tag{7.49}$$

From (7.43) and the second expression of (7.44), the first expression of (7.45) follows. By the first expression of (7.44), it holds that

$$\Pi^{\mathrm{T}}\bar{y} + \bar{z}_\Pi = \Pi^{\mathrm{T}}\bar{y} + c_\Pi - \Pi^{\mathrm{T}}\bar{y} = c_\Pi. \tag{7.50}$$

Similarly that

$$\Gamma^{\mathrm{T}}\bar{y} + \bar{z}_\Gamma = c_\Gamma. \tag{7.51}$$

Therefore, (7.46) is valid.

By (7.19), on the other hand, it is clear that (7.47) holds; and it is known from the second expression of (7.44) that (7.48) or (7.49) holds. □

Setting

$$\bar{v}_B = 0, \quad \bar{w}_B = 0, \tag{7.52}$$

$$\bar{v}_\Pi = -\bar{z}_\Pi, \; \bar{w}_\Pi = 0, \tag{7.53}$$

$$\bar{v}_\Gamma = 0, \quad \bar{w}_\Gamma = \bar{z}_\Gamma, \tag{7.54}$$

it is not difficult to verify that $(\bar{y}, \bar{v}, \bar{w})$ is a basic feasible solution to dual problem of (7.13), i.e.,

$$\begin{array}{ll} \max & b^{\mathrm{T}}y - u^{\mathrm{T}}v + l^{\mathrm{T}}w, \\ \text{s.t.} & A^{\mathrm{T}}y - v + w = c, \quad v, w, \geq 0, \end{array}$$

(see the last paragraph of Sect. 25.5). It and $\bar{x}$ satisfy complementarity condition. In this sense, $(\bar{y}, \bar{z})$ is called a dual feasible solution.

Lemma 7.6.2. *If $l_B \leq \bar{x}_B \leq u_B$ holds, then $\bar{x}$ is a basic optimal solution.*

Proof. When $l_B \leq \bar{x}_B \leq u_B$ holds, $\bar{x}$ is clearly a basic feasible solution to the (full) problem (7.13), hence the same to the local problem (7.39). By Lemma 7.6.1, it is known that $(\bar{y}, \bar{z})$ is local dual feasible, exhibiting complementarity with $\bar{x}$. Therefore, the two are local primal and dual basic optimal solutions, respectively. By Proposition 25.4.2, it is known that $\bar{x}$ is a basic optimal solution to (7.13). □

Now we will find a new dual solution to improve the objective value. To this end, define search direction

$$h = -\text{sign}(\rho_p) B^{-\mathrm{T}} e_p, \quad \sigma_{j_p} = \text{sign}(\rho_p), \tag{7.55}$$

$$\sigma_N = -N^{\mathrm{T}} h. \tag{7.56}$$

Lemma 7.6.3. *Under the preceding definition, the search direction satisfies the following conditions:*

$$B^{\mathrm{T}} h + \sigma_{j_p} e_p = 0, \quad N^{\mathrm{T}} h + \sigma_N = 0, \tag{7.57}$$

$$(b - N\bar{x}_N)^{\mathrm{T}} h + h_p \sigma_{j_p} > 0. \tag{7.58}$$

Proof. Its first half is easily verified, it is only needed to show (7.58).

From (7.55) and the second expression of (7.38), it follows that

$$\begin{aligned} h^{\mathrm{T}}(b - N\bar{x}_N) + h_p \sigma_{j_p} &= -\text{sign}(\rho_p) e_p^{\mathrm{T}} (B^{-1} b - B^{-1} N \bar{x}_N) + \text{sign}(\rho_p) h_p \\ &= -\text{sign}(\rho_p)(e_p^{\mathrm{T}} \bar{x}_B - h_p). \end{aligned}$$

Then from (7.14) and (7.40) it follows that the right-hand side of the preceding equals

$$l_{j_p} - \bar{x}_{j_p} > 0,$$

when $\rho_p > 0$, while equals

$$\bar{x}_{j_p} - u_{j_p} > 0.$$

when $\rho_p < 0$ □.

Consider the following line search scheme:

$$\hat{y} = \bar{y} + \beta h, \quad \hat{z}_{j_p} = \bar{z}_{j_p} + \beta \sigma_{j_p} = \text{sign}(\rho_p) \beta, \tag{7.59}$$

$$\hat{z}_N = \bar{z}_N + \beta \sigma_N. \tag{7.60}$$

Introduce index set

$$J = \{j \in \Gamma \mid \sigma_j < 0\} \cup \{j \in \Pi \mid \sigma_j > 0\}. \tag{7.61}$$

Assume that $J \neq \emptyset$. Then from (7.60) and sign conditions $\hat{z}_\Pi \leq 0$ and $\hat{z}_\Gamma \geq 0$, it is known that the largest possible stepsize β and pivot column index q satisfy the minimum-ratio test

$$\beta = -\bar{z}_q/\sigma_q = \min_{j \in J} -\bar{z}_j/\sigma_j \geq 0. \tag{7.62}$$

If all components of $\bar{z}_N$ are nonzero, then the solution is dual nondegenerate, hence the determined stepsize is positive.

Lemma 7.6.4. *If $J \neq \emptyset$, the new solution, determined by (7.59) and (7.60) together with (7.62), is a basic feasible solution to the local dual problem. The according objective value increases, and strictly increases if dual nondegeneracy is assumed.*

Proof. From (7.59), the first expression of (7.57) and (7.43), it is known for any $\beta \geq 0$ that

$$B^{\mathrm{T}}\hat{y} + \hat{z}_{j_p}e_p = B^{\mathrm{T}}\bar{y} + \beta(B^{\mathrm{T}}h + \sigma_{j_p}e_p) = B^{\mathrm{T}}\bar{y} = c_B. \tag{7.63}$$

From the first expression of (7.59), (7.60), the second expression of (7.57) and (7.44), it follows that

$$N^{\mathrm{T}}\hat{y} + \hat{z}_N = N^{\mathrm{T}}\bar{y} + \beta N^{\mathrm{T}}h + \bar{z}_N + \beta\sigma_N = (N^{\mathrm{T}}\bar{y} + \bar{z}_N) + \beta(N^{\mathrm{T}}h + \sigma_N) = c_N. \tag{7.64}$$

In addition, by (7.59), (7.60) and (7.57) is known that $\hat{z}_{j_p}, \hat{z}$ satisfies the sign condition at the bottom of problem (7.42), hence the new solution is basic feasible solution, associated with the objective value increasing to

$$\begin{aligned} \hat{g} &= (b - N\bar{x}_N)^{\mathrm{T}}\hat{y} + h_p\hat{z}_{j_p} \\ &= (b - N\bar{x}_N)^{\mathrm{T}}\bar{y} + \beta((b - N\bar{x}_N)^{\mathrm{T}}h + h_p\sigma_{j_p}) \\ &\geq \bar{g}, \end{aligned} \tag{7.65}$$

where the inequality comes from (7.58) and $\beta \geq 0$. In the dual nondegeneracy case, $\beta > 0$, and hence the strict inequality holds, as implies strict increase of the objective value. □

When J is empty, (7.57) is not well-defined but the following is valid.

Lemma 7.6.5. *If $J = \emptyset$, then the original problem (7.13) is infeasible.*

Proof. $J = \emptyset$ implies that

$$\sigma_\Pi \leq 0, \quad \sigma_\Gamma \geq 0. \tag{7.66}$$

Combining the preceding two expressions together with $\bar{z}_\Pi \leq 0$ and $\bar{z}_\Gamma \geq 0$ leads to

$$\hat{z}_\Pi = \bar{z}_\Pi + \beta\sigma_\Pi \leq 0, \quad \hat{z}_\Gamma = \bar{z}_\Gamma + \beta\sigma_\Gamma \geq 0, \quad \forall\, \beta > 0.$$

Similarly to the proof of Theorem 7.6.4, it can be shown that $(\hat{y}, \hat{z})$ satisfies the other constraints, with the objective value denoted again by (7.65). Thus, noting (7.58), it is known that

$$\hat{g} \to \infty \qquad \text{as} \quad \beta \to \infty,$$

Therefore, the local dual problem is unbounded. By Proposition 25.4.2, the original problem is infeasible. □

Now we need to determine a primal solution that is complementary with the dual solution, based on the local problem (7.39). For the value of basic variable x_p to change from $\bar{x}_p$ to the violated bound, it is necessary to let the value of nonbasic variable x_q change from $\bar{x}_q$ accordingly by a range, i.e.,

$$\Delta x_q = \begin{cases} -\rho_p/|\sigma_q|, \text{ if } \bar{x}_q = l_q, \\ \rho_p/|\sigma_q|, \text{ if } \bar{x}_q = u_q. \end{cases} \tag{7.67}$$

Therefore, the new values are

$$\hat{x}_B = \bar{x}_B - \Delta x_q \bar{a}_q, \quad \hat{x}_q = \bar{x}_q + \Delta x_q, \quad \hat{x}_j = \bar{x}_j, \; j \in N, \; j \neq q, \tag{7.68}$$

where $\bar{a}_q = B^{-1}a_q$, associated with the new objective value

$$\hat{f} = \bar{f} + |\Delta x_q \bar{z}_q| \geq \bar{f}. \tag{7.69}$$

Note that all components of $\hat{x}_N$ are the same as those of $\bar{x}_N$, except for $\hat{x}_q$. From the first expression of (7.68) and the second expression of (7.59), it is known that if $\rho_p > 0$, then $\hat{x}_{j_p} = l_{j_p}$ and $\hat{z} \geq 0$ hold, while if $\rho_p < 0$, then $\hat{x}_{j_p} = u_{j_p}$ and $\hat{z}_{j_p} \leq 0$ hold. Therefore, after updating basis and nonbasis by exchanging p and q, $\hat{x}$ and $(\hat{y}, \hat{z})$ exhibit complementarity, and the latter satisfies according dual feasible conditions, so that we are ready to go on the next iteration.

The overall steps are summarized into the following algorithm, a revision of Algorithm 7.5.1.

Algorithm 7.6.1 (Generalized dual simplex algorithm). Initial: $(B, N), B^{-1}$; $\bar{y}, \bar{z}, \bar{x}$ satisfying (7.38), (7.43) and (7.44). This algorithm solves the bounded-variable problem (7.13).

1. Select a pivot row index $p \in \arg\max\{|\rho_i| \mid i = 1, \cdots, m\}$, where ρ_i is defined by (7.30).
2. If $\rho_p = 0$, compute $\bar{f} = c^{\mathrm{T}}\bar{x}$, and stop.
3. Compute $\sigma_N = -N^{\mathrm{T}}h$, where $h = -\text{sign}(\rho_p)B^{-\mathrm{T}}e_p$.

4. Stop if J defined by (7.61) is empty.
5. Determine β and pivot column index q by (7.62).
6. Compute Δx_q by (7.67).
7. Compute $\bar{a}_q = B^{-1}a_q$.
8. Update $\bar{x}$ by (7.68).
9. Update $\bar{y}$, $\bar{z}_N$, $\bar{z}_{j_p}$ by (7.59) and (7.60).
10. Update B^{-1} by (3.23).
11. Update (B, N) by exchanging j_p and q.
12. Go to step 1.

Theorem 7.6.1. *Algorithm 7.6.1 generates a sequence of primal and of dual basic solutions. Assuming nondegeneracy, it terminates either at*

(i) Step 2, giving a pair of primal and dual basic optimal solutions; or at
(ii) Step 4, detecting infeasibility of the problem.

Proof. The validity comes from Lemmas 7.6.2, 7.6.4 and 7.6.5, and related discussions, made preceding Algorithm 7.6.1. □

Example 7.6.1. Solve the following problem by Algorithm 7.6.1:

$$\begin{array}{lllllllll}
\min & f = x_1 + 2x_2 - 2x_3, & & & & & & & \\
\text{s.t.} & -2x_1 + x_2 + x_3 + x_4 & & & = 0, \\
 & -x_1 - x_2 + x_3 \qquad + x_5 & & & = 0, \\
 & x_1 - x_2 - 2x_3 \qquad\qquad + x_6 & & & = 0, \\
 & 1 \le x_1 \le 5, \quad -2 \le x_2 \le \infty, \quad -3 \le x_3 \le 0, \\
 & 2 \le x_4 \le 5, \quad 0 \le x_5 \le 6, \quad -3 \le x_6 \le 0.
\end{array}$$

Answer Initial: $B = \{4, 5, 6\}$, $N = \{1, 2, 3\}$, $B^{-1} = I$, $\bar{x}_N = (1_{(-)}, -2_{(-)}, 0^{(+)})^{\mathrm{T}}$, $\bar{x}_B = (4, -1, -3)^{\mathrm{T}}$, $\bar{y} = (0, 0, 0)$, $\bar{z}_N = (1, 2, -2)^{\mathrm{T}}$, $\bar{f} = -3$.

Iteration 1:

1. $\max\{0, |0 - (-1)|, 0\} = 1$, $p = 2$, x_5 leaves the basis.
3. $h = -\text{sign}(\rho_2)B^{-\mathrm{T}}e_2 = (0, -1, 0)^{\mathrm{T}}$, $\quad \sigma_N = -N^{\mathrm{T}}h = (-1, -1, 1)^{\mathrm{T}}$.
4. $J = \{1, 2, 3\} \neq \emptyset$.
5. $\beta = \min\{-1/(-1), -2/(-1), -(-2)/1\} = 1$, $q = 1$.
6. $\Delta x_1 = \rho_2/|\sigma_1| = 1$.
7. $\bar{a}_1 = B^{-1}a_1 = (-2, -1, 1)^{\mathrm{T}}$.
8. $\bar{x}_B = \bar{x}_B - \Delta x_1 \bar{a}_1 = (4, -1, -3)^{\mathrm{T}} - (-2, -1, 1)^{\mathrm{T}} = (6, 0, -4)^{\mathrm{T}}$.
   $\bar{x}_N = \bar{x}_N + \Delta x_1 e_1 = (1, -2, 0)^{\mathrm{T}} + (1, 0, 0)^{\mathrm{T}} = (2, -2, 0)^{\mathrm{T}}$.
9. $\bar{y} = \bar{y} + \beta h = (0, -1, 0)^{\mathrm{T}} + 1 \times (0, -1, 0)^{\mathrm{T}} = (0, -2, 0)^{\mathrm{T}}$,
   $\bar{z}_N = \bar{z}_N + \beta\sigma_N = (1, 2, -2)^{\mathrm{T}} + 1(-1, -1, 1)^{\mathrm{T}} = (0, 1, -1)^{\mathrm{T}}$,
   $\bar{z}_5 = \text{sign}(\rho_2)\beta = 1$.

10. Update $B^{-1} = \begin{pmatrix} 1 & -2 & \\ & -1 & \\ & 1 & 1 \end{pmatrix}$.

11. $\bar{x}_B = (6, 2, -4)^{\mathrm{T}}$, $B = \{4, 1, 6\}$; $\bar{x}_N = (0_{(-)}, -2_{(-)}, 0^{(+)})^{\mathrm{T}}, \bar{z}_N = (1, 1, -1)^{\mathrm{T}}$, $N = \{5, 2, 3\}$.

Iteration 2:

1. $\max\{|5-6|, 0, (-3)-(-4)\} = 1$, $p = 1$, x_4 leaves the basis.
3. $h = -\mathrm{sign}(\rho_1) B^{-\mathrm{T}} e_1 = (1, -2, 0)^{\mathrm{T}}, \sigma_N = -N^{\mathrm{T}} h = (2, -3, 1)^{\mathrm{T}}$.
4. $J = \{2, 3\} \neq \emptyset$.
5. $\beta = \min\{-1/(-3), -(-1)/1\} = 1/3$, $q = 2$.
6. $\Delta x_2 = -\rho_1/|\sigma_2| = 1/3$.
7. $\bar{a}_2 = B^{-1} a_2 = (3, 1, -2)^{\mathrm{T}}$.
8. $\bar{x}_B = (6, 2, -4)^{\mathrm{T}} - (1/3)(3, 1, -2)^{\mathrm{T}} = (5, 5/3, -10/3)^{\mathrm{T}}$,
   $\bar{x}_N = \bar{x}_N + \Delta x_2 e_2 = (0, -2, 0)^{\mathrm{T}} + (0, 1/3, 0)^{\mathrm{T}} = (0, -5/3, 0)^{\mathrm{T}}$.
9. $\bar{y} = \bar{y} + \beta h = (0, -2, 0)^{\mathrm{T}}$,
   $\bar{z}_N = \bar{z}_N + \beta \sigma_N = (1, 1, -1)^{\mathrm{T}} + (1/3)(2, -3, 1)^{\mathrm{T}} = (5/3, 0, -2/3)^{\mathrm{T}}$,
   $\bar{z}_4 = \mathrm{sign}(\rho_1)\beta = -1/3$.
10. Update $B^{-1} = \begin{pmatrix} 1/3 & & \\ -1/3 & 1 & \\ 2/3 & & 1 \end{pmatrix} \begin{pmatrix} 1 & -2 & \\ & -1 & \\ & 1 & 1 \end{pmatrix} = \begin{pmatrix} 1/3 & -2/3 & \\ -1/3 & -1/3 & \\ 2/3 & -1/3 & 1 \end{pmatrix}$.
11. $B = \{2, 1, 6\}$, $N = \{5, 4, 3\}$, $\bar{x}_B = (-5/3, 5/3, -10/3)^{\mathrm{T}}$,
    $\bar{x}_N = (0_{(-)}, 5^{(+)}, 0^{(+)})^{\mathrm{T}}$,
    $\bar{z}_N = (5/3, -1/3, -2/3)^{\mathrm{T}}$.

Iteration 3:

1. $\max\{0, 0, |(-3) - (-10/3)|\} = 1/3$, $p = 3$, x_6 leaves the basis.
3. $h = -\mathrm{sign}(\rho_3) B^{-\mathrm{T}} e_3 = (-2/3, 1/3, -1)^{\mathrm{T}}, \sigma_N = -N^{\mathrm{T}} h$
   $= (-1/3, 2/3, -5/3)^{\mathrm{T}}$.
4. $J = \{1, 2\} \neq \emptyset$.
5. $2 = q \in \min\{-(5/3)/(-1/3), -(-1/3)/(2/3)\}, \beta = 1/2$,
   x_4 enters the basis.
6. $\Delta x_2 = \rho_3/|\sigma_2| = -(1/3)/(2/3) = -1/2$.
7. $\bar{a}_2 = B^{-1} a_2 = (1/3, -1/3, 2/3)^{\mathrm{T}}$.
8. $\bar{x}_B = (-5/3, 5/3, -10/3)^{\mathrm{T}} - (-1/2)(1/3, -1/3, 2/3)^{\mathrm{T}} = (-3/2, 3/2, -3)^{\mathrm{T}}$,
   $\bar{x}_N = \bar{x}_N + \Delta x_2 e_2 = (0, 5, 0)^{\mathrm{T}} + (0, -1/2, 0)^{\mathrm{T}} = (0, 9/2, 0)^{\mathrm{T}}$.
9. $\bar{y} = (0, -2, 0)^{\mathrm{T}} + (1/2)(-2/3, 1/3, -1)^{\mathrm{T}} = (-1/3, -11/6, -1/2)^{\mathrm{T}}$,
   $\bar{z}_N = \bar{z}_N + \beta \sigma_N = (5/3, -1/3, -2/3)^{\mathrm{T}} + (1/2)(-1/3, 2/3, -5/3)^{\mathrm{T}}$
   $= (3/2, 0, -3/2)^{\mathrm{T}}$,
   $\bar{z}_6 = \mathrm{sign}(\rho_3)\beta = 1/2$.
10. Update $B^{-1} = \begin{pmatrix} 1 & & -1/2 \\ & 1 & 1/2 \\ & & -3/2 \end{pmatrix} \begin{pmatrix} 1/3 & -2/3 & \\ -1/3 & -1/3 & \\ 2/3 & -1/3 & 1 \end{pmatrix} = \begin{pmatrix} & -1/2 & -1/2 \\ & -1/2 & 1/2 \\ -1 & 1/2 & -3/2 \end{pmatrix}$.

11. It is satisfied that $l_B \leq \bar{x}_B \leq u_B$. The optimal solutio and value are

$$\bar{x} = (3/2, -3/2, 0, 9/2, 0, -3)^{\mathrm{T}}, \quad \bar{f} = 3/2 + 2(-3/2) = -3/2.$$

7.7 Bound Flipping

The so-called "bound-flipping" technique can improve the effect of the generalized dual simplex method significantly. In fact, it might be the main cause for the dual simplex method to outperform its primal counterpart at present (Kirillova et al. 1979; Koberstein and Suhl 2007; Kostina 2002; Maros 2003a).

Let $(\bar{y}, \bar{z})$ be the current dual basic feasible solution and let $\bar{x}$ be the associate primal solution. Assume that a row index p has been determined by (7.31) and that a column index q determined by the minimum-ratio test (7.62). Let the nonbasic variable x_q change from the current value $\bar{x}_q$ (going up or down) toward the other bound, while keeping the other nonbasic variables unchanged. For the basic variable x_{j_p} to attain the violated bound, the value of x_q could fall either within the range between the lower and upper bounds, or beyond the other bound. In the latter case, it is favorable to adopt the "bound-flipping": fix the value of x_q on the other bound and update values of basic variables accordingly; then find a new column index q that attains the second minimum-ratio, and do the same thing again, until the value of x_{j_p} will attain the violated bound if the current value of x_q falls within the range between its lower and upper bounds. Then, a normal dual step is taken by dropping x_{j_p} from and enter x_q to the basis, and updating the primal and dual solutions. It is seen that the dual feasibility still maintains.

The bound-flipping technique is embedded in the following subalgorithm, which is called in step 10 of Algorithm 7.7.2.

Algorithm 7.7.1 (Bound-flipping subalgorithm). This algorithm provide the pivot column index q, dual stepsize β, and carries out related computations.

1. Set $j = 0, v = 0$, and compute $r_j = -\bar{z}_j/\sigma_j, \quad \forall j \in J$.
2. Set $j = j + 1$.
3. Set $v = v + \delta a_q$.
4. Set $\bar{x}_q = \begin{cases} u_q, & \text{if } \bar{x}_q = l_q, \\ l_q, & \text{if } \bar{x}_q = u_q. \end{cases}$
5. Update: $\rho_p = \rho_p - |\delta \sigma_q|$.
6. Determine q and r_q such that $r_q = \min_{j \in J} r_j$.
7. Compute Δx_q by (7.67).
8. Update: $J = J \backslash \{q\}$.
9. Go to step 13 if $J = \emptyset$.
10. Compute $\delta = \begin{cases} u_q - l_q, \text{ if } \bar{x}_q = l_q, \\ l_q - u_q, \text{ if } \bar{x}_q = u_q. \end{cases}$
11. Go to step 2 if $|\Delta x_q| \geq |\delta|$.

12. Set $\beta = r_q$.
13. Compute $u = B^{-1}v$, and update: $\bar{x}_B = \bar{x}_B - u$.
14. Return.

The following master algorithm is a slight modification of Algorithm 7.6.1.

Algorithm 7.7.2 (Generalized dual simplex algorithm: bound-flipping). Initial: (B, N), B^{-1}, $\bar{y}, \bar{z}, \bar{x}$ satisfying (7.38), (7.43) and (7.44). This algorithm solves bounded-variable problem (7.13).

1. Select row index p by (7.31) together with (7.30).
2. Stop if $\rho_p = 0$ (optimality achieved).
3. Compute σ_N by (7.56) together with (7.55).
4. Stop if J defined by (7.61) is empty (dual unbounded or primal infeasible).
5. Determine column index q by (7.62).
6. Compute Δx_q by (7.67).
7. Set $J = J \backslash \{q\}$.
8. If $J = \emptyset$, go to step 12.
9. Compute $\delta = \begin{cases} u_q - l_q, \text{ if } \bar{x}_q = l_q, \\ l_q - u_q, \text{ if } \bar{x}_q = u_q. \end{cases}$
10. If $|\Delta x_q| \geq |\delta|$, call Algorithm 7.7.1.
11. Compute $\bar{a}_q = B^{-1}a_q$.
12. Update $\bar{x}$ by (7.68).
13. Update $\bar{y}$, $\bar{z}_N$, $\bar{z}_{j_p}$ by (7.59) together with (7.60).
14. Update B^{-1} by (3.23).
15. Update (B, N) by exchanging j_p and q.
16. Go to step 1.

The bound-flipping increases computational work associated therewith, in particular, involving an additional linear system (in step 13 of Algorithm 7.7.1). This is inappreciable, however, if compared with profitable return. Since the associated dual stepsize is usually much larger than that without bound-flipping, so is the increment in objective value, especially when ρ_p is large. As a result, the number of iterations are usually decreased significantly. In fact, the bound-flipping has been unable to be omitted in current dual simplex codes.

Example 7.7.1. Solve the following problem by Algorithm 7.7.2:

$$
\begin{array}{llll}
\min & f = -x_1 + 2x_3 + 3x_4, & & \\
\text{s.t.} & -2x_1 + x_2 + x_3 + x_4 & = -2, & \\
& x_1 - x_3 + x_4 + x_5 & = 1, & \\
& x_1 - 2x_3 - 3x_4 + x_6 & = 0, & \\
& 0 \leq x_1 \leq 2, \quad -6 \leq x_2 \leq 10, & 0 \leq x_3 \leq 7, & \\
& 1 \leq x_4 \leq 5, \quad 2 \leq x_5 \leq 6, & -1 \leq x_6 \leq 6. &
\end{array}
$$

Answer Initial: $B = \{2,5,6\}$, $N = \{1,3,4\}$, $B^{-1} = I$, $\bar{x}_N = (2^{(+)}, 0_{(-)}, 1_{(-)})^{\mathrm{T}}$, $\bar{x}_B = (1,-2,1)^{\mathrm{T}}$, $\bar{y} = (0,0,0)$, $\bar{z}_N = (-1,2,3)^{\mathrm{T}}$, $\bar{f} = 1$.

Iteration 1:

1. $\max\{0, 2-(-2), 0\} = 4$, $p = 2$, x_5 leaves the basis.
3. $h = -\mathrm{sign}(\rho_2)B^{-\mathrm{T}}e_2 = (0,-1,0)^{\mathrm{T}}$, $\sigma_N = -N^{\mathrm{T}}h = (1,-1,1)^{\mathrm{T}}$.
4. $J = \{1,2\}$.
5. $\beta = \min\{-(-1)/1, -2/-1\} = 1$, $q = 1$.
6. $\Delta x_1 = -\rho_2/|\sigma_1| = -4/1 = -4$.
7. $J = J\backslash\{1\} = \{2\} \neq \emptyset$.
9. $\delta = l_1 - u_1 = 0 - 2 = -2$.
10. $|\Delta x_1| > |\delta|$, so call Algorithm 7.7.1.
    (1) $j = 0, v = 0, r_2 = -2/-1 = 2$;
    (2) $j = j + 1 = 1$;
    (3) $v = v + \delta a_1 = (-2)(-2,1,1)^{\mathrm{T}} = (4,-2,-2)^{\mathrm{T}}$;
    (4) $\bar{x}_1 = l_1 = 0$;
    (5) $\rho_2 = \rho_2 - |\delta\sigma_1| = 4 - 2 \times 1 = 2$;
    (6) $q = 2$;
    (7) $\Delta x_2 = -\rho_2/|\sigma_2| = -2/(-1) = 2$;
    (8) $J = J\backslash\{2\} = \emptyset$;
    (12) $\beta = r_2 = 2$;
    (13) $u = B^{-1}v = (4,-2,-2)^{\mathrm{T}}$;
    $\bar{x}_B = \bar{x}_B - u = (1,-2,1)^{\mathrm{T}} - (4,-2,-2)^{\mathrm{T}} = (-3,0,3)^{\mathrm{T}}$;
    (14) Return.
11. $\bar{a}_2 = B^{-1}a_2 = (1,-1,-2)^{\mathrm{T}}$.
12. $\bar{x}_B = (-3,0,3)^{\mathrm{T}} - 2(1,-1,-2)^{\mathrm{T}} = (-5,2,7)^{\mathrm{T}}$,
    $\bar{x}_N = (0,0,1)^{\mathrm{T}} + (0,2,0)^{\mathrm{T}} = (0,2,1)^{\mathrm{T}}$.
13. $\bar{y} = (0,0,0)^{\mathrm{T}} + 2(0,-1,0)^{\mathrm{T}} = (0,-2,0)^{\mathrm{T}}$,
    $\bar{z}_N = \bar{z}_N + \beta\sigma_N = (-1,2,3)^{\mathrm{T}} + 2(1,-1,1)^{\mathrm{T}} = (1,0,5)^{\mathrm{T}}$,
    $\bar{z}_{j_2} = \mathrm{sign}(\rho_2)\beta = 2$.
14. $$B^{-1} = \begin{pmatrix} 1 & 1 & \\ & -1 & \\ & -2 & 1 \end{pmatrix}.$$
16. $\bar{x}_B = (-5,2,7)^{\mathrm{T}}$, $\bar{x}_N = (0_{(-)}, 2_{(-)}, 1_{(-)})^{\mathrm{T}}$, $\bar{z}_N = (1,2,5)^{\mathrm{T}}$,
    $B = \{2,3,6\}$, $N = \{1,5,4\}$.

Iteration 2:

1. $\max\{0,0,7-6\} = 1$, $p = 3$, x_6 leaves the basis.
3. $h = -\mathrm{sign}(\rho_3)B^{-\mathrm{T}}e_3 = (0,-2,1)^{\mathrm{T}}$, $\sigma_N = -N^{\mathrm{T}}h = (1,2,5)^{\mathrm{T}}$.
4. $J = \emptyset$, hence dual unbounded or primal infeasible.

If bound-flipping had not been used, solving the preceding problem would have required much more iterations.

Chapter 8
Decomposition Method

Solving large-scale LP problems is a challenging task, putting forward high requirements on algorithms' efficiency, storage and numerical stability.

It is noticeable that the objective function and nonnegativity constraints are variable-separable, i.e., can be separated to several independent groups. If, in addition, the system of constraint equations are also variable-separable, then the large-scale problems can be handled through solving smaller problems. Let us bring up the following example:

$$\begin{array}{lllll} \min & f = (c^1)^{\mathrm{T}}x^1 + \cdots + (c^k)^{\mathrm{T}}x^k, & & & \\ \text{s.t.} & D_1x^1 & & & = h^1, \\ & & D_2x^2 & & = h^2, \\ & & & \ddots & \vdots \\ & & & D_kx^k & = h^k, \\ & & & x^1, \cdots, x^k \geq 0, & \end{array} \tag{8.1}$$

where the coefficient matrix is of block diagonal structure: orders of submatrices D_j and dimensions of subvectors $c^j, x^j, h^j,\ j = 1, \cdots, k$ are consistent, that is, each pair of c_j and x_j share a same dimension, equal to the number of columns of D_j, and the dimension of b_j is equal to the number of rows of D_j.

The feasible region of the problem is

$$P' = \{((x^1)^{\mathrm{T}}, \cdots, (x^k)^{\mathrm{T}})^{\mathrm{T}} \mid D_jx^j = h^j,\ x^j \geq 0,\ j = 1, \cdots, k\},$$

or, otherwise, denoted in the form of "Cartesian product":

$$P' = P^1 \times \cdots \times P^k; \quad P^j = \{x^j \mid D_jx^j = h^j,\ x^j \geq 0\}, \quad j = 1, \cdots, k.$$

P.-Q. PAN, *Linear Programming Computation*, DOI 10.1007/978-3-642-40754-3_8,

As variables and constraints of the problem can be separated to independent k sets, the problem comes down to solving k smaller problems, i.e.,

$$\begin{array}{ll} \min & (c^j)^{\mathrm{T}} x^j, \\ \text{s.t.} & D_j x^j = h^j, \quad x^j \geq 0, \end{array} \quad j = 1, \cdots, k.$$

If there is a basic feasible solution $\bar{x}^j$, associated objective value $\bar{f}^j$, to each of the small problems, there is a basic feasible solution, associated with according objective value, to the original problem, i.e.,

$$\bar{x} = ((\bar{x}^1)^{\mathrm{T}}, \cdots, (\bar{x}^k)^{\mathrm{T}})^{\mathrm{T}}, \quad \bar{f} = \bar{f}^1 + \cdots + \bar{f}^k. \tag{8.2}$$

The same is valid with basic optimal solutions and objective values. On the other hand, if some of the small problems is unbounded, the original is unbounded too; if required, it is not difficult to determine the associated descent extreme direction (Proposition 3.5.1).

Although such type of problems would not be encountered in practice very often, problems of partially separable structure are not unusual, as would be amenable to be solved by decomposition methods, presented in this chapter.

8.1 D-W Decomposition

Dantzig-Wolfe (D-W) decomposition (1960) partitions constraints to two sets, and handle a "master program" and a subprogram alternately to solve the original problem. The master program is expressed in terms of vertices and extreme directions of the feasible region, associated with one of the sets. Throughout solution process, the subprogram takes this set as its constraints, while its objective function varies iteration by iteration. It is solved to determine a column of the master program to enter the basis, or assert optimality of the original problem. In such a manner, the original problem is solved via handling two smaller programs. Further decomposition is possible if the set of constraints of the subprogram are of separable structure.

By partitioning the constraints to two sets, the standard LP program can be written

$$\begin{array}{ll} \min & f = c^{\mathrm{T}} x, \\ \text{s.t.} & Ax = b, \\ & Hx = h, \quad x \geq 0, \end{array} \tag{8.3}$$

where $A \in \mathcal{R}^{m \times n}$, $H \in \mathcal{R}^{m1 \times n}$, $b \in \mathcal{R}^m$, $h \in \mathcal{R}^{m1}$.

Introduce polyhedral convex set

$$P = \{x \in \mathcal{R}^n \mid Hx = h,\ x \geq 0\}.$$

Let P be nonempty, and possess the following sets of vertices and extreme directions respectively:

$$U = \{u^1, \cdots, u^s\}, \quad V = \{v^1, \cdots, v^t\},$$

According to the Representation Theorem 2.2.7, it holds that

$$P=\{x \mid x=\sum_{i=1}^{s} \alpha_i u^i + \sum_{j=1}^{t} \beta_j v^j ; \sum_{i=1}^{s} \alpha_i=1, \alpha_i \geq 0,\ i=1,\cdots,s,\ \beta_j \geq 0,\ j=1,\cdots,t\}.$$

Substituting the expression of x to $\min\{c^{\mathrm{T}}x \mid Ax = b\}$, convert (8.3) to the standard problem with respect to variables α_i, β_j, i.e.,

$$\begin{array}{lll}
\min_{\alpha_i,\beta_j} & f = \sum_{i=1}^{s} (c^{\mathrm{T}}u^i)\alpha_i + \sum_{j=1}^{t} (c^{\mathrm{T}}v^j)\beta_j, & \\
\text{s.t.} & \sum_{i=1}^{s} (Au^i)\alpha_i + \sum_{j=1}^{t} (Av^j)\beta_j = b, & \\
& \sum_{i=1}^{s} \alpha_i = 1, & \\
& \alpha_i,\ \beta_j \geq 0, \quad i = 1,\cdots,s,\ j = 1,\cdots,t, &
\end{array} \tag{8.4}$$

which is called *master program.*

The relation between feasible solutions to the master program and to the original program is determined by

$$x = \sum_{i=1}^{s} \alpha_i u^i + \sum_{j=1}^{t} \beta_j v^j, \tag{8.5}$$

corresponding to equal feasible values. Respectively, feasible and optimal solutions of the master program correspond to those of the original program. In general, however, the latter is not a basic solution even if the former is.

Let us focus on the master program side. The difficulty seems that the number $s+t$ of columns of the master program is usually far larger than the number $m+1$ of its rows, despite the number of rows would be far less than the number $m+m1$ of rows of the original program; moreover, vertices and extreme directions of P are used in expression of the coefficient matrix and costs of the master program. Fortunately, a simplex iteration only needs an entering and a leaving column available, provided that the current basis is known. As D-W composition generates an entering column in each iteration, it is sometimes called *column generation method.*

Without loss of generality, assume that the current basis matrix is

$$B = \begin{pmatrix} Au^1 & \cdots & Au^{s'} & Av^1 & \cdots & Av^{t'} \\ 1 & \cdots & 1 & 0 & \cdots & 0 \end{pmatrix}, \tag{8.6}$$

where the number of columns of B is $s' + t' = m + 1$. The basic components of the current feasible solution are defined by system

$$B(\alpha_1, \cdots, \alpha_{s'}, \beta_1, \cdots, \beta_{t'})^{\mathrm{T}} = \begin{pmatrix} b \\ 1 \end{pmatrix}.$$

Denote the feasible solution by

$$\bar{\alpha}_i \geq 0, \quad i = 1, \cdots, s'; \quad \bar{\beta}_j, \quad j = 1, \cdots, t'.$$

Then the according feasible solution and objective value of the original program is

$$\bar{x} = \sum_{i=1}^{s'} \bar{\alpha}_i u^i + \sum_{j=1}^{t'} \bar{\beta}_j v^j, \quad \bar{f} = \sum_{i=1}^{s'} (c^{\mathrm{T}} u^i) \bar{\alpha}_i + \sum_{j=1}^{t'} (c^{\mathrm{T}} v^j) \bar{\beta}_j, \tag{8.7}$$

The simplex multipliers $(\bar{y}, \bar{\gamma})$, corresponding to the first m rows and the bottom row, are defined by system

$$B^{\mathrm{T}} \begin{pmatrix} y \\ \gamma \end{pmatrix} = \hat{c}_B, \tag{8.8}$$

where $\hat{c}_B$ consists of costs of the master program, corresponding to basic columns. The reduced costs are then

$$\begin{aligned} & c^{\mathrm{T}} u^i - (Au^i)^{\mathrm{T}} \bar{y} - \bar{\gamma}, && i = 1, \cdots, s, \\ & c^{\mathrm{T}} v^j - (Av^j)^{\mathrm{T}} \bar{y}, && j = 1, \cdots, t. \end{aligned}$$

If reduced costs are all nonnegative, then optimality of the original program is achieved, and (8.7) gives its optimal solution and objective value; otherwise, some nonbasic column corresponding to a negative reduced cost is selected to enter the basis.

To avoid computing all the reduced costs, D-W decomposition solves a subprogram with P as its feasible region instead, i.e.,

$$\begin{array}{ll} \min & \zeta = (c - (A)^{\mathrm{T}} \bar{y})^{\mathrm{T}} x - \bar{\gamma}, \\ \text{s.t.} & Hx = h, \quad x \geq 0, \end{array} \tag{8.9}$$

where constant $\bar{\gamma}$ in the objective may be omitted, practically.

Let us examine the relation between the subprogram and the master and the original programs. If the feasible region P of the subprogram is empty, the master and the original programs are infeasible. Otherwise, we have the following result.

Lemma 8.1.1. *Let f^* be the optimal value of the original program. If $\bar{f}$ is the current objective value of the master program, and ζ^* is the optimal value of the subprogram, then it holds that*

$$\bar{f} + \zeta^* \leq f^* \leq \bar{f}. \tag{8.10}$$

Proof. It is clear that $f^* \leq \bar{f}$, it is only needed to show

$$\bar{f} + \zeta^* \leq f^*. \tag{8.11}$$

Let x^* be an optimal solution of the original program. Since it is also a feasible solution to the subprogram, we have

$$(c - (A)^{\mathrm{T}}\bar{y})^{\mathrm{T}}x^* - \bar{\gamma} \geq \zeta^*.$$

In view of that $B^{-1}(h^{\mathrm{T}}, 1)^{\mathrm{T}}$ is a basic solution, from the preceding expression, $Ax^* = b$ and $(\bar{y}, \bar{\gamma})$ satisfying system (8.8), it follows that

$$f^* = c^{\mathrm{T}}x^* \geq \bar{y}^{\mathrm{T}}Ax^* + \bar{\gamma} + \zeta^* = (\bar{y}^{\mathrm{T}}h + \bar{\gamma}) + \zeta^* = \hat{c}_B^{\mathrm{T}}B^{-1}\begin{pmatrix} b \\ 1 \end{pmatrix} + \zeta^* = \bar{f} + \zeta^*,$$

which gives (8.11). □

In each iteration, therefore, it is possible to estimate the optimal value of the original program by giving a lower and an upper bounds by (8.10). It holds by (8.11) that

$$\bar{f} - f^* \leq -\zeta^*.$$

If $\bar{f}$ is taken as an approximate optimal value of the original program, therefore, the according absolute error bound is $-\zeta^*$. If $\bar{f} \neq 0$, it follows further that

$$(\bar{f} - f^*)/|\bar{f}| \leq -\zeta^*/|\bar{f}|,$$

the right-hand side of which gives the relative error bound.

Theorem 8.1.1. *If the optimal value of the subprogram vanishes, optimality of the original program is achieved.*

Proof. By Lemma 8.1.1 and $\bar{f} - f^* \geq 0$, it is known that $\zeta^* \leq 0$, that is, the optimal value of the subprogram is less than or equal to 0. If it is equal to zero, then $\bar{f} \leq f^* \leq \bar{f}$, which implies that optimality of the master and hence of the original program is achieved. □

If the feasible region P of the subprogram is nonempty, consequently, solving it by the simplex method leads to one of the following two cases:

(i) A vertex optimal solution, say u^*, of the subprogram is reached.
If the optimal value vanishes, optimality of the original program is achieved, and we are done; otherwise, a column

$$w' = \begin{pmatrix} Au^* \\ 1 \end{pmatrix}. \tag{8.12}$$

is generated to enter the basis.

(ii) It is detected that the subprogram is unbounded via Proposition 3.5.1. In this case, a descent extreme direction, say $v^* \in V$, of P can be determined such that (see Proposition 3.5.1)

$$(c - (A)^{\mathrm{T}}\bar{y})^{\mathrm{T}}v^* < 0,$$

implying that the according reduced costs of the master program is negative. Then, a column

$$w' = \begin{pmatrix} Av^* \\ 0 \end{pmatrix} \tag{8.13}$$

is generated to enter the basis. To avoid repetition, other simplex steps, such as the determination of a leaving column, and etc. will not be described here.

The overall steps are summarized to the following algorithm.

Algorithm 8.1.1 (Dantzig-Wolfe Decomposition). Initial: basis matrix B of the master program, basic components $\bar{\alpha}_i$, $\bar{\beta}_j$ of the basic feasible solution and associated objective value $\bar{f}$. This algorithm solves LP problem (8.3).

1. Solve system (8.8) for $(\bar{y}, \bar{\gamma})$.
2. Call the simplex algorithm to solve subprogram (8.9).
3. If an optimal solution u^* of the subprogram is obtained, then

   (1) if the associated optimal value $\zeta^* < 0$, generate a new basic column w' by (8.12), and go to step 5;
   otherwise,

   (2) compute the optimal solution and objective value of the original program by (8.7), and stop.

4. If it is detected that the subprogram is unbounded, generate a new basic column w' by (8.13).
5. Solve system $Bw = w'$ for $\bar{w}$.

6. Stop if $\bar{w} \leq 0$. (the original program is unbounded).
7. Determine a stepsize α and leaving column by the minimum-ratio test.
8. Update $B, \bar{\alpha}_i, \bar{\beta}_j$ and $\bar{f}$.
9. Go to step 1.

Note When the optimal value ζ^* of the subprogram is close to 0, terminating the preceding algorithm gives an approximate optimal solution to the original program (Lemma 8.1.1).

It is clear that the algorithm terminates if nondegeneracy is assumed.

8.1.1 Generalization of D-W Decomposition

To generalize Algorithm 8.1.1, introduce the following concept.

Definition 8.1.1. A program consisting of $K(m+2 \leq K \leq s+t)$ columns of rank $m+1$ of the master program is a restricted subprogram.

The new reduced cost, corresponding to a leaving column, is always greater than 0 (see Proposition 3.9.1 and its proof). In each iteration of Algorithm 8.1.1, therefore, actually solved is a restricted master program with

$$K = m + 2$$

columns, consisting of basic columns and the entering column. It can be generalized to allow a restricted master program includes more columns.

Algorithm 8.1.2 (Generalized Dantzig-Wolfe decomposition). The same as Algorithm 8.1.1, except the restricted master program is solved before updating it (via determining new entering and leaving columns).

There are ways to construct an initial restricted master program. For example, take any $m+1$ independent columns to form an initial basis first. Then add new entering columns successively until the number of columns attains K; or after the number of columns is still less than K but close to computer's storage limit, add the nonbasic columns, corresponding to the smallest $K-m-1$ reduced costs.

8.2 Start-Up of D-W Decomposition

This section is devoted to the Phase-I issue of D-W decomposition.

At first, find a vertex, say u^1, of $P = \{x \in \mathcal{R}^n \mid Hx = h,\ x \geq 0\}$. If it does not exist, the original program is clearly infeasible. Then, employ u^1 and introduce

m artificial variables $\sigma_l,\ l = 1, \cdots, m$ to construct the following auxiliary Phase-I master program:

$$
\begin{aligned}
\min_{\sigma_l,\alpha_i,\beta_j}\ & w = \sum_{l=1}^{m} \sigma_l, \\
\text{s.t.}\quad & \sum_{l=1}^{m} \pm e_l \sigma_l + \sum_{i=1}^{s} (Au^i)\alpha_i + \sum_{j=1}^{t} (Av^j)\beta_j = b, \\
& \sum_{i=1}^{s} \alpha_i = 1, \\
& \sigma_l,\ \alpha_i,\ \beta_j \geq 0,\ l = 1, \cdots, m,\ i = 1, \cdots, s,\ j = 1, \cdots, t,
\end{aligned}
\tag{8.14}
$$

where for $l = 1, \cdots, m$, if $b_l - (Au^1)_l \geq 0$, $+e_l$ is taken; else, $-e_l$ taken. Thus, α_1 together with $\sigma_l,\ l = 1, \cdots, m$, constitute a set of basic variables. Basic components of the basic feasible solution are

$$
\bar{\alpha}_1 = 1, \quad \bar{\sigma}_l = |b_l - (Au^1)_l| \geq 0, \quad l = 1, \cdots, m.
$$

Then Algorithm 8.1.1 can get itself started from the initial basis consisting of columns, associated with these basic components.

Is is clear that there exists an optimal solution, corresponding to a nonnegative optimal value, to the auxiliary program: if the optimal value is greater than zero, the original program is infeasible; if, otherwise, the optimal value vanishes, a feasible solution of the master program is yielded by deleting all artificial components from the auxiliary optimal solution, resulting a basic feasible solution to the master program, as hence can be taken as an initial one for Phase-II of D-W decomposition, if artificial components are all nonbasic. In the case when there are some artificial variables being basic, additional iterations are needed to force them leave the basis. These are analogous to "following-up steps" described in Sect. 3.3.

8.3 Application of D-W Decomposition

Theoretically, D-W decomposition is able to handle general standard LP problems, and the bisection of constraints can be arbitrary. In terms of effectiveness, nevertheless, its performance varies with problems of different structures. In fact, many problems coming from practice are often of structures (partially) that are amenable to D-W decomposition.

Some models, involving block structure, may belong to this category, as illustrated by the example below.

Example 8.3.1. Solve the following program by the D-W decomposition:

$$
\begin{array}{lllllllllll}
\min & f = -x_2 + x_3 - 2x_4 - 3x_5 + x_6 - x_7 + x_8, \\
\text{s.t.} & x_1 - 3x_2 - x_3 + x_4 + 2x_5 + x_6 - x_7 - 3x_8 = 2, \\
& x_1 + 4x_3 + x_4 = 1, \\
& x_2 + x_3 - x_4 = 4, \\
& x_5 + 3x_7 - x_8 = 1, \\
& x_6 + x_7 - 3x_8 = 2, \\
& x_j \geq 0, \quad j = 1, \cdots, 8.
\end{array}
$$

Answer The first equality constraint is a coupling one, corresponding to a master program of form (8.19), while the others correspond to a subproblem of form (8.21).

The second and third equality constraints are

$$
\begin{aligned}
& x_1 + 4x_3 + x_4 = 1, \\
& x_2 + x_3 - x_4 = 4, \quad x_j \geq 0, \ j = 1, \cdots, 4,
\end{aligned}
$$

which has a basic solution $x_1 = 1, \ x_2 = 4, \ x_3 = x_4 = 0$, and the fourth and fifth are

$$
\begin{aligned}
& x_5 + 3x_7 - x_8 = 1, \\
& x_6 + x_7 - 3x_8 = 2, \quad x_j \geq 0, \ j = 5, \cdots, 8,
\end{aligned}
$$

which has a basic solution $x_5 = 1, x_6 = 2, \ x_7 = x_8 = 0$. Therefore, there is a basic feasible solution $u^1 = (1, 4, 0, 0, 1, 2, 0, 0)^{\mathrm{T}}$ to the subprogram.

Accordingly,

$$
b_1^1 - (Au^1)_1 = 2 - (1, -3, -1, 1, 2, 1, -1, -3)(1, 4, 0, 0, 1, 2, 0, 0)^{\mathrm{T}} = 2 + 7 = 9 \geq 0.
$$

Thereby, we construct an auxiliary master program of form (8.14), i.e.,

$$
\begin{array}{ll}
\min & \sigma_1, \\
\text{s.t.} & \sigma_1 - 7\alpha_1 + \sum\limits_{i=2}^{s}(Au^i)\alpha_i + \sum\limits_{j=1}^{t}(Av^j)\beta_j = 2, \\
& \sum\limits_{i=1}^{s} \alpha_i = 1, \\
& \sigma_1, \alpha_i, \beta_j \geq 0, \quad i = 1, \cdots, s, \ j = 1, \cdots, t,
\end{array}
\tag{8.15}
$$

where $A = (1, -3, -1, 1, 2, 1, -1, -3)$.

The basis matrix is $B = \begin{pmatrix} 1 & -7 \\ 0 & 1 \end{pmatrix}$. Solving

$$B(\sigma_1, \alpha_1)^{\mathrm{T}} = (2, 1)^{\mathrm{T}}$$

leads to the associated basic feasible solution, whose basic components are $\begin{pmatrix} \bar{\sigma}_1 \\ \bar{\alpha}_1 \end{pmatrix} = \begin{pmatrix} 9 \\ 1 \end{pmatrix}$, with auxiliary objective value $\bar{w} = 1$.

Phase-I: Call D-W Decomposition Algorithm 8.1.1 to solve the auxiliary master program (8.15).

Iterations 1:

1. Solving $B^{\mathrm{T}}(y, \gamma)^{\mathrm{T}} = (1, 0)^{\mathrm{T}}$ gives simplex multipliers $(\bar{y}, \bar{\gamma}) = (1, 7)$.
2. The objection function of the subprogram is

$$\begin{aligned} \zeta &= (c - (A)^{\mathrm{T}}\bar{y})^{\mathrm{T}}x - \bar{\gamma} \\ &= ((0, -1, 1, -2, -3, 1, -1, 1) - (1, -3, -1, 1, 2, 1, -1, -3))x - 7 \\ &= (-1, 2, 2, -3, -5, 0, 0, 4)x - 7. \end{aligned}$$

To be solved separately, the subprogram is decomposed to the following two smaller programs:

(i)

$$\begin{array}{ll} \min & -x_1 + 2x_2 + 2x_3 - 3x_4, \\ \text{s.t.} & x_1 + 4x_3 + x_4 = 1, \\ & x_2 + \ \ x_3 - x_4 = 4, \quad x_j \geq 0,\ j = 1, \cdots, 4. \end{array}$$

Taking basis and nonbasis $B = \{1, 2\}$, $N = \{3, 4\}$. $B^{-1} = I$, then the associated basic feasible solution is
$\bar{x}_B = (1, 4)^{\mathrm{T}} \geq 0,\ \bar{x}_N = (0, 0)^{\mathrm{T}}$,
and the reduced costs are
$\bar{z}_N = c_N - N^{\mathrm{T}}B^{-1}c_B = \begin{pmatrix} 2 \\ -3 \end{pmatrix} - \begin{pmatrix} 4 & 1 \\ 1 & -1 \end{pmatrix}^{\mathrm{T}} \begin{pmatrix} -1 \\ 2 \end{pmatrix} = \begin{pmatrix} 4 \\ 0 \end{pmatrix}$.
$\min\{4, 0\} = 0 \geq 0$. Therefore, an optimal vertex solution and optimal value are obtained, i.e.,
$(\bar{x}_1, \bar{x}_2, \bar{x}_3, \bar{x}_4)^{\mathrm{T}} = (1, 4, 0, 0)^{\mathrm{T}}, \quad \zeta_1 = 7.$

(ii)

$$\begin{array}{ll} \min & -5x_5 + +4x_8, \\ \text{s.t.} & x_5 + 3x_7 - \ \ x_8 = 1, \\ & x_6 + \ \ x_7 - 3x_8 = 2, \quad x_j \geq 0,\ j = 5, \cdots, 8. \end{array}$$

Taking basis and nonbasis $B = \{5, 6\}$, $N = \{7, 8\}$. $B^{-1} = I$,
the associated basic feasible solution is
$\bar{x}_B = (1, 2)^{\mathrm{T}} \geq 0,\ \bar{x}_N = (0, 0)^{\mathrm{T}}$,

and the reduced costs are

$$\bar{z}_N = c_N - N^{\mathrm{T}} B^{-1} c_B = \begin{pmatrix} 0 \\ 4 \end{pmatrix} - \begin{pmatrix} 3 & -1 \\ 1 & -3 \end{pmatrix}^{\mathrm{T}} \begin{pmatrix} -5 \\ 0 \end{pmatrix} = \begin{pmatrix} 15 \\ -1 \end{pmatrix}.$$

$\min\{15, -1\} = -1 < 0$, $q = 8$. Since $B^{-1}a_8 = (-1, -3)^{\mathrm{T}} \leq 0$, the subprogram is unbounded.

4. The feasible region of the subprogram has a descent extreme direction $v^1 = (0, 0, 0, 0, 1, 3, 0, 1)^{\mathrm{T}}$.
   Generate an entering column $w' = (2, 0)^{\mathrm{T}}$ by (8.13).
5. $\bar{w} = (2, 0)^{\mathrm{T}} \not\leq 0$.
7. $\alpha = \min\{9/2\}$, $p = 1$.
8. $(\bar{\sigma}_1, \bar{\alpha}_1)^{\mathrm{T}} = (9, 1)^{\mathrm{T}} - (9/2)(2, 0)^{\mathrm{T}} = (0, 1)^{\mathrm{T}}$.

$$B = \begin{pmatrix} 2 & -7 \\ 0 & 1 \end{pmatrix}. \ (\bar{\beta}_1, \bar{\alpha}_1)^{\mathrm{T}} = (9/2, 1)^{\mathrm{T}},$$

$$\begin{aligned} \bar{f} = {} & (0, -1, 1, -2, -3, 1, -1, 1)(0, 0, 0, 0, 1, 3, 0, 1)^{\mathrm{T}}(9/2) \\ & + (0, -1, 1, -2, -3, 1, -1, 1)(1, 4, 0, 0, 1, 2, 0, 0)^{\mathrm{T}} = -1/2. \end{aligned}$$

The only artificial variable σ_1 leaved the basis, hence Phase-I is finished. Phase-II: Call D-W Decomposition Algorithm 8.1.1 to solve the master program.

Iterations 2:

1. $\hat{c}_B = (c^{\mathrm{T}} v^1, c^{\mathrm{T}} u^1)^{\mathrm{T}} = (1, -5)^{\mathrm{T}}$.
   Solve $B^{\mathrm{T}}(y, \gamma)^{\mathrm{T}} = (1, -5)^{\mathrm{T}}$ for the simplex multipliers $(\bar{y}, \bar{\gamma}) = (1/2, -3/2)$.
2. The objective function of the subprogram is

$$\begin{aligned} \zeta &= (c - (A)^{\mathrm{T}} \bar{y})^{\mathrm{T}} x - \bar{\gamma} \\ &= ((0, -1, 1, -2, -3, 1, -1, 1) - (1, -3, -1, 1, 2, 1, -1, -3)(1/2))x + 3/2 \\ &= (-1/2, 1/2, 3/2, -5/2, -4, 1/2, -1/2, 5/2)x + 3/2. \end{aligned}$$

(i)

$$\begin{aligned} \min \quad & -1/2x_1 + 1/2x_2 + 3/2x_3 - 5/2x_4, \\ \text{s.t.} \quad & x_1 + 4x_3 + x_4 = 1, \\ & x_2 + \ x_3 - \ x_4 = 4, \quad x_j \geq 0, \ j = 1, \cdots, 4, \end{aligned}$$

has an optimal vertex solution $(\bar{x}_1, \bar{x}_2, \bar{x}_3, \bar{x}_4)^{\mathrm{T}} = (0, 5, 0, 1)^{\mathrm{T}}$ and optimal value $\zeta_1 = 0$.

(ii)

$$\begin{aligned} \min \quad & -4x_5 + 1/2x_6 - 1/2x_7 + 5/2x_8, \\ \text{s.t.} \quad & x_5 + 3x_7 - \ x_8 = 1, \\ & x_6 + \ x_7 - 3x_8 = 2, \quad x_j \geq 0, \ j = 5, \cdots, 8, \end{aligned}$$

has an optimal vertex solution $(\bar{x}_5, \bar{x}_6, \bar{x}_7, \bar{x}_8)^{\mathrm{T}} = (1, 2, 0, 0)^{\mathrm{T}}$ and optimal value $\zeta_1 = -3$.

Thereby, the subprogram has an optimal solution and optimal value
$u^2 = (0, 5, 0, 1, 1, 2, 0, 0)^{\mathrm{T}}, \quad \zeta^* = 0 - 3 + 3/2 = -3/2 < 0.$

3(1). $\zeta^* < 0$. Generate the entering column $w' = (-10, 1)^{\mathrm{T}}$ by (8.12).

5. $\bar{w} = (-3/2, 1)^{\mathrm{T}} \not\leq 0$.
7. $\alpha = \min\{1/1\},\ p = 2$.
8. $(\bar{\beta}_1, \bar{\alpha}_1)^{\mathrm{T}} = (9/2, 1)^{\mathrm{T}} - (-3/2, 1)^{\mathrm{T}} = (6, 0)^{\mathrm{T}}$. $B = \begin{pmatrix} 2 & -10 \\ 0 & 1 \end{pmatrix}$,

$(\bar{\beta}_1, \bar{\alpha}_2)^{\mathrm{T}} = (6, 1)^{\mathrm{T}}$, $\bar{f} = \bar{f} + \alpha\zeta^* = -1/2 - 3/2 = -2$.

Iterations 3:

1. $\hat{c}_B = (c^{\mathrm{T}}v^1, c^{\mathrm{T}}u^2)^{\mathrm{T}} = (1, -8)^{\mathrm{T}}$.
   Solve $B^{\mathrm{T}}(y, \gamma)^{\mathrm{T}} = (1, -8)^{\mathrm{T}}$ for simplex multipliers $(\bar{y}, \bar{\gamma}) = (1/2, -3)$.
2. The objective function of the subprogram is

$$\begin{aligned}\zeta &= (c - (A)^{\mathrm{T}}\bar{y})^{\mathrm{T}}x - \bar{\gamma} \\ &= ((0, -1, 1, -2, -3, 1, -1, 1) - (1, -3, -1, 1, 2, 1, -1, -3)(1/2))x + 3 \\ &= (-1/2, 1/2, 3/2, -5/2, -4, 1/2, -1/2, 5/2)x + 3.\end{aligned}$$

(i) The program is the same as that in the previous iteration, with optimal vertex solution and objective value
$(\bar{x}_1, \bar{x}_2, \bar{x}_3, \bar{x}_4)^{\mathrm{T}} = (0, 5, 0, 1)^{\mathrm{T}}, \quad \zeta_1 = 0.$
(ii) The program is the same as that in the previous iteration, with optimal vertex solution $(\bar{x}_5, \bar{x}_6, \bar{x}_7, \bar{x}_8)^{\mathrm{T}} = (1, 2, 0, 0)^{\mathrm{T}}$ and optimal value $\zeta_1 = -3$.

The optimal vertex solution and optimal value to the subprogram are

$$u^3 = (0, 5, 0, 1, 1, 2, 0, 0)^{\mathrm{T}}, \quad \zeta^* = 0 - 3 + 3 = 0.$$

3(2). $\zeta^* = 0$, an optimal solution and optimal value to the original program are obtained, i.e.,

$$\begin{aligned}x^* &= (0,\ 0,\ 0,\ 0,\ 1,\ 3,\ 0,\ 1)^{\mathrm{T}}(6) + (0,\ 5,\ 0,\ 1,\ 1,\ 2,\ 0,\ 0)^{\mathrm{T}} \\ &= (0,\ 5,\ 0,\ 1,\ 7,\ 20,\ 0,\ 0)^{\mathrm{T}}, \quad f^* = \bar{f} = -2.\end{aligned}$$

Another amenable program is of staircase structure close to block, and can be solved by the D-W decomposition in a nested manner, as is shown with the following example:

$$\begin{array}{ll} \min & f = (c^1)^{\mathrm{T}} x^1 + (c^2)^{\mathrm{T}} x^2 + (c^3)^{\mathrm{T}} x^3 + (c^4)^{\mathrm{T}} x^4, \\ \text{s.t.} & D_{11} x^1 = h^1, \\ & A_{21} x^1 + D_{22} x^2 = b^2, \\ & A_{32} x^2 + D_{33} x^3 = b^3, \\ & A_{43} x^3 + D_{44} x^4 = b^4, \\ & x^1, x^2, x^3, x^4 \geq 0, \end{array} \tag{8.16}$$

where $x^j,\ j = 1, 2, 3, 4$ are vectors of appropriate dimensions consistent with the preceding matrices.

Partition equality constraints of the above program into two parts, denoted by $\{t = 1\}$ and $\{t = 2, 3, 4\}$ respectively. Let $\{u^i\}$ and $\{v^j\}$ be respectively the vertex set and extreme direction set of the feasible region of the latter part $\{t = 2, 3, 4\}$, i.e.,

$$\begin{array}{l} A_{21} x^1 + D_{22} x^2 = b^2, \\ A_{32} x^2 + D_{33} x^3 = b^3, \\ A_{43} x^3 + D_{44} x^4 = b^4, \\ x^1, x^2, x^3, x^4 \geq 0. \end{array}$$

According to the Representation Theorem 2.2.7, the solution to the preceding can be expressed

$$x = \sum_{i=1}^{s} \alpha_i u^i + \sum_{j=1}^{t} \beta_j v^j, \quad \sum_{i=1}^{s} \alpha_i = 1, \quad \alpha_i, \beta_j \geq 0,\ i = 1, \cdots, s,\ j = 1, \cdots, t. \tag{8.17}$$

Introduce according partition

$$x = \begin{pmatrix} x^1 \\ x^2 \\ x^3 \\ x^4 \end{pmatrix}; \quad u^i = \begin{pmatrix} u^{i1} \\ u^{i2} \\ u^{i3} \\ u^{i4} \end{pmatrix}, \quad i = 1, \cdots, s; \quad v^j = \begin{pmatrix} v^{j1} \\ v^{j2} \\ v^{j3} \\ v^{j4} \end{pmatrix}, \quad j = 1, \cdots, t.$$

Substituting (8.17) to the first part $\{t = 1\}$ leads to D-W master program

$$\begin{aligned} \min_{\alpha_i, \beta_j} \quad f = & \sum_{i=1}^{s} ((c^1)^{\mathrm{T}} u^{i1} + (c^2)^{\mathrm{T}} u^{i2} + (c^3)^{\mathrm{T}} u^{i3} + (c^4)^{\mathrm{T}} u^{i4}) \alpha_i \\ & + \sum_{j=1}^{t} ((c^1)^{\mathrm{T}} v^{j1} + (c^2)^{\mathrm{T}} v^{j2} + (c^3)^{\mathrm{T}} v^{j3} + (c^4)^{\mathrm{T}} v^{j4}) \beta_j, \end{aligned}$$

$$
\begin{aligned}
\text{s.t.} \quad & \sum_{i=1}^{s} D_{11} u^{i1} \alpha_i + \sum_{j=1}^{t} D_{11} v^{j1} \beta_j = b^1, \\
& \sum_{i=1}^{s} \alpha_i = 1, \\
& \alpha_i, \beta_j \geq 0, \quad i = 1, \cdots, s, \ j = 1, \cdots, t.
\end{aligned}
$$

Assume that the current restricted master program has been solved, and $(\bar{y}_1, \bar{\gamma}_1)$ are the optimal simplex multipliers. The constructed subprogram

$$
\begin{aligned}
\min \quad & (c^1 - (D_{11})^{\mathrm{T}} \bar{y}_1)^{\mathrm{T}} x^1 + (c^2)^{\mathrm{T}} x^2 + (c^3)^{\mathrm{T}} x^3 - \bar{\gamma}_1, \\
\text{s.t.} \quad & A_{21} x^1 + D_{22} x^2 = b^2, \\
& A_{32} x^2 + D_{33} x^3 = b^3, \\
& A_{43} x^3 + D_{44} x^4 = b^4, \\
& x^1, x^2, x^3, x^4 \geq 0,
\end{aligned}
$$

just likes the original program, but the stairs are reduced by 1. Then apply the D-W decomposition again by partitioning it to two parts, $\{t = 2\}$ and $\{t = 3, 4\}$, and do as previously, until the stairs reduced to 1, and the subprogram is easily solved.

8.4 Economic Interpretation of D-W Decomposition

D-W decomposition can be interpreted differently, depending on various practical backgrounds. It is a typical instance to employ D-W decomposition to realize optimal allocation of limited resources.

Assume that a company has k plants. It is required to determine an output vector x^j for each plant j, whose production activities depend on restrictions upon available limited resources, such as manpower, facilities, materials, storage, etc., but not on inner restrictions of the other plants. In mathematical terms, x^j should satisfy constraints

$$
D_j x^j = h^j, \quad x^j \geq 0, \quad j = 1, \cdots, k,
$$

where $D_j \in \mathcal{R}^{m_j \times n_j}$ $(m_j < n_j)$ is termed consumption matrix, reflecting consumption for yielding an unit product; h^j is an available resource vector. This part of constraints are of block structure, where variables and constraints are separable. In addition, there are common resources to be shared by the k plants, as can be expressed by a set of so-called *coupling constraints*:

$$
A_1 x^1 + \cdots + A_k x^k = b,
$$

where $A_j \in \mathcal{R}^{m \times n_j}$, $j = 1, \cdots, k$.

In order to minimize the total cost of production, we are required to solve the following LP model:

$$\begin{array}{ll} \min & f = (c^1)^{\mathrm{T}} x^1 + \cdots + (c^k)^{\mathrm{T}} x^k, \\ \text{s.t.} & A_1 x^1 + \cdots + A_k x^k = b, \\ & D_j x^j = h^j, \quad x^j \geq 0, \quad j = 1, \cdots, k. \end{array} \tag{8.18}$$

Use of the D-W decomposition is advantageous if the preceding model is of large-scale, especially when there are a large number of plants associated with large consumption matrices: the company can focus on a changing small program to realize an optimal allocation of resources without going to details of inner restrictions on the plants.

For simplicity, it might be well to assume that the feasible regions

$$P^j \triangleq \{x^j \mid D_j x^j = h^j,\ x^j \geq 0\}, \quad j = 1, \cdots, k,$$

corresponding to the plants, are all bounded (actually, available resources are always limited). Using notation

$$A = (A_1, \cdots, A_k), \quad H = \begin{pmatrix} D_1 & & \\ & \ddots & \\ & & D_K \end{pmatrix}, \quad h = \begin{pmatrix} h^1 \\ \vdots \\ h^k \end{pmatrix},$$

it is clear that

$$P \triangleq \{x \in \mathcal{R}^n \mid Hx = h,\ x \geq 0\} = P^1 \times \cdots \times P^k$$

is bounded. Denote the vertex set of P by $U = \{u^1, \cdots, u^s\}$. Based on Representation Theorem 2.2.7, we obtain the following master program:

$$\begin{array}{ll} \min_{\alpha_i} & f = \sum_{i=1}^{s} (c^{\mathrm{T}} u^i) \alpha_i, \\ \text{s.t.} & \sum_{i=1}^{s} (A u^i) \alpha_i = b, \\ & \sum_{i=1}^{s} \alpha_i = 1, \\ & \alpha_i \geq 0, \quad i = 1, \cdots, s. \end{array} \tag{8.19}$$

The number of rows of the preceding is less than the number of rows of the original program, but the number of its columns would be quite large, compared with the original problem, letting alone unrealistic computation of the whole vertex set U.

Nevertheless, the company can deal with a restricted master program instead, involving only some K ($m + 2 \leq K \ll s$) columns (with rank $m + 1$) of the master program, ignoring most of inner restrictions of the plants. Now denote by $\hat{U}$ the vertex set, resulting from U by deleting vertices corresponding to the K columns.

Assume that optimality of the restricted master program was achieved and that $(\bar{y}, \bar{\gamma})$ is the optimal simplex multipliers, associated with the first m rows and the bottom row of the constraint system. In economic terms, $(\bar{y}, \bar{\gamma})$ are shadow price of the resources (Sect. 4.6), from which the reduced costs corresponding to set $\hat{U}$ follows, i.e.,

$$c^{\mathrm{T}} u^i - (Au^i)^{\mathrm{T}} \bar{y} - \bar{\gamma} = (c - (A)^{\mathrm{T}} \bar{y})^{\mathrm{T}} u^i - \bar{\gamma}, \quad u^i \in \hat{U}, \tag{8.20}$$

A negative reduced cost implies that increasing the associated nonbasic variable from 0 results in decrease of the total cost of production. Conventionally, the most negative reduced cost should be determined. The difficulty lies in that computing the reduced costs directly by (8.20) requires all vertices of P available, but the company does not know inner details of the plants.

The trick out of the difficulty is to solve the following subprogram by the simplex algorithm, i.e.,

$$\begin{array}{ll} \min & \zeta = (c - (A)^{\mathrm{T}} \bar{y})^{\mathrm{T}} x - \bar{\gamma}, \\ \text{s.t.} & x \in P, \end{array} \tag{8.21}$$

where the objective function is related to shadow price $(\bar{y}, \bar{\gamma})$.

Since P is bounded, solving (8.21) renders an optimal vertex solution, say u^*, associated with the optimal value, say ζ^*, equal to the minimum reduced cost of the restricted master program. If $\zeta^* \geq 0$, optimality of the master program, and hence the original program is achieved, so is a wanted optimal allocation scheme of resources. If, otherwise, $\zeta^* < 0$, then u^* corresponds to the wanted most negative reduced cost, hence the restricted master program may be updated by entering the master program's column, corresponding to u^* to the basis and dropping the column, determined by the minimum-ratio test, from the basis, as an iteration is complete.

It is important that subprogram (8.21) can be further decomposed to k smaller programs, i.e.,

$$\begin{array}{ll} \min & (c^j - A_j^{\mathrm{T}} \bar{y})^{\mathrm{T}} x^j - \bar{\gamma}, \\ \text{s.t.} & x^j \in P^j, \end{array} \qquad j = 1, \cdots, k.$$

If $\bar{x}^j$, $\bar{f}^j$, $j = 1, \cdots, k$, are basic optimal solutions and associated optimal values to these programs, an optimal vertex solution, of form (8.2), to subprogram (8.21) is readily available.

8.5 Benders Decomposition

The D-W decomposition presented in Sect. 8.1 converts a LP program to two smaller ones by partitioning its rows to two parts. In contrast, Benders (1962) converts a program to two smaller ones by partitioning its columns (variables) to two parts, one of which is a LP program, and the other may belong to another category. It can be used to solve large-scale mixed LP programs, especially mixed ILP programs. Subsequently, Benders decomposition is widely applied to dealing with stochastic programming programs, and is further generalized to handle nonlinear programming programs (Geoffrion 1972).

Consider

$$\begin{array}{ll} \max_{\pi,y} & f(\pi) + b^{\mathrm{T}} y, \\ \text{s.t.} & F(\pi) + A^{\mathrm{T}} y \leq c, \\ & \pi \in \Pi, \end{array} \tag{8.22}$$

where $A \in \mathcal{R}^{m\times n}$, $c \in \mathcal{R}^n$, $b \in \mathcal{R}^m$. Scalar function $f(\pi)$ and vector-valued function $F(\pi) \in \mathcal{R}^n$ and their domain $\Pi \subset \mathcal{R}^m$ will be clarified later.

For any fixed π, (8.22) is a LP program with respect to variable y. Therefore, Benders decomposition firstly deems π as a parameter to realize a partition of variables.

Lemma 8.5.1. *Program (8.22) is equivalent to the following program:*

$$\begin{array}{ll} \max_{\pi} & f(\pi) + \max_y \{b^{\mathrm{T}} y \mid A^{\mathrm{T}} y \leq c - F(\pi)\}, \\ \text{s.t.} & \pi \in \Pi \cap S, \end{array} \tag{8.23}$$

where

$$S = \{\pi \mid A^{\mathrm{T}} y + F(\pi) \leq c\}.$$

Proof. It is clear that (8.22) is infeasible, unbounded and has an optimal solution if and only if (8.23) is infeasible, unbounded and has an optimal solution accordingly.

For any fixed π, an objective value of program (8.23) is equal to the sum of $f(\pi)$ and the optimal value of the following subprogram:

$$\begin{array}{rll} D(\pi): & \max_y & b^{\mathrm{T}} y, \\ & \text{s.t.} & A^{\mathrm{T}} y \leq c - F(\pi). \end{array}$$

Let $(\bar{\pi}, \bar{y})$, $\hat{\pi}$ and $\hat{y}$ be an optimal solution to (8.22), (8.23) and $D(\hat{\pi})$, respectively. Then it is clear that $(\hat{y}, \hat{\pi})$ is a feasible solution to (8.22), and therefore

$$f(\hat{\pi}) + b^{\mathrm{T}} \hat{y} \leq f(\bar{\pi}) + b^{\mathrm{T}} \bar{y}. \tag{8.24}$$

On the other hand, it is known that

$$f(\hat{\pi}) + b^{\mathrm{T}}\hat{y} \geq f(\bar{\pi}) + \max_{y} \{b^{\mathrm{T}}y \mid A^{\mathrm{T}}y \leq c - F(\bar{\pi})\},$$

and $\bar{y}$ is clearly an optimal solution to $D(\bar{\pi})$. Therefore, it holds that

$$f(\bar{\pi}) + \max_{y} \{b^{\mathrm{T}}y \mid A^{\mathrm{T}}y \leq c - F(\bar{\pi})\} = f(\bar{\pi}) + b^{\mathrm{T}}\bar{y}.$$

Combining the previous two formulas leads to

$$f(\hat{\pi}) + b^{\mathrm{T}}\hat{y} \geq f(\bar{\pi}) + b^{\mathrm{T}}\bar{y},$$

from which and (8.24), it follows that

$$f(\hat{\pi}) + b^{\mathrm{T}}\hat{y} = f(\bar{\pi}) + b^{\mathrm{T}}\bar{y}.$$

Therefore programs (8.22) and (8.23) share the same optimal solutions and optimal value, if any. □

Thereby, program (8.22) is decomposed to two smaller programs $D(\pi)$ and (8.23), respectively involving p and q variables. Nevertheless, they are defined in implicit expressions, and difficult to handle practically. To go further, consider the dual program of $D(\pi)$, i.e.,

$$\begin{array}{ll} P(\pi): \min_x & \zeta = (c - F(\pi))^{\mathrm{T}}x, \\ \text{s.t.} & Ax = b, \quad x \geq 0. \end{array}$$

Proposition 8.5.1. *If the feasible region*

$$X = \{x \in \mathcal{R}^n \mid Ax = b,\ x \geq 0\}$$

of $P(\pi)$ is empty, then (8.22) is infeasible or unbounded.

Proof. $X = \emptyset$ means that $P(\pi)$ is infeasible for any $\pi \in \mathcal{R}^p$. According to the duality theory, program $D(\pi)$ is infeasible or unbounded, hence (8.22) is infeasible or unbounded. □

In case when $X = \emptyset$, one can solve

$$\begin{array}{ll} \min_x & (c - F(\pi))^{\mathrm{T}}x, \\ \text{s.t.} & Ax = 0, \quad x \geq 0, \end{array}$$

by the simplex algorithm to determined whether (8.22) is infeasible or unbounded, if needed. If the preceding is unbounded, it can be asserted that (8.22) is infeasible, and if optimality is achieved, (8.22) is unbounded (see discussions made in Sect. 4.2).

From now on, it is assume that $X \neq \emptyset$, whose vertex and extreme direction sets are respectively denoted by

$$U = \{u^1, \cdots, u^s\}, \quad V = \{v^1, \cdots, v^t\}.$$

Using preceding notation, we have the following result.

Lemma 8.5.2. *Given π. There is an optimal solution to program $P(\pi)$ if and only if*

$$(c - F(\pi))^{\mathrm{T}} v^j \geq 0, \quad v^j \in V. \tag{8.25}$$

Proof. It is seen that (8.25) implies inexistence of a descent unbounded direction in set X. Therefore, there exists an optimal solution to $P(\pi)$, as it is lower bounded. According Theorem 2.3.2, there is an optimal vertex solution to it.

Conversely, that there is an optimal solution to $P(\pi)$ implies that (8.23) holds, because if there is some $v^k \in V$ such that

$$(c - F(\pi))^{\mathrm{T}} v^k < 0,$$

then v^k is a descent extreme direction, hence $P(\pi)$ is unbounded below, as contradicts to the presence of an optimal solution. □

Define *dual Benders master program* by (it will be clear why prefixed by "dual")

$$\begin{array}{lll} \max_{\pi,g} & f(\pi) + g, & \\ \text{s.t.} & g \leq (c - F(\pi))^{\mathrm{T}} u^i, & u^i \in U, \\ & (c - F(\pi))^{\mathrm{T}} v^j \geq 0, & v^j \in V, \\ & & \pi \in \Pi \cap S. \end{array} \tag{8.26}$$

Theorem 8.5.1. *$(\bar{\pi}, \bar{y})$ is an optimal solution to (8.22) if and only if $\bar{y}$ is an optimal solution to $D(\bar{\pi})$ while $(\bar{\pi}, \bar{g})$ is an optimal solution to the dual Benders master program.*

Proof. It is clear that there is an optimal solution to (8.22), if and only if there is an optimal solution to the dual Benders master program.

According to Lemma 8.5.1, program (8.22) is equivalent to (8.23). On the other hand, that there is an optimal solution to program (8.22) means that so is to $D(\pi)$. By the strong dual theorem, there is an optimal solution to $P(\pi)$ with the same optimal value, i.e.,

$$\max_y \{b^{\mathrm{T}} x \mid A^{\mathrm{T}} y \leq c - F(\pi)\} = \min_x \{(c - F(\pi))^{\mathrm{T}} x \mid Ax = b,\ x \geq 0\}.$$

Substitute the preceding to (8.23) leads to

$$\begin{array}{ll} \max_\pi & f(\pi) + \min_x \{(c - F(\pi))^{\mathrm{T}} x \mid Ax = b,\ x \geq 0\}, \\ \text{s.t.} & \pi \in \Pi \cap S. \end{array} \tag{8.27}$$

If there is an optimal solution to $P(\pi)$, by Theorem 2.3.2, there is a vertex optimal solution to it. Therefore it holds that

$$\min_x \{(c - F(\pi))^{\mathrm{T}} x \mid Ax = b,\ x \geq 0\} = \min_{u^i \in U} (c - F(\pi))^{\mathrm{T}} u^i. \tag{8.28}$$

Substituting the preceding to (8.27) and applying Lemma 8.5.2 leads to

$$\begin{array}{ll} \max_{\pi} & f(\pi) + \min_{u^i \in U} (c - F(\pi))^{\mathrm{T}} u^i, \\ \text{s.t.} & (c - F(\pi))^{\mathrm{T}} v^j \geq 0, \quad v^j \in V, \\ & \pi \in \Pi \cap S, \end{array}$$

which gives (8.26). These equivalent relations ensure validity of Theorem 8.5.1. □

According to the preceding theorem, what to do next is to solve (8.26). However, it is not realistic to handle (8.26) directly, because it requires availability of all vertices and extreme directions of X, letting alone $\pi \in S$ being in implicit impression. A possible approach is to handle a relaxation program, yielding from ignoring some of the constraints. Its optimal solution is also an optimal solution to the master program if it is a feasible solution to the latter; otherwise, the relaxation program is updated by adding some constraints.

Assume that at some iteration, subsets of the vertex and extreme direction sets of X are known as

$$U' \subset U, \quad V' \subset V.$$

Define *dual Benders restricted master program*[1] by

$$\begin{array}{lll} \max_{\pi, g} & f(\pi) + g, & \\ \text{s.t.} & g \leq (c - F(\pi))^{\mathrm{T}} u^i, & u^i \in U', \\ & (c - F(\pi))^{\mathrm{T}} v^j \geq 0, & v^j \in V', \\ & & \pi \in \Pi. \end{array} \tag{8.29}$$

Theorem 8.5.2. *Let $(\bar{\pi}, \bar{g})$ be an optimal solution to the dual Benders restricted master program.*

If u^* and ζ^* are an optimal vertex solution and optimal value to the subprogram $P(\bar{\pi})$ respectively, then:

(i) If $\bar{g} > \zeta^*$, then u^* is a new vertex generated;
(ii) If $\bar{g} = \zeta^*$, then $(\bar{\pi}, \bar{g})$ is an optimal solution to the dual Benders master program.

If $P(\bar{\pi})$ is unbounded, then a new extreme direction is given.

---

[1] The term by Dantzig-Wolfe is employed here.

Proof. Since u^* and ζ^* are respectively an optimal vertex solution and optimal value to the subprogram $P(\bar{\pi})$, it holds that

$$\zeta^* = (c - F(\bar{\pi}))^{\mathrm{T}} u^* = \min\{(c - F(\bar{\pi}))^{\mathrm{T}} u^i \mid u^i \in U\}. \tag{8.30}$$

Besides, $\bar{\pi}$ satisfies

$$(c - F(\bar{\pi}))^{\mathrm{T}} v^j \geq 0, \quad v^j \in V, \tag{8.31}$$

because otherwise there is an ascent extreme direction in X, hence $P(\bar{\pi})$ is unbounded, as contradicts to the existence of an optimal solution. Moreover, it also holds that

$$\bar{\pi} \in \Pi \cap S. \tag{8.32}$$

In fact, it is clear that $\bar{\pi} \in \Pi$; and if $\bar{\pi} \notin S$, then it is derived that $D(\bar{\pi})$ is infeasible, hence $P(\bar{\pi})$ is either infeasible or unbounded, as contradicts the existence of an optimal solution to $P(\bar{\pi})$.

Since $(\bar{\pi}, \bar{g})$ is an optimal solution to (8.29), it is obtained from (8.30) that

$$\begin{aligned} f(\bar{\pi}) + \bar{g} &= f(\bar{\pi}) + \min\{(c - F(\bar{\pi}))^{\mathrm{T}} u^i \mid u^i \in U'\} \\ &\geq f(\bar{\pi}) + \min\{(c - F(\bar{\pi}))^{\mathrm{T}} u^i \mid u^i \in U\} \\ &= f(\bar{\pi}) + \zeta^*, \end{aligned}$$

hence $\bar{g} \geq \zeta^*$.

Consequently, only the following two cases arise:

(i) $\bar{g} > \zeta^*$.

In this case, from

$$\bar{g} > \zeta^* = (c - F(\bar{\pi}))^{\mathrm{T}} u^*$$

and

$$\bar{g} = \min\{(c - F(\bar{\pi}))^{\mathrm{T}} u^i \mid u^i \in U'\},$$

it follows that

$$(c - F(\bar{\pi}))^{\mathrm{T}} u^* < \min\{(c - F(\bar{\pi}))^{\mathrm{T}} u^i \mid u^i \in U'\},$$

implying that $u^* \notin U'$ is a new vertex.

(ii) $\bar{g} = \zeta^*$.

Noting (8.30), it can be obtained in this case that

$$\bar{g} = \zeta^* \leq (c - F(\bar{\pi}))^{\mathrm{T}} u^i, \quad u^i \in U,$$

which together with (8.31) and (8.32) implies that $(\bar{\pi}, \bar{g})$ is a feasible solution to the dual Benders master program. It is actually optimal since it is optimal to the relaxation program of the dual Benders master program.

If subprogram $P(\bar{\pi})$ is unbounded, on the other hand, then an extreme direction v^* is at the same time determined such that (see Proposition 3.5.1)

$$(c - F(\bar{\pi}))^{\mathrm{T}} v^* < 0.$$

Thus, from that $(\bar{\pi}, \bar{g})$ satisfies the constraints of the master program, i.e.,

$$(c - F(\bar{\pi}))^{\mathrm{T}} v^j \geq 0, \quad v^j \in V',$$

it is known that $v^* \notin V'$ is a new extreme direction. □

Proposition 8.5.2. *If the Benders restricted master program is infeasible, then program (8.22) is infeasible.*

Proof. If $\Pi = \emptyset$, (8.22) is clearly infeasible. If $\Pi \neq \emptyset$, then infeasibility of the Benders restricted master program implies that for any $\pi \in \Pi$, there is some $v^k \in V' \subset V$ such that $(c - F(\pi))^{\mathrm{T}} v^k < 0$, hence for any $\pi \in \Pi$, program $P(\pi)$ is unbounded and $D(\pi)$ is infeasible, the original program is therefore infeasible. □

The dual Benders master program (8.26) can be converted to

$$\begin{array}{lll} \max_{\pi,g} & f(\pi) + g, & \\ \text{s.t.} & (u^i)^{\mathrm{T}} F(\pi) + g \leq (u^i)^{\mathrm{T}} c, & u^i \in U, \\ & (v^j)^{\mathrm{T}} F(\pi) \qquad \leq (v^i)^{\mathrm{T}} c, & v^j \in V, \\ & & \pi \in \Pi \cap S. \end{array}$$

Accordingly, the dual Benders restricted master program (8.29) becomes

$$\begin{array}{llll} \max_{\pi,g} & f(\pi) + g, & & \\ \text{s.t.} & (u^i)^{\mathrm{T}} F(\pi) + g \leq (u^i)^{\mathrm{T}} c, & & u^i \in U', \\ & (v^j)^{\mathrm{T}} F(\pi) \qquad \leq (v^i)^{\mathrm{T}} c, & & v^j \in V', \\ & f(\pi) + g \qquad \leq M, & (*) & \pi \in \Pi, \end{array} \tag{8.33}$$

where the constraint, marked by $*$, is added to rule out possibility of unboundedness, and where M is a parameter large enough, so that such addition does not affect vertices of the original feasible region. We call the added constraint "standby", because it is active only when the program is unbounded. The parameter M can be dropped technically in real computations (see the example below).

Based on Theorem 8.5.2 and Proposition 8.5.2, we proceed somehow similarly as in Sect. 8.3 as follows:

In each iterations, a Benders restricted master program of form (8.33) is handled. If it is infeasible, the original program is infeasible; if an optimal solution $(\bar{\pi}, \bar{g})$

is obtained, then an associated subprogram $P(\bar{\pi})$ is solved. When the subprogram is unbounded, a new extreme direction is generated (Proposition 3.5.1); when an optimal vertex solution u^* and optimal value ζ^* are reached, it can be determined whether $(\bar{\pi}, \bar{g})$ is an optimal solution to the Benders master program; if it is not, a new vertex u^* is generated; if it is, then $(\bar{\pi}, \bar{y})$ is an optimal solution to the original program (8.22), where $\bar{y}$ is an optimal solution to $D(\bar{\pi})$. Of course, there is no need for solving $D(\bar{\pi})$ actually, because $\bar{y}$, as a dual solution associated with u^*, can be obtained at the same time.

In the case when optimality of (8.22) is not achieved, the restricted master program is updated by adding a constraint, corresponding to the new vertex or extreme direction, and the next iteration is carried out. In such an iteration, the restricted master program has one more constraint, as is called *row generation*. Note that subprogram $P(\bar{\pi})$ is a standard LP problem, only its objective function varies in the solution precess (with the constraints remaining unchanged).

The overall steps are summarized to the following algorithm.

Algorithm 8.5.1 (Benders Decomposition). Initial: dual Benders restricted master program of form (8.33). This algorithm solves program (8.22).

1. Solve dual Benders restricted master program.
2. Stop if it is infeasible.
3. If its optimal solution $(\bar{\pi}, \bar{g})$ is obtained, solve subprogram $P(\bar{\pi})$ by the simplex algorithm.
4. If an optimal vertex solution u^* and optimal value ζ^* to $P(\bar{\pi})$ are obtained, then:

   (1) Update (8.33) by adding the constraint, corresponding to the new vertex u^*, and go to step 1 if $\bar{g} > \zeta^*$;
   (2) Stop if $\bar{g} = \zeta^*$.

5. If subprogram $P(\bar{\pi})$ is unbounded, update (8.33) by adding the constraint, associated with the generated new extreme direction.
6. Go to step 1.

Theorem 8.5.3. *Assume that the dual Benders restricted master program either has an optimal solution or is infeasible. If it and subprogram $P(\bar{\pi})$ are solved in finitely many iterations, Algorithm 8.5.1 terminates, at either*

(i) *Step 2, detecting that the original program is infeasible; or*
(ii) *Step 4(2), achieving an optimal solution $(\bar{\pi}, \bar{y})$ to the original program, where $\bar{y}$ and u^* are complementary.*

Proof. The meaning of exit step 2 is directly from Proposition 8.5.2. According to Theorem 8.5.2, termination at step 4(2) gives an optimal solution $(\bar{\pi}, \bar{g})$ to the dual Benders master program; further, it is known by Theorem 8.5.1 that $(\bar{\pi}, \bar{y})$ is an optimal solution to the original program, where $\bar{y}$ is the dual solution, associate with u^*. Thereby, it is only needed to show the termination of Algorithm 8.5.1.

By Theorem 8.5.2, the restricted master program has one more added constraint, associated with a vertex or extreme direction per iteration. Since vertices and extreme directions in X are finite, the restricted master program must become the full master program in finitely many iterations, if the solution process does not terminate earlier. □

Although the number of constraints of the dual Benders restricted master program increases, fortunately, it hardly becomes the full dual Benders master program. Usually, the process terminates while it is very far from the latter.

8.6 Application of Benders Decomposition

As for applications, efficiency of solving the dual Benders restricted master program is crucial to Algorithm 8.5.1, as it depends on characteristic of $f(\pi)$ and the algorithm used to solve it. At present, there are quite good algorithms for the following cases, concerning (8.22):

(i) $f(\pi) = h^{\mathrm{T}}\pi,\ F(\pi) = H^{\mathrm{T}}\pi,\ h \in \mathcal{R}^m,\ H \in \mathcal{R}^{m\times n}$, and $\Pi \subset \mathcal{R}^m$ is a (bounded) polyhedron.

   In this case, the original program (8.22) is a linear program. If π is a free variable ($\Pi = \mathcal{R}^m$), in fact, (8.22) is clearly a standard dual program. As the dual restricted master program and subprogram are all linear, the simplex algorithm is a desirable tool to be used.

(ii) The same as case (i), except for adding integer requirements to components of π.

   In this case, the original program (8.22) is a mixed ILP program, and hence the dual restricted master program is a pure ILP program, except involving a free variable g. Since one inequality constraint is added to the restricted master program in each iteration, the cutting plane method (see, Sects. 10.3 and 10.5) is amenable to be employed.

(iii) $f(\pi),\ F(\pi)$ are continuously differentiable functions on a bounded and closed convex set $\Pi \subset \mathcal{R}^m$.

   In this case, the dual restricted master program is a smooth convex program, to which efficient algorithms are available (see, e.g., Sun and Yuan 2006).

Cases (ii) and (iii) are beyond the scope of this book. The following discussions are focused on case (i), related to LP.

As for how to construct an initial dual restricted master program, a direct way is to start from the feasible region

$$X = \{x \in \mathcal{R}^n \mid Ax = b,\ x \geq 0\}$$

of $P(\bar{\pi})$. If $X = \emptyset$, the original program is infeasible or unbounded (Proposition 8.5.1); otherwise, a feasible vertex, say $u^1 \in X$, can be determined. Thereby, letting

$$U' = \{u^1\}, \quad V' = \emptyset,$$

leads to an initial dual restricted master program of form (8.33).

Benders decomposition is suitable for some programs, involving block diagonal structures, such as

$$\begin{array}{ll} \min & h^{\mathrm{T}}\pi + (b^1)^{\mathrm{T}}y^1 \quad + \cdots + (b^k)^{\mathrm{T}}y^k, \\ \text{s.t.} & A_1^{\mathrm{T}}\pi + D_1^{\mathrm{T}}y^1 \quad \leq c_1, \\ & A_2^{\mathrm{T}}\pi + D_2^{\mathrm{T}}y^2 \quad \leq c^2, \\ & \vdots \\ & A_k^{\mathrm{T}}\pi + +D_k^{\mathrm{T}}y^k \leq c^k, \end{array}$$

where orders of submatrices D_i and dimensions of subvectors b^i, y^i, c^i, $i = 1, \cdots, k$ are consistent, i.e., b_i and y_i are of the same dimension, equal to the number of rows of D_i, and c_i is of dimension equal to the number of columns of D_i; π is of dimension equal to the number of rows of A_i.

If one partitions the coefficient matrix as

$$H^{\mathrm{T}} = \begin{pmatrix} A_1^{\mathrm{T}} \\ A_2^{\mathrm{T}} \\ \vdots \\ A_k^{\mathrm{T}} \end{pmatrix}, \quad A^{\mathrm{T}} = \begin{pmatrix} D_1^{\mathrm{T}} & & & \\ & D_2^{\mathrm{T}} & & \\ & & \ddots & \\ & & & D_k^{\mathrm{T}} \end{pmatrix},$$

then subprogram $P(\bar{\pi})$ can be decomposed to k smaller programs, corresponding to D_k. Let us bring up the following example.

Example 8.6.1. Solve the following program by Benders Decomposition Algorithm 8.5.1:

$$\begin{array}{lrrrrrrl} \max & 2\pi & + y_1 & + 4y_2 & + y_3 & + 2y_4, & & \\ \text{s.t.} & \pi & + \ y_1 & & & & \leq & 0, \\ & -3\pi & & + \ y_2 & & & \leq & -1, \\ & -\pi & + \ 4y_1 & + \ y_2 & & & \leq & 1, \\ & \pi & + \ y_1 & - \ y_2 & & & \leq & -2, \\ & 2\pi & & & + \ y_3 & & \leq & -3, \\ & \pi & & & & + \ y_4 & \leq & 1, \\ & -\pi & & & + \ 3y_3 & + \ y_4 & \leq & -1, \\ & -3\pi & & & - \ y_3 & - \ 3y_4 & \leq & 1. \end{array}$$

Answer Denote the program by

$$\begin{array}{ll}\max_{\pi,y} & h^{\mathrm{T}}\pi + b^{\mathrm{T}}y,\\ \text{s.t.} & H^{\mathrm{T}}\pi + A^{\mathrm{T}}y \le c,\end{array}$$

where

$h^{\mathrm{T}} = (2), b = (1,4,1,2)^{\mathrm{T}}, c = (0,-1,1,-2,-3,1,-1,1)^{\mathrm{T}}$.

$$H = (1,-3,-1,1,2,1,-1,-3), \quad A = \begin{pmatrix} 1 & 0 & 4 & 1 & 0 & 0 & 0 & 0\\ 0 & 1 & 1 & -1 & 0 & 0 & 0 & 0\\ 0 & 0 & 0 & 0 & 1 & 0 & 3 & -1\\ 0 & 0 & 0 & 0 & 0 & 1 & 1 & -3 \end{pmatrix}.$$

There is a vertex within the feasible region $X = \{x \mid Hx = b,\ x \ge 0\}$ of Subprogram $P(\pi)$, i.e.,

$u^1 = (1,4,0,0,1,2,0,0)^{\mathrm{T}}$.

$(u^1)^{\mathrm{T}}H^{\mathrm{T}}\pi = (1,4,0,0,1,2,0,0)(1,-3,-1,1,2,1,-1,-3)^{\mathrm{T}}\pi = -7\pi$.

$(u^i)^{\mathrm{T}}c = (1,4,0,0,1,2,0,0)(0,-1,1,-2,-3,1,-1,1)^{\mathrm{T}} = -5$.

The associated initial restricted master program is

$$\begin{array}{lll}\max_{\pi,g} & 2\pi + g, & \\ \text{s.t.} & -7\pi + g \le -5, & \\ & 2\pi + g \le M, & (*)\end{array}$$

where (*) marks the "standby" constraint.

Iteration 1:

1. Solve the Benders restricted master program by the simplex algorithm.

   Taking basis matrix $B = \begin{pmatrix} -7 & 2\\ 1 & 1\end{pmatrix}$, $B^{-1} = \begin{pmatrix} -1/9 & 2/9\\ 1/9 & 7/9\end{pmatrix}$ results in the dual feasible solution

$$\begin{pmatrix}\bar{\pi}\\ \bar{g}\end{pmatrix} = B^{-\mathrm{T}}\begin{pmatrix} -5\\ M\end{pmatrix} = \begin{pmatrix} 5/9 + 1/9M\\ -10/9 + 7/9M\end{pmatrix}. \tag{8.34}$$

   The associated primal solution is feasible, i.e., $B^{-1}(2,1)^{\mathrm{T}} = (0,1)^{\mathrm{T}} \ge 0$. Thereby, optimality of the restricted master program is achieved.

3. The objective function of Subprogram $P(\bar{\pi})$ is

$$\begin{aligned}(c - H^{\mathrm{T}}\bar{\pi})^{\mathrm{T}} &= (0,-1,1,-2,-3,1,-1,1)\\ &\quad -(5/9 + 1/9M)(1,-3,-1,1,2,1,-1,-3)\\ &= (-5/9 - 1/9M, 2/3 + 1/3M, 14/9 + 1/9M, -23/9 - 1/9M,\\ &\quad -37/9 - 2/9M, 4/9 - 1/9M, -4/9 + 1/9M, 8/3 + 1/3M).\end{aligned}$$

As parameter M is large enough, the terms without M in the preceding may be omitted. For simplicity, the common factor M may be omitted too; in other words, (8.34) may be replaced by

$$\begin{pmatrix} \bar{\pi} \\ \bar{g} \end{pmatrix} = B^{-\mathrm{T}} \begin{pmatrix} 0 \\ 1 \end{pmatrix}.$$

Subprogram $P(\bar{\pi})$ is decomposed to the following two program, as solved separately:

(i)

$$\begin{array}{ll} \min & -1/9x_1 + 1/3x_2 + 1/9x_3 - 1/9x_4, \\ \text{s.t.} & x_1 + 4x_3 + x_4 = 1, \\ & x_2 + \ \ x_3 - x_4 = 4, \quad x_j \geq 0, \ j = 1, \cdots, 4. \end{array}$$

Taking basis and nonbasis B_1=$\{1, 2\}$, N_1=$\{3, 4\}$. B_1^{-1}=I leading to basic feasible solution $\bar{x}_{B_1} = (1, 4)^{\mathrm{T}} \geq 0, \bar{x}_{N_1} = (0, 0)^{\mathrm{T}}$.

$$\bar{y}_1 = B_1^{-\mathrm{T}} c_{B_1} = \begin{pmatrix} -1/9 \\ 1/3 \end{pmatrix},$$

$$\bar{z}_{N_1} = c_{N_1} - N_1^{\mathrm{T}} \bar{y}_1 = \begin{pmatrix} 1/9 \\ -1/9 \end{pmatrix} \begin{pmatrix} 4 & 1 \\ 1 & -1 \end{pmatrix}^{\mathrm{T}} \begin{pmatrix} -1/9 \\ 1/3 \end{pmatrix} = \begin{pmatrix} 2/9 \\ 1/3 \end{pmatrix}.$$

$\min\{2/9, 1/3\} \geq 0$.

The reduced costs are all nonnegative, optimality is thus achieved.

The vertex optimal solution is $(\bar{x}_1, \bar{x}_2, \bar{x}_3, \bar{x}_4)^{\mathrm{T}} = (1, 4, 0, 0)^{\mathrm{T}}$, with optimal value $\zeta_1 = 11/9$.

(ii)

$$\begin{array}{ll} \min & -2/9x_5 - 1/9x_6 + 1/9x_7 + 1/3x_8, \\ \text{s.t.} & x_5 + 3x_7 - \ \ x_8 = 1, \\ & x_6 + \ \ x_7 - 3x_8 = 2, \quad x_j \geq 0, \ j = 5, \cdots, 8. \end{array}$$

Take basis and nonbasis $B_1 = \{5, 6\}$, $N_2 = \{7, 8\}$. $B_1^{-1} = I$.

The basic feasible solution is

$\bar{x}_{B_1} = (1, 2)^{\mathrm{T}} \geq 0, \ \bar{x}_{N_2} = (0, 0)^{\mathrm{T}}$.

$$\bar{y}_2 = B_1^{-\mathrm{T}} c_{B_1} = \begin{pmatrix} -2/9 \\ -1/9 \end{pmatrix},$$

$$\bar{z}_{N_2} = c_{N_2} - N_2^{\mathrm{T}} \bar{y}_2 = \begin{pmatrix} 1/9 \\ 1/3 \end{pmatrix} - \begin{pmatrix} 3 & -1 \\ 1 & -3 \end{pmatrix}^{\mathrm{T}} \begin{pmatrix} -2/9 \\ -1/9 \end{pmatrix} = \begin{pmatrix} 8/9 \\ -2/9 \end{pmatrix}.$$

The minimum reduced cost: $\min\{8/9, -2/9\} = -2/9$, $q = 8$.

$\bar{a}_8 = B_1^{-1}(-1, -3)^{\mathrm{T}} = (-1, -3)^{\mathrm{T}} \leq 0$, hence $P(\bar{\pi})$ is unbounded. The descent extreme direction is $v^* = (0, 0, 0, 0, 1, 3, 0, 1)^{\mathrm{T}}$.

5. $(v^*)^{\mathrm{T}} H^{\mathrm{T}} \pi = (0,0,0,0,1,3,0,1)(1,-3,-1,1,2,1,-1,-3)^{\mathrm{T}} \pi = 2\pi.$

$(v^*)^{\mathrm{T}} c = (0,0,0,0,1,3,0,1)(0,-1,1,-2,-3,1,-1,1)^{\mathrm{T}} = 1.$

The restricted master program is then updated to

$$\begin{aligned} \max_{\pi,g} \quad & 2\pi + g, \\ \text{s.t.} \quad & -7\pi + g \le -5, \\ & 2\pi \le 1, \\ & 2\pi + g \le M. \qquad\qquad (*) \end{aligned}$$

Iteration 2:

1. Solve the restricted master program. The basis matrix $B = \begin{pmatrix} -7 & 2 \\ 1 & 0 \end{pmatrix}$,

$B^{-1} = \begin{pmatrix} 0 & 1 \\ 1/2 & 7/2 \end{pmatrix}$, corresponds to dual feasible solution

$$\begin{pmatrix} \bar{\pi} \\ \bar{g} \end{pmatrix} = B^{-\mathrm{T}} \begin{pmatrix} -5 \\ 1 \end{pmatrix} = \begin{pmatrix} 1/2 \\ -3/2 \end{pmatrix},$$

The associated primal solution is feasible, i.e. $B^{-1}(2,1)^{\mathrm{T}} = (1, 9/2)^{\mathrm{T}} \ge 0$, hence optimality of the dual restricted master program is achieved.

3. The objective function of the subprogram $P(\bar{\pi})$ is

$$\begin{aligned} (c - H^{\mathrm{T}}\bar{\pi})^{\mathrm{T}} &= (0,-1,1,-2,-3,1,-1,1) - (1/2)(1,-3,-1,1,2,1,-1,-3) \\ &= (-1/2, 1/2, 3/2, -5/2, -4, 1/2, -1/2, 5/2). \end{aligned}$$

$P(\bar{\pi})$ can be decomposed to two programs, which are solved separately:

(i)

$$\begin{aligned} \min \quad & -1/2x_1 + 1/2x_2 + 3/2x_3 - 5/2x_4, \\ \text{s.t.} \quad & x_1 + 4x_3 + x_4 = 1, \\ & x_2 + \ \ x_3 - x_4 = 4, \quad x_j \ge 0, \ j = 1, \cdots, 4. \end{aligned}$$

(1) Take basis and nonbasis $B_1 = \{1,2\}$, $N_1 = \{3,4\}$. $B_1^{-1} = I$.
The basic feasible solution is $\bar{x}_{B_1} = (1,4)^{\mathrm{T}} \ge 0$, $\bar{x}_{N_1} = (0,0)^{\mathrm{T}}$.

$$\bar{y}_1 = B_1^{-\mathrm{T}} c_{B_1} = \begin{pmatrix} -1/2 \\ 1/2 \end{pmatrix},$$

$$\bar{z}_{N_1} = c_{N_1} - N_1^{\mathrm{T}} \bar{y}_1 = \begin{pmatrix} 3/2 \\ -5/2 \end{pmatrix} - \begin{pmatrix} 4 & 1 \\ 1 & -1 \end{pmatrix}^{\mathrm{T}} \begin{pmatrix} -1/2 \\ 1/2 \end{pmatrix} = \begin{pmatrix} 3 \\ -3/2 \end{pmatrix}.$$

The minimum reduced cost: $\min\{3, -3/2\} = -3/2,\ q = 4.$
$\bar{a}_4 = B_1^{-1}(1,-1)^{\mathrm{T}} = (1,-1)^{\mathrm{T}},\ p = 1.$
The minimum-ratio test:$\min\{1/1\} = 1,\ p = 1.$

(2) Update basis and nonbasis:

$$B_1 = \{4,2\},\ N_1 = \{3,1\}.\ B_1^{-1} = \begin{pmatrix} 1 & 0 \\ -1 & 1 \end{pmatrix}^{-1} = \begin{pmatrix} 1 & 0 \\ 1 & 1 \end{pmatrix}.$$

The basic feasible solution:

$$\bar{x}_{B_1} = B_1^{-1}(1,4)^{\mathrm{T}} = (1,5)^{\mathrm{T}} \geq 0,\ \bar{x}_{N_1} = (0,0)^{\mathrm{T}}.$$

$$\bar{y}_1 = B_1^{-\mathrm{T}} c_{B_1} = B_1^{-\mathrm{T}} \begin{pmatrix} -5/2 \\ 1/2 \end{pmatrix} = \begin{pmatrix} -2 \\ 1/2 \end{pmatrix},$$

$$\bar{z}_{N_1} = c_{N_1} - N_1^{\mathrm{T}} \bar{y}_1 = \begin{pmatrix} 3/2 \\ -1/2 \end{pmatrix} - \begin{pmatrix} 4 & 1 \\ 1 & 0 \end{pmatrix}^{\mathrm{T}} \begin{pmatrix} -2 \\ 1/2 \end{pmatrix} = \begin{pmatrix} 9 \\ 3/2 \end{pmatrix}.$$

The minimum reduced cost: $\min\{9, 3/2\} \geq 0.$
The optimal solution and associated objective value are
$(\bar{x}_1, \bar{x}_2, \bar{x}_3, \bar{x}_4)^{\mathrm{T}} = (0,5,0,1)^{\mathrm{T}}, \quad \zeta_1 = (1/2)5 - 5/2 = 0.$

(ii)

$$\begin{array}{ll} \min & -4x_5 + 1/2x_6 - 1/2x_7 + 5/2x_8, \\ \text{s.t.} & x_5 + 3x_7 - \ \ x_8 = 1, \\ & x_6 + \ \ x_7 - 3x_8 = 2, \quad x_j \geq 0,\ j = 5, \cdots, 8. \end{array}$$

Take basis and nonbasis $B_1 = \{5,6\},\ N_2 = \{7,8\}.\ B_1^{-1} = I.$
The basic feasible solution: $\bar{x}_{B_1} = (1,2)^{\mathrm{T}} \geq 0,\ \bar{x}_{N_2} = (0,0)^{\mathrm{T}}.$

$$\bar{y}_2 = B_1^{-\mathrm{T}} c_{B_1} = \begin{pmatrix} -4 \\ 1/2 \end{pmatrix},$$

$$\bar{z}_{N_2} = c_{N_2} - N_2^{\mathrm{T}} \bar{y}_2 = \begin{pmatrix} -1/2 \\ 5/2 \end{pmatrix} - \begin{pmatrix} 3 & -1 \\ 1 & -3 \end{pmatrix}^{\mathrm{T}} \begin{pmatrix} -4 \\ 1/2 \end{pmatrix} = \begin{pmatrix} 11 \\ 0 \end{pmatrix}.$$

The minimum reduced cost: $\min\{11, 0\} \geq 0.$
The optimal solution and associated objective value are
$(\bar{x}_5, \bar{x}_6, \bar{x}_7, \bar{x}_8)^{\mathrm{T}} = (1,2,0,0)^{\mathrm{T}}, \quad \zeta_2 = -4 + (1/2)2 = -3.$

Summarizing (i) and (ii) gives an optimal vertex solution and optimal objective value to the subprogram below:

$$u^* = (0,5,0,1,1,2,0,0)^{\mathrm{T}}, \quad \zeta^* = \zeta_1 + \zeta_2 = 0 + (-3) = -3. \tag{8.35}$$

4(1). $\bar{g} = -3/2 > \zeta^* = -3.$
$(u^*)^{\mathrm{T}} H^{\mathrm{T}} \pi = (0,5,0,1,1,2,0,0)(1,-3,-1,1,2,1,-1,-3)^{\mathrm{T}} \pi = -10\pi.$
$(u^*)^{\mathrm{T}} c = (0,5,0,1,1,2,0,0)(0,-1,1,-2,-3,1,-1,1)^{\mathrm{T}} = -8.$

Update the restricted master program as

$$\begin{array}{ll} \max_{\pi,g} & 2\pi + g, \\ \text{s.t.} & -7\pi + g \le -5, \\ & 2\pi \le 1, \\ & -10\pi + g \le -8, \\ & 2\pi + g \le M. \qquad\qquad (*) \end{array}$$

Iteration 3:

1. Solve the restricted master program.
   Column $(-10, 1)^{\mathrm{T}}$ enters the basis, $q = 3$.
   $B^{-1}(-10, 1)^{\mathrm{T}} = (1, -3/2)^{\mathrm{T}}$. The minimum-ratio test:min$\{1/1\} = 1,\ p = 1$.
   The basis matrix is updated as
   $B = \begin{pmatrix} -10 & 2 \\ 1 & 0 \end{pmatrix}, \ B^{-1} = \begin{pmatrix} 0 & 1 \\ 1/2 & 5 \end{pmatrix},$
   corresponding to the dual feasible solution
   $\begin{pmatrix} \bar{\pi} \\ \bar{g} \end{pmatrix} = B^{-\mathrm{T}} \begin{pmatrix} -8 \\ 1 \end{pmatrix} = \begin{pmatrix} 1/2 \\ -3 \end{pmatrix},$
   The associated primal solution is also feasible: $B^{-1}(2, 1)^{\mathrm{T}} = (1, 6)^{\mathrm{T}} \ge 0$. So optimality of the restricted master program is achieved.
3. The objective function of the subprogram $P(\bar{\pi})$ is

$$\begin{aligned} (c - H^{\mathrm{T}}\bar{\pi})^{\mathrm{T}} &= (0, -1, 1, -2, -3, 1, -1, 1) - (1/2)(1, -3, -1, 1, 2, 1, -1, -3) \\ &= (-1/2, 1/2, 3/2, -5/2, -4, 1/2, -1/2, 5/2), \end{aligned}$$

   which is the same as that in Iteration 2, hence the optimal solution and associated objective value are given by (8.35).

4(2). $\bar{g} = \zeta^* = -3$, optimality of the Benders master program is achieved. Consequently, the optimal solutio and objective value to the original program are

$$\bar{\pi} = 1/2, \quad \bar{y}^{\mathrm{T}} = (\bar{y}_1^{\mathrm{T}}, \bar{y}_2^{\mathrm{T}}) = (-2, 1/2, -4, 1/2), \qquad \bar{f} = -2.$$

8.7 Primal Benders Decomposition

The decomposition presented in Sect. 8.5 handles the dual program of a standard program. Therefore, it may be termed *dual Benders decomposition*. It is also possible to handle a standard program directly in a similar manner, resulting in a so-called *primal Benders decomposition*. This section will offer related results without detailed derivation.

Partition variables of a problem to two parts, one of which are in linear form, i.e.,

$$\begin{array}{lll} \min_{w,x} & f = f(w) + c^{\mathrm{T}}x, & \\ \text{s.t.} & F(w) + Ax = b, & \\ & w \in W,\ x \geq 0, & \end{array} \tag{8.36}$$

where $A \in \mathcal{R}^{m\times n}$, $c \in \mathcal{R}^n$, $b \in \mathcal{R}^m$. Real-valued function $f(\pi)$ and vector-valued function $F(\pi) \in \mathcal{R}^n$ and their domain $W \subset \mathcal{R}^p$ satisfy certain conditions. If $f(w)$ and $F(w)$ are all linear, and $W = \{w \in \mathcal{R}^p \mid w \geq 0\}$, the preceding is clearly a standard LP program.

Deeming w as a parameter, introduce subprogram

$$\begin{array}{ll} P(w) : \min_x & c^{\mathrm{T}}x, \\ \text{s.t.} & Ax = b - F(w), \quad x \geq 0. \end{array}$$

The dual program of the preceding is

$$\begin{array}{ll} D(w) : \max_y & \zeta = (b - F(w))^{\mathrm{T}}y, \\ \text{s.t.} & A^{\mathrm{T}}y \leq c. \end{array}$$

Assume nonempty of the feasible region of $D(w)$, i.e.,

$$Y = \{y \mid A^{\mathrm{T}}y \leq c\} \neq \emptyset,$$

whose vertex and extreme direction sets are denoted by

$$U = \{u^1, \cdots, u^s\}, \quad V = \{v^1, \cdots, v^t\}.$$

Define *primal* Benders *master program*

$$\begin{array}{llll} \min_{w,g} & f(w) + g, & & \\ \text{s.t.} & (u^i)^{\mathrm{T}}F(w) + g \geq (u^i)^{\mathrm{T}}b, & u^i & \in U, \\ & (v^j)^{\mathrm{T}}F(w) \quad\quad \geq (v^j)^{\mathrm{T}}b, & v^j & \in V, \\ & & w & \in W \cap T, \end{array} \tag{8.37}$$

where

$$T = \{w \mid F(w) + Ax = b,\ x \geq 0\}.$$

As for the relationship between the primal Benders master program and the original program (8.36), the following result arises.

Theorem 8.7.1. *$(\bar{w}, \bar{x})$ is an optimal solution to (8.36) if and only if $\bar{x}$ is an optimal solution to $P(w)$ and $(\bar{w}, \bar{g})$ is an optimal solution to the primal Benders master program.*

Thereby, one can solve program (8.36) by handling the primal Benders master program (8.37). Now turn to a relaxation program yielded from dropping some constraints. Assume that subsets of the vertex and extreme direction sets of Y are known at some iteration, i.e.,

$$U' \subset U, \quad V' \subset V.$$

Define *primal* Benders *restricted master program* as follows:

$$\begin{array}{lllll} \min_{w,g} & f(w) + g, & & & \\ \text{s.t.} & (u^i)^{\mathrm{T}} F(w) + g \geq (u^i)^{\mathrm{T}} b, & u^i & \in U', & \\ & (v^j)^{\mathrm{T}} F(w) \quad\quad\ \geq (v^j)^{\mathrm{T}} b, & v^j & \in V', & \\ & & w & \in W. & \end{array} \tag{8.38}$$

Note A "standby" constraint $f(w) + g \geq -M$ may be added to the preceding program technically, where M is a number large enough.

Theorem 8.7.2. *Let $(\bar{w}, \bar{g})$ be an optimal solution to program (8.38).*

If u^* and ζ^* are an optimal vertex solution and optimal value to subprogram $D(\bar{w})$, then:

(i) If $\bar{g} < \zeta^*$, u^* is a new vertex generated;
(ii) If $\bar{g} = \zeta^*$, $(\bar{w}, \bar{g})$ is an optimal solution to the primal Benders master program.

If $D(\bar{w})$ is unbounded, then a new extreme direction is generated.

Proposition 8.7.1. *If primal Benders restricted master program is infeasible, so is program (8.36).*

Algorithm 8.7.1 (Primal Benders Decomposition). Initial: primal Benders restricted master program of form (8.38). This algorithm solves the restricted master program (8.36).

1. Call the simplex algorithm to solve (8.38).
2. Stop if it is infeasible (the original program is infeasible).
3. If its optimal solution $(\bar{w}, \bar{g})$ is obtained, solve subprogram $D(\bar{w})$.
4. If an optimal vertex solution u^* and optimal value ζ^* to $D(\bar{w})$ are obtained, then:

   (1) update (8.38) by adding the constraint, associated with u^*, and go to step 1 if $\bar{g} < \zeta^*$.
   (2) stop if $\bar{g} = \zeta^*$ ($(\bar{w}, \bar{x})$ is an optimal solution to (8.36), where $\bar{x}$ is complementary to u^*).

5. If subprogram $D(\bar{w})$ is unbounded, update (8.38) by adding the constraint associated with the new extreme direction.
6. Go to step 1.

We are interested in linear case only. So, consider a special case of primal Benders decomposition with linear functions $f(w)$ and $F(w)$.

Letting

$$f(w) = h^{\mathrm{T}} w, \quad F(w) = Hw, \quad W = \{w \in \mathcal{R}^n \mid w \geq 0\},$$

where $H \in \mathcal{R}^{m \times n1}$, $h \in \mathcal{R}^{n1}$, then (8.36) becomes the standard LP program

$$\begin{array}{ll} \min_{w,x} & f = h^{\mathrm{T}} w + c^{\mathrm{T}} x, \\ \text{s.t.} & Hw + Ax = b, \\ & w, x \geq 0. \end{array} \tag{8.39}$$

Deeming w as a parameter, we handle subprogram

$$\begin{array}{ll} \min_{w,x} & f = c^{\mathrm{T}} x, \\ \text{s.t.} & Ax = b - Hw, \\ & x \geq 0. \end{array} \tag{8.40}$$

For simplicity, assume that set

$$Y = \{y \in \mathcal{R}^m \mid A^{\mathrm{T}} y \leq c\} \neq \emptyset \tag{8.41}$$

is bounded, whose vertex set is denoted by

$$U = \{u^1, \cdots, u^s\}. \tag{8.42}$$

Thus, the primal Benders restricted master program (8.38) becomes

$$\begin{array}{ll} \min_{w,g} & h^{\mathrm{T}} w + g, \\ \text{s.t.} & (u^i)^{\mathrm{T}} Hw + g \geq (u^i)^{\mathrm{T}} b, \quad u^i \in U'. \\ & w \geq 0. \end{array} \tag{8.43}$$

Assuming that there is an optimal solution $(\bar{w}, \bar{g})$ to the preceding program, solve associated subprogram

$$\begin{array}{ll} D(w) : \max_y & \zeta = (b - Hw)^{\mathrm{T}} y, \\ \text{s.t.} & y \in Y. \end{array}$$

If $D(w)$ has optimal value ζ^*, test for optimality of the original program by the following criterion:

$$\bar{g} = \zeta^*.$$

Another way to derive the primal Benders decomposition is to employ D-W decomposition and duality by dealing with the dual program of (8.39):

$$\begin{aligned} \max \quad & b^{\mathrm{T}} y, \\ \text{s.t.} \quad & H^{\mathrm{T}} y \le h, \\ & A^{\mathrm{T}} y \le c. \end{aligned} \tag{8.44}$$

Let Y and U be defined by (8.41) and (8.42) respectively. Express y in a convex combination of vertices in Y. Then, the D-W master program follows from (8.44), i.e.,

$$\begin{aligned} \max \quad & \sum_{i=1}^{s} (b^{\mathrm{T}} u^i)\alpha_i, \\ \text{s.t.} \quad & \sum_{i=1}^{s} (H^{\mathrm{T}} u^i)\alpha_i \le h, \\ & \sum_{i=1}^{s} \alpha_i = 1, \\ & \alpha_i \ge 0, \quad i = 1, \cdots, s. \end{aligned} \tag{8.45}$$

Then the primal Benders decomposition results from solving its dual program, which is noting but just the primal Benders restricted master program (8.43).

On the other hand, solving the dual program of the D-W master program of the original program leads to the dual Benders decomposition.

Finally, let us turn to programs with staircase structure. Some of such programs can be solved by the Benders decomposition in a nested manner. Let us take (8.16) again to show the so-called "forward nested manner" of the primal Benders decomposition.

Partition the variables to $\{x^1\}$ and $\{x^2, x^3, x^4\}$. According to the primal Benders decomposition, we need to solve the subprogram with x^1 as its parameter:

$$\begin{aligned} \min \quad & (c^2)^{\mathrm{T}} x^2 + (c^3)^{\mathrm{T}} x^3 + (c^4)^{\mathrm{T}} x^4, \\ \text{s.t.} \quad & D_{22} x^2 = b^2 - A_{21} x^1, \\ & A_{32} x^2 + D_{33} x^3 = b^3, \\ & A_{43} x^3 + D_{44} x^4 = b^4, \\ & x^2, x^3, x^4 \ge 0. \end{aligned} \tag{8.46}$$

So, we turn to solving the dual program of the preceding. It is easier to handle, as only its objective function is related to parameter x^1. In each iteration, the primal Benders restricted master program is solved to provide a parameter value $\bar{x}^1$ to the subprogram. Then (8.46) is of a staircase structure similar to that of the original program (8.16), with stairs reduced by 1. So, the primal Benders decomposition applies again. Such a process is repeated until only a single stair is left, and the program is easily solved.

Chapter 9
Interior-Point Method

As was known, the simplex method moves on the underlying polyhedron, from vertex to adjacent vertex along descent edges, until an optimal vertex is reached, or unboundedness of the problem is detected. Nevertheless, it would go through an exponential number of vertices of the polyhedron (Sect. 3.8), and even stall at a vertex forever because of cycling along degenerate edges (Sect. 3.6).

On the other hand, an optimal point can be reached directly if one goes across the interior of the polyhedron. In fact, there exists in the polyhedron a ray, emanating from any point to an optimal vertex, if any (see the last paragraph of Sect. 4.3). In other words, a single iteration is enough to solve the LP problem if the "right" descent direction is available. Although unrealistic, this idea might be the motivation of a class of so-called "interior-point methods", which moves from interior point to interior point to approach an optimal point.

Interior-point methods fall into three main categories: the potential function methods, represented by Karmarkar algorithm, the affine methods, represented by Dikin's algorithm, and the path-following methods using log barrier function. This chapter will be devoted to the most typical or efficient interior-point methods (see also Hertog and Roos 1991).

Thereafter the following basic assumptions are made on the standard LP programs:

A1: $\operatorname{rank} A = m$.
A2: $P^+ = \{x \in \mathcal{R}^n \mid Ax = b,\ x > 0\} \neq \emptyset$.
A3: $D^+ = \{(y,z) \in \mathcal{R}^m \times R^n \mid A^{\mathrm{T}}y + z = c,\ z > 0\} \neq \emptyset$.
A4: $c \notin \operatorname{range} A^{\mathrm{T}}$.

A4 is equivalent to the absence of a vector $\bar{y}$ such that $c = A^{\mathrm{T}}\bar{y}$, as means that the objective function is constant over the feasible region; and vice verse if the feasible region is nonempty. Therefore, the optimal value can only be attained on the boundary, if any.

P.-Q. PAN, *Linear Programming Computation*, DOI 10.1007/978-3-642-40754-3_9,

9.1 Karmarkar Algorithm

Karmarkar algorithm (1984) would be the most well-known interior-point algorithm, as it is the earliest one of polynomial-time complexity of the order lower than Khachiyan's ellipsoid algorithm. The basic idea behind Karmarkar algorithm inspired birth of some very efficient algorithms of such type, so as changing the state of the art of LP considerably.

Consider the so-called "Karmarkar standard problem":

$$\begin{array}{ll} \min & f = c^{\mathrm{T}} x, \\ \text{s.t.} & Ax = 0, \\ & e^{\mathrm{T}} x = 1, \quad x \geq 0, \end{array} \tag{9.1}$$

where $A \in \mathcal{R}^{m \times n}$, $c \in \mathcal{R}^n$, rank $A = m$, $m < n$, $n \geq 2$.

As a convex hull of n points $e_1, \cdots, e_n$ in the n-dimensional space, the polyhedral

$$\Gamma = \left\{ x \in \mathcal{R}^n \;\middle|\; \sum_{j=1}^{n} x_j = 1,\ x_j \geq 0 \right\}$$

is an $(n-1)$-dimensional regular simplex. It is noted that the n vertices of Γ lie in symmetric positions, and its center is e/n. The radius of its inscribed sphere and circumscribed sphere are respectively,

$$r = 1/\sqrt{n(n-1)} < 1, \quad R = \sqrt{n-1}/\sqrt{n} < 1.$$

The feasible region of problem (9.1) is the intersection of the simplex Γ and the null space, $\{x \in \mathcal{R}^n \mid Ax = 0\}$, of A. It is clear that there is an optimal solution if the feasible region is nonempty.

Concerning problem (9.1), Karmarkar makes the following two assumptions:

1. $Ae = 0$. So the center e/n of the simplex is an interior point of the feasible region.
2. The optimal solution, say x^*, satisfies $c^{\mathrm{T}} x^* = 0$, i.e. the optimal value is 0.

The preceding assumptions imply that $c^{\mathrm{T}} x > 0$ holds for all interior point x, in particular, the "center" e/n.

It is not difficult to convert a standard LP problem to Karmarkar standard problem, we will not go into details though. Interested readers are referred to related literatures, e.g., Tomlin (1987), Gay (1987) and de Ghellinck and Vial (1986).

9.1.1 Projective Transformation

Let $\bar{x}$ be an interior point, satisfying $A\bar{x} = 0$, $e^{\mathrm{T}}\bar{x} = 1$ and $\bar{x} > 0$. Denote by $\bar{X}$ the diagonal matrix, whose diagonals are components of $\bar{x}$, i.e.,

$$\bar{X} = \mathrm{diag}(\bar{x}_1, \cdots, \bar{x}_n).$$

Consider transformation

$$x' = \frac{\bar{X}^{-1}x}{e^{\mathrm{T}}\bar{X}^{-1}x} \triangleq T(x), \tag{9.2}$$

whose inverse transformation is

$$x = T^{-1}(x') = \frac{\bar{X}x'}{e^{\mathrm{T}}\bar{X}x'}. \tag{9.3}$$

$T(x)$, termed *projective transformation*, is an 1–1 mapping from Γ to Γ itself. In fact, for any $x \in \Gamma$, it holds that $x' = T(x) \in \Gamma$; conversely, for any $x' \in \Gamma$, it holds that $x = T^{-1}(x') \in \Gamma$. Under T, in particular, each vertex e_j $(j = 1, \cdots, n)$ of Γ corresponds to itself, and so does each edge; most importantly, $\bar{x}$ corresponds to the center e/n of Γ, whereas any interior point corresponds to an interior point.

Using T, (9.1) is transformed to problem

$$\begin{array}{ll} \min & f = \frac{c^{\mathrm{T}}\bar{X}x'}{e^{\mathrm{T}}\bar{X}x'}, \\ \text{s.t.} & A\bar{X}x' = 0, \\ & e^{\mathrm{T}}x' = 1, \quad x' \geq 0, \end{array} \tag{9.4}$$

which is no longer a LP problem though, because its objective function is not linear. However, it is known from (9.2) and $e^{\mathrm{T}}x = 1$ that when x is close to $\bar{x}$, the denominator in the objective function can approximately regarded as a positive constant, i.e.,

$$e^{\mathrm{T}}\bar{X}x' = \frac{e^{\mathrm{T}}x}{e^{\mathrm{T}}\bar{X}^{-1}x} = 1 \Big/ \sum_{j=1}^{n} \frac{x_j}{\bar{x}_j} \approx 1/n.$$

Karmarkar thereby employs $c^{\mathrm{T}}\bar{X}x'$ to replace the objective function approximately. Precisely, he used the following subproblem in each iteration:

$$\begin{array}{ll} \min & f = c^{\mathrm{T}}\bar{X}x', \\ \text{s.t.} & A\bar{X}x' = 0, \\ & e^{\mathrm{T}}x' = 1, \quad x' \geq 0, \end{array} \tag{9.5}$$

which and (9.4) share the same feasible region and an initial interior point e/n.

Proposition 9.1.1. *Let x^* be an optimal solution to problem (9.1). Then for any interior point $\bar{x}$, there is an optimal solution $(x^*)' = T(x^*)$, associated with objective value zero, to subproblem (9.5).*

Proof. Note that for any interior point $\bar{x} > 0$ to (9.1) and any feasible point $x' \geq 0$ to (9.4), it holds that

$$e^{\mathrm{T}}\bar{X}x' = \sum_{j=1}^{n} \bar{x}_j x'_j \geq e^{\mathrm{T}}x' \min\{\bar{x}_j \mid j = 1, \cdots, n\} > 0.$$

It is known from $(x^*)' = T(x^*)$ and Assumption 2 that $(x^*)'$ is an optimal solution to (9.4), satisfying $c^{\mathrm{T}}\bar{X}(x^*)' = 0$. In other words, there is a feasible solution $(x^*)'$, associated with objective value zero, to (9.5). Assume that $\tilde{x}'$ is its feasible solution with a negative objective value, i.e., $c^{\mathrm{T}}\bar{X}\tilde{x}' < 0$. Then it follows that

$$\frac{c^{\mathrm{T}}\bar{X}\tilde{x}'}{e^{\mathrm{T}}\bar{X}\tilde{x}'} < 0,$$

which contradicts that the optimal value of (9.4) is zero. Therefore, $(x^*)'$ is an optimal solution, associated with objective value zero, to (9.5). □

9.1.2 Karmarkar Algorithm

Assume that a descent direction has been determined. If some of components of $\bar{x}$ are close to zero, then the stepsize along the direction could be very small, and the associated improvement (decrease) in the objective function would be potty. Problem (9.5) helps get rid of such a situation, to some extent, as $\bar{x}$'s image, $\bar{x}' = T(\bar{x}) = e/n$, in x' space is now at the center of the simplex, the distance from which to each coordinate plane is the same. Of course, we do not really want to solve (9.5) itself, but only use it as a subproblem to determine a "good" search direction and associated stepsize (see, e.g., Anstreicher and Watteyne 1993; Hertog and Roos 1991; Kalantari 1990; Turner 1991).

To this end, denote the coefficient matrix of subproblem (9.5) by

$$F = \begin{pmatrix} A\bar{X} \\ \cdots \\ e^{\mathrm{T}} \end{pmatrix}. \tag{9.6}$$

Then the orthogonal projection matrix from x' space to the null of F is

$$P = I - F^{\mathrm{T}}(FF^{\mathrm{T}})^{-1}F. \tag{9.7}$$

Thus, the orthogonal projection of the objective gradient $\bar{X}c$ is

$$\Delta x = P\bar{X}c = (I - F^{\mathrm{T}}(FF^{\mathrm{T}})^{-1}F)\bar{X}c. \tag{9.8}$$

Proposition 9.1.2. *Vector Δx is nonzero, satisfying $F\Delta x = 0$ and $(\bar{X}c)^{\mathrm{T}}\Delta x > 0$.*

Proof. Assume $\Delta x = 0$. It is known from (9.8) that there is $h \in \mathcal{R}^m, h_{m+1}$ such that

$$\bar{X}c = F^{\mathrm{T}}(h^{\mathrm{T}}, h_{m+1})^{\mathrm{T}}.$$

Therefore, for any feasible solution $x' \geq 0$ to subproblem (9.5), it holds that

$$c^{\mathrm{T}}\bar{X}x' = (h^{\mathrm{T}}, h_{m+1})Fx' = h^{\mathrm{T}}A\bar{X}x' + h_{m+1}e^{\mathrm{T}}x' = h_{m+1}.$$

In other words, the feasible value is constant. Note that this is the case for $\bar{X}$, constructed from any interior point $\bar{x}$. In particular, for $\bar{x} = e/n$, subproblem's objective value at the center e/n of the simplex is $c^{\mathrm{T}}e/n^2$; on the other hand, it is known from Proposition 9.1.1 that the optimal value is zero. Thus it follows that $c^{\mathrm{T}}e/n^2 = 0$, which implies that e/n is an optimal solution, as contradicts that any interior point is not optimal. Therefore, $\Delta x \neq 0$. Then from (9.8) and $P^2 = P$ together with $P^{\mathrm{T}} = P$, both $F\Delta x = 0$ and $(\bar{X}c)^{\mathrm{T}}\Delta x > 0$ follow. □

The preceding Proposition says that $-\Delta x \neq 0$ is a descent feasible direction. Further, it is not difficult to show that $-\Delta x$, which is within the null of F, forms the largest possible angle with the objective gradient, that is, $-\Delta x$ is the steepest descent feasible vector in the null. As a result, if the stepsize from the center of the simplex along the direction is less than the radius of the inscribed sphere of the simplex, then the new interior point in x' space must lie in the interior of the feasible region, hopefully leading to a satisfactory decrease in the objective value. In precise, the new interior point in x' space is determined by

$$\hat{x}' = \frac{e}{n} - \alpha\rho\frac{\Delta x}{\|\Delta x\|}, \tag{9.9}$$

where $\alpha \in (0, 1)$, $\rho \in (0, r]$, and r is the radius of the inscribed sphere. Then mapping it to the original x-space via the inverse transformation gives

$$\hat{x} = \bar{X}\hat{x}'/e^{\mathrm{T}}\bar{X}\hat{x}'. \tag{9.10}$$

Then, an iteration is complete.

The overall steps are summarized to the following algorithm.

Algorithm 9.1.1 (Karmarkar algorithm). Given tolerance $\epsilon > 0$. Initial: $k = 1$, $\bar{x} = e/n$. This algorithm solves problem (9.1).

1. Compute $\Delta x = P\bar{X}c = (I - F^{\mathrm{T}}(FF^{\mathrm{T}})^{-1}F)\bar{X}c$, where F is defined by (9.6).
2. Compute $\hat{x}' = e/n - \alpha\rho\Delta x/\|\Delta x\|$.
3. Compute $\bar{x} = \bar{X}\hat{x}'/e^{\mathrm{T}}\bar{X}\hat{x}'$.
4. Stop if $c^{\mathrm{T}}\bar{x} < \epsilon$.
5. Set $k = k + 1$.
6. Go to step 1.

Karmarkar algorithm's inspiration is perhaps more valuable than itself, as stimulated coming out of some very successful algorithms of practical use. The idea that determining a search direction in the image space, and then transforming it back to the original space has influenced the development of interior-point methods greatly. At present, search direction of a typical interior-point method is usually a combination of a descent direction, stemming from the negative gradient, and a centering direction, generated via some transformation. Such a combination, which allows the algorithm to go farther in each iteration, is the real cause of the success of interior-point methods. As a "by-product", in addition, approaching to an optimal solution by a sequence of interior points overcomes the troublesome degeneracy encountered by the conventional simplex method.

9.1.3 Convergence

To analyze Algorithm 9.1.1, define the following *potential function* over Γ of problem (9.1):

$$f(x) = f(x,c) \triangleq n\ln(c^{\mathrm{T}}x) - \sum_{j=1}^{n} \ln x_j = \sum_{j=1}^{n} \ln \frac{c^{\mathrm{T}}x}{x_j}. \tag{9.11}$$

The corresponding potential function over Γ of problem (9.5) is then

$$f'(x') \triangleq f(x', \bar{X}c) = n\ln(c^{\mathrm{T}}\bar{X}x') - \sum_{j=1}^{n} \ln x'_j = \sum_{j=1}^{n} \ln \frac{c^{\mathrm{T}}\bar{X}x'}{x'_j}. \tag{9.12}$$

Proposition 9.1.3. *The potential function values at any point $x \in \Gamma$ and at its image x' differ by a same constant, which is dependent on $\bar{x}$.*

Proof. From (9.3), (9.11) and (9.12), it follows that

$$f(x) = f(T^{-1}(x')) = \sum_{j=1}^{n} \ln \frac{c^{\mathrm{T}}\bar{X}x'}{x'_j} - \ln(\Pi_{j=1}^{n}\bar{x}_j) = f'(x') - \ln(\Pi_{j=1}^{n}\bar{x}_j).$$

□

It is thereby seen that if a $\hat{x}'$ is determined such that $f'(\hat{x}')$ is less than $f'(e/n)$ by a constant, then $f(\hat{x})$ is accordingly less than $f(\bar{x})$ by the same amount. We will estimate

$$f'(\hat{x}') = n\ln(c^{\mathrm{T}}\bar{X}\hat{x}') - \sum_{j=1}^{n} \ln \hat{x}'_j. \tag{9.13}$$

Lemma 9.1.1. *If* Δx *and* $\hat{x}'$ *are respectively determined by (9.8) and (9.9), then*

$$\ln c^{\mathrm{T}}\bar{X}\hat{x}' \leq \ln \frac{c^{\mathrm{T}}\bar{X}e}{n} - \alpha\rho. \tag{9.14}$$

Proof. By $P^2 = P,\ P^{\mathrm{T}} = P$ and (9.8), it holds that

$$c^{\mathrm{T}}\bar{X}\Delta x = c^{\mathrm{T}}\bar{X}P\bar{X}c = c^{\mathrm{T}}\bar{X}P^2\bar{X}c = (P\bar{X}c)^{\mathrm{T}}(P\bar{X}c) = \|\Delta x\|^2 > 0, \tag{9.15}$$

which together with (9.9) gives

$$c^{\mathrm{T}}\bar{X}\hat{x}' = \frac{c^{\mathrm{T}}\bar{X}e}{n} - \alpha\rho\frac{c^{\mathrm{T}}\bar{X}\Delta x}{\|\Delta x\|} = \frac{c^{\mathrm{T}}\bar{X}e}{n} - \alpha\rho\|\Delta x\|. \tag{9.16}$$

Now consider

$$\begin{array}{ll} \min & f = c^{\mathrm{T}}\bar{X}x', \\ \text{s.t.} & A\bar{X}x' = 0, \\ & e^{\mathrm{T}}x' = 1, \\ & \|x' - e/n\| \leq R, \end{array} \tag{9.17}$$

where R is the radius of the circumscribed sphere, i.e.,

$$R = \sqrt{n-1}/\sqrt{n} < 1.$$

Noting that $-\Delta x$ is the orthogonal projection of the negative objective gradient onto the null of F, and

$$\{x \in \mathcal{R}^n \mid \|x' - e/n\| \leq R\}$$

is the spherical domain with the radius R at center of the simplex as its center, it is not difficult to show that the optimal solution is

$$x'(R) = \frac{e}{n} - R\frac{\Delta x}{\|\Delta x\|}.$$

On the other hand, since $\|x' - e/n\| \leq R$ is the circumscribed sphere of Γ, the feasible region of (9.17) includes the feasible region of subproblem (9.5), therefore the optimal value of the former is no more than the optimal value zero of the latter. So, considering (9.15) leads to

$$c^{\mathrm{T}}\bar{X}x'(R) = \frac{c^{\mathrm{T}}\bar{X}e}{n} - R\|\Delta x\| \leq 0,$$

or

$$-\|\Delta x\| \leq -\frac{c^{\mathrm{T}}\bar{X}e}{nR},$$

combining which, (9.16), $R < 1$ and $c^{\mathrm{T}}\bar{x} > 0$ gives

$$c^{\mathrm{T}}\bar{X}\hat{x}' \leq \frac{c^{\mathrm{T}}\bar{X}e}{n} - \alpha\rho\frac{c^{\mathrm{T}}\bar{X}e}{nR} = (1-\alpha\rho/R)\frac{c^{\mathrm{T}}\bar{X}e}{n} \leq (1-\alpha\rho)\frac{c^{\mathrm{T}}\bar{X}e}{n}.$$

Then, taking logarithm of the two slides of the preceding together with

$$\ln(1-\alpha\rho) \leq -\alpha\rho$$

leads to (9.14). □

Lemma 9.1.2. *If $x' \in \Gamma$ satisfies $\|x' - e/n\| \leq \alpha\rho$, then*

$$-\sum_{j=1}^{n} \ln x'_j \leq -\sum_{j=1}^{n} \ln(1/n) + \frac{(n\alpha\rho)^2}{2(1-\alpha n\rho)^2}. \tag{9.18}$$

Proof. Since $\|x' - e/n\| \leq \alpha\rho$, it holds that

$$1/n - \alpha\rho \leq x'_j \leq 1/n + \alpha\rho, \qquad j = 1, \cdots, n,$$

or

$$1 - \alpha n\rho \leq nx'_j \leq 1 + \alpha n\rho, \qquad j = 1, \cdots, n. \tag{9.19}$$

Based on Taylor formula, $\ln(nx'_j)$ can be expressed as

$$\ln(nx'_j) = \ln 1 + (nx'_j - 1) - \frac{1}{2\theta_j^2}(nx'_j - 1)^2,$$

where θ_j is between 1 and nx'_j, hence it is seen from (9.19) that $\theta_j \geq 1 - \alpha n\rho$. Thereby, the following inequality holds:

$$\ln(nx'_j) \geq (nx'_j - 1) - \frac{1}{2(1-\alpha n\rho)^2}(nx'_j - 1)^2. \tag{9.20}$$

In addition, it is known from the assumption that

$$\sum_{j=1}^{n}(nx'_j - 1) = ne^{\mathrm{T}}x' - n = 0$$

and

$$\sum_{j=1}^{n}(nx'_j - 1)^2 = \|nx' - e\|^2 = n^2\|x' - e/n\|^2 \le (\alpha n \rho)^2.$$

Then, it follows from (9.20) that

$$\sum_{j=1}^{n} \ln(nx'_j) \ge \frac{-(\alpha n \rho)^2}{2(1 - \alpha n \rho)^2},$$

which implies (9.18). □

Theorem 9.1.1. *If $\rho = 1/n < r$, $\alpha \in (0, 1/2)$, then the complexity of the number of iterations, required by Karmarkar algorithm, is $O(nL)$, where L is the binary digits of all input data.*

Proof. Combining (9.13), (9.14) and (9.18) gives

$$f'(\hat{x}') \le n \ln \frac{c^{\mathrm{T}} \bar{X} e}{n} - \alpha n \rho - \sum_{j=1}^{n} \ln(1/n) + \frac{(\alpha n \rho)^2}{2(1 - \alpha n \rho)^2} = f'(e/n) - \delta(n, \rho, \alpha), \tag{9.21}$$

where a lower bound of the decrement of the potential is

$$\delta(n, \rho, \alpha) = \alpha n \rho - \frac{(\alpha n \rho)^2}{2(1 - \alpha n \rho)^2}.$$

Substituting $\rho = 1/n$ into the preceding gives a function, only related to α, i.e.,

$$\delta(\alpha) = \alpha - \frac{\alpha^2}{2(1 - \alpha)^2} = \frac{\alpha(\alpha - 2)(\alpha - 1/2)}{(1 - \alpha)^2}, \tag{9.22}$$

to which it is clear that

$$\delta(\alpha) \begin{cases} > 0, & 0 < \alpha < 1/2, \\ = 0, & \alpha = 1/2, \\ < 0, & 1/2 < \alpha < 1. \end{cases}$$

Therefore, from the assumption of this Proposition, it follows that $\delta(\alpha) > 0$.

Let $\hat{x}$ be the original image of $\hat{x}'$, determined by (9.10). It is known by Proposition 9.1.3 that there is a constant μ, related to $\bar{x}$, such that

$$f(\bar{x}) = f'(e/n) + \mu, \quad f(\hat{x}) = f'(\hat{x}') + \mu,$$

The preceding together with (9.21) gives

$$f(\hat{x}) - f(\bar{x}) = f'(\hat{x}') - f'(e/n) \leq -\delta(\alpha),$$

hence

$$f(\hat{x}) \leq f(\bar{x}) - \delta(\alpha),$$

as holds for each iteration. Note that the initial interior point is e/n. Therefore, if $\hat{x}$ is the interior point, generated at iteration $k = 1, 2, \cdots$ (step 3 of Algorithm 9.1.1), then

$$f(\hat{x}) \leq f(e/n) - k\delta(\alpha).$$

Consequently, it holds by (9.11) that

$$n \ln(c^{\mathrm{T}}\hat{x}) - \sum_{j=1}^{n} \ln \hat{x}_j \leq n \ln(c^{\mathrm{T}}e/n) - \sum_{j=1}^{n} \ln(1/n) - k\delta(\alpha),$$

from which, $\hat{x} \in \Gamma$ and that function $\sum_{j=1}^{n} \ln x_j$ attains the highest value at e/n over Γ, it follows that

$$n \ln(c^{\mathrm{T}}\hat{x}) \leq n \ln(c^{\mathrm{T}}e/n) - k\delta(\alpha).$$

The preceding is equivalent to

$$c^{\mathrm{T}}\hat{x} \leq e^{-k\delta(\alpha)/n}(c^{\mathrm{T}}e/n). \tag{9.23}$$

Let ϵ be a given tolerance on the optimal value, then it is clear that there is a $\tau > 0$ such that

$$2^{-\tau L} < \frac{\epsilon}{c^{\mathrm{T}}e/n}. \tag{9.24}$$

Once the number k of iterations attains the smallest integer such that

$$k > (\tau/\delta(\alpha))nL,$$

then considering (9.24) gives

$$e^{-k\delta(\alpha)/n} \leq e^{-\tau L} < 2^{-\tau L} < \frac{\epsilon}{c^{\mathrm{T}}e/n},$$

which and (9.23) together lead to

$$c^{\mathrm{T}}\hat{x} < \epsilon.$$

Under the assumption, therefore, the complexity of the number of iterations, required by Karmarkar algorithm, is $O(nL)$. □

As the computational complexity of a single iteration is $O(n^3)$ for Karmarkar algorithm, the overall computational complexity is $O(n^4L)$. Karmarkar improves complexity of a single iteration to $O(n^{2.5})$, hence the overall complexity reduces to $O(n^{3.5}L)$.

In principle, it is desirable to have a larger lower bound on the decrement of the potential function. To this end, setting the derivative of (9.22) to 0 leads to its maximum $\alpha^* \approx 0.3177$ and related value $\delta(\alpha)^* \approx 0.2093$. Therefore, taking $\alpha = 0.3177$ in Karmarkar algorithm seems to be a good choice. However, it is not the case practically. As a lower *bound* on the decrement of the potential function, $\delta(\alpha)$ could be very different from the real decrement. Practice indicates that an α close to 1 often accelerates convergence, despite the polynomial complexity can not be guaranteed when $\alpha \geq 1/2$.

9.2 Affine Interior-Point Method

Because it is only amenable to the Karmarkar standard problem, Karmarkar algorithm is not very convenient for practical use. To handle the standard LP problem directly, subsequently the so-called "affine algorithm" was proposed as a variant of Karmarkar algorithm (Barnes 1986; Cavalier and Soyster 1985; Karmarkar and Ramakrishnan 1985; Vanderbei et al. 1986). However, it is found soon that the affine algorithm was proposed by Dikin as early as in 1967 (without convergence proof), but was not noted by the academic community, unfortunately.

Thereafter, concerned are the pair of the standard LP problems (4.1) and (4.2).

9.2.1 *Formulation of the Algorithm*

Let $\bar{x}$ be the current interior point. Again denote by $\bar{X}$ the diagonal matrix, whose diagonals are components of $\bar{x}$. It is clear that affine transformation

$$x' = \bar{X}^{-1}x$$

maps the positive octant to itself, and $\bar{x}$ to its "center" e. This transformation turns problem (4.1) to

$$\begin{array}{ll} \min & c^{\mathrm{T}}\bar{X}x', \\ \text{s.t.} & A\bar{X}x' = b, \quad x' \geq 0. \end{array} \tag{9.25}$$

The orthogonal projection matrix from x' space onto the null of $A\bar{X}$ is

$$P = I - \bar{X}A^{\mathrm{T}}(A\bar{X}^2A^{\mathrm{T}})^{-1}A\bar{X}, \tag{9.26}$$

and the orthogonal projection of the objective gradient $\bar{X}c$ is then

$$\Delta x' = P\bar{X}c = (I - \bar{X}A^{\mathrm{T}}(A\bar{X}^2A^{\mathrm{T}})^{-1}A\bar{X})\bar{X}c. \tag{9.27}$$

Lemma 9.2.1. *Under the basic assumption A4, it holds that* $\Delta x' \neq 0$.

Proof. Equation (9.27) can be written

$$\Delta x' = \bar{X}(c - A^{\mathrm{T}}\bar{y}),$$

where

$$\bar{y} = (A\bar{X}^2A^{\mathrm{T}})^{-1}A\bar{X}^2c.$$

Thus $\Delta x' = 0$ implies $c = A^{\mathrm{T}}\bar{y}$, as contradicts assumption A4. Therefore $\Delta x' \neq 0$. □

To determine a new interior point in x' space, take $-\Delta x'$ as a search direction:

$$\hat{x}' = e - \mu\Delta x'/\|\Delta x'\|.$$

If $\mu \in (0, 1)$ is taken, then $\hat{x}'$ lies within interior of the sphere with radius 1 at e. The original image of $\hat{x}'$ in x space is then

$$\hat{x} = \bar{x} - \mu\bar{X}\Delta x'/\|\Delta x'\|,$$

which is the "short stepsize" iteration scheme of the affine algorithm.

It is verified that the point

$$\hat{x} = \bar{x} - \alpha\bar{X}\Delta x' \tag{9.28}$$

satisfies $Ax = b$ for any α. In particular, for the short stepsize $\alpha = \mu/\|\Delta x'\| > 0$, it holds that $\hat{x} > 0$, hence the new iterate is again an interior point. Combing $P = P^2,\ P^{\mathrm{T}} = P$ and (9.27) leads to

$$c^{\mathrm{T}}\hat{x} = c^{\mathrm{T}}\bar{x} - \alpha c^{\mathrm{T}}\bar{X}P\bar{X}c = c^{\mathrm{T}}\bar{x} - \alpha\|\Delta x'\|^2 < c^{\mathrm{T}}\bar{x}, \tag{9.29}$$

as implies that the objective value strictly decreases.

It is desirable to go further without violating constraints. In fact, it turns out that letting the new interior point closer to boundary leads to bigger decrement of the objective value, and significantly enhances the efficiency of the algorithm. To this end, the following result should be noted first.

Lemma 9.2.2. *If $\Delta x' \leq 0$, the standard LP problem (4.1) is unbounded below.*

Proof. Under the assumption, $\hat{x}$ defined by (9.28) is clearly an interior point for any $\alpha > 0$. It is seen from (9.29) that

$$c^{\mathrm{T}}\hat{x} = c^{\mathrm{T}}\bar{x} - \alpha\|\Delta x'\|^2 \rightarrow -\infty \ (\alpha \rightarrow \infty).$$

Therefore, the original problem is unbounded below. □

Assume now that $\Delta x' \not\leq 0$. In this case, the largest stepsize attaining the boundary is $1/\max(\Delta x')$, hence leading to the following "long stepsize" iteration scheme:

$$\hat{x} = \bar{x} - \lambda \bar{X} \Delta x' / \max(\Delta x'), \tag{9.30}$$

where $\lambda \in (0, 1)$ is a *stepsize*.

Accompanying the original interior point $\bar{x}$, a dual estimate $(\bar{y}, \bar{z})$ is expected. If ignoring the nonnegative requirement for $\bar{z}$ but satisfying the complementarity condition as much as possible, $(\bar{y}, \bar{z})$ should be solves the following least squares problem:

$$\begin{array}{ll} \min & (1/2)\|\bar{X}z\|^2, \\ \text{s.t.} & z = c - A^{\mathrm{T}}y. \end{array}$$

It is not difficult to obtain the solution to the preceding, i.e.,

$$\bar{y} = (A\bar{X}^2A^{\mathrm{T}})^{-1}A\bar{X}^2c, \qquad \bar{z} = \bar{X}^{-1}P\bar{X}c. \tag{9.31}$$

Thus (9.27) can be written alternatively as

$$\Delta x' = \bar{X}(c - A^{\mathrm{T}}\bar{y}) = \bar{X}\bar{z}, \tag{9.32}$$

which reveals the relation between the search direction and the dual estimate. Note that $\bar{X}\bar{z}$ is noting but the associated dual gap (see Sect. 4.3).

The overall steps with long stepsize is put into the following algorithm (without generating dual estimates). It is known from (9.32) that the dual gap tens to zero as the procedure converges. Therefore, optimality is achieved approximately whenever $\|\Delta x'\|$ becomes small enough.

Algorithm 9.2.1 (Affine interior-point algorithm). Given $\lambda \in (0, 1)$, $\epsilon > 0$. Initial: interior point $\bar{x} > 0$. This algorithm solves the standard LP problem (4.1).

1. Compute $\Delta x' = P\bar{X}c = (I - \bar{X}A^{\mathrm{T}}(A\bar{X}^2A^{\mathrm{T}})^{-1}A\bar{X})\bar{X}c$.
2. Stop if $\|\Delta x'\| < \epsilon$ (optimality achieved approximately).
3. Stop if $\Delta x' \leq 0$ (unbounded below).
4. Update by $\bar{x} = \bar{x} - \lambda\bar{X}\Delta x'/\max(\Delta x')$.
5. Go to step 1.

It is seen from (9.27) that the algorithm's major work in each iteration is the computation of projection $\Delta x'$, concerning the solution of $m \times m$ system

$$(A\bar{X}^2A^{\mathrm{T}})y = A\bar{X}^2c.$$

A normal way is to compute the coefficient matrix, carry out the Cholesky factorization, and then solve the two triangular systems.

In contrast to Karmarkar algorithm, the affine algorithm is simpler, and yet performs better. In fact, the latter is the earliest interior-point algorithm found to possibly outperform the simplex method.

9.2.2 *Convergence and Start-Up*

Firstly, we state two convergence results without proofs. The first, given by Dikin (1974), is under the nondegeneracy assumption (see also Gonzaga 1990; Vanderbei and Lagarias 1990).

Theorem 9.2.1. *Assume the existence of an optimal solution to the standard problem (4.1). If all feasible solutions are nondegenerate, then the sequence of interior points, generated by (9.30)* ($\lambda \in (0, 1)$), *and sequence of dual estimates, generated by (9.31), respectively converge to relative interior points of the primal and dual optimal faces.*

If the range of λ is restricted properly, the long stepsize affine algorithm converges without the nondegeneracy assumption, as the following theorem says (Tsuchiya and Muramatsu 1995).

Theorem 9.2.2. *Assume the existence of an optimal solution to the standard problem (4.1). Then the sequence of interior points, generated by (9.30)* ($\lambda \in (0, 2/3]$), *converges to a relative interior point of the primal optimal face, and the sequence of dual estimates, generated by (9.31), converges to the relative analytic center of the dual optimal face, with asymptotic descent rate* $1 - \lambda$ *for the objective value.*

When $\lambda > 2/3$, however, convergence of the affine algorithm is not guaranteed. In fact, counter-examples have been found, indicting that $2/3$ is the largest value of λ for ensuring convergence (Hall and Vanderbei 1993; Mascarenhas 1997). As for practice, nevertheless, the case is just contrary: it is faster and more reliable to take a λ value close to 1 (e.g., $\lambda \in [0.9, 0.99]$). On the other hand, it is shown that even if $\lambda \leq 2/3$, the algorithm solves Klee-Minty problem by traversing neighborhoods of all the 2^n vertices, and hence is not of polynomial complexity (Megiddo and Shub 1989).

Now turn to the Phase-1 to provide an initial interior point to the affine algorithm.

Given any n-dimensional vector $h > 0$ such that $b - Ah \neq 0$ ($h = e$ seems to be a good choice). Introducing an artificial variable x_{n+1} and a normalized vector

$$a_{n+1} = (b - Ah)/\|b - Ah\|, \tag{9.33}$$

construct the following auxiliary problem:

$$\begin{array}{ll} \min & x_{n+1}, \\ \text{s.t.} & Ax + a_{n+1}x_{n+1} = b, \quad x, x_{n+1} \geq 0. \end{array} \tag{9.34}$$

It is clear that there is an optimal solution to the preceding problem. Moreover, $\bar{x} = h,\ \bar{x}_{n+1} = \|b - Ah\|$ is an interior point solution available. If the optimal value is strictly greater than 0, then the original problem is infeasible.

Solve the auxiliary problem by the affine algorithm with $\lambda \in (0, 2/3]$. Assume that the resulting sequence of interior points converges to x^{∞} with objective value 0, so that x^{∞} is a relative interior point of the feasible region of the original problem. Define

$$B = \{j \in A \mid x_j^{\infty} > 0\}, \quad N = \{j \in A \mid x_j^{\infty} = 0\}.$$

Then, the original problem reduces to

$$\begin{array}{ll} \min & c_B^{\mathrm{T}} x_B, \\ \text{s.t.} & B x_B = b, \quad x \geq 0, \end{array}$$

which has an interior point $x_B^{\infty} > 0$ available. Consequently, the affine algorithm can get itself started.

9.3 Dual Affine Interior-Point Method

In this section, the basic idea behind the affine method will be applied to the dual problem to derive the dual version (Adler et al. 1989).

Consider the dual problem (4.2). Assume that $(\bar{y}, \bar{z})$ is the current interior point, satisfying dual constraints. Denote by $\bar{Z}$ the diagonal matrix with components of $\bar{z}$ as its diagonals. The affine transformation $z = \bar{Z}z'$ turns the dual problem to

$$\begin{array}{ll} \max & b^{\mathrm{T}} y, \\ \text{s.t.} & A^{\mathrm{T}} y + \bar{Z}z' = c, \quad z' \geq 0, \end{array}$$

or equivalently,

$$\begin{array}{ll} \max & b^{\mathrm{T}} y, \\ \text{s.t.} & \bar{Z}^{-1} A^{\mathrm{T}} y + z' = \bar{Z}^{-1} c, \quad z' \geq 0. \end{array} \tag{9.35}$$

Note that affine transformation $z = \bar{Z}z'$ maps the positive octant of z-space to itself, and $\bar{z}$ to its "center" e. Since A is of full row rank, $\bar{Z}^{-1}A^{\mathrm{T}}$ if of full column rank.

Assume that the QR factorization of $\bar{Z}^{-1}A^{\mathrm{T}}$ is

$$\bar{Z}^{-1}A^{\mathrm{T}} = [Q_1, Q_2]\begin{pmatrix} R \\ 0 \end{pmatrix} = Q_1 R. \tag{9.36}$$

where $[Q_1, Q_2]$ is orthogonal, $Q_1 \in \mathcal{R}^{n\times m}$ and $Q_2 \in \mathcal{R}^{n\times(n-m)}$, and $R \in \mathcal{R}^{m\times m}$ is upper triangular.

Substituting (9.36) to the equality constraint of (9.35) gives

$$Q_1 R y + z' = \bar{Z}^{-1}c.$$

Premultiplying the preceding by $R^{-1}Q_1^{\mathrm{T}}$, noting $Q_1^{\mathrm{T}}Q_1 = I$ and rearranging leads to

$$y = R^{-1}Q_1^{\mathrm{T}}\bar{Z}^{-1}c - R^{-1}Q_1^{\mathrm{T}}z',$$

by which and

$$Q_1Q_1^{\mathrm{T}} + Q_2Q_2^{\mathrm{T}} = I, \quad Q_2^{\mathrm{T}}Q_2 = I,$$

program (9.35) can be deduced to the following standard form with respect to z':

$$\begin{array}{ll} \min & b^{\mathrm{T}}R^{-1}Q_1^{\mathrm{T}}z', \\ \text{s.t.} & Q_2^{\mathrm{T}}z' = Q_2^{\mathrm{T}}\bar{Z}^{-1}c,\ z' \geq 0, \end{array} \tag{9.37}$$

where constant $-b^{\mathrm{T}}R^{-1}Q_1^{\mathrm{T}}\bar{Z}^{-1}c$ was omitted from the objective function.

Proposition 9.3.1. *The orthogonal projection of the objective gradient of (9.37) to the null space of Q_2^{T} is equal to*

$$\Delta z' = \bar{Z}^{-1}A^{\mathrm{T}}(A\bar{Z}^{-2}A^{\mathrm{T}})^{-1}b. \tag{9.38}$$

Proof. By $Q_2^{\mathrm{T}}Q_1 = 0$, the orthogonal projection of the objective gradient to the null becomes

$$\Delta z' = (I - Q_2(Q_2^{\mathrm{T}}Q_2)^{-1}Q_2^{\mathrm{T}})Q_1R^{-\mathrm{T}}b = Q_1R^{-\mathrm{T}}b,$$

By $Q_1^T Q_1 = I$, the preceding is equal to

$$\Delta z' = Q_1(RR^{-1})R^{-\mathrm{T}}b = Q_1R(R^{\mathrm{T}}R)^{-1}b = Q_1R(R^{\mathrm{T}}Q_1^TQ_1R)^{-1}b,$$

which together with (9.36) gives (9.38). □

Thereby, $-\Delta z'$ is a favorable search direction in z'-space. For a new iterate to satisfy the equality constraint of (9.35), $\Delta z'$ and the associated search direction Δy in y-space should fulfil condition

$$\bar{Z}^{-1}A^{\mathrm{T}}\Delta y + \Delta z' = 0,$$

premultiplying which by $A\bar{Z}^{-1}$ and using (9.38) gives

$$\Delta y = -(A\bar{Z}^{-2}A^{T})^{-1}A\bar{Z}^{-1}\Delta z' = -(A\bar{Z}^{-2}A^{\mathrm{T}})^{-1}b. \tag{9.39}$$

Using $z = \bar{Z}z'$, vector

$$\Delta z' = -\bar{Z}^{-1}A^{\mathrm{T}}\Delta y$$

can be transformed back to z-space, i.e.,

$$\Delta z = -A^{\mathrm{T}}\Delta y. \tag{9.40}$$

The according long stepsize line search scheme is then

$$\hat{y} = \bar{y} - \lambda\beta\Delta y, \quad \hat{z} = \bar{z} - \lambda\beta\Delta z, \tag{9.41}$$

where $\lambda \in (0, 1)$, and

$$\beta = \min\{\bar{z}_j/\Delta z_j \mid \Delta z_j > 0, j = 1, \cdots, n\}. \tag{9.42}$$

In the case when $\Delta z \leq 0$, the preceding scheme is not well-defined, and the dual problem is unbounded or the primal problem is infeasible.

As for an estimate $\bar{x}$ of the associated primal solution, it can be determined as follows. If ignoring the nonnegative requirements but attempting to satisfy complementarity condition, the $\bar{x}$ should solve the following least squares problem:

$$\begin{array}{ll} \min & \|\bar{Z}x\|, \\ \text{s.t.} & A\bar{Z}^{-1}(\bar{Z}x) = b. \end{array}$$

The solution to which can be obtained by taking a look at the minimum 2-norm solution to $A\bar{Z}^{-1}w = b$ and (9.39), i.e.,

$$\bar{x} = \bar{Z}^{-2}A^{\mathrm{T}}(A\bar{Z}^{-2}A^{\mathrm{T}})^{-1}b = \bar{Z}^{-2}\Delta z.$$

Consequently, when $\bar{x} \geq 0$ and dual gap $c^{\mathrm{T}}\bar{x} - b^{\mathrm{T}}\bar{y}$ becomes sufficiently small, the procedure, generating a sequence of dual interior points, can be terminated.

The overall steps are put into the following algorithm.

Algorithm 9.3.1 (Dual affine algorithm). Given $\lambda \in (0, 1)$, $\epsilon > 0$. Initial: interior point $(\bar{y}, \bar{z})$. This algorithm solves the standard LP problem (4.1).

1. Compute $(\Delta y, \Delta z)$ by (9.39) and (9.40).
2. Stop if $\Delta z \leq 0$ (dual unbounded or primal infeasible).
3. Compute $\bar{x} = \bar{Z}^{-2}\Delta z$.
4. Stop if $\bar{x} \geq 0$ and $c^{\mathrm{T}}\bar{x} - b^{\mathrm{T}}\bar{y} < \epsilon$ (optimality achieved approximately).
5. Update $(\bar{y}, \bar{z})$ by (9.41) and (9.42).
6. Go to step 1.

The preceding algorithm starts from a dual interior point. $(\bar{y} = 0, \bar{z} = c)$ is an available one in case of $c > 0$. Otherwise, an approach analogous to that described in the last half of Sect. 9.2.2 can be employed to generate an initial dual interior point using the following auxiliary problem:

$$\begin{array}{ll} \max & -y_{m+1}, \\ \text{s.t.} & (A^{\mathrm{T}} \vdots\, c - e)\begin{pmatrix} y \\ y_{m+1} \end{pmatrix} + z = c, \quad z,\ y_{m+1} \geq 0, \end{array}$$

which can be handled by the dual affine algorithm, as it is upper bounded and has an interior point available, i.e.,

$$\bar{y} = 0, \quad \bar{y}_{m+1} = 1; \quad \bar{z} = e.$$

We point out that primal and dual affine algorithms may be derived from each another, and have similar convergence properties (Tsuchiya 1992) though the latter is superior to the former computationally.

In addition, a sequence of dual interior points may be generated alternatively by carrying out the (primal) affine algorithm. The trick is to turn to solving the standard problem in dual variable z, obtained by the so-called "dual elimination" (Sect. 25.1.3), though it is not known how well such an approach performs.

9.4 Path-Following Interior-Point Method

This type of methods perform remarkably in practice. It may be regarded as one using a "homotopy strategy", in which one creates a trivial problem, compared with a hard problem to be solved, so that there is a continuum of problems between them, and handles backwards from the trivial problem to the hard problem by solving the problems in between (exactly or approximately).

Taking the standard LP problem (4.1) as the "hard" problem, we will construct the "trivial" problem, whose optimal solution is an interior point.

As the border of the positive octant, coordinate planes $x_j = 0,\ j = 1, \cdots, n$, defining the scope of the feasible region, might well be deemed as "wall". We have seen that Karmarkar algorithm and/or the affine algorithm employe transformations to create a "centering force" to push the iterate leaving the wall. Combination of the "centering force" and the "pushing force" along the direction of the negative objective gradient gives a successful feasible downhill search direction. Another approach for generating "centering force" is via logarithmic barrier function, which was initially employed by Frisch (1955) to solve nonlinear problems. After Karmarkar algorithm emerged, the logarithmic barrier function draws high attention from the community, and is applied in the path-following method to construct the "trivial' problem.

As nonnegative constraints $x \geq 0$ are inactive for interior points, we remove them, and add logarithmic barrier terms in the objective function to construct the following problem:

$$\begin{aligned} (\mathrm{P}_{\mu}) \qquad \min \quad & f(x) = c^{\mathrm{T}}x - \mu \sum_{j=1}^{n} \ln x_j, \\ \text{s.t.} \quad & Ax = b, \end{aligned} \tag{9.43}$$

where $\mu > 0$ is called *barrier parameter*. Implying $x \geq 0$, (9.43) characterizes a class of problems with parameter $\mu > 0$. Because of the barrier terms, $f \to +\infty$ as $x_j \to 0^{+}$, so that minimization of f "rebounds" iterates from the "wall" to prevent them from leaving the interior of the feasible region.

Related to the "hard" problem by μ, the "trivial" problem (9.43) is nonlinear but not difficult to handle. The gradient and Hassian matrix of the objective function $f(x)$ are respectively

$$\nabla f(x) = c - \mu X^{-1}e, \quad \nabla^2 f(x) = \mu X^{-2},$$

where $X = \mathrm{diag}(x_1, \cdots, x_n)$. It is clear that $f(x)$ is a strict convex function over the region $x > 0$. The Lagrange function of problem (P_{μ}) is then

$$L(x, y) = c^{\mathrm{T}}x - \mu \sum_{j=1}^{n} \ln x_j - y^{\mathrm{T}}(Ax - b).$$

Therefore, point $x > 0$ is an optimal solution of (P_{μ}) if and only if there is a $y \in \mathcal{R}^m$ such that

$$Ax - b = 0, \tag{9.44}$$

$$c - \mu X^{-1}e - A^{\mathrm{T}}y = 0. \tag{9.45}$$

Using notation

$$z = c - A^{\mathrm{T}}y, \quad Z = \mathrm{diag}(z_1, \cdots, z_n).$$

the preceding conditions become the following system in variables x, y, z:

$$Ax \qquad\qquad = b, \tag{9.46}$$

$$A^{\mathrm{T}}y + z = c, \tag{9.47}$$

$$Xz = \mu e. \tag{9.48}$$

Megiddo (1989) show that:

(i) For each $\mu > 0$, problem (P_μ) either has an unique optimal solution or is unbounded below.
(ii) If there exists a solution $x > 0$ to the system for some $\mu > 0$, then so is a solution $x(\mu)$ to the system for all $\mu > 0$, and $x(\mu)$ is a continuous curve. There is a limit $\lim x(\mu)$ as $\mu \to 0^+$, which is an optimal solution to problem (P).

If the interior of the feasible region of (P) is nonempty and bounded, (9.46)–(9.48) determine a unique path $x(\mu)$ to an optimal solution. $x(\mu)$ is termed *central path* or *trajectory*. The common idea of various path-following algorithms is to find an optimal solution by following the path approximately (see also Monteiro and Adler 1989).

On the other hand, the central path can also be derived without the logarithmic barrier function. In fact, we may handle the standard LP problem directly from the optimality condition (4.15), deemed as a nonlinear system with respect to x, y, z, subject to $x, z \geq 0$. To creates iterates, which across the interior of the feasible region and approach an optimal solution on the boundary, it is only possible to satisfy the complementarity condition approximately, by using $Xz = v$ in place of $Xz = 0$, where $v > 0$ is a parameter vector. The associated solution $x(v)$ to the system is expected to tend to an optimal solution as v tends to 0. For symmetry of $x_j z_j = v_j$ for all $j = 1, \cdots, n$, it is natural to let components of v be equal by setting $v = \mu e$, as leads to (9.46)–(9.48) consequently. So, the "hard" problem is the optimality condition, while the "trivial" problem is system (9.46)–(9.48). Of course, it is not possible to solve all the systems for a continuous μ, but for some sequence $\{\mu_k\}$ instead, as will be seen in the next section.

9.4.1 Primal-Dual Method

A realization of the path-following strategy is to design a monotone descent sequence $\{\mu_k\}$ with limit 0, and for every fixed parameter $\mu = \mu_k$, $k = 1, \cdots$, solve system (9.46)–(9.48) for its solution $x(\mu_k)$. In general, it is difficult to directly obtain the exact solution to the nonlinear system (9.48). Instead, we employ the Newton method to find an approximate one. Moreover, it is not necessary to pay high cost on accuracy, since what we are interested in is not the solutions themselves but their limit.

Thus, only one Newton iteration is taken for obtaining an approximate solution, and elements of the monotone descent sequence $\{\mu_k\}$ are generated one by one.

Assume that the interior point $(\bar{x}, \bar{y}, \bar{z})$ satisfies (9.46) and (9.47), and approximately satisfies (9.48). We need to determine $(-\Delta x, -\Delta y, -\Delta z)$ such that $(\bar{x} - \Delta x,\ \bar{y} - \Delta y,\ \bar{z} - \Delta z)$ is a new approximate solution, which satisfies (9.46)–(9.48), i.e.,

$$A(\bar{x} - \Delta x) = b,$$

$$A^{\mathrm{T}}(\bar{y} - \Delta y) + (\bar{z} - \Delta z) = c,$$

$$(\bar{X} - \Delta X)(\bar{z} - \Delta z) = \mu e.$$

or equivalently,

$$A\Delta x = A\bar{x} - b, \tag{9.49}$$

$$A^{\mathrm{T}}\Delta y + \Delta z = A^{\mathrm{T}}\bar{y} + \bar{z} - c, \tag{9.50}$$

$$\bar{Z}\Delta x + \bar{X}\Delta z = \bar{X}\bar{z} + \Delta X \Delta z - \mu e. \tag{9.51}$$

where $\Delta X = diag(\Delta x_1, \ldots, \Delta x_n)$. It is known from the assumptions that the right-hand sides of (9.49) and (9.50) are equal to zero. Then, omitting the second order term $\Delta X \Delta z$ of the right-hand side of (9.51) gives a so-called "Newton equation":

$$A\Delta x = 0, \tag{9.52}$$

$$A^{\mathrm{T}}\Delta y + \Delta z = 0, \tag{9.53}$$

$$\bar{Z}\Delta x + \bar{X}\Delta z = \bar{X}\bar{z} - \mu e. \tag{9.54}$$

Introduce notation

$$D = \bar{Z}^{-1}\bar{X}. \tag{9.55}$$

Premultiplying (9.54) by $A\bar{Z}^{-1}$ and noting (9.52) leads to

$$AD\Delta z = AD\bar{z} - \mu A\bar{Z}^{-1}e,$$

and premultiplying (9.53) by $A\bar{Z}^{-1}\bar{X}$ gives

$$ADA^{\mathrm{T}}\Delta y + AD\Delta z = 0.$$

From the preceding two equations, it follows that

$$\Delta y = -(ADA^{\mathrm{T}})^{-1}A(D\bar{z} - \mu\bar{Z}^{-1}e). \tag{9.56}$$

By (9.53) and (9.54), on the other hand, it holds that

$$\Delta z = -A^{\mathrm{T}} \Delta y,$$
$$\Delta x = D(\bar{z} - \Delta z) - \mu \bar{Z}^{-1} e.$$

The preceding three expressions together determine a *Newton direction*.

Considering that the Newton point, determined along the direction by taking stepsize 1 probably violates the nonnegative conditions, instead the new iterate is defined as

$$\hat{x} = \bar{x} - \alpha \Delta x, \quad \hat{y} = \bar{y} - \alpha \Delta y, \quad \hat{z} = \bar{z} - \alpha \Delta z, \tag{9.57}$$

with the stepsize α determined by

$$\alpha = \lambda \min\{\alpha_p, \alpha_d\}, \tag{9.58}$$

where $\lambda \in (0, 1)$ (as a long stepsize performs better than a short stepsize, $\lambda = 0.99995$ might be taken), and

$$\alpha_p = \min\{\bar{x}_j / \Delta x_j \mid \Delta x_j > 0, \ \ j = 1, \cdots, n\},$$
$$\alpha_d = \min\{\bar{z}_j / \Delta z_j \mid \Delta z_j > 0, \quad z_j > 0, \ j = 1, \cdots, n\}. \tag{9.59}$$

Note that the new interior point satisfies conditions (9.46) and (9.47), but only approximately satisfies (9.48).

Since the iterate is primal and dual feasible, the iteration process can be terminated when the dual gap becomes sufficiently small, e.g., by using the following criterion:

$$\bar{x}^{\mathrm{T}} \bar{z} / (1 + |b^{\mathrm{T}} \bar{y}|) < \epsilon,$$

where $0 < \epsilon \ll 1$ is a precision tolerance. The end iterate is then regarded as an approximate optimal solution,

As a key factor, the barrier parameter μ affects the behavior of the algorithm greatly, as it determines the magnitude of the "centripetal force", compared with that of the "pushing descent force". It turns out that iterates too close to the central path are unfavorable, and degrade computational efficiency. For each given μ, it is only needed to go by one step along the Newton direction. As the bigger the μ is, the more an iterate bounced back, so it is necessary to decrease μ gradually as iterates approach the optimal solution. It is difficult to find an "ideal" way to do so, however. In this aspect, extensive research has been done, and a number of schemes have been suggested. Some schemes make the algorithm be of polynomial time complexity (e.g., Jansen et al. 1996; Kojima et al. 1989; Mizuno et al. 1993; Roos and Vial 1992 offered an unified analysis).

The first practicable scheme is found in McShane et al. (1989), which decreases μ monotonically according to the varying dual gap in solution process.

From (9.57) it is follows that

$$c^{\mathrm{T}}\hat{x} - b^{\mathrm{T}}\hat{y} = c^{\mathrm{T}}\bar{x} - b^{\mathrm{T}}\bar{y} - \delta,$$

So, the decrement of the dual gap is

$$\delta = \alpha(c^{\mathrm{T}}\Delta x - b^{\mathrm{T}}\Delta y). \tag{9.60}$$

It is known by $A(\bar{x} - \Delta x) = b$ that

$$b^{\mathrm{T}}(\bar{y} - \Delta y) = (\bar{x} - \Delta x)^{\mathrm{T}} A^{\mathrm{T}}(\bar{y} - \Delta y),$$

and by $A^{\mathrm{T}}(\bar{y} - \Delta y) + (\bar{z} - \Delta z) = c$ that

$$c^{\mathrm{T}}(\bar{x} - \Delta x) = (\bar{y} - \Delta y)^{\mathrm{T}} A(\bar{x} - \Delta x) + (\bar{z} - \Delta z)^{\mathrm{T}}(\bar{x} - \Delta x).$$

From the preceding two equalities, it follows that

$$c^{\mathrm{T}}(\bar{x} - \Delta x) - b^{\mathrm{T}}(\bar{y} - \Delta y) = (\bar{z} - \Delta z)^{\mathrm{T}}(\bar{x} - \Delta x) = \bar{z}^{\mathrm{T}}\bar{x} - \Delta z^{\mathrm{T}}\bar{x} - \bar{z}^{\mathrm{T}}\Delta x + \Delta z^{\mathrm{T}}\Delta x.$$

In addition, by (9.54) it holds that

$$\bar{z}^{\mathrm{T}}\Delta x + \bar{x}^{\mathrm{T}}\Delta z = \bar{x}^{\mathrm{T}}\bar{z} - n\mu.$$

Combining the preceding two equalities gives

$$c^{\mathrm{T}}(\bar{x} - \Delta x) - b^{\mathrm{T}}(\bar{y} - \Delta y) = n\mu + \Delta z^{\mathrm{T}}\Delta x,$$

which, by omitting the second order term, can be written

$$-(c^{\mathrm{T}}\Delta x - b^{\mathrm{T}}\Delta y) \approx -c^{\mathrm{T}}\bar{x} + b^{\mathrm{T}}\bar{y} + n\mu.$$

Substituting the preceding to (9.60) leads to

$$\delta \approx \alpha(c^{\mathrm{T}}\bar{x} - b^{\mathrm{T}}\bar{y} - n\mu).$$

In order for dual gap's decrement to be positive, therefore, μ should satisfy

$$\mu < (c^{\mathrm{T}}\bar{x} - b^{\mathrm{T}}\bar{y})/n,$$

at least. On the other hand, too big μ is not suitable, as may cause numerical instability. It might be well to take

$$\mu = (c^{\mathrm{T}}\bar{x} - b^{\mathrm{T}}\bar{y})/n^2.$$

The overall steps are summarized to the following algorithm.

Algorithm 9.4.1 (Primal-dual interior-point algorithm). Given $0 < \epsilon \ll 1$, $\lambda \in (0, 1)$. Initial: interior point $(\bar{x}, \bar{y}, \bar{z})$. This algorithm solves the standard LP problem.

1. Stop if $\bar{x}^{\mathrm{T}}\bar{z}/(1 + |b^{\mathrm{T}}\bar{y}|) < \epsilon$ (optimality achieved approximately).
2. Compute $\mu = (c^{\mathrm{T}}\bar{x} - b^{\mathrm{T}}\bar{y})/n^2$.
3. Solve system $(ADA^{\mathrm{T}})\Delta y = -A(D\bar{z} - \mu\bar{Z}^{-1}e)$.
4. Compute $\Delta z = -A^{\mathrm{T}}\Delta y,\ \Delta x = D(\bar{z} - \Delta z) - \mu\bar{Z}^{-1}e$.
5. Update $\bar{y},\ \bar{z},\ \bar{x}$ by (9.57), where α is determined by (9.58) and (9.59).
6. Go to step 1.

Speaking for itself, the so-called "interior points" so far are all feasible ones. As it is not easy to obtain such an initial one, the application of the primal-dual method has been restricted, and soon a simple variant of it had been developed.

9.4.2 *Infeasible Primal-Dual Method*

Terms "infeasible interior point" and "interior point" differ only from that the former does not necessarily satisfy primal and dual equality constraints. Infeasible primal-dual (interior-point) method is proposed by Lustig (1990) and Tanabe (1990), who modify the primal-dual (interior-point) method presented in the previous section, so that it can get itself started from an infeasible interior point. The basic idea is somehow like that behind the primal-dual simplex method (Sect. 7.1), but is realized in different framework.

Assume that the current iterate $(\bar{x}, \bar{y}, \bar{z})$ be an infeasible interior point, satisfying $\bar{x}, \bar{z} > 0$ but the residuals

$$r_p = A\bar{x} - b, \quad r_d = A^{\mathrm{T}}\bar{y} + \bar{z} - c$$

are not equal to 0, in general.

Omitting the second order term $\Delta X \Delta z$ from (9.49) to (9.51) gives the Newton equation

$$A\Delta x = r_p, \tag{9.61}$$

$$A^{\mathrm{T}}\Delta y + \Delta z = r_d, \tag{9.62}$$

$$\bar{Z}\Delta x + \bar{X}\Delta z = \bar{X}\bar{z} - \mu e. \tag{9.63}$$

Using notation D defined by (9.55), the solution to Newton equation can be written

$$\Delta y = (ADA^{\mathrm{T}})^{-1}(r_p - AD(\bar{z} - r_d) + \mu A\bar{Z}^{-1}e), \tag{9.64}$$

$$\Delta z = r_d - A^{\mathrm{T}}\Delta y, \tag{9.65}$$

$$\Delta x = D(\bar{z} - \Delta z) - \mu\bar{Z}^{-1}e. \tag{9.66}$$

In fact, the last two formulas can be derived from (9.62) and (9.63). In addition, premultiplying (9.63) by $A\bar{Z}^{-1}$ and noting (9.61) gives

$$A\bar{Z}^{-1}\bar{X}\Delta z = -r_p + A\bar{Z}^{-1}(\bar{X}\bar{z} - \mu e),$$

and premultiplying (9.62) by $A\bar{Z}^{-1}\bar{X}$ leads to

$$A\bar{Z}^{-1}\bar{X}A^{\mathrm{T}}\Delta y + A\bar{Z}^{-1}\bar{X}\Delta z = A\bar{Z}^{-1}\bar{X}r_d.$$

Then, (9.64) follows from the above two equations.

Except for different search directions, determination of the new iterate is the same as that for the primal-dual method. The infeasible primal-dual method also uses updating formula (9.57), where stepsize α is determined by (9.58)–(9.59), with $\lambda \in (0, 1)$ (usually, $\lambda = 0.99995$ is taken).

The barrier parameter, determined by Lustig et al. (1991), is

$$\mu = (c^{\mathrm{T}}\bar{x} - b^{\mathrm{T}}\bar{y} + M\gamma)/\phi(n). \tag{9.67}$$

where

$$\gamma = \|A\bar{x} - b\|/\ \|Ax^0 - b\| + \|A^{\mathrm{T}}\bar{y} + \bar{z} - c\|/\|A^{\mathrm{T}}y^0 + z^0 - c\|, \tag{9.68}$$

$$M = \xi\phi(n)\max\{\max_{j=1}^{n}|c_j|,\ \max_{i=1}^{m}|b_i|\}, \tag{9.69}$$

$$\phi(n) = \begin{cases} n^2, & \text{If } n \le 5{,}000, \\ n^{3/2}, & \text{If } n > 5{,}000. \end{cases} \tag{9.70}$$

An initial interior point (x^0, y^0, z^0) and ξ are determined by another algorithm (see discussions following Algorithm1 9.4.3).

The overall steps are put in the following algorithm.

Algorithm 9.4.2 (Infeasible primal-dual interior-point algorithm). Given $0 < \epsilon, \epsilon_p, \epsilon_d \ll 1$, $\lambda \in (0, 1)$. Initial:$(\bar{x}, \bar{y}, \bar{z})$ satisfying $\bar{x}, \bar{z} > 0$. This algorithm solves the standard LP problem.

1. Compute $r_p = A\bar{x} - b$, $r_d = A^{\mathrm{T}}\bar{y} + \bar{z} - c$.
2. Stop if $\bar{x}^{\mathrm{T}}\bar{z}/(1 + |b^{\mathrm{T}}\bar{y}|) < \epsilon$, $r_p < \epsilon_p$, $r_d < \epsilon_d$ (optimality achieved approximately).
3. Solve system $(ADA^{\mathrm{T}})\Delta y = r_p - AD(\bar{z} - r_d) + \mu A\bar{Z}^{-1}e$, where μ is determined by (9.67).
4. Compute $\Delta z = r_d - A^{\mathrm{T}}\Delta y$, $\Delta x = D(\bar{z} - \Delta z) - \mu\bar{Z}^{-1}e$.
5. Update $\bar{y}$, $\bar{z}$, $\bar{x}$ by (9.57), where α is determined by (9.58) and (9.59).
6. Go to step 1.

Theoretically, convergence of the preceding algorithm is not guaranteed, and there is no device to detect infeasibility or unboundedness either. Nevertheless, this algorithm performs well, and leads to a even better variant.

9.4.3 Predictor-Corrector Primal-Dual Method

The Newton equation (9.61)–(9.63), used in the infeasible primal-dual method, is yielded from (9.49)–(9.51) by omitting the second order term $\Delta X \Delta z$. Mehrotra's (1992) predictor-corrector primal-dual method introduces an additional correction step, involving the second order term, to improve the search direction (for simplicity, the prefix "infeasible" will be omitted).

It is noted that the left-hand side of (9.49)–(9.51) is linear in $\Delta x, \Delta y, \Delta z$, only the right-hand side involves the second order term $\Delta X \Delta z$. The system can therefore be regarded as a implicit one. Thus, one can determine a predictor solution $(\Delta x', \Delta y', \Delta z')$ to approximate its solution first, then substitute it to the right-hand side and solve the resulting system for a corrector solution.

Despite the Newton direction (9.64)–(9.66) is a clear choice for being predictor solution, Mehrotra uses the solution to the linear system below:

$$
\begin{aligned}
A\Delta x' &= r_p, \\
A^{\mathrm{T}}\Delta y' + \Delta z' &= r_d, \\
\bar{Z}\Delta x' + \bar{X}\Delta z' &= \bar{X}\bar{z},
\end{aligned}
$$

The only difference between the preceding and the Newton equation is that its right-hand side no longer involves term $-\mu e$. It can be viewed as one derived from the complementarity condition $Xz = 0$ rather than $Xs = \mu e$. So, its solution can be obtained easily from (9.64) to (9.66) by setting $\mu = 0$, i.e.,

$$\Delta y' = (ADA^{\mathrm{T}})^{-1}(r_p - AD(\bar{z} - r_d)), \tag{9.71}$$

$$\Delta z' = r_d - A^{\mathrm{T}}\Delta y', \tag{9.72}$$

$$\Delta x' = D(\bar{z} - \Delta z'). \tag{9.73}$$

Substituting $\Delta z'$ and $\Delta x'$ to the right-hand side of (9.51), we solve

$$
\begin{aligned}
A\Delta x &= r_p, \\
A^{\mathrm{T}}\Delta y + \Delta z &= r_d, \\
\bar{Z}\Delta x + \bar{X}\Delta z &= \bar{X}\bar{z} + \Delta X'\Delta z' - \mu e,
\end{aligned}
$$

to obtain a corrector solution. Only the right-hand sides of the preceding system and (9.61)–(9.63) differ, hence the derivation of the corrector solution is similar to that of (9.64)–(9.66), i.e.,

$$\Delta y = \Delta y' - (ADA^{\mathrm{T}})^{-1}(A\bar{Z}^{-1}\Delta X'\Delta z' - \mu A\bar{Z}^{-1}e), \tag{9.74}$$

$$\Delta z = r_d - A^{\mathrm{T}}\Delta y, \tag{9.75}$$

$$\Delta x = D(\bar{z} - \Delta z) + \bar{Z}^{-1}\Delta X'\Delta z' - \mu\bar{Z}^{-1}e. \tag{9.76}$$

The predictor solution may also be used to determine barrier parameter μ. Mehrotra (1992) uses the following formula:

$$\mu = (g'/\bar{x}^{\mathrm{T}}\bar{z})^2(g'/n), \tag{9.77}$$

where

$$\begin{aligned} g' &= (\bar{x} - \alpha'_p \Delta x')^{\mathrm{T}}(\bar{z} - \alpha'_d \Delta z'), \\ \alpha'_p &= 0.99995 \min\{\bar{x}_j/\Delta x'_j \mid \Delta x'_j > 0, j = 1, \cdots, n\}, \\ \alpha'_d &= 0.99995 \min\{\bar{z}_j/\Delta z'_j \mid \Delta z'_j > 0, j = 1, \cdots, n\}. \end{aligned}$$

It is noted that g' is a predicted dual gap, whereas $\bar{x}^{\mathrm{T}}\bar{z}$ is the current dual gap. Their ratio (less than 1) may be regarded as a predicted improvement at that iteration. So, a small μ is taken when the predicted improvement is significant, while a large μ is taken in the other case. This is relevant because small improvement implies that the tendency toward the "center" should be enhanced.

Mehrotra's work draws attention of the community immediately. In order to evaluate the predictor-corrector primal-dual method, compared with the pure infeasible primal-dual method, Lustig et al. (1992) conducted further numerical experiments, involving 86 Netlib standard test problems. Their numerical results indicated that in terms of iteration counts, the former outperformed the latter with 85 out of the 86 problems; and most importantly, in terms of CPU time, the former defeated the latter with 71 out of the 86, exceeding 82 %. Overall, it is well accepted that the predictor-corrector primal-dual method is superior to the infeasible primal-dual method. In order to avoid solving badly-conditioned problems and instability when iterates are close to an optimal solution, they modified the determination of the barrier parameter as follows:

Use (9.77) only when $\bar{x}^{\mathrm{T}}\bar{z} \geq 1$; otherwise, use

$$\mu = \bar{x}^{\mathrm{T}}\bar{z}/\phi(n), \tag{9.78}$$

where $\phi(n)$ is still defined by (9.70).

The overall steps are put in the following algorithm.

Algorithm 9.4.3 (Predictor-corrector primal-dual algorithm). Given $0 < \epsilon, \epsilon_p, \epsilon_d \ll 1$, $\lambda \in (0, 1)$. Initial:$(\bar{x}, \bar{y}, \bar{z})$ satisfying $\bar{x}, \bar{z} > 0$. This algorithm solves the standard LP problem.

1. Compute $r_p = A\bar{x} - b$, $r_d = A^{\mathrm{T}}\bar{y} + \bar{z} - c$.
2. Stop if $\bar{x}^{\mathrm{T}}\bar{z}/(1 + |b^{\mathrm{T}}\bar{y}|) < \epsilon$, $r_p < \epsilon_p$, $r_d < \epsilon_d$ (optimality achieved approximately).
3. Compute μ by (9.77) if $\bar{x}^{\mathrm{T}}\bar{z} \geq 1$; else, by (9.78).
4. Compute $\Delta z'$, $\Delta x'$ by (9.71)–(9.73).

5. Compute $\Delta y, \Delta z, \Delta x$ by (9.74)–(9.76).
6. Update $\bar{y}$, $\bar{z}$, $\bar{x}$ by (9.57), where α is determined by (9.58) and (9.59).
7. Go to step 1.

Now turn to how to obtain an initial point for infeasible primal-dual methods (including the predictor-corrector variant).

An initial infeasible point seems to be quite free to choose. As dispenses from satisfaction of constraint equalities, however, its quality affects algorithm's efficiency greatly. It is accepted that an initial point should be not only nearly primal and dual feasible, but also as close to the "center" as possible. It turns out that even it is close to an optimal solution, there is still troublesome with numerical difficulties if not properly centered. In view of this, Mehrotra (1992) suggests obtaining an initial point by solving a quadratic programming problem. Subsequently, Andersen et al. (1996) derive a variant, handling the following convex quadratic programming problem:

$$\begin{array}{ll} \min & c^{\mathrm{T}}x + (\eta/2)x^{\mathrm{T}}x, \\ \text{s.t.} & Ax = b, \end{array}$$

where $\eta > 0$ is a weight factor. The Lagrange function of the preceding problem is

$$L(x, w) = c^{\mathrm{T}}x + (\eta/2)x^{\mathrm{T}}x - w^{\mathrm{T}}(Ax - b).$$

An explicit expression of the solution to $\nabla L(x, w) = 0$ is

$$\bar{x} = (1/\eta)(A^{\mathrm{T}}(AA^{\mathrm{T}})^{-1}Ac - c).$$

To be an initial solution, $\bar{x}$'s components less than some positive number δ is modified to δ (e.g. $\delta = 1$). An initial dual solution $(\bar{y}, \bar{z})$ is determined similarly.

Despite it works well in practice, the above heuristic approach is scaling dependent, and does not certainly generate a well centered initial point. For Algorithm 9.4.3, moreover, it is assumed that there exists an optimal solution since there is no reliable approach to detect infeasibility or unboundedness, as is a serious drawback to practice.

9.4.4 *Homogeneous and Self-Dual Method*

The so-called "homogeneous and self-dual method" might be the most efficient interior-point method at present. It overcomes the two drawbacks of the predictor-corrector primal-dual algorithm, presented in the previous subsection, by a so-called "homogeneous and self-dual model". The model is based on "skew-symmetric self-dual artificial model", introduced by Ye et al. (1994), though it was subsequently found that Goldman and Tucker (1956b) and Tucker (1956) studied such type of models earlier. We will focus on main elements of it here in this subsection; for more details, the reader is referred to related literatures (e.g., Jansen et al. 1994; Xu and Ye 1995; Xu et al. 1996).

In stead of the original standard problem (4.1), we begin with the self-dual problem (4.9), written as

$$\begin{array}{llll} \min & c^T x - b^T y, & & \\ \text{s.t.} & \qquad Ax - b & = 0, & x \geq 0, \\ & -A^T y \qquad + c - z & = 0, & z \geq 0, \end{array} \tag{9.79}$$

which is equivalent to the pair of primal and dual standard problems (4.1) and (4.2), in the sense that they share the same optimal solutions. We attempt to modify the preceding homogeneous and self-dual problem to a better "trivial" one, whose solution is infeasible to the original.

As the expected iterates will satisfy the nonnegative but not the equality constraints, we add

$$b^T y - c^T x - \sigma = 0, \qquad \sigma \geq 0,$$

to constraints and put the preceding problem to the following problem in variables (x, y, z, σ, τ):

$$\begin{array}{lll} \min & 0, & \\ \text{s.t.} & \qquad Ax - b\tau & = 0, \\ & -A^T y \qquad + c\tau - z & = 0, \\ & b^T y - c^T x \qquad - \sigma & = 0, \\ & \qquad x, z, \sigma, \tau \geq 0, & \end{array} \tag{9.80}$$

which is related to (9.79) by σ, τ. According to Theorem 4.3.2, there exists a strictly complementary optimal solution, say $(x^*, y^*, z^*, \tau^*, \sigma^*)$, to (9.80), which can be shown to satisfy the following systems:

$$\begin{array}{ll} Xz = 0 & \tau\sigma = 0, \\ x + z > 0 & \tau + \sigma > 0, \end{array} \tag{9.81}$$

The following can be shown:

(i) Case $\tau^* > 0$. Then (x^*/τ^*) and $(y^*/\tau^*, z^*/\tau^*)$ are a pair of strictly complementary optimal solutions to (4.1) and (4.2).
(ii) Case $\tau^* = 0$. This implies $\sigma^* > 0$, and hence $b^T y^* - c^T x^* > 0$. Thus, at least one of $b^T y^* > 0$ and $-c^T x^* > 0$ holds: if $b^T y^* > 0$, the primal problem (4.1)

is infeasible; if $c^T x^* < 0$, the dual problem (4.2) is infeasible, and if $b^T y^* > 0$ and $c^T x^* < 0$, then the pair problems are both infeasible.

The wanted strictly complementary optimal solution to (9.79) can be achieved in the limit via solving a series of "trivial" problems, modified from (9.80).

To this end, introduce the feasibility and the average complementarity residual functions:

$$\begin{aligned} r_p &= b\tau - Ax, \\ r_d &= A^T y - c\tau + z, \\ r_g &= c^T x - b^T y + \sigma, \\ \mu &= (x^T z + \tau\sigma)/(n+1). \end{aligned} \tag{9.82}$$

Denote by "$\bar{\bullet}$" quantities at a current iterate $(\bar{y}, \bar{x} > 0, \bar{\tau} > 0, \bar{z} > 0, \bar{\sigma} > 0)$. Problem (9.80) is modified by putting the complementarity residual functions and logarithmic barrier terms in its objective, and laying current residuals in the place of the right-hand side of the constraints. As in the primal-dual method, the logarithmic barrier terms are for centering and keeping iterate's positiveness. The resulting problem is then

$$\begin{array}{lllll} \min & z^T x + \sigma\tau - \lambda\mu^{(0)} \sum_{j=1}^{n} (\ln x_j + \ln z_j) - \lambda\mu^{(0)} (\ln \tau + \ln \sigma), & & & \\ \text{s.t.} & \qquad\qquad Ax - b\tau & & = -\lambda \bar{r}_p, & \\ & -A^T y \qquad\qquad + c\tau - z & & = -\lambda \bar{r}_d, & \\ & b^T y - c^T x \qquad\qquad\qquad - \sigma & & = -\lambda \bar{r}_g, & \end{array} \tag{9.83}$$

where $\lambda \in [0, 1]$ is called the *path parameter*. It is note that the current iterate is feasible to (9.83).

The first optimality conditions for (9.83) is

$$\begin{array}{rl} Ax - b\tau \qquad\quad & = \lambda \bar{r}_p, \\ -A^T y \qquad + c\tau - z \qquad & = \lambda \bar{r}_d, \\ b^T y - c^T x \qquad\qquad - \sigma & = \lambda \bar{r}_g, \\ Xz \qquad & = \lambda \bar{\mu} e, \\ \tau\sigma & = \lambda \bar{\mu}, \\ x, z, \sigma, \tau > 0, & \end{array} \tag{9.84}$$

Similarly to the primal-dual method, the search direction is determined by applying Newtons's method to (9.84). Denoting by $\gamma \in [0, 1]$ a predetermined reduction rate of the path parameter, the resulting Newton equation for the search direction $(\Delta x, \Delta y, \Delta z, \Delta\sigma, \Delta\tau)$ can be written

$$\begin{pmatrix} & A & -b & & \\ -A^{\mathrm{T}} & & c & -I & \\ b^T & -c^T & & & -1 \\ & \bar{Z} & & \bar{X} & \\ & & \bar{\sigma} & & \bar{\tau} \end{pmatrix} \begin{pmatrix} \Delta y \\ \Delta x \\ \Delta \tau \\ \Delta z \\ \Delta \sigma \end{pmatrix} = \begin{pmatrix} (1-\gamma)\bar{r}_p \\ (1-\gamma)\bar{r}_d \\ (1-\gamma)\bar{r}_g \\ \gamma\bar{\mu}e - \bar{X}\bar{z} \\ \gamma\bar{\mu} - \bar{\tau}\bar{\sigma} \end{pmatrix}.$$

Setting $\gamma = 1$ gives a pure centering direction, and setting $\gamma = 0$ gives an affine direction. Such a system is solved in each iteration, as constitutes the major computational task in the method.

Once the Newton direction has been computed, a stepsize is determined by using the same formulas as in the primal-dual method, so that the new iterates is strictly positive. Of course, a stopping criteria should be also set conformably.

9.5 Notes

It is known so far that the earliest interior-point algorithm is proposed by famous mathematician von Neumann in a talk with G.B. Dantzig in 1948 (Dantzig and Thapa 2003). It is designed, without convergence proof, to generate a feasible solution to a special LP problem (which turns out to be just Karmarkar standard problem). An elegant proof was given by Dantzig in a latter to Neumann.

Originally, the logarithmic barrier function was used by Frisch (1955) to design interior-point methods for solving nonlinear programming problems. Such a nonlinear tool was applied to LP only after publication of Karmarkar algorithm (1984), inspiring a great upsurge of interior-point methods. It was utilized by the path-following method to create a search direction, a combination of a descent direction and a centering direction. In fact, the other interior-point methods presented in this chapter can also be derived alternatively in such a way (Gill et al. 1985–1986; Shanno and Bagchi 1990; for some alternatives, see Pan et al. 2006b; Zhang and Pan 2008).

To explain, we begin with the nonlinear programming problem (9.43). The gradient and Hassian matrix of its objective function at the current interior point $\bar{x}$ are, respectively,

$$\nabla f(\bar{x}) = c - \mu \bar{X}^{-1} e, \quad \nabla^2 f(\bar{x}) = \mu \bar{X}^{-2}.$$

It is noted that $\nabla^2 f(\bar{x})$ is positive definite. Setting $x = \bar{x} - \Delta x$, the problem becomes one with respect to Δx, and the latter can be approximated by the following strict convex quadratic programming problem:

$$\begin{aligned} &\min \ f(\bar{x}) - \nabla f(\bar{x})^{\mathrm{T}} \Delta x + (1/2) \Delta x^{\mathrm{T}} \nabla^2 f(\bar{x}) \Delta x, \\ &\text{s.t.} \ \ A \Delta x = 0. \end{aligned}$$

Thus, what needs to do is to find the stationary point of the following Lagrange function:

$$L(\Delta x, y) = -\nabla f(\bar{x})^{\mathrm{T}}\Delta x + (1/2)\Delta x^{\mathrm{T}}\nabla^2 f(\bar{x})\Delta x - y^{\mathrm{T}} A\Delta x,$$

where y is the Lagrange multiplier vector, by solving system

$$\mu\bar{X}^{-2}\Delta x - A^{\mathrm{T}} y = c - \mu\bar{X}^{-1}e, \tag{9.85}$$

$$A\Delta x = 0. \tag{9.86}$$

It holds by (9.85) that

$$\Delta x = (1/\mu)(\bar{X}^2 A^{\mathrm{T}} y + \bar{X}^2 c - \mu\bar{X}e), \tag{9.87}$$

substituting which to (9.86) gives

$$y = -(A\bar{X}^2 A^{\mathrm{T}})^{-1} A\bar{X}(\bar{X}c - \mu e).$$

Then, substituting the preceding to (9.87) leads to

$$\Delta x = (1/\mu)\bar{X}P(\bar{X}c - \mu e), \tag{9.88}$$

where P is the projection matrix, defined by (9.26). $-\Delta x$ is the Newton direction of problem (9.43) at $\bar{x}$. It is seen by comparing with (9.27) that the vector is just a positive combination of $\bar{X}Pe$ and the search direction $-\bar{X}\Delta x' = -\bar{X}P\bar{X}c$, used by the primal affine algorithm.

Assume that the feasible region of the original problem is nonempty. Consider problem

$$\begin{aligned} \min \ & \sum_{j=1}^{n} \ln x_j, \\ \text{s.t.} \ & Ax = b, \end{aligned}$$

to which there is an unique optimal solution that can be regarded as the center of the feasible region. It is not difficult to show that $\bar{X}Pe$ is just the Newton direction of the preceding problem at $\bar{x}$. Therefore, the search direction, defined by (9.88), reflects a common effect of a "descent force" and a "centering force". The module of the barrier parameter determines the proportion of the two tendencies. Proper choices of the parameter and stepsize lead to algorithms of polynomial-time complexity (Gonzaga 1987). On the other hand, the search direction tends to the one, used by the primal affine algorithm, as μ tends to zero. The latter is not of polynomial-time complexity.

With the aid of the same logarithmic barrier function, it is possible to derive the same search direction, used by Karmarkar algorithm, and the resulting algorithm is actually equivalent to Karmarkar algorithm if a proper barrier parameter and stepsize are taken.

As for the dual problem, construct the following nonlinear programming problem:

$$\max\ g(y) = b^{\mathrm{T}} y + \mu \sum_{j=1}^{n} \ln(c_j - a_j^{\mathrm{T}} y),$$

where a_j denotes j-indexed column of A. Introduce

$$\bar{z} = c - A^{\mathrm{T}} \bar{y}.$$

The gradient and Hassian matrix of the objective function at the current point $\bar{y}$ are respectively

$$\nabla g(\bar{y}) = b - \mu A \bar{Z}^{-1} e, \quad \nabla^2 g(\bar{y}) = -\mu A \bar{Z}^{-2} A^{\mathrm{T}}.$$

The Newton direction is then

$$\Delta y = -(A \bar{Z}^{-2} A^{\mathrm{T}})^{-1} (b/\mu + A \bar{Z}^{-1} e),$$

which is a positive combination of $-A\bar{Z}^{-1}e$ and the search direction, used by the dual affine algorithm, and the former is the Newton direction, somehow pointing to the neighborhood of the "center" of the dual feasible region of the following problem:

$$\max \sum_{j=1}^{n} \ln(c_j - a_j^{\mathrm{T}} y).$$

Another nonlinear tool applicable to interior-point method is ordinary differential equations (ODE). Let us bring up the affine algorithm. Its search direction

$$-X(I - XA^{\mathrm{T}}(AX^2A^{\mathrm{T}})^{-1}AX)Xc$$

actually determines a vector field for the feasible region. Taking it as the right-hand side, we construct the following autonomous (ODE) system:

$$\frac{\mathrm{d}x(t)}{\mathrm{d}t} = -X(I - XA^{\mathrm{T}}(AX^2A^{\mathrm{T}})^{-1}AX)Xc, \quad x(0) = \bar{x}, \tag{9.89}$$

where $\bar{x}$ is the initial interior point. To realize the dual affine algorithm, Adler et al. (1989) define an ODE trajectory, then use the power series numerical integration of first and second order to gain an optimal solution approximately. They reported remarkable numerical results.

ODE method is of general significance. For any linear or smooth nonlinear programming methods, an associated ODE system can be introduced similarly. Under mild conditions, there exists a trajectory to such a system, whose limit point is an optimal solution (Pan 1982, 1992a). More precisely, let $x(t), 0 \leq t < \gamma$ be the right-hand half of (9.89)'s trajectory, starting from $\bar{x}$, such that

$$x(t) \rightarrow x^* \qquad \text{as} \qquad t \rightarrow \gamma,$$

where x^* is the optimal solution. It is possible to solve it approximately by an available numerical integration method.

According to the author's knowledge, the earliest ODE method is proposed by Arrow and Hurwicz (1956) for solving equality constraint nonlinear programming problems. There were a number of related papers published subsequently (e.g., Abadie and Corpentier 1969; Botsaris 1974; Evtushenko 1974; Fiacco and Mccormick 1968; Tanabe 1977). A common belief seems that the computational work related to ODE is high, compared with standard methods, such as SQP, despite Brown and Bartholomew-Biggs (1987) reported very favorable numerical results with the former. Based on the author's positive experience with ODE (Pan 1982, 1992b), it is hard to understand why ODE methods fall into neglect in the academic community. After all, normal line searches are only of the first order, in certain sense. In fact, taking small stepsizes λ in Algorithm 9.2.1 amounts to carrying out numerical integration with the first order power series method or Euler method. In particular, realizing a curve search by the second order power series integration should be amenable to nonlinear programming problems. Based on this idea, recently Shi and Pan (2011) derived a "higher order" iteration scheme, and the according algorithms outperformed their conventional counterpart DFP, as well as BFGS significantly in computational experiments.

Finally, we conclude this chapter with the following remarks.

The interior-point method is so successful that some scholars believe its superiority to the simplex method for solving large-scale sparse LP problems. The homogeneous and self-dual method has already been implemented in some commercial codes, such as CPLEX (CPLEX ILOG 2007), and deemed as one of the most powerful methods. Such type of methods can be well parallelized, compared with the simplex method, parallelization of which is not very successful. Despite all these points, the interior-point method has been seriously restricted in applications due to its inherent weakness, as it provides only an approximate optimal solution, hence needs for additional "purification" in applications requiring an optimal vertex solution, and can not be "warmly" started, in contrast to the simplex methods, which can achieve optimality quickly if starts from a final basis reached in a previous solution process (see, e.g. Bixby and Saltzman 1992). Therefore, the interior-point method is unamenable to handling ILP problems, as solving such type of problems by current methods yields a large amount of correlated linear programs, which can only be left for the simplex method. As LP is mainly applied to handling integer problems, consequently it is hard to shake the domination of the simplex method (Bixby 1994, 2002; Nemhauser 1994).

At present, a race between the simplex method and the interior-point method seems to continue. No any single method is amenable to all LP problems. As a result, there are usually multiple options of methods in professional codes. To take advantages of both types, some scholars first apply the interior-point method, and then the simplex method (Bixby et al. 1992), though seeming to be too mechanically. In Chap. 24, we will present recently developed pivotal interior-point methods along this line in a natural manner.

Chapter 10
Integer Linear Programming (ILP)

The feasible region of the LP model is continuous in the sense that each variable is restricted to over a continuous interval. If variables are further restricted to integer values, it becomes an ILP model. As its feasible region consists of discrete points, ILP model differs from LP model essentially. Seeing that such type of models can be handled through a series of LP subprograms, and are so rich in practice as form a major application area of LP computation, this topic is highlighted in this chapter.

Branch-and-bound and cutting-plane methods have been principle tools for solving ILP models for more than half a century. Both of them deal with the models by solving a sequence of LP problems by simplex methods. It will be clear that only finitely many LP problems need to be solved by the former in principle, whereas the same is proved for a cutting-plane method (Gomory 1958). However, such finiteness is of theoretical value only. As well-known, the ILP problem is NP-complete (Cook 1971) and hence generally believed not to be solved in polynomial time. In practice, any of the methods alone is far from satisfaction, while the latter is even not competitive to the former in general, despite a variety of cuts suggested in the past. As a remedy, the so-called "branch-and-cut" scheme combining them is now widely used to solve ILP problems. We will not go into details here.

In the next section, basic concepts are introduced and a graphic approach to ILP is illustrated with a 2-dimensional instance. Then, the following two sections present typical branch-and-bound and cutting-plane methods, respectively. In the last two sections, new ILP methods, named *controlled- branch and controlled-cut*, are developed without any available computational results. An advanced realization of the former is delayed to Sect. 25.7.

It should be pointed out that the branch-bound and controlled-branch methods are also applicable to mixed ILP problems, in which the integrity requirement is only exerted on a part of variables. On the other hand, the presented cutting-plane and controlled-cut methods are only applicable to pure ILP problems, since they use so-called "fractional cuts".

P.-Q. PAN, *Linear Programming Computation*, DOI 10.1007/978-3-642-40754-3_10,

10.1 Introduction

Denote the largest integer no more than • by $\lfloor \bullet \rfloor$ and the smallest integer no less than • by $\lceil \bullet \rceil$.

A vector or point is called *integer* if all of its components are integer. The *feasible set* of integer points is defined by constraints of the ILP model.

It is assumed that coefficients of the objective function to be minimized are all integers. Therefore, objective values associated with integer solutions are integer.

Definition 10.1.1. LP relaxation is a LP problem, resulting from an ILP problem by dropping the integrity requirement on all or part of variables.

Clearly, it is optimal to the ILP problem if an optimal solution to its LP relaxation is integer. This is not the case in general, of course. Typically, an ILP method solves the LP relaxation by dropping the integrity requirement on all variables first. If the optimal solution is integer, then we are done. Otherwise, it solves a series of LP relaxations associated with ILP subprograms, generated by adding "valid cuts".

Definition 10.1.2. A valid cut is an inequality, satisfied by all feasible solutions to the ILP problem but violated by an optimal solution to the associated LP relaxation.

It is noted that a valid cut is closely related to an optimal solution to the LP relaxation. Adding valid cuts successively would yield iterates having more and more integer components, and eventually lead to an *integer* feasible solution. The Definition may be extended to a valid pair of cuts, used by the branch-and-bound method.

Definition 10.1.3. A *suspected-optimal* value is an integral lower bound on the optimal value of ILP problem.

Let f^* be the optimal value of the ILP problem and let $\bar{f}$ be the optimal value of the associated LP relaxation. It is clear that $f^* \geq \bar{f}$. Since f^* is integer, moreover, it holds that

$$f^* \geq \lceil \bar{f} \rceil \geq \bar{f}.$$

Therefore, $f^+ = \lceil \bar{f} \rceil$ is a suspected-optimal value. Suppose that $\bar{f}$ is integral but the associated solution is not. If the solution is an unique optimal solution, then $f^+ = \bar{f} + 1$ is a better suspected-optimal value, as it is a tighter integer lower bound on the optimal value of the ILP problem. We introduce the following special valid cut.

Definition 10.1.4. Objective cut is defined as inequality

$$c^T x \geq f^+,$$

where f^+ is a suspected-optimal value.

The use of an objective cut is the key to the proposed new methods. Such doing is motivated by the observation that f^* is often close to $\bar{f}$, or in other words, there would be a good chance for ILP problem's optimal value being equal to f^+.

10.1.1 Graphic Approach

The so-called "integer polyhedron" is defined as the convex hull of the feasible set of the ILP problem. It can be shown that its vertices are integer. Therefore, solving the ILP model is boiled down to solving the associated LP model over the integer polyhedron.

We demonstrate by solving the following 2-dimensional example graphically:

$$
\begin{array}{rrl}
 & \min & f = -2x_1 - 5x_2, \\
(ILP) & \text{s.t.} & 2x_1 + 3x_2 \leq 12, \\
 & & x_1 + x_2 \leq 5, \\
 & & x_2 \leq 3, \\
 & \text{integer} & x_1, \ x_2 \geq 0.
\end{array}
\tag{10.1}
$$

The associated LP relaxation is just problem (1.2), which was solved graphically in Sect. 2.3.1. As sketched in Fig. 10.1, the feasible region of the LP relaxation is the area, enclosed by polygon OABCD. The dots constitute the feasible set of (10.1), whose convex hull is the shaded area enclosed by polygon OAEFD. So, what should be done is just to apply the LP graphic approach to the problem over the shaded area.

In the figure, the contour line, corresponding to $-2x_1 - 5x_2 = 0$, of the objection function passes through the origin. The objective value decreases from 0 as it shifts parallel to the upper-right side. It is seen that the farthest possible position, with respect to the shaded area, is the dashed line corresponding to $-2x_1 - 5x_2 = -17$, passing through the integer vertex E. Thereby, the optimal solution to ILP problem (10.1) is $x_1^* = 1, x_2^* = 3, f^* = -17$.

As the determination of the integer polyhedron is only conceptional, however, the graphic approach is impracticable to general ILP problems.

10.2 Branch-and-Bound Method

Integer cut is such an inequality that the associated supperplane passes through an integer point in an axis and is perpendicular to the axis. For example, $x_j \leq 1$ or $x_j \geq 6$ for some j is an integer cut. If a LP relaxation with feasible region P has an optimal solution with fractional component $\bar{x}_j$, then integer cuts $x_j \leq \lfloor \bar{x}_j \rfloor$ and $x_j \geq \lceil \bar{x}_j \rceil$ are said to be a *valid pair* of cuts (associated with x_j). Adding such a pair of cuts leads to a pair of subprograms with

$$P \cap \{x \mid x_j \leq \lfloor \bar{x}_j \rfloor\} \qquad \text{and} \qquad P \cap \{x \mid x_j \geq \lceil \bar{x}_j \rceil\}$$

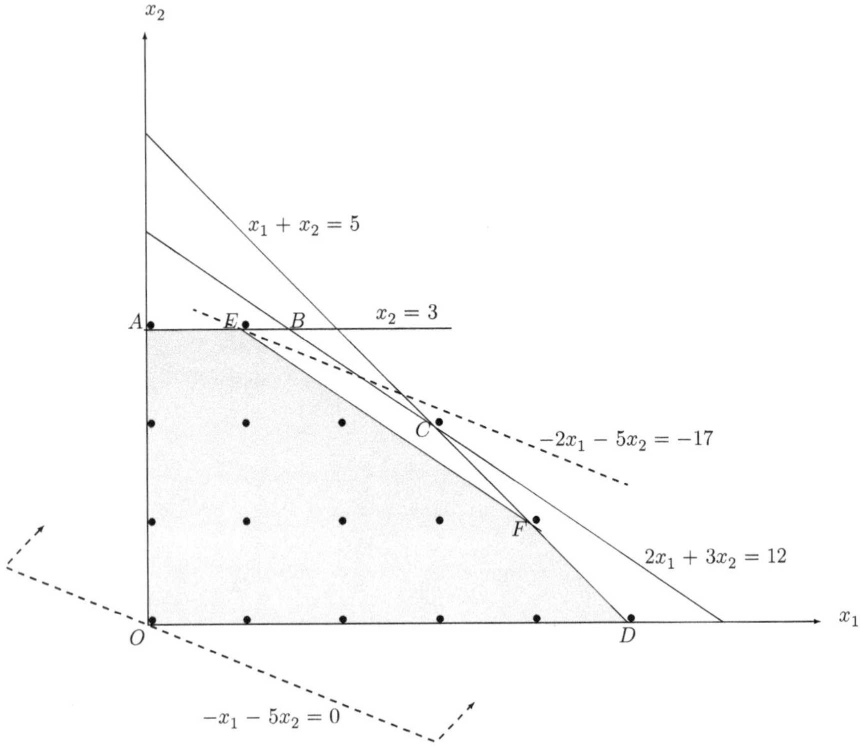

Fig. 10.1 Graphic solution to ILP problem (10.1)

as their feasible regions respectively, excluding the area $\lfloor \bar{x}_j \rfloor < x_j < \lceil \bar{x}_j \rceil$ while including all feasible integer points. The x_j is referred to as *cutting variable*.

The first branch-and-bound method was proposed by Land and Doig (1960). It adds valid pairs of cuts successively toward yielding an integer feasible solution, as described as follows:

Set an initial upper bound $\hat{f} = +\infty$ on the optimal value f^* of the ILP problem. Solve the associated LP relaxation. If the solution is integer, it is optimal to the ILP problem. In the other case, add a valid pair of cuts to decompose the ILP problem to a pair of ILP subprograms (branches): among their end solutions (if any), the one with less objective value may be declared to be optimal to the original. To this end, each ILP subprogram is treated in the same way as with the original, and so on. Such doing yields a number of branches.

A branch is deemed to be *fathomed*, meaning that it need not be branched any further because no any better feasible solution to the original ILP problem may be yielded, as in one of the following cases:

(i) The associated LP relaxation, and hence the ILP subprogram itself, is infeasible.

(ii) The solution to the associated LP relaxation is integer, and is hence feasible to the original ILP problem. If the associated objective value, say $\bar{f}$, is strictly less than the upper bound, update the latter by $\hat{f} = \bar{f}$.

(iii) It holds that $\lceil \bar{f} \rceil \geq \hat{f}$, where $\bar{f}$ is the optimal value of the associated LP relaxation.

The original ILP problem is solved when all branches of a so-called "enumerate tree" are fathomed.

There are "breadth-oriented" and "depth-oriented" strategies to develop an enumerate tree. Once a branch is found not fathomed, the former examines the nearest pending branch (in the same level), whereas the latter decomposes the branch further to two deeper branches, and examines one of them while leaving the other pending. The two strategies may be combined in some manner.

Favoured by many authors, the depth-oriented strategy always examines one of branches at a deeper level, while leaving the other branch pending, until reaching one fathomed; then, the pending branches are handled one by one, from the nearest (deepest) to the top level, until all branches are examined. This scheme would still boom a enumerate tree with unacceptable large number of branches. In fact, it could perform badly, or even fail to solve some (relatively small) ILP problems in practice. It has not been possible to find a scheme to guarantee a reasonably good enumerate tree in general.

The ILP problem (10.1) is used to show details by branch-and-bound method with depth-orientation.

(1) Set $\hat{f} = +\infty$. As was known in Sect. 2.3.1, the solution to the associated LP relaxation is $\bar{x}_1 = 1\frac{1}{2}, \bar{x}_2 = 3, \bar{f} = -18$. As component $\bar{x}_1 = 1\frac{1}{2}$ is noninteger, we add the valid pair of cuts, $x_1 \leq 1$ and $x_1 \geq 2$ to generate the following pair of ILP subprograms

$$
(ILP1)\quad \begin{array}{rl} \min & f = -2x_1 - 5x_2, \\ \text{s.t.} & 2x_1 + 3x_2 \leq 12, \\ & x_1 + x_2 \leq 5, \\ & x_2 \leq 3, \\ & x_1 \leq 1, \\ & \text{integer } x_1, x_2 \geq 0. \end{array}
\qquad
(ILP2)\quad \begin{array}{rl} \min & f = -2x_1 - 5x_2, \\ \text{s.t.} & 2x_1 + 3x_2 \leq 12, \\ & x_1 + x_2 \leq 5, \\ & x_2 \leq 3, \\ & x_1 \geq 2, \\ & \text{integer } x_1, x_2 \geq 0. \end{array}
$$

It is seen that the intersection of the feasible sets of the preceding pair of ILP subprograms is empty, whereas the union of them equals the feasible set of the original ILP problem. Figure 10.2 depicts the two feasible sets, where ILP1 and ILP2 indicate the feasible regions of the LP relaxations, associated with the subprograms. It is noted that the solution B to the LP relaxation is excluded from ILP1 $\cap$ ILP2.

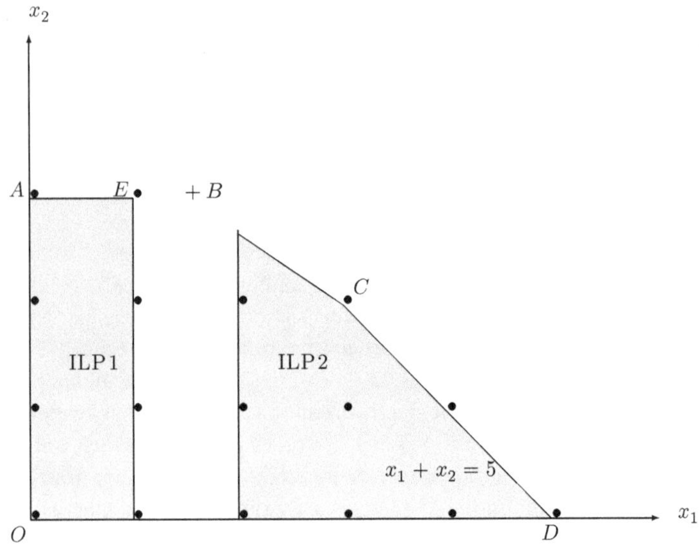

Fig. 10.2 Feasible sets for the two ILP subprograms

Thereby, solving ILP problem (10.1) is boiled down to solving subprograms (ILP1) and (ILP2), separately.

(2) Arbitrarily solve the LP relaxation associated with (ILP2) first (the dual simplex method is a proper tool – this will be clear later), and let (ILP1) pend. The solution yielded is $\bar{x}_1 = 2,\ \bar{x}_2 = 2\frac{2}{3},\ \bar{f} = -17\frac{1}{3}$. As $\bar{x}_2$ is noninteger, we use valid pair of cuts $x_2 \leq 2$ and $x_2 \geq 3$ to generate from (ILP2) a deeper level of two ILP subprograms: (ILP3) by adding $x_2 \leq 2$, and (ILP4) by $x_2 \geq 3$.

(3) Then, arbitrarily solve the LP relaxation associated with (ILP3) first, and let (ILP4) pend. The solution is $\bar{x}_1 = 3,\ \bar{x}_2 = 2,\ \bar{f} = -16$, which is integer, and hence is optimal to (ILP3) (and feasible to the original ILP problem). Thus, (ILP3) is fathomed. As $-16 < \hat{f}$, set $\bar{f} = -16$.

(4) Solve the LP relaxation associated with pending (ILP4). It is found infeasible. Therefore, (ILP4) itself is infeasible, and is hence fathomed.

The results of steps 3 and 4 together imply that (ILP2) is fathomed, with its optimal solution $\bar{x}_1 = 3,\ \bar{x}_2 = 2,\ \bar{f} = -16$.

(5) Solve the LP relaxation associated with pending (ILP1). The obtained solution is $\bar{x}_1 = 1,\ \bar{x}_2 = 3,\ \bar{f} = -17$, which is integer, and is hence optimal to (ILP1).

Finally, a comparison between the two optimal solutions to (ILP1) and (ILP2) indicates that the latter solution is optimal to the original ILP problem.

Figure 10.3 gives the enumerate tree, showing the sequence of LP relaxations (1)–(5) handled in the solution process.

The following realizes the according process via the simplex framework.

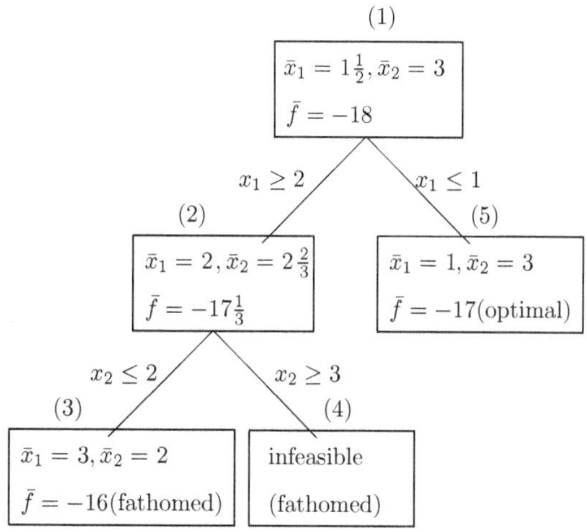

Fig. 10.3 The enumerate tree to ILP problem (10.1)

Example 10.2.1. Solve ILP problem (10.1) by the branch-and-bound method.

Answer It is solved in tableau form in accordance with the enumerate tree in Fig. 10.3. Set $\hat{f} = +\infty$. Convert the problem to the standard form by introduce slack variable x_3, x_4, x_5. The initial simplex tableau is

x_1	x_2	x_3	x_4	x_5	RHS
2	3	1			12
1	1		1		5
	1			1	3
−2	−5				

1. Call simplex Algorithm 3.2.1, yielding

x_1	x_2	x_3	x_4	x_5	RHS
1		1/2		−3/2	3/2
		−1/2	1	1/2	1/2
	1			1	3
		1		2	18

2. Insert the row, corresponding to $-x_1 + x_6 = -2$, to the preceding as the second bottom row.

x_1	x_2	x_3	x_4	x_5	x_6	RHS
1		1/2		−3/2		3/2
		−1/2	1	1/2		1/2
	1			1		3
−1					1	−2
		1		2		18

Convert the preceding to simplex tableau, and call dual simplex Algorithm 4.4.1, yielding

x_1	x_2	x_3	x_4	x_5	x_6	RHS
1					−1	2
		−1/3	1		1/3	1/3
	1	1/3			2/3	8/3
		−1/3		1	−2/3	1/3
		5/3			4/3	52/3

3. Insert the row corresponding to $x_2 + x_7 = 2$ to the preceding as the second bottom row.

x_1	x_2	x_3	x_4	x_5	x_6	x_7	RHS
1					−1		2
		−1/3	1		1/3		1/3
	1	1/3			2/3		8/3
		−1/3		1	−2/3		1/3
	1					1	2
		5/3			4/3		52/3

Convert the preceding to simplex tableau, and call dual simplex Algorithm 4.4.1, yielding

x_1	x_2	x_3	x_4	x_5	x_6	x_7	RHS
1		1/2				−3/2	3
		−1/2	1			1/2	
	1					1	2
				1		−1	1
		1/2			1	−3/2	1
		1				2	16

which gives a feasible solution to the original ILP problem (fathomed). Set upper bound $\hat{f} = -16$.

4. Insert the row, corresponding to $-x_2 + x_7 = -3$, to the end tableau of step 2 as the second bottom row.

x_1	x_2	x_3	x_4	x_5	x_6	x_7	RHS
1					−1		2
		−1/3	1		1/3		1/3
	1	1/3			2/3		8/3
		−1/3		1	−2/3		1/3
	−1					1	−3
		5/3			4/3		52/3

Convert the preceding to a simplex tableau, and call dual simplex Algorithm 4.4.1, yielding

x_1	x_2	x_3	x_4	x_5	x_6	x_7	RHS
1					−1		2
		−1/3	1		1/3		1/3
	1	1/3			2/3		8/3
		−1/3		1	−2/3		1/3
		1/3			2/3	1	−1/3
		5/3			4/3		52/3

which indicates that there is no feasible solution to the subprogram (fathomed).

5. Insert the row corresponding to $x_1 + x_6 = 1$ to the end tableau of step 1 as the second bottom row.

x_1	x_2	x_3	x_4	x_5	x_6	RHS
1		1/2		−3/2		3/2
		−1/2	1	1/2		1/2
	1			1		3
1					1	1
		1		2		18

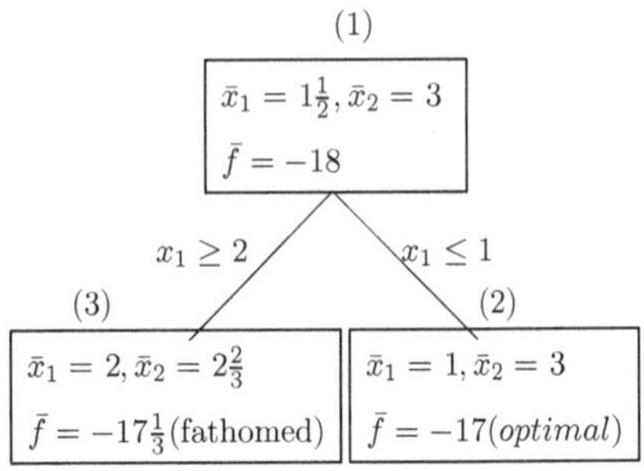

Fig. 10.4 Another enumerate tree to ILP problem (10.1)

Convert the preceding to a simplex tableau, and call dual simplex Algorithm 4.4.1, yielding

x_1	x_2	x_3	x_4	x_5	x_6	RHS
1					1	1
			1	−1	−1	1
	1			1		3
		1		−3	−2	1
				5	2	17

which give a feasible solution to the original ILP problem. A comparison between it and that of step 3 indicates that it is an optimal solution to the original.

The enumerate tree is not unique. Multiple noninteger components of a current solution implies multiple choices for a cutting variable (or valid pair of cuts), and there is two choices for an integer cut once a cutting variable is determined. Although any choice is eligible in principle, different choices may lead to very different efficiency. For instance, first handling subprogram (ILP1) rather then (ILP2) leads to another enumerate tree with less branches (see Fig. 10.4). The solution $\bar{x}_1 = 1, \bar{x}_2 = 3, \bar{f} = -17$ to the LP relaxation associated with (ILP1) is integer, and hence (ILP1) is fathomed with setting upper bound $\hat{f} = -17$. On the other hand, the solution to the LP relaxation associated with (ILP2) is reached at $\bar{x}_1 = 2, \bar{x}_2 = 2\frac{2}{3}, \bar{f} = -17\frac{1}{3}$. As the associated $\ulcorner \bar{f} \urcorner \geq \hat{f}$, (ILP2) is also fathomed. Therefore, it is asserted that $\bar{x}_1 = 1, \bar{x}_2 = 3, \bar{f} = -17$ is an optimal solution to the original ILP problem.

Consequently, the ILP problem (10.1) is solved by three calls for LP solver, while solved by five calls previously, as is a dramatic difference in efficiency. This theme will be explored in conjunction with the controlled-branch method (Sect. 10.4).

10.3 Cutting-Plane Method

The cutting-plane method presented in this section does not decompose the ILP problem, but solves a series of LP relaxations associated with modifications of the ILP problem. The modifications are made by successively adding valid cuts until reaching an optimal solution to the original ILP problem.

So, the idea of the method is quite simple. The problem is how to generate good cuts. Extensive research has been done in this respect. In this section, only so-called "fractional" cuts are generated from constraints.

Let $B = \{j_1, \cdots, j_m\}$ and $N = A \backslash B$ be optimal basis and nonbasis of the LP relaxation, respectively. Assume that the associated basic optimal solution is noninteger. Without loss of generality, consider the equation corresponding to ith row of the optimal simplex tableau, i.e.,

$$x_{j_i} + \Sigma_{j \in N} \bar{a}_{i,j} x_j = \bar{b}_i, \tag{10.2}$$

where $\bar{b}_i > 0$ is noninteger. Introduce following quantities:

$$0 \leq \alpha_{i,j} = \bar{a}_{i,j} - \lfloor \bar{a}_{i,j} \rfloor, \qquad 0 < \beta_i = \bar{b}_i - \lfloor \bar{b}_i \rfloor < 1,$$

by which, (10.2) can be written

$$x_{j_i} + \Sigma_{j \in N} (\lfloor \bar{a}_{i,j} \rfloor + \alpha_{i,j}) x_j = \lfloor \bar{b}_i \rfloor + \beta_i > 0.$$

Converting the preceding by moving the terms with integer coefficients to the left-hand side and those with fractions to the right-hand side gives

$$x_{j_i} + \Sigma_{j \in N} (\lfloor \bar{a}_{i,j} \rfloor) x_j - \lfloor \bar{b}i \rfloor = \beta_i - \Sigma_{j \in N} \alpha_{i,j} x_j.$$

For any integer feasible solution x, the left-hand side of the preceding equation is an integer, hence so is its right-hand side. In addition, since

$$\alpha_{i,j} \geq 0, \quad x_j \geq 0, \qquad \forall j \in N,$$

the right-hand is no more than $0 < \beta_i < 1$. Therefore, both sides must be an integer less than and equal to 0, i.e.,

$$\text{integer} \quad \beta_i - \Sigma_{j \in N} \alpha_{i,j} x_j \leq 0. \tag{10.3}$$

The associated basic optimal solution does not satisfy (10.3). In fact, if it is substituted to (10.3), the left-hand side becomes $\beta_i > 0$ since all its nonbasic components are zero. Thus, (10.3) is a valid cut, as, in addition, it does not exclude all integer feasible solutions.

The ILP problem is modified by adding such a cut, and the associated LP relaxation is solved then. If the yielded solution is still noninteger, a new valid cut is determined, and added to the modified ILP problem, and so on until a yielded solution is integer, and is hence optimal to the original ILP problem, or infeasibility is detected.

Example 10.3.1. Solve ILP problem (10.4) by the cutting-plane method.

Answer Call simplex Algorithm 3.2.1, yielding

x_1	x_2	x_3	x_4	RHS
1		1/8	−3/8	1/4
	1	1/8	5/8	25/4
		11/8	7/8	131/4

Arbitrarily take the first row with noninteger right-hand side. The corresponding equation is

$$x_1 + 1/8x_3 - 3/8x_4 = 1/4,$$

or equivalently

$$x_1 + 1/8x_3 + (-1 + 5/8)x_4 = 1/4.$$

The preceding can be written

$$x_1 - x_4 = 1/4 - 1/8x_3 - 5/8x_4,$$

from which the valid cut follows, i.e.,

$$1/4 - 1/8x_3 - 5/8x_4 \leq 0.$$

Insert the row corresponding to $-1/8x_3 - 5/8x_4 + x_5 = -1/4$ to the preceding tableau as the second bottom row, giving

x_1	x_2	x_3	x_4	x_5	RHS
1		1/8	−3/8		1/4
	1	1/8	5/8		25/4
		−1/8	−5/8	1	−1/4
		11/8	7/8		131/4

which is a dual feasible simplex tableau. Call the dual simplex Algorithm 4.4.1, yielding

x_1	x_2	x_3	x_4	x_5	RHS
1		1/5		−3/5	2/5
	1	0		1	6
		1/5	1	−8/5	2/5
		6/5		7/5	162/5

Arbitrarily take the third row with noninteger right-hand side. The corresponding equation is

$$1/5x_3 + x_4 - 8/5x_5 = 2/5,$$

or equivalently

$$1/5x_3 + x_4 + (-2 + 2/5)x_5 = 2/5.$$

The preceding can be written

$$x_4 - 2x_5 = 2/5 - 1/5x_3 - 2/5x_5,$$

from which the valid cut follows, i.e.,

$$2/5 - 1/5x_3 - 2/5x_5 \leq 0.$$

Insert the row corresponding to $-1/5x_3 - 2/5x_5 + x_6 = -2/5$ to the preceding tableau as the second bottom row, giving

x_1	x_2	x_3	x_4	x_5	x_6	RHS
1		1/5		−3/5		2/5
	1	0		1		6
		1/5	1	−8/5		2/5
		−1/5		−2/5	1	−2/5
		6/5		7/5		162/5

which is a dual feasible simplex tableau. Call the dual simplex Algorithm 4.4.1, yielding

x_1	x_2	x_3	x_4	x_5	x_6	RHS
1		1/2			−3/2	1
	1	−1/2			5/2	5
		1	1		−4	2
		1/2		1	−5/2	1
		1/2			7/2	31

which give an integer solution, an optimal solution to the original ILP problem:

$$\bar{x}_1 = 1, \quad \bar{x}_2 = 5, \quad \bar{f} = 31.$$

If there are multiple noninteger right-hand side of the optimal tableau to the LP relaxation, so are there multiple choices for constructing a valid cut. Although we did arbitrarily, it seems to be reasonable to form the cut that cuts deepest into the feasible region of the LP relaxation.

Let I be the set of row index corresponding to noninteger right-hand side. For any $i \in I$,

$$\beta_i / \sqrt{\Sigma_{j \in N}\, \alpha_{i,j}^2}$$

is the Euclidean distance from the associated solution to the LP relaxation to the boundary of the area defined by (10.3) (see Sect. 2.1). Unfortunately, the associated computations are cumbersome. Instead, we may use residuals approximately, as leads to determining row index i' such that

$$i' \in \arg\max_{i \in I} \beta_i .$$

10.4 Controlled-Branch Method

As was mentioned, the behavior of the branch-and-bound method is far from satisfaction, although current commercial or professional ILP codes are all rooted in it. Initially appeared unrelated and promising, the "additive" method proposed by Balas (1965) turned out to be a special case of the branch-and-bound method. As an endeavor to make progress, the method developed in this section uses the objective cut $c^T x \geq f^+$ to control the way in booming an enumerate tree. Recall that the objective cut is a valid cut, because adding it does not affect the feasible set of the ILP problem but cutting away the solution to the associated LP relaxation.

Assume that the initial LP relaxation of the ILP problem is solved, and a suspected-optimal value f^+ is determined (see the paragraph after Definition 10.1.3). Adding cut $c^T x \geq f^+$ leads to a *modified* ILP problem, whose LP relaxation has a nonempty optimal set (see Sects. 2.3 and 25.2), with objective value f^+. The set is called *valid set*, and each point in it is called *valid point*. Since an optimal solution to the initial LP relaxation is a valid point, it is optimal to the ILP problem if it is integer.

Assume now that it is noninteger. The method determines, and adds a valid pair of integer cuts to decompose the LP relaxation to two branches. Then, it solves one of them, while let the other pend. If the obtained optimal solution, if any, is a valid but noninteger point, it adds another pair of integer cuts to decompose the branch to two deeper branches, and so on, until meeting one of the following cases:

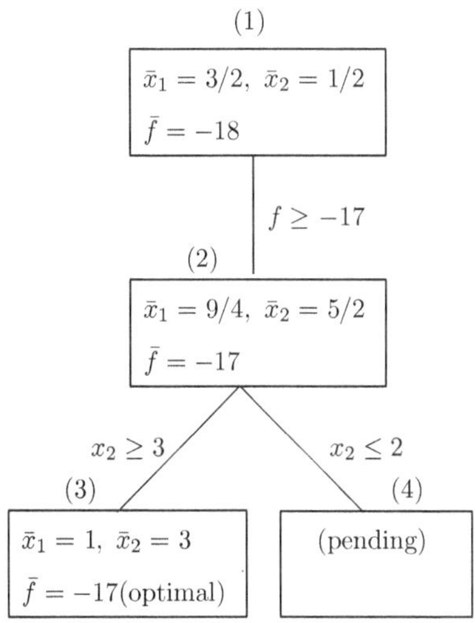

Fig. 10.5 The controlled enumerate tree for ILP problem (10.1)

(i) The optimal solution to the branch is found valid and integer, and hence is optimal to the original ILP problem.
(ii) The current branch is infeasible, as deemed fathomed.
(iii) The optimal solution to the branch is not valid, i.e., the associated objective value is strictly higher than f^+. This case is deemed "stalled", meaning that it is not branched further at the moment.

Then the method solves the pending branches one by one, from the nearest (deepest) to the top level, until all branches are examined if case (i) does not occur. This implies that there exists no valid point with the f^+. If all branches are infeasible, so is the ILP problem. Otherwise, determine the solution associated with the smallest objective value among minimums of the stalled branches. If it is integer, the solution is optimal to the original ILP problem; otherwise, the suspected-optimal value f^+ is updated in accordance with the objective value, and the new objective cut is formed, and added to the associated branch, and so on, until an optimal solution to the original ILP problem is found, or infeasibility is detected.

As for how to select a cutting variable and an integer cut to branch deeper, our scheme is based on the magnitude of their contributions to the optima value, e.g, by selecting the cutting variable x_q such that $q \in \arg\max_{j \in J} |c_j \bar{x}_j|$, where J is index set corresponding to noninteger components of the current solution.

Figure 10.5 gives the enumerate tree associated with ILP problem (10.1):

(1) The LP relaxation of the ILP problem is solved.
(2) From the optimal simplex tableau, it is known that its dual optimal solution is nondegenerate. According to Corollary 25.2.1, therefore, $\bar{x}_1 = 3/2,\ \bar{x}_2 = 1/2,\ \bar{f} = -18$ is the only optimal solution to the LP relaxation, as implies that $f^* \geq -17$. Set $f^+ = -17$ to form the initial objective cut, and add it. The solution to the resulting LP relaxation is $\bar{x}_1 = 9/4,\ \bar{x}_2 = 5/2,\ \bar{f} = -17$.

(3) As $|(-2) \times (9/4)| = 4.5 < |(-5) \times (5/2)| = 12.5$, we select x_2 as the cutting variable, and arbitrarily add $x_2 \geq 3$. The solution to the resulting LP relaxation is an integer valid point, and is hence optimal to the original ILP problem.
(4) Let the subprogram related to $x_2 \leq 2$ pend.

The following demonstrates the use of simplex methods in accordance with the numerate tree in Fig. 10.5.

Example 10.4.1. Solve ILP problem (10.1) by the controlled-branch method.

Answer The optimal simplex tableau of the LP relaxation associated with the (10.1) is cited from Example 10.2.1, i.e.,

(1)

x_1	x_2	x_3	x_4	x_5	RHS
1		1/2		−3/2	3/2
		−1/2	1	1/2	1/2
	1			1	3
		1		2	18

(2) Add the row corresponding to objective cut $2x_1+5x_2+x_6 = 17$ to the preceding as its second bottom row:

x_1	x_2	x_3	x_4	x_5	x_6	RHS
1		1/2		−3/2		3/2
		−1/2	1	1/2		1/2
	1			1		3
2	5				1	17
		1		2		18

Call dual simplex Algorithm 4.4.1, yielding

x_1	x_2	x_3	x_4	x_5	x_6	RHS
1		5/4			−3/4	9/4
		−3/4	1		1/4	1/4
	1	−1/2			1/2	5/2
		1/2		1	−1/2	1/2
		0			1	17

(3) Add $-x_2 + x_7 = -3$ to the preceding, giving

x_1	x_2	x_3	x_4	x_5	x_6	x_7	RHS
1		5/4			−3/4		9/4
		−3/4	1		1/4		1/4
	1	−1/2			1/2		5/2
		1/2		1	−1/2		1/2
	−1					1	−3
		0			1		17

Call dual simplex Algorithm 4.4.1, yielding

x_1	x_2	x_3	x_4	x_5	x_6	x_7	RHS
1					1/2	5/2	1
			1		−1/2	−3/2	1
	1					−1	3
				1		1	
		1			−1	−2	1
					1		17

which gives an integer valid point, and hence an optimal solution: $\bar{x}_1 = 1$, $\bar{x}_2 = 3$, $\bar{f} = -17$.

In the preceding Example, an optimal solution to the ILP problem is found in the first valid set. This is not the case with the following ILP problem:

$$\begin{array}{ll} \min & f = -6x_1 - 5x_2, \\ \text{s.t.} & 5x_1 + 3x_2 + x_3 \qquad = 20, \\ & -x_1 + x_2 \qquad + x_4 = \ \ 6, \\ & \text{integer } x_j \geq 0, \ j = 1, \cdots, 4. \end{array} \tag{10.4}$$

Figure 10.6 gives the enumerate tree created by the controlled-branch method, as explained as follows:

(1) The optimal solution to the associated LP relaxation is noninteger. (2) The solution to the LP relaxation, associated with the modification by adding objective cut $f \geq -32$ is valid but noninteger. (3) As $|(-6) \times (4/7)| < |(-5) \times (5\frac{5}{7})|$, cut $x_2 \geq 6$ is added, yielding a valid but noninteger point, with the branch generated by adding $x_2 \leq 5$ pending. (4) The further modification by adding cut $x_1 \geq 1$ is infeasible (fathomed). (5) Adding cut $x1 \leq 0$ to the ILP subprogram in (3), yielding an invalid solution (stalled). (6) Adding cut $x_2 \leq 5$ to the ILP subprogram in (2), yielding an invalid solution (stalled).

The yielded solution with the smallest objective value among the stalled branches is that from (6), i.e.,

$$\bar{x}_1 = 1, \ \bar{x}_2 = 5, \ \bar{f} = -31,$$

which is integer, and is hence optimal to the original ILP problem.

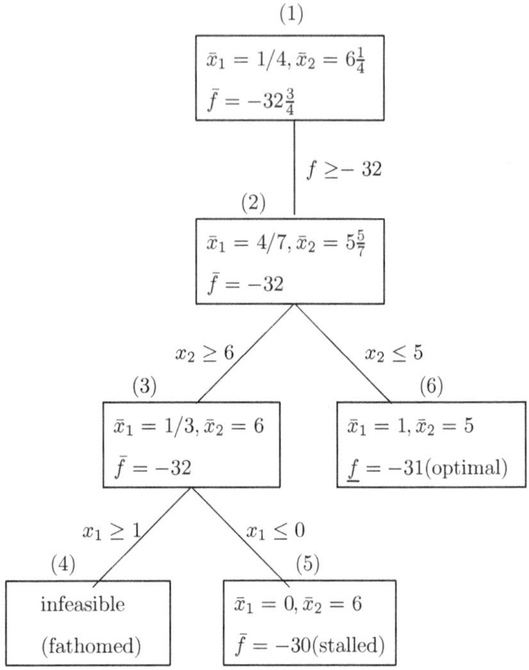

Fig. 10.6 The controlled enumerate tree for ILP problem (10.4)

Once a cutting variable is determined, what to do next is to select an integer cut from the valid pair of cuts, as is crucial to the method's efficiency. Although we took an arbitrary cut previously, it might be clever to select according to some plausible criteria.

An alternative criteria is to select one that cuts deepest into the feasible region of the LP relaxation. Assume that J is the index set of noninteger components of the associated solution $\bar{x}$. For any $j \in J$, $\alpha_j = \bar{x}_j - \llcorner \bar{x}_j \lrcorner$ is the distance from $\bar{x}$ to the boundary of $\{x \mid x_j \leq \llcorner \bar{x}_j \lrcorner\}$, and $\beta_j = \ulcorner \bar{x}_j \urcorner - \bar{x}_j$ the distance from $\bar{x}$ to the boundary of $x_j \geq \ulcorner \bar{x}_j \urcorner$ (see Sect. 2.1). Determine j' such that

$$\delta_{j'} = \max\{\max_{j \in J} \alpha_j, \max_{j \in J} \beta_j\}.$$

Then take $x_{j'}$ as the cutting variable. If $\delta_{j'} = \alpha_{j'}$, add $x_{j'} \leq \llcorner \bar{x}_{j'} \lrcorner$; otherwise, add cut $x_{j'} \geq \ulcorner \bar{x}_{j'} \urcorner$.

Based on the most-obtuse-angle heuristics, a preferable scheme is to select such an integer cut that its normal direction (pointing to the interior of the half space) forms the most obtuse angle with the negative objective gradient. More precisely, it determines index $j' = \arg\max_{j \in J} |c_j|$, and adds $x_{j'} \leq \llcorner \bar{x}_{j'} \lrcorner$ if $c_j < 0$, and $x_{j'} \geq \ulcorner \bar{x}_{j'} \urcorner$ if $c_j > 0$; in case of a tie, turn to the deepest cutting or the largest $\bar{x}_j$ in module.

10.5 Controlled-Cut Method

Assume that the optimal solution to the LP relaxation of the ILP problem is noninteger and the suspected-optimal value is f^+.

As in the controlled-branch method, the controlled-cut method firstly adds the objective cut $c^T x \geq f^+$, and solves the LP relaxation of the modified ILP problem. If it is integer, the solution is optimal to the ILP problem.

Assume now that the solution is noninteger. As in the cutting-plane method, a fractional cut is formed and added, and the modified LP relaxation is solved. If the obtained optimal solution is still noninteger, a fractional or objective cut is added, depending on whether or not the current solution is a valid point: if it is valid, i.e., associated with an objective value equal to suspected-optimal value f^+, a fractional cut is formed and added; if, otherwise, the solution is associated with an objective value strictly greater than f^+, then f^+ is updated, and the new objective cut is added. The resulting LP relaxation is solved, and so on, until an integer solution, hence optimal solution to the original ILP problem attained, or infeasibility detected.

Example 10.5.1. Solve the following ILP problem by the controlled-cut method:

$$\begin{array}{ll} \min & f = -5x_1 + x_1, \\ \text{s.t.} & -7x_1 + \ \ x_2 + x_3 \qquad\qquad = 4, \\ & \quad 2x_1 + 5x_2 \qquad\quad + x_4 \ \ = 7, \\ & \text{integer } x_j \geq 0, \ j = 1, \cdots, 4. \end{array}$$

Answer

1. Call simplex Algorithm 3.2.1, yielding

x_1	x_2	x_3	x_4	RHS
	37/2	1	7/2	57/2
1	5/2		1/2	7/2
	27/2		5/2	35/2

2. Add the row corresponding to objective cut $5x_1 - x_2 + x_5 = 17$ to the preceding as its second bottom row:

x_1	x_2	x_3	x_4	x_5	RHS
	37/2	1	7/2		57/2
1	5/2		1/2		7/2
5	−1			1	17
	27/2		5/2		35/2

After converting the preceding to the simplex tableau, call dual simplex Algorithm 4.4.1, yielding

x_1	x_2	x_3	x_4	x_5	RHS
		1	2/27	37/27	751/27
1			1/27	5/27	92/27
	1		5/27	−2/27	1/27
				1	17

3. Insert the row associated with the fractional cut $-2/27x_4 - 10/27x_5 + x_6 = -22/27$, constructed from the first row of the preceding, as the second bottom row, leading to

x_1	x_2	x_3	x_4	x_5	x_6	RHS
		1	2/27	37/27		751/27
1			1/27	5/27		92/27
	1		5/27	−2/27		1/27
			−2/27	−10/27	1	−22/27
				1		17

Call dual simplex Algorithm 4.4.1, yielding

x_1	x_2	x_3	x_4	x_5	x_6	RHS
	1	1			7/2	25
1					1/2	3
	−1			1	−5/2	2
	5		1		−1	1
	1				5/2	15

The preceding gives an integer solution, which is an optimal solution to the original ILP problem, i.e.,

$$\bar{x} = (3, 0, 25, 1)^T, \qquad f^* = -15.$$

Part II
Advanced Topics

Chapter 11
Pivot Rule

A pivot rule[1] plays a crucial role in the simplex method for solving the standard LP problem. Starting from a vertex of the feasible region, geometrically the method moves from a vertex to adjacent vertex until reaching an optimal vertex. The related "path" consists of a series of edges (which could vanish in the presence of degeneracy), joining or be joined end to end. The number of edges, termed the "length" of the path, is equal to the number of iterations taken by the simplex method. It is the pivot rule that specifies a edge to move along in each iteration, and hence determines the number of required iterations, theoretically. Indeed, the pivot rule is the characteristic or spirit of the simplex method.

Had long been used in practice, Dantzig's original rule selects a pivot column index by

$$q \in \arg\min_{j \in N} \bar{z}_j .$$

In principle, however, any index associated with a negative reduced cost is eligible to be chosen, that is, determining any column index $q \in N$ such that $\bar{z}_q < 0$. The latter rule can also guarantee non-increase of the objective value, or strict decrease under nondegeneracy. Multiple choices of a pivot column provide possibility to improve the efficiency of the simplex algorithm.

As early as at the beginning of the simplex method, Dantzig noted that the conventional rule was far from ideal, and hopefully devoted to a so-called "most-improvement rule". In fact, the decrement of objective value per iteration is not only dependent on $\bar{z}_q$ but also on stepsize α. If α is too small, the decrement could be still small, no matter how $\bar{z}_q$ is. In particular, the decrement would even vanish in the

[1]It is referred to as column pivot rule. After a pivot column is selected, the choice of a row pivot is very limited.

P.-Q. PAN, *Linear Programming Computation*, DOI 10.1007/978-3-642-40754-3_11,

presence of degeneracy. In fact, it is known from (3.9) and (3.10) that the decrement of the objective value is

$$\bar{f} - \hat{f} = -\bar{z}_q(\bar{b}_p/\bar{a}_{pq}).$$

The most-improvement rule selects q, p such that the preceding decrement is the largest possible. Although such a selection seems to be attractive, the related computational effort is too high to realize. Subsequent numerical experiments showed that the efficiency of this rule is far lower than that of the conventional rule, even though the number of required iterations is reduced (Jeroslow 1973; Kuhn and Quandt 1953).

It is desirable to pursue an "ideal" pivot rule, which specifies the shortest among paths, joining a given initial vertex to an optimal one. As a hot research topic, pivot rule has received great attention continuously since the simplex method emerged. Various pivot rules have been proposed from time to time (see, e.g., Abel 1987; Greenberg 1978; Pan 1996c; Pan and Ouiang 1993; Pan et al. 2004; Zlatev 1980). However, all efforts failed to find the "ideal" rule, and it is even unknown whether there exists such a rule that turns the simplex method to a polynomial-time one. It seems to be related to the intrinsic characteristic of the simplex framework, as Schrijver wrote (1986, pp. 141):

> The main problem seems to be that the simplex method is 'myopic', cannot see 'cut-off paths'. It does not find a path which is locally bad but globally good.

It is worse in the presence of degeneracy when the number of superplanes, meeting at a vertex, exceeds the dimension of the feasible region; consequently, some edges emanating from the vertex vanish, and the algorithm stalls for too long a time before exiting the vertex. Most of existing pivot rules, such as the conventional rule, the most-improvement rule, and the steepest-edge rule presented later, and etc., are all shown to be infinite, for which some cycling examples are found (see also Zörnig 2006).

Even so, infinite rather than finite rules are put in practical use actually. *Judging goodness of a pivot rule for the simplex method is basically a practical issue: Rule's vitality or value is only determined by its performance.* Keeping this in mind, in this chapter, more advantageous pivot rules are presented without touching their theoretical aspects, such as finiteness and etc. These rules would be the most powerful ones at present.

For simplicity, it is assumed that the current basis and nonbasis are

$$B = \{1, \ldots, m\}, \qquad N = \{m+1, \ldots, n\}. \tag{11.1}$$

that is, the basis matrix consists of the first m columns of A.

11.1 Partial Pricing

A normal approach to computing reduced costs used in pivoting is to solve system

$$B^{\mathrm{T}} y = c_B \tag{11.2}$$

first, then use the solution $\bar{y}$ to calculate reduced costs (pricing), i.e.,

$$\bar{z}_N = c_N - N^{\mathrm{T}}\bar{y}. \tag{11.3}$$

Computing all components of $\bar{z}_N$ (full pricing) by the preceding formula requires $(n - m)m$ multiplications. Thereby, a large proportion of running time is usually spent on pricing by the simplex method, especially so when $n - m \gg m$.

Such situation hastens variants of the conventional pivot rule by a so-called *partial pricing* strategy, that is, only a part of components of $\bar{z}_N$ are calculated in each iteration to determine a pivot column index. Pricing options in MINOS include *full pricing*, *sectional pricing* and *multiple pricing*, the last two of which belong to partial pricing category (see, e.g., Benichou et al. 1977; Chvatal 1983; Maros 2003b; Nocedal and Wright 1999 and etc.).

As a typical partial pricing, *sectional pricing* partitions all column indices into p (e.g., $p = 20$) sections having approximately equal number of indices. In order to handle each index equally, the m indices, associated with logical variables, are approximately equally distributed to the p sections. Started from the first index in a section, which follows the section from which the previous pivot column index is selected, pricing is done on the section (excluding basic variables) in each iteration. If a reduced cost is found in the section that is lower than a dynamically changing negative threshold, the related index is selected to enter the basis; otherwise, pricing is continued similarly with the following section, and so on. If all the p sections are priced and the minimum reduced cost is still greater than the threshold, optimality is achieved numerically if it is greater than the negative optimality tolerance (normally -10^{-6}); otherwise, the index associated with the minimum reduced cost is selected to enter the basis, and its magnitude is recorded. Thereafter, the threshold is decreased (to, e.g., 0.1 times of itself) whenever the magnitude is not sufficiently greater than (e.g., 1.1 times of) the threshold. An initial value of the threshold is set to a very large value (e.g., 10^{20}) by MINOS. So, full pricing is actually carried out in the first iteration, at least.

Multiple pricing is another partial pricing. Firstly, it determines a small set, consisting of nonbasic indices, associated with negative reduced costs. Then a pivot index is determined by some rule, e.g., by choosing an index associated with the minimum cost among them. If the cost is found is lower than some dynamically changing negative threshold, enter the according index to the basis. In the next iteration, computed are only the reduced costs associated with the rest indices in the set, from which a pivot index is determined, and so on. When all reduced costs associated with the set are greater than the negative optimal tolerance, a new set of nonbasic indices is determined, and the preceding procedure is repeated. Optimality is achieved if no such a set exists.

It is initially anticipated that partial pricing requires more iterations than full pricing because, after all, the pivot column index, determined by the latter, corresponds to a smaller cost than the former, and what is hence expected is only that the overall efficiency is improved due to the reduction of computational effort per iteration by partial pricing. Surprisingly enough, it turns out that the number of iterations, required by partial pricing, is reduced in general, so is the overall running time (see, e.g., Harris 1973; Pan 1997, 2008c). Extensive numerical experiments confirm the superiority of partial pricing to full pricing. This fact would suggest searching for better pivot rules.

11.2 Steepest-Edge Rule

In the second half of the 1980s and the first half of 1990s of the twentieth century, when interior-point methods were strongly developing, some scholars thought of such type of methods being superior to simplex methods for solving large-scale LP problems. During the same period, on the other hand, the simplex methods got active strength with numerical results on the steepest edge pivot rules (Forrest and Goldfarb 1992). Consequently, the two types of methods have been being inextricably involved in a rat race (Bixby 2002; Nemhauser 1994).

As it is assumed that the current basis matrix B consists of the first m columns of A, the $n-m$ edge directions, emanating from the current vertex, are

$$d^j = \begin{pmatrix} -B^{-1}a_j \\ e_{j-m} \end{pmatrix}, \quad j \in N, \tag{11.4}$$

where $e_{j-m} \in \mathcal{R}^{n-m}$ is the unit vector with its $(j-m)$th component 1. For each $j \in N$, the reduced cost

$$\bar{z}_j = c_j - c_B^{\mathrm{T}} B^{-1} a_j$$

is equal to the inner product of the objective gradient c and edge direction d^j, i.e.,

$$\bar{z}_j = c^{\mathrm{T}} d^j, \quad j \in N.$$

If $\bar{z}_j < 0$, then d^j is a descent edge direction (Proposition 3.5.1). Thus, Dantzig's original rule selects a column index

$$q \in \arg\min_{j \in N} \bar{z}_j,$$

associated with a descent edge direction, along which the according decrement of the objective value attains the largest for a unit increment in the entering variable.

The following pivot rule uses normalized reduced costs:

Rule 11.2.1 (Steepest-edge column rule) Select column index

$$q \in \arg\min\{\bar{z}_j / \|d^j\| \mid j \in N\}.$$

Since for any $j \in N$, $\bar{z}_j/\|d^j\|$ and the cosine of angle between c and d^j differ from the same constant factor $1/\|c\|$, the preceding rule actually selects a descent edge direction that forms the largest angle with the objective gradient, as is why the rule is prefixed by "steepest-edge". While it is attractive. computing norms $\|d^j\|, \forall j \in N$ of edge directions is too expensive to realize.

Goldfarb and Reid (1977) offered a scheme, in which $\|d^j\|^2, j \in N$ was calculated by recurrence formulas in each iteration. Further, they gave several variants of the rule and reported remarkable numerical results, obtained in extensive experiments, as received wide attention from the academic community.

The recurrence formulas can be derived based on the basic change, i.e.,

$$\hat{d}^{\mathrm{p}} = -(1/\sigma_q)d^q, \tag{11.5}$$

$$\hat{d}^j = d^j - (\sigma_j/\sigma_q)d^q, \quad j \in N,\ j \neq q, \tag{11.6}$$

where σ_j denotes entries in the pivot row, i.e.,

$$\sigma_j = a_j^{\mathrm{T}} B^{-\mathrm{T}} e_p, \quad j \in N, \tag{11.7}$$

combining which, (11.5) and (11.6) gives the following formulas:

$$\|\hat{d}^p\|^2 = (1/\sigma_q^2)\|d^q\|^2, \tag{11.8}$$

$$\|\hat{d}^j\|^2 = \|d^j\|^2 - 2(\sigma_j/\sigma_q)a_j^{\mathrm{T}}v + (\sigma_j/\sigma_q)^2\|d^q\|^2, \quad j \in N,\ j \neq q, \tag{11.9}$$

where

$$B\bar{a}_q = a_q, \quad B^{\mathrm{T}}v = \bar{a}_q. \tag{11.10}$$

In fact, only the second system of the preceding is needed to solve because $\bar{a}_q$ was already calculated. Moreover, $\bar{a}_q$ can also be employed to directly compute

$$\|d^q\|^2 = 1 + \|\bar{a}_q\|^2.$$

It is thereby seen that updating squares of norms of edge directions involves three linear systems: besides the two typical systems, another system is

$$B^{\mathrm{T}}h = e_p. \tag{11.11}$$

As Algorithm 3.5.2 also involves $B\bar{a}_q = a_q$ and (11.11), it is only needed to solve an additional system

$$B^{\mathrm{T}}v = \bar{a}_q.$$

In (11.9), in addition, only those inner products $a_j^{\mathrm{T}}v$ that correspond to nonzero σ_j is needed to calculate. Fortunately, most of σ_j often vanish in sparse computations.

Table 11.1 CPU time and ratio

Pivot rule	Dantzig	Devex	Steepest-edge	Dynamic steepest-edge	Projective steepest-edge
Hours	110.19	7.76	4.90	4.08	3.89
Time ratio	28.33	1.99	1.26	1.05	1.00

The steepest-edge rule requires more computational effort per iteration but usually much less iterations. Numerical results with this rule are listed in Table 11.1 (see the first paragraph on page 304 in conjunction with approximate steepest-edge rules).

A shortcoming is that the rule has to compute squares of norms of edge directions associated with all nonbasic indices initially. Due to the "restarting strategy" (Sect. 5.1), such computation must be carried out periodically to avoid overaccumulation of round errors. Consequently, the rule involves a considerable amount of computational effort for large-scale problems, especially when it is not possible to take advantages of computer structure (like Forrest and Goldfarb 1992), to solve linear systems efficiently. Moreover, it is clear that the partial pricing strategy is not amenable in such a context.

11.3 Approximate Steepest-Edge Rule

In this section, notations of the previous section will still be used to present three approximate variants of the steepest-edge rule.

Enlightened on the graphic approach to solving LP problems of 2-dimension, Harris (1973) proposes an idea of searching along an approximate steepest-edge. Resulting Devex rule is simple, involving the pivot row and column only, and performs remarkably in computational experiments (see also Swietanowaki 1998).

Given a set of $n - m$ column indices, termed "reference framework". For $j \in N$, assume that subvector $\hat{d}^j$ consists of components, located on the reference framework; weight t_j is endowed to the reduced cost $\bar{z}_j$, as an approximation of $\|\hat{d}^j\|$.

Rule 11.3.1 (Devex column rule) Select pivot column index

$$p \in \arg\min\{\bar{z}_j/t_j \mid j \in N\},$$

where weights t_j are determined as follows, so that the preceding rule may be regarded as selecting the steepest-edge under the reference framework.

Initially, the set of current nonbasic indices is taken as a reference framework. Set $t_j = 1$ for all indices j in it. As a result, the Devex rule coincides with the Dantzig's original rule in the first iteration.

Weights t_j are updated in subsequent iterations. It is clear that (11.5) and (11.6) are still valid if vectors involved in them are replaced by vectors, consisting of components located only on the reference framework. Thus, we have the following updating formulas:

$$\bar{t}_p = \max\{1, \|\hat{d}^q\|/|\sigma_q|\}, \tag{11.12}$$

$$\bar{t}_j = \max\{t_j, |\sigma_j/\sigma_q|\|\hat{d}^q\|\}, \quad j \in N, \ j \neq q, \tag{11.13}$$

where the last formula comes from using the larger module of vectors

$$\hat{d}^j \quad \text{and} \quad -(\sigma_j/\sigma_q)\hat{d}^q$$

to replace the module of their sum

$$\hat{d}^j - (\sigma_j/\sigma_q)\hat{d}^q.$$

Since $\bar{a}_q$ is obtained independent of weight updating in each iteration, $\|\hat{d}^q\|$ is easy to compute as d^q is available. When q does not belong to the reference framework, on the other hand, $\|\hat{d}^q\|$ could become very small even though $\|d^q\| \geq 1$. In this aspect, (11.12) ensures that all weights t_j are no less that 1. It is therefore clear that their weights never decrease even if indices stay in nonbasis for a long time.

When errors, caused by repeatedly using of the updating formulas, accumulate too high, it is necessary to determine a new reference framework and set all weights to 1 again. As weight $\bar{t}_q$ can be calculated directly, it is convenient to monitor errors in the process, which should be restarted when the calculated value differs from the updated value by a relatively large margin, e.g., when the former exceeds some times of the latter (Harris uses double).

Another scheme is to directly use $\|\hat{d}^j\|$ themselves rather than their approximation $\bar{t}_j$. Since (11.5) and (11.6) are still valid if vectors d^j in them are replaced by subvectors $\hat{d}^j$, recurrence formulas for squares of $\hat{d}^j$'s norms can be derived from (11.8) and (11.9) by replacing d^j by $\hat{d}^j$. It is still possible to directly compute

$$\|\hat{d}^q\|^2 = \delta + \|\bar{a}_q\|^2,$$

where δ is equal to 1 or 0, dependent on whether q belongs to the reference framework or not. All $\bar{a}_j$ involved in the formulas should be regarded as subvectors, corresponding to the reference framework. Pivot rules using such formulas are termed *projective steepest-edge rule*.

A further variant of the projective steepest-edge rule is to expend the reference framework when resetting it, by adding the index of the current leaving variable to the reference framework if the index is not already in it. So, only minor changes are needed in the recurrence formulas. The resulting pivot rule is termed *dynamic steepest-edge rule*.

In Table 11.1, cited are numerical results reported by Forrest and Goldfarb (1992). There were 20 test problems involved in their tests, including all 14 Netlib standard test problems that involve more than 10,000 nonzeros and 6 larger and more difficult problems collected by the authors. CPU times required by five codes based on Dantzig, Devex, the steepest-edge, dynamic steepest-edge and projective steepest-edge to solve all the 20 problems are listed in the first row of the table below. Time ratios of the first four rules to the projective steepest-edge rule are listed in the bottom row.

It is seen from the preceding that Dantzig's conventional rule is the slowest while the projective steepest-edge rule is the fastest for solving these test problems: their total time ratio is high as 28.33. However, the authors indicate that the tests are favorable to the last three rules because they take advantages of the structures of computer IBM RISC system/6000 in pricing and solving systems. These numerical results firmly establish big superiority of the steepest-edge rule and approximate steepest-edge rules, such as Devex etc, over the conventional rule.

11.4 Largest-Distance Rule

Like the steepest-edge rule or Devex rule, the so-call "largest-distance rule" is also based on normalized reduced costs, though simpler and easier to realize.

We attacks from the dual side by investigating (primal) pivot rules from the dual problem

$$\begin{array}{ll} \max & g = b^T y, \\ \text{s.t.} & A^T y \leq c, \end{array}$$

to gain some insight. In fact, a negative reduced cost

$$\bar{z}_j = c_j - a_j^{\mathrm{T}} B^{-\mathrm{T}} c_B$$

implies that the current dual solution $\bar{y} = B^{-\mathrm{T}} c_B$ violates dual constraint $a_j^{\mathrm{T}} y \leq c_j$. Entering the associated variable x_j to the basis means forcing the violated dual inequality constraint to be satisfied as a equality (binding), i.e., $a_j^{\mathrm{T}} y = c_j$. Therefore, it seems to be reasonable to enter the variable, associated with the mostly violated dual constraint, to enter the basis.

It is not surprising that the conventional rule is unsatisfactory, as it takes the nonbasic variable, associated with the most-negative reduced cost, to enter the basis, but the variable does not necessarily correspond to the mostly violated dual constraint. In fact, the signed distance of $\bar{y}$ to the boundary $a_j^{\mathrm{T}} y = c_j$ is (see Sect. 2.1)

$$d = \bar{z}_j / \|a_j\|.$$

Table 11.2 Iteration and time ratios

	Devex/SCL2		Devex/SCL1		SCL2/SCL1	
Problem	Iters	Time	Iters	Time	Iters	Time
Netlib(47)	0.26	0.33	0.29	0.34	1.09	1.03
Kennington(16)	1.75	2.18	4.95	4.33	2.84	1.98
BPMPD(17)	1.29	2.33	1.88	3.64	1.46	1.56
Average(80)	1.00	2.06	1.43	3.24	1.43	1.58

Even if $\bar{z}_j$ is the most negative, the absolute value of d may still be very small when $\|a_j\|$ is large, so that the current dual solution is not far from the boundary.

The following rule selects the variable, associated with the mostly violated dual constraint, to enter the basis (Pan 2008a).

Rule 11.4.1 (Largest-distance column rule) Select pivot column index

$$q \in \arg\min\{\bar{z}_j / \|a_j\| \mid j \in N\}.$$

The preceding actually represents a class of pivot rules if $\|a_j\|$ is regarded as any norm of a_j, though the Euclidean norm seems to be preferable.

The computation, associated with the largest-distance rule, is very simple. Norms $\|a_j\|, j = 1, \cdots, n$ of columns of the coefficient matrix remains unchanged in solution process. If these norms are calculated at the beginning, the number of divisions required by the pivot rule will not exceed $n - m$ in each iteration. It is even better to normalize columns of A before hand, so that the largest-distance rule is just the same as Dantzig's rule because any reduced cost $\bar{z}_j$ itself is equal to the signed distance of $\bar{y}$ to the boundary.

In the following, we cite numerical results reported by Pan (2008a). Tests were carried out on an IBM PC with Windows XP 2002 system, processor 1.86 GHz, 1.00 GB internal storage, about 16 digits precision, and visual Fortran 5.0 compiler. There were three codes involved:

1. Devex: Devex rule.
2. SCL1: Euclidean norm largest-distance rule.
3. SCL2: ∞-norm largest-distance rule.

MINOS 5.51 was used as a platform for the preceding three codes: only pivot rules are different. All together, there were 3 sets of 80 large-scale sparse problems were tested (see Appendix B: Tables B.1–B.3):

1. 47 (according to $m + n$) largest Netlib problems (SCRS8-STOCFOR3).
2. All 16 Kennington problems (KEN07-KEN18).
3. All 17 BPMPD problems larger than 500 KB (in compressed format) (RAT7A-DBIC1).

Table 11.2 gives iteration and CPU time ratios of these codes:

It is seen from the bottom row that the total iteration and CPU time ratios of Devex to SCL1 are 1.43 and 3.24, respectively. Therefore, SCL1 outperformed Devex significantly. It is more important that the margin of their CPU times required are larger. SCL2 also defeated Devex with CPU time ratio 2.06. Moreover, it can be expected that the largest-distance rule using partial pricing will perform better.

As for the two largest-distance rules, SCL1 outperformed SCL2 with iteration ratio 1.43 and time ratio 1.58. Therefore, the rule using Euclidean norm seems to be preferable.

Besides the remarkable performance of the largest-distance rule, the author himself was astonished by the easy realization of it: SCL1 and SCL2 were yielded from MINOS 5.51 by only inserting a single sentence for normalizing columns of the coefficient matrix in its scaling subroutine.

11.5 Nested Rule

The nested pivot rule may be classified into the partial pricing category. In particular, it would be somehow close to multiple pricing (Sect. 11.1). The key difference from the latter is that the former focuses on, and eliminate the most stubborn nonbasic indices that do not fulfil the optimality condition (in the sense within the dual tolerance).

Let $\epsilon > 0$ be dual feasibility tolerance. At the beginning of each iteration, a set N_1 of nonbasic indices is given the priority to enter the basis. Pricing is carried out on N_1 to determine a column index, associated with a negative reduced cost, by, e.g., the conventional column rule. If the associated reduced cost is less than $-\epsilon$, then the index is selected to enter the basis, and the N_1 for the next iteration is born from the current N_1 by including all its indices, associated with reduced costs less than $-\epsilon$. In the other case, do the same thing with $N_2 = N \backslash N_1$; if there is still no reduced cost less than $-\epsilon$, then optimality is achieved.

As an instance, the nested variant counterpart of the conventional rule is described below (Pan 2008b).

Rule 11.5.1 (Nested-Dantzig rule) Given a dual tolerance ϵ. Set $N_1 = N$ and $N_2 = \emptyset$.

1. Go to step 4 if $N_1 = \{j \in N_1 \mid \bar{z}_j < -\epsilon\} \neq \emptyset$.
2. Go to step 4 if $N_1 = \{j \in N_2 \mid \bar{z}_j < -\epsilon\} \neq \emptyset$.
3. Stop (optimality achieved).
4. Select a pivot column index $q \in \arg\min\{\bar{z}_j \mid j \in N_1\}$.
5. Update: $N_1 = N_1 \backslash q,\ N_2 = N \backslash N_1$.

Such a rule is prefixed by "nested" because a current set N_1, which undergoes pricing with priority, is a proper subset of its predecessor N_1. The idea behind it can be explained further as follows.

Table 11.3 Iteration and time ratios to N-Dantzig

	Dantzig		Devex		P-Dantzig		N-Devex	
Problem	Iters	Time	Iters	Time	Iters	Time	Iters	Time
Netlib(47)	5.16	5.95	1.20	1.21	4.65	4.00	1.04	0.95
Kennington(16)	5.63	5.65	5.56	5.55	3.51	2.64	1.00	0.91
BPMPD(17)	8.29	12.86	3.83	6.54	5.04	5.20	1.18	1.22
Average(80)	6.78	9.75	3.48	5.73	4.57	4.20	1.10	1.09

In the initial iteration or iterations in which N_2 is touched, a full pricing is carried out. After that, a series of nested pricing's follow, as might be called a "circle". As each N_1 is a proper subset of its predecessor in a circle, computational effort for pricing decreases monotonically iteration by iteration. In the kth iteration of a circle, moreover, the reduced costs associated N_1 are less than $-\epsilon$ all the time. Therefore, it is reasonable to enter such a "die-hard" nonbasic index to the basis.

In fact, the index set N_1, yielding from the first iteration, corresponds to nonnegativity constraints, gradients of which form acute angles with the negative reduced gradient. Based on the most-obtuse-angle heuristics, one should put forth effort to make these constraints inactive; in other words, he should give indices associated with these constraints priority to enter the basis. (see Sects. 2.5 and 5.5).

Any standard pivot rule may be modified to a nested variant. The nested steepest-edge rule can be obtained from Rule 11.5.1 by using

$$q \in \arg\min \{\bar{z}_j/\|d^j\| \mid j \in N_1\},$$

in step 4 of it instead, and the nested Devex rule obtained by using

$$q = \arg\min \{\bar{z}_j/t_j \mid j \in N_1\},$$

(for notations, see Sects. 11.2 and 11.3, respectively).

Fortunately, nested pivot rules are easy to implement. Obtained numerical results are associated with five codes (Pan 2008c):

1. Dantzig: MINOS 5.51 (full pricing option).
2. Devex: Devex rule.
3. P-Dantzig: MINOS 5.51 (default sectional pricing option).
4. N-Devex: Nested Devex rule.
5. N-Dantzig: Rule 11.5.1.

These codes are all based on MINOS 5.51 as a platform, only differing from the pivot rule used. The software and hardware environments are the same as those described in the previous section, as well as the same 80 test problems (Appendix B: Tables B.1–B.3). Table 11.3 lists total iteration and CPU time ratios of the first four codes to the fifth (N-Dantzig).

It is seen from the preceding table that for either iterations or CPU time, codes N-Dantzig and N-Devex, based on the nested pivot rules, are significantly superior to the standard ones, and N-Dantzig is the best. N-Dantzig rule outperformed Devex rule by total iteration ratio 3.48 and time ratio 5.73, and outperformed Dantzig's standard rule by total iteration ratio 6.78 and time ratio 9.75!

Further computational experiments were conducted with nested pricing. Contained in Appendix C, associated numerical results can be outlined below:

For 77 test problems, the nested-Dantzig rule defeated the steepest-edge rule by total time ratio as high as 25.22, even though the former required more iterations (total iteration ratio is 0.34). Even the largest-distance rule was not comparable with the nested-Dantzig rule. However, the nested steepest-edge rule was inferior to the steepest-edge rule, since the former restarted from scratch too many times. It should be pointed out that these tests were somehow unfair to steepest-edge rules, as they neither use recurrence formulas (thus additional two systems were solved in each iteration), nor take advantage of computer's structure to solve systems more efficiently.

In summary, nested pivot rules are superior to their standard counterparts by a large margin (except for the nested steepest-edge rule), and the nested-Dantzig rule is superior to the nested Devex rule.

11.6 Nested Largest-Distance Rule

Largest-distance rules can be improved further by incorporating the nestification tactic, as is easy to realize (Pan 2008c).

Rule 11.6.1 (Nested largest-distance column rule) The same as Rule 11.5.1, except for its step 4 using the following instead:

$$q \in \arg\min\{\bar{z}_j/\|a_j\| \mid j \in N\}.$$

Associated numerical results are as follows, where involved are the following three codes (Pan 2010):

1. Devex: Devex rule.
2. LDN1: Nested largest-distance rule with Euclidean norm.
3. LDN2: Nested largest-distance rule with ∞ norm.

The software and hardware environments are the same as those given in the previous 2 sections, with the same 80 test problems (Appendix B: Tables B.1–B.3). Table 11.4 lists total iteration and CPU time ratios.

It is seen that the nested largest-distance rules outperformed Devex rules unambiguously in terms of either iterations or CPU time. As for the two nested

Table 11.4 Iteration and time ratios

	Devex/LDN2		Devex/LDN1		LDN2/LDN1	
Problem	Iters	Time	Iters	Time	Iters	Time
Netlib(47)	1.20	1.17	1.15	1.16	0.96	0.99
Kenningt(16)	5.63	5.42	7.83	5.77	1.39	1.06
BPMPD(17)	4.32	7.43	5.43	10.17	1.26	1.37
Average(80)	3.69	6.08	4.34	7.27	1.18	1.20

ones, the nested largest-distance rules with Euclidean norm is better, as defeated Devex rule by iteration ratio 4.34 and time ratio 7.27. The margins are even larger than that between the nested-Dantzig rule and Devex rule.

In view of comparability of numerical results, reported in Sects. 11.4–11.6, we conclude that the nested largest-distance rules with Euclidean norm is much better than all commonly used rules nowadays, at least in terms of the 80 tested problems. We believe that this is the case in general.

Chapter 12
Dual Pivot Rule

In this book, pivot rules used in the dual simplex method are referred to as dual pivot rule.[1] Like in the primal simplex context, a dual pivot rule is crucial to algorithm's efficiency.

Assume that $\bar{x} = B^{-1}b \not\geq 0$. The dual Dantzig conventional rule selects a row index $p \in B$ such that $\bar{x}_p$ is the minimum among components of the basic solution. Thus, the dual objective value will increase the most possible for a unit stepsize. This rule is far from ideal, like its primal counterpart. From (4.24) and (4.25), it is known that the increment of the dual objective value will be

$$\hat{f} - \bar{f} = \bar{x}_{j_p}(\bar{z}_q/\bar{a}_{pq}).$$

According to the "most-improvement" criterion, p and q are determined such that preceding increment attains the largest possible. Unfortunately, the related computational effort is too high to be practicable, just as its primal counterpart.

This chapter will address some very promising dual pivot rules, which can be regarded as dual variants of those presented in the previous chapter.

For simplicity, assume again that the current basis matrix consists of the first m columns of A, i.e.,

$$B = \{1, \cdots, m\}, \quad N = \{m+1, \cdots, n\}.$$

[1] It is referred to as row rule. After a pivot row is selected, there is a very limited choice of a column pivot.

P.-Q. PAN, *Linear Programming Computation*, DOI 10.1007/978-3-642-40754-3_12,

12.1 Dual Steepest-Edge Rule

Consider the dual problem

$$\begin{array}{ll} \max & g = b^{\mathrm{T}} y, \\ \text{s.t.} & A^{\mathrm{T}} y \leq c. \end{array}$$

Let $\bar{y}$ be the current dual basic feasible solution, satisfying

$$B^{\mathrm{T}} \bar{y} = c_B, \tag{12.1}$$

$$N^{\mathrm{T}} \bar{y} \leq c_N. \tag{12.2}$$

Define

$$\begin{aligned} y(\beta) &= \bar{y} - \beta h^i, \\ h^i &= B^{-\mathrm{T}} e_i, \quad i = 1, \cdots, m. \end{aligned}$$

From the preceding two expressions and (12.1), for any $i \in \{1, \cdots, m\}$ and $\beta \geq 0$ it holds that

$$\begin{aligned} a_i^{\mathrm{T}} y(\beta) &= a_i^{\mathrm{T}} \bar{y} - \beta = c_i - \beta \leq c_i, \\ a_k^{\mathrm{T}} y(\beta) &= a_k^{\mathrm{T}} \bar{y} = c_k, \quad k = 1, \cdots, m, k \neq i. \end{aligned}$$

It is known that $-h^i, i = 1, \cdots, m$ is a edge direction, emanating from the vertex $\bar{y}$ of the feasible region $\{y | A^{\mathrm{T}} y \leq c\}$. The determination of row index i implies that the basic variable x_i leaves the basis, so constraint $a_i^{\mathrm{T}} y \leq c_i$ may be satisfied as a strict inequality. Since

$$-b^{\mathrm{T}} h^i = -e_i^{\mathrm{T}} B^{-1} b = -\bar{x}_i,$$

when $\bar{x}_i < 0$, the edge direction $-h^i$ forms an acute angle with the dual objective gradient b, as is an uphill direction. Therefore, the objective value will never decrease if a row index $p \in \{1, \cdots, m\}$ is selected such that $\bar{x}_p < 0$; it strictly increase if dual nondegeneracy is assumed. In particular, the following rule selects the edge that forms the smallest angle with b (Forrest and Goldfarb 1992).

Rule 12.1.1 (Dual steepest-edge row rule) Select pivot row index

$$p \in \arg\min\{\bar{x}_i / \|h^i\| \mid i = 1, \cdots, m\}.$$

Like in the primal context, practicability of the preceding rule lies in computing

$$\|h^i\|^2, \quad i = 1, \cdots, m.$$

in a recurrence manner.

Assume that x_p is selected to leave and x_q to enter the basis. Then the inverse of the basis matrix is updated by (3.28), i.e.,

$$\hat{B}^{-1} = B^{-1} - \frac{(\bar{a}_q - e_p)e_p^{\mathrm{T}}B^{-1}}{\bar{a}_{pq}},$$

where $\bar{a}_q = B^{-1}a_q$. Premultiplying the preceding expression by e_i^{T} and transposing it leads to the recurrence for edge directions, i.e.,

$$\tilde{h}^p = (1/\bar{a}_{pq})h^p, \tag{12.3}$$

$$\tilde{h}^{\mathrm{i}} = h^{\mathrm{i}} - (\bar{a}_{iq}/\bar{a}_{pq})h^{\mathrm{p}}, \quad i = 1, \cdots, m, i \neq p. \tag{12.4}$$

from which the recurrence formulas for squares of norms of edge directions follow:

$$\|\tilde{h}^{\mathrm{p}}\|^2 = (1/\bar{a}_{pq})^2\|h^{\mathrm{p}}\|^2, \tag{12.5}$$

$$\|\tilde{h}^{\mathrm{i}}\|^2 = \|h^{\mathrm{i}}\|^2 - 2(\bar{a}_{iq}/\bar{a}_{pq})u_i + (\bar{a}_{iq}/\bar{a}_{pq})^2\|h^{\mathrm{p}}\|^2, \\ i = 1, \cdots, m, i \neq p. \tag{12.6}$$

where

$$B^{\mathrm{T}}h^{\mathrm{p}} = e_p, \quad Bu = h^{\mathrm{p}}. \tag{12.7}$$

Note that $\|h^p\|^2 = (h^p)^{\mathrm{T}}h^p$ can be directly calculated, because h^{p} and $\bar{a}_q$ are obtained otherwise independently.

Computations, involved in Rule 12.1.1, is usually cheaper than that in the (primal) steepest-edge rule. They both solve an additional system (12.6), hence $Bu = h^{\mathrm{p}}$. It is noted that formula (11.9) needs to solve $B^{\mathrm{T}}v = \bar{a}_q$. Differing from (11.9), which is expensive when pivot row $e_p^{\mathrm{T}}B^{-1}N$ is dense, there is no any inner product involved in the middle term of (12.6). In practice, it is often the case of $n - m \gg m$, in which initially computing squares of norms of edge directions in the primal rule is cumbersome, compared with in the dual rule. As for taking advantage of sparsity, handling $B^{-1}a_j$ is also inferior to $B^{-\mathrm{T}}e_i$.

Now let us turn to the dual problem of form

$$\begin{array}{ll} \max & g = b^{\mathrm{T}}y, \\ \text{s.t.} & A^{\mathrm{T}}y + z = c, \quad z \geq 0. \end{array} \tag{12.8}$$

A new pivot rule can be derived if the steepest-edge direction is considered in (y, z)-space.

It is clear that the dual basic feasible solution

$$(\bar{y}, \bar{z}_N, \bar{z}_B) = (B^{-\mathrm{T}}c_B, c_N - N^{\mathrm{T}}B^{-\mathrm{T}}c_B, 0)$$

is actually the unique solution to the $(m+n) \times (m+n)$ system

$$\begin{pmatrix} B^{\mathrm{T}} & 0 & I \\ N^{\mathrm{T}} & I & 0 \\ 0 & 0 & I \end{pmatrix} \begin{pmatrix} y \\ z_N \\ z_B \end{pmatrix} = \begin{pmatrix} c_B \\ c_N \\ 0 \end{pmatrix}$$

which is, geometrically, a vertex of the polyhedron

$$\{(y, z) \in \mathcal{R}^m \times \mathcal{R}^n \mid A^{\mathrm{T}} y + z = c, z \geq 0\}.$$

The inverse of the coefficient matrix of the system is

$$\begin{pmatrix} B^{-\mathrm{T}} & 0 & -B^{-\mathrm{T}} \\ -N^{\mathrm{T}} B^{-\mathrm{T}} & I & N^{\mathrm{T}} B^{-\mathrm{T}} \\ 0 & 0 & I \end{pmatrix}$$

It is easy to verify that the last m columns of the preceding are just edge directions, emanating from the vertex, i.e.,

$$h^i = \begin{pmatrix} -B^{-\mathrm{T}} \\ N^{\mathrm{T}} B^{-\mathrm{T}} \\ I \end{pmatrix} e_i, \quad i = 1, \cdots, m. \tag{12.9}$$

Recurrence formulas of these edge directions are of the form (12.3) and (12.4), and squares of norms of them are of form (12.5) and (12.6), though (12.7) should be replaced by

$$B^{\mathrm{T}} h = e_p, \quad Bu = h + \sum_{j=m+1}^{n} \sigma_j a_j, \tag{12.10}$$

where σ_j is the jth component of the pivot row vector, i.e.,

$$\sigma = A^{\mathrm{T}} h. \tag{12.11}$$

As σ is computed while pricing, the following quantity

$$\|h^p\|^2 = h^{\mathrm{T}} h + \sigma^{\mathrm{T}} \sigma. \tag{12.12}$$

can be directly computed.

In this book, the pivot rule, based on (12.5)–(12.7), is referred to as *dual steepest-edge pivot rule I*, and that based on (12.5), (12.6) and (12.10)–(12.12) referred to as *dual steepest-edge pivot rule II*. Associated numerical results will be given in Table 12.1 in the next section, in a comparison with according approximate rules there.

12.2 Approximate Dual Steepest-Edge Rule

Harris (1973) derived a dual variant of the Devex rule, as may be regarded as an approximation of the dual steepest-edge rule II.

She constructed a "reference framework", a set of m indices of components of z. Assume that subvector $\hat{h}^{\rm i}$ consists of components, located on the reference framework, of edge direction $h^{\rm i}$. Weights $s_i, i = 1, \cdots, m$ are endowed to $\bar{x}_i$ to approximate $\|\hat{h}^{\rm i}\|$.

Rule 12.2.1 (Dual Devex row rule) Select pivot row index

$$p \in \arg\min\{\bar{x}_i/s_i | i = 1, \ldots, m\},$$

where weights s_i, $i = 1, \ldots, m$ are determined as follows, so that the rule can be regarded as an approximation of the steepest-edge rule under the reference framework.

At the beginning, index set B is taken as the reference framework, and s_i are set to 1 for all $i = 1, \ldots, m$ (see (12.9)); So, the dual Devex rule is just the same as the dual Dantzig rule in this case. Subsequently, s_i are updated iteration by iteration. Assume that σ is defined by (12.11), and components, located on the reference framework, form subvector $\hat{\sigma}$. As (12.3) and (12.4) are still valid when vectors in them are replaced by subvectors, consisting of components located on the reference framework, we have the following updates:

$$\bar{s}_p = \max\{1, \|\hat{\sigma}\|/|\bar{a}_{pq}|\},$$
$$\bar{s}_i = \max\{s_i, |\bar{a}_{iq}/\bar{a}_{pq}| \|\hat{\sigma}\|\}, \quad i = 1, \cdots, m, i \neq p,$$

the last of which comes from using the larger norm of vectors $\hat{h}^{\rm i}$ and $-(\bar{a}_{iq}/\bar{a}_{pq})h^{\rm p}$ instead of the norm of their sum (see (12.9)). In each iteration, $\bar{a}_q = B^{-1}a_q$ is computed separately. As σ is available, $\|\hat{\sigma}\|$ can be computed directly.

Similar to the (primal) Devex rule, when errors caused by repeatedly using of the updating formulas accumulate too high, it is necessary to determine a new reference framework and set all weights to 1 again. As weight $\bar{t}_q$ is calculated directly, it is convenient to monitor errors. It should be restarted when the calculated value differs from the updated value by a relatively large margin, e.g., when the former exceeds some times of the latter (Harris used double).

When errors, caused by repeatedly using of the updating formulas, accumulate too much, it is necessary to determine a new reference framework and set all weights to 1 again, just like the approximate steepest-edge rules. It is also possible to monitor errors: the process should be restarted when the calculated value of $\|\hat{\sigma}\|$ differs from the updated value s_p by a relatively large margin.

Table 12.1 CPU time and ratio

Dual pivot rule	Dantzig	Devex	Steepest-edge II	Dynamic steepest-edge	Projective steepest-edge	Steepest-edge I
Hours	177.78	67.43	12.72	10.41	7.36	6.36
Time ratio	27.95	10.60	2.00	1.64	1.16	1.00

So-called *dual projective steepest-edge rule* results from using $\|\hat{h}^i\|$ rather than $\bar{s}_i$. This rule also yields from modifying the dual steepest-edge rule under the reference framework. In fact, replacing h^i in (12.5) and (12.6) by $\hat{h}^i$ leads to recurrence formulas of $\|\hat{h}^i\|^2$, whereas u in (12.6) can be obtained by solving system

$$Bu = \sum_{j=m+1}^{n} \hat{\sigma}_j a_j.$$

A further variant of the dual steepest-edge rule results from expending the reference framework whenever it is reset, that is, by adding the pivot column index to the current reference framework if it is not in it already. Accordingly, the recurrence formulas are modified slightly. Such a variant is called *dual dynamic steepest-edge rule*.

We cite numerical results reported by Forrest and Goldfarb (1992) below. The hardware and software environments as well as 20 test problems were the same as those described for the primal case, presented in Sect. 11.3. Codes based on six dual rules: Dantzig, Devex, steepest-edge II, dynamic steepest-edge, project steepest-edge and steepest-edge I were tested. Table 12.1 gives total CPU times, required for solving all the problems, in the first row, and time ratios of the first five rules to the dual projective steepest-edge rule in the second row.

It is seen from the preceding table that the dual Dantzig rule is the slowest whereas the dual steepest-edge rule I is the fastest, with time ratio as high as 27.95, though the tests are favorable to the last four rules because they take advantages of the structures of computer IBM RISC system/6000 in pricing and solving systems.

In terms of reported numerical results, the dual rules appear to be inferior to their primal counterparts slightly (see Tables 11.1 and 12.1). However, Forrest and Goldfarb indicate that these data are unfair to the dual rules, because their primal codes were optimized while the dual ones were not; the numbers of hours required by the last four codes in Table 12.1 should be reduced by 10 % at least, though decrements of the dual Dantzig and Devex are not as so much. Therefore, the dual rules are actually efficient, compared to their primal counterparts.

The steepest-edge rule and Devex approximates as well as their dual variants are superior over the standard rules with large margins, and are widely used in many commercial codes, such as CPLEX.

12.3 Dual Largest-Distance Rule

This section derive a dual variant of the largest-distance rule (Sect. 11.4). The basic idea is to determine a leaving variable by finding an inequality constraint, mostly violated by the current solution in the reduced space.

Let $B = \{1, \cdots, m\}$, $N = A \backslash B$ be the current basis and nonbasis of the standard problem (1.8). Assume that the associated simplex tableau is dual but not primal feasible, i.e., $\bar{b} = \bar{x}_B \not\geq 0$.

For some $i \in B$, the inequality $\bar{x}_i < 0$ implies that the solution, $\bar{x}_N = 0$, in the reduced space violates constraint

$$\bar{b}_i - (w^i)^{\mathrm{T}} x_N \geq 0,$$

where $\bar{b} = B^{-1}b$ and

$$w^i = N^{\mathrm{T}} B^{-\mathrm{T}} e_i.$$

The signed distance from point $\bar{x}_N = 0$ to the boundary $\bar{b}_i - (w^i)^{\mathrm{T}} x_N = 0$ is $\bar{x}_i / \|w^i\|$, where any norm is allowed in principle though the Euclidean norm might be preferable.

The dual Dantzig conventional rule is

$$p \in \arg\min\{\bar{x}_i \,|\, i = 1, \cdots, m\},$$

which does not necessarily correspond to the mostly violated inequality constraint, since the according distance would be very small when $\|w^p\|$ is large, as the point $\bar{x}_N = 0$ may not far away from the boundary actually. So, it should not be expected that this rule performs satisfactorily.

If one determines a leaving variable that corresponds to the mostly violated inequality constraint, the following rule follows.

Rule 12.3.1 (Dual largest-distance row rule) Select a pivot row index

$$p \in \arg\min\{\bar{x}_i / \|w^i\| \quad | \, i = 1, \cdots, m\}.$$

Involving $\|w^i\|$, the preceding rule is cumbersome, compared with the (primal) largest-distance rule. Like in the primal context, however, $\|w^i\|^2$, $i = 1, \cdots, m$ can be computed recursively, so that the rule would be still practicable.

Consider the set of n-dimensional vectors

$$\sigma^i = A^{\mathrm{T}} B^{-\mathrm{T}} e_i, \quad i = 1, \cdots, m.$$

Note that w^i is an $(n - m)$-subvector of σ^i for any $i = 1, \ldots, m$. Assume that x_p leaves and x_q enters the basis, then it is not difficult to derive the following recurrence relation:

$$\tilde{\sigma}^p = (1/\bar{a}_{pq})\sigma^p,$$
$$\tilde{\sigma}^i = \sigma^i - (\bar{a}_{iq}/\bar{a}_{pq})\sigma^p, \quad i = 1, \cdots, m, i \neq p.$$

where pivot column $\bar{a}_q = B^{-1}a_q$ is available. Further, recurrence formulas of squares of their norms can be obtained, i.e.,

$$\|\tilde{\sigma}^p\|^2 = (1/\bar{a}_{pq})^2\|\sigma^p\|^2,$$
$$\|\tilde{\sigma}^i\|^2 = \|\sigma^i\|^2 - 2(\bar{a}_{iq}/\bar{a}_{pq})\sigma^{i^{\mathrm{T}}}\sigma^p + (\bar{a}_{iq}/\bar{a}_{pq})^2\|\sigma^p\|^2, \quad i = 1, \cdots, m, i \neq p,$$

combining which, $\sigma^{i^{\mathrm{T}}}\sigma^p = e_i^{\mathrm{T}}B^{-1}N\sigma^p$ and $\|\sigma^i\|^2 = \|w^i\|^2 + 1$ leads to the required formulas

$$\|\tilde{w}^p\|^2 = (1/\bar{a}_{pq})^2(\|(w^p\|^2 + 1) - 1, \tag{12.13}$$
$$\|\tilde{w}^i\|^2 = \|w^i\|^2 - 2(\bar{a}_{iq}/\bar{a}_{pq})u_i + (\bar{a}_{iq}/\bar{a}_{pq})^2(\|w^p\|^2 + 1), \quad i = 1, \cdots, m, i \neq p, \tag{12.14}$$

where

$$Bu = \sum_{j=m+1}^{n} \sigma_j a_j,$$

and σ_j is the jth component of σ^p (i.e., the $(j-m)$th component of w^p). Note that components of w^p are available entries of the pivot row, and hence $\|w^p\|^2 = (w^p)^{\mathrm{T}}w^p$ can be calculated directly.

12.4 Dual Nested Rule

This section derives a dual variant of the nested rule (Sect. 11.5). Regarding negative components of the primal basic solution as elimination target, it focuses on the most stubborn ones among them.

Let $\epsilon > 0$ be primal feasibility tolerance. At the beginning of an iteration, a set I_1 of row indices is given the priority for associated basic indies to leave the basis. A row index in I_1, associated with the smallest component of $\bar{x}_B$, is determined. If the component is less than $-\epsilon$, then the corresponding basic index is selected to leave the basis, and the I_1 for the next iteration is born from the current I_1 by including all its row indices, associated with $\bar{x}_B$'s components less than $-\epsilon$. In the other case, do the same as before with $I_2 = B \backslash I_1$; if there is still no basic components less than $-\epsilon$, then optimality is attained.

The following is a nested variant of Dantzig conventional dual rule.

Rule 12.4.1 (Dual nested row rule: Dantzig) Given primal tolerance $\epsilon > 0$. Set $I_1 = B$ and $I_2 = \emptyset$.

1. Go to step 4 if $I_1 = \{i \in I_1 \mid \bar{x}_i < -\epsilon\} \neq \emptyset$.
2. Go to step 4 if $I_1 = \{i \in I_2 \mid \bar{x}_i < -\epsilon\} \neq \emptyset$.
3. Stop (optimality achieved).
4. Select a pivot row index $p \in \arg\min \{\bar{x}_i \mid i \in I_1\}$.
5. Update: $I_1 = I_1 \backslash p,\ I_2 = B \backslash I_1$.

In the first iteration or iterations in which I_2 is touched, the preceding rule actually proceeds the same as the standard dual row rule. After such an iteration, a series of iterations with nested pivoting follow, as might be called a "circle", where each I_1 is a proper subset of its predecessor. In the kth iteration of a circle, basic components associated with I_1 are less than $-\epsilon$ all the time. It is therefore reasonable to select such a stubborn one to leave the basis.

It is possible to turn any full dual rule to a nested version. The dual nested steepest-edge rule can be obtained from Rule 12.4.1 by using

$$p \in \arg\min \{\bar{x}_i / \|h^i\| \mid i \in I_1\}$$

in step 4 of it instead, the dual nested Devex rule obtained by using

$$p \in \arg\min\{\bar{x}_i / s_i \mid i \in I_1\}$$

and the dual nested largest-distance rule by using

$$p \in \arg\min\{\bar{x}_i / \|w^i\| \mid i \in I_1\}$$

(see relevant sections for notations).

In view of the performance of their primal counterparts, it might be expected that the dual nested rules and the dual nested largest-distance rules perform satisfactorily. The dual nested rules are easy to implement, fortunately. However, there is no any computational experience at present.

Chapter 13
Simplex Phase-I Method

The tableau simplex algorithm requires a feasible simplex tableau to get itself started, whereas the revised simplex algorithm starts from a feasible basis or its inverse. The task of a Phase-I method is to provide such a starting point to Phase-II for achieving optimality. Such a two-Phase methodology has been successful in practice, compared with one-Phase, like the big-M.

Not have been used in practice, the artificial-variable method, presented in Sect. 3.3, is usually seen in textbooks only, because it is so rigid that there is no much choice for an initial basis, as the basis has to involve all the artificial variables, let alone expending problem's scale. Moreover, some cumbersome treatment has to be carried out subsequently after zero optimal value of the auxiliary program is attained but there are still artificial variables remanning basic.

This chapter will focus on practicable Phase-I methods, which are artificial variable free, or involves a single artificial variable only (for further references, see Pan 1994b; Pan and Li 2003; Pan et al. 2004).

For simplicity of exposition, the conventional pivot rule will be utilized in this chapter. Nevertheless, it is preferable to use the nested largest-distance rule instead, although all rules presented in Chap. 11 are applicable for Phase-I.

13.1 Infeasibility-Sum Method

The auxiliary objective function of this method only involves variables, associated with negative components of the current basic solution. The method is widely used in practice, as it is artificial variable free, and performs remarkably.

Assume that an initial simplex tableau of form (3.18) is available, associated with basic solution

$$\bar{x}_B = \bar{b}, \quad \bar{x}_N = 0. \tag{13.1}$$

P.-Q. PAN, *Linear Programming Computation*, DOI 10.1007/978-3-642-40754-3_13,

Introduce notation

$$I = \{i = 1, \cdots, m \mid \bar{b}_i < 0\}, \quad \bar{I} = \{1, \cdots, m\} \backslash I. \tag{13.2}$$

Assume that $I \neq \emptyset$. Construct the following auxiliary program:

$$\begin{array}{ll} \min & -\sum_{i \in I} x_{j_i}, \\ \text{s.t.} & x_B = \bar{b} - \bar{N} x_N, \\ & x_{j_i} \geq 0, \quad i \in \bar{I}; \quad x_j \geq 0, \ j \in N, \end{array} \tag{13.3}$$

where the objective function, called "infeasibility-sum", is the negative sum of variables, associated with negative components of the basic solution. Note that the constraint system of the auxiliary program is the same as that of the original problem, except the nonnegativity restriction is only imposed on variables, associated with nonnegative components of the basic solution.

It is clear that solution (13.1) is feasible to the auxiliary program. The associated feasible simplex tableau can be obtained by eliminating nonzero entries in the objective row by relevant elementary transformations. Consequently, the reduced objective function is

$$-\sum_{i \in I} x_{j_i} = w_0 + \bar{z}_N^{\mathrm{T}} x_N, \tag{13.4}$$

where

$$w_0 = -\sum_{i \in I} \bar{b}_i > 0, \quad \bar{z}_j = \sum_{i \in I} \bar{a}_{i,j}, \quad j \in N.$$

The positiveness of objective value w_0 comes from (13.2).

Theorem 13.1.1 (Infeasibility test). *If reduced costs of an auxiliary simplex tableau are all nonnegative, the original problem is infeasible.*

Proof. Assume that $\bar{z}_N \geq 0$, and $\tilde{x} \geq 0$ is a feasible solution to the original problem. Substitute $\tilde{x}$ to (13.4) results in a system, whose left-hand side is clearly less than or equal to zero, and, by $w_0 > 0$ and $\bar{z}_N \geq 0$, the right-hand side is strictly greater than zero, as is a contradiction. Therefore, the statement is valid. □

If there is a negative reduced cost, then any existing column rule applies. Assume that a column index q was determined such that

$$\bar{z}_q = \sum_{i \in I} \bar{a}_{iq} < 0. \tag{13.5}$$

As there is no nonnegativity restriction on current infeasible basic variables, the auxiliary program could be unbounded even if the original problem itself bounded.

In fact, it is the case when $\bar{I} = \emptyset$ or the following condition holds:

$$\{i \in \bar{I} \,|\, \bar{a}_{iq} > 0\} = \emptyset. \tag{13.6}$$

Thus, the conventional row rule is not applicable. To overcome this difficulty, it is modified as follows.

Rule 13.1.1 (Auxiliary row rule) Determine a row index p such that

$$\alpha = \bar{b}_p/\bar{a}_{pq} = \begin{cases} \min\{\bar{b}_i/\bar{a}_{iq} | \bar{a}_{iq} > 0, i \in \bar{I}\}, & \text{if } \{i \in \bar{I} \,|\, \bar{a}_{iq} > 0\} \neq \emptyset, \\ \max\{\bar{b}_i/\bar{a}_{iq} | \bar{a}_{iq} < 0, i \in I\}, & \text{otherwise} \end{cases} \tag{13.7}$$

It will be seen a little later the bonus, yielding from use of "max" in the second expression of (13.7).

Lemma 13.1.2. *Assume that the selected column index q satisfies (13.5). Then rule 13.1.1 determines a row index p such that the simplex tableau, resulting from the according basis change, is feasible to the auxiliary program.*

Proof. First, it can be asserted that there is a row index $t \in I$ such that $\bar{a}_{tq} < 0$; because otherwise from

$$\bar{a}_{iq} \geq 0, \quad \forall i \in I,$$

it follows that

$$\sum_{i \in I} \bar{a}_{iq} \geq 0,$$

which contradicts the assumption (13.5). Therefore, the second expression of (13.7) is always well-defined, hence so is the row rule.

Take $\bar{a}_{pq}$ as the pivot. Executing according elementary transformations leads to a new simplex tableau, whose right-hand side can be expressed in terms of its predecessor's entries (see (3.14)), i.e.,

$$\hat{b}_i = \bar{b}_i - \alpha \bar{a}_{iq}, \quad i \in \bar{I}, \, i \neq p,$$

$$\hat{b}_p = \alpha,$$

where α is determined by (13.7). It is easy to verify that these components are all nonnegative, hence the resulting simplex tableau is again feasible to the auxiliary program. □

Theorem 13.1.3. *Assume feasibility of the original problem and nondegeneracy throughout solution process. The number of basic infeasible variables strictly decreases in finitely many iterations.*

Proof. It is seen from the proof of Lemma 13.1.2 that the second expression of (13.7) is still well-defined if "max" there replaced by "min", i.e.,

$$\alpha_1 \triangleq \bar{b}_s/\bar{a}_{sq} = \min\{\bar{b}_i/\bar{a}_{iq} | \bar{a}_{iq} < 0, i \in I\} > 0. \tag{13.8}$$

If, in some iteration, the stepsize α is determined by the first expression of (13.7), and it holds that

$$\alpha \geq \alpha_1, \tag{13.9}$$

then the component, associated with row index s, of the new solution satisfies

$$\hat{x}_{j_s} = \bar{b}_s - \alpha \bar{a}_{sq} \geq 0,$$

which implies that a basic infeasible variable becomes feasible, at least.

If, in some iteration, the second expression of (13.7) is used, then, for the new solution, not only the basic components, associated with all row indices $i \in \bar{I}$, are feasible, but so are basic components, associated with row indices

$$s \in \{i \in I \mid \bar{a}_{iq} < 0\} \neq \emptyset.$$

This is of course favorable.

Assume that the number of basic infeasible variables does not strictly decreases forever. In each iteration, there is always a negative cost since the original problem is feasible (Theorem 13.1.1), and the stepsize α is determined by the first expression of (13.7), which does not satisfy (13.9). But the number of basis is finite, hence there are some bases appear infinitely, as cycling occurs. This is impossible because the objective value of the auxiliary program strictly decreases under the nondegeneracy assumption. Therefore the number of basic infeasible variables strictly decreases in finitely many iterations. □

When, after some iteration, the number of basic infeasible variables strictly decreases, but does not vanish, a new auxiliary program is constructed, whose objective function (infeasibility-sum) involves less variables than its predecessor, and iterations are carried out, until infeasibility of the original problem is detected or a feasible simplex tableau is reached. Involving a series of infeasibility-sums as objective functions in general, this method is sometimes referred to as *method with piecewise sum of infeasibilities*.

The overall steps is put in the following algorithm.

Algorithm 13.1.1 (Tableau Phase-I: infeasibility-sum). Initial: simplex tableau of form (3.18). This algorithm finds a feasible simplex tableau by handling auxiliary programs of form (13.3).

1. Stop if $\bar{b}_i \geq 0, \forall i = 1, \cdots, m$ (a feasible simplex tableau reached).
2. For all $j \in N$, compute $\bar{z}_j = \sum_{i \in I} \bar{a}_{i,j}$, where $I = \{i = 1, \cdots, m \mid \bar{b}_i < 0\}$.

3. Stop if $\bar{z}_N \geq 0$ (the original problem is infeasible).
4. Form a simplex tableau to the auxiliary program (13.3).
5. Carry out a single iteration of simplex Algorithm 3.2.1, in which Rule 13.1.1 is used for row pivoting instead.
6. Go to step 1.

Example 13.1.1. Solve the following problem by Algorithm 13.1.1 in Phase-I:

$$\begin{array}{ll} \min & f = -x_1 + x_2 - 2x_3, \\ \text{s.t.} & x_1 - 3x_2 - 2x_3 + x_4 = -4, \\ & x_1 - x_2 + 4x_3 - x_5 = 2, \\ & -3x_1 + x_2 + x_3 + x_6 = 8, \\ & x_j \geq 0, \quad j = 1, \cdots, 6. \end{array}$$

Answer Phase-I. Multiply the second equation by -1 to turn the coefficient of x_5 to 1. Setting $I = \{1, 2\}$, construct auxiliary program

$$\begin{array}{ll} \min & f = -x_4 - x_5, \\ \text{s.t.} & x_1 - 3x_2 - 2x_3 + x_4 = -4, \\ & -x_1 + x_2 - 4x_3 + x_5 = -2, \\ & -3x_1 + x_2 + x_3 + x_6 = 8, \\ & x_j \geq 0, \quad j = 1, 2, 3, 6. \end{array}$$

Its initial tableau is

x_1	x_2	x_3	x_4	x_5	x_6	RHS
1	−3	−2	1			−4
−1	1	−4		1		−2
−3	1	1			1	8
			−1	−1		

Respectively add the first and the second row to the objective row to eliminate the nonzero basic entries, yielding simplex tableau

x_1	x_2	x_3	x_4	x_5	x_6	RHS
1	−3	−2	1			−4
−1	1	−4		1		−2
−3	1	1*			1	8
	−2	−6				−6

Iteration 1:

$q = 3; \min\{8/1\} = 8, \ p = 3$. So 1 in the x_3 column and the third row is the pivot. Add $2, 4, 5$ times of row 3 to rows 1,2,4, respectively.

x_1	x_2	x_3	x_4	x_5	x_6	RHS
−5	−1		1		2	12
−13	5			1	4	30
−3	1	1			1	8
−18	4				6	42

The right-hand side of the preceding tableau is nonnegative, hence Phase-I is finished. To go to Phase-II, use coefficients of the original objective function to cover the bottom (objective) row.

x_1	x_2	x_3	x_4	x_5	x_6	RHS
−5	−1		1		2	12
−13	5			1	4	30
−3	1	1			1	8
−1	1	−2				

Add 2 times of row 3 to the bottom row to eliminate nonzero basic entries, obtaining feasible simplex tableau

x_1	x_2	x_3	x_4	x_5	x_6	RHS
−5	−1		1		2	12
−13	5			1	4	30
−3	1	1			1	8
−7	3				2	16

Carry out Phase-II from the preceding tableau. By the conventional rule, x_1 column is selected as the pivot column. All components of this column are nonpositive, as detects unboundedness of the original problem.

In practice, used is usually the following revised version of Algorithm 13.1.1.

Algorithm 13.1.2 (Phase-I: infeasibility-sum). Initial: (B, N), $B^{-1}, \bar{x}_B = B^{-1}b$. This algorithm finds a basic feasible solution to the standard LP problem by handling auxiliary programs of form (13.3).

1. Stop if $\bar{x}_B \geq 0$ (producing a basic feasible solution).
2. Construct $c_B : c_{j_i} = \begin{cases} -1, & \text{If } \bar{x}_{j_i} < 0, \\ 0, & \text{If } \bar{x}_{j_i} \geq 0, \end{cases} \quad i = 1, \cdots, m.$
3. Carry out a single iteration of Algorithm 3.5.1 (or 3.5.2), in which Rule 13.1.1 is used for row pivoting.
4. Stop if it terminates as step 3 (or step 2) (the original problem is infeasible).
5. Go to step 1.

13.2 Single-Artificial-Variable Method

Although the infeasible-sum method is artificial variable free, the row pivot rule used in it is cumbersome, compared with the normal row rule. This section describes a method, which uses the normal row rule and only involves a single artificial variable.

Let (B, N) be basis and nonbasis of the standard problem (1.8). There are two schemes for the method.

Scheme 1. The auxiliary program is based on the original data of the problem.

Given m-dimensional vector $h \geq 0$ such that $b - Bh \neq 0$. Introducing artificial variable x_{n+1} and normalized vector

$$a_{n+1} = (b - Bh)/\|b - Bh\|, \tag{13.10}$$

construct the following auxiliary program

$$\begin{array}{ll} \min & x_{n+1}, \\ \text{s.t.} & Ax + a_{n+1}x_{n+1} = b, \quad x, x_{n+1} \geq 0. \end{array} \tag{13.11}$$

It is easy to verify that

$$\bar{x}_B = h, \quad \bar{x}_N = 0, \quad \bar{x}_{n+1} = \|b - Bh\|$$

is a feasible solution to the program, though not necessarily a basic solution. For instance, setting $h = e > 0$ seems to be helpful to avoid zero stepsize, though according a_{n+1} would loose sparsity.

The constraint system of the auxiliary program is equivalent to

$$x_B = \bar{b} - \bar{a}_{n+1}x_{n+1} - \bar{N}x_N, \tag{13.12}$$

where

$$\bar{b} = B^{-1}b, \quad \bar{a}_{n+1} = B^{-1}a_{n+1}, \quad \bar{N} = B^{-1}N. \tag{13.13}$$

If $\bar{b} \geq 0$, then x_{n+1} is clearly allowed to be zero. Thus, erasing the artificial variable leads to a basic feasible solution to the original problem; otherwise, the following applies.

Proposition 13.2.1. *If $\bar{b} \not\geq 0$, row index set $\{i = 1, \cdots, m \mid \bar{a}_{i,n+1} < 0, \bar{b}_i < 0\}$ is nonempty.*

Proof. We show that if $\bar{b}_i < 0$ for some $i \in \{1, \cdots, m\}$, then $\bar{a}_{i,n+1} < 0$. Assume it is not the case, i.e.,

$$\bar{a}_{i,n+1} \geq 0.$$

Then from the preceding, (13.10) and (13.13) it is known that

$$0 \leq \bar{a}_{i,n+1} = e_i^{\mathrm{T}} B^{-1} a_{n+1} = e_i^{\mathrm{T}} B^{-1}(b - Bh)/\|b - Bh\| = (\bar{b}_i - h_i)/\|b - Bh\|,$$

hence $\bar{b}_i \geq h_i \geq 0$, as a contradiction. □

Now assume $\bar{b} \not\geq 0$. If $\bar{x}_N = 0$ is fixed and the value of $\bar{x}_{n+1}$ decreases from $\|b - Bh\|$, then $\bar{x}_B$ changes from h accordingly. If $\bar{b}_i < 0$ for some $i \in \{1, \cdots, m\}$, the according inequality constraint may block-up the value of $\bar{x}_{n+1}$ from decreasing to zero. In fact, the smallest x_{n+1} value, satisfying inequalities

$$x_B = \bar{b} - \bar{a}_{n+1} x_{n+1} \geq 0, \quad x_{n+1} \geq 0,$$

is

$$\alpha = \bar{b}_r / \bar{a}_{r,n+1} = \min\{\bar{b}_i / \bar{a}_{i,n+1} \mid \bar{a}_{i,n+1} < 0, \bar{b}_i < 0, i = 1, \cdots, m\} > 0. \quad (13.14)$$

As the value of $\bar{x}_{n+1}$ decreases, the according value of $\bar{x}_{j_r}$ hits 0 first. Let x_{j_r} leave and x_{n+1} enter the basis, then the resulting basis corresponds to a basic feasible solution to the auxiliary program, with its objective value strictly decreases. As it is lower bounded, solving the auxiliary program will render its optimal solution.

It is not difficult to show that a basic feasible solution or basis to the original problem can be obtained if the optimal value of the auxiliary program is equal to zero (x_{n+1} leaves the basis). The original problems is infeasible if the optimal value of the auxiliary program is positive.

Scheme 2. The auxiliary program is based on a canonical form of the constraint system.

Assume that the canonical form, associated with basis B, is available and that $\bar{b} = B^{-1} b \not\geq 0$. Giving m-dimensional vector $h \geq 0$, and defining

$$\bar{a}_{n+1} = (\bar{b} - h)/\|\bar{b} - h\|, \quad (13.15)$$

construct the following auxiliary program

$$\begin{array}{ll} \min & x_{n+1}, \\ \text{s.t.} & x_B = \bar{b} - \bar{a}_{n+1} x_{n+1} - \bar{N} x_N, \\ & x, x_{n+1} \geq 0. \end{array} \quad (13.16)$$

It is clear that there is a feasible solution to the program, i.e.,

$$\bar{x}_B = h, \quad \bar{x}_N = 0, \quad \bar{x}_{n+1} = \|\bar{b} - h\|.$$

Thus, determining pivot column index $q = n+1$ and row index p satisfying (13.14), one can obtain a feasible basis by performing according basis change. It is noticed that Proposition 13.2.1 is still valid in this case.

As for how to determine h, it is only required that $h \geq 0$, in principle. With respect to practice, however, it is not proper for the associated $\|\bar{a}_{n+1}\|$ being too large. A simple choice is $h = e > 0$. Another choice is setting

$$\delta_i \geq 0, \quad i = 1, \cdots, m.$$

and taking

$$h_i = \begin{cases} \delta_i, & \text{if } \bar{b}_i < \delta_i, \\ \bar{b}_i, & \text{otherwise,} \end{cases} \quad i = 1, \cdots, m.$$

where $\delta_i \geq 0, i = 1, \cdots, m$ are a set of constants, which are equal or close to zero. Consequently, (13.15) becomes

$$\bar{a}_{i,n+1} = \begin{cases} \bar{b}_i - \delta_i, & \text{if } \bar{b}_i < \delta_i \\ 0, & \text{otherwise,} \end{cases} \quad i = 1, \cdots, m. \tag{13.17}$$

We prefer the latter because it is better for maintaining sparsity of $\bar{a}_{n+1}$.

More specifically, assume availability of a simplex tableau, say (3.18), of the original problem. Inserting the artificial x_{n+1} column, we obtain the auxiliary simplex tableau below:

x_B^T	x_N^T	x_{n+1}	RHS
I	$\bar{N}$	$\bar{a}_{n+1}$	$\bar{b}$

(13.18)

where the original objective row is excluded. Determine row index r by (13.14). Then drop x_{j_r} from and enter x_{n+1} to the basis by elementary transformations, so that x_{n+1} becomes the rth basic variable. Taking the rth row as the objective row, the auxiliary program can be solved by the simplex method. If the optimal value vanishes and x_{n+1} has left the basis, a feasible simplex tableau to the original problem is obtained. If the optimal value is nonzero, the original problem is infeasible.

The steps can be summarized to the following algorithm.

Algorithm 13.2.1 (Tableau Phase-I: single-artificial-variable). Initial: a simplex tableau of form (13.18). This algorithm finds a feasible simplex tableau to the standard LP problem.

1. Select a row index $r \in \arg\min\{\bar{b}_i \mid i = 1, \cdots, m\}$.
2. Stop if $\bar{b}_r \geq 0$ (infeasible problem).
3. Determine row index $r \in \arg\min\{\bar{b}_i/\bar{a}_{i,n+1} | \bar{a}_{i,n+1} < 0, \bar{b}_i < 0, i = 1, \cdots, m\}$.
4. Carry out elementary transformation to convert $\bar{a}_{r,n+1}$ to 1, and eliminate all other nonzeros of the column.

5. Determine column index $q \in \arg\max_{j \in N} \bar{a}_{rj}$.
6. Stop if $\bar{a}_{rj} \leq 0$.
7. Determine row index $p \in I = \arg\min\{\bar{b}_i/\bar{a}_{i\,q} | \bar{a}_{i\,q} > 0;\ i = 1, \cdots, m\}$.
8. Set $p = r$ if $r \in I$.
9. Carry out elementary transformations to convert $\bar{a}_{pq}$ to 1, and eliminate all other nonzeros of the column.
10. Stop if $p = r$ (feasibility achieved).
11. Go to step 5.

It is not difficult to transform Algorithm 13.2.1 to a revised version, as is omitted here. We will see that the single-variable auxiliary program can be handled in a relevant and favorable manner in the so-called "reduced simplex" context (Sect. 15.3).

Example 13.2.1. Find a feasible simplex tableau to the following problem by Algorithm 13.2.1:

$$\begin{array}{lrcl}
\min & f = x_1 + 3x_2 - 2x_3 + 6x_4, & & \\
\text{s.t.} & -x_1 + \ x_2 - \ x_3 \qquad + x_5 & = & -1, \\
& -3x_1 + \ x_2 + 2x_3 \qquad\qquad + x_6 & = & 2, \\
& x_1 - 3x_2 - 3x_3 + x_4 \qquad\qquad & = & -4, \\
& x_j \geq 0, \quad j = 1, \cdots, 6. & &
\end{array}$$

Answer The constraint system is of a canonical form. Set $\delta_1 = \delta_2 = 0$. Introducing artificial variable x_7, construct an auxiliary program of form (13.16):

$$\begin{array}{lrcl}
\min & x_7, & & \\
\text{s.t.} & -x_1 + \ x_2 - \ x_3 \qquad + x_5 \qquad - \ x_7 & = & -1, \\
& -3x_1 + \ x_2 + 2x_3 \qquad\qquad + x_6 \qquad & = & 2, \\
& x_1 - 3x_2 - 3x_3 + x_4 \qquad\qquad - 4x_7 & = & -4, \\
& x_j \geq 0, \quad j = 1, \cdots, 7. & &
\end{array}$$

Its initial tableau is (we put the objective row of the original problem at the bottom to turn to Phase-II conveniently when Phase-I is finished).

x_1	x_2	x_3	x_4	x_5	x_6	x_7	RHS
−1	1	−1		1		−1*	−1
−3	1	2			1		2
1	−3	−3	1			−4	−4
1	3	−2	6				

Iteration 1:

1. $\min\{-1, 2, -4\} = -4 < 0$.
3. $\max\{-1/-1, -4/-4\} = 1, r = 1$.

4. Multiply row 1 by -1, then add 4 times of row 1 to row 3:

x_1	x_2	x_3	x_4	x_5	x_6	x_7	RHS
1	−1	1		−1		1	1
−3	1	2			1		2
5*	−7	1	1	−4			
1	3	−2	6				

5. $\max\{1, -1, 1, -1\} = 1 > 0,\ q = 1.$
7. $\min\{1/1, 0/5\} = 0,\ p = 3.$
8. $p \neq 1.$
9. Multiply row 3 by $1/5$, then add $-1, 3, -1$ times of row 3 to rows 1,2,4, respectively:

x_1	x_2	x_3	x_4	x_5	x_6	x_7	RHS
	2/5	4/5	−1/5	−1/5		1	1
	−16/5	13/5	3/5	−12/5	1		2
1	−7/5	1/5*	1/5	−4/5			
	22/5	−11/5	29/5	4/5			

Iteration 2:

5. $\max\{2/5, 4/5, -1/5, 1/5\} = 4/5 > 0,\ q = 3.$
7. $\min\{1/(4/5), 2/(13/5), 0/(1/5)\} = 0,\ p = 3.$
8. $p \neq 1.$
9. Multiply row 3 by 5, then add $-4/5, -13/5, 11/5$ times of row 3 to rows 1,2,4, respectively:

x_1	x_2	x_3	x_4	x_5	x_6	x_7	RHS
−4	6		−1	3		1	1
−13	15*		−2	8	1		2
5	−7	1	1	−4			
11	−11		8	−8			

Iteration 3:

5. $\max\{-4, 6, -1, 3\} = 6 > 0,\ q = 2.$
7. $\min\{1/6, 2/15\} = 2/15,\ p = 2.$
8. $p \neq 1.$
9. Multiply row 2 by $1/15$, then add $-6, 7, 11$ times of row 2 to rows 1,3,4, respectively:

x_1	x_2	x_3	x_4	x_5	x_6	x_7	RHS
6/5*			−1/5	−1/5	−2/5	1	1/5
−13/15	1		−2/15	8/15	1/15		2/15
−16/15		1	1/15	−4/15	7/15		14/15
22/15			95/15	−32/15	11/15		22/15

Iteration 4:

5. $\max\{5/6, -1/5, -1/5, -2/5\} = 5/6 > 0, q = 1.$
7. $\min\{(1/5)/(11/5)\} = 1/11,\ p = 1.$
8. $p = 1.$
9. Multiply row 1 by 5/6, then add $13/15, 16/15, -6/5, -22/15$ times of row 1 to rows 2,3,4, respectively:

x_1	x_2	x_3	x_4	x_5	x_6	x_7	RHS
1			−1/6	−1/6	−1/3	5/6	1/6
	1		−5/18	7/18	−2/9	13/18	5/18
		1	−1/9	−4/9	1/9	8/9	10/9
			61/9	−17/9	11/9	−11/9	11/9

As the artificial variable x_7 has already left the basis, the resulting is a feasible simplex tableau to the original problem.

It is noted that δ_1 and δ_2 above were all set to 0, consequently the first 3 iterations made no progress due to degeneracy. Although parameters $\delta_i, i = 1, \cdots, m$ are required to be nonnegative theoretically, therefore, they should be set to different positive values practically. Moreover, it seems to be relevant to set $\delta_i \ll 1$ such that $\|\bar{a}_{n+1}\|$ is not too large.

13.3 The Most-Obtuse-Angle Column Rule

Surprisingly enough, the conventional pivot rule used in the dual simplex algorithm was found to be useful, with favorable numerical results, for achieving primal feasibility (Pan 1994a); it is to say that the rule is itself applicable even when reduced costs are not nonnegative. This leads to a simple artificial-variable-free variant involving no reduced cost, as turns out to be a counterpart of the most-obtuse-angle row rule (Sect. 14.3).

Assume that a simplex tableau, say (3.18), is infeasible, and a pivot row index p has been selected by the conventional rule

$$p \in \arg\min\{\bar{b}_i \mid i = 1, \cdots, m\}. \tag{13.19}$$

So it holds that $\bar{b}_p < 0$.

The key of the presented Phase-I method is to use the following column rule.

Rule 13.3.1 (Most-obtuse-angle column rule) Determine pivot column index

$$q \in \arg\min_{j \in N} \bar{a}_{p\,j}. \tag{13.20}$$

In case of $\bar{a}_{pq} < 0$, after the basis change, the basic infeasible variable x_{j_p} becomes nonbasic (hence feasible), and the pth component of the right-hand side of the tableau changes from negative to positive, that is, $\hat{b}_p = \bar{b}_p/\bar{a}_{pq} > 0$.

The steps are put in the following algorithm. A generalized version of it will be derived alternatively in Sect. 20.7.

Algorithm 13.3.1 (Tableau Phase-I: the most-obtuse-angle column rule). Initial: simplex tableau of form (3.18). This algorithm finds a feasible simplex tableau to the standard LP problem.

1. Select row index $p \in \arg\min\{\bar{b}_i \mid i = 1, \cdots, m\}$.
2. Stop if $\bar{b}_p \geq 0$.
3. Determine column index $q \in \arg\min_{j \in N} \bar{a}_{p\,j}$.
4. Stop if $\bar{a}_{pq} \geq 0$.
5. Carry out elementary transformations to convert $\bar{a}_{pq}$ to 1, and eliminate all other nonzeros of the column.
6. Go to step 1.

It is noted that only the right-hand side and the pivot row of the simplex tableau are used in each iteration of the preceding Algorithm.

Theorem 13.3.1. *Assume termination of Algorithm 13.3.1. It terminates either at*

(i) Step 2, obtaining a basic feasible solution; or at
(ii) Step 4, detecting infeasibility of the problem.

Proof. When it terminates at step 2, it is clear that $\bar{b} \geq 0$, hence a feasible simplex tableau is achieved. In case of termination at step 4, it holds that $\bar{b}_p < 0$ and that $\bar{a}_{p\,j} \geq 0, \forall j \in N$. By Lemma 3.3.1, there is no feasible solution to the problem. □

The algorithm is not a monotone one, that is, the objective value would not monotonically change in solution process. Finiteness of it is hence not guaranteed even if nondegeneracy is assumed. Although Guerrero-Garcia and Santos-Palomo (2005) offer a cycling instance for this algorithm, it may be still expected that cycling hardly happens in practice.

Geometric meaning of the most-obtuse-angle column rule may be revealed by investigating in the dual space. Let $(\bar{y}, \bar{z})$ be the current dual basic solution. The $(m+n)$-dimensional vector

$$d \triangleq \begin{pmatrix} h \\ \sigma_N \\ \sigma_B \end{pmatrix} = \begin{pmatrix} -B^{-\mathrm{T}} \\ N^{\mathrm{T}} B^{-\mathrm{T}} \\ I \end{pmatrix} e_p \tag{13.21}$$

is an uphill direction. In fact, it is verified that d is in the null space of coefficient matrix of the dual constraint equalities, satisfying

$$A^{\mathrm{T}}d + \sigma = 0,$$

and forms an acute angle with the dual objective gradient, i.e.,

$$(b^{\mathrm{T}}, 0, 0)d = b^{\mathrm{T}}h = -\bar{x}_{j_p} > 0.$$

The negative σ_q implies that d forms the largest possible obtuse angle with the gradient e_{m+q} of the dual nonnegative constraint $z_q \geq 0$. Therefore, if the uphill direction is close to the dual objective gradient, then e_{m+q} tends to form the most obtuse angle with the dual objective gradient. According to the heuristic characteristic of optimal basis (Sect. 2.5), it is favorable to let the dual constraint $z_q \geq 0$ be satisfied as equality, accordingly entering x_q to the basis.

We stress that the use of the conventional dual rule in step 1 is just for simplicity, does not mean itself the best choice. In fact, it should be much better to use rules, presented in Chap. 12 instead. From a geometric point of view, in fact, the most-obtuse-angle column rule should be best matched by the dual steepest-edge rule II.

Nevertheless, Rule 13.3.1 does not employ information associated with the objective function at all. Taking into account the extent to which the current vertex violates dual constraints, we suggest the following variant.

Rule 13.3.2 (Variant of most-obtuse-angle column rule) Given constant $0 < \tau \leq 1$. Select pivot column index

$$q \in \arg\min\{\bar{z}_j \mid \bar{a}_{p\,j} \leq \tau\theta, \ j \in N\}, \qquad \theta = \min_{j \in N} \bar{a}_{p\,j} < 0$$

In order to widen the range of choice, τ should be close to 1, so that in the case when $\bar{z}_N$ has negative components, the dual constraint inequality, that is violated the most, will be satisfied as equality. It might be suitable to set $\sigma = 0.95$, or so.

An advantage of this artificial-variable-free Phase-I method lie in its good numerical stability and remarkable simplicity. Even though there are no related numerical results available at present, it seems to be promising, as its dual counterpart performs remarkably for solving large and sparse problems in computational experiments (see Sect. 14.3).

Example 13.3.1. Find a feasible simplex tableau to the following problem by Algorithm 13.3.1:

$$\begin{array}{lrl}
\min & f = 2x_1 - x_2, & \\
\text{s.t.} & 2x_1 - \ x_2 + x_3 & = -2, \\
& x_1 + 2x_2 \qquad + x_4 & = \ \ 3, \\
& -8x_1 + \ x_2 \qquad - x_4 + x_5 & = -4, \\
& x_j \geq 0, \quad j = 1, \cdots, 5. &
\end{array}$$

Answer Initial: Adding the second equality constraint to the third to eliminate x_4. Such doing results in the following simplex tableau.

x_1	x_2	x_3	x_4	x_5	RHS
2	−1*	1			−2
1	2		1		3
−7	3			1	−1
2	−1				

Iteration 1:

1. $\min\{-2, 3, -1\} = -2 < 0,\ p = 1.$
3. $\min\{2, -1\} = -1 < 0, q = 2.$
5. Multiply row 1 by −1, then add $-2, -3, 1$ times of row 1 to rows 2,3,4, respectively:

x_1	x_2	x_3	x_4	x_5	RHS
−2	1	−1			2
5		2	1		−1
−1*		3		1	−7
		−1			2

Iteration 2:

1. $\min\{2, -1, -7\} = -7 < 0,\ p = 3.$
3. $\min\{-1, 3\} = -1 < 0,\ q = 1.$
5. Add $2, 5$ times of row 3 to rows 1,2, respectively:

x_1	x_2	x_3	x_4	x_5	RHS
	1	−7			16
		17	1		34
1		−3		−1	7
		−1			2

The right-hand side of the above tableau is nonnegative, hence the obtained is a feasible simplex tableau, from which Phase-II can get itself started.

In practice, usually used is the following revised version of the preceding algorithm.

Algorithm 13.3.2 (Phase-I: most-obtuse-angle column rule). Initial: (B, N), $B^{-1}, \bar{x}_B = B^{-1}b$. This algorithm finds a basic feasible solution to the standard LP problem.

1. Select pivot row index $p \in \arg\min\{\bar{x}_{j_i} \mid i = 1, \cdots, m\}$.
2. Stop if $\bar{x}_{j_p} \geq 0$ (feasibility achieved).

3. Compute $h = B^{-T}e_p$, $\sigma_N = N^T h$.
4. Determine pivot column index $q \in= \arg\min_{j \in N} \sigma_j$.
5. Stop if $\sigma_q \geq 0$ (infeasible problem).
6. Compute $\bar{a}_q = B^{-1}a_q$ and $\alpha = \bar{x}_{j_p}/\sigma_q$.
7. Set $\bar{x}_q = \alpha$, and update:$\bar{x}_B = \bar{x}_B - \alpha\bar{a}_q$ if $\alpha \neq 0$.
8. Update B^{-1} by (3.23).
9. Update (B, N) by exchanging j_p and q.
10. Go to step 1.

13.4 Perturbation of Reduced Costs

As a result of not involving reduced costs, the feasible solution, generated by Algorithm 13.3.2, could be far from dual feasibility, and hence the number of iterations subsequently required by Phase-II would be large. In this aspect, the Phase-I method, presented in this section, seems to be advantageous (Pan 2000b,c).

It solves an auxiliary perturbed program. Let (3.18) be an infeasible simplex tableau and let $\delta_j \geq 0, \ j \in N$ be predetermined perturbation parameters. The auxiliary program results from perturbing reduced costs to nonnegative values. More precisely, introduce index set

$$J = \{j \in N | \bar{z}_j < \delta_j\}. \tag{13.22}$$

Replacing reduced costs $\bar{z}_j$ in the simplex tableau by

$$\bar{z}'_j = \begin{cases} \delta_j, & \text{if} j \in J, \\ \bar{z}_j, & \text{if} j \in N \backslash J, \end{cases} \tag{13.23}$$

leads to a dual feasible tableau, associated with a perturbed program. As a result, the dual simplex algorithm can get itself started, until optimality of the perturbed program is achieved, or dual unboundedness of it is detected.

Theorem 13.4.1. *If the auxiliary perturbed program is dual unbounded, the original problem is infeasible.*

Proof. The perturbed program has the same constraints as the original problem. Unboundedness of the former implies that there is no feasible solution to former. and hence so to the latter. □

Assume that optimality of the perturbed program is achieved. It might be well to assume that (3.18) is again the final tableau, associated with basis B. The right-hand side of it is now nonnegative. Consequently, computing $\hat{z}_N = c_N - N^T B^{-1} c_B$ (restoring reduced costs of the original problem) and covering $\bar{z}_N$ with $\hat{z}_N$ leads to a feasible simplex tableau to the original problem.

The overall steps are put into the following algorithm.

Algorithm 13.4.1 (Tableau Phase-I: reduced cost perturbation). Given perturbation parameters $\delta_j \geq 0,\ j \in N$. Initial: simplex tableau of form (3.18). This algorithm finds a feasible simplex tableau to the standard LP problem.

1. Perturb: $\bar{z}_j = \delta_j, \forall j \in \{j \in N \mid \bar{z}_j < \delta_j\}$.
2. Call the dual simplex Algorithm 4.4.1.
3. Stop if it terminates at step 3 (infeasible problem).
4. If it terminates at step 2, compute $\bar{z}_N = c_N - N^{\mathrm{T}} B^{-1} c_B$.
5. Stop (feasibility achieved).

Note There exists a perturbation-free variant of the preceding algorithm: in each iteration, one use $\bar{z}'_N$ determined by (13.23) in place of $\bar{z}_N$ for minimum-ratio test, therefore saving computations for restoring reduced costs.

Algorithm 13.4.1 performed very well in preliminary computational experiments (Pan 2000b). A reason seems to be that not all perturbed reduced costs affect the final outcome. Sometimes, reduced costs restored in step 4 of Algorithm 13.4.1 are themselves nonnegative, and hence there is no need for carrying out Phase-II. In fact, it is verified that manipulations in step 2 amount to solving the following problem

$$\begin{array}{ll} \min & b^{\mathrm{T}} y, \\ \text{s.t.} & A^{\mathrm{T}} y \leq \hat{c}, \end{array}$$

where

$$\hat{c}_j = \begin{cases} a_j^{\mathrm{T}} B^{-\mathrm{T}} c_B + \delta_j, & \text{if } j \in J, \\ \bar{z}_j, & \text{if } j \in N \backslash J. \end{cases}$$

That is to say, perturbing $\bar{z}_j < \delta_j$ to $\bar{z}_j = \delta_j$ amounts to slackening dual constraint $a_j^{\mathrm{T}} y \leq c_j$ to

$$a_j^{\mathrm{T}} y \leq a_j^{\mathrm{T}} B^{-\mathrm{T}} c_B + \delta_j = c_j + (\delta_j - c_j + a_j^{\mathrm{T}} B^{-\mathrm{T}} c_B) = c_j + (\delta_j - \bar{z}_j).$$

If slackened constraints are all inactive for dual optimal solutions, or in other words, nonnegative constraints of primal variables, associated with perturbed reduced costs, are active for primal optimal solutions, the primal and dual optimal solutions will not be affected by such perturbations at all (for detailed analysis, see Pan 2000b).

Example 13.4.1. Solve the following problem using Algorithm 13.4.1 in Phase-I:

$$\begin{array}{llllllll} \min & f = x_1 + x_2 - 3x_3, & & & & & & \\ \text{s.t.} & -2x_1 - \ \ x_2 + 4x_3 & & + x_5 & & = -4, \\ & \ \ x_1 - 2x_2 + \ \ x_3 & & & + x_6 & = \ \ 5, \\ & -x_1 \qquad\quad + 2x_3 + x_4 & & & & = -3, \\ & \qquad\qquad x_j \geq 0, \quad j = 1, \cdots, 6. & & & & & \end{array}$$

Answer

Phase-I.

Initial: There is an available simplex tableau ($B = \{5, 6, 4\}$) to this problem. Perturb negative reduced cost -3 to 0, so that it becomes dual feasible. As an illustration for implementation, we add the perturbed reduced costs at the bottom ($\delta_3 = 0$).

x_1	x_2	x_3	x_4	x_5	x_6	RHS
-2^*	-1	4		1		-4
1	-2	1			1	5
-1		2	1			-3
1	1	-3				
1	1					

Taking the bottom row as objective row, execute the dual simplex Algorithm 4.4.1. Note that the second (objective) bottom row does not take a part in pivoting.

Iteration 1:

1. $\min\{-4, 5, -3\} = -4 < 0, \ p = 1.$
3. $J = \{1, 2\} \neq \emptyset.$
4. $\min\{-1/(-2), -1/(-1)\} = 1/2, \ q = 1.$
5. Multiply row 1 by $-1/2$, then add $-1, 1, -1, -1$ times of row 1 to rows 2,3,4,5, respectively:

x_1	x_2	x_3	x_4	x_5	x_6	RHS
1	$1/2$	-2		$-1/2$		2
	$-5/2$	3		$1/2$	1	3
	$1/2$		1	$-1/2^*$		-1
	$1/2$	-1		$1/2$		-2
	$1/2$	2		$1/2$		-2

Iteration 2:

1. $\min\{2, 3, -1\} = -1 < 0, \ p = 3.$
3. $J = \{5\} \neq \emptyset.$
4. $\min\{-(1/2)/(-1/2)\} = 1, \ q = 5.$
5. Multiply row 3 by -2, then add $1/2, -1/2, -1/2, -1/2$ times of row 3 to rows 1,2,4,5, respectively:

x_1	x_2	x_3	x_4	x_5	x_6	RHS
1		−2	−1			3
	−2	3*	1		1	2
	−1		−2	1		2
	1	−1	1			−3
	1	2	1			−3

Optimality of the perturbed tableau is achieved. Deleting the perturbed (bottom) row gives a feasible simplex tableau of the original problem.

Phase-II.
Iteration 3: call the simplex Algorithm 3.2.1.
Select x_3 column as pivot column and the second row as pivot row. Multiply row 2 by $1/3$, then add $2, 1, -2$ times of row 2 to rows 1,4,5, respectively:

x_1	x_2	x_3	x_4	x_5	x_6	RHS
1	−4/3		−1/3		2/3	13/3
	−2/3	1	1/3		1/3	2/3
	−1		−2	1		2
	1/3		4/3		1/3	−7/3

The reduced costs are all nonnegative, optimality is hence achieved. Basic optimal solution and associated objective value:

$$\bar{x} = (13/3, 0, 2/3, 0, 2, 0)^{\mathrm{T}}, \qquad \bar{f} = 7/3.$$

Although the perturbation parameter was set to 0 in the preceding example, positive parameter value should be practically favorable for the sake of anti-degeneracy. It seems to be suitable to use

$$\delta_j \geq \epsilon > 0,$$

where ϵ is the feasibility tolerance ($\epsilon = 10^{-6}$, or so). Computational experiences indicate that the perturbation method is not sensitive to the magnitude of the parameters though exceeding 10^{-1} seems to be inadvisable.

Chapter 14
Dual Simplex Phase-I Method

The mission of a dual Phase-I procedure is to provide an initial dual feasible simplex tableau (or basis) to the dual simplex method.

In the dual simplex context, one can establish a dual Phase-I by introducing $n - m$ dual artificial variables, with or without penalty terms in the dual objective function, as in the primal case. We will not touch such methods, because they are not suitable for applications, just as the artificial variable method or big-M method for the primal simplex method (interested readers are referred to associated literatures, e.g., Padberg 1995). This chapter will address practicable dual Phase-I methods, which may be regarded as dual counterparts of Phase-I methods, presented in the previous chapter (for further references, see also Pan 1994d, 1995, 1996b).

For simplicity of exposition, the standard dual pivot rule will be utilized in this chapter. We stress that it is preferable to use the dual nested rule instead, although all rules presented in Chap. 12 are applicable.

14.1 Dual Infeasibility-Sum Method

This section will present a dual version of the infeasibility-sum method. Appearing in early days, the basic idea of the method draws new attention, and is put into effect in practice in recent years because of its remarkable performance (see, e.g., Koberstein and Suhl 2007; Maros 2003b).

Assume that dual basic solution

$$\bar{y} = B^{-T} c_B; \quad \bar{z}_N = c_N - N^T \bar{y}, \quad \bar{z}_B = 0 \tag{14.1}$$

is infeasible, i.e., $\bar{z}_N \not\geq 0$. Introduce index set

$$J = \{j \in N \mid \bar{z}_j < 0\}, \tag{14.2}$$

P.-Q. PAN, *Linear Programming Computation*, DOI 10.1007/978-3-642-40754-3_14,

Taking the second term of the sum of dual infeasible variables, i.e.,

$$\sum_{j\in J} z_j = \sum_{j\in J} c_j - \Big(\sum_{j\in J} a_j\Big)^{\mathrm{T}} y.$$

as objective function, we construct the following auxiliary program

$$\begin{array}{ll} \max & -(\sum_{j\in J} a_j)^{\mathrm{T}} y, \\ \text{s.t.} & B^{\mathrm{T}} y + z_B = c_B, \\ & N^{\mathrm{T}} y + z_N = c_N, \\ & z_B \geq 0; \quad z_j \geq 0, \quad j \in N/J, \end{array} \tag{14.3}$$

whose nonnegativity constraints only impose on variables, associated with nonnegative components of the dual basic solution.

Clearly, there is a basic solution to the primal program of the auxiliary program, i.e.,

$$\bar{x}_B = -B^{-1} \sum_{j\in J} a_j, \quad \bar{x}_N = 0, \tag{14.4}$$

where $\bar{x}_B$ it termed *auxiliary right-hand side.*

Theorem 14.1.1 (Dual infeasibility test). *If* $\bar{x}_B \geq 0$*, the original problem is infeasible or unbounded.*

Proof. Assume that y' is a feasible solution to the original dual problem such that

$$z' = c_N - N^{\mathrm{T}} y' \geq 0.$$

Thus, it holds that

$$\sum_{j\in J} z'_j = \sum_{j\in J} c_j - \Big(\sum_{j\in J} a_j\Big)^{\mathrm{T}} y' \geq 0. \tag{14.5}$$

On the other hand, it is known from the assumption that (14.4) gives a basic feasible solution $\bar{x}$ to the primal program of (14.3), and (14.1) gives a complementary feasible solution $(\bar{y}, \bar{z})$ of (14.3), hence the two are a pair of optimal solutions. It is clear that (y', z') is a feasible solutio to (14.3), associated with objective value no more than the optimal value, hence

$$\sum_{j\in J} z'_j \leq \sum_{j\in J} \bar{z}_j < 0,$$

where the last inequality comes from (14.2). This contradicts (14.5). Therefore, there is no feasible solution to the original dual problem when $\bar{x}_B \geq 0$. According to the duality principle, the original (primal) problem is infeasible or unbounded. □

In case when $\bar{x}_B \not\geq 0$, carry out one dual simplex iteration with a modified column pivot rule. To this end, assume that row index p has been determined such that

$$\bar{x}_{j_p} < 0. \tag{14.6}$$

Consider the following scheme for updating dual solution:

$$\hat{y} = \bar{y} - \beta B^{-\mathrm{T}} e_p, \quad \hat{z} = \bar{z} + \beta v, \quad v = A^{\mathrm{T}} B^{-\mathrm{T}} e_p, \tag{14.7}$$

where stepsize β is defined by the rule below:

Rule 14.1.1 (Auxiliary column rule) Select pivot column index q such that

$$\beta = -\bar{z}_q / v_q = \begin{cases} \min\{-\bar{z}_j / v_j \mid v_j < 0,\ j \in \bar{J}\}, & \text{if } \{j \in N \backslash J \mid v_j < 0\} \neq \emptyset, \\ \max\{-\bar{z}_j / v_j \mid v_j > 0, j \in J\}, & \text{otherwise.} \end{cases} \tag{14.8}$$

The following results are counterparts of Lemma 13.1.2 and Theorem 13.1.3 in Sect. 13.1, proofs of which are therefore omitted.

Lemma 14.1.1. *Let p be row index, satisfying (14.6). Then (14.8) is well-defined, and the associated (14.7) gives a new basic feasible dual solution to the auxiliary program.*

Theorem 14.1.2. *Assume that the original dual program is feasible. If dual feasible variables are all nondegenerate, the number of dual infeasible variables strictly decreases in finitely many iterations.*

If the number of infeasible variables reduces but does not vanish after an iteration, the process is carried out with a updated auxiliary dual program, until dual feasibility is attained or dual infeasibility or unboundedness of the original problem is detected.

The steps are put in the following algorithm.

Algorithm 14.1.1 (Dual Phase-I: Dual Infeasibility-sum). Initial: (B, N), B^{-1}, $\bar{y}, \bar{z}$. This algorithm finds a dual feasible solution by handling auxiliary programs of form (14.3).

1. If $\bar{z} \geq 0$, compute $\bar{x}_B = B^{-1}b$, and stop (dual feasibility achieved).
2. Compute $\bar{x}_B = -B^{-1} \sum_{j \in J} a_j$, where $J = \{j \in N \mid \bar{z}_j < 0\}$.
3. Stop if $\bar{x}_B \geq 0$ (the original dual problem is infeasible or unbounded).
4. Carry out a single iteration of the dual simplex Algorithm 4.5.1 in which Rule 14.1.1 is used for column pivoting instead.
5. Go to step 1.

Example 14.1.1. Find a dual basic feasible solution of the following program by Algorithm 14.1.1:

$$
\begin{array}{lrl}
\min & f = -x_1 + x_2 - 2x_3, & \\
\text{s.t.} & x_1 - 3x_2 - 2x_3 + x_4 & = -4, \\
& -x_1 + x_2 - 4x_3 \qquad + x_5 & = -2, \\
& -3x_1 + x_2 + x_3 \qquad\qquad + x_6 & = 8, \\
& x_j \geq 0, \quad j = 1, \cdots, 6. &
\end{array}
$$

Answer Taking $J = \{1, 3\}$, construct an auxiliary right-hand side $\bar{x}_B = -(-1, -5, -2)^{\mathrm{T}}$. Since components of it are all nonnegative, it is asserted that the problem is infeasible or unbounded. In fact, the problem is just the same as that in Example 13.1.1, where infeasibility of the problem is declared in Phase-II.

Example 14.1.2. Find a dual basic feasible solution to the following problem by Algorithm 14.1.1:

$$
\begin{array}{lrl}
\min & f = -5x_1 - 7x_2 + x_4, & \\
\text{s.t.} & x_1 + 2x_2 + x_3 & = 3, \\
& 2x_1 + x_2 \qquad - x_4 + x_5 & = -2, \\
& -x_1 + x_2 \qquad + x_4 \qquad + x_6 & = -1, \\
& x_j \geq 0, \quad j = 1, \cdots, 6. &
\end{array}
$$

Answer Taking $J = \{1, 2\}$ and $N \backslash J = \{3, 4, 5, 6\}$, construct the auxiliary program. $B = \{3, 5, 6\}$, $N = \{1, 2, 4\}$, $B^{-1} = I, \bar{z}_N = (-5, -7, 1)^{\mathrm{T}}$. The auxiliary right-hand side is $\bar{x}_B = (-3, -3, 0)^{\mathrm{T}}$.

Iteration 1: Carry out a single iteration of Algorithm 4.5.1, in which (14.8) is used for column pivoting instead.

1. $\min\{-3, -3, 0\} = -3, \ p = 1,$ so x_3 leaves the basis.
3. $\sigma_N = N^{\mathrm{T}} B^{-\mathrm{T}} e_p = (1, 2, 0)^{\mathrm{T}}$.
5. $\beta \ = \max\{-\bar{z}_j/\sigma_j | \sigma_j < 0, j = 1, 2, 3\} = \max\{5/1, 7/2\} = 5, q = 1,$ so x_1 enters the basis.
6. $\bar{z}_N \ = \bar{z}_N + \beta \sigma_N = (-5, -7, 1)^{\mathrm{T}} + 5 \times (1, 2, 0)^{\mathrm{T}} = (0, 3, 1)^{\mathrm{T}}, \bar{z}_{j_p} = 5.$
7. $\bar{a}_q \ = B{-}1a_q = (1, 2, -1)^{\mathrm{T}}$.
8. $\alpha \ = \bar{x}_{j_p}/\sigma_q = -3/1 = -3,$
   $\bar{x}_B = \bar{x}_B - \alpha \bar{a}_q = (-3, -3, 0)^{\mathrm{T}} - (-3) \times (1, 2, -1)^{\mathrm{T}} = (0, 3, -3)^{\mathrm{T}},$
   $\bar{x}_1 \ = \alpha = -3.$
9. $B^{-1} = \begin{pmatrix} 1 & & \\ -2 & 1 & \\ 1 & & 1 \end{pmatrix}.$
10. $B \ = \{1, 5, 6\}, \ N = \{3, 2, 4\}, \ \bar{z}_N = (5, 3, 1)^{\mathrm{T}}, \ J = \emptyset, \ \bar{x}_B = B^{-1}b = (3, -8, 2)^{\mathrm{T}}$. Dual feasibility attained.

Algorithm 14.1.1 is easily to converted to its tableau version. Note only that the pth row of the simplex tableau is $v^{\mathrm{T}} = e_p^{\mathrm{T}} B^{-1} A$, x_j column is equal to $B^{-1} a_j$, and in step 2 the following formula is used instead:

$$\bar{x}_B = -\sum_{j \in J} \bar{a}_j, \quad J = \{j \in N \,|\, \bar{z}_j < 0\}.$$

We illustrate with the preceding example.

Example 14.1.3. Find a dual feasible simplex tableau for Example 14.1.2 by the tableau version of Algorithm 14.1.1.

Answer Initial: Clearly, there is a simplex tableau to the problem, i.e.,

x_1	x_2	x_3	x_4	x_5	x_6	RHS
1*	2	1				3
2	1		−1	1		−2
−1	1		1		1	−1
−5	−7		1			

Iteration 1:

$J = \{1, 2\}$, $\bar{J} = \{4\}$, $\bar{x}_B = -((1, 2, -1)^{\mathrm{T}} + (2, 1, 1)^{\mathrm{T}}) = (-3, -3, 0)^{\mathrm{T}}$
$p = 1; \beta = \max\{-(-5)/1, -(-7)/2\} = 5, \ q = 1.$

Respectively add $-2, 1, 5$ times of row 1 to rows 2,3,4, resulting in the following tableau:

x_1	x_2	x_3	x_4	x_5	x_6	RHS
1*	2	1				3
	−3	−2	−1	1		−8
	3	1	1		1	2
	3	5	1			15

which is dual feasible.

14.2 Dual Single-Artificial-Variable Method

In this section, the idea of the single-artificial-variable method, described in Sect. 13.2, is applied to the dual problem to derive a dual Phase-I method. We introduce an artificial variable to the dual program, so that the dual simplex method can get started to produce a dual basic feasible solution to the original problem. This method had been used as dual Phase-1 for the so-called "dual projective pivot algorithm" (Pan 2005).

As in the primal context, there are two schemes:

Scheme 1. The auxiliary dual program is based on the original data of the problem.

Given any vector $v \geq 0$. Introducing artificial variable y_{m+1}, construct the following auxiliary dual program from constraints of the dual problem (4.2):

$$\begin{array}{ll} \max\ y_{m+1}, & \\ \text{s.t.}\quad A^{\mathrm{T}}y + (v-c)y_{m+1} + z = c, & z \geq 0, \end{array} \tag{14.9}$$

to which there is a feasible solution, i.e.,

$$\begin{pmatrix} \hat{y} \\ \hat{y}_{m+1} \end{pmatrix} = \begin{pmatrix} 0 \\ -1 \end{pmatrix}, \quad \begin{pmatrix} \hat{z} \\ \hat{z}_{n+1} \end{pmatrix} = \begin{pmatrix} v \\ 1 \end{pmatrix}$$

Theorem 14.2.1. *There is an optimal solution, associated with a nonpositive objective value, to the auxiliary dual program (14.9). If the optimal value is equal to zero, then the (y, z) part of its basic optimal solution is basic feasible to the original dual problem; otherwise, the latter is infeasible.*

Proof. To the auxiliary dual program (14.9), there is a feasible solution with objective value bounded above by 0, hence there is an optimal solution with nonpositive optimal value. Let $(y', y'_{m+1}, z', z'_{n+1})$ be its optimal solution, where $z'_B, z'_N \geq 0$. Assume that the optimal value is equal to zero, i.e., $y'_{m+1} = 0$. Since the optimal solution satisfies equality constraints of (14.9), it holds that

$$\begin{pmatrix} B^{\mathrm{T}} \\ N^{\mathrm{T}} \\ 0 \end{pmatrix} y' + \begin{pmatrix} z'_B \\ z'_N \\ z'_{n+1} \end{pmatrix} = \begin{pmatrix} c_B \\ c_N \\ 0 \end{pmatrix}$$

Thus, (y', z') is a basic feasible solutio to the original dual problem.

Assume now that the optimal value is less than 0, i.e., $y'_{m+1} < 0$. If the original dual problem has a feasible solution (y'', z''), then (y'', z'') together with $y''_{m+1} = z''_{n+1} = 0$ is a feasible solutio to (14.9), thereby leading to $y'_{m+1} \geq y''_{m+1} = 0$, which contradict $y'_{m+1} < 0$. Therefore, the original dual problem is infeasible if the optimal value of (14.9) is less than zero. □

According to the preceding Theorem, if the dual simplex method is used to solve the auxiliary dual problem, a dual basic feasible solution to the original problem can be obtained, if any.

Scheme 2. The auxiliary dual program is based on an canonical form of the constraint system.

Let (B, N) be basis and nonbasis of the standard LP problem, associated with a simplex tableau, say (3.18). Given any $\hat{z}_N \geq 0$. Introducing artificial variable x_{n+1}, construct the following auxiliary program:

$$\begin{array}{ll} \min & f = \bar{f} + \bar{z}_N^{\mathrm{T}} x_N, \\ \text{s.t.} & \begin{pmatrix} I & \bar{N} & 0 \\ 0 & (\hat{z}_N - \bar{z}_N)^{\mathrm{T}} & 1 \end{pmatrix} \begin{pmatrix} x_B \\ x_N \\ x_{n+1} \end{pmatrix} = \begin{pmatrix} 0 \\ 1 \end{pmatrix}, \quad x \geq 0, \ x_{n+1} \geq 1, \end{array} \tag{14.10}$$

where

$$\bar{N} = B^{-1}N, \qquad \bar{z}_N = c_N - N^{\mathrm{T}} B^{-\mathrm{T}} c_B, \qquad \bar{f} = b^{\mathrm{T}} B^{-\mathrm{T}} c_B.$$

Clearly, $(\bar{x} = 0,\ \bar{x}_{n+1} = 1)$ is a solution to the preceding program. We are interested in its dual program, i.e.,

$$\begin{array}{ll} \max & w = \bar{f} + y_{m+1}, \\ \text{s.t.} & \begin{pmatrix} z_B \\ z_N \\ z_{n+1} \end{pmatrix} = \begin{pmatrix} 0 \\ \bar{z}_N \\ 0 \end{pmatrix} - \begin{pmatrix} I & 0 \\ \bar{N}^{\mathrm{T}} & \hat{z}_N - \bar{z}_N \\ 0 & 1 \end{pmatrix} \begin{pmatrix} y \\ y_{m+1} \end{pmatrix}, \quad z, z_{n+1} \geq 0. \end{array} \tag{14.11}$$

The above is just the auxiliary dual program we want, as it is much simpler than (14.9), and has a feasible solution

$$\begin{pmatrix} \hat{y} \\ \hat{y}_{m+1} \end{pmatrix} = \begin{pmatrix} 0 \\ -1 \end{pmatrix}, \quad \begin{pmatrix} \hat{z}_B \\ \hat{z}_N \\ \hat{z}_{n+1} \end{pmatrix} = \begin{pmatrix} 0 \\ \hat{z}_N \\ 1 \end{pmatrix} \geq 0. \tag{14.12}$$

Like Theorem 14.2.1, it can be shown that if (14.11) is solved with a zero optimal value, a dual basic feasible solution to the original problem is obtained; otherwise, there is no dual feasible solution, or the original problem is infeasible or unbounded. From the last equality constraint, $z_{n+1} = -y_{m+1}$, of (14.11), it is known that max y_{m+1} and min z_{n+1} are equal, if any.

Like the dual simplex method, it is also possible to solve (14.11) via the simplex tableau. In fact, problem (14.10) has the following simplex tableau, which is not dual feasible:

x_B^{T}	x_N^{T}	x_{n+1}	f	RHS
I	$\bar{N}$			
	$(\hat{z}_N - \bar{z}_N)^{\mathrm{T}}$	1		1
	$\bar{z}_N^{\mathrm{T}}$		-1	$-\bar{f}$

Adding the $(m+1)$th row to the bottom row gives

x_B^{T}	x_N^{T}	x_{n+1}	f	RHS
I	$\bar{N}$			
	$(\hat{z}_N - \bar{z}_N)^{\mathrm{T}}$	1		1
	$\hat{z}_N^{\mathrm{T}}$	$\hat{z}_{n+1}$	-1	$-\bar{f} + 1$

(14.13)

where $\hat{z}_{n+1} = 1$. The preceding tableau, corresponding to the dual feasible solution defined by (14.12), can be taken as an auxiliary initial tableau though not a simplex one. To turn it to a simplex tableau, one more basic variable is required. To this end, determine a dual stepsize

$$\beta = \hat{z}_q/(\hat{z}_q - \bar{z}_q) = \min\{\hat{z}_j/(\hat{z}_j - \bar{z}_j) \mid \hat{z}_j - \bar{z}_j > 0,\ j \in N\}. \tag{14.14}$$

If $\beta \geq 1$, set $\beta = 1$, $q = n + 1$. Converting the entry in the $(m + 1)$th row and q column to 1, and eliminating all other nonzeros in the column by elementary transformations, so that q-indexed becomes basic. If it happens that $q = n + 1$, then deleting the $(m + 1)$th row and x_{n+1} column, and putting $\bar{b}$ in RHS column gives a dual feasible simplex tableau to the original problem. In a general case of $q \neq n + 1$, obtained is only a dual feasible tableau to the auxiliary program, which can be solved by the dual simple method.

Theoretically, any nonnegative vector $\hat{z}_N$ can be used to construct an auxiliary program. A simple way is to set $\hat{z}_N = e$. But another way would be more suitable for sparse computations: for given $\delta_j \geq 0$, $j \in N$. set

$$\hat{z}_j = \begin{cases} \delta_j, & \text{if } \bar{z}_j < \delta_j, \\ 0, & \text{otherwise,} \end{cases} \quad j \in N.$$

The associated steps can be summarized to the following algorithm.

Algorithm 14.2.1 (Tableau dual Phase-I: single-artificial-variable). Initial: auxiliary tableau of form (14.13). This algorithm finds a dual feasible tableau to the standard LP problem.

1. Determine β and q such that

$$\beta = \hat{z}_q/(\hat{z}_q - \bar{z}_q) = \min\{\hat{z}_j/(\hat{z}_j - \bar{z}_j) \mid \hat{z}_j - \bar{z}_j > 0,\ j \in N\}.$$

2. If $\beta \geq 1$, set $\beta = 1$, $q = n + 1$.
3. Convert the entry in the $(m + 1)$th row and q column to 1, and eliminate all other nonzeros in the column by elementary transformations.
4. If $q = n+1$, delete the $(m+1)$th row and x_{n+1} column, and stop (dual feasibility achieved).
5. Call the dual simplex Algorithm 4.4.1 (the $(n + 1)$-indexed column is taken as nonbasic at the beginning).

Note In the initial tableau, the first $m+1$ components of x_{n+1} column and of RHS column are equal. When carrying out the preceding algorithm, the $(n + 1)$-indexed column may be also deemed as the auxiliary right-hand side, and the right-hand side $\bar{b}$ of the original problem is filled in RHS column (but does not take a part in pivoting). In this way, once x_{n+1} enters the basis, eliminating the $(m + 1)$th row and $(n + 1)$-indexed column gives a dual feasible tableau to the original problem. The f column can be omitted, as it does not change at all in the solution process.

It is not difficult to transfer Algorithm 14.2.1 to its revised version, which is omitted here.

Example 14.2.1. Solve the following problem using Algorithm 14.2.1 as dual Phase-I:

$$\begin{array}{lrcl} \min & f = 5x_1 - 2x_2 - x_3 + x_4, & & \\ \text{s.t.} & 2x_1 - x_2 + x_3 + x_4 & = & 5, \\ & -5x_1 + 3x_2 + 2x_3 \qquad\qquad + x_5 & = & -2, \\ & -x_1 + 2x_2 - x_3 \qquad\qquad\qquad + x_6 & = & -1, \\ & x_j \geq 0, \quad j = 1, \cdots, 6. & & \end{array}$$

Answer From the first constraint equality, it follows that

$$x_4 = 5 - 2x_1 + x_2 - x_3,$$

Substitute it to the objective function to eliminate x_4:

$$f = 5 + 3x_1 - x_2 - 2x_3.$$

Setting $\hat{z}_1 = 0$, $\hat{z}_2 = \hat{z}_3 = 1$, construct the following auxiliary initial simplex tableau:

x_1	x_2	x_3	x_4	x_5	x_6	x_7	RHS
2	−1	1	1				5
−5	3	2		1			−2
−1	2	−1			1		−1
−3	1 + 1	1 + 2*				1	1
	1	1				1	−5 + 1

Phase-I: Call Algorithm 14.2.1. $m + 1 = 4$.
Iteration 1:

1. $\beta = \min\{1/(1+1), 1/(1+2)\} = 1/3 < 1, q = 3$.
3. Multiply row 4 by 1/3, then add $-1, -2, 1, -1$ times of row 4 to rows 1,2,3,5, respectively:

x_1	x_2	x_3	x_4	x_5	x_6	x_7	RHS
3	−5/3		1			−1/3	14/3
−3*	5/3			1		−2/3	−8/3
−2	8/3				1	1/3	−2/3
−1	2/3	1				1/3	1/3
1	1/3					2/3	−13/3

5. Call Algorithm 4.4.1. Note that x_7 column also represents the auxiliary right-hand side.

Iteration 2:

1. $\min\{-1/3, -2/3, 1/3, 1/3\} = -2/3 < 0, p = 2.$
4. $\min\{-1/(-3), (-2/3)/(-2/3)\} = 1/3, \ q = 1.$
5. Multiply row 2 by $-1/3$, then add $-3, 2, 1, -1$ times of row 2 to rows 1,3,4,5, respectively:

x_1	x_2	x_3	x_4	x_5	x_6	x_7	RHS
			1	1		-1^*	2
1	$-5/9$			$-1/3$		$2/9$	$8/9$
	$14/9$			$-2/3$	1	$7/9$	$10/9$
	$1/9$	1		$-1/3$		$5/9$	$11/9$
	$8/9$			$1/3$		$4/9$	$-47/9$

Iteration 3:

1. $\min\{-1, 2/9, 7/9, 5/9\} = -1 < 0, p = 1.$
4. $\min\{-(4/9)/(-1)\} = 4/9, \ q = 7.$
5. Multiply row 1 by -1, then add $-2/9, -7/9, -5/9, -4/9$ times of row 1 to rows 2,3,4,5, respectively:

x_1	x_2	x_3	x_4	x_5	x_6	x_7	RHS
			-1	-1		1	-2
1	$-5/9$		$2/9$	$-1/9$			$4/3$
	$14/9$		$7/9$	$1/9$	1		$8/3$
	$1/9$	1	$5/9$	$2/9$			$7/3$
	$8/9$		$4/9$	$7/9$			$-13/3$

Optimality of the auxiliary program is achieved with the artificial variable x_7 entering the basis. Deleting the first row and the seventh column leads to a dual feasible simplex tableau to the original problem, i.e.,

x_1	x_2	x_3	x_4	x_5	x_6	RHS
1	$-5/9$		$2/9$	$-1/9$		$4/3$
	$14/9$		$7/9$	$1/9$	1	$8/3$
	$1/9$	1	$5/9$	$2/9$		$7/3$
	$8/9$		$4/9$	$7/9$		$-13/3$

$\min\{4/3, 8/3, 7/3\} \geq 0$. Optimality is achieved, hence there is no need for dual Phase-II. The basic optimal solution and objective value are

$$\bar{x} = (4/3, 0, 7/3, 0, 0, 8/3)^{\mathrm{T}}, \quad \bar{f} = 13/3.$$

14.3 The Most-Obtuse-Angle Row Rule

This section will present a artificial-variable free dual Phase-I method, involving neither minimum-ratio test, nor reduced costs.

Assume that the current dual solution is infeasible, and a pivot column index $q \in N$ is selected (e.g., by the conventional column rule) such that

$$\bar{z}_q = c_q - c_B^{\mathrm{T}} B^{-1} a_q < 0. \tag{14.15}$$

Lemma 14.3.1. *If (14.15) holds and $B^{-1}a_q \leq 0$, then the original problem is infeasible or unbounded.*

Proof. Assume that $\bar{x}$ is a feasible solution. Define

$$d_B = -B^{-1}a_q; \quad d_q = 1; \quad d_j = 0, \quad j \in N, \; j \neq q. \tag{14.16}$$

Then vector

$$\hat{x} = \bar{x} + \alpha d$$

is a feasible solution for any $\alpha \geq 0$; in fact, it holds that

$$A\hat{x} = A\bar{x} + \alpha(Bd_B + Nd_N) = b,$$

and, since $\bar{x} \geq 0$ and $d \geq 0$, it does that $\hat{x} \geq 0$. In addition, it is known from (14.15) that vector d satisfies

$$c^{\mathrm{T}}d = (c_B^{\mathrm{T}}d_B + c_N^{\mathrm{T}}d_N) = c_q - c_B^{\mathrm{T}}\bar{a}_q = \bar{z}_q < 0, \tag{14.17}$$

hence corresponding feasible value tends to ∞ as α tends to ∞, i.e.,

$$c^{\mathrm{T}}\hat{x} = c^{\mathrm{T}}\bar{x} + \alpha c^{\mathrm{T}}d \rightarrow -\infty, \quad \alpha \rightarrow \infty,$$

as indicates lower unboundedness of the original problem. Therefore, the original problem is either infeasible, or feasible but unbounded below. □

The key of this dual Phase-I method is the following row rule (Pan 1994a, 1997).

Rule 14.3.1 (Most-obtuse-angle row rule) Select row index

$$p \in \arg\max\{\bar{a}_{i\,q} \mid i = 1, \cdots, m\}. \tag{14.18}$$

Assume that $\bar{a}_{p\,q} > 0$. Taking $\bar{a}_{p\,q}$ as the pivot, carry out according elementary transformations. As a result, the negative reduced cost, indexed by q, becomes zero, whereas the zero reduced cost, indexed by j_p, becomes positive, i.e., $-\bar{z}_q/\bar{a}_{p,q} > 0$ (see (3.13)).

These Phase-I steps can be summarized to the following algorithm.

Algorithm 14.3.1 (Dual Phase-I: the most-obtuse-angle row rule). Initial: $(B, N), B^{-1}$. This algorithm finds a dual basic feasible solution to the standard LP problem.

1. Compute $\bar{z}_N = c_N - N^{\mathrm{T}}\bar{y}$, where $\bar{y} = B^{-\mathrm{T}}c_B$.
2. Select pivot column index q such that $q \in \arg\min_{j\in N} \bar{z}_j$.
3. If $\bar{z}_q \geq 0$, compute $\bar{x}_B = B^{-1}b$, $\bar{x}_N = 0$, and stop.
4. Compute $\bar{a}_q = B^{-1}a_q$.
5. Stop if $\bar{a}_q \leq 0$.
6. Determine pivot row index $p \in \arg\max\{\bar{a}_{i\,q} \mid i = 1, \cdots, m\}$.
7. UpdateB^{-1} by (3.23).
8. Update (B, N) by exchanging j_p and q.
9. Go to step 1.

Theorem 14.3.1. *If Algorithm 14.3.1 terminates at*

(i) Step 3, then a dual basic feasible solution is obtained; or
(ii) Step 5, detecting infeasibility or lower unboundedness of the problem.

Proof. The termination at step 3 clear implies that the current solution is dual basic feasible. When terminating at from step 5, then it holds that $\bar{z}_N < 0$ and $\bar{a}_q \leq 0$; by Lemma 14.3.1, infeasibility or lower unboundedness of the problem is asserted. □

Let us investigate the most-obtuse-angle row rule geometrically. For search vector d defined by (14.16), it holds that $-\bar{a}_{p\,q} < 0$, and d forms the most obtuse angle with the gradient, e_p, of the constraint $x_{j_p} \geq 0$. In addition, (14.17) indicates that d is a downhill direction with respect to the objective function $c^{\mathrm{T}}x$. If d is close to the negative objective gradient $-c$, therefore, then e_p tends to form the most obtuse angle with $-c$. By the heuristic characteristic of optimal basis (Sect. 2.5), it is favorable to let the constraint $x_{j_p} \geq 0$ be satisfied as equality, accordingly dropping x_{j_p} from the basis.

This dual Phase-I method is advantageous for its remarkable simplicity and stability. Initially, it performed very well for solving a set of small test problems (Pan 1994a). Taking MINOS 5.3 as a platform, subsequently a code based on it outperformed MINOS 5.3 (with default options) on 48 Netlib test problems with CPU time ratio 1.37 (Pan 1997).

Koberstein and Suhl (2007) extend this method to solve more general LP problems. They reported extensive computational results, showing method's striking superiority over some commonly used dual Phase-1 methods. It is found however that the method performed unsatisfactorily with few most difficult problems, as might be due to neglecting of the current basic solution. Considering the extent to which the current vertex violating nonnegative constraints, therefore, the following variant is stated:

Rule 14.3.2 (Variant of the most-obtuse-angle row rule) Given constant $0 < \tau \leq 1$. Select pivot row index

$$p \in \arg\min\{\bar{x}_{j_i} \mid \bar{a}_{i\,q} \geq \tau\theta,\ i = 1, \cdots, m\}, \quad \theta = \max\{\bar{a}_{i\,q} \mid i = 1, \cdots, m\} > 0.$$

In case of $\tau = 1$, the preceding is just Rule 14.3.1 itself. Practically, it should be close to 1 to expend the range of choices, so that when there exist negative components of $\bar{x}_B$, the nonnegative constraint that is violated the most is satisfied as equality. It might be well to take $\sigma = 0.95$, or so.

Though Algorithm 14.3.2 uses Dantzig conventional rule for column pivoting, rules presented in Chap. 11 all apply. Although it has not been known which is the best, the most-obtuse-angle row rule seems to be best matched by the steepest-edge rule. But there are no numerical results available at this stage.

Finally, the objective value does not necessarily monotonically change in the solution process. In addition, the finiteness of the algorithm is not guaranteed even under the nondegeneracy assumption. In fact, Guerrero-Garcia and Santos-Palomo (2005) offered a cycling example. Even so, cycling has not been reported so far in practice.

Example 14.3.1. Find a dual feasible simplex basis by Algorithm 14.3.1:

$$\begin{array}{lrl} \min & f = & -5x_1 - 7x_2 + x_4, \\ \text{s.t.} & & x_1 + 2x_2 + x_3 = 3, \\ & & 2x_1 + x_2 - x_4 + x_5 = -2, \\ & & -x_1 + x_2 + x_4 + x_6 = -1, \\ & & x_j \geq 0, \quad j = 1, \cdots, 6. \end{array}$$

Answer Initial: $B = \{3, 5, 6\}$, $N = \{1, 2, 4\}$, $B^{-1} = I$.

Iteration 1:

1. $\bar{y} = B^{-\mathrm{T}} c_B = (0, 0, 0)^{\mathrm{T}}$, $\bar{z}_N = c_N - N^{\mathrm{T}} y = (-5, -7, 1)^{\mathrm{T}}$.
2. $\min\{\bar{z}_j \mid j = 1, 2, 4\} = \min\{-5, -7, 1\} = -7$, $q = 2$, hence x_2 enters the basis.
4. $\bar{a}_2 = B^{-1} a_2 = (2, 1, 1)^{\mathrm{T}}$.
6. $\max\{\bar{a}_{i\,2} \mid i = 1, 2, 3\} = \max\{2,\ 1,\ 1\} = 1$, $p = 1$, so x_3 leaves the basis.
7. Update $B^{-1} = \begin{pmatrix} 1/2 & & \\ -1/2 & 1 & \\ -1/2 & & 1 \end{pmatrix}$.
8. $B = \{2, 5, 6\}$, $N = \{1, 3, 4\}$.

Iteration 2:

1. $\bar{y} = (-7/2, 0, 0)^{\mathrm{T}}$, $\bar{z}_N = (-5, 0, 1)^{\mathrm{T}} - (-7/2, -7/2, 0)^{\mathrm{T}} = (-3/2, 7/2, 1)^{\mathrm{T}}$.
2. $\min\{-3/2, 7/2, 1\} = -3/2$, $q = 1$, so x_1 enters the basis.
4. $\bar{a}_1 = (1/2, 3/2, -3/2)^{\mathrm{T}}$.

6. $\max\{1/2, 3/2, -3/2\} = 3/2,\ p = 2$, so x_5 leaves the basis.

7. $B^{-1} = \begin{pmatrix} 1 & -1/3 & \\ 0 & 2/3 & \\ 0 & 1 & 1 \end{pmatrix} \begin{pmatrix} 1/2 & & \\ -1/2 & 1 & \\ -1/2 & & 1 \end{pmatrix} = \begin{pmatrix} 2/3 & -1/3 & \\ -1/3 & 2/3 & \\ -1 & 1 & 1 \end{pmatrix}.$

8. $B = \{2, 1, 6\},\ N = \{5, 3, 4\}$.

Iteration 3:

1. $\bar{y} = (-3, -1, 0)^{\mathrm{T}},\ \bar{z}_N = (0, 0, 1)^{\mathrm{T}} - (-1, -3, 1)^{\mathrm{T}} = (1, 3, 0)^{\mathrm{T}}$.
2. $3 = q \in \arg\min\{1, 3, 0\}$.
3. $\bar{z}_{j_3} = 0$, dual feasibility is achieved. According primal basic solution is $\bar{x}_B = B^{-1}b = (4/3, 1/3, -2)^{\mathrm{T}}, \bar{x}_N = (0, 0, 0)^{\mathrm{T}}$.

The tableau version of Algorithm 14.3.1 is of a simple form.

Algorithm 14.3.2 (Tableau Dual Phase-I: the most-obtuse-angle row rule). Initial: simplex tableau of form (3.18). This algorithm finds a dual feasible simplex tableau.

1. Select pivot column index $q \in \arg\min_{j\in N} \bar{z}_j$.
2. Stop if $\bar{z}_q \geq 0$ (dual feasibility achieved).
3. Stop if $\bar{a}_q \leq 0$ (infeasibility or lower unbounded).
4. Determine pivot row index $p \in \arg\max\{\bar{a}_{i\,q} \mid i = 1, \cdots, m\}$.
5. Carry out elementary transformations to convert $\bar{a}_{p\,q}$ to 1, and eliminate all other nonzeros of the column.
6. Go to step 1.

14.4 Perturbation of the Right-Hand Side

The basic idea of this method is simple: the right-hand side of the initial simplex tableau is perturbed to nonnegative; after the perturbed tableau is solved by the simplex method, a dual feasible simplex tableau can be obtained by recomputing the right-hand side from original data (Pan 1999a).

Assume that an initial simplex tableau is of form (3.18). Given perturbation parameter

$$\delta_i \geq 0, \quad i = 1, \cdots, m.$$

To weaken effect of degeneracy, δ_i are suggested being positive values, different from each other. The method takes

$$\bar{b}'_i = \begin{cases} \delta_i, & \text{if } \bar{b}_i \leq \delta_i, \\ \bar{b}_i, & \text{otherwise,} \end{cases} \qquad i = 1, \cdots, m \tag{14.19}$$

to replace $\bar{b}$, leading to a feasible simplex tableau.

We have the following result, the proof of which is similar to that to Theorem 7.4.1, and omitted.

Theorem 14.4.1. *If the perturbed program is unbounded, the original problem is infeasible or unbounded.*

It might be well to denote again by (3.18) the final tableau of the perturbed program, associated with basis B. Assume that the final tableau is optimal, i.e., $\bar{z}_j \geq 0, j \in N$. To restore a simplex tableau to the original problem, compute $\bar{b} = B^{-1}b$ and $\bar{f} = -c_B^{\mathrm{T}}\bar{b}$ to cover the old $\bar{b}$ and $\bar{f}$, resulting in

$$\begin{array}{c|c} \bar{A} & \bar{b} \\ \hline \bar{z}^{\mathrm{T}} & -\bar{f} \end{array}$$

which is clearly a dual feasible tableau to the original problem.

The overall steps are put in the following algorithm.

Algorithm 14.4.1 (Tableau dual Phase I: right-hand side perturbation). Given perturbation parameter $\delta_i > 0, i = 1, \cdots, m$, a simplex tableau of form (3.18). The algorithm finds a dual feasible simplex tableau to the standard LP problem.

1. Perturbation: $\bar{b}_i = \delta_i, \forall i \in \{i = 1, \cdots, m \mid \bar{b}_i \leq \delta_i\}$.
2. Call the simplex Algorithm 3.2.1.
3. Stop if termination occurs at step 3 (infeasible or unbounded).
4. Cover $\bar{b}$, $\bar{f}$ by $\bar{b} = B^{-1}b$, $\bar{f} = -c_B^{\mathrm{T}}\bar{b}$ if termination occurs at step 2.
5. Stop (dual feasibility achieved).

Note A variant of this algorithm results from not perturbing the right-hand side, but directly using $\bar{b}'$, defined by (14.19), in place of $\bar{b}$ for the minimum-ratio test, as thereby saves the restoration.

Geometrically, the perturbation of the right-hand side amounts to relaxing nonnegativity restrictions on basic variables. If these constraints are inactive at the optimal solution, then the perturbation does not matter to the determination of the optimal solution. In some cases, a LP problem can be solved by Algorithm 14.4.1 alone. In fact, the following theorem holds (for proof, see Pan 1999a).

Theorem 14.4.2. *Assume dual nondegeneracy of optimal solutions. If perturbed components of the right-hand side correspond to optimal basic variables, Algorithm 14.4.1 generates an optimal simplex tableau.*

Example 14.4.1. Find a dual feasible tableau by Algorithm 14.4.1:

$$\begin{array}{ll} \min & f = -5x_1 - 7x_2 + x_4, \\ \text{s.t.} & x_1 + 2x_2 + x_3 = 3, \\ & 2x_1 + x_2 - x_4 + x_5 = -2, \\ & -x_1 + x_2 + x_4 + x_6 = -1, \\ & x_j \geq 0, \quad j = 1, \cdots, 6. \end{array}$$

Answer Initial: A simplex tableau can be obtained from the problem. Take $\delta_2 = 1/6,\ \delta_3 = 1/12$, and put the perturbed right-hand side at the right end of the tableau.

x_1	x_2	x_3	x_4	x_5	x_6	RHS	RHS1
1	2	1				3	3
2	1		−1	1		−2	1/6
−1	1*		1		1	−1	1/12
−5	−7		1				

Call the simplex Algorithm 3.2.1.

Iteration 1:

1. $\min\{-5, -7, 1\} = -7 < 0,\ q = 2$.
3. $I = \{1, 2, 3\} \neq \emptyset$.
4. $\min\{3/2, (1/6)/1, (1/12)/1\} = 1/12,\ p = 3$.
5. Add $-2, -1, 7$ times of row 3 to rows 1,2,4, respectively:

x_1	x_2	x_3	x_4	x_5	x_6	RHS	RHS1
3		1	−2		−2	5	17/6
3*			−2	1	−1	−1	1/12
−1	1		1		1	−1	1/12
−12			8		7	−7	–

Iteration 2:

1. $\min\{-12, 8, 7\} = -12 < 0,\ q = 1$.
3. $I = \{1, 2\} \neq \emptyset$.
4. $\min\{(17/6)/3, (1/12)/3\} = 1/36, p = 2$.
5. Multiply row 2 by 1/3, then add $-3, 1, 12$ times of row 2 to rows 1,3,4, respectively:

x_1	x_2	x_3	x_4	x_5	x_6	RHS	RHS1
		1		−1	−1	6	11/4
1			−2/3	1/3	−1/3	−1/3	1/36
	1		1/3	1/3	2/3	−4/3	1/9
				4	3	−11	–

All reduced costs of the preceding tableau are nonnegative, so that a dual feasible simplex tableau can be generated by deleting the end column.

Chapter 15
Reduced Simplex Method

In this chapter and the following two chapters, some special forms of the LP problem, introduced in Sect. 25.1, will be employed to design new LP methods. In particular, this chapter will handle the so-called "reduced problem" (25.2), i.e.,

$$\begin{aligned} &\min \ x_{n+1}, \\ &\text{s.t.} \quad (A \vdots a_{n+1}) \begin{pmatrix} x \\ x_{n+1} \end{pmatrix} = b, \quad x \geq 0, \end{aligned} \tag{15.1}$$

where $a_{n+1} = -e_{m+1}$. Note that the objective variable x_{n+1} is in the place of f (thereafter the two will be regarded equal), and hence there is no sign restriction on x_{n+1}.

In the conventional simplex context, x_{n+1} appears as a dependent variable. In each iteration, variation of the basic solution as well as the value of x_{n+1} comes from variation of a chosen nonbasic variable, x_q, corresponding to a negative cost. In contrast, the key of the "reduced simplex method", presented in this chapter, is to use x_{n+1} as a special nonbasic variable, an argument, which decreases in each iteration to push the associated basic solution toward optimality.

Effectiveness of algorithms derived in this chapter is to be investigated. There are no related numerical results available at this stage.

15.1 Derivation

Consider the reduced problem (15.1). As it plays a particular role, the objective variable x_{n+1} will be separated from the set of nonbasic variables.

Assume that the constraints of (15.1) are converted to the following equivalent canonical form by a series of elementary transformations:

$$x_B = \bar{b} - \bar{N} x_N - x_{n+1} \bar{a}_{n+1} \geq 0, \quad x_N \geq 0, \tag{15.2}$$

P.-Q. PAN, *Linear Programming Computation*, DOI 10.1007/978-3-642-40754-3_15,

where $\bar{a}_{n+1} = -B^{-1}e_{m+1} \neq 0$, and

$$B = \{j_1, \cdots, j_{m+1}\}, \quad N = A \backslash B, \quad n + 1 \notin B. \tag{15.3}$$

Lemma 15.1.1. *If $\bar{a}_{n+1} \geq 0$, then problem (15.1) is infeasible or unbounded below.*

Proof. Assume that $(\bar{x}, \bar{x}_{n+1})$ is a feasible solution to (15.1), satisfying

$$\bar{x}_B = \bar{b} - \bar{N}\bar{x}_N - \bar{x}_{n+1}\bar{a}_{n+1} \geq 0. \tag{15.4}$$

Thus, for any $\alpha \geq 0$ and

$$\hat{x}_N = \bar{x}_N, \hat{x}_{n+1} = \bar{x}_{n+1} - \alpha,$$

it holds that

$$\hat{x}_B = \bar{b} - \bar{N}\hat{x}_N - \hat{x}_{n+1}\bar{a}_{n+1} = (\bar{b} - \bar{N}\bar{x}_N - \bar{x}_{n+1}\bar{a}_{n+1}) + \bar{a}_{n+1}\alpha \geq 0,$$

where the right-most inequality comes from (15.4), $\bar{a}_{n+1} \geq 0$ and $\alpha \geq 0$. This indicates that $(\hat{x}, \hat{x}_{n+1})$ is a feasible solution, and

$$\hat{x}_{n+1} \to -\infty, \quad \text{as } \alpha \to +\infty.$$

Therefore, the problem is unbounded below. □

Setting $x_N = 0$ in (15.2) leads to the following system of inequalities:

$$x_B = \bar{b} - x_{n+1}\bar{a}_{n+1} \geq 0. \tag{15.5}$$

Introduce the set of solutions to the system

$$\Phi(B) = \{x_{n+1} \mid \bar{b} - x_{n+1}\bar{a}_{n+1} \geq 0\}.$$

If this set is nonempty, then (15.5) is said *consistent*, and (15.2) is a *feasible canonical form.*

It is clear that any given $x_{n+1} = \bar{x}_{n+1}$ corresponds to a solution to (15.1), i.e.,

$$\begin{pmatrix} \bar{x}_B \\ \bar{x}_N \\ \bar{x}_{n+1} \end{pmatrix} = \begin{pmatrix} \bar{b} - \bar{x}_{n+1}\bar{a}_{n+1} \\ 0 \\ \bar{x}_{n+1} \end{pmatrix}.$$

Using the above notation, we have the following result.

Proposition 15.1.1. *$(\bar{x}, \bar{x}_{n+1})$ is a feasible solution to problem (15.1) if and only if $\bar{x}_{n+1} \in \Phi(B)$.*

Proof. Note that the constraints of (15.1) and (15.2) are equivalent. If $\bar{x}_{n+1} \in \Phi(B)$, then $(\bar{x}, \bar{x}_{n+1})$ clearly satisfies (15.2), hence is feasible. If, conversely, $(\bar{x}, \bar{x}_{n+1})$ is a feasible solution, then $\bar{x}_{n+1} \in \Phi(B)$ follows from (15.2). □

Definition 15.1.1. If, for some $p \in \{1, \cdots, m+1\}$, it holds that

$$\bar{x}_{j_p} = 0, \quad \bar{a}_{p,n+1} \neq 0, \tag{15.6}$$

then $(\bar{x}, \bar{x}_{n+1})$ is a basic solution; if, in addition,

$$\bar{x}_{j_i} \geq 0, \quad i = 1, \cdots, m+1, \tag{15.7}$$

it is a basic feasible solution. If

$$\bar{x}_{j_i} > 0, \quad \forall\, i = 1, \cdots, m+1, \quad \bar{a}_{i,n+1} < 0, \tag{15.8}$$

the basic feasible solution is said to be nondegenerate.

The preceding definitions of basic solution and basic feasible solution coincide with the same named items in the conventional simplex context. In fact, when (15.6) holds, $\bar{x}$ is just the basic solution, associated with the conventional simplex tableau, resulting from entering x_{n+1} to and dropping x_{j_p} from the basis; and if (15.7) holds, then components of the basic solution are all nonnegative, hence it is feasible. Therefore, the two will not be distinguished. It is noted however that the definition of nondegeneracy here is somewhat different from the conventional.

If $\Phi(B)$ is nonempty, it is logical to find the basic feasible solution, associated with its greatest lower bound. To this end, the following rule applies.

Rule 15.1.1 (Row rule) Assume $\bar{a}_{n+1} \not\geq 0$. Select pivot row index

$$p \in \arg\max\{\bar{b}_i/\bar{a}_{i,n+1} \mid \bar{a}_{i,n+1} < 0,\ i = 1, \cdots, m+1\}. \tag{15.9}$$

Let p be selected row index. Define

$$\begin{pmatrix} \hat{x}_B \\ \hat{x}_N \\ \hat{x}_{n+1} \end{pmatrix} = \begin{pmatrix} \bar{b} - (\bar{b}_p/\bar{a}_{p,n+1})\bar{a}_{n+1} \\ 0 \\ \bar{b}_p/\bar{a}_{p,n+1} \end{pmatrix}. \tag{15.10}$$

Using the preceding notation, we have the following result.

Lemma 15.1.2. *Assume $\Phi(B) \neq \emptyset$. If $\bar{a}_{n+1} \not\geq 0$, then $\hat{x}_{n+1}$ is its greatest lower bound, and $(\hat{x}, \hat{x}_{n+1})$ is a basic feasible solution.*

Proof. Note that condition $\bar{a}_{n+1} \not\geq 0$ ensures that (15.9) is well-defined.

Introduce notation

$$I = \{i = 1, \cdots, m+1 \mid \bar{a}_{i,n+1} < 0\}.$$

It is known from (15.9) and (15.10) that

$$\hat{x}_{n+1} \geq \bar{b}_i / \bar{a}_{i,n+1}, \quad i \in I,$$

from which it follows that

$$\bar{b}_i - \hat{x}_{n+1}\bar{a}_{i,n+1} \geq 0, \quad i \in I.$$

Note that (15.9) implies

$$\bar{a}_{p,n+1} < 0. \tag{15.11}$$

Now we show $\hat{x}_{n+1} \in \Phi(B)$. If, otherwise, it does not hold, then there is an $r \in \{1, \cdots, m+1\}$ such that

$$\bar{b}_r - \hat{x}_{n+1}\bar{a}_{r,n+1} < 0, \quad \bar{a}_{r,n+1} \geq 0.$$

There are following two cases arising:

Case (i) $\bar{a}_{r,n+1} = 0,\ \bar{b}_r < 0$. It clearly holds in this case that $\Phi(B) = \emptyset$.
Case (ii) $\bar{b}_r - \hat{x}_{n+1}\bar{a}_{r,n+1} < 0,\ \bar{a}_{r,n+1} > 0$. Then, it is known that

$$\bar{b}_p / \bar{a}_{p,n+1} = \hat{x}_{n+1} > \bar{b}_r / \bar{a}_{r,n+1}. \tag{15.12}$$

We show that

$$\bar{b}_p - x_{n+1}\bar{a}_{p,n+1} \geq 0 \tag{15.13}$$

and

$$\bar{b}_r - x_{n+1}\bar{a}_{r,n+1} \geq 0 \tag{15.14}$$

are inconsistent, as leads to $\Phi(B) = \emptyset$. In fact, for any x'_{n+1} satisfying (15.13), i.e.,

$$\bar{b}_p - x'_{n+1}\bar{a}_{p,n+1} \geq 0,$$

it follows from (15.11) and (15.12) that

$$x'_{n+1} \geq \bar{b}_p / \bar{a}_{p,n+1} > \bar{b}_r / \bar{a}_{r,n+1},$$

hence it is known by $\bar{a}_{r,n+1} > 0$ that

$$\bar{b}_r - x'_{n+1}\bar{a}_{r,n+1} < 0,$$

which indicates that x'_{n+1} does not satisfy (15.14).

Since either of the two cases leads to $\Phi(B) = \emptyset$, contradicting the assumption, it holds that $\hat{f} \in \Phi(B)$.

For any $x'_{n+1} \in \Phi(B)$, furthermore, it holds that

$$\bar{b} - x'_{n+1}\bar{a}_{n+1} \geq 0,$$

hence

$$\bar{b}_i/\bar{a}_{i,n+1} \leq x'_{n+1}, \quad i \in I,$$

which together with (15.9) and (15.10) gives

$$\hat{x}_{n+1} \leq x'_{n+1}.$$

Therefore, $\hat{x}_{n+1}$ is the greatest lower bound of $\Phi(B)$.

According the Lemma 15.1.1, on the other hand, $(\hat{x}, \hat{x}_{n+1})$ is a feasible solution. It is verified that

$$\hat{x}_{j_p} = 0. \tag{15.15}$$

Thus noting (15.11), it is known from Definition 15.1.1 that $(\hat{x}, \hat{x}_{n+1})$ is a basic feasible solution. □

After row index p determined, the following column rule is relevant.

Rule 15.1.2 (Column rule) Determine pivot column index

$$q \in \arg\min_{j \in N} \bar{a}_{p\,j}. \tag{15.16}$$

Theorem 15.1.1. *Assume $\Phi(B) \neq \emptyset$. If $\bar{a}_{pq} \geq 0$, then $(\hat{x}, \hat{x_{n+1}})$ is a basic feasible solution.*

Proof. From $\bar{a}_{p,q} \geq 0$ and (15.15), it is known that the pth row of $\bar{N}$ is nonnegative, i.e.,

$$e_p^{\mathrm{T}}\bar{N} \geq 0. \tag{15.17}$$

By Lemma 15.1.2, $(\hat{x}, \hat{f})$ is a basic feasible solution to (15.1). Assume that it is not optimal. Then there is a feasible solution, say $(\tilde{x}, \tilde{x_{n+1}})$, satisfies $\tilde{x}_{n+1} < \hat{x}_{n+1}$. Consequently, from (15.2) it follows that

$$\tilde{x}_{j_p} = \bar{b}_p - e_p^{\mathrm{T}}\bar{N}\tilde{x}_N - \bar{a}_{p,n+1}\tilde{x}_{n+1},$$

combining which, $\tilde{x}_N \geq 0$, (15.11), (15.15) and (15.17) leads to

$$\tilde{x}_{j_p} < \bar{b}_p - \bar{a}_{p,n+1}\hat{x}_{n+1} = \hat{x}_{j_p} = 0,$$

as contradicts that $\tilde{x}$ is a feasible solution. Therefore $\hat{x}$ is a basic feasible solution. □

Now assume $\bar{a}_{p,q} < 0$. Carry out the basis change by dropping j_p from and entering q to the basis. Assume that the new basis is

$$\hat{B} = \{j_1, \cdots, j_{p-1}, q, j_{p+1}, \cdots, j_{m+1}\}, \quad \hat{N} = A \backslash \hat{B},$$

where q is the pth index of $\hat{B}$. The according elementary transformations turn (15.2) to a new canonical form, setting $x_{\hat{N}} = 0$ in which gives the following system of inequalities:

$$x_{\hat{B}} = \hat{b} - x_{n+1}\hat{a}_{n+1} \geq 0. \tag{15.18}$$

Theorem 15.1.2. *Assume that the solution set* $\Phi(\hat{B}) = \{x_{n+1} \mid \hat{b} - x_{n+1}\hat{a}_{n+1} \geq 0\}$ *to (15.18) is nonempty, and that* $\hat{x}_{n+1} \in \Phi(\hat{B})$. *If* $\Phi(\hat{B})$ *is bounded below, then the largest lower bound of* $\Phi(\hat{B})$ *is less than or equal to* $\hat{x}_{n+1}$.

Proof. $Ax + x_{n+1}a_{n+1} = b$ and $x_{\hat{B}} \geq 0$ together are equivalent to

$$x_{\hat{B}} = \hat{b} - \hat{N}x_{\hat{N}} - x_{n+1}\hat{a}_{n+1} \geq 0.$$

By Lemma 15.1.2, $\hat{x}$ defined by (15.10) is a basic feasible solution, hence satisfying the preceding expression. Substituting it to the preceding and noting (15.15) gives

$$\hat{x}_{\hat{B}} = \hat{b} - \hat{x}_{n+1}\hat{a}_{n+1} \geq 0,$$

Therefore it holds that $\hat{x}_{n+1} \in \Phi(\hat{B})$. That $\Phi(\hat{B})$ is bounded below implies $\hat{a}_{n+1} \not\geq 0$, because it is unbounded below by Lemma 15.1.1, otherwise.

By Lemma 15.1.2, the greatest lower bound of $\Phi(\hat{B})$ is

$$\mu = \hat{b}_{p'}/\hat{a}_{p',n+1} = \max\{\hat{b}_i/\hat{a}_{i,n+1} \mid \hat{a}_{i,n+1} < 0,\ i = 1, \cdots, m+1\}. \tag{15.19}$$

Therefore, $\mu \leq \hat{x}_{n+1}$.

$\hat{a}_{n+1}$ can expressed in term of $\bar{a}_{n+1}$ as follows (see the first expression of (3.15)):

$$\hat{a}_{i,n+1} = \begin{cases} \bar{a}_{i,n+1} - (\bar{a}_{p,n+1}/\bar{a}_{pq})\bar{a}_{iq}, & i = 1, \cdots, m+1,\ i \neq p, \\ \bar{a}_{p,n+1}/\bar{a}_{pq}, & i = p, \end{cases}$$

Hence, from (15.15) and $\bar{a}_{pq} < 0$, it follows that

$$\hat{a}_{p,n+1} = \bar{a}_{p,n+1}/\bar{a}_{pq} > 0.$$

In addition, it is known from (15.19) that

$$\hat{a}_{p',n+1} < 0, \tag{15.20}$$

Therefore $p' \neq p$. From (15.19) and $\hat{x}$ satisfying the p'th expression of (15.18), it follows that

$$\hat{x}_{n+1} = (\hat{b}_{p'} - \hat{x}_{j_{p'}})/\hat{a}_{p',n+1} = \mu - \hat{x}_{j_{p'}}/\hat{a}_{p',n+1},$$

combining which, $p' \neq p$, (15.7) and (15.20) leads to $\mu \leq \hat{x}_{n+1}$. □

According to the preceding Theorem, such an iteration results in a new feasible canonical form, with objective value not increasing. It will be shown in Sect. 16.1 that under the nondegeneracy assumption, the objective value strictly monotonically decreases, and hence the solution process terminates in finitely many iterations, achieving optimality or detecting unboundedness of the problem.

15.2 Reduced Simplex Method

Based on the previous derivation, this section formulates the algorithm first, and then formulates its revised version.

Assume that via a series of elementary transformations, the initial tableau $(A \vdots -e_{m+1} \mid b)$ of the reduced problem (15.1) becomes

x_B^T	x_N^T	x_{n+1}	RHS
I	$\bar{N}$	$\bar{a}_{n+1}$	$\bar{b}$

(15.21)

which is termed reduced (simplex) tableau. If the system of inequalities

$$\bar{b} - x_{n+1}\bar{a}_{n+1} \geq 0$$

is consistent with respect to variable x_{n+1}, tableau (15.21) is said to be feasible.

Thereby, the overall steps described in Sect. 15.1 can be put in the following algorithm.

Algorithm 15.2.1 (Reduced simplex algorithm: tableau form). Initial: feasible reduced simplex tableau of form (15.21). This algorithm solves reduced problem (15.1).

1. Stop if $\bar{a}_{n+1} \geq 0$.

2. Determine row index p such that

$$\bar{x}_{n+1} = \bar{b}_p/\bar{a}_{p,n+1} = \max\{\bar{b}_i/\bar{a}_{i,n+1} \mid \bar{a}_{i,n+1} < 0,\ i = 1,\cdots,m+1\}.$$

3. Determine column index $q \in \arg\min_{j \in N} \bar{a}_{pj}$.
4. If $\bar{a}_{pq} \geq 0$, then compute $\bar{x}_B = \bar{b} - \bar{x}_{n+1}\bar{a}_{n+1}$, and stop.
5. Convert $\bar{a}_{pq}$ to 1, and eliminate the other nonzeros in the column by elementary transformations.
6. Go to step 1.

Theorem 15.2.1. *Assume termination of Algorithm 15.2.1. It terminates at*

(i) Step 1, detecting unboundedness of the problem; or at
(ii) Step 4, providing a basic feasible solution $\bar{x}$.

Proof. The validity is shown by Theorems 15.1.2, 15.1.1 and 15.1.1, as well as related discussions in Sect. 15.1. □

Note It is possible to start solution process directly from a conventional feasible simplex tableau. To do so, assume availability of the following conventional simplex tableau, with f replaced by x_{n+1}:

x_B^{T}	x_N^{T}	x_{n+1}	RHS
I	$\bar{N}$		$\bar{b}$
	$\bar{z}_N^{\mathrm{T}}$	-1	

where $\bar{b} \geq 0$. Starting from it, the first iteration of Algorithm 15.2.1 needs to be replaced by the following steps:

1. $\bar{a}_{n+1} = -e_{m+1} \not\geq 0$.
2. $\bar{x}_{n+1} = 0/(-1) = 0$.
3. $q \in \arg\min_{j \in N} \bar{z}_j$.
4. Optimality is achieved if $\bar{c}_q \geq 0$.
5. Carry out elementary transformations to obtain a feasible reduced tableau by taking the entry in the bottom row and x_q column as the pivot.
6. Go to step 1.

Example 15.2.1. Solve the following problem by Algorithm 15.2.1:

$$\begin{array}{ll} \min & x_{10} = -2x_1 + 4x_2 + 3x_3 - 3x_4 - 4x_5, \\ \text{s.t.} & -2x_1 - 6x_2 + 1x_3 - 3x_4 - x_5 + x_6 = 4, \\ & -x_1 - 9x_2 - 6x_3 + 2x_4 + 3x_5 + x_7 = 3, \\ & 8x_1 - 6x_2 + 3x_3 + 5x_4 + 7x_5 + x_8 = 2, \\ & 3x_1 - 2x_2 - 4x_3 - x_4 - 2x_5 + x_9 = 1, \\ & x_j \geq 0, \quad j = 1,\cdots,9. \end{array}$$

Answer The problem has the following initial tableau:

x_1	x_2	x_3	x_4	x_5	x_6	x_7	x_8	x_9	x_{10}	RHS
−2	−6	1	−3	−1	1					4
−1	−9	−6	2	3		1				3
8	−6	3	5	7			1			2
3	−2	−4	−1	−2				1		1
−2	4	3	−3	−4*					−1	

The right-hand side $(3, 7, 4, 5)^{\mathrm{T}}$ of which is nonnegative. Call Algorithm 15.2.1.
Iteration 1:

1. $\bar{a}_{10} \not\geq 0$.
2. $\bar{x}_{10} = \max\{0/(-1)\} = 0,\ p = 5$.
3. $\min\{-2, 4, 3, -3, -4\} = -4,\ q = 5$.
5. Multiply row 5 by $-1/4$, and then add $1, -3, -7, 2$ times of row 5 to rows 1,2,3,4, respectively:

x_1	x_2	x_3	x_4	x_5	x_6	x_7	x_8	x_9	x_{10}	RHS
−3/2	−7	1/4	−9/4		1				1/4	4
−5/2	−6	−15/4	−1/4			1			−3/4	3
9/2	1	33/4	−1/4*				1		−7/4	2
4	−4	−11/2	1/2					1	1/2	1
1/2	−1	−3/4	3/4	1					1/4	

Iteration 2:

1. $\bar{a}_{10} \not\geq 0$.
2. $\bar{x}_{10} = \max\{3/(-3/4), 2/(-7/4)\} = 2/(-7/4) = -8/7,\ p = 3$.
3. $\min\{9/2, 1, 33/4, -1/4\} = -1/4,\ q = 4$.
5. Multiply row 3 by -4, and then add $9/4, 1/4, -1/2, -3/4$ times of row 3 to rows 1,2,4,5, respectively:

x_1	x_2	x_3	x_4	x_5	x_6	x_7	x_8	x_9	x_{10}	RHS
−42	−16	−74			1		−9		16	−14
−7	−7	−12				1	−1		1	1
−18	−4	−33	1				−4		7	−8
13	−2	11					2	1	−3	5
14	2	24		1			3		−5	6

Table 15.1 Equivalence between the associated quantities

Quantity	Reduced	Relation	Revised reduced
Objective column	$\bar{a}_{n+1}$	=	$-B^{-1}e_{m+1}$
The righ-hand side	$\bar{b}$	=	$B^{-1}b$
Pivot row	$e_p^{\mathrm{T}}\bar{N}$	=	$e_p^{\mathrm{T}}B^{-1}N$
Pivot column	$\bar{a}_q$	=	$B^{-1}a_q$

Iteration 3:

1. $\bar{a}_{10} \not\geq 0$.
2. $\bar{x}_{10} = \max\{5/(-3), 6/(-5)\} = -6/5, p = 5.$
3. $\min\{14, 2, 24, 3\} \geq 0.$
4. $\bar{x}_{10} = -6/5,$

$$\begin{aligned}\bar{x}_B &= \bar{b} - \bar{x}_{10}\bar{a}_{n+1} = (-14, 1, -8, 5, 6)^{\mathrm{T}} - (-6/5)(16, 1, 7, -3, -5)^{\mathrm{T}} \\ &= (26/5, 11/5, 2/5, 7/5, 0)^{\mathrm{T}}. \\ B &= \{6, 7, 4, 9, 5\}.\end{aligned}$$

Basic optimal solution and according objective value are

$$\bar{x} = (0, 0, 0, 2/5, 0, 26/5, 11/5, 0, 7/5)^{\mathrm{T}}, \quad \bar{x}_{10} = -6/5.$$

Now let us derive the revised version of Algorithm 15.2.1.

Let (15.21) be the current reduced tableau, associated with basis and nonbasis matrices B, N. Premultiplying $(A\vdots - e_{m+1}|b)$ by B^{-1} gives a so-called "revised reduced tableau", as written

x_B^{T}	x_N^{T}	x_{n+1}	RHS
I	$B^{-1}N$	$B^{-1}\bar{a}_{n+1}$	$B^{-1}b$

(15.22)

Like in the conventional simplex context, reduced and revised reduced tableaus, associated with the same basis are equivalent; that is, their associated entries are equal. Based on such equivalence, it is easy to transform any reduced tableau to a revised version, and vice versa. As for the implementation of the reduced simplex method, the reduced tableau as a whole is not indispensable, and only a part of its entries are needed. Table 15.1 indicates equivalence relationship between quantities in reduce tableau (15.21) and revised reduced tableau (15.22).

Based on Table 15.1, Algorithm 15.2.1 can be revised as follows, in which $\bar{b}$ and $\bar{a}_{n+1}$ are generated recursively (see (17.13) and (17.15)).

Algorithm 15.2.2 (Reduced simplex algorithm). Initial: $(B, N), B^{-1}, \bar{b} = B^{-1}(b^{\mathrm{T}}, 0)^{\mathrm{T}}$, $\bar{a}_{n+1} = -B^{-1}e_{m+1}$, and consistent $\bar{b} - f\bar{a}_{n+1} \geq 0$. This algorithm solves the reduced problem (15.1).

1. Stop if $\bar{a}_{n+1} \geq 0$ (Unbounded).
2. Determine $\bar{x}_{n+1}$ and row index p such that

$$\bar{x}_{n+1} = \bar{b}_p/\bar{a}_{p,n+1} = \max\{\bar{b}_i/\bar{a}_{i,n+1} \mid \bar{a}_{i,n+1} < 0,\ i = 1, \cdots, m+1\}.$$

3. Compute $\sigma_N = N^{\mathrm{T}} B^{-\mathrm{T}} e_p$.
4. Determine column index $q \in \arg\min_{j \in N} \sigma_j$.
5. If $\sigma_q \geq 0$, then compute $\bar{x}_B = \bar{b} - \bar{x}_{n+1}\bar{a}_{n+1}$, and stop (optimality achieved).
6. Compute $\bar{a}_q = B^{-1}a_q$, $\nu = -\bar{a}_{p,n+1}/\sigma_q$, and $\tau = -\bar{b}_p/\sigma_q$.
7. If $\nu \neq 0$, update $\bar{a}_{n+1} = \bar{a}_{n+1} + \nu(\bar{a}_q - e_p)$.
8. If $\tau \neq 0$, update $\bar{b} = \bar{b} + \tau(\bar{a}_q - e_p)$.
9. Update B^{-1} by (3.23).
10. Update (B, N) by exchanging j_p and q.
11. Go to step 1.

It is noted that the preceding algorithm is practicable, compared with its tableau form, though the former is preferred in illustration in this book.

15.3 Reduced Phase-I: Single-Artificial-Variable

In general, an initial reduced simplex tableau is not feasible, from which the reduce simplex algorithm cannot get started. However, the algorithm can get started from a conventional feasible simplex tableau (see Note after Algorithm 15.2.1), and hence any conventional Phase-I method is applicable. In particular, the single-artificial-variable method, presented in Sect. 13.2, deserves attention, as the associated auxiliary program, involving a single artificial variable, is amenable to be solved by the reduced simplex method.

Assume $\bar{b} = B^{-1}b \not\geq 0$, $\bar{N} = B^{-1}N$. Given some m-dimensional vector $\hat{x}_B \geq 0$, and set $\bar{a}_{n+1} = \bar{b} - \hat{x}_B$. Based on the canonical form of the constraint system, an auxiliary program of form (13.16) can be constructed, i.e.,

$$\begin{aligned} &\min\ x_{n+1}, \\ &\text{s.t.}\ \ x_B = \bar{b} - \bar{a}_{n+1}x_{n+1} - \bar{N}x_N, \\ &\qquad\ x, x_{n+1} \geq 0. \end{aligned}$$

The preceding program is lower bounded, associated with the feasible solution

$$\hat{x}_B = \hat{x}_B, \quad \hat{x}_N = 0, \quad \hat{x}_{n+1} = 1.$$

As the according auxiliary tableau of form (13.18) is itself a feasible reduce simplex tableau, it can be solved by the following slight variant of the reduced simplex algorithm.

Algorithm 15.3.1 (Tableau reduced Phase-I: single-artificial-variable). Initial: reduced simplex tableau of form (13.18). $\bar{a}_{n+1} = \bar{b} - \hat{x}_B$, $\hat{x}_B \geq 0$. This algorithm finds a feasible reduced simplex tableau.

1. Determine $\bar{x}_{n+1}$ and row index p such that

$$\bar{x}_{n+1} = \bar{b}_p/\bar{a}_{p,n+1} = \max\{\bar{b}_i/\bar{a}_{i,n+1} \mid \bar{a}_{i,n+1} < 0,\ i = 1, \cdots, m\}.$$

2. Select column index $q \in \arg\min_{j \in N} \bar{a}_{pj}$.
3. Stop if $\bar{a}_{pq} \geq 0$ (infeasible problem).
4. Convert $\bar{a}_{pq}$ to 1, and eliminate the other nonzeros in the column by elementary transformations.
5. If $\bar{b} \geq 0$, restore the original objective column, and stop (feasibility achieved).
6. Go to step 1.

Example 15.3.1. Solve that following problem, using Algorithm 15.3.1 as reduced Phase-I:

$$\begin{array}{lllr}
\min & x_{10} = -2x_1 + 4x_2 + 3x_3 - 3x_4 - 4x_5, & & \\
\text{s.t.} & -2x_1 - 6x_2 + 1x_3 - 3x_4 - \ x_5 + x_6 & & = -3, \\
& -x_1 - 9x_2 - 6x_3 + 2x_4 + 3x_5 \qquad + x_7 & & = -7, \\
& 8x_1 - 6x_2 + 3x_3 + 5x_4 + 7x_5 \qquad\qquad + x_8 & & = 4, \\
& 3x_1 - 2x_2 - 4x_3 - \ x_4 - 2x_5 \qquad\qquad\qquad + x_9 & & = -5, \\
& \qquad\qquad x_j \geq 0, \quad j = 1, \cdots, 9. & &
\end{array}$$

Answer Phase-I: To turn to Phase-II conveniently, it might be well still put the original objective row at the bottom of the tableau, but which will not take a part in pivoting in Phase-I. Set $\hat{x}_B = (1, 1, 0, 1)^{\mathrm{T}}$, $\bar{a}_{10} = (-4, -8, 0, -6)^{\mathrm{T}}$, and take x_{10} column as the auxiliary objective column. Then the initial auxiliary tableau is

x_1	x_2	x_3	x_4	x_5	x_6	x_7	x_8	x_9	x_{10}	RHS
−2	−6	1	−3	−1	1				−4	−3
−1	−9*	−6	2	3		1			−8	−7
8	−6	3	5	7			1			4
3	−2	−4	−1	−2				1	−6	−5
−2	4	3	−3	−4					–	

The auxiliary program has feasible solution $\bar{x}_B = (1, 1, 0, 1)^{\mathrm{T}}$, $\bar{x}_{10} = 1$.
Phase-I: Call Algorithm 15.3.1.

Iteration 1:

1. $\max\{-3/-4, -7/-8, -5/-6\} = 7/8$, $p = 2$.
2. $\min\{-1, -9, -6, 2, 3\} = -9 < 0$, $q = 2$.
4. Multiply row 2 by $-1/9$, and then add $6, 6, 2, -4$ times of row 2 to rows 1,3,4,5, respectively:

x_1	x_2	x_3	x_4	x_5	x_6	x_7	x_8	x_9	x_{10}	RHS
−4/3		5	−13/3	−3	1	−2/3			4/3	5/3
1/9	1	2/3	−2/9	−1/3		−1/9			8/9	7/9
26/3		7	11/3	5		−2/3	1		16/3	26/3
29/9		−8/3	−13/9*	−8/3		−2/9		1	−38/9	−31/9
−22/9		1/3	−19/9	−8/3		4/9			–	−28/9

Iteration 2:

1. $\max\{(-31/9)/(-38/9)\} = 31/38,\ p = 4.$
2. $\min\{29/9, -8/3, -13/9, -8/3, -2/9\} = -8/3 < 0,\ q = 4.$
4. Multiply row 4 by $-9/13$, and then add $13/3, 2/9, -11/3, 19/94$ times of row 4 to rows 1,2,3,5, respectively:

x_1	x_2	x_3	x_4	x_5	x_6	x_7	x_8	x_9	x_{10}	RHS
−11		13		5	1			−3	14	12
−5/13	1	14/13		1/13		−1/13		−2/13	20/13	17/13
219/13		3/13		−23/13*		−16/13	1	33/13	−70/13	−1/13
−29/13		24/13	1	24/13		2/13		−9/13	38/13	31/13
−93/13		55/13		16/13		10/13		−19/13	−	25/13

Iteration 3:

1. $\max\{(-1/13)/(-70/13)\} = 1/70,\ p = 3.$
2. $\min\{219/13, 3/13, -23/13, -16/13, 33/13\} = -23/13 < 0,\ q = 5.$
4. Multiply row 3 by $-13/23$, and then add $-5, -1/13, -24/13, -16/13$ times of row 3 to rows 1,2,4,5, respectively:
5. $\bar{b} \geq 0$, Take the original objective column to overwrite the current x_{10} column, resulting in a feasible reduced tableau below:

x_1	x_2	x_3	x_4	x_5	x_6	x_7	x_8	x_9	x_{10}	RHS
842/23		314/23			1	−80/23	65/23	96/23		271/23
8/23	1	25/23				−3/23	1/23	−1/23		30/23
−219/23		−3/23		1		16/23	−13/23	−33/23		1/23
353/23		48/23	1			−26/23	24/23	45/23		53/23
105/23		101/23				−2/23*	16/23	7/23	−1	43/23

Phase-II: Call Algorithm 15.2.1.

Iteration 4:

1. $\bar{a}_{10} \not\geq 0.$
2. $\bar{x}_{10} = \max\{(43/23)/(-1)\} = -43/23,\ p = 5.$
3. $\min\{105/23, 101/23, -2/23, 16/23, 7/23\} = -2/23 < 0,\ q = 7.$
5. Multiply row 5 by $-23/2$, and then add $80/23, 3/23, -16/23, 26/23$ times of row 5 to rows 1,2,3,4, respectively:

x_1	x_2	x_3	x_4	x_5	x_6	x_7	x_8	x_9	x_{10}	RHS
−146		−162			1		−25	−8	40	−63
−13/2	1	−11/2					−1	−1/2	3/2	−3/2
27		35		1			5	1	−8	15
−44		−55	1				−8	−2	13	−22
−105/2		−101/2				1	−8	−7/2	23/2	−43/2

Iteration 5:

2. $\bar{x}_{10} = \max\{15/(-8)\} = -15/8, \ p = 3.$
3. $\min\{27, 35, 5, 1, \} = 1 \geq 0.$
4. Basic optimal solution and according objective value:

$$\bar{x} = (0, 21/16, 0, 19/8, 0, 12, 1/16, 0, 0)^{\mathrm{T}}, \quad \bar{x}_{10} = -15/8.$$

15.4 Dual Reduced Simplex Method

This section describes a dual version of Algorithm 15.2.2, still using notations in the previous two sections. To this end, firstly established are optimality conditions and related properties in the reduced simplex context.

Theorem 15.4.1. (x, x_{n+1}), *where* $x_B = \bar{b} - x_{n+1}\bar{a}_{n+1}$, $x_N = 0$, *is an optimal solution to (15.1) if the following conditions are satisfied:*

(i) $\bar{b} - x_{n+1}\bar{a}_{n+1} \geq 0, \quad \bar{b}_p - x_{n+1}\bar{a}_{p,n+1} = 0,$ *(primal feasibility)*
(ii) $e_p^{\mathrm{T}}\bar{N} \geq 0, \quad \bar{a}_{p,n+1} < 0.$ *(dual feasibility)*

Proof. The validity comes from Theorems 15.1.2, 15.1.1 and 15.1.1, as well as related discussions in Sect. 15.1. □

The pth row of the reduced simplex tableau, giving the according objective value, is called *objective row*.

Lemma 15.4.1. *Assume that*

$$e_p^{\mathrm{T}}\bar{N} \geq 0, \qquad p \in \{1, \cdots, m+1\}. \tag{15.23}$$

If $\bar{a}_{p,n+1} = 0$ *and* $\bar{b}_p < 0$, *then there is no feasible solution to (15.1).*

If $\bar{a}_{p,n+1} \neq 0$ *and* $\bar{x}_{n+1}$ *satisfies*

$$\bar{b}_p - \bar{x}_{n+1}\bar{a}_{p,n+1} = (<)\, 0,$$

then for any feasible value x'_{n+1} *(if any), the following hold:*

(i) $x'_{n+1} \leq (<)\bar{x}_{n+1}$ *when* $\bar{a}_{p,n+1} > 0.$
(ii) $x'_{n+1} \geq (>)\bar{x}_{n+1}$ *when* $\bar{a}_{p,n+1} < 0.$

Proof. The pth equality constraint of (15.2) is

$$x_p = \bar{b}_p - e_p^{\mathrm{T}}\bar{N}x_N - \bar{a}_{p,n+1}x_{n+1}.$$

Let $\tilde{x} \geq 0$ be a feasible solution, associated with objective value x'_{n+1}. Substituting it to the preceding gives

$$\tilde{x}_p = \bar{b}_p - e_p^{\mathrm{T}} \bar{N} \tilde{x}_N - \bar{a}_{p,n+1} x'_{n+1}. \tag{15.24}$$

In addition, from (15.21) and $\tilde{x} \geq 0$, it follows that

$$-e_p^{\mathrm{T}} \bar{N} \tilde{x}_N \leq 0. \tag{15.25}$$

Condition $\bar{a}_{p,n+1} = 0$ and $\bar{b}_p < 0$ together with (15.25) leads to negativity of the right-hand side of (15.24), as contradicts the nonnegative lett-hand side. Therefore, there is no feasible solution to (15.1).

(i) When $\bar{a}_{p,n+1} > 0$ and $x'_{n+1} > (\geq) \bar{x}_{n+1}$, it holds that

$$\bar{b}_p - \bar{a}_{p,n+1} x'_{n+1} < (\leq) \ \bar{b}_p - \bar{a}_{p,n+1} \bar{x}_{n+1} = (<) 0, \tag{15.26}$$

combining which and (15.25) leads to negativeness of the right-hand side of (15.24), as contradicts the nonnegative lett-hand side. Therefore, $x'_{n+1} \leq (<) \bar{x}_{n+1}$.

(ii) When $\bar{a}_{p,n+1} < 0$ and $x'_{n+1} < (\leq) \bar{x}_{n+1}$, (15.26) still holds, as also leads to negativity of the right-hand side of (15.24), leading to a contradiction. Therefore, $x'_{n+1} \geq (>) \bar{x}_{n+1}$. □

As was shown, the reduced simplex method pursues dual feasibility while maintaining primal feasibility. Conversely, pursuing primal feasibility while maintaining dual feasibility will lead to its dual version.

Assume now that the dual feasibility condition (ii) holds. Define $\bar{x}$ as follows:

$$\bar{x}_{n+1} = \bar{b}_p / \bar{a}_{p,n+1}, \quad \bar{x}_B = \bar{b}_B - \bar{x}_{n+1} \bar{a}_{n+1}, \quad \bar{x}_N = 0. \tag{15.27}$$

It is clear that $\bar{x}_{j_p} = 0$.

If $\bar{x}_B \geq 0$ holds, then the primal feasibility condition (i) is satisfied. Thus, $\bar{x}$ is an optimal solution to (15.1). Assume $\bar{x}_B \not\geq 0$. Then the following rule is applicable.

Rule 15.4.1 (Dual row rule) Select row index r such that

$$\bar{x}_{j_r} = \min\{\bar{x}_{j_i} \mid i = 1, \cdots, m+1\} < 0.$$

If $e_r^{\mathrm{T}} \bar{N} \not\geq 0$, in addition, the following rule is well-defined:

Rule 15.4.2 (Dual column rule) Select column index q such that

$$\beta = -\bar{a}_{pq} / \bar{a}_{rq} = \min\{-\bar{a}_{pj} / \bar{a}_{rj} \mid \bar{a}_{rj} < 0, \ j \in N\} \geq 0.$$

If $\beta > 0$, the reduced simplex tableau is said to be *dual nondegenerate*.

Once a pivot is determined, a basis change is executed to drop x_r from and enter x_q to the basis. It is not difficult to show that the resulting pth row still satisfies $e_p^{\mathrm{T}}\bar{N} \geq 0$. If $\bar{a}_{p,n+1} < 0$, then go to the next iteration.

The solution steps are summarized to the following algorithm.

Algorithm 15.4.1 (Dual reduced simplex algorithm: tableau form). Initial: Reduced simplex tableau of form (15.21), where $e_p^{\mathrm{T}}\bar{N} \geq 0$, $\bar{a}_{p,n+1} < 0$. This algorithm solves the reduced problem (15.1).

1. Compute $\bar{x}_{n+1} = \bar{b}_p/\bar{a}_{p,n+1}$.
2. Compute $\bar{x}_B = \bar{b} - \bar{x}_{n+1}\bar{a}_{n+1}$.
3. Select row index $r \in \arg\min\{\bar{x}_{j_i} \mid i = 1, \cdots, m+1, i \neq p\}$.
4. Stop if $\bar{x}_{j_r} \geq 0$.
5. If $J = \{j \in N \mid \bar{a}_{rj} < 0\} = \emptyset$, set $p = r$ and go to step 8.
6. Determine β and column index q such that $\beta = -\bar{a}_{pq}/\bar{a}_{rq} = \min_{j \in J} -\bar{a}_{pj}/\bar{a}_{rj}$.
7. Convert $\bar{a}_{rq}$ to 1, and eliminate the other nonzeros in the pivot column by elementary transformations.
8. Go to step 1 if $\bar{a}_{p,n+1} < 0$.
9. Stop.

Theorem 15.4.2. *Assuming dual nondegeneracy, Algorithm 15.4.1 terminates either at*

(i) Step 4, achieving a basic optimal solution $\bar{x}$; or at
(ii) Step 9, detecting infeasibility of the problem.

Proof. Termination is shown first. Assume that it does not terminate at the current iteration. It is known from steps 1 and 2 that the basic solution $\bar{x}$ is associated with objective value

$$\bar{x}_{n+1} = \bar{b}_p/\bar{a}_{p,n+1}.$$

And $\bar{x}_{j_r} < 0$ implies that

$$\bar{b}_r - \bar{x}_{n+1}\bar{a}_{r,n+1} < 0. \tag{15.28}$$

From the preceding two expressions and $\bar{a}_{p,n+1} < 0$, it follows that

$$\bar{b}_r\bar{a}_{p,n+1} - \bar{b}_p\bar{a}_{r,n+1} > 0. \tag{15.29}$$

There are the following two cases only:

(i) Passing from step 7 to step 8 to go to the next iteration. The new entry in the pth row and $n+1$ column yielded from the basis change in step 7 satisfies

$$\hat{a}_{p,n+1} = \bar{a}_{p,n+1} + \beta\bar{a}_{r,n+1} < 0. \tag{15.30}$$

and the pth component of the new right-hand side is equal to

$$\hat{b}_p \bar{b}_p + \beta \bar{b}_r. \tag{15.31}$$

In step 1 of the next iteration, the objective value calculated from the preceding two impressions is then

$$\hat{x}_{n+1} = \frac{\hat{b}_p}{\hat{a}_{p,n+1}} = \frac{\bar{b}_p + \beta \bar{b}_r}{\bar{a}_{p,n+1} + \beta \bar{a}_{r,n+1}},$$

The difference between the new and old objective values is

$$\begin{aligned} \hat{x}_{n+1} - \bar{x}_{n+1} &= \frac{\bar{b}_p \bar{a}_{p,n+1} + \beta \bar{b}_r \bar{a}_{p,n+1} - \bar{b}_p \bar{a}_{p,n+1} - \beta \bar{b}_p \bar{a}_{r,n+1}}{\bar{a}_{p,n+1}(\bar{a}_{p,n+1} + \beta \bar{a}_{r,n+1})} \\ &= \frac{\beta(\bar{b}_r \bar{a}_{p,n+1} - \bar{b}_p \bar{a}_{r,n+1})}{\bar{a}_{p,n+1}(\bar{a}_{p,n+1} + \beta \bar{a}_{r,n+1})}. \end{aligned}$$

It is known from $\bar{a}_{p,n+1} < 0$ and (15.30) that the denominator in the preceding expression is positive, whereas it is known from $\beta \geq 0$ and (15.29) that the numerator is nonnegative, therefore the objective value never decreases. Under the dual nondegeneracy assumption, the objective value strictly increases.

(ii) Passing from step 5 to step 8 to go to the next iteration. It is noted that $\bar{a}_{r,n+1} < 0$ holds in this case. The difference between the new and old objective values is

$$\frac{\bar{b}_r}{\bar{a}_{r,n+1}} - \frac{\bar{b}_p}{\bar{a}_{p,n+1}} = \frac{\bar{b}_r \bar{a}_{p,n+1} - \bar{b}_p \bar{a}_{r,n+1}}{\bar{a}_{r,n+1} \bar{a}_{p,n+1}},$$

where the denominator is clearly positive whereas, by (15.29), the numerator is also positive. Therefore, the objective value strictly increases.

If the algorithm does not terminate, then under the dual nondegeneracy assumption, the objective value strictly increases monotonically, hence no cycling occurs. This means that there are infinitely many basic solutions, as is a contradiction. Therefore, the algorithm terminates.

Note that $e_p^{\mathrm{T}} \bar{N} \geq 0$ always holds for the algorithm. By Lemma 15.4.1, optimality is achieved while termination occurs at step 4. Now assume that it occurs at step 9, hence the new entry in the pth row and $n+1$ column satisfies

$$\hat{a}_{p,n+1} = \bar{a}_{p,n+1} + \beta \bar{a}_{r,n+1} \geq 0. \tag{15.32}$$

Since $\bar{a}_{p,n+1} < 0$, in this case the β, determined in step 6, is positive. Assume there is a feasible solution, associated with objective value x'_{n+1}. It is clear that the $\bar{x}_{n+1}$, determined in step 1, satisfies

$$\bar{b}_p - \bar{x}_{n+1}\bar{a}_{p,n+1} = 0, \tag{15.33}$$

and $\bar{a}_{p,n+1} < 0$. Thus it holds by Lemma 15.4.1 that

$$x'_{n+1} \geq \bar{x}_{n+1}. \tag{15.34}$$

In case when passing through steps 7 → 8 → 9, on the other hand, it follows from (15.31), (15.32), (15.33), (15.28) and $\beta > 0$ that

$$\hat{b}_p - \bar{x}_{n+1}\hat{a}_{p,n+1} = (\bar{b}_p - \bar{x}_{n+1}\bar{a}_{p,n+1}) + \beta(\bar{b}_r - \bar{x}_{n+1}\bar{a}_{r,n+1}) = \beta(\bar{b}_r - \bar{x}_{n+1}\bar{a}_{r,n+1}) < 0. \tag{15.35}$$

If $\hat{a}_{p,n+1} > 0$, then it is know by Lemma 15.4.1 that

$$x'_{n+1} < \bar{x}_{n+1},$$

which contradicts (15.34), therefore there is no feasible solution; if $\hat{a}_{p,n+1} = 0$, then it is known by (15.35) that $\hat{b}_p < 0$; consequently, there is still no feasible solution, by Lemma 15.4.1. In case when passing through steps 5 → 8 → 9,

$$e_r^{\mathrm{T}}\bar{N} \geq 0, \quad \bar{a}_{r,n+1} \geq 0$$

and (15.28) hold. Then it can be similarly shown that there is no feasible solution. □

Example 15.4.1. Solve the following problem by Algorithm 15.4.1:

$$\begin{array}{lrl}
\min & x_{10} = & x_1 + 4x_2 + 3x_3 + 2x_4 + 9x_5, \\
\text{s.t.} & & -x_1 + 5x_2 - 4x_4 - 2x_5 + x_6 = -1, \\
& & -3x_1 - 2x_2 - 6x_3 + x_4 - x_5 + x_7 = -7, \\
& & -x_2 + 4x_3 - 6x_4 + 4x_5 + x_8 = 4, \\
& & 5x_1 + 3x_2 - 3x_3 + 3x_4 + 5x_5 + x_9 = 0, \\
& & x_j \geq 0, \quad j = 1, \cdots, 9.
\end{array}$$

Answer The initial tableau is

x_1	x_2	x_3	x_4	x_5	x_6	x_7	x_8	x_9	x_{10}	RHS
−1	5		−4	−2	1					−1
−3	−2	−6	1	−1		1				−7
	−1	4	−6	4			1			4
5	3	−3	3	5				1		
1*	4	3	2	9					−1	

Iteration 1:

To convert it to a reduced simplex tableau, take the smallest, 1, among the first 5 entries in the bottom row (in x_1 column) as pivot. Add $1, 3, -5$ times of row 5 to rows 1,2,4, respectively:

x_1	x_2	x_3	x_4	x_5	x_6	x_7	x_8	x_9	x_{10}	RHS
	9	3	−2	7	1				−1	−1
	10	3	7	26		1			−3	−7
	−1	4	−6	4			1			4
	−17	−18*	−7	−40				1	5	
1	4	3	2	9					−1	

which is a dual feasible reduced tableau with $p = 5$.

Call Algorithm 15.4.1.

Iteration 2:

1. $\bar{x}_{10} = 0/(-1) = 0$.
2. $\bar{x}_B = (-1, -7, 4, 0, 0)^{\mathrm{T}}$.
3. $\min\{-1, -7, 4, 0, 0\} = -7 < 0,\ r = 2$.
5. $J = \emptyset,\ p = 2$.
8. $\bar{a}_{2,10} = -3 < 0$.

Iteration 3:

1. $\bar{x}_{10} = (-7)/(-3) = 7/3$.
2. $\bar{x}_B = (-1, -7, 4, 0, 0)^{\mathrm{T}} - (7/3)(-1, -3, 0, 5, -1)^{\mathrm{T}} = (4/3, 0, 4, -35/3, 7/3)^{\mathrm{T}}$.
3. $\min\{4/3, 4, -35/3, 7/3\} = -35/3 < 0,\ r = 4$.
5. $J = \{2, 3, 4, 5\} \neq \emptyset$.
6. $\beta = \min\{-10/(-17), -3/(-18), -7/(-7), -26/(-40)\} = 3/18,\ q = 3$.
7. Multiply row 4 by $-1/18$, and then add $-3, -3, -4, 3$ times of row 4 to rows $1, 2, 3, 5$, respectively:

x_1	x_2	x_3	x_4	x_5	x_6	x_7	x_8	x_9	x_{10}	RHS
	37/6		−19/6*	1/3	1			1/6	−1/6	−1
	43/6		35/6	58/3		1		1/6	−13/6	−7
	−43/9		−68/9	−44/9			1	2/9	10/9	4
	17/18	1	7/18	20/9				−1/18	−5/18	
1	7/6		5/6	7/3				1/6	−1/6	

8. $\bar{a}_{2,10} = -13/6 < 0$.

Iteration 4:

1. $\bar{x}_{10} = (-7)/(-13/6) = 42/13.$
2. $\bar{x}_B = (-1, -7, 4, 0, 0)^{\mathrm{T}} - (42/13)(-1/6, -13/6, 10/9, -5/18, -1/6)^{\mathrm{T}}$
   $= (-6/13, 0, 16/39, 35/39, 7/13)^{\mathrm{T}}.$
3. $\min\{-6/13, 16/39, 35/39, 7/13\} = -6/13 < 0,\ r = 1.$
5. $J = \{4\} \neq \emptyset.$
6. $\beta = \min\{-(35/6)/(-19/6)\} = 35/19,\ q = 4.$
7. Multiply row 1 by $-6/19$, and add $-35/6, 68/9, -7/18, -5/6$ times of row 1 to rows 2, 3, 4, 5, respectively :

x_1	x_2	x_3	x_4	x_5	x_6	x_7	x_8	x_9	x_{10}	RHS
	−37/19		1	−2/19	−6/19			−1/19	1/19	6/19
	352/19			379/19	35/19	1		9/19	−47/19	−168/19
	−1,111/57			−108/19	−136/57		1	−10/57	86/57	364/57
	97/57	1		43/19	7/57			−2/57	−17/57	−7/57
1	53/19			46/19	5/19			4/19	−4/19	−5/19

8. $\bar{a}_{2,10} = -47/19 < 0.$

Iteration 5:

1. $\bar{x}_{10} = (-168/19)/(-47/19) = 168/47.$
2. $\bar{x}_B = (6/19, -168/19, 364/57, -7/57, -5/19)^{\mathrm{T}}$
   $-(168/47)(1/19, -47/19, 86/57, -17/57, -4/19)^{\mathrm{T}}$
   $= (6/47, 0, 140/141, 133/141, 23/47)^{\mathrm{T}} \geq 0.$
4. Basic optimal solution and according objective value:

$$\bar{x} = (23/47, 0, 133/141, 6/47, 0, 0, 0, 140/141, 0)^{\mathrm{T}}, \quad \bar{x}_{10} = 168/47.$$

Based on the equivalence between the reduced tableau (15.21) and the revised tableau (15.22), it is not difficult to transfer Algorithm 15.4.1 to its revision.

Algorithm 15.4.2 (Dual reduced simplex algorithm). Initial: $(B, N), B^{-1}, \bar{b} = B^{-1}b, \bar{a}_{n+1} = -B^{-1}e_{m+1}$. $\sigma_N = e_p^{\mathrm{T}} B^{-1} N \geq 0,\ \bar{a}_{p,n+1} < 0$. This algorithm solves the reduced problem (15.1).

1. Compute $\bar{x}_{n+1} = \bar{b}_p / \bar{a}_{p,n+1}$.
2. Compute $\bar{x}_B = \bar{b} - \bar{x}_{n+1}\bar{a}_{n+1}$.
3. Determine row index $r \in \arg\min\{\bar{x}_{j_i} \mid i = 1, \cdots, m+1, i \neq p\}$.
4. Stop if $\bar{x}_{j_r} \geq 0$ (optimality achieved).
5. Compute $\omega_N = N^{\mathrm{T}} B^{-\mathrm{T}} e_r$.
6. If $J = \{j \in N \mid \omega_j < 0\} = \emptyset$, set $p = r,\ \sigma_N = \omega_N$, and go to step 14.

7. Determine β and column index q such that $\beta = -\sigma_q/\omega_q = \min_{j\in J} -\sigma_j/\omega_j$.
8. Solve $B\bar{a}_q = a_q$ for $\bar{a}_q$, and compute $\nu = -\bar{a}_{r,n+1}/\omega_q$ and $\tau = -\bar{b}_r/\omega_q$
9. Update: $\bar{a}_{n+1} = \bar{a}_{n+1} + \nu(\bar{a}_q - e_r)$, where.
10. Update: $\bar{b} = \bar{b} + \tau(\bar{a}_q - e_r)$, where.
11. Update B^{-1} by (3.23) $(p = r)$.
12. Update (B, N) by exchanging j_p and q.
13. Solve $B^{\mathrm{T}}h = e_p$ and compute $\sigma_N = N^{\mathrm{T}}h$.
14. Go to step 1 if $\bar{a}_{p,n+1} < 0$.
15. Stop (infeasible problem).

It is possible to improve the dual reduced method by replacing the row Rule 15.4.1. Analogous to the dual largest-distance rule (Sect. 12.3), some rule based on how much the point $(\bar{x}_N, \bar{f})$ violates the constraints seems to be attractive as derived as follows.

Introduce a set of row vectors

$$(w^i)^{\mathrm{T}} = e_i^{\mathrm{T}} B^{-1}(N \mid a_{n+1}), \quad i = 1, \cdots, m+1.$$

For any $i = 1, \ldots, m+1$, the signed distance from point $(\bar{x}_N, \bar{f})$ to the boundary (associated with the ith row of the canonical form)

$$(B^{-1}b)_i - (w^i)^{\mathrm{T}}(x_N^T, f)^{\mathrm{T}} = 0$$

is defined by (see Sect. 2.1)

$$\bar{x}_{j_i}/\|w^i\|.$$

Rule 15.4.3 (Dual row rule: largest-distance) Select pivot row index r such that

$$\bar{x}_{j_r} = \min\{\bar{x}_{j_i}/\|w^i\| \mid i = 1, \cdots, m+1\}.$$

The recurrence formulas of $\|w^i\|^2, i = 1, \cdots, m$ are the same as (12.13) and (12.14).

Alternatively, the following approximate formulas may be used to simplify computations.

Rule 15.4.4 (Dual row rule: approximate largest-distance) Select pivot row index r such that

$$\bar{x}_{j_r} = \min\{\bar{x}_{j_i}/|\bar{a}_{i,n+1}| \mid i = 1, \cdots, m+1\}.$$

It is promising if the other rules, described in Chaps. 11 and 12, are adapted within the reduce simplex framework.

15.5 Dual Reduced Phase-I: The Most-Obtuse-Angle

Algorithm 15.4.1 requires availability of a dual feasible reduced tableau. Dual Phase-I methods presented in Chap. 14 may be applied to provide a conventional dual feasible simplex tableau of (15.1). Then, letting the objective variable x_{n+1} leave the basis gives a dual feasible reduced tableau. However, it would be more direct and effective to achieve the goal in the reduced simplex context based on the most-obtuse-angle heuristics.

The procedure can be written as follows.

Algorithm 15.5.1 (Tableau dual reduced Phase-I: the most-obtuse-angle). Initial: Reduced simplex tableau of form (15.21). This algorithm finds a dual feasible reduced simplex tableau.

1. Select pivot row index $p \in \arg\min\{\bar{a}_{i,n+1} \mid i = 1, \cdots, m+1\}$.
2. Stop if $\bar{a}_{p,n+1} \geq 0$.
3. Select pivot column index $q \in \arg\min_{j \in N} \bar{a}_{pj}$.
4. Stop if $\bar{a}_{pq} \geq 0$.
5. Convert $\bar{a}_{pq}$ to 1, and eliminate the other nonzeros in the pivot column by elementary transformations.
6. Go to step 1.

Theorem 15.5.1. *Assume finiteness of Algorithm 15.5.1. It terminates either at*

(i) Step 2, detecting infeasibility or lower unboundedness of the problem; or at
(ii) Step 4, obtaining a dual feasible reduced simplex tableau.

Proof. When it terminates at step 4, dual feasibility condition is satisfied clearly. Assume that termination occurs at step 2. If $\bar{x}$ is a feasible solution to the problem, then it satisfies

$$x_B = \bar{b} - \bar{N}x_N - x_{n+1}\bar{a}_{n+1} \geq 0, \quad x_N \geq 0.$$

Since $-\bar{a}_{n+1} \leq 0$, the preceding expression holds for all x_{n+1} satisfying $x_{n+1} \leq \bar{x}_{n+1}$, therefore the problem is unbound below. □

The preceding Algorithm is used as a dual Phase-1 procedure in the following three examples. In the first example, the infeasibility of the problem will be detected after dual Phase-1. In the second, an optimal solution will be found at the end of the dual Phase-I. In the third, an optimal solution will be achieved after the first iteration of Phase-II.

Example 15.5.1. Solve the following problem by two-phase dual reduced simplex method:

$$\begin{array}{lrcl}
\min & x_7 = x_1 - x_2 - 2x_3, & & \\
\text{s.t.} & -x_1 + \ x_2 + \ x_3 + x_4 \qquad\qquad\qquad & = & 0, \\
& x_1 - 2x_2 + \ x_3 \qquad\quad + x_5 \qquad & = & 1, \\
& x_1 + 2x_2 - 2x_3 \qquad\qquad\qquad + x_6 & = & -8, \\
& x_j \geq 0, \quad j = 1, \cdots, 6. & &
\end{array}$$

Answer Dual Phase-I: Initial tableau is

x_1	x_2	x_3	x_4	x_5	x_6	x_7	RHS
−1	1	1	1				
1	−2	1		1			1
1	2	−2			1		−8
1	−1	−2*				−1	

Iteration 1: To drop x_7 from the basis, take $p = 4$. $\min\{0, -1, -2\} = -2, q = 3$.

Multiply row 4 by $-1/2$, and add $-1, -1, 2$ times of row 4 to rows 1,2,3, respectively, obtaining the following reduced simplex tableau:

x_1	x_2	x_3	x_4	x_5	x_6	x_7	RHS
−1/2*	1/2		1			−1/2	
3/2	−5/2			1		−1/2	1
	3				1	1	−8
−1/2	1/2	1				1/2	

Phase-I: Call Algorithm 15.5.1.

Iteration 2:

1. $\min\{-1/2, -1/2, 1, 1/2\} = -1/2 < 0, \ p = 1$.
3. $\min\{-1/2, 1/2\} = -1/2, \ q = 1$.
5. Multiply row 1 by -2, and add $-3/2, 1/2$ times of row 1 to rows 2,4, respectively:

x_1	x_2	x_3	x_4	x_5	x_6	x_7	RHS
1	−1		−2			1	
	−1*		3	1		−2	1
	3				1	1	−8
		1	−1			1	

Iteration 3:

1. $\min\{1, -2, 1, 1\} = -2,\ p = 2.$
3. $\min\{-1, 3\} = -1,\ q = 2.$
5. Multiply row 2 by −1, and add 1, −3 times of row 2 to rows 1,2, respectively:

x_1	x_2	x_3	x_4	x_5	x_6	x_7	RHS
1			−5	−1		3	−1
	1		−3	−1		2	−1
			9	3	1	−5	−5
		1	−1			1	

Iteration 4:

1. $\min\{3, 2, -5, 1\} = -5,\ p = 3.$
3. $\min\{9, 3\} = 3 > 0.$
4. Dual feasibility achieved.

Dual Phase-II: Call Algorithm 15.4.1.

Iteration 5:

1. $\bar{x}_7 = (-5)/(-5) = 1.$
2. $\bar{x}_B = (-1, -1, -5, 0)^{\mathrm{T}} - 1 \times (3, 2, -5, 1)^{\mathrm{T}} = (-4, -3, 0, -1)^{\mathrm{T}},\ B = \{1, 2, 6, 3\}.$
3. $\min\{-4, -3, 0, -1\} = -4, r = 1.$
6. $\beta = \min\{-9/(-5), -3/(-1)\} = 9/5,\ q = 4.$
7. Multiply row 1 by $-1/5$, and add $3, -9, 1$ times of row 1 to rows 2,3,4, respectively:

x_1	x_2	x_3	x_4	x_5	x_6	x_7	RHS
−1/5			1	1/5		−3/5	1/5
−3/5	1			−2/5		1/5	−2/5
9/5				6/5	1	2/5	−34/5
−1/5		1		1/5		2/5	1/5

8. $\bar{a}_{3,7} = 2/5 > 0.$
9. Stop, detecting infeasibility of the problem.

Example 15.5.2. Solve the following problem by two-phase dual reduced simplex method:

$$
\begin{array}{llllllll}
\min\ x_7 = x_1 & - & 2x_2 & - & 5x_3, & & & \\
\text{s.t.} \quad -2x_1 & + & x_2 & + & x_3 & +x_4 & & = \ 1, \\
2x_1 & - & 3x_2 & + & x_3 & & +x_5 & = -1, \\
x_1 & + & 2x_2 & - & x_3 & & \quad +x_6 & = \ 2, \\
& & & & x_j \geq 0, & j = 1, \cdots, 6. & &
\end{array}
$$

Answer Initial tableau is

x_1	x_2	x_3	x_4	x_5	x_6	x_7	RHS
−2	1	1	1				1
2	−3	1		1			−1
1	2	−1			1		2
1	−2	−5				−1	

Iteration 1: $p = 4, \min\{1, -2, -5\} = -5, q = 3$.

Multiply row 4 by $-1/5$, and add $-1, -1, 1$ times of row 4 to rows 1,2,3, respectively:

x_1	x_2	x_3	x_4	x_5	x_6	x_7	RHS
−9/5*	3/5		1			−1/5	1
11/5	−17/5			1		−1/5	−1
4/5	12/5				1	1/5	2
−1/5	2/5	1				1/5	

Dual Phase-I: Call Algorithm 15.5.1.

Iteration 2:

1. $\min\{-1/5, -1/5, 1/5, 1/5\} = -1/5,\ p = 1$.
3. $\min\{-9/5, 3/5\} = -9/2, q = 1$.
5. Multiply row 1 by $-5/9$, and add $-11/5, -4/5, 1/5$ times of row 1 to rows 2,3,4, respectively:

x_1	x_2	x_3	x_4	x_5	x_6	x_7	RHS
1	−1/3		−5/9			1/9	−5/9
	−8/3*		11/9	1		−4/9	2/9
	8/3		4/9		1	1/9	22/9
	1/3	1	−1/9			2/9	−1/9

Iteration 3:

1. $\min\{1/9, -4/9, 1/9, 2/9\} = -4/9,\ p = 2.$
3. $\min\{-8/3, 11/9\} = -8/3,\ q = 2.$
5. Multiply row 2 by $-3/8$, and add $1/3, -8/3, -1/3$ times of row 2 to rows 1,3,4, respectively:

x_1	x_2	x_3	x_4	x_5	x_6	x_7	RHS
1			−17/24	−1/8		1/6	−7/12
	1		−11/24	−3/8		1/6	−1/12
			5/3	1	1	−1/3	8/3
		1	1/24	1/8		1/6	−1/12

$\min\{1/6, 1/6, -1/3, 1/6\} = -1/3, p = 3$; $\min\{5/3, 1\} > 0$, dual feasibility achieved.

Dual Phase-II: Call Algorithm 15.4.1.

Iteration 4:

1. $\bar{x}_7 = (8/3)/(-1/3) = -8.$
2. $\bar{x}_B = (-7/12, -1/12, 8/3, -1/12)^{\mathrm{T}} - (-8)(1/6, 1/6, -1/3, 1/6)^{\mathrm{T}}$
   $= (3/4, 5/4, 0, 5/4)^{\mathrm{T}} \geq 0.$
4. Basic optimal solution and according objective value:

$$\bar{x} = (3/4, 5/4, 5/4, 0, 0, 0)^{\mathrm{T}}, \quad \bar{x}_7 = -8.$$

Example 15.5.3. Solve the following problem by two-phase dual reduced simplex method:

$$\begin{array}{rl}
\min & x_9 = -2x_1 - x_2 + 2x_3 + 4x_4, \\
\text{s.t.} & x_1 - 2x_2 + 4x_3 - x_4 + x_5 = 4, \\
& 2x_1 - 3x_2 - x_3 + x_4 + x_6 = -6, \\
& x_1 + x_3 + x_4 + x_7 = 2, \\
& 2x_1 + x_2 - x_3 - 4x_4 + x_8 = -1, \\
& x_j \geq 0, \quad j = 1, \cdots, 8.
\end{array}$$

Answer Initial tableau is

x_1	x_2	x_3	x_4	x_5	x_6	x_7	x_8	x_9	RHS
1	−2	4	−1	1					4
2	−3	−1	1		1				−6
1		1	1			1			2
2	1	−1	−4				1		−1
−2*	−1	2	4					−1	

Iteration 1:

To drop objective variable x_9 from the basis, take $p = 5$; $\min\{-2, -1, 2, 4\} = -2, q = 1$.

Multiply row 5 by $-1/2$, and add $-1, -2, -1, -2$ times of row 5 to rows 1,2,3,4, respectively:

x_1	x_2	x_3	x_4	x_5	x_6	x_7	x_8	x_9	RHS
	−5/2	5	1	1				−1/2	4
	−4*	1	5		1			−1	−6
	−1/2	2	3			1		−1/2	2
		1					1	−1	−1
1	1/2	−1	−2					1/2	

Dual Phase-I: Call Algorithm 15.5.1.

Iteration 2:

1. $\min\{-1/2, -1, -1/2, -1, 1/2\} = -1, \ p = 2$.
3. $\min\{-4, 1, 5\} = -4, q = 2$.
5. Multiply row 2 by $-1/4$, and add $5/2, 1/2, -1/2$ times of row 2 to rows 1,3,5, respectively:

x_1	x_2	x_3	x_4	x_5	x_6	x_7	x_8	x_9	RHS
		35/8	−17/8	1	−5/8			1/8	31/4
	1	−1/4	−5/4		−1/4			1/4	3/2
		15/8	19/8		−1/8	1		−3/8	11/4
		1					1	−1	−1
1		−7/8	−11/8*		1/8			3/8	−3/4

Iteration 3:

1. $\min\{1/8, 1/4, -3/8, -1, 3/8\} = -1, p = 4$.
3. $\min\{1, 0, 0\} \geq 0$. Dual feasibility achieved.

Dual Phase-II: Call Algorithm 15.4.1.

Iteration 3:

1. $\bar{x}_9 \ = (-1)/(-1) = 1$.
2. $\bar{x}_B = (31/4, 3/2, 11/4, -1, -3/4)^{\mathrm{T}} - (1/8, 1/4, -3/8, -1, 3/8)^{\mathrm{T}}$
   $= (61/8, 5/4, 25/8, 0, -9/8)^{\mathrm{T}}$.
3. $\min\{61/8, 5/4, 25/8, 0, -9/8\} = -9/8 < 0, r = 5$.
5. $J \ = \{3, 4\} \neq \emptyset$.
6. $\min\{-1/(-7/8), 0/(-11/8)\} = 0, 0 \leq (-1)/(-3/8), q = 4$.
7. Multiply row 5 by $-8/11$, and add $17/8, 5/4, -19/8$ times of row 5 to rows 1,2,3, respectively:

x_1	x_2	x_3	x_4	x_5	x_6	x_7	x_8	x_9	RHS
$-17/11$		$63/11$		1	$-9/11$			$-5/11$	$98/11$
$-10/11$	1	$6/11$			$-4/11$			$-1/11$	$24/11$
$19/11$		$4/11$			$1/11$	1		$3/11$	$16/11$
		1					1	-1	-1
$-8/11$		$7/11$	1		$-1/11$			$-3/11$	$6/11$

Iteration 4:

1. $\bar{x}_9 = (-1)/(-1) = 1.$
2. $\bar{x}_B = (98/11, 24/11, 16/11, -1, 6/11)^{\mathrm{T}} - (1)(-5/11, -1/11, 3/11, -1, -3/11)^{\mathrm{T}}$
   $= (103/11, 25/11, 13/11, 0, 9/11)^{\mathrm{T}} \geq 0.$
4. Basic optimal solution and according objective value are

$$\bar{x} = (0, 25/11, 0, 9/11, 103/11, 0, 13/11, 0)^{\mathrm{T}}, \quad \bar{x}_9 = 1.$$

15.6 Notes

The reduced simplex method can be traced back to the publication of the "bisection simplex method" (Pan 1991, 1996a), which bisections an interval, including the optimal value, iteration by iteration until achieving optimality. Pan wrote (1991, p. 724).

> Finally, we indicate that justifications of ... in fact describes an approach to improving feasible solutions with dual type of canonical form,[1] in a manner similar to that in the conventional method. We are not interested in on this line though, and will develop another method...

At that time, the reduced simplex method seemed ready to come out at one's call. But unfortunately it had been overlooked by not regarding its prospects favorably until recently drawing attention again from the author.

Although there are no numerical results available at present, the method is promising for the following reasons, at least:

Firstly, while its computational effort per iteration is about the same as the conventional simplex method, a novel pivot rule is employed. Consequently, the resulting search direction corresponds to the negative reduced objective gradient as a whole (since the objective function involves a single variable), as seems to be advantageous to the conventional search direction which corresponds to a negative component only. In each iteration, as a result, the decrement in the objective variable's value is just equal to decrement in the original objective value (see also Vemuganti 2004).

---

[1]It is noting but the reduced simplex tableau.

Secondly, as was mentioned in Sect. 3.9, the numerical stability of the conventional simplex method is actual not good enough, since it could select a too small pivot in magnitude. Occasionally, it must restart from scratch to handle the troublesome case when the basis matrix is close to singularity (Sect. 5.1). In contrast, the reduced simplex method is numerically stable, since it tends to select a large pivot in magnitude. Consequently, the risk of using the restarting remedy is significantly reduced, if not avoided completely. In the stability point of view, therefore, Harris practicable row rule becomes unnecessary, although it would remain useful in the sense of the most-obtuse-angle heuristics (see Sect. 5.6).

Thirdly, a variant of the reduced simplex method shows a bright application outlook, as it seems to be a desirable framework for the implementation of the so-called "controlled-branch method" for solving ILP problems; favorably the associated LP relaxation subprograms can be handled without increasing their sizes at all (see Sect. 25.7).

Finally, the reduced simplex method would become more powerful if the basis is generalized to allow the so-called "deficient-basis" (Sect. 20.6). Moreover, a method for generating an initial deficient-basis and an associated Phase-I method can be derived using the reduced simplex framework (Sect. 20.7). These methods seems to be simple as well as efficient.

As for sparsity, on the other hand, a large amount of fill-ins could yield from transforming the conventional simplex tableau to a reduced one if nonzero costs in the original problem occupy a high proportion (see Sect. 15.2). In this case, the column, firstly selected to enter the basis, should be as sparse as possible. It would be a good idea to obtain a reduced simplex tableau directly from some crash procedure (Sect. 5.5) (excluding the objective variable from the set of basic variables).

On the other hand, some particular scaling should be applied with respect to the objective function, as the reduced simplex method would be sensitive to it.

Chapter 16
Improved Reduced Simplex Method

In contrast to the conventional simplex tableau, the reduced simplex tableau, introduced in the preceding chapter, does not explicitly offer the associated basic solution, which is computed from the last two columns of the tableau. Fortunately, a slight modification of it eliminates this "fault", leading to a more compact and practicable variant.

16.1 Variation of the Reduce Simplex Method

The key to the variant lies in derivation of an updating formula for the basic solution. To this end, let us turn to Algorithm 15.2.1.

Let (x', x'_{n+1}) be a basic feasible solution, satisfying

$$x'_B = \bar{b} - x'_{n+1}\bar{a}_{n+1} \geq 0, \quad x'_N = 0. \tag{16.1}$$

The Algorithm determines a new basic feasible solution $(\bar{x}, \bar{x}_{n+1})$ by

$$\bar{x}_B = \bar{b} - \bar{x}_{n+1}\bar{a}_{n+1}, \quad \bar{x}_N = 0, \quad \bar{x}_{n+1} = \bar{b}_p/\bar{a}_{p,n+1}, \tag{16.2}$$

where

$$p \in \arg\max\{\bar{b}_i/\bar{a}_{i,n+1} \mid \bar{a}_{i,n+1} < 0, \ i = 1, \cdots, m+1\}. \tag{16.3}$$

From Lemma 15.1.1, it is known that $\bar{x}_{n+1}$ is the greatest lower bound of the solution set $\Phi(B)$ to the inequality system (15.5). From (16.2) and (16.1), it follows that

$$\bar{x}_B = \bar{b} - \bar{x}_{n+1}\bar{a}_{n+1} = (\bar{b} - x'_{n+1}\bar{a}_{n+1}) + (x'_{n+1} - \bar{x}_{n+1})\bar{a}_{n+1} = x'_B + (x'_{n+1} - \bar{x}_{n+1})\bar{a}_{n+1}. \tag{16.4}$$

P.-Q. PAN, *Linear Programming Computation*, DOI 10.1007/978-3-642-40754-3_16,

On the other hand, it is known from (16.1) that

$$-x'_{j_i}/\bar{a}_{i,n+1} = -(\bar{b}_i - x'_{n+1}\bar{a}_{i,\,n+1})/\bar{a}_{i,n+1} = x'_{n+1} - \bar{b}_i/\bar{a}_{i,n+1},$$
$$i = 1,\cdots,m+1, \quad \bar{a}_{i,n+1} \neq 0,$$

by which it is seen that

$$\begin{aligned}&\arg\min\{-x'_{j_i}/\bar{a}_{i,n+1} \mid \bar{a}_{i,n+1} < 0,\ i = 1,\cdots,m+1\}\\ &= \arg\max\{\bar{b}_i/\bar{a}_{i,n+1} \mid \bar{a}_{i,n+1} < 0,\ i = 1,\cdots,m+1\}.\end{aligned}$$

In addition, combining the third impression of (16.2) and (16.3) gives

$$\begin{aligned}\alpha &\triangleq \min\{-x'_{j_i}/\bar{a}_{i,n+1} \mid \bar{a}_{i,n+1} < 0,\ i = 1,\cdots,m+1\}\\ &= \min\{x'_{n+1} - \bar{b}_i/\bar{a}_{i,n+1} \mid \bar{a}_{i,n+1} < 0,\ i = 1,\cdots,m+1\}\\ &= x'_{n+1} - \max\{\bar{b}_i/\bar{a}_{i,n+1} \mid \bar{a}_{i,n+1} < 0,\ i = 1,\cdots,m+1\}\\ &= x'_{n+1} - \bar{x}_{n+1} \geq 0,\end{aligned}$$

which together with (16.4) leads to

$$\bar{x}_B = x'_B + \alpha\bar{a}_{n+1}, \quad \bar{x}_{n+1} = x'_{n+1} - \alpha. \tag{16.5}$$

It is clear that if (x', x'_{n+1}) satisfies nondegeneracy condition (15.8), i.e.,

$$x'_{j_i} > 0, \quad \forall\, i = 1,\cdots,m+1, \quad \bar{a}_{i,n+1} < 0,$$

then $\alpha > 0$, hence $\bar{x}_{n+1} < x'_{n+1}$, indicating a positive decrement of the objective value.

It is noted that the new $\bar{a}_{p,n+1}$ will be positive after the basis change (by taking $\bar{a}_{pq}$ as pivot), even if $\bar{x}_{j_p} = 0$. Thus, the entering variable x_q will never leave the basis immediately in the next iteration.

Thereby, the reduced simplex tableau (15.21) can be modified to the following one, termed *improved reduced (simplex) tableau*, where there is no need for the right-hand side but $\bar{x}_B$:

x_B^{T} x_N^{T}	x_{n+1}	$\bar{x}_B$
I $\bar{N}$	$\bar{a}_{n+1}$	$\bar{x}_B$

(16.6)

To avoid accumulating errors, introduced by the recurrence formula $\bar{x}_{n+1} = x'_{n+1} - \alpha$, the final objective value may be computed directly from the original data at the end alternatively.

Based on the preceding discussions, we transfer Algorithm 15.2.1 to the following one.

Algorithm 16.1.1 (Improved reduced simplex algorithm: tableau form). Initial : improved reduced tableau of form (16.6), where $\bar{x}_B = \bar{b} - \bar{x}_{n+1}\bar{a}_{n+1} \geq 0$. This algorithm solves the reduced problem (15.1).

1. Stop if $\bar{a}_{n+1} \geq 0$.
2. Determine α and row index p such that

$$\alpha = -\bar{x}_{j_p}/\bar{a}_{p,n+1} = \min\{-\bar{x}_{j_i}/\bar{a}_{i,n+1} \mid \bar{a}_{i,n+1} < 0,\ i = 1, \cdots, m+1\}.$$

3. If $\alpha \neq 0$, add α times of x_{n+1} column to $\bar{x}_B$ column and set $\bar{x}_{n+1} = \bar{x}_{n+1} - \alpha$.
4. Determine column index $q \in \arg\min\{\bar{a}_{pj} \mid j \in N\}$.
5. Stop if $\bar{a}_{pq} \geq 0$.
6. Convert $\bar{a}_{pq}$ to 1, and eliminate the other nonzeros in the pivot column by elementary transformations.
7. Go to step 1.

Note 1 The preceding Algorithm can get started from a conventional feasible simplex tableau, using the right-hand side as the $\bar{x}_B$ column.

Note 2 Step 6 does not touch $\bar{x}_B$ column because the entry $\bar{x}_{j_p}$ of the pth row of this column vanishes.

Theorem 16.1.1. *Under the nondegeneracy assumption, Algorithm 16.1.1 terminates either at*

(i) Step 1, detecting infeasibility of the problem; or at
(ii) Step 5, offering a basic optimal solution $\bar{x}$.

Proof. It is known from the discussions preceding it that Algorithms 16.1.1 and 15.2.1 generate the same sequence of basic feasible solutions. Under the nondegeneracy assumption, the objective value decreases strictly, hence no cycling occurs. Since there are only infinitely many basic feasible solutions, therefore, it must terminate. The meanings of its exits are clear. □

Example 16.1.1. Solve following problem by Algorithm 16.1.1:

$$\begin{array}{llrl}
\min & x_{10} = -x_1 + 4x_2 + 3x_3 - 3x_4 - 2x_5, & & \\
\text{s.t.} & -x_1 - 9x_2 - 6x_3 + 2x_4 + 4x_5 + x_6 & = 1, \\
& -2x_1 - 6x_2 + x_3 - 3x_4 - x_5 + x_7 & = 4, \\
& 3x_1 - 3x_2 + 3x_3 + 5x_4 + 5x_5 + x_8 & = 5, \\
& 5x_1 - 2x_2 - 4x_3 - x_4 - 2x_5 + x_9 & = 2, \\
& x_j \geq 0, \quad j = 1, \cdots, 9. & &
\end{array}$$

Answer Initial tableau is

x_1	x_2	x_3	x_4	x_5	x_6	x_7	x_8	x_9	x_{10}	$\bar{x}_B$
−1	−9	−6	2	4	1					1
−2	−6	1	−3	−1		1				4
3	−3	3	5	5			1			5
5	−2	−4	−1	−2				1		2
−1	4	3	−3*	−2					−1	

Iteration 1: $\bar{x}_B = (1, 4, 5, 2)^{\mathrm{T}}$, $\bar{x}_{10} = 0$.

2. $\alpha = \min\{0/(-1)\} = 0$, $p = 5$.
4. $\min\{-1, 4, 3, -3, -2\} = -3$, $q = 4$.
6. Multiply row 5 by $-1/3$, and add $-2, 3, -5, 1$ times of row 5 to rows 1,2,3,4, respectively:

x_1	x_2	x_3	x_4	x_5	x_6	x_7	x_8	x_9	x_{10}	$\bar{x}_B$
−5/3	−19/3*	−4		8/3	1				−2/3	1
−1	−10	−2		1		1			1	4
4/3	11/3	8		5/3			1		−5/3	5
16/3	−10/3	−5		−4/3				1	1/3	2
1/3	−4/3	−1	1	2/3					1/3	

Iteration 2:

2. $\alpha = \min\{-1/(-2/3), -5/(-5/3)\} = -1/(-2/3) = 3/2$, $p = 1$.
3. Add $3/2$ times of x_{10} column to $\bar{x}_B$ column; $\bar{x}_{10} = -3/2$.
4. $\min\{-5/3, -19/3, -4, 8/3\} = -19/3$, $q = 2$.
6. Multiply row 1 by $-3/19$, and add $10, -11/3, 10/3, 4/3$ times of row 1 to rows 2,3,4,5, respectively:

x_1	x_2	x_3	x_4	x_5	x_6	x_7	x_8	x_9	x_{10}	$\bar{x}_B$
5/19	1	12/19		−8/19	−3/19				2/19	0
31/19		82/19		−61/19	−30/19	1			39/19	11/2
7/19		108/19		61/19	11/19		1		−39/19	5/2
118/19		−55/19		−52/19	−10/19			1	13/19	5/2
13/19		−3/19	1	2/19	−4/19				9/19	1/2

Iteration 3:

2. $\alpha = \min\{-(5/2)/(-39/19)\} = 95/78$, $p = 3$.
3. Add $95/78$ times of x_{10} column to $\bar{x}_B$ column; $\bar{x}_{10} = -3/2 - 95/78 = -106/39$.

x_1	x_2	x_3	x_4	x_5	x_6	x_7	x_8	x_9	x_{10}	$\bar{x}_B$
5/19	1	12/19		−8/19	−3/19				2/19	5/39
31/19		82/19		−61/19	−30/19	1			39/19	8
7/19		108/19		61/19	11/19		1		−39/19	0
118/19		−55/19		−52/19	−10/19			1	13/19	10/3
13/19		−3/19	1	2/19	−4/19				9/19	14/13

4. $\min\{7/19, 108/19, 61/19, 11/19\} \geq 0$.
5. Basic optimal solution and according objective value:

$$\bar{x} = (0, 5/39, 0, 14/13, 0, 0, 8, 0, 10/3)^{\mathrm{T}}, \quad \bar{x}_{10} = -106/39.$$

Based on the equivalence between the reduced tableau (15.21) and the revised tableau (15.22), the following revised version of Algorithm 16.1.1 is obtained.

Algorithm 16.1.2 (Improved Reduced Simplex Algorithm). Initial : (B, N), B^{-1}. $\bar{x}_B = \bar{b} - \bar{x}_{n+1}\bar{a}_{n+1} \geq 0$, where $\bar{b} = B^{-1}b$, $\bar{a}_{n+1} = -B^{-1}e_{m+1}$. This algorithm solves the reduced problem (15.1).

1. Stop if $\bar{a}_{n+1} \geq 0$ (unbounded problem).
2. Determine α and row index p such that

$$\alpha = -\bar{x}_{j_p}/\bar{a}_{p,n+1} = \min\{-\bar{x}_{j_i}/\bar{a}_{i,n+1} \mid \bar{a}_{i,n+1} < 0, i = 1, \cdots, m+1\}.$$

3. If $\alpha \neq 0$, update $\bar{x}_B = \bar{x}_B + \alpha\bar{a}_{n+1}$, $\bar{x}_{n+1} = \bar{x}_{n+1} - \alpha$.
4. Compute $\sigma_N = N^{\mathrm{T}}B^{-\mathrm{T}}e_p$.
5. Determine column index $q \in \arg\min_{j \in N} \sigma_j$.
6. Stop if $\sigma_q \geq 0$ (optimality achieved).
7. Compute $\bar{a}_q = B^{-1}a_q$.
8. Update $\bar{a}_{n+1} = \bar{a}_{n+1} + \nu(\bar{a}_q - e_p)$, where $\nu = -\bar{a}_{p,n+1}/\sigma_q$.
9. Update B^{-1} by (3.23).
10. Update (B, N) by exchanging j_p and q.
11. Go to step 1.

16.2 Phase-I for Improved Reduced Simplex Method

To provide a starting point, this section offers a Phase-I method for the improved reduced simplex method.

Based on the most-obtuse-angle heuristics, the following tableau algorithm results from modifying the dual Algorithm 16.3.1.

Algorithm 16.2.1 (Improved reduced Phase-I: tableau form). Initial : Improved reduced tableau of form (16.6), where $\bar{x}_B = \bar{b} - \bar{x}_{n+1}\bar{a}_{n+1}$. This algorithm finds a feasible improved reduced tableau.

1. Determine row index $p \in \arg\min\{\bar{x}_{j_i} \mid i = 1, \cdots, m+1\}$.
2. Stop if $\bar{x}_{j_p} \geq 0$ (feasibility achieved).
3. Determine column index $q \in \arg\min_{j \in N} \bar{a}_{p\,j}$.
4. If $\bar{a}_{pq} < 0$, convert $\bar{a}_{pq}$ to 1, and eliminate the other nonzeros of the column by elementary transformations, and go to step 1.
5. Stop if $\bar{a}_{p,n+1} = 0$ (infeasible problem).
6. Compute $\alpha = -\bar{x}_{j_p}/\bar{a}_{p,n+1}$.
7. If $\alpha \neq 0$, add α time of x_{n+1} column to $\bar{x}_B$ column and set $\bar{x}_{n+1} = \bar{x}_{n+1} - \alpha$.
10. Go to step 1.

Example 16.2.1. Solve the following problem using Algorithm 16.2.1 as Phase-I:

$$\begin{array}{llll}
\min & x_9 = 6x_1 + 4x_2 - 6x_3 - 4x_4, & & \\
\text{s.t.} & 3x_1 + x_2 + 3x_3 + 5x_4 + x_5 & = & 3, \\
& -4x_1 - x_2 + 5x_3 - 2x_4 + x_6 & = & 2, \\
& -6x_1 - 7x_2 - x_3 + 4x_4 + x_7 & = & -5, \\
& x_1 - 3x_2 - 2x_3 - x_4 + x_8 & = & -1, \\
& x_j \geq 0,\ j = 1, \cdots, 8. & &
\end{array}$$

Answer Phase-I: Initial tableau is

x_1	x_2	x_3	x_4	x_5	x_6	x_7	x_8	x_9	$\bar{x}_B$
3	1	3	5	1					3
−4	−1	5	−2		1				2
−6	−7	−1	4			1			−5
1	−3	−2	−1				1		−1
6	4	−6*	−4					−1	

Iteration 1: $\bar{x}_B = (3, 2, -5, -1)^{\mathrm{T}}$, $\bar{x}_9 = 0$.
Take $p = 5$, $\min\{6, 4, -6, -4\} = -6$, $q = 3$. Multiply row 5 by $-1/6$, and add $-3, -5, 1, 2$ times of row 5 to rows 1,2,3,4, respectively, to obtain an improved reduced tableau:

x_1	x_2	x_3	x_4	x_5	x_6	x_7	x_8	x_9	$\bar{x}_B$
6	3		3	1				−1/2	3
1	7/3		−16/3		1			−5/6	2
−7	−23/3*		14/3			1		1/6	−5
−1	−13/3		1/3				1	1/3	−1
−1	−2/3	1	2/3					1/6	

Phase-I: Call Algorithm 16.2.1.

Iteration 2:

1. $\min\{3, 2, -5, -1, 0\} = -5 < 0,\ p = 3.$
3. $\min\{-7, -23/3, 1/3\} = -23/3 < 0,\ q = 2.$
4. Multiply row 3 by $-3/23$, and add $-3, -7/3, 13/3, 2/3$ times of row 3 to rows 1,2,4,5, respectively:

x_1	x_2	x_3	x_4	x_5	x_6	x_7	x_8	x_9	$\bar{x}_B$
75/23			111/23	1		9/23		−10/23	24/23
−26/23			−90/23*		1	7/23		−18/23	11/23
21/23	1		−14/23			−3/23		−1/46	15/23
68/23			−53/23			−13/23	1	11/46	42/23
−9/23		1	6/23			−2/23		7/46	10/23

Iteration 3:

1. $\min\{24/23, 11/23, 15/23, 42/23, 10/23\} \geq 0.$
2. A feasible improved reduced tableau is obtained.

Phase-II: Call Algorithm 16.1.1.

Iteration 4:

1. $\bar{a}_9 \not\geq 0.$
2. $\alpha = \min\{-(24/23)/(-10/23), -(11/23)/(-18/23), -(15/23)/(-1/46)\}$
   $= (11/23)/(-18/23) = 11/18,\ p = 2.$
3. Add 11/18 times of x_9 column to $\bar{x}_B$ column.
4. $\min\{-26/23, -90/23, 7/23\} = -90/23 < 0,\ q = 4.$
6. Multiply row 2 by $-23/90$, and add $-111/23, 14/23, 53/23, -6/23$ times of row 2 to rows 1,3,4,5, respectively:

x_1	x_2	x_3	x_4	x_5	x_6	x_7	x_8	x_9	$\bar{x}_B$
28/15				1	37/30	23/30		−7/5	7/9
13/45			1		−23/90	−7/90		1/5	
49/45	1				−7/45	−8/45		1/10	23/36
163/45					−53/90	−67/90	1	7/10	71/36
−7/15		1			1/15	−1/15		1/10	19/36

Iteration 5:

1. $\bar{a}_9 \not\geq 0.$
2. $\alpha = \min\{-(7/9)/(-7/5)\} = 5/9,\ p = 1.$
3. Add 5/9 times of x_9 column to $\bar{x}_B$ column:

x_1	x_2	x_3	x_4	x_5	x_6	x_7	x_8	x_9	$\bar{x}_B$
28/15				1	37/30	23/30		−7/5	
13/45			1		−23/90	−7/90		1/5	1/9
49/45	1				−7/45	−8/45		1/10	25/36
163/45					−53/90	−67/90	1	7/10	85/36
−7/15		1			1/15	−1/15		1/10	7/12

4. $\min\{28/15, 37/30, 23/30\} \geq 0$.
5. Basic optimal solution and according objective value:

$$\bar{x} = (0, 25/36, 7/12, 1/9, 0, 0, 0, 85/36)^{\mathrm{T}}, \quad \bar{x}_9 = -7/6.$$

The following is a revision of Algorithm 16.2.1.

Algorithm 16.2.2 (Improved reduced Phase-I). Initial : (B, N), B^{-1}. $\bar{x}_B = \bar{b} - \bar{x}_{n+1}\bar{a}_{n+1}$, where $\bar{b} = B^{-1}b$, $\bar{a}_{n+1} = -B^{-1}e_{m+1}$. This algorithm finds a basic feasible solution to (15.1).

1. Determine row index $p \in \arg\min\{\bar{x}_{j_i} \mid i = 1, \cdots, m+1\}$.
2. Stop if $\bar{x}_{j_p} \geq 0$ (feasibility achieved).
3. Compute $\sigma_N = N^{\mathrm{T}} B^{-\mathrm{T}} e_p$.
4. Determine column index $q \in \arg\min_{j \in N} \sigma_j$.
5. Go to 10 if $\sigma_q < 0$.
6. Stop if $\bar{a}_{p,n+1} = 0$ (infeasible problem).
7. Compute $\alpha = -\bar{x}_{j_p}/\bar{a}_{p,n+1}$.
8. Update :$\bar{x}_B = \bar{x}_B + \alpha\bar{a}_{n+1}$, $\bar{x}_{n+1} = \bar{x}_{n+1} - \alpha$.
9. Go to step 1.
10. Compute $\bar{a}_q = B^{-1}a_q$.
11. Update :$\bar{a}_{n+1} = \bar{a}_{n+1} + \nu(\bar{a}_q - e_p)$, where $\nu = -\bar{a}_{p,n+1}/\sigma_q$.
12. Update :$\bar{x}_B = \bar{x}_B + \tau(\bar{a}_q - e_p)$, where $\tau = -\bar{x}_{j_p}/\sigma_q$.
13. Update B^{-1} by (3.23).
14. Update (B, N) by exchanging j_p and q.
15. Go to step 1

16.3 Dual Improved Reduced Simplex Method

Exploiting the improved reduced tableau of form (16.6), it is possible to pursue primal feasibility while maintaining dual feasibility, until attaining optimality. To realize this idea, we modify the dual reduced simplex Algorithm 15.4.1 as follows.

Algorithm 16.3.1 (Dual improved reduced simplex algorithm: tableau form). Initial : improved reduced tableau of form (16.6), where $e_p^{\mathrm{T}}\bar{N} \geq 0$, $\bar{a}_{p,n+1} < 0$, $\bar{x}_{n+1} = \bar{b}_p/\bar{a}_{p,n+1}$, $\bar{x}_B = \bar{b} - \bar{x}_{n+1}\bar{a}_{n+1}$. This algorithm solves the reduced problem (15.1).

1. Determine row index $r \in \arg\min\{\bar{x}_{j_i} \mid i = 1, \cdots, m+1, i \neq p\}$.
2. Stop if $\bar{x}_{j_r} \geq 0$ (optimality achieved).
3. If $J = \{j \in N \mid \bar{a}_{rj} < 0\} = \emptyset$, set $p = r$, and go to step 6.
4. Determine β and column index q such that $\beta = -\bar{a}_{pq}/\bar{a}_{rq} = \min_{j \in J} -\bar{a}_{pj}/\bar{a}_{rj}$.
5. Convert $\bar{a}_{rq}$ to 1, and eliminate the other nonzeros of the column by elementary transformations.
6. Stop if $\bar{a}_{p,n+1} \geq 0$ (infeasible problem).
7. Compute $\alpha = -\bar{x}_{j_p}/\bar{a}_{p,n+1}$.
8. If $\alpha \neq 0$, add α times of x_{n+1} column to $\bar{x}_B$ column and set $\bar{x}_{n+1} = \bar{x}_{n+1} - \alpha$.
9. Go to step 1.

Theorem 16.3.1. *Under dual nondegeneracy assumption, Algorithm 16.3.1 terminates. It does at*

(i) step 2, giving a basic optimal solution $\bar{x}$; or at
(ii) step 6, detecting infeasibility of the problem.

Proof. Comparing this algorithm with Algorithm 15.4.1, it is known that the two generates the same sequence of iterates. Thus, the result follows from Theorem 15.4.2. □

It might be well to take a closer look at the preceding Algorithm: β resulting from step 4 and the new $\bar{x}_{j_p}$ from basis change in step 5 are both nonnegative; α coming from step 7 is also nonnegative since the new $\bar{a}_{p,n+1}$ is negative. Therefore, the new objective value yielding from step 8 never decreases, and strictly increases under the dual nondegeneracy ($\beta > 0$). In case when it proceeds from step 3, moreover, α is certainly positive, hence the objective value strictly increases.

Example 16.3.1. Solve the following problem by Algorithm 16.3.1:

$$\begin{array}{llllllllllll}
\min & x_{10} = 4x_1 + 2x_2 + 3x_3 + 8x_4 + 5x_5, \\
\text{s.t.} & -x_1 + 5x_2 \quad - 4x_4 - 2x_5 + x_6 = -3, \\
& -3x_1 - 2x_2 - 6x_3 + x_4 - x_5 + x_7 = -1, \\
& \quad - x_2 + 4x_3 - 6x_4 + 4x_5 + x_8 = 5, \\
& 5x_1 + 3x_2 - 3x_3 + 3x_4 + 5x_5 + x_9 = 2, \\
& \qquad x_j \geq 0, \quad j = 1, \cdots, 9.
\end{array}$$

Answer Initial tableau:

x_1	x_2	x_3	x_4	x_5	x_6	x_7	x_8	x_9	x_{10}	$\bar{x}_B$
−1	5		−4	−2	1					−3
−3	−2	−6	1	−1		1				−1
	−1	4	−6	4			1			5
5	3	−3	3	5				1		2
4	2*	3	8	5					−1	

Iteration 1:
To convert the preceding tableau to an improved reduced one, take the smallest 2 among the first 5 entries in the bottom row (in x_2 column) as pivot. Multiply row 5 by $1/2$, and add $-5, 2, 1, -3$ times of row 5 to rows 1,2,3,4, respectively.

Take $p = 5, e_5\bar{N} \geq 0,\ \bar{a}_{5,10} = -1/2 < 0$. $\bar{x}_B = (-3, -1, 5, 2, 0)^{\mathrm{T}}$:

x_1	x_2	x_3	x_4	x_5	x_6	x_7	x_8	x_9	x_{10}	$\bar{x}_B$
−11		−15/2	−24*	−29/2	1				5/2	−3
1		−3	9	4		1			−1	−1
2		11/2	−2	13/2			1		−1/2	5
−1		−15/2	−9	−5/2				1	3/2	2
2	1	3/2	4	5/2					−1/2	

Call Algorithm 16.3.1.

Iteration 2:

1. $\min\{-3, -1, 5, 2, 0\} = -3 < 0,\ r = 1$.
3. $J\ = \{1, 3, 4, 5\} \neq \emptyset$.
4. $\beta\ = \min\{-2/(-11), -(3/2)/(-15/2), -4/-24, -(5/2)/(-29/2)\}$
   $= -4/-24 = 1/6,\ q = 4$.
5. Multiply row 1 by $-1/24$, and add $-9, 2, 9, -4$ times of row 1 to rows 2,3,4,5, respectively:

x_1	x_2	x_3	x_4	x_5	x_6	x_7	x_8	x_9	x_{10}	$\bar{x}_B$
11/24		5/16	1	29/48	−1/24				−5/48	1/8
−25/8		−93/16		−23/16	3/8	1			−1/16	−17/8
35/12		49/8		185/24	−1/12		1		−17/24	21/4
25/8		−75/16		47/16	−3/8			1	9/16	25/8
1/6	1	1/4		1/12	1/6				−1/12	−1/2

6. $\bar{a}_{5,10} = -1/12 < 0$.
7. $\alpha\ = -(-1/2)/(-1/12) = -6$.
8. Add -6 times of x_{10} column to $\bar{x}_B$ column, $\bar{x}$

x_1	x_2	x_3	x_4	x_5	x_6	x_7	x_8	x_9	x_{10}	$\bar{x}_B$
11/24		5/16	1	29/48	−1/24				−5/48	3/4
−25/8		−93/16*		−23/16	3/8	1			−1/16	−7/4
35/12		49/8		185/24	−1/12		1		−17/24	19/2
25/8		−75/16		47/16	−3/8			1	9/16	−1/4
1/6	1	1/4		1/12	1/6				−1/12	

Iteration 3:

1. $\min\{3/4, -7/4, 19/2, -1/4\} = -7/4 < 0,\ r = 2.$
3. $J = \{1, 3, 5\} \neq \emptyset.$
4. $\beta = \min\{-(1/6)/(-25/8), -(1/4)/(-93/16), -(1/12)/(-23/16)\}$
   $= -(1/4)/(-93/16) = 4/93,\ q = 3.$
5. Multiply row 2 by $-16/93$, and add $-5/16, -49/8, 75/16, -1/4$ times of row 2 to rows 1,3,4,5, respectively:

x_1	x_2	x_3	x_4	x_5	x_6	x_7	x_8	x_9	x_{10}	$\bar{x}_B$
9/31			1	49/93	−2/93	5/93			−10/93	61/93
50/93		1		23/93	−2/31	−16/93			1/93	28/93
−35/93				192/31	29/93	98/93	1		−24/31	712/93
175/31				127/31	−21/31	−25/31		1	19/31	36/31
1/31	1			2/93	17/93	4/93			−8/93	−7/93

6. $\bar{a}_{5,10} = -7/93 < 0.$
7. $\alpha = -(-7/93)/(-8/93) = -7/8.$
8. Add $-7/8$ times of x_{10} column to $\bar{x}_B$ column:

x_1	x_2	x_3	x_4	x_5	x_6	x_7	x_8	x_9	x_{10}	$\bar{x}_B$
9/31			1	49/93	−2/93	5/93			−10/93	3/4
50/93		1		23/93	−2/31	−16/93			1/93	7/24
−35/93				192/31	29/93	98/93	1		−24/31	25/3
175/31				127/31	−21/31	−25/31		1	19/31	5/8
1/31	1			2/93	17/93	4/93			−8/93	

Iteration 4:

1. $\min\{3/4, 7/24, 25/3, 5/8\} \geq 0.$
2. Basic optimal solution and according objective value:

$$\bar{x} = (0, 0, 7/24, 3/4, 0, 0, 0, 25/3, 5/8)^{\mathrm{T}}, \quad \bar{x}_{10} = 3 \times (7/24) 8 \times (3/4) = 55/8.$$

Based on the equivalence between the reduced tableau (15.21) and the revised tableau (15.22),the following revision of Algorithm 16.3.1 can be obtained.

Algorithm 16.3.2 (Dual improved reduced simplex algorithm). Initial : (B, N), B^{-1}. $\sigma = e_p^{\mathrm{T}} B^{-1} N \geq 0, \bar{a}_{p,n+1} < 0,\ \bar{x}_{n+1} = \bar{b}_p/\bar{a}_{p,n+1},\ \bar{x}_B = \bar{b} - \bar{x}_{n+1}\bar{a}_{n+1} \geq 0$, where $\bar{b} = B^{-1}b,\ \bar{a}_{n+1} = -B^{-1}e_{m+1}$. This algorithm solves the reduced problem (15.1).

1. Determine row index $r \in \arg\min\{\bar{x}_{j_i} \mid i = 1, \cdots, m+1,\ i \neq p\}$.
2. Stop if $\bar{x}_{j_r} \geq 0$ (optimality achieved).
3. Compute $\omega_N = N^{\mathrm{T}} B^{-\mathrm{T}} e_r$.
4. If $J = \{j \in N \mid \omega_j < 0\} = \emptyset$, set $p = r$, $\sigma_N = \omega_N$, and go to step 11.

5. Determine β and column index q such that $\beta = -\sigma_q/\omega_q = \min_{j\in J} -\sigma_j/\omega_j$.
6. Compute $\bar{a}_q = B^{-1}a_q$.
7. Update :$\bar{a}_{n+1} = \bar{a}_{n+1} + \nu(\bar{a}_q - e_r)$, where $\nu = -\bar{a}_{r,n+1}/\omega_q$.
8. Update :$\bar{x}_B = \bar{x}_B + \tau(\bar{a}_q - e_r)$, where $\tau = -\bar{x}_{j_r}/\omega_q$.
9. Update B^{-1} by (3.23) (p=r).
10. Update (B, N) by exchanging j_p and q.
11. Stop if $\bar{a}_{p,n+1} \geq 0$ (infeasible problem).
12. Compute $\alpha = -\bar{x}_{j_p}/\bar{a}_{p,n+1}$.
13. If $\alpha \neq 0$, update $\bar{x}_B = \bar{x}_B + \alpha\bar{a}_{n+1}$, $\bar{x}_{n+1} = \bar{x}_{n+1} - \alpha$.
14. Go to step 1.

16.4 Phase-I for the Dual Improved Reduced Simplex Method

This section offers a Phase-I method for the dual improved reduced simplex method. Based on the most-obtuse-angle heuristics, the tableau version below results from modifying the primal Algorithm 16.1.1.

Algorithm 16.4.1 (Dual improved reduced Phase-I: tableau form). Initial : improved reduced tableau of form (16.6), where $\bar{x}_B = \bar{b} - \bar{x}_{n+1}\bar{a}_{n+1}$. This algorithm finds a dual feasible improved reduced tableau.

1. Determine row index $p \in \arg\min\{\bar{a}_{i,n+1} \mid i = 1, \cdots, m+1\}$.
2. Stop if $\bar{a}_{p,n+1} \geq 0$ (infeasible or unbounded problem).
3. Compute $\alpha = -\bar{x}_{j_p}/\bar{a}_{p,n+1}$.
4. If $\alpha \neq 0$, add α times of x_{n+1} column to $\bar{x}_B$ column and set $\bar{x}_{n+1} = \bar{x}_{n+1} - \alpha$.
5. Determine column index $q \in \arg\min_{j\in N} \bar{a}_{pj}$.
6. Stop if $\bar{a}_{pq} \geq 0$ (dual feasibility achieved).
7. Convert $\bar{a}_{pq}$ to 1, and eliminate the other nonzeros in the column by elementary transformations.
8. Go to step 1.

Note Step 7 does not touch $\bar{x}_B$ column because its pth component vanishes.

Example 16.4.1. Solve the following problem using Algorithm 16.4.1 as dual Phase-I:

$$\begin{array}{llllllllll}
\min & x_9 = -4x_1 - 7x_2 + 2x_3 + 3x_4, \\
\text{s.t.} & -2x_1 + 1x_2 - 3x_3 - x_4 + x_5 & & & = -11, \\
& 3x_1 + 3x_2 + x_3 + 5x_4 & + x_6 & & = 35, \\
& 5x_1 + 4x_2 - x_3 - 2x_4 & & + x_7 & = 2, \\
& -x_1 + 6x_2 - 7x_3 + 4x_4 & & + x_8 & = 7, \\
& & & x_j \geq 0, \quad j = 1, \cdots, 8.
\end{array}$$

Answer Initial tableau:

x_1	x_2	x_3	x_4	x_5	x_6	x_7	x_8	x_9	$\bar{x}_B$
−2	1	−3	−1	1					−11
3	3	1	5		1				35
5	4	−1	−2			1			2
−1	6	−7	4				1		7
−4	−7*	2	3					−1	

where $\bar{x}_B = (-11, 35, 2, 7)^{\mathrm{T}}$, $\bar{x}_9 = 0$.
Dual Phase-I: Call Algorithm 16.4.1.

Iteration 1:

1. $\min\{0, 0, 0, 0, -1\} = -1 < 0, \ p = 5$.
3. $\alpha = 0/-1 = 0$.
5. $\min\{-4, -7, 2, 3\} = -7 < 0, \ q = 2$.
7. Multiply row 5 by $-1/7$, and add $-1, -3, -4, -6$ times of row 5 to rows 1,2,3,4, respectively:

x_1	x_2	x_3	x_4	x_5	x_6	x_7	x_8	x_9	$\bar{x}_B$
−18/7		−19/7	−4/7	1				−1/7	−11
9/7		13/7	44/7		1			−3/7	35
19/7		1/7	−2/7			1		−4/7	2
−31/7		−37/7*	46/7				1	−6/7	7
4/7	1	−2/7	−3/7					1/7	

Iteration 2:

1. $\min\{-1/7, -3/7, -4/7, -6/7, 1/7\} = -6/7 < 0, \ p = 4$.
3. $\alpha = -7/(-6/7) = 49/6$.
4. Add $47/6$ times of x_9 column to $\bar{x}_B$ column; $\bar{x}_9 = -49/6$.
5. $\min\{-31/7, -37/7, 46/7\} = -37/7 < 0, \ q = 3$.
7. Multiply row 4 by $-7/37$, and add $19/7, -13/7, -1/7, 2/7$ times of row 4 to rows 1,2,3,5, respectively:

x_1	x_2	x_3	x_4	x_5	x_6	x_7	x_8	x_9	$\bar{x}_B$
−11/37			−146/37	1			−19/37	11/37	−73/6
−10/37*			318/37		1		13/37	−27/37	63/2
96/37			−4/37			1	1/37	−22/37	−8/3
31/37		1	−46/37				−7/37	6/37	
30/37	1		−29/37				−2/37	7/37	7/6

Iteration 3:

1. $\min\{11/37, -27/37, -22/37, 6/37, 7/37\} = -27/37 < 0,\ p = 2.$
3. $\alpha = -(63/2)/(-27/37) = 259/6.$
4. Add α times of x_9 column to $\bar{x}_B$ column; $\bar{x}_9 = -49/6 - 259/6 = -154/3.$
5. $\min\{-10/37, 318/37, 13/37\} = -10/37 < 0,\ q = 3.$
7. Multiply row 2 by $-37/10$, and add $11/37, -96/37, -31/37, -30/37$ times of row 2 to rows 1,3,4,5, respectively:

x_1	x_2	x_3	x_4	x_5	x_6	x_7	x_8	x_9	$\bar{x}_B$
			−67/5	1	−11/10		−9/10	11/10	2/3
1			−159/5		−37/10		−13/10	27/10	
			412/5		48/5	1	17/5	−38/5	−85/3
		1	127/5		31/10		9/10	−21/10	7
	1		25		3		1	−2	28/3

Iteration 4:

1. $\min\{11/10, 27/10, -38/5, -21/10, -2\} = -38/5 < 0,\ p = 3.$
3. $\alpha = -(-85/3)/(-38/5) = -425/114.$
3. Add α times of x_9 column to $\bar{x}_B$ column; $\bar{x}_9 = -154/3 + 425/114 = -1{,}809/38.$
5. $\min\{412/5, 48/5, 17/5\} \geq 0.$
6. Dual feasibility is achieved:

x_1	x_2	x_3	x_4	x_5	x_6	x_7	x_8	x_9	$\bar{x}_B$
			−67/5	1	−11/10		−9/10	11/10	−261/76
1			−159/5*		−37/10		−13/10	27/10	−765/76
			412/5		48/5	1	17/5	−38/5	
		1	127/5		31/10		9/10	−21/10	1,127/76
	1		25		3		1	−2	319/19

Dual Phase-II: Call Algorithm 15.4.2.

Iteration 5:

1. $\min\{-261/76, -765/76, 1{,}127/76, 319/19\} = -765/76 < 0,\ r = 2.$
3. $J = \{4, 6, 8\} \neq \emptyset.$
4. $\beta = \min\{-(412/5)/(-159/5), -(48/5)/(-37/10), -(17/5)/(-13/10)\}$
   $= (412/5)/(159/5) = 412/159,\ q = 4.$
5. Multiply row 2 by $-5/159$, and add $67/5, -412/5, -127/5, -25$ times of row 2 to rows 1,3,4,5, respectively:

x_1	x_2	x_3	x_4	x_5	x_6	x_7	x_8	x_9	$\bar{x}_B$
−67/159				1	73/159		−56/159	−2/53	813/1,007
−5/159			1		37/318		13/318	−9/106	358/1,131
412/159					2/159	1	5/159	−32/53	−4,747/182
127/159		1			23/159		−22/159	3/53	4,311/635
125/159	1				29/318		−7/318	13/106	6,950/783

6. $\bar{a}_{3,9} = -32/53 < 0$.
7. $\alpha \ = -(-4{,}747/182)/(-32/53)$.
8. Add α times of x_9 column to $\bar{x}_B$ column:

x_1	x_2	x_3	x_4	x_5	x_6	x_7	x_8	x_9	$\bar{x}_B$
−67/159				1	73/159		−56/159	−2/53	39/16
−5/159			1		37/318		13/318	−9/106	255/64
412/159					2/159	1	5/159	−32/53	
127/159		1			23/159		−22/159	3/53	139/32
125/159	1				29/318		−7/318	13/106	229/64

Basic optimal solution and according objective value:

$$\bar{x} = (0, 229/64, 139/32, 255/64, 39/16, 0, 0, 0)^{\mathrm{T}},$$

$$\bar{x}_9 = -7 \times (229/64) + 2 \times (139/32) + 3 \times (255/64) = -141/32.$$

The following is a revision of Algorithm 16.4.1.

Algorithm 16.4.2 (Dual Improved Reduced Phase-I). Initial : $(B, N), B^{-1}$. $\bar{x}_B = \bar{b} - \bar{x}_{n+1}\bar{a}_{n+1}$, where $\bar{b} = B^{-1}b$, $\bar{a}_{n+1} = -B^{-1}e_{m+1}$. This algorithm finds a dual feasible basis to (15.1).

1. Determine row index $p \in \arg\min\{\bar{a}_{i,n+1} \mid i = 1, \cdots, m+1\}$.
2. Stop if $\bar{a}_{p,n+1} \geq 0$ (infeasible or unbounded problem).
3. Compute $\alpha = -\bar{x}_{j_p}/\bar{a}_{p,n+1}$.
4. If $\alpha \neq 0$, update $\bar{x}_B = \bar{x}_B + \alpha\bar{a}_{n+1}$, $\bar{x}_{n+1} = \bar{x}_{n+1} - \alpha$.
5. Compute $\sigma_N = N^{\mathrm{T}}B^{-\mathrm{T}}e_p$.
6. Determine column index $q \in \arg\min_{j \in N} \sigma_j$.
7. Stop if $\sigma_q \geq 0$ (dual feasibility achieved).
8. Compute $\bar{a}_q = B^{-1}a_q$.
9. Update $\bar{a}_{n+1} = \bar{a}_{n+1} + \nu(\bar{a}_q - e_p)$, where $\nu = -\bar{a}_{p,n+1}/\sigma_q$.
10. Update B^{-1} by (3.23).
11. Update (B, N) by exchanging j_p and q.
12. Go to step 1.

16.5 Bisection Reduced Simplex Method

In some applications, not only there exists an optimal solution to the LP problem, but the optimal value can be estimated. In this case, it would be possible to speed up solution process by taking advantage of information of the optimal value. If the optimal value is known in priori to be γ, e.g., after adding a new constraint $c^{\mathrm{T}}x = \gamma$, any feasible solution to the modified problem is clearly an optimal solution to the original problem. Consequently, only a Phase-I procedure is enough to solve the problem, in principle. Although this is not the case in general, it would be possible to determine some interval, including the optimal value there within.

An interval including the optimal value is referred to as *(optimal value) existence interval.* It is contracted by at least a half of the length in each iteration, until optimality achieved. This idea motivated the bisection simplex method for solving standard LP problems, as was initially expected to help bypass degenerate vertices (Pan 1991, 1994c, 1996a, Yan and Pan 2001). It may be more fitly implemented in reduced simplex context. The reader is referred to the related references for detailed theoretical results and discussions, associated with the algorithm designed in this section.

The approach to determining an existence interval is novel. Firstly, the improved reduced Phase-1 method (Sect. 16.2) is applied to generate a feasible solution, if any; the associated objective value, which is an upper bound of the optimal value, is set to μ. Then, the dual improved reduced Phase-I method (Sect. 16.4) is applied to generate a dual feasible solution, if any; the associated objective value, which a lower bound of the optimal value, is set to λ. Thus, if it holds that $\lambda = \mu$, then a pair of dual and primal basic optimal solutions is already reached, though which are not necessarily complementary. Assume that this is not the case.

Each iteration will correspond to an existence interval $[\lambda, \mu]$. It is divided by the optimal value into two sections: the left-hand and right-hand sections are comprised of dual and primal feasible values, respectively; as a common point, the optimal value itself is primal as well as dual feasible. A subalgorithm is utilized to find a primal feasible solution with the objective value $\bar{x}_{n+1} = (\lambda + \mu)/2$. The outcome will be one of the following:

(i) A primal feasible solution is found, hence $\bar{x}_{n+1}$ is a feasible objective value. Switch such value down to one, associated with a basic feasible solution. If the reached solution does not satisfy the optimality condition, the according objective value is a new upper bound on the optimal value, and set it to μ.
(ii) There is no such a primal feasible solution, hence $\bar{x}_{n+1}$ is a dual feasible value. Switch it up to a value, associated with a basic solution. If it does not satisfy the optimality condition, the according objective value is a new lower bound on the optimal value, and set it to λ.

It is clear that the resulting $[\lambda, \mu]$ is again an existence interval, contracting by a half of its length, at least.

The subalgorithm can be obtained by modifying Algorithm 16.2.1 as follows.

Algorithm 16.5.1 (Bisection subalgorithm: tableau form). Initial: improved reduced tableau of form (16.6), where $\bar{x}_B = \bar{b} - \bar{x}_{n+1}\bar{a}_{n+1}$. This procedure attempts to find a feasible solution with objective value $\bar{x}_{n+1}$, and then makes switching.

1. Determine row index $p \in \arg\min\{\bar{x}_{j_i} \mid i = 1, \cdots, m+1, i \neq p\}$.
2. Go to step 7 if $\bar{x}_{j_p} < 0$.
3. Determine α and row index p such that

$$\alpha = -\bar{x}_{j_p}/\bar{a}_{p,n+1} = \min\{-\bar{x}_{j_i}/\bar{a}_{i,n+1} \mid \bar{a}_{i,n+1} < 0, i = 1, \cdots, m+1\}.$$

4. If $\alpha > 0$, add α times of x_{n+1} column to $\bar{x}_B$ column, and set $\bar{x}_{n+1} = \bar{x}_{n+1} - \alpha$.
5. Return if $\bar{a}_{pj} \geq 0,\ \forall\ j \in N$ (optimality achieved).
6. Set $\mu = \bar{x}_{n+1}$, and return.
7. Determine column index $q \in \arg\min_{j \in N} \bar{a}_{pj}$.
8. Go to 13 if $\bar{a}_{pq} < 0$.
9. Compute $\alpha = -\bar{x}_{j_p}/\bar{a}_{p,n+1}$.
10. Add α times of x_{n+1} column to $\bar{x}_B$ column, and set $\bar{x}_{n+1} = \bar{x}_{n+1} - \alpha$.
11. Return if $\bar{x}_B \geq 0$ (optimality achieved).
12. Set $\lambda = \bar{x}_{n+1}$, and return.
13. Convert $\bar{a}_{pq}$ to 1, and eliminate the other nonzeros in the column by elementary transformations.
14. Go to step 1.

Note This subalgorithm is applicable whenever a existence interval $[\lambda, \mu]$ is available. Step 3 is always well-defined because, otherwise, $\bar{x}_p \geq 0$, hence $\bar{a}_{n+1} \not\geq 0$ leads to unboundedness of the original problem follows, as is a contradiction. Step 9 is also always well-defined because, otherwise, $\bar{a}_{p,n+1} < 0$ leads to infeasibility of the original problem, as is a contradiction.

Regarding the preceding subalgorithm, the following result is cited without proof (Pan 1991).

Theorem 16.5.1. *Assume that no cycling occurs. If the length of a existence interval is short enough, Subalgorithm 16.5.1 generates a basic optimal solution.*

Associated overall steps are put into the following master algorithm.

Algorithm 16.5.2 (Bisection reduced simplex algorithm: tableau form). Initial : improved reduced tableau of form (16.6), where $\bar{x}_B = \bar{b} - \bar{x}_{n+1}\bar{a}_{n+1}$. This algorithm solves the reduced problem (15.1).

1. Call Algorithm 16.2.1 to achieve primal feasibility:

   (1) Stop when returning from step 5 (infeasible problem);
   (2) When returning from step 2, then: if nonbasic entries in the pth row are all positive, set $\mu = \bar{x}_{n+1}$; else stop (optimality achieved).

2. Call Algorithm 16.4.1 to achieve dual feasibility:

   (1) Stop when returning from step 2 (infeasible problem);

(2) When returning from step 6: if $\bar{x}_B \not\geq 0$, set $\lambda = \bar{x}_{n+1}$; otherwise, stop (optimality achieved).

3. Stop if $\lambda = \mu$ (optimality achieved).
4. Compute $\alpha = (\mu - \lambda)/2$; if $\bar{x}_{n+1} = \lambda$, set $\alpha = -\alpha$.
5. Update $\bar{x}_{n+1} = \bar{x}_{n+1} - \alpha$, and add α times of x_{n+1} column to $\bar{x}_B$ column.
6. Call Subalgorithm 16.5.1.

   (1) Stop when returning from step 5 or 11 (optimality achieved).
   (2) Go to step 5 when returning from step 6 or 8.

The associated proof of the preceding algorithm is omitted.

Under the nondegeneracy assumption, Subalgorithm 16.5.1 either achieves optimality or contracts the existence interval by a half, at least. Based on Theorem 16.5.1, therefore, it is known that Algorithm 16.5.2 generates an optimal solution, if any, in finitely many iterations. Steps 1 and 2 generate a existence interval, or detects infeasibility or unboundedness of the problem.

Example 16.5.1. Solve the following problem by Algorithm 16.5.2:

$$
\begin{array}{lrl}
\min & x_{10} = 3x_1 - 4x_2 + 8x_3 + x_4, & \\
\text{s.t.} & x_1 + 2x_2 + x_3 + 5x_4 + x_5 & = 3, \\
& 3x_1 + 4x_2 - 7x_3 - x_4 + x_6 & = 20, \\
& -2x_1 - x_2 + x_3 + 2x_4 + x_7 & = -5, \\
& -x_1 + x_2 - 2x_3 - x_4 + x_8 & = -2, \\
& -6x_1 + 8x_2 - 16x_3 - 2x_4 + x_9 & = 50, \\
& x_j \geq 0, \quad j = 1, \cdots, 9. &
\end{array}
$$

Answer

Step 1. Call Algorithm 16.2.1 to achieve primal feasibility.
Initial tableau is

x_1	x_2	x_3	x_4	x_5	x_6	x_7	x_8	x_9	x_{10}	$\bar{x}_B$
1	2	1	5	1						3
−3	4	−7	−1		1					20
−2	−1	1	2			1				−5
−1	1	−2	−1				1			−2
−6	8	−16	−2					1		50
3	−4*	8	1						−1	

Iteration 1:
Transform the preceding to an improved reduced tableau: taking $p = 6$; $\min\{3, -4, 8, 1\} = -4$, $q = 2$; Multiplying row 6 by $-1/4$, and add $-2, -4, 1, -1, -8$ times of row 6 to rows 1,2,3,4,5, respectively:

x_1	x_2	x_3	x_4	x_5	x_6	x_7	x_8	x_9	x_{10}	$\bar{x}_B$
5/2		5	11/2	1					−1/2	3
0		1	0		1				−1	20
−11/4*		−1	7/4			1			1/4	−5
−1/4		0	−3/4				1		−1/4	−2
0		0	0					1	−2	50
−3/4	1	−2	−1/4						1/4	

Iteration 2:

1. $\min\{3, 20, -5, -2, 50, 0\} = -5$, $p = 3$.
3. $\min\{-11/4, -1, 7/4\} = -11/4, q = 1$.
4. Multiply row 3 by $-4/11$, and add $-5/2, 1/4, 3/4$ times of row 3 to rows 1,4,6, respectively:

x_1	x_2	x_3	x_4	x_5	x_6	x_7	x_8	x_9	x_{10}	$\bar{x}_B$
		45/11	78/11	1		10/11			−3/11	−17/11
		1			1				−1	20
1		4/11	−7/11			−4/11			−1/11	20/11
		1/11	−10/11			−1/11	1		−3/11	−17/11
								1	−2	50
	1	−19/11	−8/11			−3/11			2/11	15/11

Iteration 3:

1. $\min\{-17/11, 20, 20/11, -17/11, 50, 15/11\} = -17/11, \ p = 1$.
3. $\min\{45/11, 78/11, 10/11\} \geq 0$.
5. $\bar{a}_{1,10} = -3/11 < 0$.
6. $\alpha \quad = -(-17/11)/(-3/11) = -17/3$.
7. Add $-17/3$ times of x_{10} column to $\bar{x}_B$ column:

x_1	x_2	x_3	x_4	x_5	x_6	x_7	x_8	x_9	x_{10}	$\bar{x}_B$
		45/11	78/11	1		10/11			−3/11	
		1			1				−1	77/3
1		4/11	−7/11			−4/11			−1/11	7/3
		1/11	−10/11			−1/11	1		−3/11	
								1	−2	184/3
	1	−19/11	−8/11			−3/11			2/11	1/3

Iteration 4:

1. $\min\{0, 17/3, 7/3, 0, 184/3, 1/3 \geq 0$.
2. $\bar{x}_{10} = 3 \times (7/3) - 4 \times (1/3) = 17/3$, feasibility achieved.

Step 1(2). $(45/11, 78/11, 10/11) > 0$, $\mu = 17/3$.
Step 2. Call Algorithm 16.4.1 to achieve dual feasibility.

Iteration 5:

1. $\min\{-3/11, -1, -1/11, -3/11, -2, 2/11\} = -2 < 0$, $p = 5$.
3. $\alpha \ = -(184/3)/(-2) = 184/6$.
4. $\bar{x}_{10} = 17/3 - 184/6 = -25$; add $184/6$ times of x_{10} column to $\bar{x}_B$ column.
5. $\min\{0, 0, 0\} \geq 0$.
6. Dual feasibility is achieved.

Step 2(2). Basic optimal solution and according objective value:

$$\bar{x} = (7/3, 1/3, 0, 0, 0, 77/3, 0, 0, 184/3)^{\mathrm{T}}, \ \bar{x}_{10} = 17/3.$$

In the preceding example, there was no need for bisection carried out on the initial existence interval $[\lambda, \mu]$. It is of course not the case in general, though we failed to construct one requiring bisection. It would be different if tested with larger problems.

We state a revisions of Algorithms 16.5.1 and 16.5.2 below.

Algorithm 16.5.3 (Bisection subalgorithm). Initial : $(B, N), B^{-1}$. $\bar{x}_B = \bar{b} - \bar{x}_{n+1}\bar{a}_{n+1}$, where $\bar{b} = B^{-1}b$, $\bar{a}_{n+1} = -B^{-1}e_{m+1}$. This procedure attempts to find a feasible solution with objective value $\bar{x}_{n+1}$, and switch.

1. Determine row index $p \in \arg\min\{\bar{x}_{j_i} \mid i = 1, \cdots, m+1, i \neq p\}$.
2. Go to step 9 if $\bar{x}_{j_p} < 0$.
3. Determine α and row index p such that

$$\alpha = -\bar{x}_{j_p}/\bar{a}_{p,n+1} = \min\{-\bar{x}_{j_i}/\bar{a}_{i,n+1} \mid \bar{a}_{i,n+1} < 0, i = 1, \cdots, m+1\}.$$

4. If $\alpha > 0$, update $\bar{x}_B = \bar{x}_B + \alpha\bar{a}_{n+1}$, and set $\bar{x}_{n+1} = \bar{x}_{n+1} - \alpha$.
5. Compute $\sigma_N = N^{\mathrm{T}}B^{-\mathrm{T}}e_p$.
6. Determine column index $q \in \arg\min_{j \in N} \sigma_j$.
7. Return if $\sigma_q \geq 0$ (optimality achieved).
8. Set $\mu = \bar{x}_{n+1}$, and return.
9. Determine column index $q \in \arg\min_{j \in N} \sigma_j$.
10. Go to step 15 if $\sigma_q < 0$.
11. Compute $\alpha = -\bar{x}_{j_p}/\bar{a}_{p,n+1}$.
12. Update $\bar{x}_B = \bar{x}_B + \alpha\bar{a}_{n+1}$, and set $\bar{x}_{n+1} = \bar{x}_{n+1} - \alpha$.
13. Return if $\bar{x}_B \geq 0$ (optimality achieved).
14. Set $\lambda = \bar{x}_{n+1}$, and return.

15. Compute $\bar{a}_q = B^{-1}a_q$.
16. Update $\bar{a}_{n+1} = \bar{a}_{n+1} + \nu(\bar{a}_q - e_p)$, where $\nu = -\bar{a}_{p,n+1}/\sigma_q$.
17. Update B^{-1} by (3.23).
18. Update (B, N) by exchanging j_p and q.
19. Go to step 1.

Algorithm 16.5.4 (Bisection reduced simplex Algorithm). Initial : $(B, N), B^{-1}$. $\bar{x}_B = \bar{b} - \bar{x}_{n+1}\bar{a}_{n+1}$, where $\bar{b} = B^{-1}b$, $\bar{a}_{n+1} = -B^{-1}e_{m+1}$. This algorithm solves the reduced problem (15.1).

1. Call Algorithm 16.2.2 to achieve primal feasibility:
   (1) Stop when returning from step 6 (infeasible problem);
   (2) When returning from step 2, then: if $\sigma_N < 0$, set $\mu = \bar{x}_{n+1}$; else stop (optimality achieved).
2. Call Algorithm 16.4.2 to achieve dual feasibility:
   (1) Stop when returning from step 2 (unbounded primal problem);
   (2) When returning from step 7, then: if $\bar{x}_B < 0$, set $\lambda = \bar{x}_{n+1}$; else stop (optimality achieved).
3. Stop if $\lambda = \mu$ (optimality achieved).
4. Compute $\alpha = (\mu - \lambda)/2$; set $\alpha = -\alpha$ if $\bar{x}_{n+1} = \lambda$.
5. Update $\bar{x}_{n+1} = \bar{x}_{n+1} - \alpha$; $\bar{x}_B = \bar{x}_B + \alpha\bar{a}_{n+1}$.
6. Call Subalgorithm 16.5.3.
   (1) Stop when returning from step 7 or 13 (optimality achieved).
   (2) Go to step 8 when returning from step 8 or 14.

Chapter 17
D-Reduced Simplex Method

Consider the so-called "D-reduced problem" (25.3), i.e.,

$$\begin{array}{ll} \min & c^{\mathrm{T}}x, \\ \text{s.t.} & Ax = e_r, \quad x \geq 0. \end{array} \tag{17.1}$$

The associated dual problem is of an objective function involving a single variable. A so-called "D-reduced simplex method" will be developed to solve the preceding problem.

As it will play a specially role in the presented method, the r-indexed row will be separated off to handle. Therefore, we introduce the following column and row index sets:

$$B = \{j_1, \cdots, j_{m-1}\}, \quad N = A \backslash B, \quad R = \{1, \cdots, m\} \backslash \{r\} = \{i_1, \cdots, i_{m-1}\}. \tag{17.2}$$

17.1 D-Reduced Simplex Tableau

Let $I \in \mathcal{R}^{(m-1)\times(m-1)}$ be the unit matrix. Assume that the initial tableau of problem (17.1) is converted to the following so-called "D-reduced (simplex) tableau" by the Gaussian-Jordan elimination with row and column exchanges:

x_B^{T}	x_N^{T}	f	RHS
I	$\bar{N}_R$		
	$\bar{\omega}_N^{\mathrm{T}}$		1
	$\bar{z}_N^{\mathrm{T}}$	-1	$-\bar{f}$

(17.3)

where the mth (r-indexed) row is called *datum row*. Of course, a D-reduced tableau represents the original problem itself, just as a conventional simplex tableau, though now the RHS column alone does not give the basic solution explicitly.

P.-Q. PAN, *Linear Programming Computation*, DOI 10.1007/978-3-642-40754-3_17,

It is noted that the datum row represents equation

$$\bar{\omega}_N^{\mathrm{T}} x_N = 1. \tag{17.4}$$

whereas the bottom (objective) row represents equation

$$\bar{z}_N^{\mathrm{T}} x_N - f = -\bar{f}. \tag{17.5}$$

Lemma 17.1.1. *If nonbasic entries in the datum row are no more than zero, there is no feasible solution.*

Proof. Since $\bar{\omega}_N \leq 0$, if $\hat{x} \geq 0$ is a feasible solution, then it follows that the left side of (17.4) is less than or equal to 0, as contradicts the right-hand side 1. Therefore, there is no feasible solution. □

Denote by $\bar{a}_q(R)$ the subvector of $\bar{a}_q$, associated with row index set R.

Definition 17.1.1. If it holds that

$$\bar{\omega}_q \neq 0, \quad \bar{z}_q = 0, \tag{17.6}$$

the qth column is a datum column of the D-reduced tableau (17.3).

Proposition 17.1.1. *The D-reduced tableau with the qth datum column gives the basic solution*

$$\bar{x}_B = -\bar{x}_q \bar{a}_q(R), \qquad \bar{x}_q = 1/\bar{\omega}_q, \quad \bar{x}_j = 0, \; q \neq j \in N, \tag{17.7}$$

associated with objective value $\bar{f}$.

Proof. The D-reduced tableau (17.3) corresponds to problem

$$\begin{aligned} \max \; & f = \bar{f} + \bar{z}_N^{\mathrm{T}} x_N, \\ \text{s.t.} \; & x_B + \bar{N}_R x_N = 0, \\ & \bar{\omega}_N^{\mathrm{T}} x_N = 1, \quad x_B, x_N \geq 0. \end{aligned} \tag{17.8}$$

Substituting $\bar{x}_j = 0, \; q \neq j \in N, \; \bar{x}_q = 1/\bar{\omega}_q$ to the equality constraint gives (17.7). Carrying out elementary transformations, by taking $\bar{\omega}_q$ as pivot, will convert (17.3) to a conventional simplex tableau, associated with a basic solution, which is just that given by (17.7). If $\bar{z}_q = 0$, the bottom row remains unchanged, and hence the according objective value of the solution is equal to $\bar{f}$. □

It is seen that a D-reduced tableau could give multiple basic solutions, as its datum column would be specified differently.

D-reduced tableau satisfying

$$\bar{z}_q = 0, \quad \bar{\omega}_q > 0; \quad \bar{a}_{iq} \leq 0, \quad i \in R, \tag{17.9}$$

is called *feasible*, and that satisfying

$$\bar{z}_N \geq 0, \tag{17.10}$$

is called *dual feasible*.

Lemma 17.1.2. *A feasible D-reduced tableau corresponds to a basic feasible solution. A dual feasible D-reduced tableau satisfying (17.6) corresponds to a dual basic feasible solution.*

Proof. Consider the conventional simplex tableau, resulting from the basis change by taking $\bar{\omega}_q \neq 0$ as pivot. If (17.9) holds, then the basic solution $\bar{x} \geq 0$, defined by (17.7), is clearly a feasible solution. If (17.6) and (17.10) both hold, the tableau corresponds to a dual basic feasible solution $\bar{z}_B = 0,\ \bar{z}_N \geq 0$. □

Lemma 17.1.3. *A D-reduced tableau, satisfying*

$$\bar{\omega}_q > 0, \quad \bar{a}_{iq} \leq 0, \quad i \in R, \tag{17.11}$$

$$\bar{z}_q = 0, \quad \bar{z}_N \geq 0, \tag{17.12}$$

corresponds to a pair of primal and dual basic optimal solutions.

Proof. Under the assumption, conditions (17.9) and (17.10) are both satisfied. By Lemma 17.1.2, the D-reduced tableau corresponds to a primal and a dual basic feasible solution, which are complementary. □

Lemma 17.1.4. *Let (17.3) be a feasible D-reduced tableau. If, for some $q' \in N$, it holds that*

$$\bar{z}q' < 0, \quad \bar{a}_{i,q'} \leq 0,\ i \in R, \quad \bar{\omega}_{q'} \leq 0,$$

then the problem is lower unbounded.

Proof. The feasibility of the D-reduced tableau implies satisfaction of (17.9) for some $q \in N$. Carrying out elementary transformations, with $\bar{\omega}_q > 0$ as pivot, will lead to a feasible simplex tableau with the sigh of the q' column remaining unchanged; there is no change to the objective row, and hence to $\bar{z}_{q'} < 0$. Therefore the problem is lower unbounded (Lemma 3.2.2). □

Lemma 17.1.5. *If a D-reduced tableau satisfies*

$$\bar{z}_q < 0; \quad \bar{a}_{iq} \leq 0, \quad i \in R, \tag{17.13}$$

then the problem has no feasible solution, or has no dual feasible solution with objective value $\bar{f}$.

Proof. Consider the dual problem, represented by D-reduced tableau (17.3), i.e.,

$$\begin{aligned}
&\max\ \bar{f} + g, \\
&\text{s.t.}\ \begin{pmatrix} I & \\ \bar{N}_R^{\mathrm{T}} & \bar{\omega}_N \end{pmatrix} \begin{pmatrix} y \\ g \end{pmatrix} + \begin{pmatrix} z_B \\ z_N \end{pmatrix} = \begin{pmatrix} 0 \\ \bar{z}_N \end{pmatrix}, \quad z_B,\ z_N \geq 0,
\end{aligned} \tag{17.14}$$

where the objective function involves a single variable g. Setting $g = 0$ turns this problem to

$$\begin{aligned}
&\max\ \bar{f}, \\
&\text{s.t.}\ \begin{pmatrix} I \\ \bar{N}_R^{\mathrm{T}} \end{pmatrix} y + \begin{pmatrix} z_B \\ z_N \end{pmatrix} = \begin{pmatrix} 0 \\ \bar{z}_N \end{pmatrix}, \quad z_B,\ z_N \geq 0,
\end{aligned} \tag{17.15}$$

where the objective value equals $\bar{f}$. On the other hand, deleting the mth row of the tableau gives a (conventional) simplex tableau of (17.15). Since (17.13) holds, by Lemma 14.3.1 the problem either has a feasible solution or is lower unbounded. If there is no feasible solution, the D-reduced tableau has no feasible solution; if lower unbounded, then (17.15) has no feasible solution, hence there is no dual feasible solution, associated with objective value $\bar{f}$. □

A D-reduced tableau corresponds to a dual problem with objective function involving a single variable. It will be seen a little later that the datum row corresponds to the dual reduced gradient, and basis changes carried out on the D-reduced tableau only touch rows associated with R, not objective value $\bar{f}$. Desired changes in objective value will be realized by adding appropriate times of the datum row to the bottom row.

Alternatively, an initial D-reduced tableau may be obtained from a (conventional) simplex tableau as follows. If $\bar{b}_r$ of the right-hand side is nonzero, multiplying the r-indexed row by $1/\bar{b}_r$, and then adding appropriate times of the row to the other rows respectively gives a D-reduced tableau with the r-indexed row being datum row. The j_r-indexed column can be taken as datum column, as it becomes nonbasic, i.e.,

$$\hat{a}_{r,j_r} = 1/\bar{b}_r; \quad \hat{a}_{i,j_r} = -\bar{b}_i/\bar{b}_r,\ i \in R,$$

corresponding to zero reduced cost. Further, if the conventional tableau is feasible, i.e., $\bar{b} \geq 0$, then the resulting is a feasible D-reduced tableau.

17.2 Dual D-Reduced Simplex Method

Starting from a dual feasible D-reduced tableau, the method presented in this section generates a sequence of such tableaus to achieve primal feasibility, and obtain a pair of primal and dual basic optimal solutions.

Let (17.3) be a dual feasible D-reduced tableau and let (17.10) hold. It is noted by Preposition 17.1.1 that the objective value of the basic solution $\bar{x}$ is $\bar{f}$, which is equal to the opposite number of the south-eastern entry of the tableau.

If entries in the datum row are all less than or equals 0, the problem has no feasible solution (Lemma 17.1.1). This is usually not the case, and hence the following rule is well-defined.

Rule 17.2.1 (Dual column rule) Determine β and column index q such that

$$\beta = \bar{z}_q/\bar{\omega}_q = \min\{\bar{z}_j/\bar{\omega}_j \mid \bar{\omega}_j > 0,\ j \in N\} \geq 0, \tag{17.16}$$

where β is termed *dual stepsize*. If $\beta \neq 0$, add $-\beta$ times of the datum row to the bottom row to convert $\bar{z}_q$ to 0, then apply the following row rule.

Rule 17.2.2 (Dual row rule) Select row index

$$p \in \arg\max_{i \in R} \bar{a}_{iq}.$$

Since $\bar{\omega}_q > 0$, an optimal D-reduced tableau is already obtained if $\bar{a}_{pq} \leq 0$ (by taking column q as datum column); otherwise, convert it to 1, and eliminate the other nonzeros of column q by elementary transformations. Thus, an iteration is complete; consequently, column q enters the basis, whereas column j_p leaves the basis, becoming a new datum column.

The resulting tableau remains dual feasible. In fact, Rule 17.2.2 ensures a largest possible stepsize β subject to dual constraints. Now the south-eastern entry of the tableau becomes $-\bar{f} + \beta$, so the increment of the objective value is just $\beta \geq 0$. If $\beta > 0$, the D-reduced tableau is said to be dual nondegenerate, a case in which the objective value strictly increases. In degenerate case of $\beta = 0$, the objective value will remain unchanged.

Repeat the preceding steps until optimality achieved or infeasibility is detected. It is not difficult to show that, the leaving column j_p will not enter the basis immediately, and the entering column q not leave the basis immediately either. In addition, the basis change will not touch the objective value, whereas only adding some times of the datum row to the bottom row changes it. As the f column does not vary in the process at all, it can be omitted.

The overall steps can be put into the following model.

Algorithm 17.2.1 (Dual D-reduced simplex algorithm: tableau form). Initial: dual feasible D-reduced tableau of form (17.3). This algorithm solves the D-reduced problem (17.1).

1. Stop if $J = \{j \in N \mid \bar{\omega}_j > 0\} = \emptyset$.
2. Determine β and column index q such that $\beta = \bar{z}_q/\bar{\omega}_q = \min_{j \in J} \bar{z}_j/\bar{\omega}_j$.
3. If $\beta > 0$, add $-\beta$ times of datum row to the bottom row.
4. Determine row index $p \in \arg\max_{i \in R} \bar{a}_{i\,q}$.
5. If $\bar{a}_{p\,q} \leq 0$, compute $\bar{x}_q = 1/\bar{\omega}_q,\ \bar{x}_B = -\bar{x}_q\bar{a}_q(R)$, and stop.

6. Convert $\bar{a}_{pq}$ to 1, and eliminate the other nonzeros in the column by elementary transformations.
7. Go to step 1.

Theorem 17.2.1. *Under dual nondegeneracy assumption, Algorithm 17.2.1 terminates either at*

(i) Step 1, detecting infeasibility of the problem; or at
(ii) Step 5, giving a basic optimal solution.

Proof. The proofs of termination under dual nondegeneracy is the same as in the conventional simplex context. Meanings of the exits come from Lemmas 17.1.1 and 17.1.3, as well as discussion preceding the Algorithm □

The preceding Algorithm may be viewed as solving the dual problem through the tableau. To explain, return to the dual problem (17.14).

It is clear that there is the following solution to the dual problem:

$$y = 0, \quad g = 0,$$
$$z_B = 0, \quad z_N = \bar{z}_N,$$

which is feasible when $\bar{z}_N \geq 0$. The reduced gradient of the objective function is given by

$$\Delta y = 0, \quad \Delta g = 1,$$

along which the objective value in the reduced space increases the fastest. The associated vector in z space is then

$$\Delta z_B = 0, \quad \Delta z_N = -\bar{\omega}_N.$$

This leads to the following update:

$$\hat{z}_N = \bar{z}_N - \beta\bar{\omega}_N, \quad \hat{g} = \beta. \tag{17.17}$$

It is noted that the dual stepsize $\beta \geq 0$ equals the according increment of the objective value. If $\bar{\omega}_N \not\leq 0$, the greatest possible value, maintaining dual feasibility, is

$$\beta = \bar{z}_q/\bar{\omega}_q = \min\{\bar{z}_j/\bar{\omega}_j \mid \bar{\omega}_j > 0, \ j \in N\}. \tag{17.18}$$

It is then seen that the dual D-reduced simplex algorithm is a dual variant of the reduced simplex algorithm, as it actually handles the dual problem by latter's train of thought. The sense of such doing lies in that the resulting search direction in the former corresponds to the whole dual reduced gradient, while, in contrast, the search direction in the (conventional) dual simplex algorithm corresponds to a single component of the dual reduced gradient only.

Example 17.2.1. Solve the following problem by Algorithm 17.2.1:

$$\begin{array}{lrrrrrrrl}
\min & 3x_1 & +5x_2 & +2x_3 & +4x_4, & & & & \\
\text{s.t.} & 2x_1 & +5x_2 & -3x_3 & -4x_4 & +x_5 & & & = -2, \\
 & -4x_1 & -3x_2 & +6x_3 & -2x_4 & & +x_6 & & = 4, \\
 & -x_1 & -4x_2 & -x_3 & +2x_4 & & & +x_7 & = -5, \\
 & & & & x_j \geq 0, & & j = 1, \cdots, 7. & &
\end{array}$$

Answer Initial tableau

x_1	x_2	x_3	x_4	x_5	x_6	x_7	RHS
2	5	−3	−4	1			−2
−4	−3	6	−2		1		4
−1	−4	−1	2			1	−5*
3	5	2	4				

Iteration 1:
$\max\{|-2|, |4|, |-5|\} = 5,\ r = 3$, take the third row as datum row. Multiply row 3 by $-1/5$, and add $2, -4$ times of row 3 to rows 1,2, respectively, resulting in the following D-reduced tableau:

x_1	x_2	x_3	x_4	x_5	x_6	x_7	RHS
12/5	33/5*	−13/5	−24/5	1		−2/5	
−24/5	−31/5	26/5	−2/5		1	4/5	
1/5	4/5	1/5	−2/5			−1/5	1
3	5	2	4				

which is dual feasible, as the bottom line is nonnegative.
Call Algorithm 17.2.1.

Iteration 2:

1. $J = \{1, 2, 3\} \neq \emptyset$.
2. $\beta = \min\{3/(1/5), 5/(4/5), 2/(1/5)\} = 5/(4/5),\ q = 2$.
3. Add $-5/(4/5)$ times of row 3 to the bottom row.
4. $\max\{33/5, -31/5\} = 33/5 > 0,\ p = 1$.
6. Multiply row 1 by $5/33$, and add $31/5, -4/5$ times of row 1 to rows 2,3, respectively:

x_1	x_2	x_3	x_4	x_5	x_6	x_7	RHS
4/11	1	−13/33	−8/11	5/33		−2/33	
−28/11		91/33*	−54/11	31/33	1	14/33	
−1/11		17/33	2/11	−4/33		−5/33	1
7/4		3/4	13/2			5/4	−25/4

Iteration 3:

1. $J = \{3, 4\} \neq \emptyset$.
2. $\beta = \min\{(3/4)/(17/33), (13/2)/(2/11)\} = (3/4)/(17/33)$, $q = 3$.
3. Add $-(3/4)/(17/33)$ times of row 3 to the bottom row.
4. $\max\{-13/33, 91/33\} = 91/33 > 0$, $p = 2$.
6. Multiply row 2 by $33/91$, and add $13/33, -17/33$ times of row 2 to rows 1,3, respectively:

x_1	x_2	x_3	x_4	x_5	x_6	x_7	RHS
	1		−10/7	2/7	1/7		
−12/13		1	−162/91	31/91	33/91	2/13	
5/13			100/91	−27/91	−17/91	−3/13	1
32/17			106/17	3/17		25/17	−131/17

Iteration 4:

1. $J = \{1, 4\} \neq \emptyset$.
2. $\beta = \min\{(32/17)/(5/13), (106/17)/(100/91)\} = (32/17)/(5/13)$, $q = 1$.
3. Add $(32/17)/(5/13)$ times of row 3 to the bottom row:

x_1	x_2	x_3	x_4	x_5	x_6	x_7	RHS
	1		−10/7	2/7	1/7		
−12/13		1	−162/91	31/91	33/91	2/13	
5/13			100/91	−27/91	−17/91	−3/13	1
			6/7	57/35	32/35	13/5	−63/5

4. $\max\{-12/13, 0\} \leq 0$.
5. $\bar{x}_1 = 13/5$. The basic optimal solution and according objective value:

$$\bar{x} = (13/5, 0, 12/5, 0, 0, 0, 0)^{\mathrm{T}}, \quad \bar{f} = 63/5.$$

Further, we give the revised version of Algorithm 17.2.1. Denote by B_R the submatrix consisting of entries associated with B, R (similarly below). Assume that row $r = m$ is datum row. It is not difficult to derive the following revised tableau equivalent to the D-reduced tableau (17.3):

x_B^{T}	x_N^{T}	f	RHS
I	$B_R^{-1}N_R$		
	$e_r^{\mathrm{T}}N - e_r^{\mathrm{T}}BB_R^{-1}N_R$		1
	$c_N^{\mathrm{T}} - c_B^{\mathrm{T}}B_R^{-1}N_R$	-1	

(17.19)

Assume the column q enters the basis. Introduce $\bar{a}_q(R) = B_R^{-1}a_q(R)$, where $a_q(R)$ is a_q's subvector associated with row index set R. The formula for updating the inverse of the basis matrix is similar to (3.23) i.e.,

$$\hat{B}_R^{-1} = E_s B_R^{-1}, \tag{17.20}$$

where

$$E_p = \begin{pmatrix} 1 & & -\bar{a}_{i_1\,q}/\bar{a}_{i_s,q} & & \\ & \ddots & \vdots & & \\ & & -\bar{a}_{i_{s-1}\,q}/\bar{a}_{p,q} & & \\ & & 1/\bar{a}_{i_s,q} & & \\ & & -\bar{a}_{i_{s+1}\,q}/\bar{a}_{i_s,q} & & \\ & & \vdots & \ddots & \\ & & -\bar{a}_{i_{m-1},q}/\bar{a}_{i_s,q} & & 1 \end{pmatrix} \tag{17.21}$$

Based on equivalence between (17.3) and (17.19), a revised version of Algorithm 17.2.1 is stated as follows.

Algorithm 17.2.2 (Dual D-reduced simplex algorithm). Initial: B, R, N defined by (17.2). B_R^{-1} and $\bar{z}_N \geq 0$. This algorithm solves the D-reduced problem (17.1).

1. Compute $\bar{\omega}_N = N^{\mathrm{T}}e_m - N_R^{\mathrm{T}}B_R^{-\mathrm{T}}B^{\mathrm{T}}e_m$.
2. Stop if $J = \{j \in N \mid \bar{\omega}_j > 0\} = \emptyset$ (dual unbounded problem).
3. Determine β and column index q such that $\beta = \bar{z}_q/\bar{\omega}_q = \min_{j\in J} \bar{z}_j/\bar{\omega}_j$.
4. If $\beta > 0$, update by $\bar{z}_N = \bar{z}_N - \beta\bar{\omega}_N$.
5. Compute $\bar{a}_q(R) = B_R^{-1}a_q(R)$.
6. Select $p \in \arg\max_{i\in R}\, \bar{a}_{i\,q}$ $(p = i_s)$.
7. If $\bar{a}_{p,q} \leq 0$, compute

$$\bar{x}_q = 1/\bar{\omega}_q, \quad \bar{x}_B = -\bar{x}_q\bar{a}_q(R), \quad \bar{f} = c_q\bar{x}_q + c_B^{\mathrm{T}}\bar{x}_B,$$

8. Stop (optimality achieved).
9. Update B_R^{-1} by (17.20).
10. Update (B, N) by exchanging j_s and q.
11. Go to step 1.

17.3 Dual D-Reduced Phase-I

The algorithms, described in the previous section, have to start from a dual feasible tableau or solution. If $\bar{z}_N = c_N - N_R^{\mathrm{T}} B_R^{-\mathrm{T}} c_B \geq 0$, such a starting point is readily available; otherwise, there is a need for dual Phase-I procedure to provide it. An initial dual feasible tableau for Algorithm 17.2.1 can be obtained by the following modification of D-reduced simplex Algorithm 17.5.1.

Algorithm 17.3.1 (Dual D-reduced Phase-I: tableau form). Initial: D-reduced tableau of form (17.3). This algorithm finds a dual feasible D-reduced tableau.

1. Determine column index $q \in \arg\min_{j \in N} \bar{z}_j$.
2. Stop if $\bar{z}_q \geq 0$.
3. If $\bar{\omega}_q \neq 0$, add $\beta = -\bar{z}_q / \bar{\omega}_q$ times of the datum row to the bottom row.
4. Determine row index $p \in \arg\max_{i \in R} \bar{a}_{i\,q}$.
5. Go to 7 if $\bar{a}_{p\,q} > 0$.
6. Stop if $\bar{\omega}_q = 0$.
7. Convert $\bar{a}_{p\,q}$ to 1, and eliminate the other nonzeros in the column by elementary transformations.
8. Go to step 1.

Theorem 17.3.1. *Assume termination of Algorithm 17.3.1. It terminates either at*

(i) Step 2, giving a dual feasible D-reduced tableau; or at
(ii) Step 6, detecting infeasibility or lower unboundedness.

Proof. Validity of the meaning of exit 2 is clear. Assume that it returns from exit 6. Then $\bar{z}_q < 0$, $\bar{a}_q \leq 0$ and $\bar{\omega}_q = 0$ hold. If the problem is feasible, by Lemma 17.1.1 there is $q' \in N$ such that $\bar{\omega}_{q'} > 0$. Taking $\bar{\omega}_{q'}$ as pivot, a (conventional) simplex tableau is obtained by carrying out elementary transformations, without changing the q column. As components of the column are positive, the problem is infeasible or lower unbounded, by Lemma 14.3.1. □

Example 17.3.1. Solve following problem using Algorithm 17.3.1 as Phase-I:

$$\begin{array}{llllllllll}
\min & -3x_1 & -\,5x_2 & +\,4x_3 & +\,8x_4, & & & & & \\
\text{s.t.} & 6x_1 & -\,2x_2 & +\,3x_3 & +\,2x_4 & +\,x_5 & & & = & 2, \\
& -2x_1 & +\,4x_2 & -\,x_3 & -\,8x_4 & & +\,x_6 & & = & -4, \\
& 4x_1 & +\,5x_2 & +\,2x_3 & +\,x_4 & & & +\,x_7 & = & 6, \\
& & & & x_j \geq 0, & j = 1, \cdots, 7. & & & &
\end{array}$$

Answer Initial tableau

x_1	x_2	x_3	x_4	x_5	x_6	x_7	RHS
6	−2	3	2	1			2
−2	4	−1	−8		1		−4*
4	5	2	1			1	6
−3	−5	4	8				

Iteration 1:
Arbitrarily take row 2 to be as the datum row. Multiply the row by $-1/4$, and add $-2, -6$ times of it to rows 1,3, respectively:

x_1	x_2	x_3	x_4	x_5	x_6	x_7	RHS
5		5/2	−2	1	1/2		
1/2	−1	1/4	2		−1/4		1
1	11*	1/2	−11		3/2	1	
−3	−5	4	8				

Dual Phase-I: Call Algorithm 17.3.1. $r = 2$.

Iteration 2:

1. $\min\{-3, -5, 4, 8\} = -5 < 0,\ q = 2$.
3. Add $\beta = -(-5)/(-1) = -5$ times of row 2 to the bottom row.
4. $\max\{0, 11\} = 11 > 0,\ p = 3$.
7. Add $1/11$ times of row 3 to the bottom row:

x_1	x_2	x_3	x_4	x_5	x_6	x_7	RHS
5*		5/2	−2	1	1/2		
13/22		13/44	1		−5/44	1/11	1
1/11	1	1/22	−1		3/22	1/11	
−11/2		11/4	−2		5/4		−5

Iteration 3:

1. $\min\{-11/2, 11/4, -2, 5/4, 0\} = -11/2 < 0,\ q = 1$.
3. Add $\beta = -(-11/2)/(13/22)$ times of row 2 to the bottom row.
4. $\max\{5, 1/11\} = 5 > 0,\ p = 1$.
7. Multiply row 1 by $1/5$, and add $-13/22, -1/11$ times of row 1 to rows 2,3, respectively:

x_1	x_2	x_3	x_4	x_5	x_6	x_7	RHS
1		1/2	−2/5	1/5	1/10		
			68/55	−13/110	−19/110	1/11	1
	1		−53/55	−1/55	7/55	1/11	
		11/2	95/13		5/26	11/13	56/13

Iteration 4:

1. $\min\{11/2, 95/13, 0, 5/26, 11/13\} \geq 0,\ q = 1$.
2. Dual feasibility is achieved.

Dual Phase-II: Call Algorithm 17.2.1: $r = 2,\ q = 2$.

Iteration 5:

1. $J = \{4, 7\}$.
2. $\beta = \max\{-(95/13)/(68/55), -(11/13)/(1/11)\} = -(95/13)/(68/55)$, q=4.
3. Add β times of row 2 to the bottom row.

x_1	x_2	x_3	x_4	x_5	x_6	x_7	RHS
1		1/2	−2/5	1/5	1/10		
			68/55	−13/110	−19/110	1/11	1
	1		−53/55	−1/55	7/55	1/11	
		11/2		95/136	165/136	21/68	−109/68

4. $\max\{-2/5, -53/55\} \leq 0$.
5. Take $\bar{x}_4 = 55/68$. Basic optimal solution and according objective value:

$$\bar{x} = (11/34, 53/68, 0, 55/68, 0, 0, 0, 1)^{\mathrm{T}}, \quad \bar{f} = 109/68.$$

Based on equivalence between the reduced tableau (17.3) and the revised tableau (17.19), the tableau Algorithm 17.3.1 can be turn to the following revision.

Algorithm 17.3.2 (Dual D-reduced Phase-I). Initial: B, R, N defined by (17.2). B_R^{-1}; $\bar{z}_N = c_N - N_R^{\mathrm{T}} B_R^{-\mathrm{T}} c_B$. This algorithm finds a dual feasible D-reduced basis.

1. Determine column index $q \in \arg\min_{j \in N} \bar{z}_j$.
2. Return if $\bar{z}_q \geq 0$ (dual feasibility achieved).
3. Compute $\bar{\omega}_N = N^{\mathrm{T}} e_m - N_R^{\mathrm{T}}(B_R^{-\mathrm{T}} B e_m)$.
4. If $\bar{\omega}_q \neq 0$, update $\bar{z}_N = \bar{z}_N + \beta\bar{\omega}_N$, where $\beta = -\bar{z}_q/\bar{\omega}_q$.
5. Compute $\bar{a}_q(R) = B_R^{-1} a_q(R)$.
6. Determine $p \in \arg\max_{i \in R}\ \bar{a}_{i\,q}$ $(p = i_s)$.
7. Go to 9 if $\bar{a}_{p,q} > 0$.
8. Return if $\bar{\omega}_q = 0$ (infeasible or lower unbounded).
9. Update B_R^{-1} by (17.20).
10. Update (B, N) by exchanging j_s and q.
11. Go to step 1.

There are no any numerical results with the algorithms, described in this section. It seems to be better to obtain a (conventional) dual feasible simplex tableau (basis) first, and then to convert it to a D-reduced tableau (basis).

17.4 Dual D-Reduced Phase-I: Single-Artificial-Variable

This section describes another dual D-reduced Phase-I method, which may be regarded as a variant of the dual single-artificial-variable method (Sect. 14.2), where a (conventional) dual feasible tableau is generated via solving auxiliary dual program (14.9) or (14.11). As the objective function only involves a single variable, in fact, the auxiliary programs are amenable to be solved by the dual D-reduced method.

A tableau version will be derived by handling (14.11). To this end, turn to an initial auxiliary tableau of form (14.13) (f column is omitted), i.e.,

x_B^{T}	x_N^{T}	x_{n+1}	RHS
I	$\bar{N}$		
	$(\hat{z}_N - \bar{z}_N)^{\mathrm{T}}$	1	1
	$\hat{z}_N^{\mathrm{T}}$	$\hat{z}_{n+1}$	$-\bar{f}+1$

(17.22)

where $\hat{z}_N \geq 0$ is some given vector, and $\hat{z}_{n+1} = 1$.

As the tableau itself is a dual feasible D-reduced tableau with the $(m+1)$th row as the datum row, Algorithm 17.2.1 is applicable. As the artificial variable x_{n+1} is nonbasic at the beginning, we set

$$N := N \cup \{n+1\}.$$

When optimality of the auxiliary program is achieved, x_{n+1} enters the basis ($\hat{z}_{n+1}$ becomes 0), then delete the $(m+1)$th row and x_{n+1} column to gain a (conventional) dual feasible tableau, which can be further transformed to a dual feasible D-reduced tableau; otherwise, it is detected that there exists no dual feasible solution.

The last column of (17.22) can be omitted in operations, as the first $m+1$ components of the last two columns are the same (associated with e_{m+1}), and not touched by subsequent row elementary transformations at all; that is to say, x_{n+1} column will also play the role of the right-hand side column of the auxiliary program. In the following example, the right-hand side $\bar{b}$ of the original problem is put in RHS column, so that it is convenient to turn to Phase-II when Phase-I is finished.

Example 17.4.1. Solve Example 17.3.1 by the two-phase dual D-reduced method, by starting Phase-I from a single-artificial-variable auxiliary tableau of form (17.22).

Answer Taking $\hat{z}_1 = \hat{z}_2 = 1, \hat{z}_3 = \hat{z}_4 = 0$, construct auxiliary initial tableau

x_1	x_2	x_3	x_4	x_5	x_6	x_7	x_8	RHS
6	−2	3	2	1				2
−2	4	−1	−8		1			−4
4	5*	2	1			1		6
1 + 3	1 + 5	−4	−8				1	1
1	1						1	1

Dual Phase-I: Call Algorithm 17.2.1. The artificial variable x_8 column is also deemed as the right-hand side of the auxiliary program. The fourth row is datum row: $r = m + 1 = 4$.

Iteration 1:

1. $J = \{1, 2, 8\} \neq \emptyset$.
2. $\beta = \max\{-1/4, -1/6, -1/1\} = -1/6,\ q = 2$.
3. Add $-1/6$ times of row 4 to the bottom row.
4. $\max\{-2, 4, 5\} = 5 > 0,\ p = 3$.
6. Multiply row 3 by $1/5$, and add $2, -4 - 6$ times of row 3 to rows 1,2,4, respectively:

x_1	x_2	x_3	x_4	x_5	x_6	x_7	x_8	RHS
38/5		19/5	12/5	1		2/5		22/5
−26/5		−13/5	−44/5		1	−4/5		−44/5
4/5	1	2/5	1/5			1/5		6/5
−4/5		−32/5	−46/5			−6/5	1	−31/5
1/3		2/3	4/3				5/6	5/6

Iteration 2:

1. $J = \{8\} \neq \emptyset$.
2. $\beta = \max\{-(5/6)/1\} = -5/6,\ q = 8$.
3. Add $-5/6$ times of row 4 to the bottom row.

x_1	x_2	x_3	x_4	x_5	x_6	x_7	x_8	RHS
38/5		19/5	12/5	1		2/5		22/5
−26/5		−13/5	−44/5		1	−4/5		−44/5
4/5	1	2/5	1/5			1/5		6/5
−4/5		−32/5	−46/5			−6/5	1	−31/5
1		6	9			1		6

4. $\max\{0, 0, 0\} \leq 0$.
5. Delete the fourth row and x_8 column:

x_1	x_2	x_3	x_4	x_5	x_6	x_7	RHS
38/5		19/5	12/5	1		2/5	22/5
−26/5		−13/5	−44/5		1	−4/5	−44/5*
4/5	1	2/5	1/5			1/5	6/5
1		6	9			1	6

Iteration 3:
Take row $r = 2$ as datum row. Multiply the row by $-5/44$, and add $-22/5, -6/5$ times of it to rows 1,3, respectively, resulting in a dual feasible D-reduced tableau of the original problem:

x_1	x_2	x_3	x_4	x_5	x_6	x_7	RHS
5*		5/2	−2	1	1/2		
13/22		13/44	1		−5/44	1/11	1
1/11	1	1/22	−1		3/22*	1/11	
1		6	9			1	6

Dual Phase-II: Call Algorithm 17.2.1. $r = 2$.

Iteration 4:

1. $J = \{1, 3, 4, 7\} \neq \emptyset$.
2. $\beta = \max\{-1/(13/22), -6/(13/44), -9/1, -1/(1/11)\} = -1/(13/22) \neq 0$, $q = 1$.
3. Add $-22/13$ times of row 2 to the bottom row.
4. $\max\{5, 1/11\} = 5 > 0,\ p = 1$.
6. Multiply row 1 by $1/5$, and add $-13/22, -1/11$ times of row 1 to rows 2,3, respectively:

x_1	x_2	x_3	x_4	x_5	x_6	x_7	RHS
1		1/2	−2/5	1/5	1/10		
			68/55	−13/110	−19/110	1/11	1
	1		−53/55	−1/55	7/55	1/11	
		11/2	95/13		5/26	11/13	56/13

Iteration 5:

1. $J = \{4, 7\} \neq \emptyset$.
2. $\beta = \max\{-(95/13)/(68/55), -(11/13)/(1/11)\} = -(95/13)/(68/55)$, q=4.

3. Add $-(95/13)/(68/55)$ row 2 to the bottom row:

x_1	x_2	x_3	x_4	x_5	x_6	x_7	RHS
1		1/2	−2/5	1/5	1/10		
			68/55	−13/110	−19/110	1/11	1
	1		−53/55	−1/55	7/55	1/11	
		11/2		95/136	165/136	21/68	−109/68

5. Take $\bar{x}_4 = 55/68$. Basic optimal solution and according objective value are

$$\bar{x} = (11/34, 53/68, 0, 55/68, 0, 7, 1)^{\mathrm{T}}, \quad \bar{f} = 109/68.$$

17.5 D-Reduced Simplex Method

Starting from a feasible D-reduced tableau, the so-called "D-reduced simplex algorithm", described in this section, proceeds toward dual feasibility, iteration by iteration, while maintaining primal feasibility.

Let (17.3) be a feasible D-reduced tableau with column q as datum column, satisfying conditions (17.6) and (17.9). Column pivoting is carried out first.

Rule 17.5.1 (Column rule) Select column index

$$q' \in \arg\min\{\bar{z}_j \mid j \in N,\ j \neq q\}. \tag{17.23}$$

If $\bar{z}_{q'} \geq 0$, it is clear that optimality is already achieved. Now assume that

$$\bar{z}_{q'} < 0. \tag{17.24}$$

According to whether

$$\bar{a}_{iq'} \leq 0, \quad i \in R \tag{17.25}$$

holds or not, one of the following two cases is handled:

(i) Condition (17.25) holds.

If $\bar{\omega}_{q'} \leq 0$, the problem is lower unbounded (Lemma 17.1.4); If $\bar{\omega}_{q'} > 0$, then set $q = q'$, compute

$$\beta = -\bar{z}_q/\bar{\omega}_q. \tag{17.26}$$

and add β times of the datum row to the bottom row, so that $\bar{z}_q$ is converted to zero. Taking column q as the new datum column, it is clear that the D-reduced tableau is still feasible, thus an iteration is complete. As $\beta > 0$, the objective value $\bar{f}$, associated with the tableau, strictly decreases.

(ii) Condition (17.25) does not hold.
Then the following rule is well-defined:

Rule 17.5.2 (Row rule) Determine row index p such that

$$\alpha = -\bar{a}_{pq}/\bar{a}_{pq'} = \min\{-\bar{a}_{iq}/\bar{a}_{iq'} \mid \bar{a}_{iq'} > 0,\ i \in R\}. \tag{17.27}$$

It is known from (17.9) and (17.27) that

$$\bar{a}_{pq} \leq 0, \quad \bar{a}_{pq'} > 0. \tag{17.28}$$

Therefore, $\alpha \geq 0$. When the strict inequality in (17.9) holds, the D-reduced tableau is said to be *nondegenerate*. In this case, $\alpha > 0$.

Lemma 17.5.1. *Assume that (17.24), (17.6) and (17.9) hold, and q' and p are determined by (17.23) and (17.27), respectively. Then the problem is lower unbounded if*

$$\bar{\omega}_{q'} < 0, \quad \alpha \geq -\bar{\omega}_q/\bar{\omega}_{q'}. \tag{17.29}$$

Proof. Multiply the datum row of the D-reduced tableau by $1/\bar{\omega}_q$, then for $i \in R$, add $-\bar{a}_{iq}$ times of the datum row to the ith row. The resulting is a (conventional) feasible simplex tableau. The q' column's component at the datum row of this tableau is

$$\tilde{\omega}_{q'} = \bar{\omega}_{q'}/\bar{\omega}_q < 0, \tag{17.30}$$

where the inequality comes from the first expression of (17.6) and of (17.29). Besides, the other components of q' column are

$$\tilde{a}_{iq'} = \bar{a}_{iq'} - \bar{a}_{iq}(\bar{\omega}_{q'}/\bar{\omega}_q), \quad i \in R. \tag{17.31}$$

If $\bar{a}_{iq'} \leq 0,\ i \in R$, then from (17.9), (17.30) and (17.31), it is known that $\tilde{a}_{iq'} \leq 0$; and if $\bar{a}_{iq'} > 0,\ i \in R$, from the seconde expression of (17.29) and (17.27), it follows that

$$-\bar{a}_{iq}/\bar{a}_{iq'} \geq -\bar{\omega}_q/\bar{\omega}_{q'} > 0, \quad i \in R, \tag{17.32}$$

where the strict inequality is from the first expression of (17.6) and of (17.29). From $\bar{a}_{iq'} > 0$ and (17.29), it follows that $\bar{a}_{iq} < 0$ and

$$-\bar{a}_{iq'}/\bar{a}_{iq} \leq -\bar{\omega}_{q'}/\bar{\omega}_q,$$

multiply which by $-\bar{a}_{iq} > 0$ and combining the resulting expression with (17.31) gives $\tilde{a}_{iq'} \leq 0$. Thus, it was shown that the q' column of the feasible simplex tableau is less than or equals zero. By Lemma 3.2.2, the problem is lower unbounded. □

Now assume that (17.29) does not hold. Execute elementary transformations on the feasible D-reduced tableau to convert $\bar{a}_{pq'}$ to 1, and eliminate the other nonzeros in the column (including those in the bottom row). Thus, x_{j_p} leaves and $x_{q'}$ enters the basis. As $p \in R$, the objective value $\bar{f}$ remains unchanged. If the q-indexed entry of the resulting bottom row is nonzero, add appropriate times of the datum row to the bottom row to eliminate it. For the resulting tableau, the following result stands good.

Proposition 17.5.1. *Assume that (17.24), (17.6) and (17.9) hold, and q' and p are determined by (17.23) and (17.27), respectively. If $\bar{\omega}_{q'} \geq 0$, or*

$$\bar{\omega}_{q'} < 0, \quad \alpha < -\bar{\omega}_q/\bar{\omega}_{q'}, \tag{17.33}$$

then the new tableau is a feasible D-reduced tableau, with column q as the datum column. The according objective value never increases, and strictly decreases under nondegeneracy assumption.

Proof. Mark entries of the new tableau by ˆ. Validity of the following inequality will be shown first:

$$\hat{a}_{iq} \leq 0, \quad i \in R. \tag{17.34}$$

The p ($\neq r$)th component of the q column of the new tableau is

$$\hat{a}_{pq} = \bar{a}_{pq}/\bar{a}_{pq'} \leq 0, \tag{17.35}$$

where the inequality follows from (17.9) and the first expression of (17.28). The other components of the column are

$$\hat{a}_{iq} = \bar{a}_{iq} - \bar{a}_{iq'}(\bar{a}_{pq}/\bar{a}_{pq'}), \quad i \in R,\ i \neq p. \tag{17.36}$$

For $i \in R, i \neq p$, if $\bar{a}_{iq'} \leq 0$, then by (17.9), (17.35) and (17.36), it is known that $\hat{a}_{iq} \leq 0$; and if $\bar{a}_{iq'} > 0$, by (17.27) and (17.36), it is known that $\hat{a}_{iq} \leq 0$. Therefore, (17.34) hold.

On the other hand, the q column's component at the datum row of the new tableau is

$$\hat{\omega}_q = \bar{\omega}_q - \bar{\omega}_{q'}(\bar{a}_{pq}/\bar{a}_{pq'}). \tag{17.37}$$

If $\bar{\omega}_{q'} \geq 0$, from the first expression of (17.6), (17.35) and (17.37), it follows that

$$\hat{\omega}_q > 0. \tag{17.38}$$

If, otherwise, (17.33) holds, from the second expression of (17.6) and (17.27), it follows that

$$-\bar{a}_{pq}/\bar{a}_{pq'} < -\bar{\omega}_q/\bar{\omega}_{q'}.$$

Multiplying the preceding by $\bar{\omega}_{q'} < 0$ gives

$$-\bar{\omega}_{q'}(\bar{a}_{pq}/\bar{a}_{pq'}) > -\bar{\omega}_q,$$

combining which and (17.37) leads to (17.38).

Note that $\bar{z}_q = 0$. After a basis change by taking $\bar{a}_{pq'}$ as pivot, the bottom row's entry at the q column equals

$$\nu = \bar{z}_{q'}\alpha \leq 0.$$

where the inequality follows from (17.27) and (17.35). If $\nu = 0$, then it is known by (17.34) and (17.38) that resulting is a feasible D-reduced tableau with q column as the datum one, and the according objective value does not increase. If nondegeneracy is assumed, then $\alpha > 0$ and $\nu < 0$. Add $-\nu/\hat{\omega}_q$ times of the datum row of the current tableau to the bottom row to eliminate its entry at the q column, the right-hand side entry at the bottom row becomes

$$-(\bar{f} + \nu/\hat{\omega}_q).$$

Thus, from $\nu < 0$ and $\hat{\omega}_q > 0$, it is known that the according objective value strictly decreases. □

The preceding steps are repeated until an optimal tableau obtained or lower unboundedness of the problems detected. The overall steps are put into the following algorithm.

Algorithm 17.5.1 (D-reduced simplex algorithm: tableau form). Initial: feasible D-reduced tableau of form (17.3) with q-indexed column as the datum one. This algorithm solves D-reduced problem (17.1).

1. Select column index $q' \in \arg\min\{\bar{z}_j \mid j \in N,\ j \neq q\}$.
2. If $\bar{z}_{q'} \geq 0$, set $\bar{x}_q = 1/\bar{\omega}_q,\ \bar{x}_B = -\bar{x}_q\bar{a}_q(R),\ i \in R$, and stop.
3. Go to step 6 if $I = \{i \in R \mid \bar{a}_{i\,q'} > 0\} \neq \emptyset$.
4. Stop if $\bar{\omega}_{q'} \leq 0$.
5. Set $q = q'$, and go to step 9.
6. Determine α and row index p such that $\alpha = -\bar{a}_{p\,q}/\bar{a}_{p\,q'} = \min\{-\bar{a}_{i\,q}/\bar{a}_{i\,q'} \mid \bar{a}_{i\,q'} > 0,\ i \in R\}$.
7. Stop if $\bar{\omega}_{q'} < 0$ and $\alpha \geq -\bar{\omega}_q/\bar{\omega}_{q'}$.
8. Convert $\bar{a}_{p\,q'}$ to 1 and eliminate the other nonzeros in the column by elementary transformations.
9. Add $\beta = -\bar{z}_q/\bar{\omega}_q$ times of the datum row to the bottom row.
10. Go to step 1.

Theorem 17.5.1. *Under nondegeneracy assumption, Algorithm 17.5.1 terminates either at*

(i) Step 2, giving a basic optimal solution; or at
(ii) Step 4 or 7, detecting lower unboundedness of the problem.

Proof. The proof of the termination of the Algorithm under nondegeneracy assumption is the same as that in the conventional context. Meanings of its exits follow from Lemmas 17.1.3, 17.1.4 and 17.5.1, as well as the discussions preceding the Algorithm. □

Example 17.5.1. Solve the following problem by Algorithm 17.5.1:

$$\begin{aligned} \min\ & 5x_1 - 4x_2 + 7x_3 - x_4, \\ \text{s.t.}\ & -3x_1 - 2x_2 + 5x_3 - 4x_4 + x_5 = 4, \\ & -6x_1 - 4x_2 + 3x_3 + 2x_4 + x_6 = 2, \\ & x_1 + 5x_2 - 4x_3 + x_4 + x_7 = 3, \\ & x_j \geq 0, \quad j = 1, \cdots, 7. \end{aligned}$$

Answer Initial tableau

x_1	x_2	x_3	x_4	x_5	x_6	x_7	RHS
−3	−2	5	−4	1			4*
−6	−4	3	2		1		2
1	5	−4	1			1	3
5	−4	7	−1				

Iteration 1:
$\max\{|4|, |2|, |3|\} = 4,\ r = 1$, Take the first row as pivot row. Multiply the row by $1/4$, and add $-2, -3$ times of it to rows 2,3, respectively:

x_1	x_2	x_3	x_4	x_5	x_6	x_7	RHS
−3/4	−1/2	5/4	−1	1/4			1
−9/2	−3	1/2	4	−1/2	1		
13/4	13/2*	−31/4	4	−3/4		1	
5	−4	7	−1				

x_5 becomes a nonbasis variable. For the tableau, $\bar{a}_{1,5} = 1/4 > 0,\ \bar{a}_{2,5} = -1/2 \leq 0,\ \bar{a}_{3,5} = -3/4 \leq 0$, hence it is a feasible D-reduced tableau.

Call Algorithm 17.5.1. $r = 1,\ q = 5$.

Iteration 2:

1. $\min\{5, -4, 7, -1\} = -4 < 0,\ q' = 2$.
3. $I \quad = \{3\} \neq \emptyset$.
6. $\alpha \quad = \min\{-(-3/4)/(13/2)\} = 3/26, p = 3$.
7. $\bar{a}_{1,2} = -1/2 < 0,\ \alpha = 3/26 < -(1/4)/(-1/2) = 1/2$.

8. Multiply row 3 by $2/13$, and add $1/2, 3, 4$ times of row 3 to rows 1,2,4, respectively:

x_1	x_2	x_3	x_4	x_5	x_6	x_7	RHS
$-1/2$		$17/26$	$-9/13$	$5/26$		$1/13$	1
-3		$-40/13$	$76/13$	$-11/13$	1	$6/13$	
$1/2$	1	$-31/26$	$8/13$	$-3/26$		$2/13$	
7		$29/13$	$19/13$	$-6/13$		$8/13$	

9. Add $\beta = -(-6/13)/(5/26)$ times of row 1 to the bottom row:

x_1	x_2	x_3	x_4	x_5	x_6	x_7	RHS
$-1/2$		$17/26$	$-9/13$	$5/26$		$1/13$	1
-3		$-40/13$	$76/13$*	$-11/13$	1	$6/13$	
$1/2$	1	$-31/26$	$8/13$	$-3/26$		$2/13$	
$29/5$		$19/5$	$-1/5$			$4/5$	$12/5$

Iteration 3:

1. $\min\{29/5, 19/5, -1/5, 4/5\} = -1/5 < 0,\ q' = 4.$
3. $I \;\; = \{2, 3\} \neq \emptyset.$
6. $\alpha \;\; = \min\{-(-11/13)/(76/13), -(-3/26)/(8/13)\}$
   $= -(-11/13)/(76/13) = 11/76,\ p = 2.$
7. $\bar{a}_{1,4} = -9/13 < 0, \alpha = 11/76 < -(5/26)/(-9/13) = 5/18.$
8. Multiply row 2 by $13/76$, and add $9/13, -8/13, 1/5$ times of row 2 to rows 1,3,4, respectively:

x_1	x_2	x_3	x_4	x_5	x_6	x_7	RHS
$-65/76$		$11/38$		$7/76$	$9/76$	$5/38$	1
$-39/76$		$-10/19$	1	$-11/76$	$13/76$	$3/38$	
$31/38$	1	$-33/38$		$-1/38$	$-2/19$	$2/19$	
$433/76$		$351/95$		$-11/380$	$13/380$	$31/38$	$12/5$

9. Add $\beta = -(-11/380)/(7/76)$ times of row 1 to the bottom row.

x_1	x_2	x_3	x_4	x_5	x_6	x_7	RHS
$-65/76$		$11/38$		$7/76$	$9/76$	$5/38$	1
$-39/76$		$-10/19$	1	$-11/76$	$13/76$	$3/38$	
$31/38$	1	$-33/38$		$-1/38$	$-2/19$	$2/19$	
$38/7$		$53/14$			$1/14$	$6/7$	$19/7$

Iteration 4:

1. $\min\{38/7, 53/14, 1/14, 6/7\} \geq 0$.
2. Set $\bar{x}_5 = 76/7$. Basic optimal solution and according objective value:

$$\bar{x} = (0, 2/7, 0, 11/7, 76/7, 0, 0, 1)^{\mathrm{T}}, \quad \bar{f} = -19/7.$$

A revised version of Algorithm 17.5.1 can be obtained by deriving the according search direction using the inverse of the basis matrix, as done for derivation of the (conventional) revised simplex algorithm. But, for simplicity, here we directly apply the equivalence between the D-reduced tableau (17.3) and revised tableau (17.19).

Denoting a_q's subvector, associated with R, by $a_q(R)$, and so on, a revised version of Algorithm 17.5.1 can be described as follows.

Algorithm 17.5.2 (D-reduced simplex algorithm). Initial: B, R, N defined by (17.2). B_R^{-1}; $\bar{z}_N = c_N - N_R^{\mathrm{T}} B_R^{-\mathrm{T}} c_B$, $\bar{\omega}_N = N^{\mathrm{T}} e_m - N_R^{\mathrm{T}}(B_R^{-\mathrm{T}} B e_m)$. $\bar{a}_q(R) = B_R^{-1} a_q(R) \leq 0$, $\bar{\omega}_q > 0$, $\bar{z}_q = 0$. This algorithm solves the D-reduced problem (17.1).

1. Determine column index $q' \in \arg\min\{\bar{z}_j \mid j \in N,\ j \neq q\}$.
2. If $\bar{z}_{q'} \geq 0$, compute

$$\bar{x}_q = 1/\bar{\omega}_q, \ \bar{x}_B = -\bar{x}_q \bar{a}_q(R), \qquad \bar{f} = c_q \bar{x}_q + c_B^{\mathrm{T}} \bar{x}_B,$$

3. Stop (optimality achieved).
4. Compute $\bar{a}_{q'}(R) = B_R^{-1} a_{q'}(R)$.
5. Go to step 7 if $I = \{i \in R \mid \bar{a}_{i\,q} > 0\} \neq \emptyset$.
6. Stop if $\bar{\omega}_{q'} \leq 0$ (infeasible problem).
7. Set $q = q'$, and go to step 12.
8. Determine α and p such that $\alpha = -\bar{a}_{p\,q}/\bar{a}_{p\,q'} = \min_{i \in I} -\bar{a}_{i\,q}/\bar{a}_{i\,q'}$ $(p = i_s)$.
9. Stop if $\bar{\omega}_{q'} < 0$ and $\alpha \geq -\bar{\omega}_q/\bar{\omega}_{q'}$ (unbounded problem).
10. Compute $\sigma_N = N_R^{\mathrm{T}} B_R^{-\mathrm{T}} e_s$.
11. Update $\bar{\omega}_N = \bar{\omega}_N + \nu \sigma_N$, $\bar{\omega}_{j_p} = \nu$, where $\nu = -\bar{\omega}_{q'}/\sigma_{q'}$.
12. Update $\bar{z}_N = \bar{z}_N + \beta_1 \sigma_N$, $\bar{z}_{j_p} = \beta_1$, where $\beta_1 = -\bar{z}_{q'}/\sigma_{q'}$.
13. Update $\bar{z}_N = \bar{z}_N + \beta_2 \bar{\omega}_N$, where $\beta_2 = -\bar{z}_q/\bar{\omega}_q$.
14. Update B_R^{-1} by (17.20).
15. Update (B, N) by exchanging j_s and q.
16. Go to step 1.

It has not been known at present how the method, derived in this section, performs, though it seems to be inferior to the dual D-reduced simplex method (Sect. 17.2).

17.6 D-Reduced Phase-I: The Most-Obtuse-Angle

As usual, the D-reduced simplex method, described in the previous section, is actually only a Phase-II procedure. Any conventional Phase-I method can be sightly modified to generate a start point (Chap. 13). Dropping the initial condition $\bar{z}_N \geq 0$, in addition, the dual Algorithm 17.2.1 can be directly used to achieve primal feasibility.

According to the analysis made at the last half of Sect. 17.2, the reduce gradient of the dual objective function corresponds to the negative datum row in z-space. Based on the most-obtuse-angle heuristics, the dual nonnegativity constraint (e.g., $z_q \geq 0$), whose gradient and the negative datum row form the most obtuse angle, should be binding, that is, the according index be chosen to enter the basis. Based on this idea, the following algorithm can be obtained by modifying the dual Algorithm 17.2.1.

Algorithm 17.6.1 (Tableau D-reduced Phase-I: the most-obtuse-angle). Initial: D-reduced tableau of form (17.3). This algorithm finds a feasible D-reduced tableau.

1. Determine column index $q \in \arg\max_{j \in N} \bar{\omega}_j$.
2. Stop if $\bar{\omega}_q \leq 0$.
3. Add $\beta = -\bar{z}_q/\bar{\omega}_q$ times of the datum row to the bottom row.
4. Determine row index $p \in \arg\max\{\bar{a}_{iq} \mid i \in R\}$.
5. Stop if $\bar{a}_{p,q} \leq 0$.
6. Convert $\bar{a}_{p\,q}$ to 1, and eliminate the other nonzeros in the column by elementary transformations.
7. Go to step 1.

Theorem 17.6.1. *Assume termination of Algorithm 17.6.1. It terminates at*

(i) Step 2, detecting infeasibility of the problem; or at
(ii) Step 5, giving a feasible D-reduced tableau.

Proof. It follows from Lemmas 17.1.1 and 17.2.1. □

Example 17.6.1. Solve following problem using Algorithm 17.6.1 as Phase-I:

$$
\begin{array}{ll}
\min & f = -x_1 + 5x_2 - 3x_3 - x_4, \\
\text{s.t.} & \quad x_1 + 1x_2 + 2x_3 + 5x_4 + x_5 \qquad\qquad\qquad = \ \ 3, \\
& -7x_1 + 3x_2 + 4x_3 - \ x_4 \qquad + x_6 \qquad\qquad = -5, \\
& -2x_1 - \ x_2 + \ x_3 - \ x_4 \qquad\qquad + x_7 \qquad = -2, \\
& -5x_1 - 6x_2 + 8x_3 - 2x_4 \qquad\qquad\qquad + x_8 = \ \ 6, \\
& \qquad\qquad\qquad x_j \geq 0, \quad j = 1, \cdots, 8.
\end{array}
$$

Answer Initial tableau

x_1	x_2	x_3	x_4	x_5	x_6	x_7	x_8	RHS
1	1	2	5	1				3*
−7	3	4	−1		1			−5
−2	−1	1	−1			1		−2
−5	−6	8	−2				1	6
−1	5	−3	−1					

Iteration 1:
Arbitrarily take the first row to be as the datum row. Multiply the row by $1/3$, and add $5, 2, -6$ times of it to rows 2,3,4, respectively, yielding the following D-reduced tableau:

x_1	x_2	x_3	x_4	x_5	x_6	x_7	x_8	RHS
1/3	1/3	2/3	5/3	1/3				1
−16/3	14/3	22/3	22/3*	5/3	1			
−4/3	−1/3	7/3	7/3	2/3		1		
−7	−8	4	−12	−2			1	
−1	5	−3	−1	1/5				

Phase-I: Call Algorithm 17.6.1. $r = 1$.

Iteration 2:

1. $\max\{1/3, 1/3, 2/3, 5/3, 1/3\} = 5/3 > 0,\ q = 4$.
3. Add $\beta = -(-1)/(5/3) = 3/5$ times of row 1 to the bottom row.
4. $\max\{22/3, 7/3, -12\} = 22/3 > 0,\ p = 2$.
6. Multiply row 2 by $3/22$, and add $-5/3, -7/3, 12$ times of row 2 to rows 1,3,4, respectively:

x_1	x_2	x_3	x_4	x_5	x_6	x_7	x_8	RHS
17/11	−8/11	−1		−1/22	−5/22			1
−8/11	7/11	1	1	5/22	3/22			
4/11*	−20/11			3/22	−7/22	1		
−173/11	−4/11	16		8/11	18/11		1	
−4/5	26/5	−13/5		1/5				3/5

Iteration 3:

1. $\max\{17/11, -8/11, -1, -1/22, -5/22\} = 17/11 > 0,\ q = 1$.
3. Add $\beta = -(-4/5)/(17/11)$ times of row 1 to the bottom row.
4. $\max\{-8/11, 4/11, -173/11\} = 4/11 > 0\ , p = 3$.

6. Multiply row 3 by $11/4$, and add $-17/11, 8/11, 173/11$ times of row 3 to rows 1,2,4, respectively:

x_1	x_2	x_3	x_4	x_5	x_6	x_7	x_8	RHS
	7	−1		−5/8	9/8	−17/4		1
	−3	1	1	1/2	−1/2	2		
1	−5			3/8	−7/8	11/4		
	−79	16		53/8	−97/8	173/4	1	
	82/17	−53/17		3/17	−2/17			19/17

Iteration 4:

1. $\max\{7, -1, -1, -5/8, 9/8, -17/4\} = 7 > 0,\ q = 2.$
3. Add $\beta = -(82/17)/7$ times of row 1 to the bottom row:

x_1	x_2	x_3	x_4	x_5	x_6	x_7	x_8	RHS
	7	−1		−5/8	9/8	−17/4		1
	−3	1*	1	1/2	−1/2	2		
1	−5			3/8	−7/8	11/4		
	−79	16		53/8	−97/8	173/4	1	
		−17/7		17/28	−25/28	41/14		3/7

4. $\max\{-3, -5, -79\} \leq 0.$
5. Feasibility is achieved.

Phase-II: Call Algorithm 17.5.1. $r = 1,\ q = 2.$

Iteration 5:

1. $\min\{-17/7, 17/28, -25/28, 41/14\} = -17/7 < 0,\ q' = 3.$
3. $I \quad = \{2, 4\} \neq \emptyset.$
6. $\alpha \quad = \min\{-(-3)/1, -(-79/16)\} = 3,\ p = 2.$
7. $\bar{a}_{1,3} = -1 < 0,\ \alpha = 3 < -7/(-1) = 7.$
8. Add $1, -6, 17/7$ times of row 2 to rows 1,4,5, respectively.
9. Add $-(-51/7)/4$ times of row 1 to the bottom row:

x_1	x_2	x_3	x_4	x_5	x_6	x_7	x_8	RHS
	4		1	−1/8	5/8	−9/4		1
	−3	1	1	1/2	−1/2	2		
1	−5			3/8	−7/8	11/4		
	−31		−16	−11/8	−33/8	45/4	1	
			17/4	51/32	−31/32	59/16		9/4

Iteration 6:

1. $\min\{17/4, 51/32, -31/32, 59/16\} = -31/32 < 0,\ q' = 6.$
3. $I = \emptyset.$
4. $5/8 > 0.$
5. $q = 6.$
9. Add $-(-31/32)/(5/8)$ times of row 1 to the bottom row:

x_1	x_2	x_3	x_4	x_5	x_6	x_7	x_8	RHS
	4		1	−1/8	5/8	−9/4		1
	−3	1	1	1/2	−1/2	2		
1	−5			3/8	−7/8	11/4		
	−31		−16	−11/8	−33/8	45/4	1	
	31/5		29/5	7/5		1/5		19/5

Iteration 7:

1. $\min\{31/5, 29/5, 7/5, 1/5\} \geq 0.$
2. Basic optimal solution and according objective value are

$$\bar{x} = (7/5, 0, 4/5, 0, 0, 8/5, 0, 33/5)^{\mathrm{T}}, \quad \bar{f} = -19/5.$$

Based on equivalence between the reduced tableau (17.3) and the revised tableau (17.19), a revised version of Algorithm 17.6.1 can be stated, as matches Algorithm 17.5.2.

Algorithm 17.6.2 (D-reduced Phase-I). Initial: B, R, N defined by (17.2). B_R^{-1}; $\bar{z}_N = c_N - N_R^{\mathrm{T}} B_R^{-\mathrm{T}} c_B \geq 0$. This algorithm finds a feasible D-reduced basis.

1. Compute $\bar{\omega}_N = N^{\mathrm{T}} e_m - N_R^{\mathrm{T}}(B_R^{-\mathrm{T}} B e_m)$.
2. Determine column index $q \in \arg\max_{j \in N} \bar{\omega}_j$.
3. Stop if $\bar{\omega}_q \leq 0$ (infeasible problem).
4. Update $\bar{z}_N = \bar{z}_N + \beta \bar{\omega}_N$, where $\beta = -\bar{z}_q / \bar{\omega}_q$.
5. Compute $\bar{a}_q(R) = B_R^{-1} a_q(R)$.
6. Determine $p \in \arg\max_{i \in R} \bar{a}_{i\,q}$ $(p = i_s)$.
7. Stop if $\bar{a}_{p,q} \leq 0$ (feasible D-reduced basis B_R obtained).
8. Update B_R^{-1} by (17.20).
9. Update (B, N) by exchanging j_s and q.
10. Go to step 1.

17.7 Dual Bisection D-Reduced Simplex Method

In Sect. 16.5, the bisection reduced simplex algorithm is described, where an existence interval $[\lambda, \mu]$, containing the optimal value, is contracted by at least a half of the length in each iteration, until reaching an optimal solution. Following the same train of thought, Shen and Pan (2006) proposed a dual bisection simplex algorithm. In this section, we will present a variant based on the D-reduced simplex framework fitly.

Let $[\lambda, \mu]$ be an existence interval. If a dual feasible solution is found with objective value $\bar{f} = (\lambda + \mu)/2$, the value is switched up to one associated with a dual basic feasible solution; if the latter does not satisfy the optimality condition, the according objective value is a new lower bound on the optimal value, and hence set it to λ. If there exists no dual feasible solution with objective value $\bar{f} = (\lambda + \mu)/2$, the value must be primal feasible. The value is switched down to one associated with a dual basic solution; if the latter does not satisfy the optimality condition, the according value is a new upper bound on the optimal value, and the value is then set to μ. Consequently, the resulting $[\lambda, \mu]$ is again an existence interval, contracted by a half of its length, at least.

The contracting procedure is realized by following algorithm.

Algorithm 17.7.1 (Dual subalgorithm: tableau form). Initial: D-reduced tableau of form (17.3). This algorithm finds a dual feasible D-reduced tableau with objective value $\bar{f}$, and does switching.

1. Determine column index $q \in \arg\min_{j \in N} \bar{z}_j$.
2. Go to step 7 if $\bar{z}_q < 0$.
3. Determine β and column index q such that $\beta = -\bar{z}_q/\bar{\omega}_q = \max\{-\bar{z}_j/\bar{\omega}_j \mid \bar{\omega}_j > 0,\ j \in N\}$.
4. If $\beta \neq 0$, add β times of row r to the bottom row.
5. Return if $\bar{a}_{i\,q} \leq 0, i \in R$ (optimality achieved).
6. Set $\lambda = \bar{f}$, and return.
7. Determine row index $p \in \arg\max_{i \in R} \bar{a}_{i\,q}$.
8. Go to step 12 if $\bar{a}_{p\,q} > 0$.
9. Add $\beta = -\bar{z}_q/\bar{\omega}_q$ times of row r to the bottom row.
10. Return if $\bar{z}_N \geq 0$ (optimality achieved).
11. Set $\mu = \bar{f}$, and return.
12. Convert $\bar{a}_{p\,q}$ to 1, and eliminate the other nonzeros in the column by elementary transformations.
13. Go to step 1.

Theorem 17.7.1. *Given an existence interval $[\lambda, \mu]$, and $\bar{f} = (\lambda + \mu)/2$. Algorithm 17.7.1 returns from*

(i) Step 5 or 10, gives a basic optimal solution; or from
(ii) Step 6, gives a new left end of the existence interval; of from
(iii) Step 11, gives a new right end of it.

In one of cases (ii) and (iii), the existence interval contracts by a half of length, at least.

Proof. It is noted that there exists an optimal solution to the problem under the assumption, and $\bar{f}$ is either primal or dual feasible if it is not optimal. If returning from step 5, then not only dual feasibility is achieved, but for q, determined at step 3, it holds that $\bar{\omega}_q > 0$ and

$$\bar{a}_{iq} \leq 0, \quad i \in R. \tag{17.39}$$

thus primal feasibility is also achieved, therefore so is optimality. When returning from step 6, dual feasibility is achieved, hence $\bar{f}$ is a new lower bound of the optimal value. If returning from step 10, then optimality is attained because in addition to dual feasibility achieved, from $\bar{z}_q < 0$ and (17.39) and Lemma 17.1.5, it is asserted that primal feasibility is achieved. In the case when returning from step 11, primal feasibility is achieved and $\bar{f}$ is a new upper bound of the optimal value. It is clear that the resulting existence interval contracts by a half of the length, at least. □

The master algorithm uses the D-reduced Phase-I Algorithm 17.6.1 to produce an upper bound μ of the optimal value, and the dual D-reduced Phase-I Algorithm (Sect. 17.4) to produce a lower bound λ. Then it call Subalgorithm 17.7.1 in each iteration to bisection the interval, until optimality is achieved. The overall steps are put into the following algorithm.

Algorithm 17.7.2 (Bisection D-reduced simplex algorithm: tableau form). Initial: D-reduced tableau of form (17.3). This algorithm solves the D-reduced problem (17.1).

1. Call Algorithm 17.6.1 to pursue primal feasibility:
   (1) Stop if returning from (2) (infeasible problem);
   (2) Returning from (5): if $\bar{z}_N < 0$, set $\mu = \bar{f}$; else stop (optimality achieved).
2. Call Algorithm 17.3.1 to pursue dual feasibility:
   (1) Stop if returning from step 6 (infeasible or lower unbounded problem);
   (2) If returning from step 2, determine β and column index q such that

$$\beta = -\bar{z}_q/\bar{\omega}_q = \max\{-\bar{z}_j/\bar{\omega}_j \mid \bar{\omega}_j > 0, \ j \in N\}.$$

3. If $\beta \neq 0$, add β times of the datum row to the bottom row.
4. Stop if $\bar{\omega}_q > 0$; $\bar{a}_{iq} \leq 0, i \in R$ (optimality achieved).
5. Set $\lambda = \bar{f}$.
6. Stop if $\lambda = \mu$ (optimality achieved).
7. Add $\beta = (\lambda - \mu)/2$ times of the datum row to the bottom row.

8. Call Algorithm 17.7.1.
   (1) If returning from step 6, set $\beta = (\lambda - \mu)/2$ and go to step 7;
   (2) If returning from step 11, set $\beta = (\mu - \lambda)/2$ and go to step 7;
   (3) Stop if returning from step 5 or 10 (optimality achieved).

Proofs about meanings of the exits of the preceding algorithm are omitted (interested readers is referred to Yuan Shen and P.-Q. Pan 2006)

Example 17.7.1. Solve following problem by Algorithm 17.7.2:

$$
\begin{array}{lrcl}
\min & f = 10/7 - 5x_1 + 7x_2 - 6x_3 + 5x_4, & & \\
\text{s.t.} & 2x_1 + 5x_2 - 3x_3 + 5x_4 + x_5 & = & 1, \\
& -4x_1 - 3x_2 + 8x_3 - 2x_4 \quad + x_6 & = & -1, \\
& x_1 - x_2 + x_3 + 6x_4 \quad + x_7 & = & 6, \\
& -7x_1 + 3x_2 + 4x_3 - 2x_4 \quad + x_8 & = & -2, \\
& x_j \geq 0, \quad j = 1, \cdots, 8. & &
\end{array}
$$

Answer Initial tableau

x_1	x_2	x_3	x_4	x_5	x_6	x_7	x_8	RHS
2	5	−3	5	1				1
−4	−3	8	−2		1			−1
1	−1	1	6			1		6
−7	3	4	−2				1	−2*
−5	7	−6	5					−10/7

Iteration 1:
Arbitrarily take the fourth row as the datum row. Multiply the row by $-1/2$, and add $-1, 1, -6$ times of it to rows 1,2,3, respectively, resulting in D-reduced tableau:

x_1	x_2	x_3	x_4	x_5	x_6	x_7	x_8	RHS
−3/2	13/2	−1	4	1			1/2	
−1/2	−9/2	6	−1		1		−1/2	
−20	8	13				1	3	
7/2	−3/2	−2	1				−1/2	1
−5	7	−6	5					−10/7

Step 1. Call Algorithm 17.6.1 to pursue primal feasibility. $r = 4$.
Iteration 2:

1. $\max\{7/2, -3/2, -2, 1, -1/2\} = 7/2 > 0,\ q = 1$.
3. Add $\beta = -(-5)/(7/2) = 10/7$ times of row 4 to the bottom row.
4. $\max\{-3/2, -1/2, -20\} = -1/2 \leq 0$.
5. Primal feasibility achieved:

x_1	x_2	x_3	x_4	x_5	x_6	x_7	x_8	RHS
−3/2	13/2	−1	4	1			1/2	
−1/2	−9/2	6	−1		1		−1/2	
−20	8	13*				1	3	
7/2	−3/2	−2	1				−1/2	1
	34/7	−62/7	45/7				−5/7	

Step 1(2) $\mu = 0$.
Step 2. Call Algorithm 17.3.1 to pursue dual feasibility: $r = 4$.
Iteration 3:

1. $\min\{0, 34/7, -62/7, 45/7, -5/7\} = -62/7 < 0,\ q = 3$.
3. Add $\beta = -(-62/7)/(-2)$ times of row 4 to the bottom row.
4. $\max\{-1, 6, 13\} = 13 > 0,\ p = 3$.
7. Multiply row 3 by $1/13$, and add $1, -6, 2$ times of row 3 to rows 1,2,4, respectively:

x_1	x_2	x_3	x_4	x_5	x_6	x_7	x_8	RHS
−79/26	185/26		4	1		1/13	19/26	
227/26*	−213/26		−1		1	−6/13	−49/26	
−20/13	8/13	1				1/13	3/13	
11/26	−7/26		1			2/13	−1/26	1
−31/2	23/2		2				3/2	−31/7

Iteration 4:

1. $\min\{-31/2, 23/2, 2, 0, 3/2\} = -31/2 < 0,\ q = 1$.
3. Add $\beta = -(-31/2)/(11/26)$ times of row 4 to the bottom row.
4. $\max\{-79/26, 227/26, -20/13\} = 227/26 > 0,\ p = 2$.
7. Multiply row 2 by $26/227$, and add $79/26, 20/13, -11/26$ times of row 2 to rows 1,3,4, respectively:

x_1	x_2	x_3	x_4	x_5	x_6	x_7	x_8	RHS
	968/227		829/227	1	79/227	−19/227	17/227	
1	−213/227		−26/227		26/227	−12/227	−49/227	
	−188/227	1	−40/227		40/227	−1/227	−23/227	
	29/227		238/227		−11/227	40/227	12/227	1
	18/11		425/11			62/11	1/11	2,480/77

Iteration 5:

1. $\min\{18/11, 425/11, 0, 62/11, 1/11\} \geq 0$.
2. Returned.

Step 2(2) $\beta = \max\{-(18/11)/(29/227), -(425/11)/(238/227), -(62/11)/(40/227), -(1/11)/(12/227)\} = -(1/11)/(12/227)$, $q = 8$.
Step 3. Add $\beta = -(1/11)/(12/227)$ times of row 4 to the bottom row:

x_1	x_2	x_3	x_4	x_5	x_6	x_7	x_8	RHS
	968/227		829/227	1	79/227	−19/227	17/227	
1	−213/227		−26/227		26/227	−12/227	−49/227	
	−188/227	1	−40/227		40/227	−1/227	−23/227	
	29/227		238/227		−11/227	40/227	12/227	1
	17/12		221/6		1/12	16/3		2,561/84

Step 4. $\bar{a}_{4,8} = 12/227 > 0, \bar{a}_{1,8} = 17/227 > 0$.
Step 5. $\lambda = -2{,}561/84$.
Step 7. Add $\beta = (-2{,}561/84 - 0)/2 = -2{,}561/168$ times of row 4 to the bottom row:

x_1	x_2	x_3	x_4	x_5	x_6	x_7	x_8	RHS
	968/227		829/227	1	79/227	−19/227	17/227*	
1	−213/227		−26/227		26/227	−12/227	−49/227	
	−188/227	1	−40/227		40/227	−1/227	−23/227	
	29/227		238/227		−11/227	40/227	12/227	1
	−491/925		1,814/87		3,164/3,849	1,583/598	−1,432/1,777	2,561/168

Step 8. Call Algorithm 17.7.1.
Iteration 6:

1. $\min\{-491/925, 1,814/87, 3,164/3,849, 1,583/598, -1,432/1,777\} = -1{,}432/1{,}777 < 0,\ q = 8$.
7. $\max\{17/227, -49/227, -23/227\} = 17/227 > 0,\ p = 1$.
12. Multiply row 1 by 227/17, and add $49/227, 23/227, -12/227, 1,432/1,777$ times of row 1 to rows 2,3,4, respectively:

x_1	x_2	x_3	x_4	x_5	x_6	x_7	x_8	RHS
	968/17		829/17	227/17	79/17	−19/17	1	
1	193/17		177/17	49/17	19/17	−5/17		
	84/17	1	81/17	23/17	11/17	−2/17		
	−49/17		−26/17	−12/17	−5/17	4/17		1
	5,488/121		5,293/88	2,561/238	717/157	1,247/714		2,561/168

Iteration 7:

1. $\min\{5{,}488/121, 5{,}293/88, 2{,}561/238, 717/157, 1{,}247/714\} \geq 0$.
3. $\beta = \max\{-(1{,}247/714)/(4/17)\}$, $q = 7$.
4. Add $-(1{,}247/714)/(4/17)$ times of row 4 to the bottom row:

x_1	x_2	x_3	x_4	x_5	x_6	x_7	x_8	RHS
	968/17		829/17	227/17	79/17	−19/17	1	
1	193/17		177/17	49/17	19/17	−5/17		
	84/17	1	81/17	23/17	11/17	−2/17		
	−49/17		−26/17	−12/17	−5/17	4/17		1
	267/4		143/2	16	27/4			219/28

5. $\bar{a}_q(R) = (-19/17, -5/17, -2/17)^{\mathrm{T}} \leq 0$. Setting $\bar{x}_8 = 17/4$, the basic optimal solution and according objective value are

$$\bar{x} = (5/4, 0, 1/2, 0, 0, 0, 17/4, 19/4, 1)^{\mathrm{T}}, \quad \bar{f} = -219/28.$$

Chapter 18
Criss-Cross Simplex Method

Methods perform very differently when solving a same problem. It is common that a problem that is solved slowly by the simplex method would be solved fast by the dual simplex method; and vise versa. Consequently, LP packages often include multiple options to be chosen, as it seems to be impossible to predetermine which method would be better to solve a given problem. Pursuing a method with features of both the primal and dual methods, scholars have attempted to combine primal and dual simplex methods for years. A class of resulting variants may be described by "criss-cross" because of their switching between primal and dual iterations. The primal-dual algorithm (Sect. 7.1) and the self-dual parametric simplex algorithm (Sect. 7.2) may be also classified into this category.

Assume that the standard LP problem (1.8) has the following simplex tableau:

x_B^{T}	x_N^{T}	f	RHS
I	$\bar{N}$		$\bar{b}$
	$\bar{z}_N^T$	-1	$-\bar{f}$

(18.1)

Initially, Talacko and Rockefeller (1960), Balinski and Gomory (1963), and Llewellyn (1964) suggested switching between primal and dual steps under a certain mechanism, but not supported by numerical results.

Later, S. Zionts (1969) proposed a so-called "criss-cross" algorithm, carrying out the primal and dual simplex iterations alternately. In a primal iteration, a pair of column and row are selected by the primal pivot rule, then an according basis change is carried out; in a dual iteration, a pair of row and column are selected by the dual pivot rule, then an according basis change is carried out; in case when the minimum-ratio is infinite in row (column) selection, it turns to a dual (primal) iteration, and so on. If no any iteration can be carried out, it is asserted that there exists no optimal solution; otherwise, if primal (dual) feasibility is achieved first, then the primal (dual) simplex algorithm is executed subsequently to achieve optimality. So, the algorithm is actually a primal or dual Phase-I method, depending on which type of feasibility is attained first. Just like the standard simplex algorithm, finiteness of Zionts algorithm is not guaranteed.

P.-Q. PAN, *Linear Programming Computation*, DOI 10.1007/978-3-642-40754-3_18,

A decade later or more, Chang (1979), Terlaky (1985), and Wang (1987) proposed a purely combinatorial criss-cross variant independently. Their algorithm performs a primal or a dual iteration based on the following rule: it determines the smallest among indices, associated with all negative components of $\bar{x}_B = \bar{b}$ and $\bar{z}_N$. If the determined index is associated with $\bar{z}_N$ ($\bar{x}_B = \bar{b}$), then a primal (dual) iterations is carried out. Shown to be finite, the algorithm uses a criteria simpler than that of Zionts algorithm to detect nonexistence of an optimal solution.

Unfortunately, the performance of all the above algorithms are far inferior to the simplex algorithm. Also, Roos (1990) offered an instance for which the number of iterations required by the finite rule is exponential in the number of variables and constraints.

In this chapter, we present efficient or promising criss-cross simplex variants. Like the simplex algorithm, finiteness of them is not guaranteed; even the objective value may not change monotonically in solution process. But it can still be expected that cycling rarely happens in practice.

18.1 Most-Obtuse-Angle Criss-Cross Method

As mentioned previously, Zionts method performs primal and dual iterations alternately. A fault of such doing is that while performing the primal iteration, reduced costs would already be nearly nonnegative while negative components of the right-hand side are large in magnitude, in other words, a dual iteration should be more relevant; and vice versa.

The criss-cross method (Yan and Pan 2009) applies the most-obtuse-angle column rule (Sect. 13.3) and row rule (Sect. 14.3) in the primal and dual iterations, respectively. Therefore, no any minimum-ratio test is involved. It does not mechanically perform primal and dual iterations alternately, but instead, depending on strength of current primal and dual infeasibility. To this end, a predetermined threshold value is used to control switching between the two sides, so that a primal (dual) iteration is taken if the primal (dual) infeasibility is in some sense stronger than dual (primal) infeasibility.

A slight variant of the algorithm is formulated below, which achieves optimality in a single Phase, and involves no threshold.

Algorithm 18.1.1 (Most-obtuse-angle criss-cross algorithm: tableau form). Initial: the simplex tableau of form (18.1). This algorithm solves the standard LP problem.

1. Determine column index $q \in \arg\min_{j\in N} \bar{z}_j$.
2. Determine row index $p \in \arg\min\{\bar{b}_i \mid i = 1,\cdots,m\}$.
3. Compute $\mu = \min\{\bar{z}_q, \bar{b}_p\}$.
4. Stop if $\mu \geq 0$ (optimality achieved).
5. If $\mu = \bar{z}_q$, then:

   (1) redetermine row index $p \in \arg\max\{\bar{a}_{i\,q} \mid i = 1,\cdots,m\}$;
   (2) stop if $\bar{a}_{p\,q} \leq 0$ (infeasible or unbounded problem);

else

(3) redetermine column index $q \in \arg\min_{j \in N} \bar{a}_{pj}$;

(4) stop if $\bar{a}_{pq} \geq 0$ (infeasible problem).

6. Convert $\bar{a}_{pq}$ to 1, and eliminate the other nonzeros in the column by elementary transformations.
7. Go to step 1.

Note As step 3 involves comparison between the right-hand side and reduced costs, it is important to scale data before hand. If the data are not well scaled, it would be better to set some suitable threshold value $\tau > 0$, and use

$$\mu = \min\{\bar{z}_q, \tau \bar{b}_p\}$$

in step 3 instead.

For simplicity of description, we used conventional primal and dual rules in steps 1 and 2 in Algorithm 18.1.1, respectively. In practice, however, it would be much better to use the nested largest-distance Rule 11.6.1 and the dual nested Rule 12.4.1, instead. In addition, although the most-obtuse-angle row and column rules are respectively used in step 5(1) and 5(3), it should be better to utilize their variants, Rules 14.3.2 and 13.3.2.

Cited from Yan and Pan (2009), the following numerical results roughly reflect the behavior of Algorithm 18.1.1. The experiments were preliminary, only involving the 25 smallest standard NETLIB problems (Appendix B: Table B.4: Problems AFIRO-SCAGR25). The code, programmed in Fortran without exploiting sparse structure, was compared with MINOS 5.51 only in terms of the number of required iterations: it outperformed MINOS 5.51 with total iteration ratio 1.24. As there was no minimum-ratio test involved in each iteration by the most-obtuse-angle criss-cross algorithm, the margin between it and the simplex algorithm should be even larger with respect to CPU time.

Example 18.1.1. Solve the following problem by Algorithm 18.1.1:

$$\begin{array}{ll} \min & f = -2x_1 - x_2, \\ \text{s.t.} & x_1 - x_2 + x_3 = 2, \\ & x_1 + 2x_2 + x_4 = 8, \\ & -x_1 - x_2 + x_5 = -3, \\ & x_j \geq 0, \quad j = 1, \cdots, 5. \end{array}$$

Answer Initial simplex tableau:

x_1	x_2	x_3	x_4	x_5	RHS
1	−1	1			2
1	2		1		8
−1*	−1			1	−3
−2	−1				

Iteration 1:

1. $\min\{-2, -1\} = -2 < 0,\ q = 1.$
2. $\min\{2, 8, -3\} = -3 < 0,\ p = 3.$
3. $\mu = \min\{-2, -3\} = -3.$
5. Take $p = 3$, and carry out the dual iteration.
   5(3) $\min\{-1, -1\} = -1, q = 1.$
6. Multiply row 3 by -1, and then add $-1, -1, 2$ times of row 3 to rows 1,2,4, respectively:

x_1	x_2	x_3	x_4	x_5	RHS
	-2	1		1*	-1
	1		1	1	5
1	1			-1	3
	1			-2	6

Iteration 2:

1. $\min\{1, -2\} = -2 < 0,\ q = 5.$
2. $\min\{-1, 5, 3\} = -1 < 0,\ p = 1.$
3. $\mu = \min\{-2, -1\} = -2.$
5. Take $q = 5$, and carry out the primal iteration.
   5(1) $\max\{1, 1, -1\} = 1,\ p = 1.$
6. Add $-1, 1, 2$ times of row 1 to rows 2,3,4, respectively:

x_1	x_2	x_3	x_4	x_5	RHS
	-2	1		1	-1
	3*	-1	1		6
1	-1	1			2
	-3	2			4

Iteration 3:

1. $\min\{-3, 2\} = -3 < 0,\ q = 2.$
2. $\min\{-1, 6, 2\} = -1 < 0,\ p = 1.$
3. $\mu = \min\{-3, -1\} = -3.$
5. Take $q = 2$, and carry out the primal iteration.
   5(1) $\max\{-2, 3, -1\} = 3,\ p = 2.$
6. Multiply row 2 by $1/3$, and then add $2, 1, 3$ times of row 2 to rows 1,3,4, respectively:

x_1	x_2	x_3	x_4	x_5	RHS
		$1/3$	$2/3$	1	3
	1	$-1/3$	$1/3$		2
1		$2/3$	$1/3$		4
		1	1		10

The preceding simplex tableau is optimal. The basic optimal solution and optimal value:

$$\bar{x} = (4, 2, 0, 0, 3)^{\mathrm{T}}, \qquad \bar{f} = -10.$$

The following is the revised version of Algorithm 18.1.1, where Rules 14.3.2 13.3.2 are used in steps 5(3) and 5(6), respectively.

Algorithm 18.1.2 (The most-obtuse-angle criss-cross algorithm). Given $\tau > 0$. Initial: $(B, N), B^{-1}$, $\bar{x}_B = B^{-1}b$, $\bar{z}_N = c_N - N^{\mathrm{T}} B^{-T} c_B$ and $\bar{f} = c_B^{\mathrm{T}} \bar{x}_B$. This algorithm solves the standard LP problem.

1. Determine $q \in \arg\min_{j \in N} \bar{z}_j$.
2. Determine $p \in \arg\min\{\bar{x}_{j_i} \mid i = 1, \cdots, m\}$.
3. Compute $\mu = \min\{\bar{x}_{j_p}, \bar{z}_q\}$.
4. Stop if $\mu \geq 0$ (optimality achieved).
5. If $\mu = \bar{z}_q$, then:
   (1) compute $\bar{a}_q = B^{-1}a_q$;
   (2) stop if $\bar{a}_q \leq 0$, (infeasible or unbounded problem);
   (3) redetermine row index
   $p \in \arg\min\{\bar{x}_{j_i} \mid \bar{a}_{iq} \geq \tau\theta,\ i = 1, \cdots, m\}$, where $\theta = \max\{\bar{a}_{iq} \mid i = 1, \cdots, m\} > 0$.
   else
   (4) compute $\sigma_N = N^{\mathrm{T}} B^{-T} e_p$;
   (5) stop if $\sigma_N \geq 0$ (infeasible problem);
   (6) redetermine column index
   $q \in \arg\min\{\bar{z}_j \mid \bar{a}_{pj} \leq \tau\theta,\ j \in N\}$, where $\theta = \min_{j \in N} \bar{a}_{pj} < 0$.
6. Update $\bar{x}_B = \bar{x}_B + \alpha \bar{a}_q$, and set $\bar{x}_q = \alpha$, $\bar{f} = \bar{f} - \alpha \bar{z}_q$, where $\alpha = -\bar{x}_{j_p} / \bar{a}_{pq}$.
7. Update $\bar{z}_N = \bar{z}_N + \beta \sigma_N$, $\bar{z}_{j_p} = \beta$, $\bar{f} = \bar{f} - \beta \bar{x}_{j_p}$, where $\beta = -\bar{z}_q / \bar{a}_{pq}$.
8. Update B^{-1} by (3.23).
9. Update (B, N) by exchanging j_p and q.
10. Go to step 1.

18.2 Perturbation Simplex Method

Although degeneracy hardly causes cycling in practice, it often leads to stalling (i.e., staying at some vertex for too long a time before exiting it) and hence degrades the conventional simplex algorithm's efficiency. To get rid of this drawback, perturbation algorithms (Pan 1999a, 2000b) are designed, which can get started from any basis.

The basic idea of the perturbation algorithms is closely related to sensitivity analysis. In fact, perturbing the current primal (dual) basic solution amounts to accordingly perturbing the right-hand side b (costs c). In case when such doing is not large enough for changing the optimal basis, that is, the perturbed problem and the original problem have the same optimal basis, then the optimal solution to the original problem can be readily obtained after an optimal basis to the perturbed problem reached. Even this is not the case, a subsequent solution process would achieve optimality easier.

The following algorithm can be obtained by combining the primal and dual Phase-I perturbation approaches (Sects. 13.4 and 14.4).

Algorithm 18.2.1 (Perturbation simplex algorithm 1: tableau form). Initial: a simplex tableau of form (18.1). Perturbation parameters $\delta_j \geq 0,\ j = 1,\cdots,n$ and $\zeta_i \geq 0,\ i = 1,\ldots,m$. This algorithm solve the standard LP problem.

1. Call the simplex Algorithm 3.2.1: before the minimum-ratio test, set
   $\bar{b}_i = \zeta_i,\ i \in \{i = 1,\ldots,m \mid \bar{b}_i < \zeta_i\}$.
2. Stop if returning from step 3 (infeasible or unbounded problem).
3. If returning from step 2, then:

   (1) set $\bar{b} = B^{-1}b$;
   (2) stop if $\bar{b} \geq 0$ (optimality achieved).

4. Call the dual simplex Algorithm 4.4.1: before the minimum-ratio test, set
   $\bar{z}_j = \delta_j,\ j \in \{j \in N \mid \bar{z}_j < \delta_j\}$.
5. Stop if returning from step 3 (infeasible problem).
6. If returning from step 2, then

   (1) set $\bar{z}_N = c_N - N^T B^{-1} c_B$;
   (2) stop if $\bar{z}_N \geq 0$ (optimality achieved).

7. Go to step 1.

The resetting of the right-hand side in step 1 is to ensure primal feasibility as well as nondegeneracy in primal simplex iterations. At the beginning of step 4, dual feasibility is already achieved in principle, and the resetting of the reduced costs is to ensure dual nondegeneracy in dual simplex iterations, as is designed for solving large-scale and highly degenerate problems. In our preliminary tests, there are only fewer switches between primal and dual iterations required.

The finiteness of the algorithm is not guaranteed, as the objective value does not strictly monotonically changes in the primal and dual simplex iterations. However, we believe that it is more difficult to construct cycling instances to perturbation algorithms, if possible; moreover, there would be less possibility for the solution process stalling. Based on our experience, we believe that degeneracy may not be a problem if the search direction used is close to the direction of the negative objective gradient.

The following algorithm is a variant of the preceding algorithm. It appears to be simpler and attractive, as the perturbation parameters are just used in place of the reduced costs (components of the right-hand side) for the primal (dual) minimum-ratio test, if relevant. (see Note after Algorithm 13.4.1).

Algorithm 18.2.2 (Perturbation simplex algorithm 2: tableau form). Initial: a simplex tableau of form (18.1). Perturbation parameters $\delta_j \geq 0,\ j = 1,\cdots,n$. This algorithm solve the standard LP problem.

1. Call the simplex Algorithm 3.2.1: for the minimum-ratio test, use δ_{j_i} in place of $\bar{b}_i$ if $\bar{b}i < \delta_{j_i},\ i = 1,\ldots,m$.
2. Stop if returning from step 3 (infeasible or unbounded problem).
3. Stop if $\bar{b} \geq 0$, when returning from step 2. (optimality achieved).
4. Call the dual simplex Algorithm 4.4.1: for the minimum-ratio test, use δ_j in place of $\bar{z}_j$ if $\bar{z}_j < \delta_j\ j \in N$.
5. Stop if returning from step 3 (infeasible problem).
6. Assume returning from step 2. Stop if $\bar{z}_N \geq 0$ (optimality achieved).
7. Go to step 1.

Preliminary computational experiments have been done with Algorithms 18.2.1 vs. the conventional two-phase simplex algorithm (code RSA; see Notation). There are more than 100 test problems involved, but these problems are small, involving up to 22 decision variables and inequality constraints. In the tests, the conventional code requires iterations about 2–3 times of those by Algorithms 18.2.1 (Pan 1999a, 2000b). Therefore, the perturbation algorithm is much more efficient than the conventional, at least for small problems.

It might be possible to improve the perturbation algorithms by having the called primal (dual) algorithm terminate easier, when optimality is yet far away. To do so, primal (dual) feasibility tolerance should vary dynamically in solution process, which should be larger while primal (dual) infeasibility is strong, and reduce subsequently until some predetermined value reached. In addition, the termination criteria of the called primal (dual) simplex algorithm should also vary conformably.

Example 18.2.1. Solve following problem by Algorithm 18.2.2:

$$
\begin{array}{lrcl}
\min & f = -2x_1 - 5x_2 + 4x_3, & & \\
\text{s.t.} & -2x_1 + 3x_2 - 5x_3 + x_4 & = & -17, \\
 & -4x_1 + 8x_2 - 2x_3 \qquad + x_5 & = & 1, \\
 & 7x_1 - 4x_2 - x_3 \qquad\qquad + x_6 & = & -6, \\
 & x_j \geq 0, \quad j = 1,\cdots,6. & &
\end{array}
$$

Answer Set $\delta_j = 1,\ j = 1,\ldots,6$. Initial simplex tableau:

x_1	x_2	x_3	x_4	x_5	x_6	RHS
−2	3	−5	1			−17
−4	8*	−2		1		1
7	−4	−1			1	−6
−2	−5	4				

1. Call simplex Algorithm 3.2.1.

Iteration 1:

1. $\min\{-2, -5, 4\} = -5 < 0,\ q = 2$.
4. In the minimum-ratio test, $\bar{b}_1$ is replaced by 1 ($\bar{b}_1 = -17 < 1$): $\min\{1/3, 1/8\} = 1/8, p = 2$.
5. Multiply row 2 by $1/8$, and then add $-3, 4, 5$ times of row 2 to rows 1,3,4, respectively:

x_1	x_2	x_3	x_4	x_5	x_6	RHS
$-1/2$		$-17/4$	1	$-3/8$		$-139/8$
$-1/2$	1	$-1/4$		$1/8$		$1/8$
5^*		-2		$1/2$	1	$-11/2$
$-9/2$		$11/4$		$5/8$		$5/8$

Iteration 2:

1. $\min\{-9/2, 11/4, 5/8\} = -9/2 < 0,\ q = 1$.
4. $\bar{b}_3 = -11/2 < 1$ is replaced by 1:$\min\{1/5\} = 1/5,\ p = 3$.
5. Multiply row 3 by $1/5$, and then add $1/2, 1/2, 9/2$ times of row 3 to rows 1,2,4, respectively:

x_1	x_2	x_3	x_4	x_5	x_6	RHS
		$-89/20^*$	1	$-13/40$	$1/10$	$-717/40$
	1	$-9/20$		$7/40$	$1/10$	$-17/40$
1		$-2/5$		$1/10$	$1/5$	$-11/10$
		$19/20$		$43/40$	$9/10$	$-173/40$

1. $\min\{19/20, 43/40, 9/10\} \geq 0$,
3. Returned from 2.
4. Call dual simplex Algorithm 4.4.1.

Iteration 3:

1. $\min\{-717/40, -17/40, -11/10\} = -717/40 < 0,\ p = 1$.
4. $\min\{-(19/20)/(-89/20), -(43/40)/(-13/40)\}$
   $= -(19/20)/(-89/20) = 19/89, q = 3$.
5. Multiply row 1 by $-20/89$, and then add $9/20, 2/5, -19/20$ times of row 1 to rows 2,3,4, respectively:

x_1	x_2	x_3	x_4	x_5	x_6	RHS
		1	$-20/89$	$13/178$	$-2/89$	$717/178$
	1		$-9/89$	$37/178$	$8/89$	$247/178$
1			$-8/89$	$23/178$	$17/89$	$91/178$
			$19/89$	$179/178$	$82/89$	$-1{,}451/178$

Iteration 4:

1. $\min\{717/178, 247/178, 91/178\} \geq 0$.
6. Returned from 2. $\bar{z}_N = (19/89, 179/178, 82/89)^{\mathrm{T}} \geq 0$.

Basic optimal solution and optimal value:

$$\bar{x} = (91/178, 247/178, 717/178, 0, 0, 0)^{\mathrm{T}}, \quad \bar{f} = 1{,}451/178.$$

The preceding problem was solved by calling the primal and dual simplex algorithm once each. For large-scale and highly degenerate problems, modification amounts in the minimum-ratio tests are usually large, and there would be more calls required.

Perturbation parameters δ_j are required to be nonnegative (e.g., $\delta_j = 1$ could be taken in the preceding example). In practice, however, it seems to be relevant to set them different from each other, and small (e.g., no more than 10^{-3}); they should become smaller and smaller while approaching optimality.

We formulate the revised version of Algorithm 18.2.2 below.

Algorithm 18.2.3 (Perturbation simplex algorithm 2). Given perturbation parameter $\delta_j \geq 0, \; j = 1, \ldots, n$. Initial: (B, N), B^{-1}, $\bar{x}_B = B^{-1}b$, $\bar{z}_N = c_N - N^{\mathrm{T}} B^{-T} c_B$. This algorithm solve the standard LP problem.

1. Call simplex Algorithm 3.5.1: for the minimum-ratio test, use δ_{j_i} in place of $\bar{x}_{j_i}$ if $\bar{x}_{j_i} < \delta_{j_i}, \; i = 1, \ldots, m$.
2. Stop if returning from step 5 (infeasible or unbounded problem).
3. Stop if $\bar{x}_B \geq 0$, when returning from step 3. (optimality achieved).
4. Call the dual simplex Algorithm 3.5.2: for the minimum-ratio test, use δ_j in place of $\bar{z}_j$ if $\bar{z}_j < \delta_j, \; j \in N$.
5. Stop if returning from step 4 (infeasible problem).
6. Stop if $\bar{z}_N \geq 0$, when returning from step 2. (optimality achieved).
7. Go to step 1.

18.3 Criss-Cross Reduced Simplex Method

The same idea behind the most-obtuse-angle criss-cross method (Sect. 18.1) may be implemented in the reduced simplex context. The resulting algorithm can be obtained by combining Algorithms 16.4.1 and 16.2.1.

Algorithm 18.3.1 (Criss-cross reduced simplex algorithm: tableau form). Initial: improved reduced tableau of form (16.6), where $\bar{x}_B = \bar{b} - \bar{x}_{n+1}\bar{a}_{n+1}$. This algorithm solves the reduced problem (15.1).

1. Determine row index $p_1 \in \arg\min\{\bar{a}_{i,n+1} \mid i = 1, \cdots, m+1\}$.
2. Stop if $\bar{a}_{p_1,n+1} \geq 0$ (infeasible or unbounded problem).

3. Determine row index $p_2 \in \arg\min\{\bar{x}_{j_i} \mid i = 1, \cdots, m+1\}$.
4. Compute $\mu = \min\{\bar{a}_{p_1, n+1}, \bar{x}_{j_p}\}$.
5. If $\mu = \bar{a}_{p_1, n+1}$, then:

   (1) compute $\alpha = -\bar{x}_{j_{p_1}} / \bar{a}_{p_1, n+1}$;
   (2) if $\alpha \neq 0$, add α times of x_{n+1} column to $\bar{x}_B$ column and set $\bar{x}_{n+1} = \bar{x}_{n+1} - \alpha$;
   (3) determine column index $q \in \arg\min_{j \in N} \bar{a}_{p_1, j}$;
   (4) assuming $\bar{a}_{p_1, q} \geq 0$. Stop if $\bar{x}_B \geq 0$ (optimality achieved); else go to step 3;
   (5) convert $\bar{a}_{p_1, q}$ to 1, and eliminate the other nonzeros in the column by elementary transformations;
   else
   (6) determine column index $q \in \arg\min_{j \in N} \bar{a}_{p_2, j}$;
   (7) if $\bar{a}_{p_2, q} < 0$, convert $\bar{a}_{p_2, q}$ to 1, and eliminate the other nonzeros in the column by elementary transformations, then go to step 1;
   (8) stop if $\bar{a}_{p_2, n+1} \geq 0$ (infeasible problem);
   (9) compute $\alpha = -\bar{x}_{j_{p_2}} / \bar{a}_{p_2, n+1}$;
   (10) add α times of x_{n+1} column to $\bar{x}_B$ column.

6. Go to step 1.

Example 18.3.1. Solve the following problem by Algorithm 18.3.1:

$$
\begin{array}{llr}
\min & x_9 = 2x_1 + 7x_2 - 4x_3 + 5x_4, & \\
\text{s.t.} & -3x_1 + 3x_2 + 5x_3 + \ x_4 + x_5 & = \ 13/30, \\
& \ x_1 - 2x_2 - 1x_3 - 3x_4 \qquad + x_6 & = -11/30, \\
& -6x_1 + \ x_2 - 4x_3 \quad 7x_4 \qquad\qquad + x_7 & = \ 1/2, \\
& -4x_1 - 5x_2 + 2x_3 + \ x_4 \qquad\qquad\qquad + x_8 & = -17/30, \\
& \qquad\qquad x_j \geq 0, \quad j = 1, \cdots, 8. &
\end{array}
$$

Answer Initial tableau:

x_1	x_2	x_3	x_4	x_5	x_6	x_7	x_8	x_9	RHS
−3	3	5	1	1					13/30
1	−2	−1	−3		1				−11/30
−6	1	−4	7			1			1/2
−4	−5	2	1				1		−17/30
2	7	−4*	5					−1	

Iteration 1: convert the preceding to a reduced tableau. $\bar{x}_9 = 0$. Set $p = 5$, $\min\{2, 7, -4, 5\} = -4 < 0$, $q = 3$.

Multiply row 5 by $-1/4$, then add $-5, 1, 4, -2$ times of row 5 to rows 1,2,3,4, respectively:

x_1	x_2	x_3	x_4	x_5	x_6	x_7	x_8	x_9	RHS
−1/2*	47/4		29/4	1				−5/4	13/30
1/2	−15/4		−17/4		1			1/4	−11/30
−8	−6		2			1		1	1/2
−3	−3/2		7/2				1	−1/2	−17/30
−1/2	−7/4	1	−5/4					1/4	

Iteration 2:

1. $\min\{-5/4, 1/4, 1, -1/2, 1/4\} = -5/4 < 0,\ p_1 = 1.$
3. $\min\{13/30, -11/30, 1/2, -17/30, 0\} = -17/30,\ p_2 = 4.$
4. $\mu = \min\{-5/4, -17/30\} = -5/4.$
5. $\mu = -5/4.$

   (1) $\alpha = -(13/30)/(-5/4) = 26/75$;
   (2) add 26/75 times of x_9 column to RHS column;
   (3) $\min\{-1/2, 47/4, 29/4\} = -1/2 < 0,\ q = 1$;
   (5) multiply row 1 by −2, then add $-1/2, 8, 3, 1/2$ times of row 1 to rows 2,3,4,5, respectively:

x_1	x_2	x_3	x_4	x_5	x_6	x_7	x_8	x_9	RHS
1	−47/2		−29/2	−2				5/2	
	8		3	1	1			−1	−7/25
	−194		−114	−16		1		21	127/150
	−72		−40	−6			1	7	−37/50
	−27/2	1	−17/2	−1				3/2	13/150

Iteration 3:

1. $\min\{5/2, -1, 21, 7, 3/2\} = -1 < 0,\ p_1 = 2.$
3. $\min\{0, -7/25, 127/150, -37/50, 13/150\} = -37/50,\ p_2 = 4.$
4. $\mu = \min\{-1, -37/50\} = -1.$
5. $\mu = -1.$

   (1) $\alpha = -(-7/25)/(-1) = -7/25$;
   (2) add −7/25 times of x_9 column to RHS column:

x_1	x_2	x_3	x_4	x_5	x_6	x_7	x_8	x_9	RHS
1	−47/2		−29/2	−2				5/2	−7/10
	8		3	1	1			−1	
	−194*		−114	−16		1		21	−151/30
	−72		−40	−6			1	7	−27/10
	−27/2	1	−17/2	−1				3/2	−1/3

(3) $\min\{8,3,1\} \geq 0$;
(4) $\bar{x}_B \not\geq 0$.

Iteration 4:

3. $\min\{-7/10, 0, -151/30, -27/10, -1/3\} = -151/30,\ p_2 = 3$.
4. $\mu = \min\{-1, -151/30\} = -151/30$.
5. $\mu \neq -1$.

(6) $\min\{-194, -114, -16\} = -194 < 0,\ q = 2$;
(7) multiply row 3 by $-1/194$, then add $47/2, -8, 72, 27/2$ times of row 3 to rows 1,2,4,5, respectively:

x_1	x_2	x_3	x_4	x_5	x_6	x_7	x_8	x_9	RHS
1			−67/97	−6/97		−47/388		−17/388	−306/3,389
			−165/97	33/97	1	4/97		−13/97	−302/1,455
	1		57/97	8/97		−1/194		−21/194	151/5,820
			224/97	−6/97		−36/97*	1	−77/97	−807/970
		1	−55/97	11/97		−27/388		15/388	197/11,640

Iteration 5:

1. $\min\{-17/388, -13/97, -21/194, -77/97, 15/388\} = -77/97 < 0,\ p_1 = 4$.
3. $\min\{-306/3{,}389,\ -302/1{,}455,\ 151/5{,}820,\ -807/970,\ 197/11{,}640\}$ $= -807/970,\ p_2 = 4$.
4. $\mu = \min\{-77/97, -807/970\} = -807/970$.
5. $\mu \neq -77/97$.

(6) $\min\{224/97, -6/97, -36/97\} = -36/97 < 0,\ q = 7$;
(7) multiply row 4 by $-97/36$, then add $47/388, -4/97, 1/194, 27/388$ times of row 4 to rows 1,2,3,5, respectively:

x_1	x_2	x_3	x_4	x_5	x_6	x_7	x_8	x_9	RHS
1			−13/9	−1/24			−47/144	31/144	29/160
			−13/9*	1/3	1		1/9	−2/9	−3/10
	1		5/9	1/12			−1/72	−7/72	3/80
			−56/9	1/6		1	−97/36	77/36	269/120
		1	−1	1/8			−3/16	3/16	83/480

Iteration 6:

1. $\min\{31/144, -2/9, -7/72, 77/36, 3/16\} = -2/9 < 0,\ p_1 = 2$.
3. $\min\{29/160, -3/10, 3/80, 269/120, 83/480\} = -3/10,\ p_2 = 2$.
4. $\mu = \min\{-2/9, -3/10\} = -3/10$.
5. $\mu \neq -2/9$.

(6) $\min\{-13/9, 1/3, 1/9\} = -13/9 < 0,\ q = 4$;
(7) multiply row 2 by $-9/13$, then add $13/9, -5/9, 56/9, 1$ times of row 2 to rows 1,3,4,5, respectively:

x_1	x_2	x_3	x_4	x_5	x_6	x_7	x_8	x_9	RHS
1				−3/8	−1		−7/16	7/16	77/160
			1	−3/13	−9/13		−1/13	2/13	27/130
	1			11/52	5/13		3/104	−19/104	−81/1,040
				−33/26	−56/13	1	−165/52	161/52	5,513/1,560
		1		−11/104	−9/13		−55/208	71/208	475/1,248

Iteration 7:

1. $\min\{7/16, 2/13, -19/104, 161/52, 71/208\} = -19/104 < 0,\ p_1 = 3$.
3. $\min\{77/160, 27/130, -81/1{,}040, 5{,}513/1{,}560, 475/1{,}248\} = -81/1{,}040$, $p_2 = 3$.
4. $\mu = \min\{-19/104, -81/1{,}040\} = -19/104$.
5. $\mu = -19/104$.

(1) $\alpha = -(-81/1{,}040)/(-19/104) = -81/190$;
(2) add $-81/190$ times of x_9 column to RHS column;

x_1	x_2	x_3	x_4	x_5	x_6	x_7	x_8	x_9	RHS
1				−3/8	−1		−7/16	7/16	28/95
			1	−3/13	−9/13		−1/13	2/13	27/190
	1			11/52	5/13		3/104	−19/104	
				−33/26	−56/13	1	−165/52	161/52	631/285
		1		−11/104	−9/13		−55/208	71/208	67/285

(3) $\min\{11/52, 5/13, 3/104\} \geq 0$;
(4) Optimal solution and optimal value are:

$$\bar{x} = (28/95, 0, 67/285, 27/190, 0, 0, 631/285, 0)^{\mathrm{T}},$$

$$\bar{x}_9 = 2(28/95) - 4(67/285) + 5(27/190) = 41/114.$$

18.4 Perturbation Reduced Simplex Method

The same idea behind perturbation method (Sect. 18.2) can also be realized in the reduced simplex context. Utilizing Algorithms 16.1.1 and 16.3.1 gives the following algorithm.

Algorithm 18.4.1 (Perturbation reduced simplex algorithm: tableau form). Given perturbation parameters $\delta_j = 1,\ j = 1,\cdots,n$. Initial: improved reduced tableau of form (16.6), where $\bar{x}_B = \bar{b} - \bar{f}\bar{a}_{n+1}$. This algorithm solves the reduced problem (15.1).

1. Call the improved reduced simplex Algorithm 16.1.1: for the minimum-ratio test, use δ_{j_i} in place of $\bar{x}_{j_i}$ if $\bar{x}_{j_i} < \delta_{j_i},\ i = 1,\ldots,m$, though α is computed by the original formula.
2. Stop if returning from step 1 (infeasible problem).
3. If returning form step 5, then stop if $\bar{x}_B \geq 0$ (optimality achieved).
4. Call the improved dual reduced simplex Algorithm 16.3.1: for the minimum-ratio test, use δ_j in place of $\bar{a}_{pj}$ if $\bar{a}_{pj} < \delta_j$, though β is computed by the original formula.
5. Stop if returning from step 6 (infeasible problem).
6. Assume returning from step 2. Stop if $\bar{a}_{pj} \geq 0,\ j \in N$ (optimality achieved).
7. Go to step 1.

Example 18.4.1. Solve the following problem by Algorithm 18.4.1:

$$
\begin{array}{lrcl}
\min & x_9 = -2x_1 + 7x_2 - 5x_3 - 3x_4, & & \\
\text{s.t.} & -x_1 + 4x_2 - 5x_3 - 2x_4 + x_5 & = & 12, \\
 & x_1 - 3x_2 + 3x_3 \quad 5x_4 \qquad + x_6 & = & 2, \\
 & 7x_1 - 6x_2 - x_3 - 4x_4 \qquad\qquad + x_7 & = & -15, \\
 & -3x_1 + x_2 + 2x_3 - x_4 \qquad\qquad\qquad + x_8 & = & -2, \\
 & x_j \geq 0, \quad j = 1,\cdots,8. & &
\end{array}
$$

Answer Set perturbation parameters $\delta_j = 1,\ j = 1,\cdots,8$.

1. Call Algorithm 16.1.1. Start from the initial tableau: $\bar{f} = 0$.

x_1	x_2	x_3	x_4	x_5	x_6	x_7	x_8	x_9	RHS
−1	4	−5	−2	1					12
1	−3	3	5		1				2
7	−6	−1	−4			1			−15
−3	1	2	−1				1		−2
−2	7	−5*	−3					−1	

Iteration 1:

1. $\bar{a}_9 = -e_5 \not\geq 0$.
2. $\alpha = 0/(-1) = 0,\ p = 5$.
4. $\min\{-2, 7, -5, -3\} = -5 < 0,\ q = 3$.
6. Multiply row 5 by $-1/5$, then add $5, -3, 1, -2$ times of row 5 to rows 1,2,3,4, respectively.

x_1	x_2	x_3	x_4	x_5	x_6	x_7	x_8	x_9	RHS
1	−3		1	1				1	12
−1/5	6/5		16/5		1			−3/5	2
37/5	−37/5		−17/5			1		1/5	−15
−19/5*	19/5		−11/5				1	−2/5	−2
2/5	−7/5	1	3/5					1/5	

Iteration 2:

2. Determine row index by replacing the right-hand side term −2 by 1: $\min\{-2/(-3/5), -1/(-2/5)\} = 5/2,\ p = 4, \alpha = -(-2)/(-2/5) = -5$.
3. Add −5 times of x_9 column to RHS column.
4. $\min\{-19/5, 19/5, -11/5\} = -19/5 < 0,\ q = 1$.
6. Multiply row 4 by −5/19, then add $-1, 1/5, -37/5, -2/5$ times of row 4 to rows 1,2,3,5, respectively:

x_1	x_2	x_3	x_4	x_5	x_6	x_7	x_8	x_9	RHS
	−2		8/19	1			5/19	17/19	7
	1		63/19		1		−1/19	−11/19	5
			−146/19*			1	37/19	−11/19	−16
1	−1		11/19				−5/19	2/19	
	−1	1	7/19				2/19	3/19	−1

Iteration 3:

2. Determine row index by replacing the right-hand side term −16 by 1:

$$\min\{-5/(-11/19), -1/(-11/19)\} = 19/11,\ p = 3;$$

$$\alpha = -(-16)/(-11/19) = -304/11;$$

3. Add −304/11 times of x_9 column to RHS column.
4. $\min\{0, -146/19, 37/19\} = -146/19 < 0,\ q = 4$.
6. Multiply row 3 by −19/146, then add $-8/19, -63/19, -11/19, -7/19$ times of row 3 to rows 1,2,4,5, respectively:

x_1	x_2	x_3	x_4	x_5	x_6	x_7	x_8	x_9	RHS
	−2			1		4/73	27/73	63/73	−195/11
	1				1	63/146	115/146	−121/146	21
			1			−19/146	−37/146	11/146	
1	−1					11/146	−17/146	9/146	−32/11
	−1	1				7/146	29/146	19/146	−59/11

Iteration 4:

2. $\alpha = \min\{-21/(-121/146)\} = 3{,}066/121,\ p = 2.$
3. Add $3{,}066/121$ times of x_9 column to RHS column.
4. $\min\{1, 63/146, 115/146\} \geq 0.$
5. Returning with $\bar{x}_B \not\geq 0$.

x_1	x_2	x_3	x_4	x_5	x_6	x_7	x_8	x_9	RHS
	-2			1		$4/73$	$27/73$	$63/73$	$501/121$
	1				1	$63/146$	$115/146$	$-121/146$	
			1			$-19/146$	$-37/146$	$11/146$	$21/11$
1	-1					$11/146$	$-17/146$	$9/146$	$-163/121$
	-1^*	1				$7/146$	$29/146$	$19/146$	$-250/121$

4. Call Algorithm 16.3.1.

Iteration 5:

1. $\min\{501/121, 0, 21/11, -163/121, -250/121\} = -250/121 < 0, r = 5.$
3. $J = \{2\} \neq \emptyset.$
4. $\min\{-1/(-1)\} = 1, q = 2.$
5. Multiply row 5 by -1, then add $2, -1, 1$ times of row 5 to rows 1,2,4, respectively:

x_1	x_2	x_3	x_4	x_5	x_6	x_7	x_8	x_9	RHS
		-2		1		$-3/73$	$-2/73$	$44/73$	$91/11$
		1			1	$35/73$	$72/73$	$-51/73$	$-250/121$
			1			$-19/146$	$-37/146$	$11/146$	$21/11$
1		-1				$2/73$	$-23/73$	$-5/73$	$87/121$
	1	-1				$-7/146$	$-29/146$	$-19/146$	$250/121$

6. $\bar{a}_{2,9} = -51/73 < 0.$
7. $\alpha \quad = -(-250/121)/(-51/73) = -1{,}943/657.$
8. Add $-1{,}943/657$ times of x_9 column to RHS column.

x_1	x_2	x_3	x_4	x_5	x_6	x_7	x_8	x_9	RHS
		-2		1		$-3/73$	$-2/73$	$44/73$	$331/51$
		1			1	$35/73$	$72/73$	$-51/73$	
			1			$-19/146$	$-37/146$	$11/146$	$86/51$
1		-1				$2/73$	$-23/73$	$-5/73$	$47/51$
	1	-1				$-7/146$	$-29/146$	$-19/146$	$125/51$

Iteration 6:

1. $\min\{331/51, 0, 86/51, 47/51, 125/51\} \geq 0$.
6. Returning from step 2, and $\bar{a}_{2j} \geq 0,\ j \in N$. Optimal solution and optimal value:

$$\bar{x} = (47/51, 125/51, 0, 86/51, 331/51, 0, 0, 0)^{\mathrm{T}},$$
$$\bar{x}_9 = (-2, 7, -3)(47/51, 125/51, 86/51)^{\mathrm{T}} = 523/51.$$

Although the preceding problem was solved by calling the improved primal and dual reduced simplex algorithms once each, there would be more calls required for solving large-scale and highly degenerate problems.

18.5 Mixed Two-Phase Method

The conventional two-phase primal (dual) simplex method utilizes primal (dual) iterations in both Phase-I and Phase-II. As was mentioned in Chaps. 13 and 14, nevertheless, a primal rule could be used to attain dual feasibility whereas a dual rule to attain primal feasibility. As differing from conventional ones, in other words, it is possible to solve problems by dual (primal) Phase-I and primal (dual) Phase-II. There are multiple ways to construct such mixed two-phase methods. In this section, this idea will be illustrated by examples in the reduced simplex context only.

Primal-dual Scheme. The primal reduced simplex algorithm is used for Phase-I whereas the dual reduced simplex algorithm for Phase-II. When primal feasibility happens to be achieved in a Phase-I iteration, the problem is solved without Phase-II at all, as is illustrated in the following example.

Example 18.5.1. Solve following problem by the Primal-dual Scheme:

$$\begin{array}{llr}
\min & x_8 = -3x_1 - 2x_2 + 4x_3 + 2x_4, & \\
\text{s.t.} & x_1 + 2x_2 - x_3 + 2x_4 + x_5 & = 4, \\
& 2x_1 + 3x_3 - 4x_4 + x_6 & = 1, \\
& -x_1 + 2x_2 - 3x_4 + x_7 & = -3, \\
& x_j \geq 0, \quad j = 1, \cdots, 7. &
\end{array}$$

Answer Initial tableau is

x_1	x_2	x_3	x_4	x_5	x_6	x_7	x_8	RHS
1	2	−1	2	1				4
2		3	−4		1			1
−1	2		−3			1		−3
−3*	−2	4	2				−1	

Phase-I: Call reduced simplex Algorithm 15.2.1.

Iteration 1:

1. $\bar{a}_8 \not\geq 0$.
2. $\bar{x}_8 = 0, p = 4$.
3. $\min\{-3, -2, 4, 2\} = -3,\ q = 1$.
5. Multiply row 4 by $-1/3$, and add $-1, -2, 1$ times of row 4 to rows 1,2,3, respectively:

x_1	x_2	x_3	x_4	x_5	x_6	x_7	x_8	RHS
	4/3	1/3	8/3	1			−1/3	4
	−4/3	17/3	−8/3*		1		−2/3	1
	8/3	−4/3	−11/3			1	1/3	−3
1	2/3	−4/3	−2/3				1/3	

which is a reduced simplex tableau.

Iteration 2:

1. $\bar{a}_8 \not\geq 0$.
2. $\bar{x}_8 = \max\{4/(-1/3), 1/(-2/3)\} = -3/2, p = 2$.
   As $x_7 = -3 - (-3/2)(1/3) = -5/2 < 0$, the associated basic solution is infeasible; but Algorithm 15.2.1 is still carried out.
3. $\min\{-4/3, 17/3, -8/3\} = -8/3,\ q = 4$.
5. Multiply row 2 by $-3/8$, and add $-8/3, 11/3, 2/3$ times of row 2 to rows 1,3,4, respectively:

x_1	x_2	x_3	x_4	x_5	x_6	x_7	x_8	RHS
		6		1	1		−1	5
	1/2	−17/8	1		−3/8		1/4	−3/8
	9/2	−73/8			−11/8	1	5/4	−35/8
1	1	−11/4			−1/4		1/2	−1/4

Iteration 3:

1. $\bar{a}_8 \not\geq 0$.
2. $\bar{x}_8 = \max\{5/(-1)\} = -5,\ p = 1$.
3. $\min\{0, 6, 1\} \geq 0$.
4. $x_B = (5, -3/8, -35/8, -1/4)^{\mathrm{T}} - (-5)(-1, 1/4, 5/4, 1/2)^{\mathrm{T}} = (0, 7/8, 15/8, 9/4)^{\mathrm{T}} \geq 0$.

Optimality is achieved. Basic optimal solution and optimal value:

$$\bar{x} = (9/4, 0, 0, 7/8, 0, 0, 15/8)^{\mathrm{T}}, \qquad \bar{x}_8 = -5.$$

It is noted that the problem was solved by the Primal-dual Scheme without Phase II.

Dual-primal scheme. The dual reduced simplex algorithm is used for Phase-I whereas the primal reduced simplex algorithm for Phase-II. When dual feasibility happens to be achieved in some Phase-I iteration, the problem is solved without Phase-II at all.

Example 18.5.2. Solve the Example 18.5.1 by the Dual-primal Scheme.

Answer Iteration 1: The reduced simplex tableau (the second one in the Example) is obtained, i.e.,

x_1	x_2	x_3	x_4	x_5	x_6	x_7	x_8	RHS
	4/3	1/3	8/3	1			−1/3	4
	−4/3	17/3	−8/3		1		−2/3	1
	8/3	−4/3	−11/3*			1	1/3	−3
1	2/3	−4/3	−2/3				1/3	

Phase-I: Call dual reduced simplex Algorithm 15.4.1.

$$\min\{-1/3, -2/3, 1/3, 1/3\} = -2/3, \quad p = 2.$$

Iteration 2:

1. $\bar{x}_8 \ = 1/(-2/3) = -3/2.$
2. $\bar{x}_B \ = \ (4, 1, -3, 0)^{\mathrm{T}} - (-3/2)(-1/3, -2/3, 1/3, 1/3)^{\mathrm{T}} \ = \ (7/2, 0, -5/2, 1/2)^{\mathrm{T}}.$
3. $\min\{7/2, 0, -5/2, 1/2\} = -5/2 < 0,\ r = 3.$
5. $J \ = \{3, 4\} \neq \emptyset.$
6. $\min\{-(17/3)/(-4/3), (8/3)/(-11/3)\} = -8/11,\ q = 4.$
7. Multiply row 3 by $-3/11$, and add $-8/3, 8/3, 2/3$ times of row 3 to rows 1,2,4, respectively:

x_1	x_2	x_3	x_4	x_5	x_6	x_7	x_8	RHS
	36/11	−7/11		1		8/11	−1/11	20/11
	−36/11*	73/11			1	−8/11	−10/11	35/11
	−8/11	4/11	1			−3/11	−1/11	9/11
1	2/11	−12/11				−2/11	3/11	6/11

8. $\bar{a}_{2,8} = -10/11 < 0.$

Iteration 3:

1. $\bar{x}_8 \ = (35/11)/(-10/11) = -35/10.$
2. $\bar{x}_B \ = (20/11, 35/11, 9/11, 6/11)^{\mathrm{T}} - (-35/10)(-1/11, -10/11, -1/11, 3/11)^{\mathrm{T}} = (3/2, 0, 1/2, 3/2)^{\mathrm{T}} \geq 0.$
4. Primal feasibility achieved.

Phase-II: Call reduced simplex Algorithm 15.2.1.

Iteration 4:

2. $\bar{x}_8 = \max\{(20/11)/(-1/11), (35/11)/(-10/11), (9/11)/(-1/11)\}$
   $= (35/11)/(-10/11) = -7/2,\ p = 2.$
3. $\min\{-36/11, 73/11, -8/11\} = -36/11,\ q = 2.$
5. Multiply row 2 by $-11/36$, and add $-36/11, 8/11, -2/11$ times of row 2 to rows 1,3,4, respectively:

x_1	x_2	x_3	x_4	x_5	x_6	x_7	x_8	RHS
		6		1	1		−1	5
	1	−73/36			−11/36	2/9	5/18	−35/36
		−10/9	1		−2/9	−1/9	1/9	1/9
1		−13/18			1/18	−2/9	2/9	13/18

Iteration 5:

2. $\bar{x}_8 = \max\{5/(-1)\} = -5,\ p = 1.$
3. $\min\{6, 1, 0\} \geq 0.$
4. $\bar{x}_B = (5, -35/36, 1/9, 13/18)^{\mathrm{T}} - (-5)(-1, 5/18, 1/9, 2/9)^{\mathrm{T}}$
   $= (0, 5/12, 2/3, 11/6)^{\mathrm{T}} \geq 0.$

The basic optimal solution and associated objective value:

$$\bar{x} = (11/6, 5/12, 0, 2/3, 0, 0, 0)^{\mathrm{T}}, \quad \bar{x}_8 = -5.$$

This problem was solved by the Dual-Primal Scheme via 4 iterations and 3 basis changes, as is slightly inefficient, compared to the Primal-dual Scheme. It seems to be impossible to know which is more efficient for solving a given problem before hands.

Chapter 19
Generalizing Reduced Simplex Method

In Sects. 7.3–7.7, the general LP problem in form (7.11) was converted to the bounded-variable problem (7.13), and the latter was then solved by a generalized primal or dual simplex method. In this chapter, the primal and dual reduced simplex methods will be generalized to solve the bounded-variable problem. Some Phase-I procedures will be developed to provide a starting point to get them started.

19.1 Generalized Reduced Simplex Method

The objective value will be again deemed as a variable, and denoted by x_{n+1}. Placing the objective function $x_{n+1} = c^{\mathrm{T}}x$ in the constraint part, the bounded-variable problem (7.13) can be transformed to the following form, equivalently:

$$\begin{array}{ll} \min & x_{n+1}, \\ \text{s.t.} & A\begin{pmatrix} x \\ x_{n+1} \end{pmatrix} = b, \quad l \leq x \leq u, \end{array} \tag{19.1}$$

where $A \in \mathcal{R}^{(m+1)\times(n+1)}$, $b \in \mathcal{R}^{m+1}$, rank $A = m+1$, $m < n$; Both l and u are given vectors (may include infinite components). As its objective function involves a single variable x_{n+1}, this problem will be referred to as *reduced bounded-variable problem*. Note that x_{n+1} is a free variable.

In this section, the reduced simplex method will be generalized to solve problem (19.1).

Assume that the current basis and nonbasis are

$$B = \{j_1, \cdots, j_{m+1}\}, \quad N = A\backslash B, \; n+1 \notin B.$$

Then, the constraints of (19.1) are equivalent to the following canonical form:

$$l_B \leq x_B = \bar{b} - \bar{N}x_N - \bar{a}_{n+1}x_{n+1} \leq u_B, \quad l_N \leq x_N \leq u_N,$$

P.-Q. PAN, *Linear Programming Computation*, DOI 10.1007/978-3-642-40754-3_19,

where

$$\bar{b} = B^{-1}b, \quad \bar{N} = B^{-1}N, \quad \bar{a}_{n+1} = B^{-1}a_{n+1}.$$

If $(\bar{x}, \bar{x}_{n+1})$ satisfies

$$\bar{x}_j = l_j \text{ or } u_j, \quad j \in N, \tag{19.2}$$

$$l_B \leq \bar{x}_B = \bar{b} - \bar{N}\bar{x}_N - \bar{a}_{n+1}\bar{x}_{n+1} \leq u_B, \tag{19.3}$$

it is called *basic feasible solution*; if, further, $l_B < \bar{x}_B < u_B$, it is said to be nondegenerate. It is clear that Lemma 7.4.1 is valid in this context, that is, if the bounded-variable problem (7.13) has a feasible solution, it has a basic feasible solution, and if it has an optimal solution, then it has a basic optimal solution.

Consider the following line search scheme:

$$\hat{x}_B = \bar{x}_B + \alpha\bar{a}_{n+1}, \quad \hat{x}_{n+1} = \bar{x}_{n+1} - \alpha, \quad \hat{x}_N = \bar{x}_N, \tag{19.4}$$

where the stepsize α is determined by the rule below.

Rule 19.1.1 (Row rule) Select stepsize α and row index p such that

$$\alpha = \alpha_p = \min\{\alpha_i \mid i = 1, \cdots, m+1\}, \tag{19.5}$$

where

$$\alpha_i = \begin{cases} (u_{j_i} - \bar{x}_{j_i})/\bar{a}_{i,n+1}, & \text{if } \bar{a}_{i,n+1} > 0, \\ (l_{j_i} - \bar{x}_{j_i})/\bar{a}_{i,n+1}, & \text{if } \bar{a}_{i,n+1} < 0, \\ \infty, & \text{if } \bar{a}_{i,n+1} = 0, \end{cases} \quad i = 1, \cdots, m+1. \tag{19.6}$$

Lemma 19.1.1. *If $(\bar{x}, \bar{x_{n+1}})$ is a basic feasible solution, then the solution $(\hat{x}, \hat{x}_{n+1})$, defined by (19.4), is a basic feasible solution, associated with the objective value not increasing, or strictly decreasing if $\bar{x}$ is nondegenerate.*

Proof. Note that $\hat{x}_N = \bar{x}_N$. For any real number α, the new solution given by (19.4) satisfies the equality constraints of (19.1); from $(\bar{x}, \bar{x}_{n+1})$ satisfying the equality constraint, in fact, it follows that

$$\begin{aligned} A(\hat{x}^{\mathrm{T}}, \hat{x}_{n+1})^{\mathrm{T}} &= B\hat{x}_B + N\hat{x}_N + \hat{x}_{n+1}a_{n+1} \\ &= B(\bar{x}_B + \alpha\bar{a}_{n+1}) + N\bar{x}_N + (\bar{x}_{n+1} - \alpha)a_{n+1} \\ &= B\bar{x}_B + N\bar{x}_N + \bar{x}_{n+1}a_{n+1} \\ &= b. \end{aligned}$$

On the other hand, the α determined by (19.5) is clearly nonnegative such that the new solution $\bar{x}$ satisfies bound constraints $l \leq \hat{x} \leq u$. Therefore $(\hat{x}, \hat{x}_{n+1})$ is a basic feasible solution. When $\bar{x}$ is nondegenerate, in addition, it is clear that $\alpha > 0$, and hence

$$\hat{x_{n+1}} = \bar{x}_{n+1} - \alpha < \bar{x}_{n+1},$$

which implies that the objective value decreases. □

It is noted that decrement of the objective value is equal to the stepsize. In fact, the α, given by (19.6), is the largest possible stepsize to maintain feasibility of $\hat{x}$. In practice, the problem should be deemed as unbounded if α is too large. Note that the j_p-indexed component of the new solution is equal to the related upper or lower bound, i.e.,

$$\hat{x}_{j_p} = \begin{cases} u_{j_p}, \text{ if } \bar{a}_{p,n+1} > 0, \\ l_{j_p}, \text{ if } \bar{a}_{p,n+1} < 0. \end{cases} \tag{19.7}$$

Introduce

$$\sigma_N = N^{\mathrm{T}} B^{-\mathrm{T}} e_p. \tag{19.8}$$

and set

$$J = \{j \in \Gamma \mid \bar{a}_{p,n+1}\sigma_j > 0\} \cup \{j \in \Pi \mid \bar{a}_{p,n+1}\sigma_j < 0\}, \tag{19.9}$$

where Γ and Π are defined by (7.19). Then the following is valid for the new solution.

Lemma 19.1.2. *If J is empty, the new solution $(\hat{x}, \hat{x}_{n+1})$, defined by (19.4), is a basic optimal solution.*

Proof. By (19.5) and (19.6) it is known that $\bar{a}_{p,n+1} \neq 0$. From the pth equation of the canonical form $x_B = \bar{b} - \bar{N}x_N - \bar{a}_{n+1}x_{n+1}$, it follows that

$$x_{n+1} = (\bar{b}_p - \sigma_N^{\mathrm{T}} x_N - x_{j_p})/\bar{a}_{p,n+1}. \tag{19.10}$$

Assuming that (x', x'_{n+1}) is any feasible solution, it holds that

$$l_j \leq x'_j \leq u_j, \quad j = 1, \cdots, n. \tag{19.11}$$

If $\bar{a}_{p,n+1} > 0$, then $J = \emptyset$ implies that

$$\sigma_\Gamma \leq 0, \quad \sigma_\Pi \geq 0. \tag{19.12}$$

Thus, from the first expression of (19.12) and $x'_\Gamma \geq l_\Gamma = \bar{x}_\Gamma$, it follows that

$$\sigma_\Gamma^{\mathrm{T}} x'_\Gamma \leq \sigma_\Gamma^{\mathrm{T}} \bar{x}_\Gamma,$$

and from the second expression of (19.12) and $x'_\Pi \leq u_\Pi = \bar{x}_\Pi$, it follows that

$$\sigma_\Pi^{\mathrm{T}} x'_\Pi \leq \sigma_\Pi^{\mathrm{T}} \bar{x}_\Pi.$$

Combining the preceding two expressions and $\bar{a}_{p,n+1} > 0$ gives

$$-(\sigma_N^{\mathrm{T}} x'_N)/\bar{a}_{p,n+1} \geq -(\sigma_N^{\mathrm{T}} \bar{x}_N)/\bar{a}_{p,n+1} = -(\sigma_N^{\mathrm{T}} \hat{x}_N)/\bar{a}_{p,n+1}. \tag{19.13}$$

On the other hand, from (19.7), (19.11) and $\bar{a}_{p,n+1} > 0$, it follows that

$$-x'_{j_p}/\bar{a}_{p,n+1} \geq -u_{j_p}/\bar{a}_{p,n+1} = -\hat{x}_{j_p}/\bar{a}_{p,n+1}. \tag{19.14}$$

Therefore, by $(\hat{x}, \hat{x}_{n+1})$ and (x', x'_{n+1}) satisfying (19.10), (19.13) and (19.14), it holds that

$$x'_{n+1} = (\bar{b}_p - \sigma_N^{\mathrm{T}} x'_N - x'_{j_p})/\bar{a}_{p,n+1} \geq (\bar{b}_p - \sigma_N^{\mathrm{T}} \hat{x}_N - \hat{x}_{j_p})/\bar{a}_{p,n+1} = \hat{x}_{n+1}.$$

Therefore, by considering Lemma 19.1.1, it is known that $\hat{x}$ is a basic optimal solution.

It can be shown analogously that the statement is valid if $\bar{a}_{p,n+1} < 0$. □

When $J \neq \emptyset$, the following rule is well-defined.

Rule 19.1.2 (Column rule) Select column index

$$q \in \arg\max_{j \in J} |\sigma_j|. \tag{19.15}$$

Then after the associated basis change, when j_p left from and q entered to the basis, an iteration is complete. It is clear that the resulting basis corresponds to the new solution $(\hat{x}, \hat{x})$, given by (19.4) together with (19.5) and (19.6).

The overall steps can be put into the following algorithm.

Algorithm 19.1.1 (Generalized reduced simplex algorithm). Initial: (B, N), B^{-1}; $\bar{a}_{n+1} = -B^{-1}e_{m+1}$; basic feasible solution $\bar{x}$. This algorithm solves the reduced bounded-variable problem (19.1).

1. Determine stepsize α and row index p by (19.5) and (19.6).
2. If $\alpha \neq 0$, update: $\bar{x}_B = \bar{x}_B + \alpha \bar{a}_{n+1}, \bar{x}_{n+1} = \bar{x}_{n+1} - \alpha$
3. Compute $\sigma_N = N^{\mathrm{T}} B^{-\mathrm{T}} e_p$.
4. Stop if J, defined by (19.9), is empty.
5. Determine column index q by (19.15).
6. Compute $\bar{a}_q = B^{-1} a_q$ and $\nu = -\bar{a}_{p,n+1}/\sigma_q$.
7. If $\nu \neq 0$, update $\bar{a}_{n+1} = \bar{a}_{n+1} + \nu(\bar{a}_q - e_p)$.

8. Update B^{-1} by (3.23).
9. Update (B, N) by exchanging j_p and q.
10. Go to step 1.

Theorem 19.1.1. *Algorithm 19.1.1 produces a sequence of basic feasible solutions. Assuming degeneracy, it terminates at step 4, offering a basic optimal solution.*

Proof. The validity comes from Lemmas 19.1.1, 19.1.2 and the discussions preceding Algorithm 19.1.1.

Example 19.1.1. Solve the following problem by Algorithm 19.1.1:

$$\begin{array}{ll}\min & x_9 = -x_1 - 3x_2 - 2x_3 + 4x_4 - 5x_5,\\ \text{s.t.} & -2x_1 + x_2 - 3x_3 - x_4 + x_5 + x_6 = -52,\\ & 3x_1 - 3x_2 + x_3 + 5x_4 - 2x_5 + x_7 = 26,\\ & x_1 - 6x_2 - 2x_3 - 4x_4 + 6x_5 + x_8 = 7,\\ & 2 \le x_1 \le 10, \quad -5 \le x_2 \le 12, \quad 1 \le x_3 \le 15,\\ & -4 \le x_4 \le 27, \quad -3 \le x_5 \le 14, \quad -9 \le x_6 \le 18,\\ & -7 \le x_7 \le 13, \quad -10 \le x_8 \le 21.\end{array}$$

Given a feasible solution: $\bar{x} = (2, -5, 15, -4, -3, 1, 4, 7)^{\mathrm{T}}$, $\bar{x}_9 = -18$.

Answer The initial tableau is

x_1	x_2	x_3	x_4	x_5	x_6	x_7	x_8	x_9	RHS
−2	1	−3	−1	1	1				−52
3	−3	1	5	−2		1			26
1	−6	−2	−4	6			1		7
−1*	−3	−2	4	−5				−1	

Iteration 1: Convert the preceding to a reduced tableau by taking $p = 5$, $q = 1$.

Multiply row 5 by -1, then add $2, -3, -1, 1$ times of row 5 to rows 1,2,3,4, respectively:

x_1	x_2	x_3	x_4	x_5	x_6	x_7	x_8	x_9	RHS
	7	1	−9	11	1			2	−52
	−12	−5	17	−17		1		−3	26
	−9	−4		1			1	−1	7
1	3	2	−4	5				1	

Taking the left-side of the preceding tableau as original data, call Algorithm 19.1.1.

Initial: $B = \{6, 7, 8, 1\}$, $N = \{2, 3, 4, 5\}$, $B^{-1} = I$, $\bar{x}_N = (-5_{(-)}, 15^{(+)}, -4_{(-)}, -3_{(-)})^{\mathrm{T}}$, $\bar{x}_B = (1, 4, 7, 2)^{\mathrm{T}}$, $\bar{x}_9 = -18$. $\bar{a}_9 = (2, -3, -1, 1)^{\mathrm{T}}$. The $\bar{x}$ is a basic feasible solution.

Iteration 2:

1. $\alpha \ = \min\{(18-1)/2, (-7-4)/(-3), (-10-7)/(-1), (10-2)/(1)\}$
   $= \min\{17/2, 11/3, 17, 8\} = 11/3,\ p = 2.$
2. $\bar{x}_B = (1, 4, 7, 2)^{\mathrm{T}} + (11/3)(2, -3, -1, 1)^{\mathrm{T}} = (25/3, -7, 10/3, 17/3)^{\mathrm{T}},$
   $\bar{x}_9 = -18 - 11/3 = -65/3.$
3. $\sigma_N = N^{\mathrm{T}} B^{-\mathrm{T}} e_2 = (-12, -5, 17, -17)^{\mathrm{T}}.$
4. $J \ = \{2, 5\} \neq \emptyset.$
5. $\max\{|-12|, |-17|\} = 17,\ q = 5.$
6. $\bar{a}_5 \ = B^{-1} a_5 = (11, -17, 1, 5)^{\mathrm{T}}.$
7. $\nu \ = -(-3)/(-17) = -3/17,$
   $\bar{a}_{n+1} = (2, -3, -1, 1)^{\mathrm{T}} + (-3/17)(11, -17-1, 1, 5)^{\mathrm{T}}$
   $= (1/17, 3/17, -20/17, 2/17)^{\mathrm{T}}.$
8. $$B^{-1} = \begin{pmatrix} 1 & 11/17 & & \\ & -1/17 & & \\ & 1/17 & 1 & \\ & 5/17 & & 1 \end{pmatrix}.$$
9. $\bar{x}_B = (25/3, -3, 10/3, 17/3)^{\mathrm{T}},\ B = \{6, 5, 8, 1\},$
   $\bar{x}_N = (-5_{(-)}, 15^{(+)}, -4_{(-)}, -7_{(-)})^{\mathrm{T}},\ N = \{2, 3, 4, 7\}.$

Iteration 3:

1. $\alpha \ = \min\{(18-(25/3))/(1/17), (14-(-3))/(3/17), (-10-(10/3))/$
   $(-20/17), (10-(17/3))/(2/17)\} = \min\{493/3, 289/3, 34/3, 221/6\}$
   $= 34/3,\ p = 3.$
2. $\bar{x}_B = (25/3, -3, 10/3, 17/3)^{\mathrm{T}} + (34/3)(1/17, 3/17, -20/17, 2/17)^{\mathrm{T}}$
   $= (9, -1, -10, 7)^{\mathrm{T}},\ \bar{x}_9 = -65/3 - 34/3 = -99/3.$
3. $\sigma_N = N^{\mathrm{T}} B^{-\mathrm{T}} e_3 = (-165/17, -73/17, 1, 1/17)^{\mathrm{T}}.$
4. $J \ = \{2\} \neq \emptyset.$
5. $\max\{|-165/17|\|\} = 165/17,\ q = 2.$
6. $\bar{a}_2 \ = B^{-1} a_2 = (-13/17, 12/17, -165/17, -9/17)^{\mathrm{T}}.$
7. $\nu \ = -(-20/17)/(-165/17) = -4/33,$
   $\bar{a}_{n+1} = (1/17, 3/17, -20/17, 2/17)^{\mathrm{T}} + (-4/33)(-13/17, 12/17, -165/17-1,$
   $-9/17)^{\mathrm{T}} = (5/33, 1/11, 4/33, 2/11)^{\mathrm{T}}.$
8. $$B^{-1} = \begin{pmatrix} 1 & & -13/165 & \\ & 1 & 4/55 & \\ & & -17/165 & \\ & & -3/55 & 1 \end{pmatrix} \begin{pmatrix} 1 & 11/17 & & \\ & -1/17 & & \\ & 1/17 & 1 & \\ & 5/17 & & 1 \end{pmatrix}$$
   $$= \begin{pmatrix} 1 & 106/165 & -13/165 & \\ & -3/55 & 4/55 & \\ & -1/165 & -17/165 & \\ & 16/55 & -3/55 & 1 \end{pmatrix}.$$

9. $\bar{x}_B = (9, -1, -5, 7)^{\mathrm{T}}, \ B = \{6, 5, 2, 1\},$
$\bar{x}_N = (-10_{(-)}, 15^{(+)}, -4_{(-)}, -7_{(-)})^{\mathrm{T}}, N = \{8, 3, 4, 7\}.$

Iteration 4:

1. $\alpha \ = \min\{(18 - 9)/(5/33), (14 - (-1))/(1/11), (12 - (-5))/(4/33), (10 - 7)/(2/11)\}$
$= \min\{297/5, 165, 561/4, 33/2\} = 33/2, \ p = 4.$
2. $\bar{x}_B = (9, -1, -5, 7)^{\mathrm{T}} + (33/2)(5/33, 1/11, 4/33, 2/11)^{\mathrm{T}}$
$= (23/2, 1/2, -3, 10)^{\mathrm{T}}, \bar{x}_9 = -99/3 - 33/2 = -297/6.$
3. $\sigma_N = N^{\mathrm{T}} B^{-\mathrm{T}} e_4 = (-3/55, 42/55, 52/55, 16/55)^{\mathrm{T}}.$
4. $J \ = \{4, 7\} \neq \emptyset.$
5. $\max\{|52/55|, |16/55|\} = \{52/55, 16/55\} = 52/55, \ q = 4.$
6. $\bar{a}_4 \ = B^{-1} a_4 = (317/165, -51/55, -17/165, 52/55)^{\mathrm{T}}.$
7. $\nu \ = -(2/11)/(52/55) = -5/26,$
$\bar{a}_{n+1} = (5/33, 1/11, 4/33, 2/11)^{\mathrm{T}} + (-5/26)(317/165, -51/55, -17/165, 52/55 - 1)^{\mathrm{T}} = (-17/78, 7/26, 11/78, 5/26)^{\mathrm{T}}.$
8.
$$B^{-1} = \begin{pmatrix} 1 & & & -317/156 \\ & 1 & & 51/52 \\ & & 1 & 17/156 \\ & & & 55/52 \end{pmatrix} \begin{pmatrix} 1 & 106/165 & -13/165 & \\ & -3/55 & 4/55 & \\ & -1/165 & -17/165 & \\ & 16/55 & -3/55 & 1 \end{pmatrix}$$
$$= \begin{pmatrix} 1 & 2/39 & 5/156 & -317/156 \\ & 3/13 & 1/52 & 51/52 \\ & 1/39 & -17/156 & 17/156 \\ & 4/13 & -3/52 & 55/52 \end{pmatrix}.$$
9. $\bar{x}_B = (23/2, 1/2, -3, -4)^{\mathrm{T}}, \ B = \{6, 5, 2, 4\},$
$\bar{x}_N = (-10_{(-)}, 15^{(+)}, 10^{(+)}, -7_{(-)})^{\mathrm{T}}, N = \{8, 3, 1, 7\}.$

Iteration 5:

1. $\alpha \ = \min\{(-9 - 23/2)/(-17/78), (14 - (1/2))/(7/26), (12 - (-3))/(11/78), (27 - (-4))/(5/26)\}$
$= \min\{1{,}599/17, 351/7, 1{,}170/11, 806/5\} = 351/7, \ p = 2.$
2. $\bar{x}_B = (23/2, 1/2, -3, -4)^{\mathrm{T}} + (351/7)(-17/78, 7/26, 11/78, 5/26)^{\mathrm{T}}$
$= (4/7, 14, 57/14, 79/14)^{\mathrm{T}}, \bar{x}_9 = -297/6 - 351/7 = -1395/14.$
3. $\sigma_N = N^{\mathrm{T}} B^{-\mathrm{T}} e_2 = (1/52, 19/26, 51/52, 3/13)^{\mathrm{T}}.$
4. $J \ = \{8, 7\} \neq \emptyset.$
5. $\max\{|1/52|, |3/13|\} = 3/13 > 0, \ q = 7.$
6. $\bar{a}_7 \ = B^{-1} a_7 = (2/39, 3/13, 1/39, 4/13)^{\mathrm{T}}.$
7. $\nu \ = -(7/26)/(3/13) = -7/6,$
$\bar{a}_{n+1} = (-17/78, 7/26, 11/78, 5/26)^{\mathrm{T}} + (-7/6)(2/39, 3/13 - 1, 1/39, 4/13)^{\mathrm{T}}$
$= (-5/18, 7/6, 1/9, -1/6)^{\mathrm{T}}.$

8.
$$B^{-1} = \begin{pmatrix} 1 & -2/9 & & \\ & 13/3 & & \\ & -1/9 & 1 & \\ & -4/3 & & 1 \end{pmatrix} \begin{pmatrix} 1 & 2/39 & 5/156 & -317/156 \\ & 3/13 & 1/52 & 51/52 \\ & 1/39 & -17/156 & 17/156 \\ & 4/13 & -3/52 & 55/52 \end{pmatrix}$$
$$= \begin{pmatrix} 1 & & 1/36 & -9/4 \\ & 1 & 1/12 & 17/4 \\ & & -1/9 & \\ & & -1/12 & -1/4 \end{pmatrix}.$$

9. $\bar{x}_B = (4/7, -7, 57/14, 79/14)^{\mathrm{T}},\ B = \{6, 7, 2, 4\},$
$\bar{x}_N = (-10_{(-)}, 15^{(+)}, 10^{(+)}, 14^{(+)})^{\mathrm{T}},\ N = \{8, 3, 1, 5\}.$

Iteration 6:

1. $\alpha = \min\{(-9 - 4/7)/(-5/18), (13 - (-7))/(7/6), (12 - (57/14))/(1/9), (-3 - (79/14))/(-1/6)\}$
$= \min\{1, 206/35, 120/7, 999/14, 363/7\} = 120/7,\ p = 2.$
2. $\bar{x}_B = (4/7, -7, 57/14, 79/14)^{\mathrm{T}} + (120/7)(-5/18, 7/6, 1/9, -1/6)^{\mathrm{T}}$
$= (-88/21, 13, 251/42, 39/14)^{\mathrm{T}}, \bar{x}_9 = -1395/14 - 120/7 = -1635/14.$
3. $\sigma_N = N^{\mathrm{T}} B^{-\mathrm{T}} e_2 = (1/12, 19/6, 17/4, 13/3)^{\mathrm{T}}.$
4. $J = \{8\} \neq \emptyset.$
5. $\max\{|1/12|\} = 1/12 > 0,\ q = 8.$
6. $\bar{a}_8 = B^{-1} a_8 = (1/36, 1/12, -1/9, -1/12)^{\mathrm{T}}.$
7. $\nu = -(7/6)/(1/12) = -14,$
$\bar{a}_{n+1} = (-5/18, 7/6, 1/9, -1/6)^{\mathrm{T}} + (-14)(1/36, 1/12 - 1, -1/9, -1/12)^{\mathrm{T}}$
$= (-2/3, 14, 5/3, 1)^{\mathrm{T}}.$

8.
$$B^{-1} = \begin{pmatrix} 1 & -4/13 & & \\ & 12 & & \\ & 4/3 & 1 & \\ & 1 & & 1 \end{pmatrix} \begin{pmatrix} 1 & & 1/36 & -9/4 \\ & 1 & 1/12 & 17/4 \\ & & -1/9 & \\ & & -1/12 & -1/4 \end{pmatrix}$$
$$= \begin{pmatrix} 1 & -4/13 & 1/468 & -185/52 \\ & 12 & 1 & 51 \\ & 4/3 & & 17/3 \\ & 1 & & 4 \end{pmatrix}.$$

9. $\bar{x}_B = (-88/21, -10, 251/42, 39/14)^{\mathrm{T}},\ B = \{6, 8, 2, 4\},$
$\bar{x}_N = (13^{(+)}, 15^{(+)}, 10^{(+)}, 14^{(+)})^{\mathrm{T}},\ N = \{7, 3, 1, 5\}.$

Iteration 7:

1. $\alpha = \min\{(-9 - (-88/21))/(-2/3), (21 - (-10))/14, (12 - (251/42))/(5/3), (27 - (39/14))/1\}$
$= \min\{101/14, 31/14, 253/70, 339/14\} = 31/14,\ p = 2.$
2. $\bar{x}_B = (-88/21, -10, 251/42, 39/14)^{\mathrm{T}} + (31/14)(-2/3, 14, 5/3, 1)^{\mathrm{T}}$
$= (-17/3, 21, 29/3, 5)^{\mathrm{T}}, \bar{x}_9 = -1635/14 - 31/14 = -119.$

3. $\sigma_N = N^{\mathrm{T}} B^{-\mathrm{T}} e_2 = (1, 166, 51, 52)^{\mathrm{T}}$.
4. $J = \emptyset$. The basic optimal solution and optimal value :

$$\bar{x} = (10, 29/3, 15, 5, 14, -17/3, 13, 21)^{\mathrm{T}},$$

$$\bar{x}_9 = (-1, -3, -2, 4, -5)(10, 29/3, 15, 5, 14)^{\mathrm{T}} = -119.$$

Finally, we formulate the tableau version of Algorithm 19.1.1 based on Table 15.1.

Algorithm 19.1.2 (Generalized reduced simplex algorithm: tableau form). Initial: improved reduced tableau of form (16.6), associated with $(\bar{x}, \bar{x}_{n+1})$ with (19.2) and (19.3) satisfied. This algorithm solves the reduced bounded-variable problem (19.1).

1. Determine stepsize α and row index p by (19.5) and (19.6).
2. If $\alpha \neq 0$, add α times of x_{n+1} column to $\bar{x}_B$ column $\bar{x}_{n+1} = \bar{x}_{n+1} - \alpha$.
3. Stop if $J = \{j \in \Gamma \mid \bar{a}_{p,n+1}\bar{a}_{pj} > 0\} \cup \{j \in \Pi \mid \bar{a}_{p,n+1}\bar{a}_{pj} < 0\} = \emptyset$ (optimality achieved).
4. Determine column index $q \in \arg\max_{j \in J} |\bar{a}_{pj}|$.
5. Convert $\bar{a}_{pq}$ to 1, and eliminate the other nonzeros in the column by elementary transformations (see the Note below).
6. Go to step 1.

Note Manipulations in step 5 does not touch $\bar{x}_B$ column but replace its pth component by $\bar{x}_q$.

When executing the preceding tableau algorithm, the solution $\bar{x}$, not just $\bar{x}_B$, should be traced iteration by iteration, e.g., via the tableau used in the next section.

19.2 Generalized Reduced Phase-I

As it is easy to turn the reduced and conventional simplex tableaus from each to another, a Phase-I procedure for the generalized simplex method can be utilized to get the generalized reduced method started, and vice versa. On the other hand, it is not difficult to generalize Phase-I procedures for the standard LP problem (Chap. 13) to get the generalized reduced method started.

Based on the most-obtuse-angle heuristics, the generalized dual reduced simplex method can be modified to a Phase-I procedure. To this end, it is only needed to modify steps 4 and 13 of Algorithm 19.6.1, as well as meanings of the exits.

Algorithm 19.2.1 (Generalized reduced Phase-I algorithm: tableau form). Initial: D-reduced tableau of form (19.6), q column as the datum column, associated with solution $\bar{x}$. This algorithm finds a feasible D-reduced tableau to the bounded-variable problem (19.6).

1. Determine ρ_q by (19.28).
2. Go to step 8 if $\rho_q = 0$.
3. Stop if N_1 defined by (19.29) is empty (infeasible problem).
4. Determine β and column index q' such that $q' \in \arg\max_{j\in N_1} |\bar{\omega}_j|$, $\beta = -\bar{z}_{q'}/\bar{\omega}_{q'}$.
5. If $\beta \neq 0$, add β times of the datum row to the bottom row.
6. Update $\bar{x}_q, \bar{x}_{q'}$ and $\bar{x}_B$ by (19.38), (19.40) and (19.41).
7. Set $q = q'$, and go to step 1.
8. Compute $\rho_{j_t}, t = 1, \cdots, m-1$ by (19.28), and determine s by (19.43).
9. If $\rho_{j_s} = 0$, compute $\bar{f} = c^{\mathrm{T}}\bar{x}$, and stop (feasibility achieved).
10. Go to step 13 if $\bar{a}_{i_s,q} = 0$.
11. Convert $\bar{a}_{i_s,q}$ to 1, and eliminate the other nonzeros in the column by elementary transformations.
12. Set $q = j_s$, and go to step 3.
13. Determine β and column index q' such that $q' \in \arg\max_{j\in N_2} |\bar{a}_{i_s,j}|$, $\beta = -\bar{z}_{q'}/\bar{a}_{i_s,q'}$.
14. Update $\bar{x}_{j_s}, \bar{x}_q$ and $\bar{x}_B$ by (19.47)–(19.49).
15. Go to step 1.

Example 19.2.1. Solve the following problem by the two-phase generalized reduced simplex algorithm, using Algorithm 19.2.1 in Phase-I:

$$\begin{array}{ll} \min & f = -7x_1 + 6x_2 + 3x_3 - 4x_4, \\ \text{s.t.} & -2x_1 - 3x_2 + 7x_3 - 5x_4 + x_5 = 2, \\ & 3x_1 - 5x_2 - 2x_3 - x_4 + x_6 = -15, \\ & -2x_1 - 4x_2 + 3x_3 - 6x_4 + x_7 = 0, \\ & -5x_1 + 6x_2 - 3x_3 + 3x_4 + x_8 = 7, \\ & -7 \le x_1 \le 6, \quad -3 \le x_2 \le 9, \quad -5 \le x_3 \le 12, \\ & -4 \le x_4 \le 11, \quad -7 \le x_5 \le 5, \quad -8 \le x_6 \le 14, \\ & -7 \le x_7 \le 7, \quad -6 \le x_8 \le 10. \end{array}$$

Answer Phase-I: Initial tableau is

x_1	x_2	x_3	x_4	x_5	x_6	x_7	x_8	RHS
−2	−3	7	−5	1				2*
3	−5	−2	−1		1			−15
−2	−4	3	−6			1		
−5	6	−3	3				1	7
−7	6	3	−4					

Iteration 1:
Convert the preceding to a D-reduced tableau.
Set $\bar{x}_N = (-7_{(-)}, 9^{(+)}, 12^{(+)}, -4_{(-)})^{\mathrm{T}}$ $(N = \{1,2,3,4\})$.
$\bar{x}_B = b - N\bar{x}_N = (-89, 71, -38, -34)^{\mathrm{T}}$ $(B = \{5,6,7,8\})$.
$\rho_5 = -7 - (-89) = 82, \rho_6 = 14 - (71) = -57,\ \rho_7 = -7 - (-38) = 31.$
$\rho_8 = -6 - (-34) = 28.\ \max\{|82|, |-57|, |31|, |28|\} = 82,\ p = 1.$

To enter RHS column to the basis, multiply row 1 by $1/2$, then add $15, -7$ times of it to rows 2,4, respectively:

	x_1	x_2	x_3	x_4	x_5	x_6	x_7	x_8	RHS
	-1	$-3/2$	$7/2$	$-5/2$	$1/2$				1
	-12	$-55/2$	$101/2$	$-77/2$	$15/2$	1			
	-2	-4	3	-6			1		
	2	$33/2$	$-55/2$	$41/2$	$-7/2$			1	
	-7	6	3	-4					
u	6	9	12	11	5	14	7	10	
$\bar{x}$	-7	9	12	-4	-89	71	-38	-34	
l	-7	-3	-5	-4	-7	-8	-7	-6	

The first row is its datum row ($r = 1$); x_5 is its datum column ($q = 5$). $\bar{x}_B = (71, -38, -34)^{\mathrm{T}}(B = \{6, 7, 8\})$. $\bar{x}_N = (-7_{(-)}, 9^{(+)}, 12^{(+)}, -4_{(-)}, -89)^{\mathrm{T}}$ ($N = \{1, 2, 3, 4, 5\}$).

Phase-I: Call Algorithm 19.2.1:

Iteration 2:

1. $\rho_5 = -7 - (-89) = 82 \neq 0$.
3. $\text{sign}(\rho_5 \bar{a}_{1,5}) = 1,\ N_1 = \{1, 3, 4\} \neq \emptyset$.
4. $\max\{|-1|, |7/2|, |-5/2|\} = 7/2, q' = 3,\ \beta = -3/(7/2) = -6/7$.
5. add $-6/7$ times of row 1 to row 5.
6. $\bar{x}_5 = -89 + 82 = -7$; $\bar{x}_3 = 12 - ((1/2)/(7/2))82 = 2/7$;

$$\begin{aligned}\bar{x}_B(r) &= (71, -38, -34)^{\mathrm{T}} - 82(-1/2, 15/2, 0, -7/2)^{\mathrm{T}} \\ &\quad - (2/7 - 12)(-7/2, 101/2, 3, -55/2)^{\mathrm{T}} \\ &= (1, 333/7, -20/7, -484/7)^{\mathrm{T}}(B(r) = \{6, 7, 8\}).\end{aligned}$$

7. $q = 3$.

	x_1	x_2	x_3	x_4	x_5	x_6	x_7	x_8	RHS
	-1	$-3/2$	$7/2$	$-5/2$	$1/2$				1
	-12	$-55/2$	$101/2$	$-77/2$	$15/2$	1			
	-2	-4	3	-6			1		
	2	$33/2$	$-55/2^{*}$	$41/2$	$-7/2$			1	
	$-43/7$	$51/7$		$-13/7$	$-3/7$				$6/7$
u	6	9	12	11	5	14	7	10	
$\bar{x}$	-7	9	$2/7$	-4	-7	$333/7$	$-20/7$	$-484/7$	
l	-7	-3	-5	-4	-7	-8	-7	-6	

Iteration 3:

1. $\rho_3 = 0$.
8. $\rho_6 = 14 - 333/7 = -235/7 < 0,\ \rho_7 = 0,$
   $\rho_8 = -6 - (-484/7) = 442/7 > 0;\ s = 3,\ i_s = 4.$

10. $\bar{a}_{4,3} = -55/2 \neq 0$.
11. multiply row 4 by $-2/55$, then add $-7/2, -101/2, -3$ times of row 4 to rows 1,2,3, respectively.
12. $q = 8$.

	x_1	x_2	x_3	x_4	x_5	x_6	x_7	x_8	RHS
	$-41/55$	$3/5$		$6/55$	$3/55$			$7/55$	1
	$-458/55$	$14/5$		$-47/55$	$59/55$	1		$101/55$	
	$-98/55$	$-11/5$		$-207/55$	$-21/55$		1	$6/55$	
	$-4/55$	$-3/5$	1	$-41/55$	$7/55$			$-2/55$	
	$-43/7$	$51/7$		$-13/7$	$-3/7$				$6/7$
u	6	9	12	11	5	14	7	10	
$\bar{x}$	-7	9	$2/7$	-4	-7	$333/7$	$-20/7$	$-484/7$	
l	-7	-3	-5	-4	-7	-8	-7	-6	

Iteration 4:

3. $\operatorname{sign}(\rho_8 \bar{a}_{1,8}) = 1,\ N_1 = \{1, 2\} \neq \emptyset$.
4. $\max\{|-41/55|, |-3/5|\} = 41/55, q' = 1,\ \beta = -(-43/7)/(-41/55)$.
5. add β times of row 1 to row 5.
6. $\bar{x}_8 = -484/7 + 442/7 = -6;\ \bar{x}_1 = -7 - ((7/55)/(-41/55))(442/7)$
   $= 155/41$.
   $$\begin{aligned}\bar{x}_B(r) = (333/7, -20/7, 2/7)^{\mathrm{T}} &- (442/7)(-7/55, 101/55, 6/55, -2/55)^{\mathrm{T}} \\ &- (155/41 - (-7))\,(41/55, -458/55, -98/55, -4/55)^{\mathrm{T}} \\ = (1, 877/41, 388/41, 138/41)^{\mathrm{T}}\ &(B = \{6, 7, 3\}).\end{aligned}$$
7. $q\ = 1$.

	x_1	x_2	x_3	x_4	x_5	x_6	x_7	x_8	RHS
	$-41/55$	$3/5$		$6/55$	$3/55$			$7/55$	1
	$-458/55$*	$14/5$		$-47/55$	$59/55$	1		$101/55$	
	$-98/55$	$-11/5$		$-207/55$	$-21/55$		1	$6/55$	
	$-4/55$	$-3/5$	1	$-41/55$	$7/55$			$-2/55$	
		$96/41$		$-113/41$	$-36/41$			$-43/41$	$373/41$
u	6	9	12	11	5	14	7	10	
$\bar{x}$	$\frac{155}{41}$	9	$\frac{138}{41}$	-4	-7	$\frac{877}{41}$	$\frac{388}{41}$	-6	
l	-7	-3	-5	-4	-7	-8	-7	-6	

Iteration 5:

1. $\rho_1 = 0$.
8. $\rho_6 = 14 - 877/41 = -303/41 < 0,\ \rho_7 = 7 - 388/41 = -101/41,\ \rho_3 = 0;$
   $s\ = 1,\ i_s = 2$.
10. $\bar{a}_{2,1} = -458/55 \neq 0$.

11. multiply row 2 by $-55/458$, then add $41/55, 98/55, 4/55$ times of row 2 to rows 1,3,4, respectively.
12. $q = 6$.

	x_1	x_2	x_3	x_4	x_5	x_6	x_7	x_8	RHS
		80/229		85/458	−19/458	−41/458		−17/458	1
	1	−77/229		47/458	−59/458	−55/458		−101/458	
		−641/229		−820/229	−140/229	−49/229	1	−65/229	
		−143/229	1	−169/229	27/229	−2/229		−12/229	
		96/41		−113/41	−36/41			−43/41	373/41
u	6	9	12	11	5	14	7	10	
$\bar{x}$	$\frac{155}{41}$	9	$\frac{138}{41}$	−4	−7	$\frac{877}{41}$	$\frac{388}{41}$	−6	
l	−7	−3	−5	−4	−7	−8	−7	−6	

Iteration 6:

3. $\text{sign}(\rho_6 \bar{a}_{1,6}) = 1,\ N_1 = \{2, 5, 8\} \neq \emptyset$.
4. $\max\{|-80/229|, |19/458|, |17/458|\}{=}80/229, q'{=}2,\ \beta{=}-(96/41)/(80/229)$.
5. add β times of row 1 to row 5.
6. $\bar{x}_6 = 877/41 - 303/41 = 14;$
   $\bar{x}_2 = 9 - ((-41/458)/(80/229))(-303/41) = 1{,}137/160;$
   $\bar{x}_B = (155/41, 388/41, 138/41)^{\mathrm{T}} - (-303/41)(41/458, -55/458, -49/229, -2/229)^{\mathrm{T}} - (1{,}137/160-9)(-80/229, -77/229, -641/229, -143/229)^{\mathrm{T}}$
   $= (1, 361/160, 413/160, 339/160)^{\mathrm{T}} (B = \{1, 7, 3\})$.
7. $q = 2$.

	x_1	x_2	x_3	x_4	x_5	x_6	x_7	x_8	RHS
		80/229		85/458	−19/458	−41/458		−17/458	1
	1	−77/229		47/458	−59/458	−55/458		−101/458	
		−641/229		−820/229	−140/229	−49/229	1	−65/229	
		−143/229	1	−169/229	27/229	−2/229		−12/229	
				−4	−3/5	3/5		−4/5	491/205
u	6	9	12	11	5	14	7	10	
$\bar{x}$	$\frac{361}{160}$	$\frac{1{,}137}{160}$	$\frac{339}{160}$	−4	−7	14	$\frac{413}{160}$	−6	
l	−7	−3	−5	−4	−7	−8	−7	−6	

Iteration 7:

1. $\rho_2 = 0$;
8. $\rho_1 = \rho_7 = \rho_3 = 0$; feasibility achieved.

The RHS column leaves and the datum column enters the basis ($p = 1, q = 2$): multiply row 1 by $229/80$, then add $77/229, 641/229, 143/229$ times of row 1 to row 2,3,4, respectively:

Overwrite the RHS column by the f column.

	x_1	x_2	x_3	x_4	x_5	x_6	x_7	x_8	f
		1		17/32	−19/160	−41/160		−17/160	
	1			9/32	−27/160	−33/160		−41/160	
				−67/32	−151/160	−149/160	1	−93/160	
			1	−13/32	7/160	−27/160		−19/160	
				−4*	−3/5	3/5		−4/5	−1
u	6	9	12	11	5	14	7	10	
$\bar{x}$	$\frac{361}{160}$	$\frac{1{,}137}{160}$	$\frac{339}{160}$	−4	−7	14	$\frac{413}{160}$	−6	
l	−7	−3	−5	−4	−7	−8	−7	−6	

Iteration 8:
Take $p = 5$. $\max\{|-4|, |-3/5|, |3/5|, |-4/5|\} = 4, q = 4$.

Multiply row 5 by $-1/4$, then add $17/32, -9/32, 67/32, 13/32$ times of row 5 to rows 1,2,3,4, respectively:

and add $\bar{x}_B$ column, yielding a feasible improved reduced tableau (x_9 and f columns are equal):

	x_1	x_2	x_3	x_4	x_5	x_6	x_7	x_8	x_9	$\bar{x}_B$
		1			127/640	113/640		17/80	17/128	1,137/160
	1				−27/128	−21/128		−5/16	−9/128	361/160
					−403/640	−797/640*	1	−13/80	67/128	413/160
			1		67/640	−147/640		−3/80	13/128	339/160
				1	3/20	−3/20		1/5	1/4	−4
u	6	9	12	11	5	14	7	10		
$\bar{x}$	$\frac{361}{160}$	$\frac{1{,}137}{160}$	$\frac{339}{160}$	−4	−7	14	$\frac{413}{160}$	−6		
l	−7	−3	−5	−4	−7	−8	−7	−6		

Phase-II: Call Algorithm 19.1.2.

Iteration 9:

1. $\alpha_1 = (-3 - 1{,}137/160)/(-17/128) = 6{,}468/85$,
   $\alpha_2 = (-7 - 361/160)/(-9/128) = 5{,}924/45$,
   $\alpha_3 = (7 - 413/160)/(67/128) = 2{,}828/335$,
   $\alpha_4 = (12 - 339/160)/(13/128) = 6{,}324/65$,
   $\alpha_5 = (11 - (-4))/(1/4) = 60,\ p = 3,\ \alpha = 2{,}828/335$.
3. Add α times of x_9 column to the $\bar{x}_B()$ column.
4. $J = \{6\} \neq \emptyset$.
6. $\max\{|-797/640|\} = 797/640, q = 6$.
7. Multiply row 3 by $-640/797$, then add $-113/640, 21/128, 147/640, 3/20$ times of row 3 to rows 1,2,4,5, respectively:
   (Do not touch $\bar{x}_B$ column, except for its third component replace by $\bar{x}_6 = 14$).

	x_1	x_2	x_3	x_4	x_5	x_6	x_7	x_8	x_9	$\bar{x}_B$
		1			87/797		113/797	151/797*	165/797	401/67
	1				−102/797		−105/797	−232/797	−111/797	557/335
					403/797	1	−640/797	104/797	−335/797	14
			1		176/797		−147/797	−6/797	4/797	997/335
				1	180/797		−96/797	175/797	149/797	−633/335
u	6	9	12	11	5	14	7	10		
$\bar{x}$	$\frac{557}{335}$	$\frac{401}{67}$	$\frac{997}{335}$	$-\frac{633}{335}$	−7	14	7	−6		
l	−7	−3	−5	−4	−7	−8	−7	−6		

Iteration 10:

1. $\alpha_1 = (-3 - 401/67)/(-165/797) = 13{,}758/317$,
   $\alpha_2 = (-7 - 557/335)/(-111/797) = 35{,}516/571$,
   $\alpha_3 = (-8 - 14)/(-335/797) = 7{,}537/144$,
   $\alpha_4 = (12 - 997/335)/(4/797) = 219{,}357/122$,
   $\alpha_5 = (11 - (-633/335))/(149/797) = 8{,}963/130$, $p = 1$, $\alpha = 13{,}758/317$.
3. Add α times of x_9 column to the $\bar{x}_B$ column.
4. $J = \{5, 8\} \neq \emptyset$.
6. $\max\{|-87/797|, |-151/797|\} = 151/797, q = 8$.
7. Multiply row 1 by 797/151, then add $232/797, -104/797, 6/797, -175/797$ times of row 1 to rows 2,3,4,5, respectively:
   (Do not touch $\bar{x}_B$ column, except for its first component replace by $\bar{x}_8 = -6$).

	x_1	x_2	x_3	x_4	x_5	x_6	x_7	x_8	x_9	$\bar{x}_B$
		797/151			−87/151		−113/151	1	−165/151	−6
	1	−232/151			6/151		13/151		27/151	−241/55
		104/151			65/151	1	−136/151		−85/151	−140/33
		−6/151	1		34/151		−27/151		2/151	527/165
		175/151		1	15/151		−43/151		−8/151	1,027/165
u	6	9	12	11	5	14	7	10		
$\bar{x}$	$-\frac{241}{55}$	−3	$\frac{527}{165}$	$\frac{1{,}027}{165}$	−7	$-\frac{140}{33}$	7	−6		
l	−7	−3	−5	−4	−7	−8	−7	−6		

Iteration 11:

1. $\alpha_1 = (10 - (-6))/(165/151) = 2{,}416/165$,
   $\alpha_2 = (6 - (-241/55))/(27/151) = 2{,}845/49$,
   $\alpha_3 = (-8 - (-140/33))/(-85/151) = 5{,}981/896$,
   $\alpha_4 = (12 - 527/165)/(2/151) = 4{,}654/7$,
   $\alpha_5 = (-4 - 1{,}027/165)/(-8/151) = 11{,}000/57$, $p = 3$, $\alpha = 5{,}981/896$.

3. Add α times of x_9 column to the $\bar{x}_B$ column:

	x_1	x_2	x_3	x_4	x_5	x_6	x_7	x_8	x_9	$\bar{x}_B$
		797/151			−87/151		−113/151	1	−165/151	22/17
	1	−232/151			6/151		13/151		27/151	−271/85
		104/151			65/151	1	−136/151		−85/151	−8
		−6/151	1		34/151		−27/151		2/151	279/85
		175/151		1	15/151		−43/151		−8/151	499/85
u	6	9	12	11	5	14	7	10		
$\bar{x}$	$-\frac{271}{85}$	−3	$\frac{279}{85}$	$\frac{499}{85}$	−7	−8	7	$\frac{22}{17}$		
l	−7	−3	−5	−4	−7	−8	−7	−6		

4. $J = \emptyset$. The optimal solution and optimal value:

$$\bar{x} = (-271/85, -3, 279/85, 499/85, -7, -8, 7, 22/17)^{\mathrm{T}},$$
$$\bar{f} = (-7, 6, 3, -4)(-271/85, -3, 279/85, 499/85)^{\mathrm{T}} = -792/85.$$

19.3 Generalized Reduced Phase-I: Single-Artificial-Variable

The generalized Phase-I procedure described in Sect. 19.2 is based on D-reduced simplex framework. In this section, another generalized Phase-I procedure, based on the reduced simplex framework, is derived by generalizing the single-artificial-variable method (Sect. 15.3).

Assume that $\bar{x}$ satisfies

$$\bar{x}_j = l_j \text{ or } u_j, \quad j \in N, \tag{19.16}$$

but not

$$l \le \bar{b} - \bar{N}\bar{x}_N \le u.$$

For given $\bar{x}_B$ such that $l \le \bar{x}_B \le u$, set the artificial column

$$\bar{a}_{n+1} = \bar{b} - \bar{N}\bar{x}_N - \bar{x}_B, \quad \bar{x}_{n+1} = 1. \tag{19.17}$$

Construct an auxiliary improved reduced tableau of form (16.6) (where three additional rows are inserted to give $u, \bar{x}$ and l). It is clear this tableau is feasible, from which the generalized reduced simplex method can get itself started. If the optimal objective value vanishes, then a feasible improved reduced tableau of the original problem is obtained, and hence it is ready to go to Phase-II.

The following algorithm results from slightly modifying Algorithm 19.1.2 to suit the special requirement of the zero optimal value of the auxiliary program.

Algorithm 19.3.1 (Tableau generalized reduced Phase-I: single-artificial-variable). Initial: an improved reduced tableau of form (16.6). $\bar{x}, \bar{a}_{n+1}, \bar{x}_{n+1}$ satisfying (19.16) and (19.17). This algorithm find a feasible improved reduced tableau for the bounded-variable problem (19.1).

1. Determine stepsize α and row index p by (19.5) and (19.6).
2. If $\alpha > \bar{x}_{n+1}$, set $\alpha = \bar{x}_{n+1}$.
3. Set $\bar{x}_{n+1} = \bar{x}_{n+1} - \alpha$.
4. If $\alpha \neq 0$, add α times of x_{n+1} column to the $\bar{x}_B$.
5. Stop if $\bar{x}_{n+1} = 0$ (feasibility achieved).
6. Stop if

$$J = \{j \in \Gamma \mid \bar{a}_{p,n+1}\bar{a}_{pj} > 0\} \cup \{j \in \Pi \mid \bar{a}_{p,n+1}\bar{a}_{pj} < 0\} = \emptyset,$$

   (infeasible problem).
7. Determine column index $q \in \arg\max_{j\in J} |\bar{a}_{pj}|$.
8. Convert $\bar{a}_{pq}$ to 1, and eliminate the other nonzeros in the column by elementary transformations,
   (without touching $\bar{x}_B$ column, only its pth component is replace by $\bar{x}_q$).
9. Go to step 1.

Example 19.3.1. Solve Example 19.2.1 by the two-phase generalized reduced simplex algorithm, using Algorithm 19.3.1 in Phase-I.

Answer Set $\bar{x}_N = (-7_{(-)}, 9^{(+)}, 12^{(+)}, -4_{(-)})^{\mathrm{T}}$ $(N = \{1,2,3,4\})$.

Given $\bar{x}_B = (0,0,0,0)^{\mathrm{T}} (B = \{5,6,7,8\})$. $\bar{a}_9 = (2,-15,0,7)^{\mathrm{T}} - (91,-86,38,41)^{\mathrm{T}} = (-89,71,-38,-34)^{\mathrm{T}}$. $\bar{x}_9 = 1$.

Form the initial auxiliary tableau:

	x_1	x_2	x_3	x_4	x_5	x_6	x_7	x_8	x_9	$\bar{x}_B$
	−2	−3	7*	−5	1				−89	
	3	−5	−2	−1		1			71	
	−2	−4	3	−6			1		−38	
	−5	6	−3	3				1	−34	
	−7	6	3	−4						
u	6	9	12	11	5	14	7	10	1	
$\bar{x}$	−7	9	12	−4					1	
l	−7	−3	−5	−4	−7	−8	−7	−6	0	

Phase-I: Call Algorithm 19.3.1.

Iteration 1:

1. $\alpha = \min\{(-7-0)/(-89), (14-0)/(71), (-7-0)/(-38), (-6-0)/(-34)\};$
   $= \min\{7/89, 14/71, 7/38, 3/17\} = 7/89, \ p = 1.$
2. $7/89 < 1$.
3. $\bar{x}_9 = 1 - 7/89 = 82/89 \neq 0$.

4. add $7/89$ times of x_9 column to the $\bar{x}_B$ column.
5. $J = \{1,3,4\} \neq \emptyset$.
6. $\max\{|-2|, |7|, |-5|\} = 7$, $q = 3$.
7. multiply row 1 by $1/7$, then add $2, -3, 3, -3$ times of row 1 to rows 2,3,4,5, respectively:
   (without touching $\bar{x}_B$ column, only its first component is replaced by $\bar{x}_3 = 12$).

	x_1	x_2	x_3	x_4	x_5	x_6	x_7	x_8	x_9	$\bar{x}_B$
	−2/7	−3/7	1	−5/7	1/7				−89/7	12
	17/7	−41/7		−17/7	2/7	1			319/7	497/89
	−8/7	−19/7		−27/7	−3/7		1		1/7	−266/89
	−41/7*	33/7		6/7	3/7			1	−505/7	−238/89
	−43/7	51/7		−13/7	−3/7					
u	6	9	12	11	5	14	7	10	1	
$\bar{x}$	−7	9	12	−4	−7	497/89	−266/89	−238/89	82/89	
l	−7	−3	−5	−4	−7	−8	−7	−6	0	

Iteration 2:

1. $\alpha = \min\{(-5-12)/(-89/7), (14-497/89)/(319/7), (7-(-266/89))/(1/7), (-6-(-238/89))/(-505/7)\}$
   $= \min\{119/89, 306/1{,}657, 6{,}223/89, 107/2{,}321\} = 107/2{,}321$, $p = 4$.
2. $107/2{,}321 < 82/89$.
3. $\bar{x}_9 = 82/89 - 107/2{,}321 = 442/505$.
4. add $107/2{,}321$ times of x_9 column to the $\bar{x}_B$ column.
6. $J = \{1,2\} \neq \emptyset$.
7. $\max\{|-41/7|, |33/7|\} = 41/7$, $q = 1$.
8. multiply row 1 by $-7/42$, then add $2/7, -17/7, 8/7, 43/7$ times of row 1 to rows 2,3,4,5, respectively:
   (without touching $\bar{x}_B$ column, only its fourth component is replaced by $\bar{x}_1 = -7$).

	x_1	x_2	x_3	x_4	x_5	x_6	x_7	x_8	x_9	$\bar{x}_B$
		−27/41	1	−31/41	5/41			−2/41	−377/41	5,764/505
		−160/41*		−85/41	19/41	1		17/41	642/41	3,881/505
		−149/41		−165/41	−21/41		1	−8/41	583/41	−1,506/505
	1	−33/41		−6/41	−3/41			−7/41	505/41	−7
		96/41		−113/41	−36/41			−43/41		
u	6	9	12	11	5	14	7	10	1	
$\bar{x}$	−7	9	$\frac{5{,}764}{505}$	−4	−7	$\frac{3{,}881}{505}$	$-\frac{1{,}506}{505}$	−6	$\frac{442}{505}$	
l	−7	−3	−5	−4	−7	−8	−7	−6	0	

Iteration 3:

1. $\alpha = \min\{(-5 - 5{,}764/505)/(-377/41), (14 - 3{,}881/505)/(642/41),$
$(7 - (-1{,}506/505))/(583/41), (6 - (-7))/(505/41)\}$
$= \min\{1{,}171/656, 221/548, 245/349, 533/505\} = 221/548,\ p = 2.$
2. $221/548 < 442/505.$
3. $\bar{x}_9 = 442/505 - 221/548 = 101/214 \neq 0.$
4. add $221/548$ times of x_9 column to the $\bar{x}_B$.
6. $J = \{2, 5, 8\} \neq \emptyset.$
7. $\max\{|-160/41|, |19/41|, |17/41|\} = 160/41,\ q = 2.$
8. multiply row 2 by $-41/160$, then add $27/41, 149/41, 33/41, -96/41$ times of row 2 to rows 1,3,4,5, respectively:
   (without touching $\bar{x}_B$ column, only its second component is replaced by $\bar{x}_2{=}9$).

	x_1	x_2	x_3	x_4	x_5	x_6	x_7	x_8	x_9	$\bar{x}_B$
			1	−13/32	7/160	−27/160		−19/160	−947/80	1,649/214
		1		17/32	−19/160	−41/160		−17/160	−321/80	9
				−67/32	−151/160	−149/160	1	−93/160	−29/80	3,834/1,393
	1			9/32	−27/160	−33/160		−41/160	727/80	−3,169/1,559
				−4	−3/5	3/5		−4/5		
u	6	9	12	11	5	14	7	10	1	
$\bar{x}$	$-\frac{3{,}169}{1{,}559}$	9	$\frac{1{,}649}{214}$	−4	−7	14	$\frac{3{,}834}{1{,}393}$	−6	$\frac{101}{214}$	
l	−7	−3	−5	−4	−7	−8	−7	−6	0	

Iteration 4:

1. $\alpha = \min\{(-5 - 1{,}649/214)/(-947/80), (-3 - 9)/(-321/80),$
$(-7 - 3{,}834/1{,}393)/(-29/80), (6 - (-3{,}169/1{,}559))/(727/80)\}$
$= \min\{3{,}498/3{,}259, 320/107, 3{,}605/134, 5{,}803/6{,}565\} = 5{,}803/6{,}565,\ p{=}2.$
2. $\alpha = 101/214.$
3. $\bar{x}_9 = 442/505 - 221/548 = 101/214 \neq 0.$
4. Add $101/214$ times of x_9 column to the $\bar{x}_B$ column.
5. Feasibility is achieved. Cover the artificial column by x_9 column of the original problem, giving the following feasible tableau.

	x_1	x_2	x_3	x_4	x_5	x_6	x_7	x_8	x_9	$\bar{x}_B$
			1	−13/32	7/160	−27/160		−19/160		339/160
		1		17/32	−19/160	−41/160		−17/160		1,137/160
				−67/32	−151/160	−149/160	1	−93/160		4,845/1,877
	1			9/32	−27/160	−33/160		−41/160		4,605/2,041
				−4*	−3/5	3/5		−4/5	−1	
u	6	9	12	11	5	14	7	10		
$\bar{x}$	$\frac{4{,}605}{2{,}041}$	$\frac{1{,}137}{160}$	$\frac{339}{160}$	−4	−7	14	$\frac{4{,}845}{1{,}877}$	−6		
l	−7	−3	−5	−4	−7	−8	−7	−6		

Take $p = 5$. $J = \{4, 5, 6, 8\} \neq \emptyset$. $\max\{|-4|, |-3/5|, |3/5|, |-4/5|\} = 4$, $q = 4$. Multiply row 5 by $-1/4$, then add $13/32, -17/32, 67/32, -9/32$ times of row 5 to rows 1,2,3,4, respectively (does not touch $\bar{x}_B$ column), leading to the following feasible reduced tableau:

	x_1	x_2	x_3	x_4	x_5	x_6	x_7	x_8	x_9	$\bar{x}_B$
			1		67/640	−147/640		−3/80	13/128	339/160
		1			−127/640	−113/640		−17/80	−17/128	1,137/160
					−403/640	−797/640*	1	−13/80	67/128	4,845/1,877
	1				−27/128	−21/128		−5/16	−9/128	4,605/2,041
				1	3/20	−3/20		1/5	1/4	−4
u	6	9	12	11	5	14	7	10		
$\bar{x}$	$\frac{4,605}{2,041}$	$\frac{1,137}{160}$	$\frac{339}{160}$	−4	−7	14	$\frac{4,845}{1,877}$	−6		
l	−7	−3	−5	−4	−7	−8	−7	−6		

Phase-II: Call Algorithm 19.1.2.

Iteration 5:

1. $\alpha = \min\{(12 - 339/160)/(13/128), (-3 - 1{,}137/160)/(-17/128), (7 - 4{,}845/1{,}877)/(67/128), (-7 - 4{,}605/2{,}041)/(-9/128), (11 - (-4))/(1/4)\}$
$= \min\{6{,}324/65, 6{,}468/85, 3{,}191/378, 5{,}924/45, 60\} = 3{,}191/378$, $p = 3$.
3. Add $3{,}191/378$ times of x_9 column to the $\bar{x}_B$ column.
4. $J = \{6\} \neq \emptyset$.
5. $\max\{|-797/640|\} = 797/640$, $q = 6$.
6. Multiply row 3 by $-640/797$, then add $147/640, 113/640, 21/128, 3/20$ times of row 3 to row 1,2,4,5, respectively:
(without touching $\bar{x}_B$ column, only its third component is replaced by $\bar{x}_6 = 14$).

	x_1	x_2	x_3	x_4	x_5	x_6	x_7	x_8	x_9	$\bar{x}_B$
			1		176/797		−147/797	−6/797	4/797	997/335
		1			−87/797		−113/797	−151/797*	−165/797	401/67
					403/797	1	−640/797	104/797	−335/797	14
	1				−102/797		−105/797	−232/797	−111/797	1,252/753
				1	180/797		−96/797	175/797	149/797	−2,857/1,512
u	6	9	12	11	5	14	7	10		
$\bar{x}$	$\frac{1,252}{753}$	$\frac{401}{67}$	$\frac{997}{335}$	$-\frac{2,857}{1,512}$	−7	14	7	−6		
l	−7	−3	−5	−4	−7	−8	−7	−6		

Iteration 6:

1. $\alpha = \min\{(12 - 997/335)/(4/797), (-3 - 401/67)/(-165/797),$
   $(-8 - 14)/(-335/7{,}797), (-7 - 1{,}252/753)/(-111/797),$
   $(11 - (-2{,}857/1{,}512))/(149/97)\}$
   $= \min\{219{,}357/7{,}122, 13{,}758/317, 7{,}537/144, 33{,}028/531, 11{,}514/167\}$
   $= 13{,}758/317, \ p = 2.$
3. Add 13,758/317 times of x_9 column to the $\bar{x}_B$ column.
4. $J = \{5, 8\} \neq \emptyset$.
5. $\max\{|-87/797|, |-151/797|\} = 151/797, \ q = 8.$
6. Multiply row 2 by $-797/151$, then add $6/797, -104/797, 232/797, -175/797$ times of row 2 to rows 1,3,4,5, respectively:
   (without touching $\bar{x}_B$ column, only its second component is replaced by $\bar{x}_8 = -6$).

	x_1	x_2	x_3	x_4	x_5	x_6	x_7	x_8	x_9	$\bar{x}_B$
		−6/151	1		34/151		−27/151		2/151	527/165
		−797/151			87/151		113/151	1	165/151	−6
		104/151			65/151	1	−136/151		−85/151	−140/33
	1	−232/151			6/151		13/151		27/151	−241/55
		175/151		1	15/151		−43/151		−8/151	1,027/165
u	6	9	12	11	5	14	7	10		
$\bar{x}$	$-\frac{241}{55}$	−3	$\frac{527}{165}$	$\frac{1{,}027}{165}$	−7	$-\frac{140}{33}$	7	−6		
l	−7	−3	−5	−4	−7	−8	−7	−6		

Iteration 7:

1. $\alpha = \min\{(12 - 527/165)/(2/151), (10 - (-6))/(165/151),$
   $(-8 - (-140/33))/7(-85/151), (6. - (-241/55))/(27/151),$
   $(-4 - 1{,}027/165)/(-8/151)\}$
   $= \min\{4{,}654/7, 2{,}416/165, 75{,}981/896, 2{,}845/49, 11{,}000/57\}$
   $= 5{,}981/896, \ p = 3.$
3. Add 5,981/896 times of x_9 column to the $\bar{x}_B$ column.

	x_1	x_2	x_3	x_4	x_5	x_6	x_7	x_8	x_9	$\bar{x}_B$
		−6/151	1		34/151		−27/151		2/151	279/85
		−797/151			87/151		113/151	1	165/151	22/17
		104/151			65/151	1	−136/151		−85/151	−8
	1	−232/151			6/151		13/151		27/151	−271/85
		175/151		1	15/151		−43/151		−8/151	499/85
u	6	9	12	11	5	14	7	10		
$\bar{x}$	$-\frac{271}{85}$	−3	$\frac{279}{85}$	$\frac{499}{85}$	−7	−8	7	$\frac{22}{17}$		
l	−7	−3	−5	−4	−7	−8	−7	−6		

4. $J = \emptyset$. The basic optimal solution and optimal value:

$$\bar{x} = (-271/85, -3, 279/85, 499/85, -7, -8, 7, 22/17)^{\mathrm{T}},$$

$$\bar{x}_9 = (-7, 6, 3, -4)(-271/85, -3, 279/85, 499/85)^{\mathrm{T}} = -792/85.$$

19.4 Generalized Dual Reduced Simplex Method

Generalized reduced Algorithm 19.1.2 pursues dual feasibility while maintaining primal feasibility. Assume that an improved reduced simplex tableau is dual feasible, i.e., for some row index p, it holds that

$$J = \{j \in \Gamma \mid \bar{a}_{p,n+1}\bar{a}_{pj} > 0\} \cup \{j \in \Pi \mid \bar{a}_{p,n+1}\bar{a}_{p\,j} < 0\} = \emptyset \qquad (19.18)$$

and $\bar{x}_{j_p} = l_{j_p}$ if $\bar{a}_{p,n+1} < 0$ whereas $\bar{x}_{j_p} = u_{j_p}$ if $\bar{a}_{p,n+1} > 0$. The following algorithm is its dual version, derived via pursuing primal feasibility while maintaining dual feasibility.

Algorithm 19.4.1 (Generalized dual reduced simplex algorithm: tableau form). Initial: dual feasible improved reduced tableau of form (16.6). This algorithm solves the bounded-variable problem (19.1).

1. Select row index $r \in \arg\max\{|\rho_i| \mid i = 1, \cdots, m+1,\ i \neq p\}$, where ρ_i is defined by (25.48).
2. Stop if $\rho_r = 0$ (optimality achieved).
3. Set $p = r$, and go to step 7 if

$$J = \{j \in \Gamma \mid \mathrm{sign}(\rho_r)\bar{a}_{r\,j} < 0\} \cup \{j \in \Pi \mid \mathrm{sign}(\rho_r)\bar{a}_{r\,j} > 0\} = \emptyset.$$

4. Determine index q such that $\beta = |\bar{a}_{p\,q}/\bar{a}_{r\,q}| = \min_{j \in J} |\bar{a}_{p\,j}/\bar{a}_{r\,j}|$.
5. Convert $\bar{a}_{r\,q}$ to 1, and eliminate the other nonzeros of the column by elementary transformations (without touching the $\bar{x}_B$ column except for its rth component replaced by the value of $\bar{x}_q$).
6. Update $\bar{x}_B = \bar{x}_B - \rho_r \bar{a}_{j_r}$, $\bar{x}_{j_r} = \bar{x}_{j_r} + \rho_r$.
7. Compute $\alpha = -\rho_p/\bar{a}_{p,n+1}$.
8. Stop if $\alpha < 0$ (dual unbounded or infeasible problem).
9. If $\alpha \neq 0$, add $-\alpha$ times of x_{n+1} column to $\bar{x}_B$ column, and set $\bar{x}_{n+1} = \bar{x}_{n+1} + \alpha$.
10. Go to step 1.

Practically, the bound flipping tactic should be incorporated into Algorithm 19.4.1, with its revised version formed therewith.

Example 19.4.1. Solve the following problem by Algorithm 16.3.1:

$$
\begin{array}{rl}
\min & x_8 = x_1 - 2x_2 + 5x_3 + x_4, \\
\text{s.t.} & 3x_1 - 2x_2 + x_3 - 4x_4 + x_5 = -3, \\
& 6x_1 + 3x_2 - 3x_3 + 6x_4 + x_6 = 4, \\
& -5x_1 - 4x_2 - 6x_3 + 2x_4 + x_7 = -7, \\
& 0 \le x_1 \le 10, \quad 1 \le x_2 \le 10, \quad -20 \le x_3 \le 5, \\
& -4 \le x_4 \le 7, \quad 0 \le x_5, x_6, x_7 \le 15.
\end{array}
$$

Answer $B = \{5,6,7\}$, $N = \{1,2,3,4\}$. Set $\bar{x}_N = (0_{(-)}, 10^{(+)}, -20_{(-)}, -4_{(-)})^{\mathrm{T}}$, then

$$
\bar{x}_B = b - N x_N = \begin{pmatrix} -3 \\ 4 \\ -7 \end{pmatrix} - \begin{pmatrix} 3 & -2 & 1 & -4 \\ 6 & 3 & -3 & 6 \\ -5 & -4 & -6 & 2 \end{pmatrix} \begin{pmatrix} 0 \\ 10 \\ -20 \\ -4 \end{pmatrix} = \begin{pmatrix} 21 \\ -62 \\ -79 \end{pmatrix},
$$

$\bar{x}_8 = (1, -2, 5, 1)(0, 10, -20, -4)^T = -124.$

Initial tableau:

	x_1	x_2	x_3	x_4	x_5	x_6	x_7	x_8	$\bar{x}_B$
	3	−2	1	−4	1				21
	6	3	−3	6		1			−62
	−5	−4	−6	2			1		−79
	1*	−2	5	1				−1	124
u	10	10	5	7	15	15	15	$+\infty$	
$\bar{x}$	0	10	−20	−4	21	−62	−79	−124	
l	0	1	−20	−4	0	0	0	$-\infty$	

Iteration 1: Turn the preceding to an improved reduced tableau by taking $\bar{a}_{4,1}$ at the pivot.

Add $-3, -6, 5$ times of row 4 to rows 1,2,3, respectively (without touching the $\bar{x}_B$ column but its fourth component is replaced by the value, 0, of $\bar{x}_1$).

	x_1	x_2	x_3	x_4	x_5	x_6	x_7	x_8	$\bar{x}_B$
		4	−14	−7	1			3	21
		15	−33	0		1		6	−62
		−14	19	7			1	−5	−79
	1	−2	5	1				−1	0
u	10	10	5	7	15	15	15	$+\infty$	
$\bar{x}$	0	10	−20	−4	21	−62	−79	−124	
l	0	1	−20	−4	0	0	0	$-\infty$	

which is a dual feasible reduced tableau with $p = 4$, $p = 4$.

Iteration 2:

1. $\rho_1 = 15 - 21 = -6$, $\rho_2 = 0 - (-62) = 62$, $\rho_3 = 0 - (-79) = 79$. $\max\{|-6|, |62|, |79|\} = 79$, $r = 3$.

3. $J \ = \emptyset,\ p = 3.$
7. $\alpha \ = -79/(-5) > 0.$
9. $\bar{x}_B = (21, -62, -79, 0)^T - (79/5)(3, 6, -5, -1)^T = (-132/5, -784/5, 0, 79/5)^T$
   $\bar{x}_8 \ = -124 + 79/5 = -541/5.$

	x_1	x_2	x_3	x_4	x_5	x_6	x_7	x_8	$\bar{x}_B$
		4	−14	−7	1			3	−132/5
		15	−33*	0		1		6	−784/5
		−14	19	7			1	−5	0
	1	−2	5	1				−1	79/5
u	10	10	5	7	15	15	15	$+\infty$	
$\bar{x}$	79/5	10	−20	−4	−132/5	−784/5	0	−541/5	
l	0	1	−20	−4	0	0	0	$-\infty$	

Iteration 3:

1. $\rho_1 \ = 0 - (-132/5) = 132/5,\ \rho_2 = 0 - (-784/5) = 784/5,$
   $\rho_4 \ = 10 - 79/5 = -29/5.$
   $\max\{|132/5|, |784/5|, |-29/5|\} = 784/5,\ r = 2.$
3. $J \ = \{2, 3\}.$
4. $\min\{|(-14)/15|, |19/(-33)|\} = 19/33,\ q = 3.$
5. Multiply row 2 by $-1/33$, and add $14, -19, -5$ times of row 2 to rows 1,3,4, respectively
   (without touching the $\bar{x}_B$ column but its second component is replaced by the value, -20, of $\bar{x}_3$).
6. $\bar{x}_B = (-132/5, -20, 0, 79/5)^T - (784/5)(-14/33, -1/33, 19/33, 5/33)^T$
   $= (1{,}324/33, -2{,}516/165, -14{,}896/165, -1{,}313/165)^T, \bar{x}_6 = 0.$
7. $\alpha \ = -(14{,}896/165)/(-17/11) = 5{,}199/89 > 0).$
9. $\bar{x}_B = (1{,}324/33, -2{,}516/165, -14{,}896/165, -1{,}313/165)^T - (5{,}199/89)(5/11, -2/11, -17/11, -1/11)^T = (502/37, -1{,}828/395, 0, -1{,}771/669)^T,$
   $\bar{x}_8 \ = -541/5 + 5{,}199/89 = -846/17.$

	x_1	x_2	x_3	x_4	x_5	x_6	x_7	x_8	$\bar{x}_B$
		−26/11		−7	1	−14/33		5/11	$\frac{502}{37}$
		−5/11	1	0		−1/33		−2/11	$-\frac{1{,}828}{395}$
		−59/11		7		19/33	1	−17/11	0
	1	3/11*		1		5/33		−1/11	$-\frac{1{,}771}{669}$
u	10	10	5	7	15	15	15	$+\infty$	
$\bar{x}$	$-\frac{1{,}771}{669}$	10	$-\frac{1{,}828}{395}$	−4	$\frac{502}{37}$	0	0	$-\frac{846}{17}$	
l	0	1	−20	−4	0	0	0	$-\infty$	

Iteration 4:

1. $\rho_1, \rho_2 = 0, \rho_4 = 0 - (-1{,}771/669) = 1{,}771/669,$
   $\max\{|-1{,}771/669|\},\ r = 4.$
3. $J = \{2\}.$
4. $\min\{|(3/11)/(-59/11)|\},\ q = 2.$
5. Multiply row 4 by 11/3, and add $26/11, 5/11, 59/11$ times of row 4 to rows 1,2,3, respectively
   (without touching the $\bar{x}_B$ column but its fourth component is replaced by the value, 10, of $\bar{x}_2$).
6. $\bar{x}_B = (502/37, -1{,}828/395, 0, 10)^T - (1{,}771/669)(26/3, 5/3, 59/3, 11/3)^T$
   $= (-8{,}822/941, -3{,}851/426, -15{,}046/289, 589/2{,}007)^T, \bar{x}_1 = 0.$
7. $\alpha\ = -(38{,}992/749)/(-10/3) = 19{,}319/1{,}237 > 0).$
9. $\bar{x}_B = (-8{,}822/941, -3{,}851/426, -15{,}046/289, 589/2{,}007)^T - (19{,}319/1{,}237)$
   $(-1/3, -1/3, -10/3, -1/3)^T = (-15{,}504/3{,}721, -2{,}166/565, 0, 11/2)^T,$
   $\bar{x}_8\ = -846/17 + 19{,}319/1{,}237 = -205/6.$

	x_1	x_2	x_3	x_4	x_5	x_6	x_7	x_8	$\bar{x}_B$
	26/3			5/3	1	8/9		$-1/3$	$-\frac{97{,}125}{25{,}337}$
	5/3		1	5/3		2/9		$-1/3$	$-\frac{2{,}166}{565}$
	59/3			80/3		32/9	1	$-10/3$	0
	11/3	1		11/3		5/9		$-1/3$	$\frac{11}{2}$
u	10	10	5	7	15	15	15	$+\infty$	
$\bar{x}$	0	$\frac{11}{2}$	$-\frac{2{,}166}{565}$	-4	$-\frac{97{,}125}{25{,}337}$	0	0	$-\frac{205}{6}$	
l	0	1	-20	-4	0	0	0	$-\infty$	

Iteration 5:

1. $\rho_1\ = 97{,}125/25{,}337,\ \rho_2, \rho_4 = 0. \max\{|97{,}125/25{,}337|\},\ r = 1.$
3. $J\ = \emptyset,\ p = 1.$
7. $\alpha\ = -(97{,}125/25{,}337)/(-1/3) = 325{,}662/26{,}053 > 0.$
9. $\bar{x}_B = (-15{,}504/3{,}721, -2{,}166/565, 0, 11/2)^T - (325{,}662/26{,}053)$
   $(-1/3, -1/3, -10/3, -1/3)^T = (0, 1/3, 125/3, 29/3)^T,$
   $\bar{x}_8\ = -205/6 + 325{,}662/26{,}053 = -65/3.$

	x_1	x_2	x_3	x_4	x_5	x_6	x_7	x_8	$\bar{x}_B$
	26/3			5/3	1	8/9		$-1/3$	0
	5/3		1	5/3		2/9		$-1/3$	1/3
	59/3			80/3*		32/9	1	$-10/3$	125/3
	11/3	1		11/3		5/9		$-1/3$	29/3
u	10	10	5	7	15	15	15	$+\infty$	
$\bar{x}$	0	29/3	1/3	-4	0	0	125/3	$-65/3$	
l	0	1	-20	-4	0	0	0	$-\infty$	

Iteration 6:

1. $\rho_2, \rho_2, \rho_4 = 0,\ \rho_3 = 15 - 125/3 = -80/3,\ r = 3.$
3. $J\{1, 4, 6\} \neq \emptyset.$
4. $\min\{|(26/3)/(59/3)|, |(5/3)/(80/3)|, |(8/9)/(32/9)|\} = (5/3)/(80/3),$ $q = 4.$
5. Multiply row 3 by 3/80, and add $-5/3, -5/3, -11/3$ times of row 3 to rows 1,2,4, respectively (without touching the $\bar{x}_B$ column but its third component is replaced by the value, -4, of $\bar{x}_4$).
6. $\bar{x}_B = (0, 1/3, -4, 29/3)^T - (-80/3)(-1/16, -1/16, 3/80, -11/80)^T$ $= (-5/3, -4/3, -3, 6)^T, \bar{x}_7 = 125/3 - 80/3 = 15.$
7. $\alpha = -(5/3)/(-1/8) = 40/3 > 0).$
9. $\bar{x}_B = (-5/3, -4/3, -3, 6)^T - (40/3)(-1/8, -1/8, -1/8, 1/8)^T = (0, 1/3, - 4/3, 13/3)^T, \bar{x}_8 = -65/3 + 40/3 = -25/3.$

	x_1	x_2	x_3	x_4	x_5	x_6	x_7	x_8	$\bar{x}_B$
	119/16				1	2/3	−1/16	−1/8	0
	7/16		1			0	−1/16	−1/8	1/3
	59/80			1		2/15	3/80	−1/8	−4/3
	77/80	1				1/15	−11/80	1/8	13/3
u	10	10	5	7	15	15	15	$+\infty$	
$\bar{x}$	0	13/3	1/3	−4/3	0	0	15	−25/3	
l	0	1	−20	−4	0	0	0	$-\infty$	

Iteration 7:

1. $\rho_2, \rho_3, \rho_4 = 0.$
2. Terminated with optimal solution and objective value:

$$\bar{x} = (0, 13/3, 1/3, -4/3, 0, 0, 15)^T,\ x_8 = -25/3.$$

19.5 Generalized Dual Reduced Phase-I

When components of l and u are all finite, it is, in principle, possible to set values of nonbasic variables such that the tableau is dual feasible. Numerically, however, too large bound value is unacceptable, let alone magnitude of some components of bounds may be infinite. Therefore, it is still required to provide an initial dual feasible tableau or solution for the generalized dual reduced method.

A choice is to directly modify the reduced simplex Algorithm 19.1.2, ignoring the requirement for the initial reduced tableau being feasible, as follows.

Algorithm 19.5.1 (Generalized dual reduced Phase-I: tableau form). Initial: improved reduced tableau of form (16.6) and solution $\bar{x}$. This algorithm finds a dual feasible tableau to the bounded-variable problem (19.1).

1. Determine stepsize α and row index p by (19.5) and (19.6).
2. If $\alpha \neq 0$, add α times of x_{n+1} column to the $\bar{x}_B$ column.
3. Stop if $J = \{j \in \Gamma \mid \bar{a}_{p,n+1}\bar{a}_{pj} > 0\} \cup \{j \in \Pi \mid \bar{a}_{p,n+1}\bar{a}_{pj} < 0\} = \emptyset$ (dual feasibility achieved).
4. Determine column index $q \in \arg\max_{j\in J} |\bar{a}_{pj}|$.
5. Convert $\bar{a}_{pq}$ to 1, and eliminate the other nonzeros in the column by elementary transformations.
   (without touching $\bar{x}_B$ column, only its pth component is replace by $\bar{x}_q$).
6. Go to step 1.

Note Termination at step 3 gives a generalized dual feasible improved reduced tableau. The tableau can be easily converted to a conventional dual feasible tableau, used as a starting point for Algorithm 7.6.1: let x_{j_p} column leave and x_{n+1} column enter the basis, and change the sign of entries of the pth row to be as the objective function row; or, otherwise, is converted to a dual feasible D-reduced tableau used for getting Algorithm 19.6.1 started.

Example 19.5.1. Find a dual feasible solution of the following problem by Algorithm 19.5.1:

$$
\begin{array}{lllllllllll}
\min & x_8 = 4x_1 - 5x_2 + 2x_3 + x_4, \\
\text{s.t.} & 3x_1 & - & 4x_2 & + & x_3 & - & 5x_4 & + x_5 & & = -3, \\
& -2x_1 & - & 6x_2 & - & 3x_3 & + & 6x_4 & & + x_6 & = 26, \\
& -6x_1 & + & 2x_2 & - & 5x_3 & + & 3x_4 & & + x_7 & = 21, \\
& \multicolumn{10}{l}{-\infty \le x_1 \le 9, \quad -7 \le x_2 \le +\infty, \quad -1 \le x_3 \le 12,} \\
& \multicolumn{10}{l}{0 \le x_4 \le 8, \quad 0 \le x_5 \le 10, \quad -5 \le x_6 \le 6,} \\
& \multicolumn{10}{l}{0 \le x_7 \le +\infty.}
\end{array}
$$

Answer Set $\bar{x}_N = (9^{(+)}, -7_{(-)}, -1_{(-)}, 0_{(-)})^{\mathrm{T}}$ $(N = \{1,2,3,4\})$. $\bar{x}_B = b - N\bar{x}_N = (-3, 26, 21)^{\mathrm{T}} - (54, 27, -63, 69)^{\mathrm{T}} = (-57, -1, 84)^{\mathrm{T}}$ $B = \{5, 6, 7\}$.

Initial: improved reduced tableau (where the RHS column is listed to be easier to go to Phase-II):

	x_1	x_2	x_3	x_4	x_5	x_6	x_7	x_8	RHS	$\bar{x}_B$
	3	−4	1	−5	1				−3	−57
	−2	−6	−3	6		1			26	−1
	−6	2	−5	3			1		21	84
	4	−5*	2	1				−1		
u	9	∞	12	8	10	6	∞			
$\bar{x}$	9	−7	−1	0	−57	−1	84			
l	$-\infty$	−7	−1	0	0	−5	0			

Iteration 1: Turn the preceding to a reduced tableau.
Take $p = 4$, $J - \{1, 2\} \neq \emptyset$. $\max\{|4|, |-5|\} = 5$, $q = 2$.

Multiply row 4 by $-1/5$, then add $4, 6, -2$ times of row 4 to rows 1,2,3, respectively
(without touching $\bar{x}_B$ column, only its fourth component is replaced by $\bar{x}_2 = 7$).

	x_1	x_2	x_3	x_4	x_5	x_6	x_7	x_8	RHS	$\bar{x}_B$
	$-1/5$		$-3/5$	$-29/5$	1			$4/5$	-3	-57
	$-34/5^*$		$-27/5$	$24/5$		1		$6/5$	26	-1
	$-22/5$		$-21/5$	$17/5$			1	$-2/5$	21	84
	$-4/5$	1	$-2/5$	$-1/5$				$1/5$		-7
u	9	∞	12	8	10	6	∞			
$\bar{x}$	9	-7	-1	0	-57	-1	84			
l	$-\infty$	-7	-1	0	0	-5	0			

Call Algorithm 19.5.1.

Iteration 2:

1. $\alpha = \min\{(10-(-57))/(4/5), (6-(-1))/(6/5),$
$(0-84)/(-2/5), (\infty-(-7))/(1/5)\} = 35/6, \ p = 2.$
2. Add $35/6$ times of x_8 column to the $\bar{x}_B$ column.
3. $J = \{1, 4\} \neq \emptyset$.
4. $\max\{|-34/5|, |24/5|\} = 34/5, \ q = 1.$
5. Multiply row 2 by $-5/34$, then add $1/5, 22/5, 4/5$ times of row 2 to rows 1,3,4, respectively
(Do not touch $\bar{x}_B$ column, except the second component of which is replace by $\bar{x}_1 = 9$):

	x_1	x_2	x_3	x_4	x_5	x_6	x_7	x_8	RHS	$\bar{x}_B$
			$-15/34$	$-101/17$	1	$-1/34$		$13/17$	$-64/17$	$-157/3$
	1		$27/34$	$-12/17$		$-5/34$		$-3/17$	$-65/17$	9
			$-12/17^*$	$5/17$		$-11/17$	1	$-20/17$	$71/17$	$245/3$
		1	$4/17$	$-13/17$		$-2/17$		$1/17$	$-52/17$	$-35/6$
u	9	∞	12	8	10	6	∞			
$\bar{x}$	9	$-35/6$	-1	0	$-157/3$	6	$245/3$			
l	$-\infty$	-7	-1	0	0	-5	0			

Iteration 3.

1. $\alpha = \min\{(10-(-157/3))/(13/17), (-\infty-9)/(-3/17),$
$(0-245/3)/(-20/17), (\infty-(-35/6))/(1/17)\} = 833/12, \ p = 3.$
2. Add $833/12$ times of x_8 column to the $\bar{x}_B$ column.
3. $J = \{3\} \neq \emptyset$.
4. $\max\{|-12/17|\}, \ q = 3.$

5. Multiply row 3 by $-17/12$, then add $15/34, -27/34, -4/17$ times of row 3 to rows 1,2,4, respectively:
(without touching $\bar{x}_B$ column, only its third component is replaced by $\bar{x}_3 = -1$).

	x_1	x_2	x_3	x_4	x_5	x_6	x_7	x_8	RHS	$\bar{x}_B$
				$-49/8$	1	$3/8$	$-5/8$	$3/2$*	$-51/8$	$3/4$
	1			$-3/8$		$-7/8$	$9/8$	$-3/2$	$7/8$	$-13/4$
			1	$-5/12$		$11/12$	$-17/12$	$5/3$	$-71/12$	-1
		1		$-2/3$		$-1/3$	$1/3$	$-1/3$	$-5/3$	$-7/4$
u	9	∞	12	8	10	6	∞			
$\bar{x}$	$-13/4$	$-7/4$	-1	0	$3/4$	6	0			
l	$-\infty$	-7	-1	0	0	-5	0			

Iteration 4:

1. $\alpha = \min\{(10 - 3/4)/(3/2), (-\infty - (-13/4))/(-3/2), (12 - (-1))/(5/3), (-7 - (-7/4))/(-1/3)\} = 37/6,\ p = 1.$
2. Add $37/6$ times of x_8 column to the $\bar{x}_B$ column.
3. $J = \emptyset$.
   Dual feasibility is achieved. Convert the end tableau to a conventional dual feasible one: delete $\bar{x}_B$ column, and taking the entry at the first row and f column as a pivot to carry out elementary transformations (such that the pivot becomes -1). See the resulting tableau below (whose fourth row is now the objective row, which was the previous first row).

	x_1	x_2	x_3	x_4	x_5	x_6	x_7	x_8	RHS	$\bar{x}_B$
	1			$-13/2$	1	$-1/2$	$1/2$		$-11/2$	$-25/2$
			1	$115/18$	$-10/9$	$1/2$	$-13/18$		$7/6$	$167/18$
		1		$-73/36$	$2/9$	$-1/4$	$7/36$		$-37/12$	$-137/36$
				$49/12$	$-2/3$	$-1/4$	$5/12$	-1	$17/4$	10
u	9	∞	12	8	10	6	∞			
$\bar{x}$	$-25/2$	$-137/36$	$167/18$	0	10	6	0			
l	$-\infty$	-7	-1	0	0	-5	0			

$\rho_1, \rho_2, \rho_3 = 0$. The basic optimal solution and optimal value:

$$\bar{x} = (-25/2, -137/36, 167/18, 0, 10, 6, 0)^{\mathrm{T}},$$

$$\bar{x}_8 = 4(-25/2) - 5(-137/36) + 2(167/18) = -149/12.$$

The following algorithm is a variant of Algorithm 19.5.1, which matches Algorithm 7.5.2.

Algorithm 19.5.2 (Tableau generalized dual reduced Phase-I algorithm: a variant of the most-obtuse-angle rule). Given $0 < \tau \leq 1$. Initial: improved reduced tableau of form (16.6) and solution $\bar{x}$. This algorithm finds a dual feasible tableau to the bounded-variable problem (19.1).

The same as Algorithm 19.5.1, except its step 1 replaced by

1. Determine stepsize α and row index p such that

$$\alpha = \alpha_p = \min\{\alpha_i \mid |\bar{a}_{i,n+1}| \geq \tau\theta,\ i = 1, \cdots, m+1\},$$

where α_i is defined by (19.6).

19.6 Generalized Dual D-Reduced Simplex Method

Consider the D-reduced bounded-variable problem below:

$$\begin{array}{ll} \min & f = c^{\mathrm{T}} x, \\ \text{s.t.} & Ax = e_m, \quad l \leq x \leq u. \end{array}$$

In this section, the dual D-reduced simplex method (Sect. 17.2) is generalized to solve this problem. A tableau version will be derived first, and then its revised version will be formulated.

The key point is to update the objective row based on the datum row as far as possible.

To trace solution $\bar{x}$, three additional rows are listed at the bottom of tableau (17.3), i.e.,

	x_B^{T}	x_N^{T}	f	RHS
	I	$\bar{N}_R$		
		$\bar{w}_N^{\mathrm{T}}$		1
		$\bar{z}_N^{\mathrm{T}}$	-1	$-\bar{f}$
u	u_B^{T}	u_N^{T}		
$\bar{x}$	$\bar{x}_B^{\mathrm{T}}$	$\bar{x}_N^{\mathrm{T}}$		
l	l_B^{T}	l_N^{T}		

where the associated sets B, N, R are defined by

$$\begin{array}{ll} B = \{j_1, \cdots, j_{m-1}\}, & N = A \backslash B, \\ R = \{1, \cdots, m\} \backslash \{r\} = \{i_1, \cdots, i_{m-1}\}, & \end{array} \tag{19.19}$$

$r \notin R$; q column is the datum column, satisfying

$$\bar{\omega}_q \neq 0, \quad \bar{z}_q = 0. \tag{19.20}$$

Let $\bar{x}$ be the basic solution, defined by (17.7), such that

$$\bar{x}_j = l_j \text{ or } u_j, \quad j \in N, \ j \neq q \tag{19.21}$$

$$\bar{x}_B + \bar{N}_R \bar{x}_N = 0, \quad \sum_{j \in N} \bar{\omega}_j \bar{x}_j = 1. \tag{19.22}$$

If x_q enters the basis, $\bar{x}$ is clearly just the basic solution, associated with the new basis in the conventional sense, with objective value

$$\bar{f} = \bar{z}_N^{\mathrm{T}} \bar{x}_N - \bar{z}_{n+1}. \tag{19.23}$$

Define column index set

$$\Gamma = \{\, j \in N, j \neq q \mid \bar{x}_j = l_j\}, \quad \Pi = \{\, j \in N, j \neq q \mid \bar{x}_j = u_j\}. \tag{19.24}$$

It is clear that $\Gamma \cup \Pi = N \backslash \{q\}$, $\Gamma \cap \Pi = \emptyset$.

The D-reduced tableau, satisfying the following conditions:

$$\bar{z}_\Gamma \geq 0, \quad \bar{z}_\Pi \leq 0, \tag{19.25}$$

is called dual feasible D-reduced tableau. If components of l and u are all finite, it is always possible to set the nonbasic variables to relevant bounds such that any D-reduced tableau is dual feasible.

Proposition 19.6.1. *The objective value $\bar{f}$ of the dual feasible D-reduced tableau is a lower-bound of all feasible values.*

Proof. Let x' be an arbitrary feasible solution. Then, it is known from (19.24) that

$$\bar{x}_\Gamma = l_\Gamma \leq x'_\Gamma, \quad x'_\Pi \leq u_\Pi = \bar{x}_\Pi,$$

which together with (19.25) gives

$$\bar{z}_\Gamma^{\mathrm{T}} \bar{x}_\Gamma \leq \bar{z}_\Gamma^{\mathrm{T}} x'_\Gamma, \quad \bar{z}_\Pi^{\mathrm{T}} x'_\Pi \geq \bar{z}_\Pi^{\mathrm{T}} \bar{x}_\Pi, \quad \bar{z}_{n+1} + \bar{z}_N^{\mathrm{T}} x'_N \geq \bar{z}_{n+1} + \bar{z}_N^{\mathrm{T}} \bar{x}_N = \bar{f}. \tag{19.26}$$

This indicates that the $\bar{f}$ is a lower-bound on all feasible values, if there exits a feasible solution. □

Provided that $\bar{x}$ is dual feasible, if, in addition, it satisfies

$$l_q \leq \bar{x}_q \leq u_q, \quad l_B \leq \bar{x}_B \leq u_B, \tag{19.27}$$

then it is clearly a basic optimal solution. In the other case, a sequence of dual feasible tableaus can be generated such that the related objective value increases monotonically, until reaching a basic optimal solution or detecting primal infeasibility.

Introduce the "boundary violation amounts" as follows:

$$\rho_j = \begin{cases} l_j - \bar{x}_j, & \text{if } \bar{x}_j < l_j, \\ u_j - \bar{x}_j, & \text{if } \bar{x}_j > u_j, \\ 0, & \text{if } l_j \le \bar{x}_j \le u_j, \end{cases} \qquad j \in \{1, \cdots, n\}. \tag{19.28}$$

There will be the following two cases to be dealt with:

(1) $\bar{x}_q$ violates boundary, i.e., $\rho_q \neq 0$.

Define set

$$N_1 = \{j \in \Gamma \mid \mathrm{sign}(\rho_q \bar{\omega}_q)\,\bar{\omega}_j < 0\} \cup \{j \in \Pi \mid \mathrm{sign}(\rho_q \bar{\omega}_q)\,\bar{\omega}_j > 0\}, \tag{19.29}$$

where the sign function $\mathrm{sign}(t)$ is defined by (7.14).

Lemma 19.6.1. *Assume $\rho_q \neq 0$. If N_1 is empty, there no feasible solution exists.*

Proof. In view of (19.24), the second expression of (19.22) can be written

$$\bar{\omega}_q \bar{x}_q = 1 - \left(\sum_{j \in \Gamma} \bar{\omega}_j l_j + \sum_{j \in \Pi} \bar{\omega}_j u_j \right). \tag{19.30}$$

Assume that x' is a feasible solution, hence it holds that

$$l_j \le x'_j \le u_j \tag{19.31}$$

and that

$$\bar{\omega}_q x'_q = 1 - \left(\sum_{j \in \Gamma} \bar{\omega}_j x'_j + \sum_{j \in \Pi} \bar{\omega}_j x'_j \right). \tag{19.32}$$

Consider the case of $\rho_q \bar{\omega}_q > 0$. In this case, $N_1 = \emptyset$ implies

$$\bar{\omega}_j \ge 0, \ \forall\, j \in \Gamma, \quad \bar{\omega}_j \le 0, \ \forall\, j \in \Pi,$$

combining which and (19.31) leads to

$$\sum_{j \in \Gamma} \bar{\omega}_j x'_j + \sum_{j \in \Pi} \bar{\omega}_j x'_j \ge \sum_{j \in \Gamma} \bar{\omega}_j l_j + \sum_{j \in \Pi} \bar{\omega}_j u_j.$$

Then from the preceding, (19.30) and (19.32), it follows that

$$\begin{aligned}\bar{\omega}_q x'_q &= 1 - \left(\sum_{j\in\Gamma} \bar{\omega}_j x'_j + \sum_{j\in\Pi} \bar{\omega}_j x'_j\right) \\ &\le 1 - \left(\sum_{j\in\Gamma} \bar{\omega}_j l_j + \sum_{j\in\Pi} \bar{\omega}_j u_j\right) \\ &= \bar{\omega}_q \bar{x}_q.\end{aligned}$$

If $\bar{\omega}_q > 0,\ \rho_q > 0$, the preceding gives $x'_q \le \bar{x}_q < l_q$, as contradicts that x' is a feasible solution. If $\bar{\omega}_q < 0,\ \rho_q < 0$, on the other hand, it gives $x'_q \ge \bar{x}_q > u_q$, also contradicting that x' is a feasible solution.

In the case of $\rho_q \bar{\omega}_q < 0$, a contradiction can be derived similarly. Therefore, there no feasible solution exists. □

Now assume that $N_1 \ne \emptyset$. Determine β_1 and column index q' such that

$$\beta_1 = |\bar{z}_{q'}/\bar{\omega}'_q| = \min_{j\in N_1} |\bar{z}_j/\bar{\omega}_j|, \quad \theta_1 = \mathrm{sign}(\rho_q \bar{\omega}_q)\beta_1. \tag{19.33}$$

If $\theta_1 \ne 0$, the associated D-reduced tableau is said to be *dual nondegenerate*.

Add θ_1 times of the datum row to the bottom row. It might be well to denote the resulting tableau again by (19.6). Then it is clear that

$$\bar{z}_q = \theta_1 \bar{\omega}_q, \tag{19.34}$$

and that $\bar{\omega}_{q'} \ne 0,\ \bar{z}_{q'} = 0$, where q' column is taken as the datum column.

Let us determine the corresponding solution $\hat{x}$ and the associated objective value $\hat{f}$. Since $(\bar{x}, \bar{f})$ and $(\hat{x}, \hat{f})$ both are solutions to the system, presented by the new tableau, it holds that

$$\bar{x}_B = -\bar{N}_R \bar{x}_N, \quad \hat{x}_B = -\bar{N}_R \hat{x}_N, \tag{19.35}$$

$$\sum_{j\in N} \bar{\omega}_j \bar{x}_j = 1, \quad \sum_{j\in N} \bar{\omega}_j \hat{x}_j = 1, \tag{19.36}$$

$$\bar{f} = \bar{z}_N^{\mathrm{T}} \bar{x}_N - \bar{z}_{n+1}, \quad \hat{f} = \bar{z}_N^{\mathrm{T}} \hat{x}_N - \bar{z}_{n+1}. \tag{19.37}$$

Denote by $\bar{a}_j(R)$ the jth column of $\bar{N}_R$ for any $j \in N$.

Set $\hat{x}_q$ to the violated bound, i.e.,

$$\hat{x}_q = \bar{x}_q + \rho_q. \tag{19.38}$$

Besides $\hat{x}_q$ and $\hat{x}_{q'}$, let the other nonbasic components of $\hat{x}$ be equal to those of $\bar{x}$ correspondingly, i.e.,

$$\hat{x}_j = \bar{x}_j, \quad j \in N,\ j \ne q', j \ne q. \tag{19.39}$$

From (19.39) and subtracting the two equalities of (19.36), it follows that

$$\bar{\omega}_q(\hat{x}_q - \bar{x}_q) + \bar{\omega}_{q'}(\hat{x}_{q'} - \bar{x}_{q'}) = 0.$$

combining which and (19.38) gives

$$\hat{x}_{q'} = \bar{x}_{q'} - (\bar{\omega}_q/\bar{\omega}_{q'})\rho_q. \tag{19.40}$$

From subtracting the two equalities of (19.35) together with (19.38) and (19.39), the updating formula follows, i.e.,

$$\hat{x}_B = \bar{x}_B - \rho_q \bar{a}_q(R) - (\hat{x}_{q'} - \bar{x}_{q'})\bar{a}_{q'}(R). \tag{19.41}$$

It is easily verified that such a resulting $\hat{x}$ is a dual feasible solution, associated with the new tableau.

Noting $\bar{z}_{q'} = 0$, on the other hand, from (19.39), (19.38) and (19.33) together with subtracting the two equalities of (19.37), it follows that

$$\hat{f} = \bar{f} + \rho_q \bar{z}_q = \bar{f} + |\rho_q \bar{\omega}_q|\beta_1 \geq \bar{f}. \tag{19.42}$$

Therefore, the objective value does not decreases, and strictly increases when the D-reduced tableau is dual nondegenerate, i.e., $\beta_1 > 0$.

Set $q = q'$, and repeat the preceding steps. Under dual nondegeneracy assumption, the procedure terminates in finitely many iterations, finally satisfying $l_q \leq \bar{x}_q \leq u_q$, or detecting nonexistence of a feasible solution. Note that such an iteration involves no basis change, but only updates the objective row while maintaining the other part unchanged.

(2) $\bar{x}_q$ is within the boundary, i.e., $\rho_q = 0$.

Determine row index

$$s \in \arg\max \{|\rho_{j_t}| \mid t = i, \cdots, m-1\}. \tag{19.43}$$

If $\rho_{j_s} = 0$, (19.27) clearly holds, and hence optimality is achieved.

Now assume that $\rho_{j_s} \neq 0$. Note that $\rho_{j_s} > 0$ implies that $\bar{x}_{j_s}$ violates the lower-bound, and $\rho_{j_s} < 0$ implies that it violates the upper-bound. Only the following two cases arise:

(i) $\bar{a}_{i_s,q} \neq 0$.

Taking $\bar{a}_{i_s,q}$ as pivot, convert it to 1, and eliminate the other nonzeros in the column by elementary transformations. Thus, j_s column leaves from and q column enters to the basis. Note that the bottom row remains unchanged since $\bar{z}_q = 0$. The bottom component of j_sth column is zero. Since the old $\bar{\omega}_q$ is nonzero, the new $\bar{\omega}_{j_s}$ must be nonzero, therefore j_sth column becomes the new datum column, so set $q = j_s$. As the solution $\bar{x}$ is unchanged, it is clear the new tableau maintains dual feasibility, though now the component, associated with the new datum column, violates the boundary. Consequently, it becomes case (1) via such an iteration.

(ii) $\bar{a}_{i_s,q} = 0$.

Using set

$$N_2 = \{j \in \Gamma \| \operatorname{sign}(\rho_{j_s})\, \bar{a}_{i_s,j} < 0\} \cup \{j \in \Pi \mid \operatorname{sign}(\rho_{j_s})\, \bar{a}_{i_s,j} > 0\}, \tag{19.44}$$

determine column index q' and β_2 such that

$$\beta_2 = |\bar{z}_{q'}/\bar{a}_{i_s,q'}| = \min_{j \in N_2} |\bar{z}_j/\bar{a}_{i_s,j}|, \quad \theta_2 = \operatorname{sign}(\rho_{j_s})\beta_2. \tag{19.45}$$

Then, convert $\bar{a}_{i_s,q'}$ to 1, and eliminate the other nonzeros in the column by elementary transformations. The basis and nonbasis, associated with the resulting tableau, are

$$B = \{j_1, \cdots, j_{s-1}, q', j_{s+1}, \cdots, j_{m-1}\}, \quad N = N \backslash q' \cup \{j_s\}. \tag{19.46}$$

Note that in this case, the according elementary transformations do not change the qth column (which remains the datum column).

Similar to case (1), the $\hat{x}$, associated with the resulting tableau, can be obtained from (19.35) and (19.37) as follow. Set the nonbasic component $\hat{x}_{j_s}$ to the violated bound, i.e.,

$$\hat{x}_{j_s} = \bar{x}_{j_s} + \rho_{j_s}. \tag{19.47}$$

Let all nonbasic components of $\hat{x}$ be the same as those of $\bar{x}$, except for the q and j_s-indexed ones. Thus, from (19.36) it follows that

$$\bar{\omega}_q(\hat{x}_q - \bar{x}_q) + \bar{\omega}_{j_s}(\hat{x}_{j_s} - \bar{x}_{j_s}) = 0.$$

So the following updating formula follows

$$\hat{x}_q = \bar{x}_q - (\bar{\omega}_{j_s}/\bar{\omega}_q)\rho_{j_s}. \tag{19.48}$$

Then, from (19.35) it is obtained that

$$\hat{x}_B = \bar{x}_B - \rho_{j_s}(\bar{a}_{j_s}(R) - (\hat{x}_q - \bar{x}_q)\bar{a}_q(R)). \tag{19.49}$$

It is clear that the new tableau is dual feasible for $\hat{x}$.

From that $\bar{z}_{j_s} = \theta_2$ holds for the new tableau, (19.37) and (19.45), it follows that

$$\hat{f} = \bar{f} + \rho_{j_s}\bar{z}_{j_s} = \bar{f} + \rho_{j_s}\theta_2 = \bar{f} + |\rho_{j_s}|\beta_2 \geq \bar{f}, \tag{19.50}$$

which implies that the objective value does not decrease, and strictly increases when the D-reduced tableau is dual nondegenerate.

The overall steps can be summarized to the following algorithm, where the objective value is not updated by (19.50) iteration by iteration, but is computed at the end of the algorithm instead.

Algorithm 19.6.1 (Generalized dual D-reduced simplex algorithm: tableau form). Initial: dual feasible D-reduced tableau of form (19.6), qth column is the datum column, associated with solution $\bar{x}$. This algorithm solves the D-reduced bounded-variable problem (19.6).

1. Go to step 7 if ρ_q defined by (19.28) is 0.
2. Stop if N_1 defined by (19.29) is empty.
3. Determine β_1 and column index q' by (19.33).
4. If $\beta_1 \neq 0$, add β_1 times of the datum row to the bottom row.
5. Update $\bar{x}_q, \bar{x}_{q'}$ and $\bar{x}_B$ by (19.38), (19.40) and (19.41).
6. Set $q = q'$, and go to step 1.
7. Determine s by (19.43).
8. If $\rho_{j_s} = 0$, compute $\bar{f} = c^{\mathrm{T}}\bar{x}$, and stop.
9. Go to step 12 if $\bar{a}_{i_s,q} = 0$.
10. Convert $\bar{a}_{i_s,q}$ to 1, and eliminate the other nonzeros in the column by elementary transformations.
11. Set $q = j_s$, and go to step 2.
12. Determine column index q' by (19.45).
13. Convert $\bar{a}_{i_s,q'}$ to 1, and eliminate the other nonzeros in the column by elementary transformations.
14. Update $\bar{x}_{j_s}, \bar{x}_q$ and $\bar{x}_B$ by (19.47)–(19.49).
15. Go to step 1.

Theorem 19.6.1. *Algorithm 19.6.1 creates a sequence of basic feasible solutions. Under the dual nondegeneracy assumption, it terminates either at*

(i) Step 2, detecting infeasibility of the problem; or at
(ii) Step 8, giving a basic optimal solution.

Proof. The validity comes from Lemma 19.6.1 and the analysis preceding Algorithm 19.6.1.

Example 19.6.1. Solve the following problem by Algorithm 19.6.1:

$$\begin{array}{llllllllll}
\min & f = 2x_1 - 4x_2 - 7x_3 + 5x_4, \\
\text{s.t.} & -3x_1 & + \; x_2 & + \; 3x_3 & + \; 5x_4 & + \; x_5 & & & = & 5, \\
& -6x_1 & + \; 2x_2 & - \; 4x_3 & - \; 4x_4 & & + \; x_6 & & = & -20, \\
& 4x_1 & - \; x_2 & - \; 2x_3 & + \; 3x_4 & & & + \; x_7 & = & -8, \\
& \multicolumn{9}{l}{-10 \le x_1 \le 10, \quad -8 \le x_2 \le 20, \quad 1 \le x_3 \le 15,} \\
& \multicolumn{9}{l}{-10 \le x_4 \le 17, \quad 0 \le x_5 \le 15, \quad -1 \le x_6 \le 12,} \\
& \multicolumn{9}{l}{-8 \le x_7 \le 15.}
\end{array}$$

Answer Initial tableau:

x_1	x_2	x_3	x_4	x_5	x_6	x_7	RHS
−3	1	3	5	1			5
−6	2	−4	−4		1		−20*
4	−1	−2	3			1	−8
2	−4	−7	5				

Iteration 1: Convert the preceding to a D-reduced tableau.
Take $\bar{x}_N = (-10_{(-)}, 20^{(+)}, 15^{(+)}, -10_{(-)})^{\mathrm{T}}$ $(N = \{1,2,3,4\})$.
$\bar{x}_B = b - N\bar{x}_N = (-40, -100, 112)^{\mathrm{T}}$ $(B = \{5,6,7\})$.
$\rho_5 = 0 - (-40) = 40, \rho_6 = (-1) - (-100) = 99,$
$\rho_7 = 15 - 112 = -97.$ $\max\{|-40|, |-99|, |97|\} = 99,\ p = 2.$
Enter the right-hand side to the basis: multiply row 2 by 1/20, then add 5, −8 times of row 2 to rows 1,3, respectively:

	x_1	x_2	x_3	x_4	x_5	x_6	x_7	RHS
	−9/2	3/2	2	4	1	1/4		
	3/10	−1/10	1/5	1/5		−1/20		1
	32/5	−9/5	−2/5	23/5		−2/5	1	
	2	−4	−7	5				
u	10	20	15	17	15	12	15	
$\bar{x}$	−10	20	15	−10	−40	−100	112	
l	−10	−8	1	−10	0	−1	−8	

The second row is the datum row, and x_6 column is the datum column. $\bar{x}_B = (-40, 112)^{\mathrm{T}}(B = \{5,7\})$.

$\bar{x}_N = (-10_{(-)}, 20^{(+)}, 15^{(+)}, -10_{(-)}, -100)^{\mathrm{T}}$ $(N = \{1,2,3,4,6\})$. $\bar{f} = -255$
The resulting tableau is then dual feasible.

Call Algorithm 19.6.1: $r = 2,\ q = 6$.

Iteration 2:

1. $\rho_6 = -1 - (-100) = 99 > 0.$
2. $\mathrm{sign}(\rho_6 \bar{a}_{2,6}) = -1,\ N_1 = \{1,2,4\} \neq \emptyset.$
3. $\beta_1 = \min\{|2/(3/10)|, |-4/(-1/10)|, |5/(1/5)|\} = 20/3,$
   $q' = 1,\ \theta_1 = -1(20/3) = -20/3.$
4. Add −20/3 times of row 2 to row 4.
5. $\bar{x}_6 = -100 + 99 = -1.\ \bar{x}_1 = -10 - ((-1/20)/(3/10))99 = 13/2.$
   $\bar{x}_B = (-40, 112)^{\mathrm{T}} - 99(1/4, -2/5)^{\mathrm{T}} - (13/2 - (-10))(-9/2, 32/5)^{\mathrm{T}}$
   $= (19/2, 46)^{\mathrm{T}},\ B = \{5,7\}.$
6. $q = 1.$

	x_1	x_2	x_3	x_4	x_5	x_6	x_7	RHS
	$-9/2$	$3/2$	2	4	1	$1/4$		
	$3/10$	$-1/10$	$1/5$	$1/5$		$-1/20$		1
	$32/5^*$	$-9/5$	$-2/5$	$23/5$		$-2/5$	1	
		$-10/3$	$-25/3$	$11/3$		$1/3$		$-20/3$
u	10	20	15	17	15	12	15	
$\bar{x}$	$13/2$	20	15	-10	$19/2$	-1	46	
l	-10	-8	1	-10	0	-1	-8	

Iteration 3:

1. $\rho_1 \; = 0.$
7. $\rho_5 \; = 0, \rho_7 = 15 - (46) = -31.\; i_3 = 3,\; j_3 = 7.$
9. $\bar{a}_{3,1} = 32/5 \neq 0.$
10. multiply row 3 by $5/32$, then add $9/2, -3/10$ times of row 3 to rows 1,2, respectively.
11. $q \; = 7.$

	x_1	x_2	x_3	x_4	x_5	x_6	x_7	RHS
		$15/64$	$55/32$	$463/64$	1	$-1/32$	$45/64$	
		$-1/64$	$7/32$	$-1/64$		$-1/32$	$-3/64$	1
	1	$-9/32$	$-1/16$	$23/32$		$-1/16$	$5/32$	
		$-10/3$	$-25/3$	$11/3$		$1/3$		$-20/3$
u	10	20	15	17	15	12	15	
$\bar{x}$	$13/2$	20	15	-10	$19/2$	-1	46	
l	-10	-8	1	-10	0	-1	-8	

Iteration 4:

1. $\rho_7 \; = 15 - 46 = -30 < 0.$
2. $\mathrm{sign}(\rho_7 \bar{a}_{2,7}) = 1,\; N_1 = \{3, 4, 6\} \neq \emptyset.$
3. $\beta_1 \; = \min\{|(-25/3)/(7/32)|, |(11/3)/(-1/64)|, |(1/3)/(-1/32)|\} = 32/3,$ $q' = 6,\; \theta_1 = -(-1)(32/2) = 32/3.$
4. Add $32/3$ times of row 2 to row 4.
5. $\bar{x}_7 = 46 + (-31) = 15; \bar{x}_6 = -1 - ((-3/64)/(-1/32))(-31) = 91/2;$ $\bar{x}_B = (19/2, 13/2)^{\mathrm{T}} - (-31)(45/64, 5/32)^{\mathrm{T}} - (91/2 - (-1))(-1/32, -1/16)^{\mathrm{T}}$ $= (131/4, 57/4)^{\mathrm{T}},\; B = \{5, 1\}.$
6. $q \; = 6.$

	x_1	x_2	x_3	x_4	x_5	x_6	x_7	RHS
		15/64	55/32	463/64	1	−1/32	45/64	
		−1/64	7/32	−1/64		−1/32	−3/64	1
	1	−9/32	−1/16	23/32		−1/16	5/32	
		−7/2	−6	7/2			−1/2	4
u	10	20	15	17	15	12	15	
$\bar{x}$	57/4	20	15	−10	131/4	91/2	15	
l	−10	−8	1	−10	0	−1	−8	

Iteration 5:

1. $\rho_6 = 12 - 91/2 = -67/2 < 0$.
2. $\text{sign}(\rho_6 \bar{a}_{2,6}) = 1,\ N_1 = \{3, 4\} \neq \emptyset$.
3. $\beta_1 = \min\{|(-6)/(7/32)|, |(7/2)/(-1/64)|\} = 192/7,\ q' = 3,\ \theta_1 = 192/7$.
4. Add $192/7$ times of row 2 to row 4.
5. $\bar{x}_6 = 91/2 + (-67/2) = 12;$
   $\bar{x}_3 = 15 - ((-1/32)/(7/32))(-67/2) = 143/14;$
   $\bar{x}_B = (131/4, 57/4)^{\mathrm{T}} - (-67/2)(-1/32, -1/16)^{\mathrm{T}} - (143/14 - (15))$
   $(55/32, -1/16)^{\mathrm{T}} = (559/14, 83/7)^{\mathrm{T}}, B = \{5, 1\}$.
6. $q = 3$.

	x_1	x_2	x_3	x_4	x_5	x_6	x_7	RHS
		15/64	55/32*	463/64	1	−1/32	45/64	
		−1/64	7/32	−1/64		−1/32	−3/64	1
	1	−9/32	−1/16	23/32		−1/16	5/32	
		−55/14		43/14		−6/7	−25/14	220/7
u	10	20	15	17	15	12	15	
$\bar{x}$	83/7	20	143/14	−10	559/14	12	15	
l	−10	−8	1	−10	0	−1	−8	

Iteration 6:

1. $\rho_3 = 0$.
7. $\rho_5 = 15 - 559/14 = -349/14,\ \rho_1 = 10 - 83/7 = -13/7.\ i_1 = 1,\ j_1 = 5$.
9. $\bar{a}_{1,3} = 55/32 \neq 0$.
10. Add $-7/32, 1/16$ times of row 1 to rows 2,3, respectively:
11. $q = 5$.

	x_1	x_2	x_3	x_4	x_5	x_6	x_7	RHS
		3/22	1	463/110	32/55	−1/55	9/22	
		−1/22		−103/110*	−7/55	−3/110	−3/22	1
	1	−3/11		54/55	2/55	−7/110	2/11	
		−55/14		43/14		−6/7	−25/14	220/7
u	10	20	15	17	15	12	15	
$\bar{x}$	83/7	20	143/14	−10	559/14	12	15	
l	−10	−8	1	−10	0	−1	−8	

Iteration 7:

1. $\rho_5 = 15 - 559/14 < 0$.
2. $\text{sign}(\rho_5 \bar{a}_{2,5}) = 1,\ N_1 = \{4\} \neq \emptyset$.
3. $\beta_1 = \min\{|(43/14)/(103/110)|\} = 2{,}365/721, q' = 4,\ \theta_1 = 2{,}365/721.$
4. Add 2,365/721 times of row 2 to row 4.
5. $\bar{x}_5 = 559/14 + (-349/14) = 15;$

$$\bar{x}_4 = -10 - ((-7/55)/(-103/110))(-349/14) = -681/103;$$
$$\begin{aligned}\bar{x}_B &= (143/14, 83/7)^{\mathrm{T}} - (-349/14)(32/55, 2/55)^{\mathrm{T}} \\ &\quad - (-681/103 - (-10))(463/110, 54/55)^{\mathrm{T}} \\ &= (1{,}077/103, 972/103)^{\mathrm{T}}, \quad B = \{3, 1\}.\end{aligned}$$

6. $q = 4$.

	x_1	x_2	x_3	x_4	x_5	x_6	x_7	RHS
		3/22	1	463/110	32/55	−1/55	9/22	
		−1/22		−103/110	−7/55	−3/110	−3/22	1
	1	−3/11		54/55	2/55	−7/110	2/11	
		−420/103			−43/103	−195/206	−230/103	3,575/103
u	10	20	15	17	15	12	15	
$\bar{x}$	972/103	20	1,077/103	−681/103	15	12	15	
l	−10	−8	1	−10	0	−1	−8	

Iteration 8:

1. $\rho_4 = 0$.
7. $\rho_3 = \rho_1 = 0$. The optimal solution and optimal value are

$$\begin{aligned}\bar{x} &= (972/103, 20, 1{,}077/103, -681/103, 15, 12, 15)^{\mathrm{T}}, \\ \bar{f} &= (2, -4, -7, 5)(972/103, 20, 1{,}077/103, -681/103)^{\mathrm{T}} = -17{,}240/103.\end{aligned}$$

Based on the equivalence between the D-reduced tableau (17.3) and the revised tableau (17.19), the revised version of Algorithm 19.6.1 can be stated as follows.

Algorithm 19.6.2 (Generalized dual D-reduced simplex algorithm). Initial: $B,\ R,\ N,\ B_R^{-1},\ \bar{z}_N = c_N - N_R^{\mathrm{T}} B_R^{-\mathrm{T}} c_B.\ \bar{x},\ \bar{f}$ satisfying (19.21) and (19.25). This algorithm solves the D-reduced bounded-variable problem (19.6).

1. Go to step 8 if ρ_q, determined by (19.28), is equal to 0.
2. Compute $\bar{\omega}_N = N^{\mathrm{T}} e_m - N_R^{\mathrm{T}}(B_R^{-\mathrm{T}} B e_m)$.
3. Stop if $N_1 = \{j \in \Gamma \mid \mathrm{sign}(\rho_q \bar{\omega}_q)\bar{\omega}_j < 0\} \cup \{j \in \Pi \mid \mathrm{sign}(\rho_q \bar{\omega}_q)\bar{\omega}_j > 0\} = \emptyset$, (infeasible problem).
4. Determine θ_1 and column index q' such that $\beta_1 = |\bar{z}_{q'}/\bar{\omega}_{q'}| = \min_{j \in N_1} |\bar{z}_j/\bar{\omega}_j|$, $\theta_1 = \mathrm{sign}(\rho_q \bar{\omega}_q)\beta$.
5. If $\theta_1 \neq 0$, update $\bar{z}_N = \bar{z}_N + \theta_1 \bar{\omega}_N$.
6. Update $\bar{x}_q,\ \bar{x}_{q'}$ and $\bar{x}_B$ by (19.38), (19.40) and (19.41).
7. Set $q = q'$, and go to step 1.
8. Determine s by (19.43).
9. If $\rho_{j_s} = 0$, compute $\bar{f} = c^{\mathrm{T}}\bar{x}$, and stop (optimality achieved).
10. Compute $\sigma_N = N_R^{\mathrm{T}} B_R^{-\mathrm{T}} e_s$.
11. Go to step 15 if $\sigma_q = 0$.
12. Compute $\bar{a}_q(R) = B_R^{-1} a_q(R)$.
13. Update B_R^{-1} by (17.20).
14. Set $q = j_s$, and go to step 2.
15. Determine column index q' such that $|\bar{z}_{q'}/\sigma_{q'}| = \min_{j \in N_2} |\bar{z}_j/\sigma_j|$.
16. Compute $\bar{a}_{q'}(R) = B_R^{-1} a_{q'}(R)$.
17. Update B_R^{-1} by (17.20).
18. Update $\bar{x}_{j_s}, \bar{x}_{q'}$ and $\bar{x}_B$ by (19.47)–(19.49).
19. Update (B, N) by exchanging j_s and q'.
20. Go to step 1.

Chapter 20
Deficient-Basis Method

As was known, the nondegeneracy assumption is entirely contrary to reality. When the simplex method is used to solve large-scale sparse problems, degeneracy or even high degeneracy almost always presents. Therefore, finiteness of the simple method is not guaranteed, and there are indeed few cycling instances (Sect. 3.6).

As a common belief, on the other hand, stalling caused by degeneracy degrades performance of the conventional simplex method seriously, as a huge number of iterations may stay at a vertex for too long a time before exiting it. In fact, stalling has long been thought as a headache for solving large-scale and highly degenerate problems, where vanished basic components occupy a large proportion (Hattersley and Wilson 1988; Megiddo 1986b; Ryan and Osborne 1988).

To overcome drawbacks related to degeneracy, the conventional basis will be generalized to allow deficient case. In the following two chapters, primal and dual deficient-basis methods will be developed. It will be shown that it is possible to take advantage of high degeneracy to reduce computational work significantly (Pan 1998b, 1999b, 2000a).

Consider the standard LP problem (1.8). Conventionally, the coefficient matrix A is assumed to be of full row rank. In this chapter and thereafter, the generalized assumption is made that b belongs to the column space of A, i.e.,

$$b \in \text{range}(B),$$

or equivalently, system $Ax = b$ is consistent.

20.1 Deficient-Basis and Tableau

The concept "basis" has been the center of the simplex method for years. Since it is a submatrix consisting of m linear independent columns of A, however, it is unavoidable for some of basic solution's components to vanish, whenever the

P.-Q. PAN, *Linear Programming Computation*, DOI 10.1007/978-3-642-40754-3_20,

right-hand side of the problem belongs to a proper subset of range(B). To overcome this difficulty, we redefine the basis concept.

Definition 20.1.1. Basis (matrix) is such a nonsingular square submatrix of A that the subspace spanned by the associated columns includes b.

It is clear that there exists a basis of such type under the assumption of $b \in$ range(A), and that a basis is actually a standard (full) basis if its order equals m, whereas it is said to be *deficient* if the order is less than m. In addition, the associated basic solution will have no zero components, if the deficient basis yields from dropping columns corresponding to zero components of a standard basic solution.

Columns (rows), associated with a basis, are said to be *basic*, and the other columns (rows) are *nonbasic*. Variables (components) are said to be *basic* and nonbasic, conformably.

Let k be the order of a basis. Denote by B and R the index sets of basic columns and rows, respectively by

$$B = \{j_1, \cdots, j_k\}, \quad \text{and} \quad R = \{i_1, \cdots, i_k\}. \tag{20.1}$$

Denote the associated nonbasic column and row index sets, respectively by

$$N = A \backslash B, \quad \text{and} \quad R' = \{1, \cdots, m\} \backslash R. \tag{20.2}$$

Once B, R are given, so are N, R'. Note that R' is empty when $k = m$ whereas N is always nonempty. Although $1 \leq k \leq m$ is usually assumed, it is sometimes useful and easy to include the case $k = 0$, in which B and R are both empty.

Without confusion, the preceding set notations will also be utilized to denote the submatrices or vectors, consisting of the associated columns and rows; e.g., the submatrix consisting of basic columns is still denoted by B, and as well as

B_R $\in \mathcal{R}^{k\times k}$: basis, consisting of entries in basic columns and rows.
$B_{R'}$ $\in \mathcal{R}^{(m-k)\times k}$: submatrix consisting of entries in basic columns and nonbasic rows.
N_R $\in \mathcal{R}^{k\times(n-k)}$: submatrix consisting of entries in nonbasic columns and basic rows.
$N_{R'}$ $\in \mathcal{R}^{(m-k)\times(n-k)}$: submatrix consisting of entries in nonbasic columns and rows.
b_R $\in \mathcal{R}^{k}$: subvector consisting of basic components (in basic rows) of the right-hand side.
$b_{R'}$ $\in \mathcal{R}^{m-k}$: subvector consisting of nonbasic components (in nonbasic rows) of the right-hand side.
$a_q(R)$ $\in \mathcal{R}^{k}$: the q-indexed column of N_R.
$a_q(R')$ $\in \mathcal{R}^{m-k}$: the q-indexed column of $N_{R'}$.

Since $b \in \text{range}(B)$, an initial tableau of the standard LP problem, say

$$
\begin{array}{cc|c|c}
x_B^{\mathrm{T}} & x_N^{\mathrm{T}} & f & \text{RHS} \\
\hline
B_R & N_R & & b_R \\
B_{R'} & N_{R'} & & b_{R'} \\
\hline
c_B^{\mathrm{T}} & c_N^{\mathrm{T}} & -1 &
\end{array}
\tag{20.3}
$$

can be converted to an equivalent form by elementary transformations with row and columns exchanges, i.e.,

$$
\begin{array}{cc|c|c}
x_B^{\mathrm{T}} & x_N^{\mathrm{T}} & f & \text{RHS} \\
\hline
I & \bar{N}_R & & \bar{b}_R \\
 & \bar{N}_{R'} & & \\
\hline
 & \bar{z}_N^{\mathrm{T}} & -1 & -\bar{f}
\end{array}
\tag{20.4}
$$

The preceding is referred to as *deficient-basis tableau*, where the unit matrix I corresponds to basis B_R, featured by basic columns and rows.

$\bar{z}_N^{\mathrm{T}}$ is called *reduced costs (coefficients)*. A solution to $Ax = b$, given by

$$\bar{x}_B = \bar{b}_R, \quad \bar{x}_N = 0,$$

is still called *basic solution*, as in case of the coefficient matrix of full row rank, it is always possible to expand a deficient basis to a full one, associated with the same solution, by entering some nonbasic columns. Note that the j_t-indexed basic variable $\bar{x}_{j_t}$ is in the i_t-indexed row of the tableau. Thus,

$$\bar{x}_{j_t} = \bar{b}_{i_t}, \quad t = 1, \cdots, k.$$

The solution is feasible if $\bar{x}_B \geq 0$, and optimal if, in addition, the associated objective value attains the minimum value over the feasible region. The tableau is said to be feasible or optimal, conformably.

There are results similar to those in the conventional simplex context.

Lemma 20.1.1. *Assume that the deficient-basis tableau (20.4) is feasible. If its reduced costs are all nonnegative, then it gives a pair of primal and dual basic optimal solutions.*

Proof. It is clear that the tableau gives a pair of complementary primal and dual solutions, which are primal and dual feasible respectively, as satisfy the optimality conditions. □

20.2 Deficient-Basis Method: Tableau Form

Let us now discuss about how to transform a feasible deficient-basis tableau to a new one while maintaining feasibility, and thereby derive a tableau version of the so-called "deficient-basis method".

Assume that the current deficient-basis tableau of form (20.4) is feasible, i.e., $\bar{b}_R \geq 0$, and that $\bar{z}_N \not\geq 0$. We select a pivot column index q such that $\bar{z}_q < 0$, e.g., by the counterpart of Dantzig conventional rule below:

$$q \in \arg\min\{\bar{z}_j \mid j \in N\}. \tag{20.5}$$

Thus, it holds that $\bar{z}_q < 0$.

Denote the q-indexed columns of $\bar{N}_R$ and $\bar{N}_{R'}$ by $\bar{a}_q(R)$ and $\bar{a}_q(R')$, respectively.

The determination of a pivot row index depends on whether column $\bar{a}_q(R')$ vanishes, i.e., whether a_q belongs to the column space of B. There are the following two cases arising:

(i) Rank-increasing iteration: $R' \neq \emptyset$ and $\bar{a}_q(R') \neq 0$.

In this case, it is not allowed for $\bar{x}_q$ to be taken on any nonzero value while keeping the other nonbasic components of the current solution unchanged, as such doing violates some of the equality constraints, associated with nonbasic rows. To detor the difficulty, determine row index

$$p \in \arg\max_{i \in R'} |\bar{a}_{i,q}|. \tag{20.6}$$

Then take $\bar{a}_{pq}$ as a pivot to carry out elementary transformations, so that the q-indexed nonbasic column and p-index nonbasic row both becomes basic while the basic columns and rows remain basic. Thus, the rank of the basis increases by 1, and the iterations is complete. It is noted that the right-hand side remains unchanged.

(ii) Rank-remaining iteration: $R' = \emptyset$ or $\bar{a}_q(R') = 0$.

Increase $\bar{x}_q$ from zero to decrease the objective value while keeping the other nonbasic components of the current solution feasible. In the following case, such decreasing will be unlimited.

Lemma 20.2.1. *Assume that the deficient-basis tableau is feasible and that $R' = \emptyset$ or $\bar{a}_q(R') = 0$. If $\bar{z}_q < 0$ and $\bar{a}_q(R) \leq 0$, the original problem is lower unbounded.*

Proof. The proof is an analogue to that of Lemma 3.2.2. □

If, otherwise, $\bar{a}_q(R) \not\leq 0$ does not hold, then there is a blocking variable, associated with α and s such that

$$\alpha = \bar{b}_{i_s}/\bar{a}_{i_s,q} = \min\{\bar{b}_{i_t}/\bar{a}_{i_t,q} \mid \bar{a}_{i_t,q} > 0,\ t = 1, \cdots, k\} \geq 0, \tag{20.7}$$

where stepsize α will be used to update the basic feasible solution. Taking $\bar{a}_{i_s\,q}$ as the pivot, carry out according elementary transformations. Consequently, the j_s-index column leaves and q-index column enters the basis, while the basic rows remain basic. So, the rank of the basis remains, and an iteration is complete. It is noted such doing does not touch nonbasic rows at all (including entries in the right-hand side).

As $\bar{z}_q$ is eliminated, both cases change components of the dual solution, locating at the bottom row. Since $\bar{b}_{i_s} = 0$, a rank-increasing iteration does not change tableau's right-hand side, hence the solution and objective value. On the other hand, a rank-remaining iteration changes these quantities, in general; in fact, it is known from (3.13) that the objective value becomes

$$\hat{f} = \bar{f} + \alpha \bar{z}_q \leq \bar{f},$$

where the right-most inequality results from (20.7) and $\bar{z}_q < 0$. Therefore, the objective value does not increase, and in case of nondegeneracy ($\bar{b}_R > 0$), α is positive, hence the objective value strictly decreases.

The overall procedure can be summarized as follows.

Algorithm 20.2.1 (Deficient-basis algorithm: tableau form). Initial: feasible tableau of form (20.4) with $B,\ N,\ R,\ R',\ 1 \leq k \leq m$. This algorithm solves the standard LP problem.

1. Determine column index $q \in \arg\min_{j \in N} \bar{z}_j$.
2. Stop if $\bar{z}_q \geq 0$.
3. Go to step 7 if $R' = \emptyset$ or $\bar{a}_q(R') = 0$.
4. Determine $p \in \arg\max_{i \in R'} |\bar{a}_{i\,q}|$.
5. Convert $\bar{a}_{p\,q}$ to 1, and eliminate the other nonzeros in the column by elementary transformations.
6. Set $k = k + 1$, bring q from N to B and p from R' to R, and go to step 1.
7. Stop if $\bar{a}_q(R) \leq 0$.
8. Determine s by (20.7).
9. Convert $\bar{a}_{i_s,q}$ to 1, and eliminate the other nonzeros in the column by elementary transformations.
10. Update (B, N) by exchanging j_s and q.
11. Go to step 1.

Theorem 20.2.1. *Assume that rank-remaining iterations are all nondegenerate. Algorithm 20.2.1 terminates either at*

(i) step 2, giving a basic optimal solution; or at
(ii) step 7, detecting lower unboundedness.

Proof. The proof on termination of Algorithm 20.2.1 is similar to that in the conventional case. The meanings of its exits come from Lemmas 20.1.1 and 20.2.1 together with the related discussions preceding Algorithm 20.2.1. □

Example 20.2.1. Solve the following problem by Algorithm 20.2.1:

$$\begin{array}{ll}\min & f = -2x_1 + 3x_2 - 4x_3 + x_4 + 2x_7 - 3x_8,\\ \text{s.t.} & -x_1 + x_2 + x_3 - x_4 + x_5 + x_7 - 2x_8 = 2,\\ & 2x_1 - x_2 + x_3 + x_4 + x_6 - x_7 + x_8 = 1,\\ & x_1 - 2x_2 - x_4 - 3x_8 = 0,\\ & -x_1 + x_2 + x_4 + x_8 = 0,\\ & 2x_1 - x_2 + x_4 - x_8 = 0,\\ & x_j \geq 0, \quad j = 1, \cdots, 8.\end{array}$$

Answer $k = 2, B = \{5, 6\},\ R = \{1, 2\},\ N = \{1, 2, 3, 4, 7, 8\},\ R' = \{3, 4, 5\}$.
Initial feasible deficient-basis tableau:

x_1	x_2	x_3	x_4	x_5	x_6	x_7	x_8	RHS
−1	1	1	−1	1		1	−2	2
2	−1	1*	1		1	−1	1	1
1	−2		−1				−3	
−1	1		1				1	
2	−1		1				−1	
−2	3	−4	1			2	−3	

Iteration 1:

1. $\min\{-2, 3, -4, 1, 2, -3\} = -4 < 0,\ q = 3$.
3. $\bar{a}_q(R') = 0$.
7. $\bar{a}_q(R) \not\leq 0$.
8. $\min\{2/1, 1/1\} = 1,\ s = 2,\ i_s = 2$.
9. add $-1, 4$ times of row 2 to rows 1,6, respectively.
10. $B = \{5, 3\},\ N = \{1, 2, 4, 6, 7, 8\}$.

x_1	x_2	x_3	x_4	x_5	x_6	x_7	x_8	RHS
−3	2		−2	1	−1	2*	−3	1
2	−1	1	1		1	−1	1	1
1	−2		−1				−3	
−1	1		1				1	
2	−1		−2				−1	
6	−1		5		4	−2	1	4

Iteration 2:

1. $\min\{6, -1, 5, 4, -2, 1\} = -2 < 0,\ q = 7$.
3. $\bar{a}_q(R') = 0$.
7. $\bar{a}_q(R) \not\leq 0$.
8. $\min\{1/2\} = 1/2,\ s = 1,\ i_s = 1$.
9. Multiply row 2 by $1/2$, then add $1, 2$ times of row 2 to rows 2,6, respectively.
10. $B = \{7, 3\},\ N = \{1, 2, 4, 5, 6, 8\}$.

x_1	x_2	x_3	x_4	x_5	x_6	x_7	x_8	RHS
−3/2	1		−1	1/2	−1/2	1	−3/2	1/2
1/2		1		1/2	1/2		−1/2	3/2
1	−2		−1				−3*	
−1	1		1				1	
2	−1		−2				−1	
3	1		3	1	3		−2	5

Iteration 3:

1. $\min\{3, 1, 3, 1, 3, -2\} = -2 < 0,\ q = 8.$
3. $\bar{a}_q(R') \neq 0;$
4. $\max\{|-3|, |1|, |-1|\} = 3,\ p = 3.$
5. Multiply row 3 by $-1/3$, then add $-1, 1, 2$ times of row 3 to rows 4,5,6, respectively.
6. $k = 3,\ B = \{7, 3, 8\},\ R = \{1, 2, 3\},\ N = \{1, 2, 4, 5, 6\},\ R' = \{4, 5\}.$

x_1	x_2	x_3	x_4	x_5	x_6	x_7	x_8	RHS
-2	2		$-1/2$	$1/2$	$-1/2$	1		$1/2$
$1/3$	$1/3$	1	$1/6$	$1/2$	$1/2$			$3/2$
$-1/3$	$2/3$		$1/3$				1	
$-2/3$	$1/3$		$2/3$					
$5/3$	$-1/3$		$-5/3$					
$7/3$	$7/3$		$11/3$	1	3			5

Iteration 4:

1. $\min\{7/3, 7/3, 11/3, 1, 3\} \geq 0.$
2. The basic optimal solution and optimal value are:

$$\bar{x} = (0, 0, 3/2, 0, 0, 0, 1/2, 0)^{\mathrm{T}}, \qquad \bar{f} = -5.$$

20.3 Deficient-Basis Method

In this section, we derive the revised version of Algorithm 20.2.1.

A deficient basis characterizes an iteration, just as a standard basis in the conventional simplex context. All entries in a deficient-basis tableau can be expressed using the inverse of the deficient-basis matrix.

Let B_R be the basis matrix, associated with sets B, R, and let B_R^{-1} be its inverse. Then, it is known that

$$\bar{b}_{R'} = b_{R'} - B_{R'} B_R^{-1} b_R = 0.$$

Thus, the revised tableau, corresponding to the deficient-basis tableau (20.4), is

x_B^{T}	x_N^{T}	f	RHS
I	$B_R^{-1} N_R$		$B_R^{-1} b_R$
	$N_{R'} - B_{R'} B_R^{-1} N_R$		
	$c_N^{\mathrm{T}} - c_B^{\mathrm{T}} B_R^{-1} N_R$	-1	$-c_B^{\mathrm{T}} B_R^{-1} b_R$

(20.8)

Table 20.1 Equivalence between the associated quantities

Quantity	Tableau	Relation	Revised tableau
Objective row	$\bar{z}_N$	$=$	$c_N^{\mathrm{T}} - c_B^{\mathrm{T}} B_R^{-\mathrm{T}} N_R$
Basic components of pivot column	$\bar{a}_q(R)$	$=$	$B_R^{-1} a_q(R)$
Nonbasic components of pivot column	$\bar{a}_q(R')$	$=$	$a_q(R') - B_{R'} B_R^{-1} a_q(R)$
Basic components of the right-hand side	$\bar{b}_R$	$=$	$B_R^{-1} b_R$
Nonbasic entries of the pth basic row	$e_p^{\mathrm{T}} \bar{N}_R$	$=$	$e_p^{\mathrm{T}} B_R^{-1} N_R$

Let $q \in N$ be the pivot column index selected by (20.5). The following tableau offers correspondence between wanted quantities in the preceding tableau and tableau (20.4) (see Table 20.1):

As for updating the inverse B_R^{-1} of the deficient basis, the following two types of iterations should be treated separately:

(i) Rank-increasing iteration: $R' \neq \emptyset$ and $\bar{a}_q(R') = a_q(R') - B_{R'} B_R^{-1} a_q(R) \neq 0$.

Assume that pivot row index $p \in R'$ was determined by (20.6). The new basis matrix (increasing rank by 1) is of form

$$\hat{B}_{\hat{R}} = \left(\begin{array}{c|c} B_R & a_q(R) \\ \hline e_p^{\mathrm{T}} B & a_{pq} \end{array}\right) \begin{array}{c} k \\ 1 \end{array} \qquad (20.9)$$

$$\begin{array}{cc} k & 1 \end{array}$$

where $e_p^{\mathrm{T}} B$ denotes the pth row of B, a_{pq} denotes the pth component of a_q, and $a_q(R)$ corresponds to basic components of a_q. It is easy to verify that the inverse is of the following form:

$$\hat{B}_{\hat{R}}^{-1} = \left(\begin{array}{c|c} U & v \\ \hline d^{\mathrm{T}} & \tau \end{array}\right) \begin{array}{c} k \\ 1 \end{array} \qquad (20.10)$$

$$\begin{array}{cc} k & 1 \end{array}$$

where

$$\begin{aligned} h^{\mathrm{T}} &= e_p^{\mathrm{T}} B, \\ \tau &= 1/(a_{pq} - h^{\mathrm{T}} \bar{a}_q(R)), \\ v &= -\tau \bar{a}_q(R), \\ d^{\mathrm{T}} &= -\tau h^{\mathrm{T}} B_R^{-1}, \\ U &= B_R^{-1} - v h^{\mathrm{T}} B_R^{-1}. \end{aligned}$$

(ii) Rank-remaining iteration: $R' = \emptyset$ or $\bar{a}_q(R') = a_q(R') - B_{R'}B_R^{-1}a_q(R) = 0$.

Essentially, the updating in this case is the same as in the conventional simplex context, only B_R^{-1} is now of order less than m, in general. Assume that $s \in \{1, \cdots, k\}$ was determined by (20.7). The updating formula is similar to (3.23), i.e.,

$$\hat{B}_R^{-1} = E_s B_R^{-1}, \tag{20.11}$$

where

$$E_s = \begin{pmatrix} 1 & & -(\bar{a}_q(R))_1/(\bar{a}_q(R))_s & & \\ & \ddots & \vdots & & \\ & & -(\bar{a}_q(R))_{s-1}/(\bar{a}_q(R))_s & & \\ & & 1/(\bar{a}_q(R))_s & & \\ & & -(\bar{a}_q(R))_{s+1}/(\bar{a}_q(R))_s & & \\ & & \vdots & \ddots & \\ & & -(\bar{a}_q(R))_k/(\bar{a}_q(R))_s & & 1 \end{pmatrix}$$

Based on Table 20.1, Algorithm 20.2.1 can be revised as follows.

Algorithm 20.3.1 (Deficient-basis algorithm). Initial: B, R, $1 \le k \le m$, B_R^{-1}; $\bar{x}_B = B_R^{-1}b_R \ge 0$. This algorithm solves the standard LP problem.

1. Compute $\bar{z}_N = c_N - N_R^{\mathrm{T}}\bar{y}$, where $\bar{y} = B_R^{-\mathrm{T}}c_B$.
2. Determine pivot column index $q \in \arg\min_{j\in N} \bar{z}_j$.
3. If $\bar{z}_q \ge 0$, compute $\bar{f} = c_B^{\mathrm{T}}\bar{x}_B$, and stop (optimality achieved).
4. Compute $\bar{a}_q(R) = B_R^{-1}a_q(R)$.
5. Go to step 9 if $R' = \emptyset$ or $\bar{a}_q(R') = a_q(R') - B_{R'}\bar{a}_q(R) = 0$.
6. Determine row index $p \in \arg\max_{i\in R'} |\bar{a}_{i\,q}|$.
7. Update B_R^{-1} by (20.10).
8. Set $k = k + 1$, bring q from N to B and p from R' to R, and go to step 1.
9. Stop if $\bar{a}_q(R) \le 0$ (lower unbounded problem).
10. Determine s and α such that
    $\alpha = \bar{x}_{i_s}/\bar{a}_{i_s,q} = \min\{\bar{x}_{i_t}/\bar{a}_{i_t,q} \mid \bar{a}_{i_t,q} > 0,\ t = 1, \cdots, k\}$.
11. If $\alpha \ne 0$, set $\bar{x}_q = \alpha$ and update $\bar{x}_B = \bar{x}_B - \alpha\bar{a}_q(R)$.
12. Update B_R^{-1} by (20.11).
13. Update (B, N) by exchanging j_s and q.
14. Go to step 1.

The preceding algorithm involves solution of two linear systems in each iteration (in steps 1 and 4, respectively), which are generally smaller than those in the conventional simplex context, especially for large-scale and highly degenerate problems. Moreover, a deficient basis is potentially better conditioned than a

Table 20.2 Iteration and time ratios of MINOS 5.3 to PDBSA 1.0

Problem	Iterations	Time	% degen
Small (20)	1.09	1.50	1.06
Medium (15)	0.96	1.30	0.85
Large (15)	1.08	1.41	0.87
Average (50)	1.07	1.41	0.93
Kennington (8)	1.51	2.07	0.61
BPMPD (4)	1.64	2.05	0.63
Average (12)	1.52	2.06	0.63

conventional one. More importantly, the associated search direction appears to be better than that in the conventional context. These advantages are conformed by computational experiments, as reported below:

Originally, the rectangular matrix B is defined as the "deficient basis", as is not compact, compared with B_R. Numerical results, obtained from the related experiments, are cited here, as these results still roughly reflect the behavior of Algorithm 20.3.1 (Pan 1998b, 2008b).

These experiments were conducted on a Pentium II 550E PC, with Windows 98 operating system, 256 MB RAM and about 16 decimal point precision. Visual Fortran 5.0 compiler was used.

There were two codes involved in the comparison:

1. MINOS 5.3.
2. PDBSA 1.0: is based on an algorithm close to Algorithm 20.3.1

Taking MINOS 5.3 as a platform, the second code used Phase-I procedure the same as MINOS 5.3 (also with the same default options).

Total 65 standard test problems without bounds and ranges sections were tested, which are classified into two groups: the first group includes 50 Netlib test problems (Appendix B: Table B.4), classified to 3 sets according to increasing $m + n$: Small, Medium and Large. The set Small includes 20 problems, Medium and Large include 15 each. The second group involves 15 larger problems, including 8 Kennington problems (Appendix B: Table B.5), 4 BPMPD (Appendix B: Table B.6) and the largest 3 Netlib problems (QAP12, MAROS-R7 and STOCFOR3).

Iteration and time ratios of MINOS 5.3 to PDBSA 1.0 are listed in the following table (see Table 20.2), where the last column gives degenerate iterations percentage ratio. The largest 3 Nelib problems are not included, because MINOS 5.3 failed to solve them.

At first, it is seen from Table 20.2 that the new code clearly outperformed MINOS 5.3, overall: time ratios for the first and the second groups are 1.41 and 2.06, respectively. It seems that the superiority of the new code is greater with large-scale problems.

Secondly, the new code appeared to be more stable and reliable, as it solved all the 65 test problems, including QAP12, MAROS-R7 and STOCFOR3, which MINOS 5.3 failed to solve.

It is noted that, the time ratios are significantly larger than iteration ratios. This is the case because systems solved per iteration by the new code are small, compare with those solved by MINOS 5.3, and the computational efforts are reduced, consequently.

Finally, it is seen from the last column that except for the smallest 20 problems, degenerate iteration ratios are less than 1, that is, the proportion of degenerate iterations yielded by the new code is larger than that by MINOS 5.3. Even so, the new code was still an unambiguous winner. Forrest and Goldfarb (1992) are right: algorithm's efficiency does not rest on the proportion of degenerate iterations.

Example 20.3.1. Solve Example 20.2.1 by Algorithm 20.3.1.

Answer Initial: $k = 2$, $B = \{5, 6\}$, $R = \{1, 2\}$. $N = \{1, 2, 3, 4, 7, 8\}$, $R' = \{3, 4, 5\}$. $B_R = I$. $\bar{x}_B = (2, 1)^{\mathrm{T}} \geq 0$.

Iteration 1:

1. $\bar{z}_N = (-2, 3, -4, 1, 2, -3)^{\mathrm{T}}$.
2. $\min\{-2, 3, -4, 1, 2, -3\} = -4 < 0$, $q = 3$.
4. $\bar{a}_q(R) = \begin{pmatrix} 1 & \\ & 1 \end{pmatrix} \begin{pmatrix} 1 \\ 1 \end{pmatrix} = \begin{pmatrix} 1 \\ 1 \end{pmatrix}$.
5. $\bar{a}_q(R') = \begin{pmatrix} 0 \\ 0 \\ 0 \end{pmatrix} - \begin{pmatrix} 0 & 0 \\ 0 & 0 \\ 0 & 0 \end{pmatrix} \begin{pmatrix} 1 \\ 1 \end{pmatrix} \neq 0$.
9. $\bar{a}_q(R) \nleq 0$.
10. $\alpha = \min\{2/1, 1/1\} = 1$, $s = 2$, $i_s = 2$.
11. $\bar{x}_3 = 1$, $\bar{x}_B = (2, 1)^{\mathrm{T}} - 1(1, 1)^{\mathrm{T}} = (1, 0)^{\mathrm{T}}$.
12. $B_R^{-1} = \begin{pmatrix} 1 & -1 \\ & 1 \end{pmatrix} \begin{pmatrix} 1 & \\ & 1 \end{pmatrix} = \begin{pmatrix} 1 & -1 \\ & 1 \end{pmatrix}$.
13. $B = \{5, 3\}$, $N = \{1, 2, 4, 6, 7, 8\}$.

Iteration 2:

1. $\bar{y} = (0, -4)^{\mathrm{T}}$,
   $\bar{z}_N = (-2, 3, 1, 0, 2, -3)^{\mathrm{T}} - (-8, 4, -4, -4, 4, -4)^{\mathrm{T}} = (6, -1, 5, 4, -2, 1)^{\mathrm{T}}$.
2. $\min\{6, -1, 5, 4, -2, 1\} = -2 < 0$, $q = 7$.
4. $\bar{a}_q(R) = \begin{pmatrix} 1 & -1 \\ & 1 \end{pmatrix} \begin{pmatrix} 1 \\ -1 \end{pmatrix} = \begin{pmatrix} 2 \\ -1 \end{pmatrix}$.
5. $\bar{a}_q(R') = \begin{pmatrix} 0 \\ 0 \\ 0 \end{pmatrix} - \begin{pmatrix} 0 & 0 \\ 0 & 0 \\ 0 & 0 \end{pmatrix} \begin{pmatrix} 2 \\ -1 \end{pmatrix} \neq 0$.
9. $\bar{a}_q(R) \nleq 0$.
10. $\alpha = \min\{1/2\} = 1/2$, $s = 1$, $i_s = 1$.
11. $\bar{x}_3 = 1/2$, $\bar{x}_B = (1, 1)^{\mathrm{T}} - (1/2)(2, -1)^{\mathrm{T}} = (0, 3/2)^{\mathrm{T}}$.
12. $B_R^{-1} = \begin{pmatrix} 1/2 & 0 \\ 1/2 & 1 \end{pmatrix} \begin{pmatrix} 1 & -1 \\ & 1 \end{pmatrix} = \begin{pmatrix} 1/2 & -1/2 \\ 1/2 & 1/2 \end{pmatrix}$.
13. $B = \{7, 3\}$, $N = \{1, 2, 4, 5, 6, 8\}$.

Iteration 3:

1. $\bar{y} = (-1, -3)^{\mathrm{T}}$,
   $\bar{z}_N = (-2, 3, 1, 0, 0, -3)^{\mathrm{T}} - (-5, 2, -2, -1, -3, -1)^{\mathrm{T}} = (3, 1, 3, 1, 3, -2)^{\mathrm{T}}$.
2. $\min\{3, 1, 3, 1, 3, -2\} = -2 < 0,\ q = 8$.
4. $\bar{a}_q(R) = \begin{pmatrix} 1/2 & -1/2 \\ 1/2 & 1/2 \end{pmatrix} \begin{pmatrix} -2 \\ 1 \end{pmatrix} = \begin{pmatrix} -3/2 \\ -1/2 \end{pmatrix}$.
5. $\bar{a}_q(R') = \begin{pmatrix} -3 \\ 1 \\ -1 \end{pmatrix} - \begin{pmatrix} 0 & 0 \\ 0 & 0 \\ 0 & 0 \end{pmatrix} \begin{pmatrix} -3/2 \\ -1/2 \end{pmatrix} = \begin{pmatrix} -3 \\ 1 \\ -1 \end{pmatrix} \neq 0$.
6. $\max\{|-3|, |1|, |-1|\} = 3,\ s = 3,\ i_s = 3$.
7. $\tau = (-3 - (0, 0)(-3/2, -1/2)^{\mathrm{T}})^{-1} = -1/3$.
   $v = -(-1/3)(-3/2, -1/2)^{\mathrm{T}} = (-1/2, -1/6)^{\mathrm{T}},\ d^{\mathrm{T}} = (0, 0)$.
   $$U = \begin{pmatrix} 1/2 & -1/2 \\ 1/2 & 1/2 \end{pmatrix} - \begin{pmatrix} -1/2 \\ -1/6 \end{pmatrix} (0, 0) = \begin{pmatrix} 1/2 & -1/2 \\ 1/2 & 1/2 \end{pmatrix}.$$
   $$\hat{B}_{\hat{R}}^{-1} = \begin{pmatrix} 1/2 & -1/2 & -1/2 \\ 1/2 & 1/2 & -1/6 \\ 0 & 0 & -1/3 \end{pmatrix}.$$
8. $k = 3,\ B = \{7, 3, 8\},\ R = \{1, 2, 3\},\ B_R^{-1} = \hat{B}_{\hat{R}}^{-1},\ \bar{x}_8 = 0$.

Iteration 4:

1. $\bar{y} = (-1, -3, 2/3)^{\mathrm{T}}$,
   $\bar{z}_N = (-2, 3, 1, 0, 0)^{\mathrm{T}} - (-13/3, 2/3, -8/3, -1, -3)^{\mathrm{T}} = (7/3, 7/3, 11/3, 1, 3)^{\mathrm{T}} \geq 0$.
3. The basic optimal solution and optimal value:
   $$\bar{x} = (0, 0, 3/2, 0, 0, 0, 1/2, 0)^{\mathrm{T}}, \quad \bar{f} = (-4, 2)(3/2, 1/2)^{\mathrm{T}} = -5.$$
   which coincides with the result in Example 20.2.1.

20.4 Phase-I: Single-Artificial-Variable

Standard Phase-I methods presented in Chap. 13 are applicable (or after slightly modified) in the deficient-basis context. It is also possible to design new Phase-I methods by taking advantage of deficient basis, as the method presented in this section (Pan and Pan 2001).

Introducing an artificial variable x_{n+1}, associated with column the same as the right-hand side. The resulting auxiliary program is then

$$\begin{aligned} \min\ & f = x_{n+1}, \\ \text{s.t.}\ & Ax + bx_{n+1} = b, \quad x, x_{n+1} \geq 0, \end{aligned} \tag{20.12}$$

which has the initial tableau below:

x^{T}	x_{n+1}	f	RHS
A	b		b
	1	-1	

Take the x_{n+1} column as the first basic column. Assuming $b_r \neq 0$, take the rth row as the first basic row, i.e.,

$$B = \{n + 1\}, \quad R = \{r\}.$$

Thus,

$$B = \{n + 1\}, \quad N = A, \quad R = \{r\}, \quad R' = \{1, \cdots, m\} \backslash R.$$

Carry out elementary transformations to turn the r-th component of the x_{n+1} column to 1, and eliminate the other nonzeros in the column. By row and column exchanges, the resulting deficient-basis tableau is of form

x_B^{T}	x_N^{T}	f	RHS
1	$\bar{N}_R$		1
	$\bar{N}_{R'}$		
	$-\bar{N}_R$	-1	-1

(20.13)

where $\bar{N}_R \in \mathcal{R}^{1\times n}$, $\bar{N}_{R'} \in \mathcal{R}^{(m-1)\times n}$ are associated with the auxiliary feasible solution $\bar{x}_{n+1} = 1$, $\bar{x}_N = 0$ and objective value 1. It can be solved by the deficient-basis Algorithm 20.2.1. If the optimal value of the auxiliary program vanishes, a feasible solution to the original problem is obtained (when x_{n+1} leaves the basis); otherwise, there is no feasible solution to the original problem.

It is important to maintain sparsity of the deficient-basis tableau (20.13). Although, in theory, any row associated with a nonzero component of the right-hand side can be taken as the first basic row, a clever choice is to select a row having the minimum number of nonzeros, because a dense basic row would cause a lot of fill-ins. Also, when the right-hand side is dense, the deficient-basis tableau would become dense, as consequently looses the value of the deficient-basis method.

Practically, the f column can be omitted from the initial tableau. The x_{n+1} column can also be omitted because it is the same as the RHS column (except for the bottom row). The objective row of the original problem may be placed in the tableau, as it is convenient to go to Phase-II after Phase-I finishes. In addition, the auxiliary objective row may be omitted too, as it differs from the basic row associated with x_{n+1} just by a sign (except for entries in x_{n+1} column), though this is not done with the following example, where it is placed at the bottom.

Example 20.4.1. Solve the following problem by the two-phase deficient-basis algorithm:

$$
\begin{array}{llrl}
\min & f = -3x_1 + 2x_2 - x_3 + 3x_5 + 2x_7 - 6x_8, & & \\
\text{s.t.} & -5x_1 + x_2 + 2x_3 + x_4 + x_5 - 1x_7 - 3x_8 & = & -15, \\
& -3x_2 + x_3 + 3x_4 - x_5 - 2x_6 + 2x_7 + x_8 & = & -11, \\
& -3x_3 + 5x_4 - 3x_5 + x_6 - 3x_7 + 2x_8 & = & 2, \\
& + x_3 + x_5 - x_8 & = & 0, \\
& - x_3 - x_5 + 2x_8 & = & 0, \\
& x_j \geq 0, \quad j = 1, \cdots, 8. & &
\end{array}
$$

Answer Phase-I: Auxiliary initial tableau (the RHS column also represents the x_9 column)

x_1	x_2	x_3	x_4	x_5	x_6	x_7	x_8	RHS(x_9)
−5	1	2	1	1		−1	−3	−15
	−3	1	3	−1	−2	2	1	−11
		−3	5	−3	1	−3	2	2
		1		1			−1	
		−1		−1			2	
−3	2	−1		3		2	−6	
								1

Iteration 1:
Multiply row 1 by $-1/15$, then add $11, -2, -1$ times of row 2 to rows 2,3,7, respectively:

x_1	x_2	x_3	x_4	x_5	x_6	x_7	x_8	RHS(x_9)
1/3	−1/15	−2/15	−1/15	−1/15		1/15	1/5	1
11/3*	−56/15	−7/15	34/15	−26/15	−2	41/15	16/5	
−2/3	2/15	−41/15	77/15	−43/15	1	−47/15	8/5	
		1		1			−1	
		−1		−1			2	
−3	2	−1		3		2	−6	
−1/3	1/15	2/15	1/15	1/15		1/15	−1/5	

The preceding auxiliary tableau is a feasible deficient-basis tableau:

$$s = 1, \quad B = \{9\}, \quad R = \{1\}, \quad N = \{1,2,3,4,5,6,7,8\}, \quad R' = \{2,3,4,5\}.$$

Phase-I: Call Algorithm 20.2.1.

Iteration 2:

1. $\min\{-1/3, 1/15, 2/15, 1/15, 1/15, 1/15, -1/5\} = -1/3 < 0, \ q = 1.$
3. $\bar{a}_q(R') = (11/3, -2/3, 0, 0)^T \neq 0.$

4. $\max\{|11/3|, |-2/3|\} = 11/3,\ i_s = 2$.
5. Multiply row 2 by 3/11, then add $-1/3, 2/3, 3, 1/3$ times of row 2 to rows 1,3,6,7, respectively.
6. $k = 2,\ B = \{9, 1\},\ R = \{1, 2\},\ N = \{2, 3, 4, 5, 6, 7, 8\},\ R' = \{3, 4, 5\}$.

x_1	x_2	x_3	x_4	x_5	x_6	x_7	x_8	RHS(x_9)
	3/11	−1/11	−3/11	1/11	2/11	−2/11	−1/11	1
1	−56/55	−7/55	34/55	−26/55	−6/11	41/55	48/55	
	−6/11*	−31/11	61/11	−35/11	7/11	−29/11	24/11	
		1		1			−1	
		−1		−1			2	
	−58/55	−76/55	102/55	87/55	−18/11	233/55	−186/55	
	−3/11	1/11	3/11	−1/11	−2/11	2/11	1/11	

Iteration 3:

1. $\min\{-3/11, 1/11, 3/11, -1/11, -2/11, 2/11, 1/11\} = -3/11 < 0,\ q = 2$.
3. $\bar{a}_q(R') = (-6/11, 0, 0)^T \neq 0$.
4. $\max\{|-6/11|, 0, 0\} = 6/11,\ i_s = 3$.
5. Multiply row 3 by $-11/6$, then add $-3/11, 56/55, 58/55, 3/11$ times of row 3 to rows 1,2,6,7, respectively.
6. $k = 3,\ B = \{9, 1, 2\},\ R = \{1, 2, 3\},\ N = \{3, 4, 5, 6, 7, 8\},\ R' = \{4, 5\}$.

x_1	x_2	x_3	x_4	x_5	x_6	x_7	x_8	RHS(x_9)
		−3/2	5/2*	−3/2	1/2	−3/2	1	1
1		77/15	−146/15	82/15	−26/15	17/3	−16/5	
	1	31/6	−61/6	35/6	−7/6	29/6	−4	
		1		1			−1	
		−1		−1			2	
		61/15	−133/15	116/15	−43/15	28/3	−38/5	
		3/2	−5/2	3/2	−1/2	3/2	−1	

Iteration 4:

1. $\min\{3/2, -5/2, 3/2, -1/2, 3/2, -1\} = -5/2 < 0,\ q = 4$.
3. $\bar{a}_q(R') = (0, 0)^T$.
7. $\bar{a}_q(R) \not\leq 0$.
8. $\alpha = \min\{1/(5/2)\} = 2/5,\ s = 1,\ i_s = 1$. x_9 leaves the basis.
9. Multiply row 1 by 2/5, then add $146/15, 61/6, 133/15, 5/2$ times of row 1 to rows 2,3,6,7, respectively.
10. $B = \{4, 1, 2\},\ N = \{3, 5, 6, 7, 8, 9\}$.

x_1	x_2	x_3	x_4	x_5	x_6	x_7	x_8	RHS(x_9)
		$-3/5$	1	$-3/5$	$1/5$	$-3/5$	$2/5$	$2/5$
1		$-53/75$		$-28/75$	$16/75$	$-13/75$	$52/75$	$292/75$
	1	$-14/15$		$-4/15$	$13/15$	$-19/15$	$1/15$	$61/15$
		1		1			-1	
		-1		-1			2^*	
		$-94/75$		$181/75$	$-82/75$	$301/75$	$-304/75$	$266/75$
								1

As the artificial x_9 column has already become nonbasic, the preceding tableau is feasible. Delete the x_9 column and the bottom row, then turn to Phase-II.

Phase-II: Call Algorithm 20.2.1.

Iteration 5:

1. $\min\{-94/75, 181/75, -82/75, 301/75, -304/75\} = -304/75 < 0,\ q = 8.$
3. $\bar{a}_q(R') = (-1, 2)^T \neq 0.$
4. $\max\{|-1|, |2|\} = 2,\ i_s = 5.$
5. Multiply row 5 by $1/2$, then add $-2/5, -52/75, -1/15, 1, 304/75$ times of row 5 to rows 1,2,3,4,6, respectively.
6. $k = 4,\ B = \{4, 1, 2, 8\},\ R = \{1, 2, 3, 5\},\ N = \{3, 5, 6, 7\},\ R' = \{4\}.$

x_1	x_2	x_3	x_4	x_5	x_6	x_7	x_8	RHS
		$-2/5$	1	$-2/5$	$1/5$	$-3/5$		$2/5$
1		$-9/25$		$-2/75$	$16/75$	$-13/75$		$292/75$
	1	$-9/10$		$-7/30$	$13/15$	$-19/15$		$61/15$
		$1/2^*$		$1/2$				
		$-1/2$		$-1/2$			1	
		$-82/25$		$29/75$	$-82/75$	$301/75$		$266/75$

Iteration 6:

1. $\min\{-82/25, 29/75, -82/75, 301/75\} = -82/25 < 0,\ q = 3.$
3. $\bar{a}_q(R') \neq 0.$
4. $i_s = 4.$
5. Multiply row 4 by 2, then add $2/5, 9/25, 9/10, 1/2, 82/25$ times of row 4 to rows 1,2,3,5,6, respectively.
6. $k = 5,\ B = \{4, 1, 2, 8, 3\},\ R = \{1, 2, 3, 5, 4\},\ N = \{5, 6, 7\},\ R' = \emptyset.$

x_1	x_2	x_3	x_4	x_5	x_6	x_7	x_8	RHS
			1		$1/5^*$	$-3/5$		$2/5$
1				$1/3$	$16/75$	$-13/75$		$292/75$
	1			$2/3$	$13/15$	$-19/15$		$61/15$
		1		1				
							1	
				$11/3$	$-82/75$	$301/75$		$266/75$

Iteration 7:

1. $\min\{11/3, -82/75, 301/75\} = -82/75 < 0,\ q = 6.$
3. $R' = \emptyset.$
7. $\bar{a}_q(R) \nleq 0.$
8. $\alpha = \min\{1/(20/87)\} = 87/20,\ s = 1,\ i_s = 1.$
9. Multiply row 1 by 5, then add $-16/75, -13/15, 82/75$ times of row 1 to rows 2,3,6, respectively.
10. $B = \{6, 1, 2, 8, 3\},\ N = \{4, 5, 7\}.$

x_1	x_2	x_3	x_4	x_5	x_6	x_7	x_8	RHS
			5		1	−3		2
1			−16/15	1/3		7/15		52/15
	1		−13/3	2/3		4/3		7/3
		1		1				
							1	
			82/15	11/3		11/15		86/15

Iteration 8:

1. $\min\{82/15, 11/3, 11/15\} \geq 0.$
2. The basic optimal solution and optimal value:

$$\bar{x} = (52/15, 7/3, 0, 0, 0, 2, 0, 0)^{\mathrm{T}}, \quad \bar{f} = -86/15.$$

20.5 On Implementation

The deficient-basis algorithm may be regarded as a variant of the conventional simplex method. According implementation issues, such as the LU factorization of the basis matrix, L^{-1} stored in factors form, LU update in the rank-remaining iteration, and etc., are analogous to those, presented in Chap. 5. Besides all, one should pay close attention to take advantage of basis deficiency to reduce computational work. In this section, only the following two aspects are discussed.

20.5.1 Initial Basis

It is known for the standard LP problem (1.8) that

$$\text{rank } A = \text{rank } (A \mid b) \leq m.$$

The LU factorization of an initial basis matrix can be created by a crash procedure like that presented in Sect. 5.5. In addition to balancing sparsity and stability, however, the resulting initial basis should be of order as low as possible. To this end, some variant of Markowitz Rule 5.3.1 would be utilized.

Initially, set $B = \emptyset,\ R = \emptyset$. Suppose, for some k, Gauss elimination has been carried out on matrix $(A \mid b)$ such that some k columns (excluding b) of the latter are upper triangular; in other words, k basic columns and rows have been determined, and the according $\bar{B}_R$ is an upper triangular k-order submatrix of $(\bar{A} \mid \bar{b})$.

If $k = m$, or if $k < m$ but the following row index set is empty,

$$I' = \{i \in R' \mid |\bar{b}_i| > \epsilon_0\}, \tag{20.14}$$

where $\epsilon_0 > 0$ is a threshold for numerical 0 (see Sect. 5.1), then the LU factorization of an k-order basis matrix has already been obtained. Otherwise, the following rule is utilized to determine the $(k+1)$th basic column and row.

Rule 20.5.1. Given constant $0 < \sigma \leq 1$. Select column index q and row index p such that

$$(r_p - 1)(c_q - 1) = \min\{(r_i - 1)(c_j - 1) \mid |\bar{a}_{ij}| \geq \sigma\theta,\ i \in I',\ j \in N\}, \tag{20.15}$$

where r_i denotes the number of nonzeros in the ith row and c_j the number of nonzeros in the jth column, and

$$\theta = \max\{|\bar{a}_{ij}| \mid i \in I',\ j \in N\}. \tag{20.16}$$

Note that the preceding rule fails if

$$\{(i,j) \in I' \times N \mid |\bar{a}_{ij}| \geq \sigma\theta\} = \emptyset.$$

In this case, $Ax = b$ may be regarded numerically inconsistent.

20.5.2 LU Updating in Rank-Increasing Iteration

Updating LU factors in a rank-remaining iteration is essentially the same as that in the conventional simplex context (Sect. 5.4). In contrast, updating LU factors in a rank-increasing iteration is simpler.

Let $L^{-1}B_R = U$ be the LU factorization of B_R. From (20.9), it follows that

$$\left(\begin{array}{c|c} L^{-1} & \\ \hline & 1 \end{array}\right) \hat{B}_{\hat{R}} = \left(\begin{array}{c|c} U & L^{-1}a_q(R) \\ \hline e_{i_s}^{\mathrm{T}} B & a_{i_s q} \end{array}\right).$$

which is upper triangular, except for the bottom row. Premultiply a series of Gauss matrices $G_1, \cdots, G_k$ to eliminate entries of the bottom row (except for the diagonal), it becomes an upper triangular matrix $\tilde{U}$. The LU factorization of the new basis matrix is then

$$\tilde{L}^{-1}\hat{B}_{\hat{R}} = \tilde{U}, \quad \tilde{L}^{-1} = G_k \cdots G_1 \left(\begin{array}{c|c} L^{-1} & \\ \hline & 1 \end{array}\right).$$

20.6 Deficient-Basis Reduced Method

In this section, the deficient basis will be embedded to the (improved) reduced method (Chap. 16) to solve large-scale and highly degenerate problems. Thereby, the related search direction corresponds to the entire objective gradient, and the method is stable, compared with the standard deficient-basis method presented previously.

We are concerned with the reduced problem (25.2), i.e.,

$$\begin{array}{ll} \min & x_{n+1}, \\ \text{s.t.} & [A \vdots a_{n+1}] \begin{pmatrix} x \\ x_{n+1} \end{pmatrix} = b, \quad x \geq 0. \end{array} \tag{20.17}$$

Assume that B, R are basic columns and rows, denoted by (20.1) (the row index $m + 1$ is the rth element of R); and N, R' are nonbasic columns and rows, denoted by (20.2). It's initial tableau is then

x_B^{T}	x_N^{T}	x_{n+1}	RHS
B_R	N_R	$-e_r$	b_R
$B_{R'}$	$N_{R'}$		$b_{R'}$

(20.18)

where B_R is the deficient basis, $r \in \{1, \cdots, k\}$, and $-e_r \in \mathcal{R}^k$ denotes the basic row part of $a_{n+1} = -e_{m+1}$. Note that x_{n+1} column is put alone as a special nonbasic column. Let $\bar{x}_{n+1}$ be given. Setting $x_{n+1} = \bar{x}_{n+1}$ and $\bar{x}_N = 0$, it can be determined from the first k rows of (20.18) that

$$\bar{x}_B = B_R^{-1} b_R + \bar{x}_{n+1} B_R^{-1} e_r, \tag{20.19}$$

which determines a solution $\bar{x}$ to $Ax = b$, corresponding objective value $\bar{x}_{n+1}$. Then, if

$$B_{R'} B_R^{-1} e_r = 0, \qquad b_{R'} - B_{R'} B_R^{-1} b_R = 0,$$

then tableau (20.18) can be transformed to the so-called "(improved) deficient-basis reduced tableau" of form

x_B^{T}	x_N^{T}	x_{n+1}	$\bar{x}_B$
I	$\bar{N}_R$	$\bar{w}$	$\bar{x}_B$
	$\bar{N}_{R'}$		

(20.20)

If $\bar{x}_B \geq 0$, then (20.20) is a feasible tableau, and hence ready to carry out a rank-increasing or rank-remaining iteration (see Sect. 20.2). Note that rank-increasing iteration does not change both x_{n+1} column and $\bar{x}_B$ column.

Based on preceding discussions, integrating steps of Algorithms 16.1.1 and 20.2.1 gives the following algorithm.

Algorithm 20.6.1 (Deficient-basis reduced algorithm: tableau form). Initial: feasible deficient-basis reduced tableau of form (20.20) with $B,\ R,\ N,\ R',\ 1 \leq k \leq m+1$. $\bar{x}_B,\ \bar{x}_{n+1}$. This algorithm solves the reduced problem (20.17).

1. Stop if $\bar{w} \geq 0$ (unbounded problem).
2. Determine α and s such that $\alpha = -\bar{x}_{j_s}/\bar{w}_s = \min\{-\bar{x}_{j_t}/\bar{w}_t \mid \bar{w}_t < 0,\ t = 1,\cdots,k\}$.
3. If $\alpha \neq 0$, update $\bar{x}_B = \bar{x}_B + \alpha\bar{w},\ \bar{x}_{n+1} = \bar{x}_{n+1} - \alpha$.
4. Determine column index $q \in \arg\min_{j \in N}\ \bar{a}_{i_s,j}$.
5. Stop if $\bar{a}_{i_s,q} \geq 0$ (optimality achieved).
6. Go to step 10 if $R' = \emptyset$ or $\bar{a}_q(R') = 0$.
7. Determine row index $p \in \arg\max_{i \in R'} |\bar{a}_{pq}|$.
8. Convert $\bar{a}_{pq}$ to 1, and eliminate the other nonzeros in the column by elementary transformations.
9. Set $k = k+1$, bring q from N to B and p from R' to R, and go to step 4.
10. Convert $\bar{a}_{i_s,q}$ to 1, and eliminate the other nonzeros in the column by elementary transformations.
11. Update (B, N) by exchanging j_s and q.
12. Go to step 1.

Note $\bar{x}_B$ column is not changed in steps 8 and 10.

Example 20.6.1. Solve the following problem by Algorithm 20.6.1:

$$
\begin{array}{llrrrrrrrl}
\min & x_8 = 3x_2 + 2x_4 + x_5 - 4x_6 - 2x_7, \\
\text{s.t.} & x_1 & - & 2x_2 & & + 3x_4 & + x_5 & - x_6 & - 2x_7 & = 4, \\
& & & 4x_2 & + x_3 & - 5x_4 & & + 2x_6 & + 3x_7 & = 12, \\
& & & 6x_2 & & & - 2x_5 & & & = 0, \\
& & & -5x_2 & & & & & - x_7 & = 0, \\
& & & x_2 & & & - 3x_5 & & + 4x_7 & = 0, \\
& & & & & & x_j \geq 0, & j = 1,\cdots,7.
\end{array}
$$

Answer Initial: $k = 2,\ B = \{1, 3\},\ R = \{1, 2\}$, and the feasible deficient-basis tableau is

x_1	x_2	x_3	x_4	x_5	x_6	x_7	x_8	RHS
1	−2		3	1	−1	−2		4
	4	1	−5		2	−3		12
	6			−2				
	−5					−1		
	1			−3		4		
	3		2	1	−4*	−2	−1	

$\bar{z}_6 = -4 < 0,\ \bar{a}_6(R') = (0, 0, 0)^{\mathrm{T}}$, Take $q = 6$. Multiply row 6 by $-1/4$, then add $1, -2$ times of row 6 to rows 1,2, respectively, obtaining the feasible deficient-basis reduced tableau below:

x_1	x_2	x_3	x_4	x_5	x_6	x_7	x_8	$\bar{x}_B$
1	−11/4		5/2	3/4		−3/2	1/4	4
	11/2	1	−4*	1/2		−4	−1/2	12
	6			−1/2				
	−5					−1		
	1			−3		4		
	−3/4		−1/2	−1/4	1	1/2	1/4	

Call Algorithm 20.6.1: $k = 3,\ B = \{1, 3, 6\},\ R = \{1, 2, 6\}.\ N = \{2, 4, 5, 7\},\ R' = \{3, 4, 5\}$.

Iteration 1: $\bar{x}_B = (4, 12, 0)^T,\ \bar{x}_8 = 0$

1. $\bar{w} = (14, -1/2, 1/4)^{\mathrm{T}} \not\geq 0$.
2. $\alpha = \min\{-(12)/(-1/2)\} = 24,\ s = 2$.
3. Add 24 times of x_8 column to $\bar{x}_B$ column; $\bar{x}_8 = -24$.
4. $\min\{11/2, -4, 1/2, -4\} = -4 < 0,\ q = 4$.
6. $\bar{a}_q(R') = 0$.
10. Multiply row 2 by $-1/4$, then add $-5/2, 1/2$ times of row 2 to rows 1,6, respectively.
11. $B = \{1, 4, 6\},\ R = \{1, 2, 6\},\ N = \{2, 3, 5, 7\},\ R' = \{3, 4, 5\}$.

x_1	x_2	x_3	x_4	x_5	x_6	x_7	x_8	$\bar{x}_B$
1	11/16	5/8		17/16		−4	−1/16	10
	−11/8	−1/4	1	−1/8		1	1/8	
	6			−1/2				
	−5					−1		
	1			−3		4*		
	−23/16	−1/8		−5/16	1	1	5/16	6

Iteration 2:

1. $\bar{w} = (-1/16, 1/8, 5/16)^{\mathrm{T}} \not\geq 0$.
2. $\alpha = \min\{-(10)/(-1/16)\} = 160,\ s, i_s, j_s = 1$.
3. Add 160 times of x_8 column to $\bar{x}_B$ column; $\bar{x}_8 = -24 - 160 = -184$.
4. $\min\{11/16, 5/8, 17/16, -4\} = -4 < 0,\ q = 7$.
6. $\bar{a}_q(R') = (0, -1, 4)^T \neq 0$.
7. $\max\{0, |-1|, |4|\} = 4,\ p = 5$.
8. Multiply row 5 by $1/4$, then add $4, -1, 1, -1$ times of row 5 to rows 1,2,4,6, respectively.
9. $k = 4,\ B = \{1, 6, 4, 7\},\ R = \{1, 2, 6, 5\},\ N = \{2, 3, 5\},\ R' = \{3, 4\}$.

x_1	x_2	x_3	x_4	x_5	x_6	x_7	x_8	$\bar{x}_B$
1	27/16	5/8		−31/16			−1/16	
	−13/8	−1/4	1	5/8			1/8	20
	6			−1/2*				
	−19/4			−3/4				
	1/4			−3/4		1		
	−27/16	−1/8		7/16	1		5/16	56

Iteration 3:

4. $\min\{27/16, 5/8, -31/16\} = -31/16 < 0,\ q = 5$.
6. $\bar{a}_q(R') = (-2, 3/4)^T \neq 0$.
7. $\max\{|-1/2|, |-3/4|\} = 3/4,\ p = 4$.
8. Multiply row 4 by $-4/3$, then add $31/16, -5/8, 1/2, 3/4, -7/16$ times of row 4 to rows 1,2,3,5,6, respectively.
9. $k = 5,\ B = \{1, 6, 4, 7, 5\},\ R = \{1, 2, 4, 5, 6\},\ N = \{2, 3\},\ R' = \{3\}$.

x_1	x_2	x_3	x_4	x_5	x_6	x_7	x_8	$\bar{x}_B$
1	335/24	5/8					−1/16	
	−67/12	−1/4	1				1/8	20
	55/6							
	19/3			1				
	5					1		
	−107/24	−1/8			1		5/16	56

Iteration 4:

4. $\min\{335/24, 5/8\} \geq 0$.
5. The basic optimal solution and optimal value:

$$\bar{x} = (0, 0, 0, 20, 0, 56, 0)^{\mathrm{T}}, \qquad \bar{x}_8 = -184.$$

As for the revised version of Algorithm 20.6.1, it posses features of both deficient-basis and reduced simplex approaches; e.g., the updates (20.10) and (20.11) of the inverse of the basis still apply. Besides, updating of $\bar{x}_B$ column is the same as in Algorithm 20.6.1, and updating x_{n+1} column is by the first expression of (3.15).

Assume that B_R is the deficient basis, then the revised deficient-basis reduced tableau is

$$\begin{array}{cc|c|c} x_B^{\mathrm{T}} & x_N^{\mathrm{T}} & x_{n+1} & \bar{x}_B \\ \hline I & B_R^{-1}N_R & -B_R^{-1}e_r & \bar{x}_B \\ & N_{R'} - B_{R'}B_R^{-1}N_R & & \\ \hline \end{array} \tag{20.21}$$

Based on equivalence between (20.20 and (20.21), it is easy to write the revised version of Algorithm 20.6.1.

Algorithm 20.6.2 (Deficient-basis reduced algorithm). Initial: $1 \le k \le m + 1$, B, R, B_R^{-1}, $\bar{x}_{n+1}$; $\bar{w} = -B_R^{-1}e_r$, $\bar{x}_B = B_R^{-1}b_R - \bar{x}_{n+1}\bar{w} \ge 0$. This algorithm solves the reduced problem.

1. Stop if $\bar{w} \ge 0$ (unbounded problem).
2. Determine α and s such that $\alpha = -\bar{x}_{j_s}/\bar{w}_s = \min\{-\bar{x}_{j_t}/\bar{w}_t \mid \bar{w}_t < 0,\ t = 1,\cdots,k\}$.
3. If $\alpha \ne 0$, update $\bar{x}_B = \bar{x}_B + \alpha\bar{w}$, $\bar{x}_{n+1} = \bar{x}_{n+1} - \alpha$.
4. Compute $\sigma_N = N_R^{\mathrm{T}}B_R^{-\mathrm{T}}e_s$.
5. Determine column index $q \in \arg\min_{j\in N}\sigma_j$.
6. Stop if $\sigma_q \ge 0$ (optimality achieved).
7. Compute $\bar{a}_q(R) = B_R^{-1}a_q(R)$.
8. Go to step 13 if $R' = \emptyset$ or $\bar{a}_q(R') = a_q(R') - B_{R'}\bar{a}_q(R) = 0$.
9. Determine $p \in \arg\max_{i\in R'} |(\bar{a}_{i,q}|$.
10. Update B_R^{-1} by (20.10).
11. Set $k = k + 1$, $\bar{x}_q = 0$, $\bar{a}_{k,n+1} = 0$, and bring q from N to B and p from R' to R.
12. Go to step 4.
13. Compute $\bar{\alpha} = \bar{a}_{i_s,n+1}/\bar{a}_{i_s,q}$.
14. If $\alpha \ne 0$, set $\bar{w}_s = \bar{\alpha}$, and update $\bar{w} = \bar{w} - \bar{\alpha}\bar{a}_q(R)$.
15. Update B_R^{-1} by (20.11).
16. Update (B, N) by exchanging j_s and q.
17. Go to step 1.

Example 20.6.2. Solve the following problem by the preceding algorithm:

$$\begin{array}{lrl} \min & x_8 = & -3x_1 + x_4 + 4x_5 - 2x_6 + 2x_7, \\ \text{s.t.} & -3x_1 & + x_3 - x_4 - 2x_5 + 4x_6 - 2x_7 = 9, \\ & & + x_2 + 4x_4 + 3x_5 - 3x_6 - 5x_7 = 2, \\ & 6x_1 & - 6x_4 - 2x_5 = 0, \\ & & + 4x_7 = 0, \\ & x_1 & - x_4 - x_7 = 0, \\ & & x_j \ge 0, \quad j = 1,\cdots,7. \end{array}$$

Answer The reduced problem is associated with the following tableau:

x_1	x_2	x_3	x_4	x_5	x_6	x_7	x_8	RHS
−3		1	−1	−2	4	−2		9
	1		4	3	−3	−5		2
6			−6	−2				
						4		
1			−1			−1		
−3			1	4	−2	2	−1	

Initial: The preceding is a feasible deficient-basis tableau. A feasible reduced (deficient) basis is yielded by transforming the bottom row to a basic row and x_6 column to a basic column, i.e., $k = 3$, $B = \{3, 2, 6\}$, $R = \{1, 2, 6\}$, $N = \{1, 4, 5, 7\}$, $R' = \{3, 4, 5\}$.

$$B_R = \begin{pmatrix} 1 & & 4 \\ & 1 & -3 \\ & & -2 \end{pmatrix}, \quad B_R^{-1} = \begin{pmatrix} 1 & & 2 \\ & 1 & -3/2 \\ & & -1/2 \end{pmatrix},$$

$$\bar{a}_{n+1} = -B_R^{-1} e_3 = \begin{pmatrix} -2 \\ 3/2 \\ 1/2 \end{pmatrix}. \quad \bar{x}_8 = 0, \quad \bar{x}_B = B_R^{-1} b_R = \begin{pmatrix} 9 \\ 2 \\ 0 \end{pmatrix} \geq 0.$$

Iteration 1:

1. $\bar{w} = (-2, 3/2, 1/2)^{\mathrm{T}} \not\leq 0$.
2. $\alpha = \min\{-9/(-2)\} = 9/2, \ s = 1$.
3. $\bar{x}_B = (9, 2, 0)^{\mathrm{T}} + (9/2)(-2, 3/2, 1/2)^{\mathrm{T}} = (0, 35/4, 9/4)^{\mathrm{T}}, \ \bar{x}_8 = -9/2$.
4. $$\sigma_N = \begin{pmatrix} -3 & -1 & -2 & -2 \\ & 4 & 3 & -5 \\ -3 & 1 & 4 & 2 \end{pmatrix}^{\mathrm{T}} \begin{pmatrix} 1 & & 2 \\ & 1 & -3/2 \\ & & -1/2 \end{pmatrix}^{\mathrm{T}} \begin{pmatrix} 1 \\ 0 \\ 0 \end{pmatrix} = \begin{pmatrix} -9 \\ 1 \\ 6 \\ 2 \end{pmatrix}.$$
5. $\min\{-9, 1, 6, 2\} = -9 < 0, \ q = 1$.
7. $$\bar{a}_q(R) = \begin{pmatrix} 1 & & 2 \\ & 1 & -3/2 \\ & & -1/2 \end{pmatrix} \begin{pmatrix} -3 \\ 0 \\ -3 \end{pmatrix} = \begin{pmatrix} -9 \\ 9/2 \\ 3/2 \end{pmatrix}.$$
8. $$\bar{a}_q(R') = \begin{pmatrix} 6 \\ 0 \\ 1 \end{pmatrix} - \begin{pmatrix} 0 & 0 & 0 \\ 0 & 0 & 0 \\ 0 & 0 & 0 \end{pmatrix} \begin{pmatrix} 6 \\ -3 \\ -2 \end{pmatrix} = \begin{pmatrix} 6 \\ 0 \\ 1 \end{pmatrix} \neq 0.$$
9. $\max\{|6|, 0, |1|\} = 6, \ p = 3$.
10. $\tau = (6 - (0, 0, 0)(-9, 9/2, 3/2)^{\mathrm{T}})^{-1} = 1/6$.
 $v = -(1/6)(-9, 9/2, 3/2)^{\mathrm{T}} = (3/2, -3/4, -1/4)^{\mathrm{T}}$.

$$d^{\mathrm{T}} = -(1/6)(0,0,0)\begin{pmatrix}1 & & 2\\ & 1 & -3/2\\ & & -1/2\end{pmatrix} = (0,0,0).$$

$$U = \begin{pmatrix}1 & & 2\\ & 1 & -3/2\\ & & -1/2\end{pmatrix} - \begin{pmatrix}3/2\\ -3/4\\ -1/4\end{pmatrix}(0\ 0\ 0) = \begin{pmatrix}1 & & 2\\ & 1 & -3/2\\ & & -1/2\end{pmatrix}.$$

$$\hat{B}_{\hat{R}}^{-1} = \begin{pmatrix}1 & & 2 & 3/2\\ & 1 & -3/2 & -3/4\\ & & -1/2 & -1/4\\ & & & 1/6\end{pmatrix}.$$

11. $k = 4,\ B = \{3,2,6,1\},\ R = \{1,2,6,3\},\ N = \{4,5,7\},$
$R' = \{4,5\}, B_R^{-1} = \hat{B}_{\hat{R}}^{-1}.\ \bar{x}_1 = 0,\ \bar{a}_{4,n+1} = 0.$

Iteration 2:

4. $$\sigma_N = \begin{pmatrix}-1 & -2 & -2\\ 4 & 3 & -5\\ 1 & 4 & 2\\ -6 & -2 & \end{pmatrix}^{\mathrm{T}} \begin{pmatrix}1 & & 2 & 3/2\\ & 1 & -3/2 & -3/4\\ & & -1/2 & -1/4\\ & & & 1/6\end{pmatrix}^{\mathrm{T}} \begin{pmatrix}1\\0\\0\\0\end{pmatrix} = \begin{pmatrix}-8\\3\\2\end{pmatrix}.$$

5. $\min\{-8,3,2\} = -13/2 < 0,\ q = 4.$

7. $$\bar{a}_q(R) = \begin{pmatrix}1 & & 2 & 3/2\\ & 1 & -3/2 & -3/4\\ & & -1/2 & -1/4\\ & & & 1/6\end{pmatrix}\begin{pmatrix}-1\\4\\1\\-6\end{pmatrix} = \begin{pmatrix}-8\\7\\1\\-1\end{pmatrix}.$$

8. $$\bar{a}_q(R') = \begin{pmatrix}0\\-1\end{pmatrix} - \begin{pmatrix}0&0&0&0\\0&0&0&1\end{pmatrix}\begin{pmatrix}-8\\7\\1\\-1\end{pmatrix} = \begin{pmatrix}0\\0\end{pmatrix}.$$

13. $\bar{\alpha} = -2/-8 = 1/4.$

14. $\bar{w} = (-2,3/2,1/2,0)^{\mathrm{T}} - (1/4)(-8,7,1,-1)^{\mathrm{T}} = (0,-1/4,1/4,1/4)^{\mathrm{T}},$
$\bar{w}_1 = 1/4.$

15. $$\hat{B}_{\hat{R}}^{-1} = \begin{pmatrix}-1/8 & & & \\ 7/8 & 1 & & \\ 1/8 & & 1 & \\ -1/8 & & & 1\end{pmatrix}\begin{pmatrix}1 & & 2 & 3/2\\ & 1 & -3/2 & -3/4\\ & & -1/2 & -1/4\\ & & & 1/6\end{pmatrix} = \begin{pmatrix}-1/8 & & -1/4 & -3/16\\ 7/8 & 1 & 1/4 & 9/16\\ 1/8 & & -1/4 & -1/16\\ -1/8 & & -1/4 & -1/48\end{pmatrix}.$$

16. $B = \{4,2,6,1\},\ N = \{3,5,7\},\ B_R^{-1} = \hat{B}_{\hat{R}}^{-1}.$

Iteration 3:

1. $\bar{a}_{n+1} = (1/4,-1/4,1/4,1/4)^{\mathrm{T}} \not\geq 0.$
2. $\alpha = \min\{-(35/4)/(-1/4)\} = 35,\ s = 2.$
3. $\bar{x}_B = (0,35/4,9/4,0)^{\mathrm{T}} + 35(1/4,-1/4,1/4,1/4)^{\mathrm{T}} = (35/4,0,11,35/4)^{\mathrm{T}},$
$\bar{x}_8 = -9/2 - 35 = -79/2.$

$$4.\ \sigma_N = \begin{pmatrix} 1 & -2 & -2 \\ & 3 & -5 \\ & 4 & 2 \\ & -2 & \end{pmatrix}^{\mathrm{T}} \begin{pmatrix} -1/8 & & -1/4 & -3/16 \\ 7/8 & 1 & 1/4 & 9/16 \\ 1/8 & & -1/4 & -1/16 \\ -1/8 & & -1/4 & -1/48 \end{pmatrix}^{\mathrm{T}} \begin{pmatrix} 0 \\ 1 \\ 0 \\ 0 \end{pmatrix} = \begin{pmatrix} 7/8 \\ 9/8 \\ -25/4 \end{pmatrix}.$$

5. $\min\{7/8, 9/8, -25/4\} = -25/4 < 0,\ q = 7.$

$$7.\ \bar{a}_q(R) = \begin{pmatrix} -1/8 & & -1/4 & -3/16 \\ 7/8 & 1 & 1/4 & 9/16 \\ 1/8 & & -1/4 & -1/16 \\ -1/8 & & -1/4 & -1/48 \end{pmatrix}^{\mathrm{T}} \begin{pmatrix} -2 \\ -5 \\ 2 \\ 0 \end{pmatrix} = \begin{pmatrix} -1/4 \\ -25/4 \\ -3/4 \\ -1/4 \end{pmatrix}.$$

$$8.\ \bar{a}_q(R') = \begin{pmatrix} 4 \\ -1 \end{pmatrix} - \begin{pmatrix} 0 & 0 & 0 & 0 \\ -1 & 0 & 0 & 1 \end{pmatrix} \begin{pmatrix} -1/4 \\ -25/4 \\ -3/4 \\ -1/4 \end{pmatrix} = \begin{pmatrix} 4 \\ -1 \end{pmatrix}.$$

9. $\max\{|4|, |-1|\} = 4,\ p = 4.$

10. $\tau = (4-0)^{-1} = 1/4,$

$v = -(1/4)(-1/4, -25/4, -3/4, -1/4)^{\mathrm{T}} = (1/16, 25/16, 3/16, 1/16)^{\mathrm{T}}.$

$d^{\mathrm{T}} = -(1/4)(0, 0, 0, 0)B_R^{-1} = (0, 0, 0, 0).$

$U = B_R^{-1} - (1/16, 25/16, 3/16, 1/16)^{\mathrm{T}}(0, 0, 0, 0) = B_R^{-1}.$

$$\hat{B}_{\hat{R}}^{-1} = \begin{pmatrix} -1/8 & & -1/4 & -3/16 & 1/16 \\ 7/8 & 1 & 1/4 & 9/16 & 25/16 \\ 1/8 & & -1/4 & -1/16 & 3/16 \\ -1/8 & & -1/4 & -1/48 & 1/16 \\ & & & & 1/4 \end{pmatrix}.$$

11. $k = 5,\ B = \{4, 2, 6, 1, 7\},\ R = \{1, 2, 6, 3, 4\},\ N = \{3, 5\},\ R' = \{5\}.$

$B_R^{-1} = \hat{B}_{\hat{R}}^{-1}.\ \bar{x}_7 = 0,\ \bar{a}_{5,n+1} = 0.$

Iteration 4:

$$4.\ \sigma_N = \begin{pmatrix} 1 & -2 \\ & 3 \\ & 4 \\ & -2 \\ & \end{pmatrix}^{\mathrm{T}} \begin{pmatrix} -1/8 & & -1/4 & -3/16 & 1/16 \\ 7/8 & 1 & 1/4 & 9/16 & 25/16 \\ 1/8 & & -1/4 & -1/16 & 3/16 \\ -1/8 & & -1/4 & -1/48 & 1/16 \\ & & & & 1/4 \end{pmatrix}^{\mathrm{T}} \begin{pmatrix} 0 \\ 1 \\ 0 \\ 0 \\ 0 \end{pmatrix} = \begin{pmatrix} 7/8 \\ 9/8 \end{pmatrix}.$$

5. $\min\{7/8, 9/8\} \geq 0.$

6. The basic optimal solution and optimal value:

$$\bar{x} = (35/4, 0, 0, 35/4, 0, 11, 0)^{\mathrm{T}}, \qquad \bar{x}_8 = -79/2.$$

20.6.1 Starting-Up

The preceding method requires a feasible deficient-basis reduced tableau (or solution) to get itself started. Such a reduced tableau may be converted from a feasible deficient-basis tableau, which can be generated by a single-artificial-variable based approach (see Sects. 20.4 and 20.7).

Assume that a feasible deficient-basis tableau is available. If column index set

$$J = \{j \in N \mid \bar{z}_j < 0\}$$

is empty, then optimality is already attained, and we are done. In the other case, determine a pivot column index

$$q = \arg\min_{j \in J} \bar{z}_j .$$

Taking the largest component of $\bar{a}_q(R')$ in module as pivot, perform a rank-increasing iteration if $R' \neq \emptyset$ and $\bar{a}_q(R') \neq 0$.

These steps are repeated until $J = \emptyset$ or $R' = \emptyset$ or $\bar{a}_q(R') = 0$. In the latter two cases, a wanted tableau can be created by performing elementary transformations by taking $\bar{z}_q$ as pivot (before that, $\bar{x}_B$ column should be in place of RHS column and the value of the south-east corner be set to zero).

Alternatively, the index $q \in J$ may be selected such that $\bar{a}_q(R')$ has the minimum number of nonzero components, if any.

20.7 Phase-I: The Most-Obtuse-Angle Column Rule

As was seen in Sect. 20.4, a feasible deficient-basis tableau can be created by solving the auxiliary program (20.12) using the deficient-basis method. Since the program is itself in reduced form, it is more attractive to use instead the deficient-basis reduced method, presented in the previous section. Since the program has equal x_{n+1} column and the right-hand side, the associated computation can be simplified considerably.

To explain, put the initial tableau of (20.12) in the following form:

x^{T}	x_{n+1}	RHS
A	b	b

Take $\bar{x} = 0,\ \bar{x}_{n+1} = 1$ as the initial solution. Assume that the current deficient-basis tableau is

x_B^{T}	x_N^{T}	x_{n+1}	RHS
I	$\bar{N}_R$	$\bar{b}_R$	$\bar{b}_R$
	$\bar{N}_{R'}$		

where $B = \{j_1, \cdots, j_k\},\ R = \{i_1, \cdots, i_k\}$. It represents equalities

$$x_B + \bar{N}_R x_N = -\bar{b}_R x_{n+1} + \bar{b}_R, \qquad \bar{N}_{R'} x_N = 0.$$

If set

$$S = \{t \mid \bar{b}_{i_t} < 0,\ t = 1, \ldots, k\}$$

is empty, then $\bar{x}_{n+1}$ can be decreased from 1 until reaching 0, as leads to a feasible solution to the original problem, i.e.,

$$\bar{x}_B = \bar{b}_R \geq 0, \qquad \bar{x}_N = 0.$$

If S is nonempty, the deficient-basis reduced method determines a row index

$$s \in \arg\min_{t \in S} -\bar{x}_{j_t}/\bar{b}_{i_t}.$$

In our case, there is usually a tie in selection of s, since $\bar{x}_{j_t} = 0,\ \forall\, t \in S$. As the according stepsize vanishes, no solution updating should occur. For such a highly degenerate case, we use the following rule instead:

$$s \in \arg\min_{t \in S}\ \bar{b}_{i_t}.$$

This rule is not only stable, but also advantageous in the sense of the most-obtuse-angle heuristics.

Thereby, the preceding tableau is simplified to

x_B^{T}	x_N^{T}	RHS
I	$\bar{N}_R$	$\bar{b}_R$
	$\bar{N}_{R'}$	

(20.22)

The subsequent steps are then the same as those in Algorithm 20.6.1. The resulting algorithm turns out to be a generalized version of Algorithm 13.3.1.

Algorithm 20.7.1 (Deficient-basis Phase-I: the most-obtuse-angle column rule). Initial: deficient-basis tableau of form (20.22) with $B,\ R,\ N,\ R',\ 1 \leq k \leq m$. This algorithm finds a feasible deficient-basis tableau.

The same as Algorithm 20.6.1 except for its steps 1–5 replaced by

1. Determine $s \in \arg\min\{\bar{b}_{i_t} \mid t = 1, \ldots, k\}$.
2. Stop if $\bar{b}_{i_s} \geq 0$ (feasibility achieved).
3. Do nothing.
4. Determine $q \in \arg\min_{j \in N}\ \bar{a}_{i_s, j}$.
5. Stop if $\bar{a}_{i_s, q} \geq 0$ (infeasible problem).

To get the preceding Algorithm started, the procedure described in Sect. 20.5.1 may be utilized for creating an initial deficient-basis. For matching the Algorithm, however, nothing would be compared with the following.

Algorithm 20.7.2 (Initial basis: tableau form). Initial: $(\bar{A} = A \mid \bar{b} = b),\ B = \emptyset,\ R = \emptyset, N = A,\ R' = \{1, \cdots, m\},\ k = 0$. This algorithm finds a deficient-basis tableau.

1. Determine $p \in \arg\max_{i \in R'} |\bar{b}_i|$.
2. Stop if $|\bar{b}_p| = 0$ (deficient-basis attained).
3. Stop if $J = \{j \mid \bar{a}_{p\,j}\bar{b}_p > 0,\ j \in N\} = \emptyset$ (infeasible problem).
4. Determine $q \in \arg\max_{j \in J} |\bar{a}_{p\,j}|$.
5. Convert $\bar{a}_{p\,q}$ to 1, and eliminate the other nonzeros in the column by elementary transformations.
6. Set $k = k + 1$, bring q from N to B and p from R' to R.
7. Go to step 1 if $R' \neq \emptyset$.
8. Stop (full basis attained).

Example 20.7.1. Solve the following problem :

$$\begin{array}{rrrrrrrrrl} \min & f = 7x_1 & +\ 2x_2 & +\ 3x_3 & -\ 2x_4 & & & +\ 9x_7 & -\ 2x_8, & \\ \text{s.t.} & -5x_1 & -\ \ x_2 & -\ 5x_3 & +\ 2x_4 & +\ \ x_5 & +\ \ x_6 & -\ 3x_7 & -\ 3x_8 & = -9, \\ & 2x_1 & +\ 2x_2 & -\ 2x_3 & +\ \ x_4 & +\ 3x_5 & -\ \ x_6 & +\ \ x_7 & -\ 2x_8 & = -9, \\ & 5x_1 & -\ 3x_2 & & -\ 3x_4 & +\ 5x_5 & -\ 3x_6 & +\ 2x_7 & +\ \ x_8 & = \ \ 2, \\ & & 3x_2 & & +\ \ x_4 & & & +\ 3x_7 & & = \ \ 0, \\ & & & & & & & -\ 4x_7 & & = \ \ 0. \end{array}$$

Answer

(1) Call Algorithm 20.7.2 to determine an initial basis. The equality constraints can be represented by the following tableau:

x_1	x_2	x_3	x_4	x_5	x_6	x_7	x_8	RHS
−5	−1	−5	2	1	1	−3	−3	−9
2	2	−2	1	3	−1	1	−2	−9
5	−3		−3	5	−3	2	1	2
	3		1			3		
						−4		

corresponding to $B = \emptyset,\ N = \{1, \cdots, 8\},\ R = \emptyset,\ R' = \{1, \cdots, 5\},\ k = 0$.

Iteration 1:

1. $\max \{|-9|, |-9|, |2|\} = 9,\ p = 1$.
4. $\max \{|-5|, |-1|, |-5|, |-3|, |-3|\} = 5,\ q = 1$.
5. Multiply row 1 by $-1/5$, then add $-2, -5$ times of row 1 to rows 2,3, respectively.
6. $k = 1,\ B = \{1\},\ N = \{2, \cdots, 8\},\ R = \{1\},\ R' = \{2, \cdots, 5\}$.

x_1	x_2	x_3	x_4	x_5	x_6	x_7	x_8	RHS
1	1/5	1	−2/5	−1/5	−1/5	3/5	3/5	9/5
	8/5	−4	9/5	17/5	−3/5	−1/5	−16/5	−63/5
	−4	−5	−1	6	−2	−1	−2	−7
	3		1			3		
						−4		

Iteration 2:

1. $\max\{|-63/5|, |-7|\} = 63/5,\ p = 2.$
4. $\max\{|-4|, |-3/5|, |-1/5|, |-16/5|\} = 4,\ q = 3.$
5. Multiply row 2 by $-1/4$, then add $-1, 5$ times of row 2 to rows 1,3, respectively.
6. $k = 2,\ B = \{1, 3\},\ N = \{2, 4, \cdots, 8\},\ R = \{1, 2\},\ R' = \{3, 4, 5\}.$

x_1	x_2	x_3	x_4	x_5	x_6	x_7	x_8	RHS
1	3/5		1/20	13/20	−7/20	11/20	−1/5	−27/20
	−2/5	1	−9/20	−17/20	3/20	1/20	4/5	63/20
	−6		−13/4	7/4	−5/4	−3/4	2	35/4
	3		1			3		
						−4		

Iteration 3:

1. $\max\{|35/4|\},\ p = 3.$
4. $\max\{|7/4|, |2|\} = 2,\ q = 8.$
5. Multiply row 3 by $1/2$, then add $1/5, -4/5$ times of row 3 to rows 1,2, respectively.
6. $k = 3,\ B = \{1, 3, 8\},\ N = \{2, 4, 5, 6, 7\},\ R = \{1, 2, 3\},\ R' = \{4, 5\}.$

x_1	x_2	x_3	x_4	x_5	x_6	x_7	x_8	RHS
1			−11/40	33/40	−19/40	19/40		−19/40
	2	1	17/20	−31/20	13/20	7/20		−7/20
	−3		−13/8	7/8	−5/8	−3/8	1	35/8
	3		1			3		
						−4		

A deficient-basis attained

(2) Call Algorithm 20.7.1 to achieve feasibility.

Iteration 4:

1. $\min\{-19/40, -7/20, 35/8\} = -19/40,\ s = 1.$
3. $\min\{-11/40, -19/40\} = -19/40,\ q = 6.$
6. $\bar{a}_q(R') = 0.$

10. Multiply row 1 by $-40/19$, then add $-13/20, 5/8$ times of row 1 to rows 2,3, respectively.
11. $k = 3,\ B = \{6, 3, 8\},\ N = \{1, 2, 4, 5, 7\},\ R = \{1, 2, 3\},\ R' = \{4, 5\}$.

x_1	x_2	x_3	x_4	x_5	x_6	x_7	x_8	RHS
$-40/19$			$11/19$	$-33/19$	1	-1		1
$26/19$	2	1	$9/19$	$-8/19$		1		-1
$-25/19$	-3		$-24/19$	$-4/19$		-1	1	5
	3		1			3		
						-4		

Iteration 5:

1. $\min\{1, -1, 5\} = -1,\ s = 2$.
3. $\min\{-8/19\},\ q = 5$.
6. $\bar{a}_q(R') = 0$.
10. Multiply row 2 by $-19/8$, then add $33/19, 4/19$ times of row 2 to rows 1,3, respectively.
11. $k = 3,\ B = \{6, 5, 8\},\ N = \{1, 2, 3, 4, 7\},\ R = \{1, 2, 3\},\ R' = \{4, 5\}$,

Feasibility is attained: $\bar{x}_B = (41/8, 19/8, 11/2)^T,\ \bar{x}_N = 0$. Add the objective row at the bottom and $f = x_9$ column, and add 2 times row 3 to the bottom line to transform it to a feasible deficient-basis tableau:

x_1	x_2	x_3	x_4	x_5	x_6	x_7	x_8	x_9	RHS
$-31/4$	$-33/4$	$-33/8$	$-11/8$		1	$-41/8$			$41/8$
$-13/4$	$-19/4$	$-19/8$	$-9/8$	1		$-19/8$			$19/8$
-2	-4	$-1/2$	$-3/2$			$-3/2$	1		$11/2$
	3		1			3			
						-4			
3	-6	2	-5			6		-1	11

with solution

$$\bar{x} = (0, 0, 0, 0, 19/8, 41/8, 0, 11/2, 0)^T, \qquad \bar{x}_9 = -11. \tag{20.23}$$

(3) Convert the preceding to a reduced tableau by the approach described in Sect. 20.6.1.

Iteration 6:
$J = \{2, 4\},\ \min\{-6, -5\} = -6,\ q = 2,\ \bar{a}_q(R') \neq 0$.
Multiply row 4 by $1/3$, then add $33/4, 19/4, 4, 6$ times of row 6 to rows 1,2,3,6, respectively.

$k = 4,\ B = \{6, 5, 8, 2\},\ N = \{1, 3, 4, 7\},\ R = \{1, 2, 3, 4\},\ R' = \{5\}.$

x_1	x_2	x_3	x_4	x_5	x_6	x_7	x_8	x_9	RHS
−31/4		−33/8	11/8		1	25/8			41/8
−13/4		−19/8	11/24	1		19/8			19/8
−2		−1/2	−1/6			5/2	1		11/2
0	1	0	1/3			1			
0		0				−4			
3		2	−3			12		−1	11

Iteration 7:
Put $\bar{x}_B$ column in place of RHS column (by zeroing the value 11 at the south-east corner).
$J = \{4\},\ q = 4,\ \bar{a}_q R' = 0.$
Multiply row 6 by $-1/3$, then add $-11/8, -11/24, 1/6, -1/3$ times of row 6 to rows 1,2,3,4, respectively.

x_1	x_2	x_3	x_4	x_5	x_6	x_7	x_8	x_9	$\bar{x}_B$
−51/8		−77/24			1	69/8		−11/24	41/8
−67/24		−149/72		1		101/24		−11/72	19/8
−13/6		−11/18				11/6	1	1/18	11/2
1/3	1	2/9				7/3		−1/9	
0		0				−4		0	
−1		−2/3	1			−4		1/3	

$k = 5,\ B = \{6, 5, 8, 2, 4\},\ N = \{1, 3, 7\},\ R = \{1, 2, 3, 4, 6\},\ R' = \{5\}.$

(4) Call Algorithm 20.6.1.

Iteration 7:

1. $\bar{w} = (-11/24, -11/72, 1/18, -1/9, 1/3)^T \not\geq 0.$
2. $\alpha = 0,\ s = 4.$
4. $\min\{1/3, 2/9, 7/3\} \geq 0.$
5. Optimality achieved by solution (20.23).

Chapter 21
Dual Deficient-Basis Method

This chapter will attack the standard LP problem from the dual side. To achieve optimality, the method presented in the previous chapter proceeds toward dual feasibility while maintaining primal feasibility. In this chapter, will derived is its dual version, achieving primal feasibility while maintaining dual feasibility (Pan 1998a, 2004, 2005). In addition, some promising methods are developed by combining the deficient-basis and D-reduced methods.

21.1 Dual Deficient-Basis Method: Tableau Form

Assume that deficient-basis tableau (20.4) is dual feasible, satisfying $\bar{z}_j \geq 0$. If $\bar{b} \geq 0$, optimality is already achieved; else, determine row index i_s such that

$$i_s \in \arg\min_{i_t \in R} \bar{b}_{i_t} < 0. \tag{21.1}$$

Such doing will let the basic infeasible variable x_{j_p} leave the basis, hence become feasible.

Lemma 21.1.1. *Assume that $\bar{z}_N \geq 0$ and $\bar{b}_{i_s} < 0$. If column index set*

$$J = \{j \in N \mid \bar{a}_{i_s\,j} < 0\}$$

is empty, then problem is infeasible.

Proof. It is seen that since $\bar{b}_{i_s} < 0$, the i_sth row of the tableau actually gives an upper-hill direction, with respect to the dual objective function. If the entries of this row are all nonnegative, then the stepsize taken along this direction is allowed arbitrarily large while maintaining dual feasibility. This means that the dual problems of (1.8) is unbounded, hence there is no feasible solution to (1.8). □

P.-Q. PAN, *Linear Programming Computation*, DOI 10.1007/978-3-642-40754-3_21,

Now assume that $J \neq \emptyset$. The largest possible stepsize that can be taken is

$$\beta = -\bar{z}_q/\bar{a}_{i_s q} = \min\{-\bar{z}_j/\bar{a}_{i_s j} \mid \bar{a}_{i_s j} < 0,\ j \in J\} \geq 0, \tag{21.2}$$

where column index q corresponds to a dual constraint, broking increasing of the dual objective value. Adding $\beta \geq 0$ times of the i_sth row to the bottom (objective) row updates the dual solution. Consequently, $\bar{z}_q$ vanishes, and the south-east entry of the tableau becomes

$$-\hat{f} = -\bar{f} + \beta\bar{b}_{i_s} \leq -\bar{f},$$

which means that the objective value $\hat{f}$ does not decrease, and strictly increases under the dual nondegeneracy assumption ($\beta > 0$).

There will be the following two cases to be handled differently:

(i) Rank-remaining iteration: $R' = \emptyset$ or $\bar{a}_q(R') = 0$.

Convert $\bar{a}_{i_s,q}$ to 1, and eliminate the other nonzeros in the column by elementary transformations. Consequently, the j_s-indexed basic column becomes nonbasic, and the q-indexed nonbasic column becomes basic, while the basic rows remain basic. So, the rank of the basis remains. Such doing turns $\bar{b}_{i_s}$ to positive, but does not touch nonbasic rows at all (including entries in the right-hand side).

(ii) Rank-increasing iteration: $R' \neq \emptyset$ and $\bar{a}_q(R') \neq 0$.

Determine row index $p \in \arg\max_{i \in R'} |\bar{a}_{i\,q}|$. Convert $\bar{a}_{p,q}$ to 1, and eliminate the other nonzeros in the column by elementary transformations. Thus, while all basic columns and rows remain basic, the q-indexed nonbasic column and p-indexed nonbasic row becomes basic. Hence the rank of the basis matrix increases by 1. Since the right-hand side remains unchanged, the same row index i_s will be selected in the next iteration, until encountering a rank-remaining iteration to turn $\bar{b}_{i_s}$ positive.

The associated steps can be summarized to the following algorithm.

Algorithm 21.1.1 (Dual deficient-basis algorithm: tableau form). Initial : a dual feasible deficient-basis tableau of form (20.4) with $B,\ N,\ R,\ R',\ 1 \leq k \leq m$. This algorithm solves the standard LP problem.

1. Determine s such that $i_s \in \arg\min_{i_t \in R} \bar{b}_{i_t}$.
2. Stop if $\bar{b}_{i_s} \geq 0$.
3. Stop if $J = \{j \in N \mid \bar{a}_{i_s j} < 0\}$.
4. Determine column index q and stepsize β by (21.2).
5. If $\beta \neq 0$, add β times of the i_sth row to the bottom row.
6. Go to step 10 if $R' = \emptyset$ or $\bar{a}_q(R') = 0$.
7. Determine row index $p \in \arg\max_{i \in R'} |\bar{a}_{i\,q}|$.

8. Convert $\bar{a}_{pq}$ to 1, and eliminate the other nonzeros in the column by elementary transformations.
9. Set $k = k + 1$, bring q from N to B and p from R' to R, and go to step 3.
10. Convert $\bar{a}_{i_s q}$ to 1, and eliminate the other nonzeros in the column by elementary transformations.
11. Update (B, N) by exchanging j_s and q.
12. Go to step 1.

Theorem 21.1.1. *Under the dual nondegenerate assumption, Algorithm 21.1.1 terminates either at*

(i) Step 2, reaching a basic optimal solution; or at
(ii) Step 3, detecting infeasibility of the problem.

Proof. The proof of finiteness of algorithm is similar to that in the conventional simplex context. The meanings of its exits come from Lemmas 20.1.1 and 21.1.1, as well as discussions preceding the algorithm. □

Example 21.1.1. Solve the following problem by the Algorithm 21.1.1:

$$\begin{array}{llllllllll}
\min & f = 4x_3 + x_4 + 3x_5 + x_6 + 3x_7 + 2x_8, \\
\text{s.t.} & x_1 \quad + 2x_3 - x_4 + x_5 - x_6 - 3x_7 - 3x_8 = -10, \\
& \quad + x_2 + x_3 + 2x_4 - x_5 + x_6 - 2x_7 + x_8 = -7, \\
& \quad - 3x_3 - x_4 - 3x_5 - x_6 + 2x_8 = 0, \\
& \quad + x_3 + x_4 + x_5 + 3x_6 - x_8 = 0, \\
& \quad - x_3 + x_4 - x_5 - 2x_6 + 2x_8 = 0, \\
& \quad x_j \geq 0, \quad j = 1, \cdots, 8.
\end{array}$$

Answer $s = 2, B = \{1, 2\}, R = \{1, 2\}. N = \{3, 4, 5, 6, 7, 8\}, R' = \{3, 4, 5\}$. The initial feasible dual deficient-basis tableau is

x_1	x_2	x_3	x_4	x_5	x_6	x_7	x_8	RHS
1		2	−1	1	−1	−3	−3	−10
	1	1	2	−1	1	−2	1	−7
		−3	−1	−3	−1		2*	
		1	1	1	3		−1	
		−1	1	−1	−2		2	
		4	1	3	1	3	2	

Iteration 1:

1. $\min\{-10, -7\} = -10 < 0,\ s = 1,\ i_s = 1$.
3. $J = \{4, 6, 7, 8\} \neq \emptyset$.
4. $\beta = \min\{-1/(-1), -1/(-1), -3/(-3), -2/(-3)\} = 2/3,\ q = 8$.

5. Add $2/3$ times of row 1 to the bottom row.
6. $\bar{a}_q(R') = (2, -1, 2)^T \neq 0$.
7. $\max\{|2|, |-1|, |2|\} = 2,\ p = 3$.
8. Multiply row 3 by $1/2$, then add $3, -1, 1, -2$ times of row 3 to rows 1,2,4,5, respectively.
9. $k = 3,\ B = \{1, 2, 8\},\ r = \{1, 2, 3\},\ N = \{3, 4, 5, 6, 7\},\ R' = \{4, 5\}$.

x_1	x_2	x_3	x_4	x_5	x_6	x_7	x_8	RHS
1		−5/2	−5/2	−7/2	−5/2	−3		−10
	1	5/2	5/2	1/2	3/2	−2		−7
		−3/2	−1/2	−3/2	−1/2		1	
		−1/2	1/2	−1/2	5/2			
		2	2*	2	−1			
2/3		16/3	1/3	11/3	1/3	1		−20/3

Iteration 2:

3. $J = \{3, 4, 5, 6, 7\} \neq \emptyset$.
4. $\beta = \min\{-(16/3)/(-5/2), -(1/3)/(-5/2), -(11/3)/(-7/2), -(1/3)/(-5/2), -1/-3\} = 2/15,\ q = 4$.
5. Add $2/15$ times of row 1 to the bottom row.
6. $\bar{a}_q(R') = (-1/2, 2)^T \neq 0$.
7. $\max\{|1/2|, |2|\} = 2,\ p = 5$.
8. Multiply row 5 by $1/2$, then add $5/2, -5/2, 1/2, -1/2$, times of row 5 to rows 1,2,3,4, respectively.
9. $k = 4,\ B = \{1, 2, 8, 4\},\ R = \{1, 2, 3, 5\},\ N = \{3, 5, 6, 7\},\ R' = \{4\}$.

x_1	x_2	x_3	x_4	x_5	x_6	x_7	x_8	RHS
1				−1	−15/4	−3		−10
	1			−2	11/4	−2		−7
		−1		−1	−3/4		1	
		−1		−1	11/4*			
		1	1	1	−1/2			
4/5		5		16/5		3/5		−8

Iteration 3:

3. $J = \{5, 6, 7\} \neq \emptyset$.
4. $\beta = \min\{-(16/5)/(-5/2), 0/(-15/4), -(3/5)/(-3)\} = 0,\ q = 6$.
6. $\bar{a}_q(R') = (11/4) \neq 0$.
7. $\max\{|11/4|\} = 11/4,\ p = 4$.

8. Multiply row 4 by $4/11$, then add $15/4, -11/4, 3/4, 1/2$ times of row 4 to rows 1,2,3, respectively.
9. $k = 5,\ B = \{1, 2, 8, 4, 6\},\ R = \{1, 2, 3, 5, 4\},\ N = \{3, 5, 7\},\ R' = \emptyset$.

x_1	x_2	x_3	x_4	x_5	x_6	x_7	x_8	RHS
1		−15/11		−26/11		−3*		−10
	1	1		−1		−2		−7
		−14/11		−14/11			1	
		−4/11		−4/11	1			
		9/11	1	9/11				
4/5		5		16/5		3/5		−8

Iteration 4:

3. $J = \{3, 5, 7\} \neq \emptyset$.
4. $\beta = \min\{-5/(-15/11), -(16/5)/(-26/11), -(3/5)/-3\} = 1/5,\ q = 7$.
5. Add $1/5$ times of row 1 to the bottom row.
6. $R' = \emptyset$.
10. Multiply row 1 by $-1/3$, then add $2, -3/5$ times of row 1 to rows 2,6, respectively.
11. $B = \{7, 2, 8, 4, 6\},\ N = \{1, 3, 5\}$.

x_1	x_2	x_3	x_4	x_5	x_6	x_7	x_8	RHS
−1/3		5/11		26/33		1		10/3
−2/3*	1	21/11		19/33				−1/3
		−14/11		−14/11			1	
		−4/11		−4/11	1			
		9/11	1	9/11				
1		52/11		30/11				−10

Iteration 5:

1. $\min\{10/3, -1/3\} = -1/3 < 0,\ s = 2,\ i_s = 2$.
3. $J = \{1\} \neq \emptyset$.
4. $\beta = \min\{-1/(-2/3)\} = 3/2,\ q = 1$.
5. Add $3/2$ times of the second row to the bottom row.
6. $R' = \emptyset$.
10. Multiply row 2 by $-3/2$, then add $1/3, -1$ times of row 2 to rows 1,6, respectively.
11. $B = \{7, 1, 8, 4, 6\},\ N = \{2, 3, 5\}$.

x_1	x_2	x_3	x_4	x_5	x_6	x_7	x_8	RHS
	−1/2	−1/2		1/2		1		7/2
1	−3/2	−63/22		−19/22				1/2
		−14/11		−14/11			1	
		−4/11		−4/11	1			
		9/11	1	9/11				
	3/2	167/22		79/22				−21/2

Iteration 6:

1. $\min\{7/2, 1/2\} \geq 0$.
2. The basic optimal solution and optimal value:

$$\bar{x} = (1/2, 0, 0, 0, 0, 0, 7/2, 0)^{\mathrm{T}}, \ \bar{f} = 21/2,$$

21.2 Dual Deficient-Basis Method

As all quantities involved in Algorithm 21.1.1 can be found in Table 20.1. It is a easy task to write its revised version.

Algorithm 21.2.1 (Dual deficient-basis algorithm). Initial :$B,\ N,\ R,\ R',\ 1 \leq k \leq m$. B_R^{-1}; $\bar{z}_N = c_N - N_R^{\mathrm{T}} B_R^{-\mathrm{T}} c_B \geq 0$; $\bar{z}_B = 0$; $\bar{x}_B = B_R^{-1} b_R$. This algorithm solves the standard LP problem.

1. Determine s such that $\bar{x}_{j_s} = \min\{\bar{x}_{j_t} \mid t = 1, \cdots, k\}$.
2. If $\bar{x}_{j_s} \geq 0$, compute $\bar{f} = c_B^{\mathrm{T}} \bar{x}_B$, and stop (optimality achieved).
3. Compute $\sigma_N = N_R^{\mathrm{T}} B_R^{-\mathrm{T}} e_s$.
4. Stop if $J = \{j \in N \mid \sigma_j < 0\} = \emptyset$ (infeasible problem).
5. Determine β and column index q such that $\beta = -\bar{z}_q/\sigma_q = \min_{j \in J} -\bar{z}_j/\sigma_j$.
6. If $\beta \neq 0$, update $\bar{z}_N = \bar{z}_N + \beta \sigma_N$, $\bar{z}_{j_s} = \bar{z}_{j_s} + \beta$.
7. Compute $\bar{a}_q(R) = B_R^{-1} a_q(R)$.
8. Go to step 12 if $R' = \emptyset$ or $\bar{a}_q(R') = a_q(R') - B_{R'} \bar{a}_q(R) = 0$.
9. Determine row index $p \in \arg\max_{i \in R'} |\bar{a}_{i\,q}|$.
10. Update B_R^{-1} by (20.10).
11. Update $k = k + 1$, $\bar{x}_q = 0$, bring q from N to B and p from R' to R, and go to step 3.
12. Compute $\alpha = \bar{x}_{j_s}/\bar{a}_{i_s, q}$.
13. If $\alpha \neq 0$, set $\bar{x}_q = \alpha$, and update $\bar{x}_B = \bar{x}_B - \alpha \bar{a}_q(R)$.
14. Update B_R^{-1} by (20.11).
15. Update (B, N) by exchanging j_s and q.
16. Go to step 1.

Table 21.1 Ratio of MINS 5.51 to RDPPA 1.10

Problem	Iteration counts	Time	% degen
Small (20)	1.02	1.56	1.08
Medium (15)	1.10	1.06	0.89
Large (15)	1.44	1.25	1.07
Average (50)	1.37	1.24	1.02
Kennington (8)	5.68	2.83	0.30
BPMPD (5)	3.08	1.41	0.92
Netlib (2)	0.94	1.01	1.80
Average (15)	4.58	2.14	0.47

The preceding Algorithm needs a dual feasible tableau (solution) to get itself started. The (conventional) dual simplex Phase-I methods (Chap. 14) are applicable to this purpose, almost without any modification.

Originally proposed was an analogue to Algorithm 21.2.1, which uses the rectangular matrix B as deficient basis (Pan 2005). As they roughly reflect the behavior of Algorithm 21.2.1, the numerical results reported in the paper are cited below.

The computational experiments were carried out on a Pentium III 550E PC with Windows 98 operating system, 256 MB inner storage and about 16 decimal precision. Visual Fortran 5.0 compiler was used. There were following two codes involved:

1. MINOS 5.51.
2. RDPPA 1.10.

The second code was developed by modifying MINOS 5.5.1 as less as possible. The test set of 65 standard LP problems were the same as those in Sect. 20.3, classified to the same sets with negligible difference.

Table 21.1 lists iterations and time ratios (MINOS 5.51/RDPPA 1.10), in which the last column gives degenerate iterations percentage ratios.

It is seen from Table 21.1 that the new code significantly outperformed MINOS 5.51 overall. For the first group of problems, the time ratio is 1.24, and for the second group, the time ratio reaches 2.14. So, the superiority margin is higher with large-scale sparse problems. Comparing these results with those listed in Sect. 20.3, we conclude that the dual deficient-basis method appears to be better than its primal counterpart.

From the last column of Table 21.1, it is seen that the degenerate iteration ratio for the first group is 1.02, very close to 1, whereas that for the second group of larger problems is significantly less than 1. Overall, the proportion of degenerate iterations is high with the new method. Therefore, it is not true that degeneracy itself degrades algorithm's efficiency.

Example 21.2.1. Solve the following problem by Algorithm 21.2.1:

$$
\begin{array}{llllllllll}
\min & f = 4x_3 + x_4 + 3x_5 + x_6 + 5x_7 + 2x_8, \\
\text{s.t.} & x_1 \quad + 2x_3 - x_4 + x_5 - x_6 - 3x_7 - 3x_8 = -9, \\
& \quad + x_2 + x_3 + 2x_4 - x_5 + x_6 - 2x_7 + x_8 = -4, \\
& \quad - 3x_3 \quad - 3x_5 \quad + 2x_8 = 0, \\
& \quad + x_3 \quad + x_5 \quad - x_8 = 0, \\
& \quad - x_3 \quad - x_5 \quad + 2x_8 = 0, \\
& \qquad x_j \geq 0, \quad j = 1, \cdots, 8.
\end{array}
$$

Answer Initial :$s - 2$, $B = \{1,2\}$, $R = \{1,2\}$. $N = \{3,4,5,6,7,8\}$, $R' = \{3,4,5\}$. $B_R = I$. $\bar{z}_N = (4,1,3,1,5,2)^{\mathrm{T}}$. $\bar{z}_B = 0$, $\bar{x}_B = (-9,-4)^{\mathrm{T}}$.

Iteration 1:

1. $\min\{-9,-4\} = -9 < 0$, $s = 1$.
3. $\sigma = (2,-1,1,-1,-3,-3)^{\mathrm{T}}$.
4. $J = \{4,6,7,8\} \neq \emptyset$.
5. $\beta = \min\{-1/(-1), -1/(-1), -5/(-3), -2/(-3)\} = 2/3$, $q = 8$.
6. $\bar{z}_N = (4,1,3,1,5,2)^{\mathrm{T}} + (2/3)(2,-1,1,-1,-3,-3)^{\mathrm{T}}$
   $= (16/3, 1/3, 11/3, 1/3, 3, 0)^{\mathrm{T}}$.
   $N = \{3,4,5,6,7,8\}$, $\bar{z}_1 = 0 + 2/3 = 2/3$.
7. $\bar{a}_q(R) = (-3,1)^{\mathrm{T}}$.
8. $\bar{a}_q(R') = (2,-1,2)^{\mathrm{T}} \neq 0$.
9. $\max\{|2|, |-1|, |2|\} = 2$, $p = 3$.
10. $p = 3$, $\tau = (2 - (0,0)(-3,1)^{\mathrm{T}})^{-1} = 1/2$.
    $v = -(1/2)(-3,1)^{\mathrm{T}} = (3/2, -1/2)^{\mathrm{T}}$, $d^{\mathrm{T}} = (0,0)$.
    $$U = \begin{pmatrix} 1 & \\ & 1 \end{pmatrix} - \begin{pmatrix} 3/2 \\ 1/2 \end{pmatrix} (0 \;\; 0) = \begin{pmatrix} 1 & \\ & 1 \end{pmatrix}.$$
    $$\hat{B}_{\hat{R}}^{-1} = \begin{pmatrix} 1 & & 3/2 \\ & 1 & -1/2 \\ & & 1/2 \end{pmatrix}.$$
11. $k = 3$, $B = \{1,2,8\}$, $R = \{1,2,3\}, \bar{x}_8 = 0$, $N = \{3,4,5,6,7\}$,
    $R' = \{4,5\}$.

Iteration 2:

3. $s = 1$, $$\sigma_N = \begin{pmatrix} 2 & -1 & 1 & -1 & -3 \\ 1 & 2 & -1 & 1 & -2 \\ -3 & & -3 & 0 & \end{pmatrix}^{\mathrm{T}} \begin{pmatrix} 1 \\ 0 \\ 3/2 \end{pmatrix} = (-5/2, -1, -7/2, -1, -3)^{\mathrm{T}}.$$
4. $J = \{3,4,5,6,7\} \neq \emptyset$.
5. $\beta = \min\{-(16/3)/(-5/2), -(1/3)/(-1), -(11/3)/(-7/2), -(1/3)/(-1),$
   $-(3)/(-3)\} = 1/3$, $q = 4$.
6. $\bar{z}_N = (16/3, 1/3, 11/3, 1/3, 3)^{\mathrm{T}} + (1/3)(-5/2, -1, -7/2, -1, -3)^{\mathrm{T}}$
   $= (9/2, 0, 5/2, 0, 2)^{\mathrm{T}}$, $N = \{3,4,5,6,7\}$, $\bar{z}_1 = 2/3 + 1/3 = 1$.

7. $\bar{a}_q(R) = \begin{pmatrix} 1 & & 3/2 \\ & 1 & -1/2 \\ & & 1/2 \end{pmatrix} \begin{pmatrix} -1 \\ 2 \\ 0 \end{pmatrix} = \begin{pmatrix} -1 \\ 2 \\ 0 \end{pmatrix}$,

8. $\bar{a}_q(R') = \begin{pmatrix} 0 \\ 0 \end{pmatrix} - \begin{pmatrix} 0 & 0 & -1 \\ 0 & 0 & 2 \end{pmatrix} \begin{pmatrix} -1 \\ 2 \\ 0 \end{pmatrix} = \begin{pmatrix} 0 \\ 0 \end{pmatrix}$.

12. $\alpha = (-9)/(-1) = 9$.

13. $\bar{x}_4 = 9,\ \bar{x}_B = (-9, -4, 0)^{\mathrm{T}} - 9(-1, 2, 0)^{\mathrm{T}} = (0, -22, 0)^{\mathrm{T}}$.

14. $B_R^{-1} = \begin{pmatrix} -1 & & \\ 2 & 1 & \\ & & 1 \end{pmatrix} \begin{pmatrix} 1 & & 3/2 \\ & 1 & -1/2 \\ & & 1/2 \end{pmatrix} = \begin{pmatrix} -1 & & -3/2 \\ 2 & 1 & 5/2 \\ & & 1/2 \end{pmatrix}$.

15. $B = \{4, 2, 8\},\ N = \{1, 3, 5, 6, 7\}$.

Iteration 3:

1. $\min\{9, -22, 0\} = -22 < 0,\ s = 2$.

3. $\sigma_N = \begin{pmatrix} 1 & 2 & 1 & -1 & -3 \\ & 1 & -1 & 1 & -2 \\ & -3 & -3 & & \end{pmatrix}^{\mathrm{T}} \begin{pmatrix} 2 \\ 1 \\ 5/2 \end{pmatrix} = (2, -5/2, -13/2, -1, -8)^{\mathrm{T}}$.

4. $J = \{3, 5, 6, 7\} \neq \emptyset$.

5. $\beta = \min\{-(9/2)/(-5/2), -(5/2)/(-13/2), 0/(-1), -2/(-8)\} = 0$,
$q = 6$.

7. $\bar{a}_q(R) = \begin{pmatrix} -1 & & -3/2 \\ 2 & 1 & 5/2 \\ & & 1/2 \end{pmatrix} \begin{pmatrix} -1 \\ 1 \\ 0 \end{pmatrix} = \begin{pmatrix} 1 \\ -1 \\ 0 \end{pmatrix},\ \bar{a}_q(R') = \begin{pmatrix} 0 \\ 0 \end{pmatrix}$.

8. $\bar{a}_q(R') = 0$.

12. $\alpha = (-22)/(-1) = 22$.

13. $\bar{x}_B = (9, -22, 0)^{\mathrm{T}} - 22(1, -1, 0)^{\mathrm{T}} = (-13, 0, 0)^{\mathrm{T}},\ \bar{x}_6 = 22$.

14. $B_R^{-1} = \begin{pmatrix} 1 & 1 & \\ & -1 & \\ & & 1 \end{pmatrix} \begin{pmatrix} -1 & & -3/2 \\ 2 & 1 & 5/2 \\ & & 1/2 \end{pmatrix} = \begin{pmatrix} 1 & 1 & 1 \\ -2 & -1 & -5/2 \\ & & 1/2 \end{pmatrix}$.

15. $B = \{4, 6, 8\},\ N = \{1, 2, 3, 5, 7\}$.

Iteration 4:

1. $\min\{-13, 22, 0\} = -13 < 0,\ s = 1$.

3. $\sigma_N = \begin{pmatrix} 1 & & 2 & 1 & -3 \\ & 1 & 1 & -1 & -2 \\ & & -3 & -3 & \end{pmatrix}^{\mathrm{T}} \begin{pmatrix} 1 \\ 1 \\ 1 \end{pmatrix} = (1, 1, 0, -3, -5)^{\mathrm{T}}$.

4. $J = \{5, 7\} \neq \emptyset$.

5. $\beta = \min\{-(5/2)/(-3), -2/(-5)\} = 2/5,\ q = 7$.

6. $\bar{z}_N = ((1, 0, 9/2, 5/2, 2)^{\mathrm{T}} + (2/5)(1, 1, 0, -3, -5)^{\mathrm{T}}$
$= (7/5, 2/5, 9/2, 13/10, 0)^{\mathrm{T}}, N = \{1, 2, 3, 5, 7\}$,
$\bar{z}_4 = 0 + 2/5 = 2/5$.

7. $\bar{a}_q(R) = \begin{pmatrix} 1 & 1 & 1 \\ -2 & -1 & -5/2 \\ & & 1/2 \end{pmatrix} \begin{pmatrix} -3 \\ -2 \\ 0 \end{pmatrix} = \begin{pmatrix} -5 \\ 8 \\ 0 \end{pmatrix}, \ \bar{a}_q(R') = \begin{pmatrix} 0 \\ 0 \end{pmatrix}.$
8. $\bar{a}_q(R') = 0$.
12. $\alpha = (-13)/(-5) = 13/5$.
13. $\bar{x}_B = (-13, 22, 0)^{\mathrm{T}} - (13/5)(-5, 8, 0)^{\mathrm{T}} = (0, 6/5, 0)^{\mathrm{T}}, \ \bar{x}_7 = 13/5$.
14. $B_R^{-1} = \begin{pmatrix} -1/3 & & \\ -2/3 & 1 & \\ & & 1 \end{pmatrix} \begin{pmatrix} 1 & 1 & 1 \\ -2 & -1 & -5/2 \\ & & 1/2 \end{pmatrix} = \begin{pmatrix} -1/3 & -1/3 & -1/3 \\ -8/3 & -5/3 & -19/6 \\ & & 1/2 \end{pmatrix}.$
15. $B = \{7, 6, 8\}, \ N = \{1, 2, 3, 4, 5\}$.

Iteration 5:

1. $\min\{13/5, 6/5, 0\} \geq 0$.
2. The basic optimal solution and optimal value:

$$\bar{x} = (0, 0, 0, 0, 0, 6/5, 13/5, 0)^{\mathrm{T}}, \quad \bar{f} = (1, 5)(6/5, 13/5)^{\mathrm{T}} = 71/5.$$

21.3 Dual Deficient-Basis D-Reduced Method: Tableau Form

In this section, the deficient basis will be incorporated to the dual D-reduced simplex method (Sect. 17.2). The resulting method is stable, compared with the dual deficient-basis method, presented in the previous section, and the related search direction corresponds to the entire dual objective gradient.

Consider the D-reduced problem (17.1). Assume that the order of the deficient basis equals $1 \leq k \leq m - 1$ and that the r-indexed row is the datum row ($1 \leq r \leq m$). Sets B, R, N, R' are defined by

$$B = \{j_1, \cdots, j_k\}, \quad \{r\} \notin R = \{i_1, \cdots, i_k\}, \qquad N = A \backslash B,$$
$$R' = \{1, \cdots, m\} \backslash (R \cup \{r\}). \tag{21.3}$$

The initial tableau of (17.1) is

x_B^{T}	x_N^{T}	f	RHS
B_R	N_R		
$B_{R'}$	$N_{R'}$		
ω_B^{T}	ω_N^{T}		1
c_B^{T}	c_N^{T}	-1	

(21.4)

which can be converted to the following form by a series of elementary transformations:

x_B^{T}	x_N^{T}	f	RHS
I	$\bar{N}_R$		
	$\bar{N}_{R'}$		
	$\bar{\omega}_N^{\mathrm{T}}$		1
	$\bar{z}_N^{\mathrm{T}}$	-1	

(21.5)

The preceding is called *deficient-basis D-reduced tableau*, where the mth row is the datum row, indexed by $r \notin R, R'$.

Assume that the current tableau is dual feasible, i.e., $\bar{z}_N \geq 0$. If entries of the datum row are no more than zero, then there is no feasible solution to the original problem (Lemma 17.1.1). If this is not the case, determined β and column index q such that

$$\beta = \bar{z}_q/\bar{\omega}_q = \min\{\bar{z}_j/\bar{\omega}_j \mid \bar{\omega}_j > 0,\ j \in N\} \leq 0.$$

Add $-\beta$ times of the datum row to the bottom row to turn $\bar{z}_q$ to 0. Then the south-east entry of the tableau becomes $-\beta$, hence the largest possible increment of the objective value just equals $\beta \geq 0$. Under the nondegeneracy assumption, $\beta > 0$, hence the objective value strictly increases.

The determination of pivot row index depends on whether $\bar{a}_q(R')$ vanishes.

(i) Rank-remaining iteration: $R' = \emptyset$ or $\bar{a}_q(R') = 0$.

Determine $s \in \arg\max\{\bar{a}_{i_t,q} \mid t = 1, \ldots, k\}$. If $\bar{a}_{i_s,q} \leq 0$, it is clear by $\bar{\omega}_q > 0$ that the tableau corresponds to a basic feasible solution, and hence optimality is achieved (taking the q-indexed column as the datum column). If $\bar{a}_{i_s,q} > 0$, take it as the pivot to make a basis change. In such an iteration, the rank of the basis remains unchanged.

(ii) Rank-increasing iteration: $R' \neq \emptyset$ and $\bar{a}_q(R') \neq 0$.

Determine row index $p \in \arg\max_{i \in R'} |\bar{a}_{i\,q}|$. Take $\bar{a}_{p\,q}$ as the pivot to make a basis change. Then the rank of the basis increases to $k+1$.

In either iterations, the elementary transformations do not touch the datum row, the bottom row and the right-hand side; so the resulting tableau is again a dual feasible D-reduced tableau. Repeat these steps until optimality achieved, or infeasibility of the problem detected.

The overall steps are summarized into the following algorithm.

Algorithm 21.3.1 (Dual deficient-basis D-reduced algorithm 1: tableau form). Initial : dual feasible deficient-basis D-reduced tableau of form (21.5) with $B,\ R,\ N,\ R',\ 1 \leq k \leq m-1$. This algorithm solves the D-reduced problem (17.1).

1. Stop if $J = \{j \in N \mid \bar{\omega}_j > 0\} = \emptyset$ (infeasible problem).
2. Determine β and column index q such that $\beta = \bar{z}_q/\bar{\omega}_q = \min_{j \in J} \bar{z}_j/\bar{\omega}_j$.
3. If $\beta \neq 0$, add $-\beta$ times of the datum row to the bottom row.
4. Go to step 8 if $R' = \emptyset$ or $\bar{a}_q(R') = 0$.

5. Determine row index $p \in \arg\max_{i \in R'} |\bar{a}_{i\,q}|$.
6. Convert $\bar{a}_{pq}$ to 1, and eliminate the other nonzeros in the column by elementary transformations.
7. Set $k = k + 1$, bring q from N to B and p from R' to R, and go to step 1.
8. Determine $s \in \arg\max\{\bar{a}_{i_t,q} \mid t = 1, \ldots, k\}$.
9. If $\bar{a}_{i_s,q} \leq 0$, compute $\bar{x}_q = 1/\bar{\omega}_q$, $\bar{x}_B = -\bar{x}_q \bar{a}_q(R)$, and stop (optimality achieved).
10. Convert $\bar{a}_{i_s,q}$ to 1, and eliminate the other nonzeros in the column by elementary transformations.
11. Update (B, N) by exchanging j_s and q.
12. Go to step 1.

Example 21.3.1. Solve the following problem by Algorithm 21.3.1:

$$
\begin{array}{llrcl}
\min & f = 3x_4 + 2x_5 + x_6 + 5x_7 + 6x_8, & & & \\
\text{s.t.} & x_1 \qquad\qquad - 6x_4 + x_5 + 4x_6 + 6x_7 & & = & 9, \\
& \quad + x_2 \qquad + 2x_4 \qquad\quad - 7x_6 - 8x_7 + x_8 & & = & -6, \\
& \qquad\quad + x_3 \qquad - 4x_5 \qquad\qquad\qquad + 2x_8 & & = & 0, \\
& \qquad\qquad\quad + 3x_4 + 2x_5 \qquad\quad + 3x_7 & & = & 3, \\
& \qquad\qquad\quad - x_4 + 2x_5 - 4x_6 - 6x_7 + 3x_8 & & = & 2, \\
& \qquad\qquad\qquad x_j \geq 0, \quad j = 1, \ldots, 8. & & &
\end{array}
$$

Answer Initial tableau is

x_1	x_2	x_3	x_4	x_5	x_6	x_7	x_8	RHS
1			−6	1	4	6		9
	1		2		−7	−8	1	−6
		1		−4			2	0
			3	2		3	0	3
			−1	2	−4	−6	3	2*
			3	2	1	5	6	0

Multiply row 5 by 1/2, then add $-9, 6, -3$ times of row 5 to rows 1,2,4, respectively:

x_1	x_2	x_3	x_4	x_5	x_6	x_7	x_8	RHS
1			−3/2	−8	22	33	−27/2	
	1		−1	6	−19	−26	10	
		1		−4			2	
			9/2	−1*	6	12	−9/2	
			−1/2	1	−2	−3	3/2	1
			3	2	1	5	6	

Iteration 1: $k = 3$, $B = \{1, 2, 3\}$, $N = A \backslash B$, $R = \{1, 2, 3\}$, $R' = \{4\}$.

1. $J = \{5, 8\}$.
2. $\beta = \min\{2/1, 6/(3/2)\} = 2/1$, $q = 5$.
3. Add -2 times of row 5 to the bottom row.
4. $\bar{a}_q(R') \neq 0$.
5. $\max\{|-1|\}$, $p = 4$.
6. Multiply row 4 by -1, then add $8, -6, 4, -1$ times of row 4 to rows 1,2,3,5, respectively:
7. $k = 4$, $B = \{1, 2, 3, 5\}$, $N = A \backslash B$, $R = \{1, 2, 3, 4\}$, $R' = \emptyset$.

x_1	x_2	x_3	x_4	x_5	x_6	x_7	x_8	RHS
1			−75/2		−26	−63	45/2	
	1		26*		17	46	−17	
		1	−18		−24	−48	20	
			−9/2	1	−6	−12	9/2	
			4		4	9	−3	1
			4		5	11	3	−2

Iteration 2:

1. $J = \{4, 6, 7\}$.
2. $\beta = \min\{4/4, 5/4, 11/9)\} = 1$, $q = 4$.
3. Add -1 times of row 5 to the bottom row.
4. $R' = \emptyset$.
8. $\max\{-75/2, 26, -18, -9/2\} = 26 > 0$, $s = 2$.
10. Multiply row 2 by $1/26$, then add $75/2, 18, 9/2, -4$ times of row 2 to rows 1,3,4,5, respectively:
11. $k = 4$, $B = \{1, 4, 3, 5\}$, $N = A \backslash B$, $R = \{1, 2, 3, 4\}$, $R' = \emptyset$.

x_1	x_2	x_3	x_4	x_5	x_6	x_7	x_8	RHS
1	75/52				−77/52	87/26	−105/52	
	1/26		1		17/26*	23/13	−17/26	
	9/13	1			−159/13	−210/13	107/13	
	9/52			1	−159/52	−105/26	81/52	
	−2/13				18/13	25/13	−5/13	1
	1/9					11/18	113/18	−67/18

Iteration 3:

1. $J = \{6, 7\}$.
2. $\beta = 0$, $q = 6$.

4. $R' = \emptyset$.
8. $\max\{-77/52, 17/26, -159/13, -159/52\} = 17/26 > 0,\ s = 2$.
10. Multiply row 2 by 26/17, then add $77/52, 159/13, 159/52, -18/13$ times of row 2 to rows 1,3,4,5, respectively:
11. $k = 4,\ B = \{1, 4, 3, 5\},\ N = A\backslash B,\ R = \{1, 2, 3, 4\},\ R' = \emptyset$.

x_1	x_2	x_3	x_4	x_5	x_6	x_7	x_8	RHS
1	26/17		77/34			125/17	−7/2	
	1/17		26/17		1	46/17	−1	
	24/17	1	318/17			288/17	−4	
	6/17		159/34	1		72/17	−3/2	
	−4/17		−36/17			−31/17	1	1
	1/9					11/18	113/18	−67/18

Iteration 4:

1. $J = \{8\}$.
2. $\beta = 113/18,\ q = 8$.
4. $R' = \emptyset$.
8. $\max\{-7/2, -1, -4, -3.2\} = -1 \le 0,\ s = 2$.
9. $\bar{x}_8 = 1,\ \bar{x}_B = (7/2, 1, 4, 3/2)^T$. The basic optimal solution:

$$\bar{x} = (7/2, 0, 4, 0, 3/2, 1, 0, 1)^T, \qquad \bar{f} = 10.$$

It might be accepted that if the same approach is used, the search direction determined within a subspace of high dimension has better quality than that within a subspace of lower dimension. Therefore, it should be attractive to get Algorithm 21.3.1 started from $k = 0$. The according sets are

$$B, R = \emptyset, \quad N = A, \quad R' = \{1, \cdots, r-1, r+1, \cdots, m\}, \tag{21.6}$$

and the basis matrix B_R may be regarded as of 0-order. Subsequent steps are the same as those in the usual case. The according tableau is called 0-*order D-reduced tableau*; if it is still denoted by tableau (21.5), I and $\bar{N}_R$ should be viewed as empty sets, $\bar{N}_{R'}$ the submatrix of A after the r-indexed (datum) row deleted. If the bottom row satisfies $\bar{z}_N \ge 0$, it is called 0-order dual feasible D-reduced tableau. Initially, therefore, a series of rank-increasing iterations are usually performed under condition $\bar{a}_q(R') \ne 0$.

Example 21.3.2. Solve the following problem by Algorithm 21.3.1, starting from the 0-order D-reduced tableau:

$$
\begin{array}{lrrrrrrrcr}
\min & f = 2x_1 & + x_2 & + 2x_3 & + x_4 & + 4x_5 & + 5x_6 & + 2x_7, & & \\
\text{s.t.} & & + 2x_2 & & & - 3x_5 & - 4x_6 & + 6x_7 & = & 0, \\
& -2x_1 & & - 4x_3 & & & + x_6 & - x_7 & = & -1, \\
& & + 3x_2 & & - 2x_4 & + 4x_5 & & & = & 8, \\
& -3x_1 & & + 2x_3 & - 4x_4 & & - 2x_6 & & = & 0, \\
& & 2x_2 & - x_3 & & + 2x_5 & & - 5x_7 & = & -1, \\
& & & & & x_j \geq 0, & j = 1, \cdots, 7. & & &
\end{array}
$$

Answer Initial tableau:

x_1	x_2	x_3	x_4	x_5	x_6	x_7	RHS
	2			−3	−4	6	
−2		−4			1	−1	−1
	3		−2	4			8
−3		2	−4		−2		
	2	−1		2		−5	−1
2	1	2	1	4	5	2	

Convert the preceding to the 0-order D-reduced tableau:$\max\{|-1|, |8|, |-1|\} = 8$, $r = 3$, Take the row 3 as the datum row (which also involves minimum nonzeros). Multiply the datum row by $1/8$, then add it to row 2 as well as row 5:

x_1	x_2	x_3	x_4	x_5	x_6	x_7	RHS
	2			−3	−4	6	
−2	3/8	−4	−1/4	1/2	1	−1	
	3/8		−1/4	1/2			1
−3		2	−4		−2		
	19/8*	−1	−1/4	5/2		−5	
2	1	2	1	4	5	2	

where $k = 0,\ B = \emptyset,\ R = \emptyset,\ N = \{1, \cdots, 7\},\ R' = \{1, 2, 4, 5\}$, datum row $r = 3$.

Iteration 1:

1. $J = \{2, 5\} \neq \emptyset$.
2. $\beta = \min\{1/(3/8), 4/(1/2)\} = 8/3,\ q = 2$.
3. Add $-8/3$ times of row 3 to the bottom row.
4. $\bar{a}_q(R') \neq 0$.
5. $\max\{|2|, |3/8|, |3/8|, |19/8|\} = 19/8,\ p = 5$.
6. Multiply row 5 by $8/19$, then add $-2, -3/8, -3/8, -1$ times of row 5 to rows 1,2,3,6, respectively.
7. $k = 1,\ B = \{2\},\ R = R \cup \{5\} = \{5\},\ N = \{1, 3, 4, 5, 6, 7\},\ R' = \{1, 2, 4\}$.

x_1	x_2	x_3	x_4	x_5	x_6	x_7	RHS
		16/19	4/19	−97/19	−4	194/19*	
−2		−73/19	−4/19	2/19	1	−4/19	
		3/19	−4/19	2/19		15/19	1
−3		2	−4		−2		
	1	−8/19	−2/19	20/19		−40/19	
2		2	5/3	8/3	5	2	−8/3

Iteration 2:

1. $J = \{3,5,7\} \neq \emptyset$.
2. $\beta = \min\{2/(3/19), (8/3)/(2/19), 2/(15/19)\} = 38/15,\ q = 7$.
3. Add $-38/15$ times of row 3 to the bottom row.
4. $\bar{a}_q(R') \neq 0$.
5. $\max\{|194/19|, |-4/19|\} = 194/19,\ p = 1$.
6. Multiply row 1 by$19/194$, then add $4/19, -15/19, 40/19, -2$ times of row 1 to rows 2,3,4,6, respectively.
7. $k = 2,\ B = \{2,7\},\ R = \{5,1\},\ N = \{1,3,4,5,6\},\ R' = \{2,4\}$.

x_1	x_2	x_3	x_4	x_5	x_6	x_7	RHS
		8/97	2/97	−1/2	−38/97	1	
−2		−371/97	−20/97		89/97		
		9/97	−22/97	1/2	30/97		1
−3		2	−4		−2		
	1	−24/97	−6/97		−80/97		
2		8/5	11/5	12/5	5		−26/5

Iteration 3:

1. $J = \{3,5,6\} \neq \emptyset$.
2. $\beta = \min\{(8/5)/(9/97), (12/5)/(1/2), 5/(30/97)\} = 24/5,\ q = 5$.
3. Add $-24/5$ times of row 3 to the bottom row:

x_1	x_2	x_3	x_4	x_5	x_6	x_7	RHS
		8/97	2/97	−1/2	−38/97	1	
−2		−371/97	−20/97		89/97		
		9/97	−22/97	1/2	30/97		1
−3		2	−4		−2		
	1	−24/97	−6/97		−80/97		
2		112/97	319/97		341/97		−10

4. $\bar{a}_q(R') = 0$.
8. $\max\{-1/2, 0\} = 0 \leq 0$.

9. $\bar{x}_5 = 2, \bar{x}_B = (0, 1)^{\mathrm{T}}$. The basic optimal solution and optimal value:

$$\bar{x} = (0, 0, 0, 0, 2, 0, 1)^{\mathrm{T}}, \qquad \bar{f} = 10.$$

After pivot column index q is determined, there may be choices for taking either $p \in R'$ or $i_s \in R$ as the pivot row index. In doing so, Algorithm 21.3.1 gives the priority to R'. An alternative is to focus on numerical stability as follows. Define

$$\zeta = \max\{\zeta_1, \zeta_2\}, \quad \zeta_1 = |\bar{a}_{p\,q}| = \max_{i \in R'} |\bar{a}_{i\,q}| \quad \zeta_2 = \bar{a}_{i_s,q} = \max\{\bar{a}_{i_t,q} \mid t = 1, \ldots, k\}$$

(where ζ_1 or ζ_2 is set to zero if R' or R is empty). Then, if $\zeta = \zeta_1$, take $p \in R'$; if $\zeta = \zeta_2$, take $i_s \in R$.

Nevertheless, it seems to be preferable to keep the rank of the basis as low as possible. To this end, the following variant gives the priority to R rather than R'.

Algorithm 21.3.2 (Dual deficient-basis D-reduced algorithm 2: tableau form). Initial: dual feasible deficient-basis D-reduced tableau of form (21.5) with $B,\ R,\ N,\ R', 1 \leq k \leq m-1$. This algorithm solves the D-reduced problem (17.1).

1. Stop if $J = \{j \in N \mid \bar{\omega}_j > 0\} = \emptyset$ (infeasible problem).
2. Determine column index q and β such that $\beta = \bar{z}_q/\bar{\omega}_q = \min_{j \in J} \bar{z}_j/\bar{\omega}_j$.
3. If $\beta \neq 0$, add $-\beta$ times of the datum row to the bottom row.
4. Determine $s \in \arg\max\{\bar{a}_{i_t,q} \mid t = 1, \ldots, k\}$.
5. Go to step 8 if $\bar{a}_{i_s,q} \leq 0$.
6. Convert $\bar{a}_{i_s,q}$ to 1, and eliminate the other nonzeros in the column by elementary transformations.
7. Update (B, N) by exchanging j_s and q, and go to step 1.
8. If $R' = \emptyset$ or $\bar{a}_q(R') = 0$, compute $\bar{x}_q = 1/\bar{\omega}_q,\ \bar{x}_B = -\bar{x}_q\bar{a}_q(R)$, and stop (optimality achieved).
9. Determine row index $p \in \arg\max_{i \in R'} |\bar{a}_{i\,q}|$.
10. Convert $\bar{a}_{p\,q}$ to 1, and eliminate the other nonzeros in the column by elementary transformations.
11. Set $k = k + 1$, bring q from N to B and p from R' to R.
12. Go to step 1

Example 21.3.3. Solve the problem in Example 21.3.1 by Algorithm 21.3.2:

$$\begin{array}{rl}
\min & f = 2x_1 + x_2 - 2x_3 + x_4 - 3x_5 + x_6 + 2x_7 - 2x_8, \\
\text{s.t.} & 2x_1 - 3x_2 + x_3 - 6x_4 + x_5 + 4x_6 + 6x_7 = 0, \\
& 5x_1 + 6x_3 - 2x_4 + 10x_6 + 8x_7 - x_8 = 0, \\
& -x_1 + 8x_2 - 4x_5 + 2x_8 = 0, \\
& 4x_1 - 2x_3 + 5x_4 + 8x_5 - 4x_6 + 3x_7 = 1, \\
& x_j \geq 0, \quad j = 1, \ldots, 8.
\end{array}$$

Answer Begin with the seconde tableau in the Answer, i.e.,

x_1	x_2	x_3	x_4	x_5	x_6	x_7	x_8	RHS
1			−3/2	−8	22	33	−27/2	
	1		−1	6	−19	−26	10	
		1	0	−4	0	0	2	
			9/2	−1	6	12	−9/2	
			−1/2	1	−2	−3	3/2	1
			3	2	1	5	6	

Iteration 1: $k = 3$, $B = \{1, 2, 3\}$, $N = A\backslash B$, $R = \{1, 2, 3\}$, $R' = \{4\}$.

1. $J = \{5, 8\}$.
2. $\beta = \min\{2/1, 6/(3/2)\} = 2/1$, $q = 5$.
3. Add −2 times of row 5 to the bottom row.
4. $\max\{-8, 6, -4\} = 6, s = 2$.
5. $\bar{a}_{i_s,q} > 0$
6. Multiply row 2 by 1/6, then add $8, 4, 1, -1$ times of row 2 to rows 1,3,4,5, respectively:
7. $B = \{1, 5, 3\}$, $N = A\backslash B$, $R = \{1, 2, 3\}$, $R' = \{4\}$.

x_1	x_2	x_3	x_4	x_5	x_6	x_7	x_8	RHS
1	4/3		−17/6		−10/3	−5/3	−1/6	
	1/6		−1/6	1	−19/6	−13/3	5/3	
	2/3	1	−2/3		−38/3	−52/3	26/3	
	1/6		13/3		17/6	23/3	−17/6	
	−1/6		−1/3		7/6	4/3	−1/6	1
	0		4		5	11	3	−2

Iteration 2:

1. $J = \{6, 7\}$.
2. $\beta = \min\{5/(7/6), 11/(4/3))\} = 30/7$, $q = 6$.
3. Add −30/7 times of row 5 to the bottom row.
4. $\max\{-10/3, -19/3, -38/3\} \leq 0$.
8. $\bar{a}_q(R') \neq 0$.
9. $\max\{|17/6|\}$, $p = 4$.
10. Multiply row 4 by 6/17, then add $10/3, 19/6, 38/3, -7/6$ times of row 4 to rows 1,2,3,5, respectively:
11. $k = 4$, $B = \{1, 5, 3, 6\}$, $N = A\backslash B$, $R = \{1, 2, 3, 4\}$, $R' = \emptyset$.

x_1	x_2	x_3	x_4	x_5	x_6	x_7	x_8	RHS
1	26/17		77/34			125/17	−7/2	
	6/17		159/34	1		72/17	−3/2	
	24/17	1	318/17			288/17	−4	
	1/17		26/17		1	46/17	−1	
	−4/17		−36/17			−31/17	1	1
	5/7		38/7			37/7	26/7	−44/7

Iteration 3:

1. $J = \{8\}$.
2. $\beta = 26/7$, $q = 8$.
3. Add $-26/7$ times of row 5 to the bottom row.
4. $\max\{-7/2, -3/2, -4, -1\} \leq 0$.
8. $R' = \emptyset$. $\bar{x}_8 = 1$, $\bar{x}_B = (7/2, 3/2, 4, 1)^T$. The optimal solution and objective value:

$$\bar{x} = (7/2, 0, 4, 0, 3/2, 1, 0, 1)^T, \qquad \bar{f} = 44/7 + 26/7 = 10.$$

21.4 Dual Deficient-Basis D-Reduced Method

Assume that the r-indexed row is the datum row. Denote by B_R the basis matrix, associated with basic column and row index sets B and R, and so on. It is not difficult to show that the deficient-basis D-reduced tableau (21.5) is equivalent to the following revised tableau:

x_B^{T}	x_N^{T}	f	RHS
I	$B_R^{-1}N_R$		
	$N_{R'} - B_{R'}B_R^{-1}N_R$		
	$e_m^{\mathrm{T}}N - e_m^{\mathrm{T}}BB_R^{-1}N_R$		1
	$c_N^{\mathrm{T}} - c_B^{\mathrm{T}}B_R^{-1}N_R$	−1	

(21.7)

Based on such an equivalence, Algorithms 21.3.1 and 21.3.2 can be revised. But only former's revision is formulated here.

Algorithm 21.4.1 (Dual deficient-basis D-reduced algorithm 1). Initial: B, R, N, R', $1 \leq k \leq m-1$. B_R^{-1}, $\bar{z}_N \geq 0$. This algorithm solves the D-reduced problem (17.1).

1. Compute $\bar{\omega}_N = N^{\mathrm{T}}e_m - N_R^{\mathrm{T}}B_R^{-T}B^{\mathrm{T}}e_m$.
2. Stop if $J = \{j \in N \mid \bar{\omega}_j > 0\} = \emptyset$ (infeasible problem).
3. Determine column index q and β such that $\beta = \bar{z}_q/\bar{\omega}_q = \min_{j\in J} \bar{z}_j/\bar{\omega}_j$.

4. If $\beta \neq 0$, update $\bar{z}_N = \bar{z}_N - \beta\bar{\omega}_N$.
5. Compute $\bar{a}_q(R) = B_R^{-1}a_q(R)$.
6. Go to step 10 if $R' = \emptyset$ or $\bar{a}_q(R') = a_q(R') - B_{R'}\bar{a}_q(R) = 0$.
7. Determine row index $p \in \arg\max_{i\in R'} |\bar{a}_{i\,q}|$.
8. Update B_R^{-1} by (20.10).
9. Set $k = k+1$, bring q from N to B and p from R' to R, and go to step 1.
10. Determine $s \in \arg\max\{\bar{a}_{i_t,q} \mid t = 1,\ldots,k\}$.
11. If $\bar{a}_{i_s,q} \leq 0$, compute $\bar{x}_q = 1/\bar{\omega}_q,\ \bar{x}_B = -\bar{x}_q\bar{a}_q(R),\ \bar{f} = c_q\bar{x}_q + c_B^{\mathrm{T}}\bar{x}_B$, and stop (optimality achieved).
12. Update B_R^{-1} by (17.20).
13. Update (B, N) by exchanging j_s and q.
14. Go to step 1.

The Algorithm can get itself started if $\bar{z}_N = c_N - N_R^{\mathrm{T}}B_R^{-\mathrm{T}}c_B \geq 0$. In the other case, methods, presented in Sect. 17.3 or Chap. 14, can be utilized to provide an initial dual feasible solution, if slightly modified.

We point out that the so-called "reduced dual elimination" (Sect. 25.1.4) is a convenient tool to convert the dual problem to the D-reduced form. The resulting problem can be solved by the dual method presented in this section by generating a sequence of primal feasible solutions.

Example 21.4.1. Solve the following problem by Algorithm 21.4.1:

$$\begin{array}{lrl}
\min & f = & x_1 + 2x_3 + 3x_4 + 5x_6 + 2x_8, \\
\text{s.t.} & & x_1 + x_5 - 2x_6 + 3x_8 = 1, \\
& & -2x_1 - 4x_3 + 3x_4 - 3x_6 + x_7 - x_8 = 3, \\
& & -x_3 + 4x_8 = 0, \\
& & -2x_1 + x_2 + 2x_3 - 4x_4 + 4x_6 = -2, \\
& & 3x_1 + 2x_6 - 2x_8 = 0, \\
& & x_j \geq 0, \quad j = 1,\cdots,8.
\end{array}$$

Answer The initial tableau is

x_1	x_2	x_3	x_4	x_5	x_6	x_7	x_8	RHS
1				1	−2		3	1
−2		−4	3		−3	1	−1	3
		−1					4	
−2	1	2	−4		−4			−2*
3					2		−2	
1		2	3		5		2	

which is a dual feasible deficient-basis tableau with $B = \{5, 7, 2\},\ R = \{1, 2, 4\}$.

Convert the preceding to a D-reduced tableau: It might be well to take row 3 is the datum row. Multiply row 3 by $-1/3$, then add $-4, 6$ times of row 3 to rows 1,2, respectively:

x_1	x_2	x_3	x_4	x_5	x_6	x_7	x_8	RHS
	1/2	1	−2	1			3	
−5	3/2	−1	−3		3	1	−1	
		−1					4	
1	−1/2	−1	2		−2			1
3					2		−2	
1		2	3		5		2	

For this tableau, $k = 2,\ B = \{5, 7\},\ R = \{1, 2\},\ N = \{1, 2, 3, 4, 6, 8\},\ R' = \{3, 5\}$, the datum row: $r = 4$. $B_R = I$. $\bar{z}_N = (1, 0, 2, 3, 5, 2)^{\mathrm{T}} \geq 0$.

Iteration 1:

1. $\bar{\omega}_N = (1, -1/2, -1, 2, -2, 0)^{\mathrm{T}}$.
2. $J \ = \{1, 4\}$.
3. $\beta \ = \min\{1/1, 3/2\} = 1,\ q = 1$.
4. $\bar{z}_N \ = (1, 0, 2, 3, 5, 2)^{\mathrm{T}} - (1, -1/2, -1, 2, -2, 0)^{\mathrm{T}} = (0, 1/2, 3, 1, 7, 2)^{\mathrm{T}}$.
5. $\bar{a}_q(R) \ = (0, -5)^{\mathrm{T}}$.
6. $\bar{a}_q(R') = (0, 3)^{\mathrm{T}} \neq 0$.
7. $\max\{|3|\} = 3,\ p = 5$.
8. $\tau \ = 1/3, v = -(1/3)(0, -5)^{\mathrm{T}} = (0, 5/3)^{\mathrm{T}}, d^{\mathrm{T}} = (0, 0), U = I$.

$$\hat{B}_{\hat{R}}^{-1} = \begin{pmatrix} 1 & & \\ & 1 & 5/3 \\ & & 1/3 \end{pmatrix}.$$

9. $k \ = 3,\ B = \{5, 7, 1\},\ R = \{1, 2, 5\}.N = \{2, 3, 4, 6, 8\},\ R' = \{3\}$.

Iteration 2:

1. $$\bar{\omega}_N = (-1/2, -1, 2, -2, 0)^{\mathrm{T}} - \begin{pmatrix} 1/2 & 1 & -2 & & 3 \\ 3/2 & -1 & -3 & 3 & -1 \\ & & & 2 & -2 \end{pmatrix}^{\mathrm{T}} \begin{pmatrix} 1 & & \\ & 1 & 5/3 \\ & & 1/3 \end{pmatrix}^{\mathrm{T}} \begin{pmatrix} 0 \\ 0 \\ 1 \end{pmatrix}$$
$$= (-1/2, -1, 2, -8/3, 2/3)^{\mathrm{T}}.$$
2. $J \ = \{4, 8\}$.
3. $\beta \ = \min\{1/2, 2/(2/3)\} = 1/2,\ q = 4$.
4. $\bar{z}_N \ = (1/2, 3, 1, 7, 2)^{\mathrm{T}} - (1/2)(-1/2, -1, 2, -8/3, 2/3)^{\mathrm{T}}$
$= (3/4, 7/2, 0, 25/3, 5/3)^{\mathrm{T}}$.
5. $$\bar{a}_q(R) = \begin{pmatrix} 1 & & \\ & 1 & 5/3 \\ & & 1/3 \end{pmatrix} \begin{pmatrix} -2 \\ -3 \\ 0 \end{pmatrix} = \begin{pmatrix} -2 \\ -3 \\ 0 \end{pmatrix}.$$
6. $\bar{a}_q(R') = 0$.

10. $\max\{-2, -3, 0\} \leq 0$.
11. $\bar{x}_4 = 1/2,\ \bar{x}_B = -(1/2)(-2, -3, 0)^{\mathrm{T}} = (1, 3/2, 0)^{\mathrm{T}}$. The basic optimal solution and optimal value:

$$\bar{x} = (0, 0, 0, 1/2, 1, 0, 3/2, 0)^{\mathrm{T}}, \quad \bar{f} = 3/2.$$

21.5 Deficient-Basis D-Reduced Gradient Method: Tableau Form

It is noted that the search direction $-\bar{\omega}_N$, used in Algorithms 21.3.1 or 21.4.1, is independent of the current dual iterate. It is imaginable that if the current iterate is close to boundary, the stepsize would become very small, so that the according change in the dual objective value is negligible. In order to "go further", in this section the current iterate is exploited to modify the search direction, so that it is not only uphill with respect to the dual objective, but also has a "centering" tendency to leave the nearby boundary, pointing to interior of the dual feasible region. Such a trick has been used to design interior-point methods in Chap. 9.

Let sets B, R, N, R' be defined by (20.1) and (20.2). The according D-reduced tableau (21.5) represents the following D-reduced problem:

$$\begin{aligned} &\max\ y_r, \\ &\text{s.t.}\ \begin{pmatrix} I & 0 & 0 \\ \bar{N}_R^{\mathrm{T}} & \bar{N}_{R'}^{\mathrm{T}} & \bar{\omega}_N \end{pmatrix} \begin{pmatrix} y_R \\ y_{R'} \\ y_r \end{pmatrix} + \begin{pmatrix} z_B \\ z_N \end{pmatrix} = \begin{pmatrix} 0 \\ \bar{z}_N \end{pmatrix}, \ z_B, z_N \geq 0. \end{aligned} \tag{21.8}$$

Let $(\bar{z}_N \geq 0,\ \bar{z}_B = 0)$ be the current dual feasible solution. Setting $0 < \delta \ll 1$, introduce notation

$$\tilde{z}_N^{-1} = \tilde{Z}_N^{-1} e, \quad \tilde{z}_j = \begin{cases} \bar{z}_j, & \bar{z}_j \geq \delta, \\ \delta, & \bar{z}_j < \delta, \end{cases} \quad j \in N, \tag{21.9}$$

where $\tilde{Z}_N$ is the diagonal matrix, whose diagonal entries are components of $\tilde{z}_N$.

In order to create a "centering" direction, we maximize the following auxiliary function:

$$\max\ (\tilde{z}_N^{-1})^{\mathrm{T}} z_N,$$

subject to constraints of (21.8). From the equality constraints, it follows that

$$(\tilde{z}_N^{-1})^{\mathrm{T}} z_N = (n - k) - (\bar{\omega}_N^{\mathrm{T}} \tilde{z}_N^{-1}) y_r - (\bar{N}_R \tilde{z}_N^{-1})^{\mathrm{T}} y_R - (\bar{N}_{R'} \tilde{z}_j^{-1})^{\mathrm{T}} y_{R'}.$$

where the first term is constant. But, we do not want that the objective variable y_r be involved in the maximization of the auxiliary function. By constraints of (21.8), it is known that $y_R + z_B = 0$, hence if the value of z_B does not change, the related value of y_R does not either. Thus, by neglecting the first three terms of the auxiliary function, the problem comes down to the maximization of the following objective:

$$\max \quad -(\bar{N}_{R'}\bar{z}_j^{-1})^{\mathrm{T}} y_{R'},$$

subject to constraints of (21.8). The reduced gradient of the preceding program is

$$\Delta y_{R'} = -\bar{N}_{R'}\bar{z}_N^{-1}. \tag{21.10}$$

If $\Delta y_R,\ \Delta y_r = 0$ is taken, the according direction in z-space is then

$$\begin{aligned}\Delta z_N &= -\bar{N}_{R'}^{\mathrm{T}}\Delta y_{R'} = \bar{N}_{R'}^{\mathrm{T}}\bar{N}_{R'}\bar{z}_N^{-1},\\ \Delta z_B &= 0.\end{aligned}$$

Now it is logical to take

$$-\tilde{\omega}_N = -\bar{\omega}_N + \mu\bar{N}_{R'}^{T}\bar{N}_{R'}\bar{z}_N^{-1}, \tag{21.11}$$

$$-\tilde{\omega}_B = 0 \tag{21.12}$$

rather than $-\bar{\omega}_N$ as the search direction, where parameter $\mu \geq 0$ is a weight of the modification.

Assume that $\bar{a}_j(R'),\ j \in N$, is the subvector, associated with R', of $\bar{a}_j$. The vector $-\tilde{\omega}_N$, defined by (21.11) is of the following property.

Lemma 21.5.1. *Given $\mu > 0$. Assume for some $k \in N$ it holds that*

$$\bar{z}_k = 0, \quad \bar{a}_k(R') \neq 0, \quad \text{and} \quad \bar{z}_j \neq 0, \quad \forall\, j \in N,\ j \neq k.$$

Then $-\tilde{\omega}_k > 0$ if δ is sufficiently small.

Proof. From (21.9) and (21.11), it follows that

$$\begin{aligned}-\tilde{\omega}_k &= -\bar{\omega}_k + \mu(\bar{a}_k(R')^{\mathrm{T}}\bar{N}_{R'})\bar{z}_N^{-1}\\ &= -\bar{\omega}_k + \mu\bar{a}_k(R')^{\mathrm{T}}\bar{a}_k(R')/\delta + \mu\sum_{j\in N, j\neq k}\bar{a}_k(R')^{\mathrm{T}}\bar{a}_j(R')/\bar{z}_j.\end{aligned} \tag{21.13}$$

by which, it is known that when δ is small enough, the sign of $-\tilde{\omega}_k$ coincides with that of

$$\mu\|\bar{a}_k(R')\|^2/\delta > 0.$$

□

In addition, the following result can be shown similarly.

Lemma 21.5.2. *Given $\mu > 0$ and $N' \in N$. Assume that*

$$\bar{z}_j = 0, \quad 0 \le \bar{a}_j(R') \ne 0, \ \forall\ j \in N' \quad \text{and} \quad \bar{z}_j \ne 0, \ \forall\ j \in N \backslash N'.$$

Then $-\tilde{\omega}_k > 0$ if δ is sufficiently small.

The preceding Lemma says that under certain conditions, a sufficiently small δ enable the component, associated with $\bar{z}_j = 0$, of the new search direction $-\tilde{\omega}$ is strictly greater than zero to get rid of dual degeneracy's effect.

It is seen from (21.11) and (21.12) that such a search direction yields from multiplying rows of R' of the D-reduced tableau by according components of $\mu\Delta y_{R'}$, and adding the resulting rows to the datum row. Since the associated components of the right-hand side are zero, the whole right-hand side remains unchanged. Besides, the bottom row is not touched by these manipulations. Starting from such a D-reduced tableau, proceeded is a so-called "outer iteration", consisting of a series of "inner" D-reduced rank-increasing iterations, and terminating whenever $R' = \emptyset$ or $\bar{a}_q(R') = 0$; if the optimality conditions are not satisfied yet, i.e., $\bar{z}_q = 0, \ \bar{\omega}_q > 0$ but $\bar{a}_q(R) \not\le 0$, the next outer iteration is carried out.

An outer iteration may start from any dual feasible deficient-basis D-reduced tableau, including the 0-order tableau, though it seems to be favorable to take the D-reduced tableau associated with sets

$$B = B\backslash\{j_t \mid t \in T\}, \quad R = R\backslash\{i_t \mid t \in T\}, \quad T = \{t = 1, \cdots, k \mid \bar{a}_{i_t,q} > 0\}. \tag{21.14}$$

Such doing will force rows and columns, associated with negative components of the current primal feasible solution, to leave the basis. Another scheme to be chosen is to let a part of them leave the basis, e.g., the rows and columns associated with negative components of larger magnitude. In fact, under the assumption of Lemma 21.5.2, $-\bar{\omega}_{j_t} > 0$ holds for all $t \in T$ if δ is small enough. Thus, $\bar{z}_{j_t} = 0$ does not matter with the determination of stepsize and the related index. However, in case when $\bar{z}_N$ is degenerate, a positive stepsize is still not guaranteed, if $\bar{a}_q(R) \not\le 0$ and it holds for some $j \in N$ that $\bar{z}_j = 0, \ \bar{\omega}_j > 0$.

The according algorithm can be obtained by modifying Algorithm 21.3.1.

Algorithm 21.5.1 (Deficient-basis D-reduced gradient algorithm: tableau form). Given $\mu > 0$; $\delta > 0$. Initial : dual feasible deficient-basis D-reduced tableau of form (21.5), associated with $B, \ R, \ N, \ R', \ 1 \le k \le m-1$. This algorithm solves the D-reduced problem (17.1).

1. Cover nonbasic components of the datum row by $\bar{\omega}_N^{\mathrm{T}} = \bar{\omega}_N^{\mathrm{T}} - \mu(\bar{z}_N^{-1})^{\mathrm{T}} \bar{N}_{R'}^{\mathrm{T}} \bar{N}_{R'}$.
2. Stop if $J = \{j \in N \mid \bar{\omega}_j > 0\} = \emptyset$ (infeasible problem).
3. Determine column index q and β such that $\beta = \bar{z}_q/\bar{\omega}_q = \min_{j \in J} \bar{z}_j/\bar{\omega}_j$.
4. If $\beta \ne 0$, add $-\beta$ times of the datum row to the bottom row.
5. Go to step 9 if $R' = \emptyset$ or $\bar{a}_q(R') = 0$.
6. Determine row index $p \in \arg\max_{i \in R'} |\bar{a}_{i\,q}|$.

7. Convert $\bar{a}_{pq}$ to 1, and eliminate the other nonzeros in the column by elementary transformations.
8. Set $k = k + 1$, bring q from N to B and p from R' to R, and go to step 2.
9. Determine $s \in \arg\max\{\bar{a}_{i_t,q} \mid t = 1, \ldots, k\}$.
10. If $\bar{a}_{i_s,q} > 0$, update B, R by (21.14), and go to step 1.
11. Compute $\bar{x}_q = 1/\bar{\omega}_q$, $\bar{x}_{j_t} = -\bar{x}_q \bar{a}_{i_t,q}$, $t = 1, \cdots, k$.
12. Stop (optimality achieved).

Note There are two layers of iterations in the preceding Algorithm. The inner one consists of steps 2–7.

Example 21.5.1. Solve the following problem by Algorithm 21.5.1:

$$\begin{array}{lrrrrrrl}
\min & f = x_1 & + 2x_2 & + x_3 & + 7x_4 & + 3x_5 & + x_6, & \\
\text{s.t.} & -3x_1 & & - x_3 & & - 3x_5 & + 2x_6 & = -1, \\
& -2x_1 & - x_2 & + x_3 & - 6x_4 & + 6x_5 & - x_6 & = -2, \\
& & + 3x_2 & - 2x_3 & & + 8x_5 & + 4x_6 & = 0, \\
& -6x_1 & & & + 9x_4 & - 3x_5 & & = 3, \\
& & & & x_j \geq 0, & j = 1, \cdots, 6. & &
\end{array}$$

Answer $\mu = 1$ and $\delta = 1/10$ are taken. Initial tableau:

x_1	x_2	x_3	x_4	x_5	x_6	RHS
−3		−1		−3	2	−1
−2	−1	1	−6	6	−1	−2
	3	−2		8	4	
−6			9	−3		3*
1	2	1	7	3	1	

Convert the preceding to a D-reduced tableau:max$\{|-1|, |-2|, |0|, |3|\} = 3$, $r = 4$, take row 4 as the datum row.

Multiply row 4 by $1/3$, then add $1, 2$ times of row 4 to rows 1,2, respectively:

x_1	x_2	x_3	x_4	x_5	x_6	RHS
−5		−1	3	−4	2	
−6	−1	1		4	−1	
	3	−2		8	4	
−2			3	−1		1
1	2	1	7	3	1	

Iteration 1:

$k = 0$. $N = \{1, \cdots, 6\}$, $R' = \{1, 2, 3\}$.

1. $\bar{N}_{R'} = \begin{pmatrix} -3 & & -1 & & -3 & -5 & 2 \\ -2 & -1 & 1 & -6 & 6 & -2 & -1 \\ & 3 & -2 & & 8 & 3 & 4 \end{pmatrix}.$

   Set $\bar{\omega}_N^{\mathrm{T}} = (-2, 0, 0, 3, -1, 0) - (1, 1/2, 1, 1/7, 1/3, 1)\bar{N}_{R'}^{\mathrm{T}}\bar{N}_{R'}$
   $= (-1{,}208/21, -71/3, 529/42, 124/7, -345/7, -841/42)$
   to as the left part of row 4:

x_1	x_2	x_3	x_4	x_5	x_6	RHS
−5		−1	3	−4	2	
−6	−1	1*		4	−1	
	3	−2		8	4	
−1,208/21	−71/3	529/42	124/7	−345/7	−841/42	1
1	2	1	7	3	1	

2. $J = \{3, 4\} \neq \emptyset$.
3. $\beta = \min\{1/(529/42), 7/(124/7)\} = 1/(529/42),\ q = 3$.
4. Add $-\beta$ times of the datum row to the bottom row.
5. $\bar{a}_q(R') \neq 0$.
6. $\max\{|-1|, |1|, |-2|\} = 2,\ p = 2$.
7. Add $1, 2, -529/42, -1$ times of row 2 to rows 1,3,4,5, respectively.
8. $B = \{3\},\ R = \{2\},\ N = \{1, 2, 4, 5, 6\},\ R' = \{1, 3\}$.

x_1	x_2	x_3	x_4	x_5	x_6	RHS
−11	−1		3		1	
−6	−1	1		4	−1	
−12*	1			16	2	
379/21	−155/14		124/7	−299/3	−52/7	1
2,945/529	2,052/529		2,959/529	159/23	1,370/529	−42/529

Iteration 2:

2. $J = \{1, 4\} \neq \emptyset$.
3. $\beta = \min\{(2{,}945/529)/(379/21), (2{,}959/529)/(124/7)\}$
   $= (2{,}945/529)/(379/21),\ q = 1$.
4. Add $-\beta$ times of the datum row to the bottom row.
5. $\bar{a}_q(R') \neq 0$.
6. $\max\{|-11|, |-12|\} = 12,\ p = 3$.
7. Multiply row 3 by $-1/12$, and add $11, 6, -379/21, -2{,}945/529$ times of row 3 to rows 1,2,4,5, respectively.
8. $B = \{3, 1\},\ R = \{2, 3\},\ N = \{2, 4, 5, 6\},\ R' = \{1\}$.

x_1	x_2	x_3	x_4	x_5	x_6	RHS
	−23/12		3*	−44/3	−5/6	
	−3/2	1		−4	−2	
1	−1/12			−4/3	−1/6	
	−2,411/252		124/7	−4,763/63	−557/126	1
	5,529/758		49/379	6,477/172	1,850/379	−147/379

Iteration 3:

2. $J = \{4\} \neq \emptyset$.
3. $\beta = (49/379)/(124/7)$, $q = 4$.
4. Add $-\beta$ times of the datum row to the bottom row.
5. $\bar{a}_q(R') \neq 0$.
6. $\max\{|3|\}$, $p = 1$.
7. Multiply row 3 by 1/3, then add −124/7, −49/379 times of rows 4,5, respectively.
8. $B = \{3, 1, 4\}$, $R = \{2, 3, 1\}$, $N = \{2, 5, 6\}$, $R' = \emptyset$.

x_1	x_2	x_3	x_4	x_5	x_6	RHS
	−23/36		1	−44/9	−5/18	
	−3/2	1		−4	−2	
1	−1/12			−4/3	−1/6	
	7/4			11	1/2	1
	1,760/239			3,477/91	1,307/266	−49/124

Iteration 4:

2. $J = \{2, 5, 6\} \neq \emptyset$.
3. $\beta = \min\{(1,760/239)/(7/4), (3,477/91)/11, (1,307/266)/(1/2)\}$
 $= (3{,}477/91)/11$, $q = 5$.
4. Add $-\beta$ times of the datum row to the bottom row.

x_1	x_2	x_3	x_4	x_5	x_6	RHS
	−23/36		1	−44/9	−5/18	
	−3/2	1		−4	−2	
1	−1/12			−4/3	−1/6	
	7/4			11	1/2	1
	2,027/1,577				629/198	−383/99

5. $\bar{a}_q(R') = 0$.
9. $\max\{-44/9, -4, -4/3\} \leq 0$.
11. $\bar{x}_5 = 1/11$, $\bar{x}_B = (1/11)(4, 4/3, 44/9)^{\mathrm{T}} = (4/11, 4/33, 4/9)^{\mathrm{T}}$.

12. The basic optimal solution and optimal value:

$$\bar{x} = (4/33, 0, 4/11, 4/9, 1/11, 0)^{\mathrm{T}}, \qquad \bar{f} = 383/99.$$

Example 21.5.2. Solve the following D-reduced problem by Algorithm 21.5.1:

$$\begin{array}{rl} \min & f = 3x_4 + x_6 + 2x_7, \\ \text{s.t.} & x_1 \qquad\qquad - 2x_4 + 5x_5 - 5x_6 + 2x_7 = 0, \\ & \quad + x_2 \qquad - 3x_4 + 3x_5 - 2x_6 + x_7 = 0, \\ & \qquad + x_3 + x_4 - 4x_5 + 6x_6 - 3x_7 = 0, \\ & \qquad\qquad - 6x_4 \qquad + 3x_6 + x_7 = 0, \\ & \qquad\qquad + x_4 + 4x_5 + 2x_6 - x_7 = 1, \\ & \qquad\qquad x_j \geq 0, \quad j = 1, \cdots, 7. \end{array}$$

Answer $\mu = 1$ and $\delta = 1/10$ are taken. Initial tableau:

x_1	x_2	x_3	x_4	x_5	x_6	x_7	RHS
1			−2	5	−5	2	
	1		−3	3	−2	1	
		1	1	−4	6	−3	
			−6	0	3	1	
			1	4	2	−1	1
			3	0	1	2	

Row 5 is the datum row: $r = 5$. $B = \{1, 2, 3\}$, $R = \{1, 2, 3\}$, $N = \{4, 5, 6, 7\}$, $R' = \{4\}$.

Column 5 is the datum: $\bar{a}_{1,5} = 5 < 0$, $\bar{a}_{2,5} = 3 > 0$, $T = \{1, 2\}$,

Reset $B = \{3\}$, $R = \{3\}$, $N = \{1, 2, 4, 5, 6, 7\}$, $R' = \{1, 2, 4\}$.

Iteration 1:

1. $\bar{N}_{R'} = \begin{pmatrix} 1 & & -2 & 5 & -5 & 2 \\ & 1 & -3 & 3 & -2 & 1 \\ & & -6 & & 3 & 1 \end{pmatrix}.$

Cover nonbasic components of row 5 by

$$\begin{aligned} \bar{\omega}_N^{\mathrm{T}} &= (0, 0, 1, 4, 2, -1) - (10, 10, 1/3, 10, 1, 1/2)\bar{N}_{R'}^{\mathrm{T}}\bar{N}_{R'} \\ &= (-166/3, -75/2, 1{,}399/6, -2{,}311/6, 2{,}095/6, -452/3). \end{aligned}$$

x_1	x_2	x_3	x_4	x_5	x_6	x_7	RHS
1			−2	5	−5*	2	
	1		−3	3	−2	1	
		1	1	−4	6	−3	
			−6	0	3	1	
−166/3	−75/2		1,399/6	−2,311/6	2,095/6	−452/3	1
			3	0	1	2	

2. $J = \{4, 6\} \neq \emptyset$.
3. $\beta = \min\{3/(1{,}399/6), 1/(2{,}095/6)\} = 6/2{,}095,\ q = 6$.
4. Add $-\beta$ times of the datum row to the bottom row.
5. $\bar{a}_q(R') \neq 0$.
6. $\max\{|-5|, |-2|, |3|\} = 5,\ p = 1$.
7. Multiply row 1 by $-1/5$, then add $2, -6, -3, -2{,}095/6, -1$ times of row 1 to rows 2,3,4,5,6, respectively.
8. $B = \{3, 6\},\ R = \{3, 1\},\ N = \{1, 2, 4, 5, 7\},\ R' = \{2, 4\}$.

x_1	x_2	x_3	x_4	x_5	x_6	x_7	RHS
−1/5			2/5	−1	1	−2/5	
−2/5	1		−11/5	1		1/5	
6/5		1	−7/5	2		−3/5	
3/5*			−36/5	3		11/5	
29/2	−75/2		187/2	−36		−11	1
332/2,095	45/419		1,397/599	1,102/999		5,094/2,095	−6/2,095

Iteration 2:

2. $J = \{1, 4\} \neq \emptyset$.
3. $\beta = \min\{(332/2{,}095)/(29/2), (1{,}397/599)/(187/2)\}$
   $= (332/2{,}095)/(29/2), q = 1$.
4. Add $-\beta$ times of the datum row to the bottom row.
5. $\bar{a}_q(R') \neq 0$.
6. $\max\{|-2/5|, |3/5|\} = 3/5,\ p = 4$.
7. Multiply row 4 by 5/3, then add $1/5, 2/5, -6/5, -29/2, -332/2{,}095$ times of row 4 to rows 1,2,3,5,6, respectively.
8. $B = \{3, 6, 1\},\ R = \{3, 1, 4\},\ N = \{2, 4, 5, 7\},\ R' = \{2\}$.

x_1	x_2	x_3	x_4	x_5	x_6	x_7	RHS
			−2		1	1/3	
	1		−7*	3		5/3	
		1	13	−4		−5	
1			−12	5		11/3	
	−75/2		535/2	−217/2		−385/6	1
	15/29		38/29	217/145		74/29	−2/145

Iteration 3:

2. $J = \{4\} \neq \emptyset$.
3. $\beta = (38/29)/(535/2),\ q = 4$.
4. Add $-\beta$ times of the datum row to the bottom row.
5. $\bar{a}_q(R') \neq 0$.

6. $\max\{|-7|\}$, $p = 2$.
7. Multiply row 2 by $-1/7$, then add $2, -13, 12, -535/2, -38/29$ times of row 2 to rows 1,3,4,5,6 respectively:
8. $B = \{3, 6, 1, 4\}$, $R = \{3, 1, 4, 2\}$, $N = \{2, 5, 7\}$, $R' = \emptyset$.

x_1	x_2	x_3	x_4	x_5	x_6	x_7	RHS
	−2/7			−6/7	1	−1/7	
	−1/7		1	−3/7		−5/21	
	13/7	1		11/7		−40/21	
1	−12/7			−1/7		17/21	
	5/7			43/7		−10/21	1
	75/107			217/107		920/321	−2/107

Iteration 4:

2. $J = \{2, 5\} \neq \emptyset$.
3. $\beta = \min\{(75/107)/(5/7), (217/107)/(43/7)\} = (217/107)/(43/7)$, $q = 5$.
4. Add $-\beta$ times of the datum row to the bottom row:

x_1	x_2	x_3	x_4	x_5	x_6	x_7	RHS
	−2/7			−6/7	1	−1/7	
	−1/7		1	−3/7		−5/21	
	13/7	1		11/7		−40/21	
1	−12/7			−1/7		17/21	
	5/7			43/7		−10/21	1
	20/43					130/43	−15/43

5. $\bar{a}_q(R') = 0$.
9. $\max\{11/7, -6/7, -1/7, -3/7\} = 11/7 > 0$, $s = 1$.
10. $T = \{1\}$, $B = \{6, 1, 4\}$, $R = \{1, 4, 2\}$, $N = \{2, 3, 5, 7\}$, $R' = \{3\}$.

Iteration 5:

1. $\bar{N}_{R'} = (13/7, 1, 11/7, -40/2)$. Cover nonbasic components of row 5 by
$$\bar{\omega}_N^{\mathrm{T}} = (5/7, 0, 43/7, -10/21) - (43/20, 10, 10, 43/130)\bar{N}_{R'}^{\mathrm{T}}\bar{N}_{R'} = (-22{,}540/423, -12{,}067/415, -7{,}554/191, 15{,}649/285).$$

x_1	x_2	x_3	x_4	x_5	x_6	x_7	RHS
	−2/7			−6/7	1	−1/7	
	−1/7		1	−3/7		−5/21	
	13/7	1		11/7		−40/21*	
1	−12/7			−1/7		17/21	
	−22,540/423	−12,067/415		−7,554/191		15,649/285	1
	20/43					130/43	−15/43

2. $J = \{7\} \neq \emptyset$.
3. $\beta = (130/43)/(15{,}649/285)$, $q = 7$.
4. Add $-\beta$ times of the datum row to the bottom row.
5. $\bar{a}_q(R') \neq 0$.
6. $\max\{|-40/21|\}$, $p = 3$.
7. Multiply row 3 by $-21/40$, then add $1/7, 5/21, -17/21, -15{,}649/285, -130/43$ times of row 3 to rows 1,2,4,5,6, respectively.
8. $B = \{6, 1, 4, 7\}$, $R = \{1, 4, 2, 3\}$, $N = \{2, 3, 5\}$, $R' = \emptyset$.

x_1	x_2	x_3	x_4	x_5	x_6	x_7	RHS
	$-17/40$	$-3/40$		$-39/40$	1		
	$-3/8$	$-1/8$	1	$-5/8$			
	$-39/40$	$-21/40$		$-33/40$		1	
1	$-37/40$	$17/40$		$21/40$			
	$100{,}463/401{,}851$	$-19{,}713/78{,}851$		$23/4$			1
	$1{,}397/411$	$658/411$		$2{,}624/1{,}205$			$-767/1{,}899$

Iteration 6:

2. $J = \{2, 5\} \neq \emptyset$.
3. $\beta = \min\{(1{,}397/411)/(100{,}463/401{,}851), (2{,}624/1{,}205)/(23/4)\}$
$= (2{,}624/1{,}205)/(23/4)$, $q = 5$.
4. Add $-\beta$ times of the datum row to the bottom row.

x_1	x_2	x_3	x_4	x_5	x_6	x_7	RHS
	$-17/40$	$-3/40$		$-39/40$	1		
	$-3/8$	$-1/8$	1	$-5/8$			
	$-39/40$	$-21/40$		$-33/40$		1	
1	$-37/40$	$17/40$		$21/40$			
	$100{,}463/401{,}851$	$-19{,}713/78{,}851$		$23/4$			1
	$76/23$	$39/23$					$-18/23$

5. $\bar{a}_q(R') = 0$.
9. $\max\{-39/40, 21/40, -5/8, -33/40\} = 21/40 > 0$, $s = 2$.
10. $T = \{2\}$, $B = \{6, 4, 7\}$, $R = \{1, 2, 3\}$, $N = \{1, 2, 3, 5\}$, $R' = \{4\}$.

Iteration 7:

x_1	x_2	x_3	x_4	x_5	x_6	x_7	RHS
	$-17/40$	$-3/40$		$-39/40$	1		
	$-3/8$	$-1/8$	1	$-5/8$			
	$-39/40$	$-21/40$		$-33/40$		1	
1	$-37/40*$	$17/40$		$21/40$			
$-3{,}379/222$	$4{,}571/319$	$-3{,}823/569$		$-3{,}498/1{,}561$			1
	$76/23$	$39/23$					$-18/23$

2. $J = \{2\} \neq \emptyset$.
3. $\beta = (76/23)/(4{,}571/319)$, $q = 2$.
4. Add $-\beta$ times of the datum row to the bottom row.
5. $\bar{a}_q(R') \neq 0$.
6. $\max\{|-37/40|\}$, $p = 4$.
7. Multiply row 4 by $-40/37$, then add $17/40, 3/8, 39/40, -4{,}571/319, -76/23$ times of row 4 to rows 1,2,3,5,6, respectively.
8. $B = \{6, 4, 7, 2\}$, $R = \{1, 2, 3, 4\}$, $N = \{1, 3, 5\}$, $R' = \emptyset$.

x_1	x_2	x_3	x_4	x_5	x_6	x_7	RHS
−17/37		−10/37		−45/37	1		
−15/37		−11/37	1	−31/37			
−39/37		−36/37		−51/37		1	
−40/37	1	−17/37		−21/37			
3,957/14,641		−793/5,868		218/37			1
2,471/704		6,039/1,861		910/1,761			−997/984

Iteration 8:

2. $J = \{1, 5\} \neq \emptyset$.
3. $\beta = \max\{(2{,}471/704)/(3{,}957/14{,}641), (910/1{,}761)/(218/37)\}$
   $= (910/1{,}761)/(218/37)$, $q = 5$.
4. Add $-\beta$ times of the datum row to the bottom row.

x_1	x_2	x_3	x_4	x_5	x_6	x_7	RHS
−17/37		−10/37		−45/37	1		
−15/37		−11/37	1	−31/37			
−39/37		−36/37		−51/37		1	
−40/37	1	−17/37		−21/37			
3,957/14,641		−793/5,868		218/37			1
380/109		355/109					−120/109

5. $\bar{a}_q(R') = 0$.
9. $\max\{-45/37, -31/37, -51/37, -21/37\} \leq 0$.
11. $\bar{x}_5 = 37/218$, $\bar{x}_B = (37/218)(45/37, 31/37, 51/37, 21/37)^{\mathrm{T}}$
    $= (45/218, 31/218, 51/218, 21/218)^{\mathrm{T}}$.
12. The basic optimal solution and optimal value:

$$\bar{x} = (0, 21/218, 0, 31/218, 37/218, 45/218, 51/218)^{\mathrm{T}}, \quad \bar{f} = 120/109.$$

21.6 Deficient-Basis D-Reduced Gradient Method

It is not easy to convert the tableau Algorithm 21.5.1 to the revised version because of the difficulty to update B_R^{-1} in step 10. The tableau Algorithm itself seems to be of higher practical value. However, the situation will be different if each outer iterations starts from the 0-order tableau instead. This idea is embodied in the following algorithm.

Algorithm 21.6.1 (Deficient-basis D-reduced gradient algorithm). Given $\mu > 0; \delta > 0$. Initial: $\bar{z} \geq 0$. This algorithm solves the D-reduced problem (17.1).

1. Set $k = 0$ and give B, R, N, R' by (21.6).
2. Compute $\omega_N = N^{\mathrm{T}} e_m - \mu N_{R'}^{\mathrm{T}} N_{R'} \bar{z}_N^{-1})$, and set $\bar{\omega}_N = \omega_N$.
3. Stop if $J = \{j \in N \mid \bar{\omega}_j > 0\} = \emptyset$ (infeasible problem).
4. Determine column index q and θ such that $\theta = \bar{z}_q / \bar{\omega}_q = \min_{j \in J} \bar{z}_j / \bar{\omega}_j$.
5. If $\theta \neq 0$, update $\bar{z}_N = \bar{z}_N - \theta \bar{\omega}_N$.
6. Compute $\bar{a}_q(R) = B_R^{-1} a_q(R),\ \bar{a}_q(R') = a_q(R') - B_{R'} \bar{a}_q(R)$.
7. Go to step 12 if $R' = \emptyset$ or $\bar{a}_q(R') = 0$.
8. Determine row index $p \in \arg\max_{i \in R'} |\bar{a}_{i\,q}|$.
9. Update B_R^{-1} by (20.10).
10. Set $k = k + 1$, bring q from N to B and p from R' to R,
11. Compute $\bar{\omega}_N = \omega_N - N_R^{\mathrm{T}} (B_R^{-T} \omega_B)$, and go to step 3.
12. Determine $s \in \arg\max\{\bar{a}_{i_t, q} \mid t = 1, \ldots, k\}$.
13. Go to step 1 if $\bar{a}_{i_s, q} > 0$.
14. Compute $\bar{x}_q = 1/\bar{\omega}_q,\ \bar{x}_B = -\bar{x}_q \bar{a}_q(R),\ \bar{f} = c_q \bar{x}_q + c_B^{\mathrm{T}} \bar{x}_B$.
15. Stop (optimality achieved).

Note This Algorithm contains steps 3–11 as its inner steps.

An alternative way is to utilize, to a larger extent, current information to form the datum row r. The first iteration starts from

$$\omega = A^{\mathrm{T}} e_m - \mu A_{R'}^{\mathrm{T}} \bar{A}_{R'} \bar{z}^{-1}, \tag{21.15}$$

where $A_{R'}$ denotes the matrix, resulting from A by deleting row r. Assume that at the end of some subsequent inner iterations, the basic column and row index sets are B and R, not satisfying the optimality condition; the datum row $\bar{\omega}$ is

$$\bar{\omega}_B = 0, \quad \bar{\omega}_N = \omega_N - N_R^{\mathrm{T}} (B_R^{-\mathrm{T}} \omega_B),$$

and the basic row matrix $\bar{A}_R$ is

$$\bar{B}_R = I, \quad \bar{N}_R = B_R^{-1} N_R.$$

Then the datum row $\hat{\omega}$ for the next outer iteration is formed by combining the preceding two items as follows:

$$\begin{aligned}\hat{\omega}_B &= \bar{\omega}_B - \mu \bar{B}_R^{\mathrm{T}} \bar{A}_R \bar{z}^{-1} = -\mu(\bar{z}_B^{-1} + B_R^{-1} N_R \bar{z}_N^{-1}),\\ \hat{\omega}_N &= \bar{\omega}_N - \mu \bar{N}_R^{\mathrm{T}} \bar{A}_R \bar{z}^{-1}\\ &= \omega_N - N_R^{\mathrm{T}} B_R^{-\mathrm{T}} \omega_B - \mu \bar{N}_R^{\mathrm{T}}(\bar{z}_B^{-1} + B_R^{-1} N_R \bar{z}_N^{-1})\\ &= \omega_N - N_R^{\mathrm{T}} B_R^{-\mathrm{T}}(\omega_B - \hat{\omega}_B).\end{aligned}$$

Thereby, we have the following variant of Algorithm 21.6.1.

Algorithm 21.6.2 (Variant of the dual reduced gradient algorithm). The same as Algorithm 21.6.1, except for steps 1 and 2 replaced, respectively, by

1. Compute ω by (21.15), and set $\bar{\omega} = \omega$.
2. Set $\bar{\omega} = \omega,\ k = 0;\ B, R, N, R'$ is determined by (21.6).
   and steps 13–15 replaced by
13. If $\bar{a}_{p,q} \leq 0$, compute $\bar{x}_q = 1/\bar{\omega}_q,\ \bar{x}_B = -\bar{x}_q \bar{a}_q(R),\ \bar{f} = c_q \bar{x}_q + c_B^{\mathrm{T}} \bar{x}_B$, and stop (optimality achieved).
14. Compute $v = -\bar{z}_B^{-1} - B_R^{-1} N_R \bar{z}_N^{-1}$.
15. Compute $\omega_N = \omega_N - N_R^{\mathrm{T}} B_R^{-T}(\omega_B - v),\ \omega_B = v$.
16. Go to step 1.

It is noted that the preceding variant involves solution of additional two systems. The behavior of it is not known yet so far.

21.7 Notes

The deficient-basis methods, presented in the previous two chapters might still be viewed as variants of the simplex method. However, the solution process can not be interpreted as a movement from simplex to simplex. As A is not of full row rank in general, in fact, there is no correspondence between feasible deficient-basis tableaus and vertices of the feasible region, though in case when A is of full row rank, it is possible to expand a deficient basis to a full one by entering some nonbasic columns to it, and hence associated with a vertex.

In Pan's original work (1998–2008; see also Guerrero-Garcia and Santos-Palomo 2009), the conventional basis is generalized to a submatrix B of A, whose column space includes b. As a result, deficient basis is rectangle rather than square. Besides $\bar{x}_N = 0$, in that context, conventional $\bar{x}_B = B^{-1}b$ was essentially replaced by $\bar{x}_B = B^{+}b$, where B^{+} denotes the Moore-Penrose inverse (Pan and Ouyang 1994; also see, e.g., He and Sun 1991); as more compact and efficient, it is replaced by $x_B = B_R^{-1} b_R$ in current context.

The orthogonal transformation was used for triangularization of the basis in some of the original deficient-basis algorithms. An advantage of such doing is that the resulting search direction is an *orthogonal* projection of the negative objective gradient onto the associated null space. Thereby, such a direction forms the smallest angle with the negative objective gradient among vectors in the null space, as is attractive, compared with the reduced gradient approach. Unfortunately, the orthogonal transformation may lead to a large amount of fill-ins, and may be suitable only for solving dense problems. This is why the associated quite remarkable computational results have not been cited so far in this book, despite many contents in the previous two chapters can be handled based on the orthogonal transformation, otherwise.

The deficient or full basis together with a pivot rule for basis change plays a central role in the simplex-like methods. In particular, we all realize that the rule is crucial to the success of such type of methods. It should be desirable to utilize some analogue to rules presented in Chaps. 11 and 12, preferably the primal (dual) nested largest-distance rule, though analogues to Dantzig's original rule were employed in previous chapters for sake of simplicity (see, e.g., Pan et al. 2006a).

As for the bounded-variable problem (7.13), the concept of basis can be generalized as follows.

Definition 21.7.1. Let B_R be the nonsingular square submatrix consisting of entries associated with B, R, and let $\bar{x}_N$ be on bound. B_R is a basis (matrix) if the range space of B includes $b - N\bar{x}_N$.

It is noted that such a basis is closely related to values taken on by nonbasic variables. Based on this definition, it is not difficult to generalize the deficient-basis algorithms established previously, though we will not go into details here.

Chapter 22
Face Method

Like the simplex method, the so-called "face method", presented in this chapter, also involves pivot choice. However, it does not use elementary but orthogonal transformations to determine a search direction, and generates iterates that are not necessarily vertices, but boundary points. It proceeds from face to face in the feasible region, until reaching an optimal face together with a pair of primal and dual optimal solutions. Therefore, in some sense, the face method may be regarded as a generalization of the simplex method, which proceeds from 0-dimensional face (vertex) to 0-dimensional face (vertex).

In this chapter, the face method is developed first. Then its Phase-I procedure is described. A generalized version of the face method and a variant based on affine-scaling are derived.

We are concerned with the reduced problem (15.1), i.e.,

$$\begin{array}{ll} \min & x_{n+1}, \\ \text{s.t.} & Ax = b, \quad x_j \geq 0, \; j = 1, \cdots, n, \end{array} \tag{22.1}$$

where $A \in \mathcal{R}^{(m+1)\times(n+1)}$, $b \in \mathcal{R}^{m+1}$, rank $A = m + 1$, $m < n$, $a_{n+1} = Ae_{n+1} = -e_{m+1}$. The assumption on the rank of A will be dropped later.

Assume that $B \in \mathcal{R}^{(m+1)\times k}$, $N \in \mathcal{R}^{(m+1)\times(n-k+1)}$ is a partition of A, where

$$\text{rank } B = m + 1, \quad m + 1 \leq k \leq n + 1.$$

Respectively, B and N are called *face matrix* and *nonface matrix*; the associated variables are called *face variables* and *nonface variables*. Without confusion, denote the associated index sets by B and N, respectively, i.e.,

$$B = \{1, \cdots, k - 1, n + 1\}, \quad N = \{1, \cdots, n + 1\} \backslash B.$$

P.-Q. PAN, *Linear Programming Computation*, DOI 10.1007/978-3-642-40754-3_22,

To simplify computation, thereafter it is always arranged such that the kth (last) column of the face matrix equals the $(n+1)$th (last) column of A, i.e.,

$$Be_k = -e_{m+1}. \tag{22.2}$$

The face, associated with the face matrix B, is defined by

$$P_B = \{x \in A \mid Bx_B = b;\ x_s \geq 0,\ s = 1, \cdots, k-1,\ x_N = 0\}. \tag{22.3}$$

A point on the face is called *face point*. Thus, an 0-dimensional face has an unique face point, vertex.

Definition 22.1. A face is level face if the objective value is constant over it. A level face is an *optimal face* if the related objective value equals the optimal value.

The largest optimal face, i.e., that of the highest dimension, is the optimal set (Sect. 2.3).

22.1 Face Method Using Cholesky Factorization

Assume that $\bar{x}$ is the current face point on P_B, i.e., $\bar{x} \in P_B$. As $\bar{x}$ is uniquely determined by $\bar{x}_B$, the two will not be strictly distinguished hereafter; for simplicity, $\bar{x}_B \in P_B$ is often used in place of $\bar{x} \in P_B$.

22.1.1 The Steepest Downhill

We determine a search direction firstly. Consider the following subprogram:

$$\begin{array}{ll} \min & x_{n+1}, \\ \text{s.t.} & Bx_B = b, \\ & x_j \geq 0, \quad j \in B, j \neq n+1, \end{array} \tag{22.4}$$

which actually minimizes the objective over the face P_B.

Introduce $k \times k$ orthogonal projection matrix, i.e.,

$$P = I - B^{\mathrm{T}}(BB^{\mathrm{T}})^{-1}B. \tag{22.5}$$

Then, the orthogonal projection of the objective gradient, e_k, of the subprogram onto the null space of B is

$$\Delta_B = Pe_k = e_k - B^{\mathrm{T}}\bar{y}, \quad BB^{\mathrm{T}}\bar{y} = -e_{m+1}, \tag{22.6}$$

and $-\Delta_B$ is a desirable search direction. In fact, we have the following result.

Proposition 22.1.1. *Vector* Δ_B *satisfies* $B\Delta_B = 0$.

Proof. By (22.6), it is easy to verify validity of the equality. □

Proposition 22.1.2. *The following statements are equivalent:*

$$(i)\ \Delta_B \neq 0; \qquad (ii)\ \Delta_k > 0; \qquad (iii)\ e_k \notin \mathrm{range}(B^{\mathrm{T}}).$$

Proof. Premultiply the first expression of (22.6) by Δ_B^{T}. Then combining the result and Proposition 22.1.1 gives

$$\Delta_B^{\mathrm{T}}\Delta_B = \Delta_k.$$

So, statements (i) and (ii) are equivalent. If $\Delta_B = 0$, it is known by the first expression of (22.6) that $e_k = B^{\mathrm{T}}\bar{y}$, i.e., $e_k \in \mathrm{range}(B^{\mathrm{T}})$, hence (iii) results in (i). Conversely, if $e_k \in \mathrm{range}(B^{\mathrm{T}})$, then there is some vector u such that

$$e_k = B^{\mathrm{T}}u. \tag{22.7}$$

Premultiplying the preceding by B together with (22.2) gives

$$-e_{m+1} = BB^{\mathrm{T}}u,$$

which, the second expression of (22.6) and B of full row rank lead to $u = \bar{y}$. Thus, it yields from (22.7) that $e_k = B^{\mathrm{T}}\bar{y}$, substituting which to the first expression of (22.6) gives $\Delta_B = 0$. Therefore, it is shown that (i) and (iii) are equivalent. □

Proposition 22.1.3. *If* $\Delta_B \neq 0$, *then it holds that:*

$$e_k^{\mathrm{T}}\Delta_B = \Delta_k > 0, \tag{22.8}$$

$$-e_k^{\mathrm{T}}\Delta_B/\|\Delta_B\| \leq e_k^{\mathrm{T}}v/\|v\|, \quad \forall\, 0 \neq v \in \mathrm{Null}(B). \tag{22.9}$$

Proof. Expression (22.8) can be derived directly from Proposition 22.1.2. Note that for P defined by (22.5), it holds that

$$Pv = v \neq 0, \quad \forall\, 0 \neq v \in \mathrm{Null}(B). \tag{22.10}$$

It is obtained by Cauchy inequality that

$$(-Pe_k)^{\mathrm{T}}(Pv) \leq \|Pe_k\|\|Pv\|,$$

multiplying which by $1/\|Pv\|$ gives

$$(-Pe_k)^{\mathrm{T}}(Pv)/\|Pv\| \leq \|Pe_k\|,$$

i.e.,

$$(-Pe_k)^{\mathrm{T}}(Pv)/\|Pv\| \leq (Pe_k)^{\mathrm{T}}(Pe_k)/\|Pe_k\|.$$

Combining the preceding expression, $P^2 = P$, $P^{\mathrm{T}} = P$, (22.10) and $\Delta_B = Pe_k$ gives (22.9). □

According to Proposition 22.1.3, vector $-\Delta_B \neq 0$ forms the most-obtuse-angle in the null space of B with the objective gradient e_k of the subprogram, and is hence a steepest downhill. In this sense, taking the vector as a search direction is the "best" choice.

It is noted that the computation of Δ_B mainly lies on solving $(m + 1) \times (m + 1)$ system

$$BB^{\mathrm{T}}y = -e_{m+1}. \tag{22.11}$$

Since B is of full row rank, there exists the Cholesky factorization (Golub and Van Loan 1989)

$$BB^{\mathrm{T}} = LL^{\mathrm{T}}, \tag{22.12}$$

where L is nonsingular lower triangular. Thereby, system (22.11) becomes

$$LL^{\mathrm{T}}y = -e_{m+1}, \tag{22.13}$$

which can be put into the following two triangular systems:

$$Lv = -e_{m+1}, \qquad L^{\mathrm{T}}y = v.$$

The solution to the first system is readily available, i.e., $v = -(1/\nu)e_{m+1}$, where ν is the $(m + 1)$th diagonal of L. Consequently, there is only one triangular system left to be solved, i.e.,

$$L^{\mathrm{T}}y = -(1/\nu)e_{m+1} \tag{22.14}$$

22.1.2 Updating Solution

The new iterate is defined by the following line search scheme:

$$\hat{x}_B = \bar{x}_B - \alpha\Delta_B, \tag{22.15}$$

where α is a stepsize to be determined. If index set

$$J = \{j \in B \mid \Delta_j > 0, j \neq n + 1\}$$

is nonempty, determine α and p such that

$$\alpha = \bar{x}_p/\Delta_p = \min_{j \in J} \bar{x}_j/\Delta_j \geq 0, \tag{22.16}$$

which is the largest possible stepsize for the new iterate remaining within the feasible region. If $\bar{x}_j = 0$ for some $j \in B,\ j \neq n+1$, then face point $\bar{x}$ is said to be *degenerate*. In this case, stepsize α could vanish, and the resulting iterate is actually not "new" but the same as the old. Note that whether $\bar{x}$ is degenerate only depends upon the first $k-1$ components of $\bar{x}_B$.

Lemma 22.1.1. *Assume that $\bar{x}_B \in P_B$ and $\Delta_B \neq 0$.*

(i) If $J \neq \emptyset$, then $\hat{x}_B \in P_B$ is a boundary point. The objective value does not increase, and strictly decreases if $\bar{x}_B$ is nondegenerate.
(ii) If $J = \emptyset$, the original problem is lower unbound.

Proof. (i) $\Delta_B \neq 0$ and $J \neq \emptyset$ together ensure that (22.16) is well-defined. As $\bar{x}_B \in P_B$, it is known from Proposition 22.1.1 that $\hat{x}_B \in P_B$; further, it is clear that $\hat{x}_p = 0$, hence $\hat{x}$ is a boundary point, associated with the new objective value

$$\hat{x}_{n+1} = \bar{x}_{n+1} - \alpha\Delta_k. \tag{22.17}$$

It is known from (22.16) that $\alpha \geq 0$. Besides, from $\Delta_B \neq 0$ and Proposition 22.1.2, it follows that $\Delta_k > 0$, hence $\hat{x}_{n+1} \leq \bar{x}_{n+1}$. If $\bar{x}_B$ is nondegenerate, then $\alpha > 0$, leading to $\hat{x}_{n+1} < \bar{x}_{n+1}$.

(ii) When $J = \emptyset$, it is clear that $\hat{x}_B \in P_B$ holds for any $\alpha > 0$. It is known by (22.17) that

$$\hat{x}_{n+1} \to -\infty, \qquad \text{as} \qquad \to \infty.$$

Therefore the objective value is lower unbounded over P_B, hence over the feasible region. □

22.1.3 Face Contraction

Assume that $\Delta_B \not\geq 0$ and the new iterate $\hat{x}_B$ was determined by (22.15) with $\hat{x}_p = 0$. Updating B and N accordingly gives the face and nonface index sets, associated with the new iterate, i.e.,

$$\breve{B} = B\backslash\{p\}, \quad \breve{N} = N \cup \{p\}.$$

The associated face matrix $\breve{B} \in \mathcal{R}^{(m+1)\times(k-1)}$ results from the old B by dropping the p-indexed column. Then we have the following result.

Proposition 22.1.4. *If* rank $B = m + 1$, *then* rank $\breve{B} = m + 1$.

Proof. It is clear that rank $\breve{B} \leq m + 1$. It is known by (22.16) that $\Delta_p \neq 0$, and by Proposition 22.1.1 that

$$\sum_{j \in B} \Delta_j a_j = B\Delta_B = 0,$$

thus

$$a_p = -(1/\Delta_p) \sum_{j \in B, j \neq p} \Delta_j a_j.$$

The above equality implies that $a_p \in$ range $\breve{B}$. Thereby, rank $\breve{B} < m + 1$ leads to rank $B < m + 1$, as contradicts the assumption. Therefore rank $\breve{B} = m + 1$. □

The preceding Proposition implies that the new face matrix $\breve{B}$ is again of full row rank, and hence there exists the Cholesky factorization

$$\breve{B}\breve{B}^{\mathrm{T}} = \breve{L}\breve{L}^{\mathrm{T}}.$$

Proposition 22.1.5. *WW^T is independent of the ordering of the columns of W for any matrix W.*

Proof. The validity is verified by the fact that

$$(WQ^T)(WQ^T)^T = W(Q^TQ)W^T = WW^T$$

holds for any column permutation Q^T. □

Proposition 22.1.6. *If $Lu = a_p$, then $1 - \|u\|^2 > 0$.*

Proof. According to Proposition 22.1.5, we may assume that $B = (\breve{B}, a_p)$, and hence

$$BB^T = \breve{B}\breve{B}^T + a_p a_p^T,$$

so that

$$\breve{B}\breve{B}^T = BB^T - a_p a_p^T = LL^T - a_p a_p^T. \tag{22.18}$$

It is clear that there exists $v \neq 0$ such that $LL^Tv = a_p$, which with (22.18) gives

$$\breve{B}\breve{B}^T = LL^T - LL^Tvv^TLL^T.$$

From $u = L^T v$ and the positiveness of $\breve{B}\breve{B}^T$, it follows that

$$v^T \breve{B}\breve{B}^T v = u^T u - (u^T u)^2 = u^T u(1 - u^T u) > 0,$$

which implies $1 - \|u\|^2 > 0$. □

Saunders (1972) offers a procedure to obtain the new factor from its predecessor, as is put into the following algorithm.

Algorithm 22.1.1 (Subalgorithm: contraction). Initial: the Cholesky factor L of BB^{T}. This algorithm finds the Cholesky factor $\breve{L}$ of $\breve{B}\breve{B}^{\mathrm{T}}$.

1. Solves lower triangular system $Lu = a_p$.
2. Compute $\gamma = \sqrt{1 - \|u\|^2}$.
3. Determine Givens rotations $J_{m+1}, \cdots, J_1$, where J_i is in the $(m+2, i)$-plane, such that

$$J_1 \cdots J_{m+1} \begin{pmatrix} u \\ \gamma \end{pmatrix} = \begin{pmatrix} 0 \\ 1 \end{pmatrix}.$$

4. Compute $J_1 \cdots J_{m+1} \begin{pmatrix} L^{\mathrm{T}} \\ 0 \end{pmatrix} = \begin{pmatrix} \breve{L}^{\mathrm{T}} \\ a_p^{\mathrm{T}} \end{pmatrix}$, to obtain $\breve{L} \in \mathcal{R}^{(m+1)\times(m+1)}$.
5. Return.

It is noted that $J_1 \cdots J_{m+1}$ in step 3 of the preceding are well-defined.
To see the output of the Subalgorithm is what we want, set

$$J_1 \cdots J_{m+1} \triangleq Q, \qquad Q \begin{pmatrix} L^{\mathrm{T}} \\ 0 \end{pmatrix} \triangleq \begin{pmatrix} \breve{L}^{\mathrm{T}} \\ s^T \end{pmatrix},$$

where Q is orthogonal, $\breve{L}$ is lower-triangular, and s is a vector. Then, it follows that

$$\begin{pmatrix} u^{\mathrm{T}} & \gamma \\ L & 0 \end{pmatrix} \begin{pmatrix} u & L^{\mathrm{T}} \\ \gamma & 0 \end{pmatrix} = \begin{pmatrix} u^{\mathrm{T}} & \gamma \\ L & 0 \end{pmatrix} Q^T Q \begin{pmatrix} u & L^{\mathrm{T}} \\ \gamma & 0 \end{pmatrix} = \begin{pmatrix} 0 & 1 \\ \breve{L} & s \end{pmatrix} \begin{pmatrix} 0 & \breve{L}^{\mathrm{T}} \\ 1 & s^T \end{pmatrix}$$

so that

$$u^{\mathrm{T}} u + \gamma^2 = 1, \qquad Lu = s, \qquad LL^{\mathrm{T}} = \breve{L}\breve{L}^{\mathrm{T}} + ss^{\mathrm{T}}.$$

The second equality of the preceding and $Lu = a_p$ together lead to $s = a_p$, and hence from the third equality and (22.18), it follows that

$$\breve{B}\breve{B}^T = LL^T - a_p a_p^T = \breve{L}\breve{L}^{\mathrm{T}}.$$

It is noted that a_p is recomputed as the vector s, and thus a non-trivial discrepancy between s and a_p could imply significant numerical error in L, as provides us with some sort of numerical check.

22.1.4 Optimality Test

In case of $\Delta_B = 0$, the search direction defined by (22.6) is useless. In particular, it is the case when $k = m + 1$, and hence the projection matrix is null.

Lemma 22.1.2. *Assume that Δ_B is defined by (22.6). If Δ_B vanishes, then P_B is a level face; and vice versa if $0 < \bar{x}_B \in P_B$.*

Proof. Assume that $\bar{y}$ satisfies the second expression of (22.6). From $\Delta_B = 0$ and the first expression of (22.6), it follows that

$$e_k = B^{\mathrm{T}}\bar{y}.$$

Transpose the two sides of the preceding equality, and premultiply the result by any $x_B \in P_B$ to obtain

$$e_k^{\mathrm{T}}x_B = \bar{y}^{\mathrm{T}}Bx_B,$$

combining which and $Bx_B = b$ gives

$$x_{n+1} = \bar{y}^{\mathrm{T}}b.$$

This implies that the objective value is constant over P_B, and hence P_B is a level face.

Now assume that P_B is a level face and that $0 < \bar{x}_B \in P_B$ with $\Delta_B \neq 0$. In case when $\Delta_B \leq 0$, it is known by (ii) of Lemma 22.1.1 that the objective value is lower unbounded over P_B, as contradicts that P_B is a level face; when $\Delta_B \not\leq 0$, since x_B is nondegenerate, it is known from (i) of Lemma 22.1.1 that the objective value, associated with $\hat{x}_B \in P_B$ given by (22.15) and (22.16), strictly decreases, as also contradicts that P_B is a level face. Therefore, Δ_B vanishes. □

Now consider the dual problem of (22.1)

$$\begin{aligned} &\max\ (b^{\mathrm{T}}, 0)^{\mathrm{T}}y, \\ &\text{s.t.}\ \begin{pmatrix} B^{\mathrm{T}} \\ N^{\mathrm{T}} \end{pmatrix} y + \begin{pmatrix} z_B \\ z_N \end{pmatrix} = \begin{pmatrix} e_k \\ 0 \end{pmatrix}, \quad z_B,\ z_N \geq 0,\ z_{n+1} = 0. \end{aligned} \tag{22.19}$$

Lemma 22.1.3. *Assume that $\bar{x}_B \in P_B$ and $\Delta_B = 0$. If $\bar{z}_N = -N^{\mathrm{T}}\bar{y} \geq 0$, then $(\bar{x}_B, \bar{x}_N)$ and $(\bar{z}_B, \bar{z}_N, \bar{y})$ are a pair of primal and dual optimal solutions, where $\bar{x}_N = 0,\ \bar{z}_B = 0$, and P_B is an optimal face to (22.1).*

Proof. $(\bar{x}_B, \bar{x}_N)$ $(\bar{x}_N = 0)$ is clearly a feasible solution to (22.1). On the other hand, combining (22.6) and $\Delta_B = 0$ gives $B^{\mathrm{T}}\bar{y} = e_k$. Thereby, it is easy to verify that $(\bar{z}_B, \bar{z}_N, \bar{y})$ is a feasible solution to (22.19), satisfying the complementarity condition, and hence are a pair of primal and dual optimal solutions. By Lemma 22.1.2, $\Delta_B = 0$ implies that P_B is a level face. It includes the optimal solution, hence is an optimal face. □

According to Proposition 22.1.2, $\Delta_B = 0$ if and only if $\Delta_k = 0$. Therefore condition $\Delta_B = 0$ may be replaced by $\Delta_k = 0$ simply.

22.1.5 *Face Expansion*

Assume $\Delta_B = 0$ and $\bar{z}_N = -N^{\mathrm{T}}\bar{y} \not\geq 0$. In this case, optimality of the level face P_B can not be asserted, but a column index q can be determined such that

$$q \in \arg\min_{j \in N} \bar{z}_j.$$

Accordingly, update the face and nonface index sets by

$$\hat{B} = B \cup \{q\}, \quad \hat{N} = N \backslash \{q\}. \tag{22.20}$$

As a result, the number of columns of the face matrix will increase by 1.

Thus, the q-indexed column is added to the face matrix (as its second last column). Denote the resulting new face matrix by $\hat{B} \in \mathcal{R}^{(m+1)\times(k+1)}$, which is still of full row rank. The according updating of the Cholesky factor is quite simple (Golub 1965), as stated as follows.

Algorithm 22.1.2 (Subalgorithm: expansion). Initial: the Cholesky factor L of BB^{T}. This algorithm finds Cholesky factor $\hat{L}$ of $\hat{B}\hat{B}^{\mathrm{T}}$.

Determine Givens rotations $J_1, \cdots, J_{m+1}$, where J_i is in the $(i, m+2)$-plane, such that

$$J_{m+1} \cdots J_1 \begin{pmatrix} L^{\mathrm{T}} \\ a_q^{\mathrm{T}} \end{pmatrix} = \begin{pmatrix} \hat{L}^{\mathrm{T}} \\ 0 \end{pmatrix},$$

which gives $\hat{L} \in \mathcal{R}^{(m+1)\times(m+1)}$.

The validity of the preceding Subalgorithm is verified by premultiplying the above equality by its transpose:

$$\hat{L}\hat{L}^{\mathrm{T}} = LL^{\mathrm{T}} + a_q a_q^{\mathrm{T}} = BB^{\mathrm{T}} + a_q a_q^{\mathrm{T}} = \hat{B}\hat{B}^{\mathrm{T}}.$$

Another scheme, seeming to be attractive, is to use

$$\hat{B} = B \cup \{j \in N \mid \bar{z}_j < 0\}, \quad \hat{N} = A \backslash \hat{B}$$

rather than (22.20), though adding two columns to the face matrix will complicate updating of the Cholesky factor.

22.1.6 Face Algorithm

Any A's submatrix $B \in \mathcal{R}^{(m+1)\times k}, m+1 \le k \le n+1$ of full row rank can be taken as an initial face matrix. But we prefer the largest one, i.e., $k = n+1,\ B = A$.

Now drop the assumption that A is of full row rank. Assume that an orthogonal matrix $[Q_1, Q_2] \in \mathcal{R}^{(n+1)\times(n+1)}$ and a permutation $\Pi \in \mathcal{R}^{(m+1)\times(m+1)}$ have been found such that

$$A^{\mathrm{T}}\Pi = [Q_1, Q_2]\begin{pmatrix} R & T \\ 0 & 0 \end{pmatrix},$$

where $R \in \mathcal{R}^{r\times r}$ is nonsingular upper triangular. Then $L = R^{\mathrm{T}}$ is the initial Cholesky factor. In fact, rank $A = r$, and the last $m - r + 1$ equations of $\Pi^{\mathrm{T}}Ax = \Pi^{\mathrm{T}}b$ is redundant; if the original notation is still used to denote the problem, resulting from dropping these redundant rows, then $A^{\mathrm{T}} = Q_1 R$, and the wanted Cholesky factorization is

$$AA^{\mathrm{T}} = R^{\mathrm{T}}Q_1^{\mathrm{T}}Q_1 R = LL^{\mathrm{T}}. \tag{22.21}$$

The overall steps can be put into the following algorithm.

Algorithm 22.1.3 (Face algorithm). Initial: $(B, N),\ m+1 \le k \le n+1$ (B includes $n+1$ as its last index), feasible solution $\bar{x}$ and Cholesky factor L of BB^{T}. This algorithm solves the reduced problem (22.1).

1. Solves upper triangular system $L^{\mathrm{T}}\bar{y} = -(1/\nu)e_{m+1}$, where ν is the $(m+1)$th diagonal of L.
2. If $k > m+1$, compute $\Delta_B = e_k - B^{\mathrm{T}}\bar{y}$.
3. Go to step 10 if $k = m+1$ or $\Delta_B = 0$.
4. Stop if $J = \{j \in B \mid \Delta_j > 0,\ j \ne n+1\} = \emptyset$.
5. Determine stepsize α and index p such that $\alpha = \bar{x}_p/\Delta_p = \min_{j\in J}\bar{x}_j/\Delta_j$.
6. If $\alpha \ne 0$, update: $\bar{x}_B = \bar{x}_B - \alpha\Delta_B$.
7. Update (B, N) by braining p from B to N.
8. Update L by Subalgorithm 22.1.1.
9. Set $k = k-1$, and go to step 1.
10. Compute $\bar{z}_N = -N^{\mathrm{T}}\bar{y}$.
11. Stop if $\bar{z}_N \ge 0$.
12. Determine index $q \in \arg\min_{j\in N}\bar{z}_j$.
13. Update L by Subalgorithm 22.1.2.

14. Bring q from N to B as its second last index.
15. Set $k = k + 1$.
16. Go to step 1.

Theorem 22.1.1. *Under the nondegeneracy assumption, Algorithm 22.1.3 terminates either at*

(i) Step 4, detecting lower unboundedness of (22.1); or at
(ii) Step 11, giving an optimal face together with a pair of primal and dual optimal solutions.

Proof. Under the nondegeneracy assumption, the proof of the finiteness of Algorithm 22.1.3 is the same as for the simplex algorithm. The meanings of the exits come from Lemmas 22.1.1 and 22.1.3. □

In a contraction iteration, the face algorithm solves two triangular systems (including one involved in updating L by Subalgorithm 22.1.1), whereas it solve only one in an expansion iteration, in contrast to the simplex algorithm that solves four. Therefore, the former involves less computational effort than the latter per iteration. Moreover, an explicit representation of L suffices, and there is no need for any storage of the Givens rotations, used to create the Cholesky factorization initially and to update Cholesky factors subsequently.

The dimension of the end optimal face produced by the Algorithm should be more than 0, in general, though it would not be necessarily the optimal set.

Preliminary computational experiments were conducted. Compiled using the Visual FORTRAN 5.0, test codes were run under a Windows XP system Home Edition Version 2002 on an IBM PC with an Intel(R) Pentium(R) processor 1.00 GB of 1.86 GHz memory, and about 16 digits of precision. The following two dense codes in FORTRAN 77 are compared:

1. RSA: The simplex algorithm (see Notation).
2. FALP: Face Algorithm 22.1.3, supported by Algorithm 22.2.1 being as Phase-I.

Test set of problems are the 26 smallest Netlib problems (Appendix B: Table B.4: AFIRO-DEGEN2). As code RSA failed to solve problem DEGEN2 within 10,000 iterations, only total ratios of RSA to FALP on the other 25 problems are given below:

total iterations ratio: 1.51, total time ratio: 4.91.

It is indeed impressive that the face algorithm outperformed the simplex algorithm with time ratio near 5 (see Appendix D for details).

Example 22.1.1. Solve the following problem by Algorithm 22.1.3:

$$\begin{array}{lrrrrrl}
\min & f = 2x_1 & & - x_3 & & - x_5, & \\
\text{s.t.} & 2x_1 & & - x_3 & & & = 1, \\
 & x_1 & & - x_3 & + 3x_4 & & = 0, \\
 & & - x_2 & - x_3 & & - 2x_5 & = -2, \\
 & & & x_j \geq 0, & j = 1, \cdots, 5. & &
\end{array}$$

Answer It is noted that there is a feasible solution $(1, 1, 0, 1, 1)^{\mathrm{T}}$ with objective value 1 to the preceding problem. At first, convert the problem to the following reduced form:

$$
\begin{array}{lrrrrrrl}
\min & f, & & & & & & \\
\text{s.t.} & 2x_1 & & -\ x_3 & & & & = \ \ 2, \\
 & x_1 & & -\ x_3 & +\ 3x_4 & & & = \ \ 4, \\
 & & -\ x_2 & -\ x_3 & & -\ 2x_5 & & = -4, \\
 & 2x_1 & & -\ x_3 & & -\ \ x_5 & -\ f & = \ \ 0, \\
 & & & x_j \geq 0, & & j = 1, \cdots, 5. & &
\end{array}
$$

Initial: $k = 6$, $B = \{1, 2, 3, 4, 5, 6\}$, $N = \emptyset$, initial feasible solution $\bar{x}_B = (1, 1, 0, 1, 1, 1)^{\mathrm{T}}$. Respectively, the face matrix B and the Cholesky factor of BB^{T} are

$$
B = \begin{pmatrix} 2 & & -1 & & & \\ 1 & & -1 & 3 & & \\ & -1 & & -1 & -2 & \\ 2 & & -1 & & -1 & -1 \end{pmatrix},
$$

$$
L = \begin{pmatrix} 2{,}889/1{,}292 & & & \\ 1{,}292/963 & 549/181 & & \\ & -181/183 & 679/303 & \\ 2{,}889/1{,}292 & & 606/679 & 848/773 \end{pmatrix}.
$$

Iteration 1:

1. $\nu \ = 848/773, \bar{y} = (213/278, 15/139, 46/139, -231/278)^{\mathrm{T}}$.
2. $\Delta_B = (0, 0, 0, 0, 0, 1)^{\mathrm{T}} - (-3/139, -46/139, -6/139, -1/139, 47/278,$ $231/278)^{\mathrm{T}} = (3/139, 46/139, 6/139, 1/139, -47/278, 47/278)^{\mathrm{T}} \neq 0$.
4. $J \ = \{1, 2, 3, 4\} \neq \emptyset$.
5. $\alpha \ = \min\{1/(3/139), 1/(46/139), 0/(6/139), 1/(1/139)\} = 0, \ p = 3$.
7. $B \ = \{1, 2, 4, 5, 6\}, \ N = \{3\}$.
8. $L \ = \begin{pmatrix} 2 & & & \\ 1 & 3 & & \\ & -1 & 2{,}889/1{,}292 & \\ 2 & & 2{,}584/2{,}889 & 505/461 \end{pmatrix}.$
9. $k \ = 5$.

Iteration 2:

1. $\nu \ = 505/461, \bar{y} = (7/9, 1/9, 1/3, -5/6)^{\mathrm{T}}$.
2. $\Delta_B = (0, 0, 0, 0, 1)^{\mathrm{T}} - (0, -1/3, 0, 131{,}910/791{,}459, 5/6)^{\mathrm{T}}$ $= (0, 1/3, 0, -131{,}910/791{,}459, 1/6)^{\mathrm{T}} \neq 0$.
4. $J \ = \{2\} \neq \emptyset$.

5. $\alpha \;= \min\{1/(1/3)\} = 3,\ p = 2.$
6. $\bar{x}_B \;= (1,1,1,1,1)^{\mathrm{T}} - 3(0,1/3,0,-131,910/791,459,1/6)^{\mathrm{T}}$
   $\;= (1,0,1,3/2,1/2)^{\mathrm{T}}.$
7. $B \;= \{1,4,5,6\},\ N = \{2,3\}.$
8. $$L = \begin{pmatrix} 2 & & & \\ 1 & 3 & & \\ & -1 & 2 & \\ 2 & & 1 & 1 \end{pmatrix}.$$
9. $k \;= 4.$

Iteration 3:

1. $\nu \;= 1,\ \bar{y} = (11/12,1/6,1/2,-1)^{\mathrm{T}}.$
2. $\Delta_B = (0,0,0,0,1)^{\mathrm{T}} - (0,-1/3,0,131{,}910/791{,}459,5/6)^{\mathrm{T}}$
   $= (0,1/3,0,-131{,}910/791{,}459,1/6)^{\mathrm{T}} \neq 0.$
10. $\bar{z}_N \;= (1/2,1/12)^{\mathrm{T}} \geq 0.$
11. The basic optimal solution and optimal value:

$$\bar{x} \;= (1,0,0,1,3/2)^{\mathrm{T}}, \qquad \bar{f} = 1/2.$$

22.2 Face Phase-I: Single-Artificial-Variable

Any Phase-I method can be utilized to provide a feasible solution to get the face algorithm started. This section consider a Phase-I method involving only a single artificial variable.

Assume that $Ae \neq b$. The auxiliary program used is as follows:

$$\begin{array}{ll} \min & x_{n+1}, \\ \text{s.t.} & Ax + a_{n+1}x_{n+1} = b, \quad x, x_{n+1} \geq 0, \end{array}$$

where $a_{n+1} = b - Ae$.

At the beginning of Phase-I, set

$$B = \{1,\cdots,n,n+1\}, \quad N = \emptyset. \tag{22.22}$$

Thus, the initial auxiliary feasible solution possesses all its components 1, and the auxiliary program can be solved by a slight modification of Algorithm 22.1.3:

In step 1, two triangular systems are solved to obtain $\bar{y}$. In step 5, $n+1$ is taken as a candidate for selection of row index p. Once it leaves B (while $\bar{x}_{n+1}$ becomes 0), a feasible solution is obtained. In addition, step 4 should be dropped because the auxiliary program is certainly lower bounded.

The steps are put in the following algorithm.

Algorithm 22.2.1 (Face Phase-I: single-artificial-variable). Initial: $B = \{1, \cdots, n, n+1\}$, $N = \emptyset$, $k = n+1$; $\bar{x} = e$, $\bar{x}_{n+1} = 1$; $BB^{\mathrm{T}} = LL^{\mathrm{T}}$. This algorithm finds a feasible solution to (22.1).

The same as Algorithm 22.1.3, except for its steps 1–6 replaced by

1. Solve $LL^{\mathrm{T}}\bar{y} = -b$.
2. If $k > m+1$, compute $\Delta_B = e_k - B^{\mathrm{T}}\bar{y}$.
3. Go to step 10 if $k = m+1$ or $\Delta_B = 0$.
4. Determine α and index p such that $\alpha = \bar{x}_p/\Delta_p = \min\{\bar{x}_j/\Delta_j \mid \Delta_j > 0, j \in B\}$.
5. If $\alpha \neq 0$, update: $\bar{x}_B = \bar{x}_B - \alpha\Delta_B$.
6. Stop if $\bar{x}_{n+1} = 0$ (feasibility achieved).

Example 22.2.1. Solve the following problem by the two-phase face algorithm:

$$\begin{array}{lrcrcrcrcrcr}
\min & x_1 & + & 2x_3 & + & 3x_4, & & & & & & \\
\text{s.t.} & 2x_1 & - & x_2 & - & 2x_3 & - & x_4 & & & = & -1, \\
& & + & 2x_2 & & & - & 2x_4 & + & x_5 & = & 1, \\
& x_1 & - & 2x_2 & & & & & - & x_5 & = & -2, \\
& & & x_j \geq 0, & & j = 1, \cdots, 5. & & & & & &
\end{array}$$

Answer Phase-I: Call Algorithm 22.2.1.

Initial: $k = 6$, $B = \{1,2,3,4,5,6\}$, $N = \emptyset$, $\bar{x} = e$, $a_6 = (1,0,0)^{\mathrm{T}}$. B and the Cholesky factor of BB^{T}:

$$B = \begin{pmatrix} 2 & -1 & -2 & -1 & & 1 \\ & 2 & & -2 & 1 & \\ 1 & -2 & & & -1 & \end{pmatrix}, \quad L = \begin{pmatrix} -1{,}257/379 & & \\ & -3 & \\ -2{,}394/1{,}985 & 5/3 & -2{,}804/2{,}109 \end{pmatrix}.$$

Iteration 1:

1. $\bar{y} = (29/175, -4/35, -36/175)^{\mathrm{T}}$.
2. $\Delta_B = (-22/175, -3/175, 58/175, -11/175, -16/175, 146/175)^{\mathrm{T}} \neq 0$.
4. $\alpha = \min\{1/(58/175), 1/(146/175)\} = 175/146$, $p = 6$.
5. $\bar{x}_B = (1,1,1,1,1,1)^{\mathrm{T}} - (175/146)(-22/175, -3/175, 58/175, -11/175, -16/175, 146/175)^{\mathrm{T}}$
   $= (84/73, 149/146, 44/73, 157/146, 81/73, 0)^{\mathrm{T}}$.
6. $\bar{x}_6 = 0$, return with
   $\bar{f} = (1,0,2,3,0)(84/73, 149/146, 44/73, 157/146, 81/73)^{\mathrm{T}} = 815/146$.

Phase-II: Call Algorithm 22.1.3.

Initial: $k = 6$, $B = \{1,2,3,4,5,6\}$, $N = \emptyset$, $\bar{x} = (84/73, 149/146, 44/73, 157/146, 81/73, 815/146)^{\mathrm{T}}$. B and the Cholesky factor of BB^{T}:

$$B = \begin{pmatrix} 2 & -1 & -2 & -1 & & \\ & 2 & & -2 & 1 & \\ 1 & -2 & & & -1 & \\ 1 & & 2 & 3 & 0 & -1 \end{pmatrix}, \quad L = \begin{pmatrix} -721/228 & & & \\ & -3 & & \\ -721/570 & 5/3 & -740/581 & \\ 721/456 & 2 & 905/3{,}458 & -392/135 \end{pmatrix}.$$

Iteration 2:

1. $\bar{y} \;= (-61/1{,}231, -114/1{,}231, -30/1{,}231, -146/1{,}231)^{\mathrm{T}}$.
2. $\Delta_B = (298/1{,}231, 107/1{,}231, 170/1{,}231, 149/1{,}231, 84/1{,}231, 1{,}085/1{,}231)^{\mathrm{T}}$.
4. $J \;\neq \emptyset$.
5. $\alpha \;= \min\{(84/73)/(298/1{,}231), (149/146)/(107/1{,}231)(44/73)/(170/1{,}231), (157/146)/(149/1{,}231), (81/73)/(84/1{,}231)(815/146)/(1{,}085/1{,}231)\}$ $= 1{,}305/299,\; p = 3$.
6. $\bar{x}_B = (84/73, 149/146, 44/73, 157/146, 81/73, 815/146)^{\mathrm{T}} - (1{,}305/299)(298/1{,}231, 107/1{,}231, 170/1{,}231, 149/1{,}231, 84/1{,}231, 1{,}085/1{,}231)^{\mathrm{T}}$ $= (1{,}211/12{,}867, 109/170, 0, 93/170, 69/85, 15{,}366/8{,}855)^{\mathrm{T}}$.
7. $B \;= \{1, 2, 4, 5, 6\},\; N = \{3\}$.
8. $$L \;= \begin{pmatrix} -2{,}158/881 & & & \\ & -3 & & \\ -1{,}762/1{,}079 & 5/3 & 963/1{,}292 & \\ 881/2{,}158 & 2 & -2{,}889/1{,}292 & 524/387 \end{pmatrix}.$$
9. $k \;= 5$.

Iteration 3:

1. $\bar{y} \;= (1, -14/11, -18/11, -6/11)^{\mathrm{T}}$.
2. $\Delta_B = (2/11, 3/11, 1/11, -4/11, 83{,}204/183{,}049)^{\mathrm{T}}$.
4. $J \;\neq \emptyset$.
5. $\alpha \;= \min\{(1{,}211/12{,}867)/(2/11), (109/170)/(3/11), (93/170)/(1/11)(15{,}366/8{,}855)/(83{,}204/183{,}049)\}$ $= 1{,}203/2{,}324,\; p = 1$.
6. $\bar{x}_B = (1{,}211/12{,}867, 109/170, 93/170, 69/85, 15{,}366/8{,}855)^{\mathrm{T}} - (1{,}203/2{,}324)(2/11, 3/11, 1/11, -4/11, 83{,}204/183{,}049)^{\mathrm{T}}$ $= (0, 1/2, 1/2, 1, 3/2)^{\mathrm{T}}$.
7. $B \;= \{2, 4, 5, 6\},\; N = \{1, 3\}$.
8. $$L \;= \begin{pmatrix} 1{,}393/985 & & & \\ & -3 & & \\ 1{,}393/985 & 5/3 & -1{,}121/2{,}378 & \\ -2{,}378/1{,}121 & 2 & 985/1{,}393 & -1 \end{pmatrix}.$$
9. $k \;= 4$.

Iteration 4:

1. $\bar{y} \;= (0, -3/2, -3/2, -1)^{\mathrm{T}}$.
3. $k \;= 4$.
10. $\bar{z}_N = N^{\mathrm{T}}\bar{y} = (5/2, 2)^{\mathrm{T}} \geq 0$.
11. A 0-dimensional optimal face is obtained. The basic optimal solution and optimal value:

$$\bar{x} = (0, 1/2, 0, 1/2, 1, 3/2)^{\mathrm{T}}, \quad \bar{x}_6 = 3/2.$$

22.3 Generalizing the Face Method

In this section, we generalize Algorithm 22.1.3 to solve the bounded-variable reduced problem, i.e.,

$$
\begin{aligned}
&\min\ x_{n+1},\\
&\text{s.t.}\ \ Ax = b,\\
&\qquad l_j \le x_j \le u_j, \quad j = 1,\cdots,n,
\end{aligned}
\tag{22.23}
$$

where A, b are the same as in (22.1). The formula for computing the search direction Δ_B and the iteration scheme are also applicable here. As bound constraints are added, however, (22.16) for determining the stepsize α and index p has to be replaced by

$$\alpha = \alpha_p = \min\{\alpha_j \mid j \in B, j \ne n+1\}, \tag{22.24}$$

where

$$\alpha_j = \begin{cases} (\bar{x}_j - u_j)/\Delta_j, & \text{if } \Delta_j < 0, \\ (\bar{x}_j - l_j)/\Delta_j, & \text{if } \Delta_j > 0, \\ \infty, & \text{if } \Delta_j = 0, \end{cases} \quad j \in B,\ j \ne n+1. \tag{22.25}$$

In addition, the optimality should be tested by examining whether the following set is empty:

$$J' = \{j \in \Gamma \mid \bar{z}_j < 0\} \cup \{j \in \Pi \mid \bar{z}_j > 0\} \tag{22.26}$$

where Γ and Π are defined by (7.19).

The algorithm can be formulated as follows.

Algorithm 22.3.1 (Generalized face algorithm). Initial: (B, N), $m+1 \le k \le n+1$, feasible solution $\bar{x}$ and Cholesky factor L of BB^{T}. This algorithm solves bounded-variable reduced problem (22.23).

The same as Algorithm 22.1.3, except for steps 4,5,11 and 12, replaced respectively by

4. Compute α_j by (22.25).
5. Determine stepsize α and index p by (22.24).
11. Stop if J', defined by (22.26), is empty (optimality achieved).
12. Determine index $q \in \arg\max_{j \in J'} |\bar{z}_j|$.

Example 22.3.1. Solve the following problem by Algorithm 22.3.1:

$$
\begin{aligned}
\min\ \ & f = x_1 - x_3 + 2x_4, \\
\text{s.t.}\ \ & +3x_1 + x_2 - 5x_3 + x_4 = 3, \\
& -2x_1 + 6x_3 + x_5 = 5, \\
& -6x_1 - 3x_2 + 3x_3 - 2x_5 = -11, \\
& 1 \le x_1 \le 9, \quad -2 \le x_2 \le 8, \\
& 0 \le x_3 \le 12, \quad 0 \le x_4 \le 10, \quad 0 \le x_5 \le 15.
\end{aligned}
$$

Answer There is a feasible solution $(1,2,1,3,1)^{\mathrm{T}}$ with objective value 6.
Initial:

$$k=6, \quad B=\{1,2,3,4,5,6\}, \quad N=\emptyset,$$

feasible solution $\bar{x} = (1,2,1,3,1,6)^{\mathrm{T}}$. Face matrix B and the Cholesky factor of BB^{T}:

$$B = \begin{pmatrix} 3 & 1 & -5 & 1 & & \\ -2 & & 6 & & 1 & \\ -6 & -3 & 3 & & -2 & \\ 1 & & -1 & 2 & & -1 \end{pmatrix},$$

$$L = \begin{pmatrix} -6 & & & \\ 6 & -2{,}889/1{,}292 & & \\ 6 & 2{,}279/637 & 549/181 & \\ -5/3 & -2{,}584/2{,}889 & 1{,}267/915 & -1{,}001/816 \end{pmatrix}.$$

Iteration 1:

1. $\nu = -1{,}001/816, \bar{y} = (772/623, 468/623, 27/89, -414/623)^{\mathrm{T}}$.
2. $\Delta_B = (24/89, -205/623, 71/623, 8/89, -90/623, 209/623)^{\mathrm{T}} \neq 0$.
4. $\alpha_{1,2,3,4,5} = ((1-1)/(24/89), (8-2)/(-205/623), (1-0)/(71/623),$
   $(3-0)/(8/89), (1-10)/(-90/623))$.
5. $\alpha = 0,\ p=1$.
6. $\bar{x}_B = (1,2,1,3,1,6)^{\mathrm{T}}$.
7. $B = \{2,3,4,5,6\},\ N=\{1\}$.
8. $L = \begin{pmatrix} -1{,}351/260 & & & \\ 1{,}351/234 & -1{,}057/552 & & \\ 1{,}351/390 & 1{,}057/506 & -330/139 & \\ -260/193 & -843/908 & -2{,}128/1{,}401 & -251/249 \end{pmatrix}.$
9. $k = 5$.

Iteration 2:

1. $\nu = -251/249, \bar{y} = (124/63, 220/189, 17/27, -62/63)^{\mathrm{T}}$.
2. $\Delta_B = (-5/63, -1/63, 0, 2/21, 1/63)^{\mathrm{T}} \neq 0$.
4. $\alpha_{2,3,5} = ((2-8)/(-5/63), (1-12)/(-1/63), (1-0)/(2/21))$.
5. $\alpha = 21/2,\ p=5$.
6. $\bar{x}_B = (1,2,1,3,1,6)^{\mathrm{T}} + (21/2)(5/63, 1/63, 0, -2/21, -1/63)^{\mathrm{T}}$
   $= (17/6, 7/6, 3, 0, 35/6)^{\mathrm{T}}$.
7. $B = \{2,3,4,6\},\ N=\{1,5\}$.
8. $L = \begin{pmatrix} -1{,}351/260 & & & \\ 1{,}351/234 & -1{,}762/1{,}079 & & \\ 1{,}351/390 & 1{,}079/881 & 2{,}378/1{,}121 & \\ -260/193 & -749/688 & 1{,}393/985 & -1 \end{pmatrix}.$
9. $k = 4$.

Iteration 3:

1. $\nu = -1$, $\bar{y} = (124/63, 220/189, 17/27, -62/63)^{\mathrm{T}}$.
2. $k = 4$.
10. $\bar{z}_N = (4/3, 1/6)^{\mathrm{T}}$.
11. $J' = \emptyset$. The basic optimal solution and optimal value:

$$\bar{x} = (1, 17/6, 7/6, 3, 0, 35/6)^{\mathrm{T}}, \quad \bar{x}_6 = 35/6.$$

22.4 Affine Face Method

It is seen from (22.15) and (22.16) that in case when $\bar{x}_B$ is degenerate, α would vanish, and hence leading to a zero stepsize. It is more than that. It is known from (22.17) that the difference between the new and old objective values is

$$\hat{x}_{n+1} - \bar{x}_{n+1} = -\alpha \Delta_k.$$

Therefore, even though the first $k-1$ components of $\bar{x}_B$ are positive in principle, the stepsize would still be so small that the decrement of the objective value is negligible if some components are too close to zero. Actually, this depends on relative magnitudes of components. To avoid too small stepsizes, the affine interior-point method (Sect. 9.2) exploits affine transformation to attain a relative equilibrium between components. In this section, the affine transformation is employed to modify the search direction, used in the face method, so that it is not only a downhill but also points to the interior of the feasible region in some sense.

Assume that $\bar{x}_B$ is nondegenerate. Denote by $\bar{X}_B$ the $k \times k$ diagonal matrix, whose diagonals are components of vector $\bar{x}_B$, except for the kth diagonal 1, i.e.,

$$\bar{X}_B = \mathrm{diag}(\bar{x}_1, \cdots, \bar{x}_{k-1}, 1). \tag{22.27}$$

Consider affine transformation

$$x_B = \bar{X}_B x'_B. \tag{22.28}$$

This transformation is invertible, and determines an 1–1 correspondence between $\bar{x}_B$ and

$$\bar{x}'_B = (\underbrace{1, \cdots, 1}_{k-1}, \bar{x}_{n+1})^{\mathrm{T}}.$$

Despite some of the first $k - 1$ components of $\bar{x}_B$ may be close to zero, the corresponding components of $\bar{x}'_B$ are all equal to 1, so that a line search along any direction in $\bar{x}'_B$-space will not lead to too small stepsize.

By carrying out transformation (22.28), the subprogram (22.4) is converted to

$$\begin{aligned} &\min\ u_{n+1}, \\ &\text{s.t.}\ \ (B\bar{X}_B)x'_B = b, \\ &\qquad u_j \geq 0, \quad j \in B,\ j \neq n+1. \end{aligned}$$

The projection matrix onto the null space of $B\bar{X}_B$ is then

$$P = I - \bar{X}_B B^{\mathrm{T}}(B\bar{X}_B^2 B^{\mathrm{T}})^{-1} B\bar{X}_B. \tag{22.29}$$

Thereby, the orthogonal projection of the negative objective gradient onto the null space is $-Pe_k$, corresponding to vector

$$\Delta_B = -\bar{X}_B P e_k$$

in x_B-space. Thereby, combining

$$\bar{X}_B e_k = e_k,\ Be_k = -e_{m+1},$$

and (22.29) gives that search direction below:

$$\Delta_B = -e_k + \bar{X}_B^2 B^{\mathrm{T}}\bar{y}, \quad (B\bar{X}_B^2 B^{\mathrm{T}})\bar{y} = -e_{m+1}. \tag{22.30}$$

If the Cholesky factorization $B\bar{X}_B^2 B^{\mathrm{T}} = LL^{\mathrm{T}}$ is available, the $\bar{y}$ can be obtained by solving an upper triangular system only, i.e.,

$$L^{\mathrm{T}}\bar{y} = -(1/\nu)e_{m+1},$$

where ν is the $(m+1)$the diagonal of L.

If $\bar{x}$ is degenerate, nevertheless, transformation (22.28) is not invertible, and becomes useless. This difficulty may be overcome as follows. Let δ be a given small positive number. Define

$$\hat{x}_j = \begin{cases} \bar{x}_j, & \text{If } \bar{x}_j \geq \delta, \\ 1, & \text{If } \bar{x}_j < \delta, \end{cases} \quad j \in B,\ j \neq n+1, \tag{22.31}$$

Then construct an affine transformation with $\hat{x}_B$ in place of $\bar{x}_B$, i.e., by using

$$\bar{X}_B = \mathrm{diag}(\hat{x}_1, \cdots, \hat{x}_{k-1}, 1). \tag{22.32}$$

instead of (22.27).

The difference between the resulting algorithm and the face algorithm lies in search direction only. The following algorithm is obtained by slightly modifying Algorithm 22.1.3.

Algorithm 22.4.1 (Affine face algorithm). Given $0 < \delta \ll 1$. Initial: feasible solution $\bar{x}$. This algorithm solves the reduced problem (22.1).

1. Set $k = n+1$, $B = A$, $N = \emptyset$ and compute Cholesky factorization $B\bar{X}_B^2 B^{\mathrm{T}} = LL^{\mathrm{T}}$, where $\bar{X}_B$ is defined by (22.32).
2. Solves $L^{\mathrm{T}}\bar{y} = -(1/\nu)e_{m+1}$, where ν is the $(m+1)$th diagonal of L.
3. Compute $\Delta_B = -e_k + \bar{X}_B^2 B^{\mathrm{T}}\bar{y}$.
4. Go to step 11 if $\Delta_B = 0$.
5. Stop if $J = \{j \in B \mid \Delta_j < 0,\ j \neq n+1\} = \emptyset$ (lower unbounded problem).
6. Determine stepsize α and index p such that $\alpha = -\bar{x}_p/\Delta_p = \min_{j\in J} -\bar{x}_j/\Delta_j$.
7. If $\alpha \neq 0$, update: $\bar{x}_B = \bar{x}_B + \alpha\Delta_B$.
8. Update (B, N) by bringing p from B to N.
9. Update L by Algorithm 22.1.1.
10. Set $k = k-1$; if $k > m+1$, go to step 2.
11. Compute $\bar{z}_N = -N^{\mathrm{T}}\bar{y}$.
12. Stop if $\bar{z}_N \geq 0$ (optimality achieved).
13. Go to step 1.

Note This Algorithm contains steps 2–10 as its inner steps.

To reduce effects of degeneracy in practice, it seems to be favorable to replace (22.31) alternatively by

$$\hat{x}_j = \begin{cases} \bar{x}_j, & \text{if } \bar{x}_j \geq \delta_j, \\ \delta_j, & \text{if } \bar{x}_j < \delta_j, \end{cases} \qquad j \in B,\ j \neq n+1.$$

where $\delta_j > 0$, $n+1 \neq j \in B$ are different small numbers conformable with magnitude of the first $k-1$ components of $\bar{x}_B$.

Example 22.4.1. Solve the following problem by Algorithm 22.4.1, starting from feasible solution $(0,0,3,2,4)^{\mathrm{T}}$:

$$\begin{array}{llllllll}
\min & x_1 & +5x_2 & -2x_4. & & & & \\
\text{s.t.} & & +3x_2 & +\ x_3 & & -\ x_5 & = & -1, \\
 & -3x_1 & -\ x_2 & & +3x_4 & -2x_5 & = & -2, \\
 & x_1 & -2x_2 & +2x_3 & & +\ x_5 & = & 10, \\
 & & x_j \geq 0, & j = 1,\cdots,5. & & & &
\end{array}$$

Answer Take $\delta = 10^{-6}$. Initial: feasible solution $\bar{x} = (0,0,3,2,4,-4)^{\mathrm{T}}$.

Iteration 1:

1. $k = 6$, $B = \{1,2,3,4,5,6\}$, $N = \emptyset$. $\bar{X}_B = \mathrm{diag}(1,1,3,2,4,1)^{\mathrm{T}}$. Face matrix B and the Cholesky factor of $B\bar{X}_B^2 B^{\mathrm{T}}$ are respectively,

$$B = \begin{pmatrix} & 3 & 1 & & -1 & \\ -3 & -1 & & 3 & -2 & \\ 1 & -2 & 2 & & 1 & \\ 1 & 5 & & -2 & & -1 \end{pmatrix},$$

$$L = \begin{pmatrix} -2{,}449/420 & & & \\ -5{,}809/1{,}168 & 3{,}869/419 & & \\ 769/1{,}121 & -4{,}313/1{,}346 & -3{,}394/499 & \\ -1{,}402/545 & -14{,}233/2{,}934 & 2{,}753/822 & -1{,}941/1{,}519 \end{pmatrix}.$$

2. $\bar{y} = (827/1{,}382, -1{,}205/2{,}826, -269/892, -2{,}737/4{,}469)^{\mathrm{T}}$.
3. $\Delta_B = (451/1{,}235, -487/2{,}051, -35/822, -635/2{,}923, -576/763, -1{,}732/4{,}469)^{\mathrm{T}} \neq 0$.
5. $J = \{2, 3, 4, 5\} \neq \emptyset$.
6. $\alpha = 0,\ p = 2$.
8. $B = \{1, 3, 4, 5, 6\},\ N = \{2\}$.
9. $L = \begin{pmatrix} -5 & & & \\ -32/5 & 3{,}019/366 & & \\ -2/5 & -2{,}682/589 & -4{,}040/713 & \\ & -1{,}653/505 & 3{,}573/1{,}456 & 2{,}273/2{,}022 \end{pmatrix}.$
10. $k = 5$.

Iteration 2:

2. $\bar{y} = (1{,}977/2{,}944, -1{,}095/2{,}176, -73/213, -1{,}517/1{,}917)^{\mathrm{T}}$.
3. $\Delta_B = (80/213, -80/639, 560/1{,}917, -80/639, -400/1{,}917)^{\mathrm{T}} \neq 0$.
5. $J = \{3, 5\} \neq \emptyset$.
6. $\alpha = \min\{-3/(-80/639), -4/(-80/639)\} = 1{,}917/80,\ p = 3$.
7. $\bar{x}_B = (0, 3, 2, 4, -4)^{\mathrm{T}} + (1{,}917/80)(80/213, -80/639, 560/1{,}917, -80/639, -400/1{,}917)^{\mathrm{T}} = (9, 0, 9, 1, -9)^{\mathrm{T}}$.
8. $B = \{1, 4, 5, 6\},\ N = \{2, 3\}$.
9. $L = \begin{pmatrix} -4 & & & \\ -8 & -2{,}207/329 & & \\ 4 & 1{,}292/2{,}889 & 2{,}584/2{,}889 & \\ & 1{,}292/321 & -2{,}584/2{,}889 & -1 \end{pmatrix}.$
10. $k = 4 = m + 1$.
11. $\bar{z}_N = (1{,}097/1{,}456, 80/5{,}751)^{\mathrm{T}} \geq 0$.
12. The 0-dimension optimal face (basic optimal solution) and optimal value:

$$\bar{x} = (9, 0, 0, 9, 1, -9)^{\mathrm{T}}, \qquad \bar{x}_6 = -9.$$

22.5 Notes

The quality of an initial partition (B, N) seems to be important to the face algorithm. Although initially setting $B = A$ would be a convenient choice, a small B should be favorable, ideally containing the right (unknown optimal) face columns. Therefore, it might be advisable to determine an initial B, whose cardinality is no more than $2m$, especially for large-scale problems with $n \gg m$, by using a crash procedure based on the most-obtuse-angle heuristics, balanced with sparsity considerations (Sect. 5.5).

As was reported, the associated computational results with the face algorithm are remarkable though preliminary. This outcome is not surprising because it uses the steepest downhill as search direction, and involves less computational effort per iteration mainly due to fewer linear systems involved than the simplex method.

Based on the "dual elimination" (Sect. 25.1.3), it is possible to design another type of method in the following way. By a series of elementary transformations with row exchanges, the standard dual problem (4.2) can be deduced to a problem of form (25.6), i.e.,

$$\begin{array}{ll} \min & -g = -\bar{g} + \bar{x}^{\mathrm{T}} z, \\ \text{s.t.} & G_2 z = d_2, \quad z \geq 0, \end{array} \tag{22.33}$$

together with $Uy + G_1 z = d_1$. After the preceding standard problem is converted to the reduced form, face Algorithm 22.1.3 and affine face Algorithm 22.4.1 become applicable to achieve optimality by generating a sequence of *dual* feasible solutions.

Gill and Murray (1973) use the Cholesky factorization $BB^T = LL^T$, where B is the usual square basis, to improve the numerical stability of the simplex algorithm. Based on the fact that BB^T and hence L is independent of the ordering of the columns of B, Saunders (1972) proposes an approach to keep L sparse throughout the solution process. It initially carries out row and column permutations to turn the coefficient matrix A to a lower-triangular form as much as possible, and then uses the resulting row permutation in subsequent iterations. This approach is clearly applicable in the context of the face method.

Now let us turn to a just published result by Zhang et al. (2013), an alternative approach to the face algorithm with favorable computational results. It updates $(BB^T)^{-1}$ by the Sherman-Morrison formula to create the search directions. The approach appears to be attractive for solving large-scale sparse problems, as it bypasses updating of Cholesky factors, and hence involves no orthogonal transformations, except for the initial iteration.

In the following introduction, we will modify it sightly by computing the search directions more directly.

Let $B = (a_{j_1}, \cdots, a_{j_k})$ be the face matrix. It is noted that BB^T is invariant to B's column permutation.

Assume that B resulted from adding column a_{j_t}, $t \in \{1, \cdots, k\}$ to the predecessor face matrix B', in a face expansion iteration. Then it holds that

$$BB^T = \sum_{r=1,\cdots,k} a_{j_r} a_{j_r}^T = \left(\sum_{\substack{r=1,\cdots,k \\ r \neq t}} a_{j_r} a_{j_r}^T \right) + a_{j_t} a_{j_t}^T = B' B'^T + a_{j_t} a_{j_t}^T.$$

By Sherman-Morrison formula (See, e.g., Golub and Van Loan 1989), it follows that

$$(BB^T)^{-1} = (B'B'^T)^{-1} - \frac{(B'B'^T)^{-1}a_{j_t}a_{j_t}^T(B'B'^T)^{-1}}{1 + a_{j_t}^T(B'B'^T)^{-1}a_{j_t}} = (B'B'^T)^{-1} - \frac{hh^T}{1 + a_{j_t}^T h}, \tag{22.34}$$

where

$$h = (B'B'^T)^{-1}a_{j_t}.$$

If, otherwise, B is yielded from B' by dropping column a_{j_t} in a face contraction iteration, it is easily known that

$$(BB^T)^{-1} = (B'B'^T)^{-1} + \frac{hh^T}{1 - a_{j_t}^T h}, \tag{22.35}$$

Let B_1 be the initial face matrix. For iterations $l = 2, 3, \cdots$, denote again by a_{j_t} the column entering or leaving the face matrix B_{l-1}. Based on (22.34) and (22.35), the updating formula of $(B_l B_l^T)^{-1}$ can be written uniformly, i.e.,

$$(B_l B_l^T)^{-1} = (B_{l-1} B_{l-1}^T)^{-1} - \text{sign}(\eta_{l-1})\frac{h_{l-1}h_{l-1}^T}{1 + \eta_{l-1}}, \tag{22.36}$$

where

$$h_{l-1} = (B_{l-1} B_{l-1}^T)^{-1}a_{j_t}, \qquad \eta_{l-1} = \begin{cases} a_{j_t}^T h_{l-1} > 0 \text{ if } a_{j_t} \text{ enters} \\ -a_{j_t}^T h_{l-1} < 0 \text{ if } a_{j_t} \text{ leaves} \end{cases}$$

From the preceding and the second formula of (22.6), the update formula follows:

$$\bar{y}_l = -(B_l B_l^T)^{-1}e_{m+1} = \bar{y}_{l-1} + \text{sign}(\eta_{l-1})\frac{h_{l-1,m+1}h_{l-1}}{1 + \eta_{l-1}}, \tag{22.37}$$

where $h_{l-1,m+1}$ denotes the $(m+1)$th component of h_{l-1}. Assume that the first $l-1$ iterations were carried out, and hence $\{h_i, \eta_i\}, i = 1, \cdots, l-1$ were available. After $\bar{y}_l$ is obtained via (22.37), consequently we are able to compute the search direction by the first formula of (22.6), i.e.,

$$\Delta_{B_l} = e_k - B_l^{\mathrm{T}}\bar{y}_l. \tag{22.38}$$

Successively using (22.36) for $l, \cdots, 2$ leads to

$$(B_l B_l^T)^{-1} = (B_{l-2} B_{l-2}^T)^{-1} - \text{sign}(\eta_{l-2})\frac{h_{l-2}h_{l-2}^T}{1+\eta_{l-2}} - \text{sign}(\eta_{l-1})\frac{h_{l-1}h_{l-1}^T}{1+\eta_{l-1}}$$

$$\vdots$$

$$= (B_1 B_1^T)^{-1} - \sum_{i=1}^{l-1} \text{sign}(\eta_i)\frac{h_i h_i^T}{1+\eta_i}.$$

If a_q is selected to enter the face matrix at the current iteration, thereby, we can compute

$$h_l = (B_l B_l^T)^{-1} a_q = (B_1 B_1^T)^{-1} a_q - \sum_{i=1}^{l-1} \text{sign}(\eta_{l-1})\frac{(h_{l-1}^T a_q)h_{l-1}}{1+\eta_{l-1}} \tag{22.39}$$

$$\eta_l = \begin{cases} a_q^T h_l > 0 \text{ if } a_q \text{ enters} \\ -a_q^T h_l < 0 \text{ if } a_q \text{ leaves} \end{cases} \tag{22.40}$$

so that (h_i, η_i) can be produced for $i = 1, 2, \cdots$, iteration by iteration, using the same initial Cholesky factorization

$$B_1 B_1^T = (LL^T).$$

Chapter 23
Dual Face Method

The same idea of the face method can be applied to the dual problem to derive a dual variant. The resulting method seems to be even more efficient than its primal counterpart.

23.1 Dual Face Method Using Cholesky Factorization

Using notation

$$A := \begin{pmatrix} A & -b \\ 0 & 1 \end{pmatrix}, \quad y := \begin{pmatrix} y \\ y_{m+1} \end{pmatrix}, \quad z := \begin{pmatrix} z \\ z_{n+1} \end{pmatrix}, \quad c := \begin{pmatrix} c \\ 0 \end{pmatrix},$$

we turn the standard dual problem to

$$\begin{array}{ll} \max & y_{m+1}, \\ \text{s.t.} & A^{\mathrm{T}} y + z = c, \quad z \geq 0, \end{array} \tag{23.1}$$

where $A \in \mathcal{R}^{(m+1)\times(n+1)}$, $c \in \mathcal{R}^{(n+1)}$, rank $A = m + 1$, $m < n$, and

$$Ae_{n+1} = (-b^{\mathrm{T}}, 1)^{\mathrm{T}}, \quad e_{m+1}^{\mathrm{T}} A = e_{n+1}^{\mathrm{T}}. \tag{23.2}$$

Assume that this problem is feasible.

Let (B, N) be a partition of matrix A, where $B \in \mathcal{R}^{(m+1)\times k}$ and

$$\text{rank } B = k, \qquad 1 \leq k \leq m + 1.$$

So, B is of full column rank. The related variables are respectively termed *dual face* and *dual nonface* variables, and matrices termed *dual face* and *dual nonface* matrices.

P.-Q. PAN, *Linear Programming Computation*, DOI 10.1007/978-3-642-40754-3_23,

As $z_{n+1} = 0$ is fixed throughout the solution process, it will be possible to keep the $(n + 1)$th (last) column of A to be as the kth (last) column of B to simplify computation, as was in the face algorithm. Thus, it holds that

$$Be_k = (-b^{\mathrm{T}}, 1)^{\mathrm{T}}, \quad e_{m+1}^{\mathrm{T}} B = e_k^{\mathrm{T}}, \quad e_{m+1}^{\mathrm{T}} N = 0. \tag{23.3}$$

Partition (B, N) corresponds to dual face

$$D_N = \{(y, z) \in \mathcal{R}^m \times \mathcal{R}^n \mid A^{\mathrm{T}} y + z = c, \ z_B = 0, \ z_N \geq 0\}. \tag{23.4}$$

A point on a dual face is termed (dual) face point. 0-dimensional face ($k = m + 1$) has a unique dual face point – vertex. A face is said to be *(dual) level face* if the (dual) objective value is constant over it, and said to be *dual optimal* if the constant is equal to the optimal value.

For simplicity, thereafter $(y, z_N) \in D_N$ will often be used in place of $(y, z) \in D_N$.

23.1.1 The Steepest Uphill

It might be well to assume that $B = \{1, \cdots, k-1, n+1\}$. Consider the following subprogram:

$$\begin{array}{ll} \max & y_{m+1}, \\ \text{s.t.} & B^{\mathrm{T}} y = c_B, \\ & N^{\mathrm{T}} y + z_N = c_N, \quad z_N \geq 0, \end{array} \tag{23.5}$$

whose feasible region is face D_N.

Using $(m + 1) \times (m + 1)$ projection

$$P = I - B(B^{\mathrm{T}} B)^{-1} B^{\mathrm{T}}, \tag{23.6}$$

the objective gradient, e_{m+1}, of the subprogram (23.5) is projected to the null of B^{T}, yielding the following search direction:

$$\Delta y = P e_{m+1} = e_{m+1} - B \bar{x}_B, \quad B^{\mathrm{T}} B \bar{x}_B = e_k, \tag{23.7}$$

$$\Delta z_N = -N^{\mathrm{T}} \Delta y. \tag{23.8}$$

Proposition 23.1.1. *For Δy defined by (23.7), it holds that $B^{\mathrm{T}} \Delta y = 0$.*

Proof. By (23.7), it is easy to verify the validity of this equation. □

Proposition 23.1.2. *The following statements are equivalent*

$$(i)\ \Delta y \neq 0; \qquad (ii)\ (\Delta y)_{m+1} > 0; \qquad (iii)\ e_{m+1} \notin \text{range}(B).$$

Proof. Premultiplying the fist expression of (23.7) by Δy^{T}, then from Proposition 23.1.1 it follows that

$$0 \leq \Delta y^{\mathrm{T}} \Delta y = (\Delta y)_{m+1},$$

which implies equivalence of (i) and (ii). If $\Delta y = 0$, then it is known by the fist expression of (23.7) that $e_{m+1} = B\bar{x}_B$, hence $e_{m+1} \in \text{range}(B)$, so (iii) implies (i). Conversely, if $e_{m+1} \in \text{range}(B)$, then there is some vector u such that

$$e_{m+1} = Bu. \tag{23.9}$$

Premultiply the preceding equality by B^{T}, then from (23.3) it follows that

$$e_k = B^{\mathrm{T}} Bu,$$

combining which and the second expression of (23.7) and B of full column rank gives $u = \bar{x}_B$. Thus, (23.9) implies

$$e_{m+1} = B\bar{x}_B.$$

Substituting the preceding equality to the fist expression of (23.7) leads to $\Delta y = 0$. So (i) implies (iii). Further, it can be asserted that (i) and (iii) are equivalent. □

Proposition 23.1.3. *If $\Delta y \neq 0$, then*

$$e_{m+1}^{\mathrm{T}} \Delta y = (\Delta y)_{m+1} > 0, \tag{23.10}$$

$$e_{m+1}^{\mathrm{T}} \Delta y / \|\Delta y\| \geq e_{m+1}^{\mathrm{T}} u / \|u\|, \quad \forall\, 0 \neq u \in \text{Null}(B^{\mathrm{T}}). \tag{23.11}$$

Proof. Equality (23.10) directly follows from Proposition 23.1.2. For P defined by (23.6), it holds that

$$Pu = u \neq 0, \quad \forall\, 0 \neq u \in \text{Null}(B^{\mathrm{T}}). \tag{23.12}$$

It is known from Cauchy inequality that

$$\|Pe_{m+1}\| \|Pu\| \geq (Pe_{m+1})^{\mathrm{T}} (Pu),$$

Multiplying the preceding inequality by $1/\|Pu\|$ renders

$$\|Pe_{m+1}\| \geq (Pe_{m+1})^{\mathrm{T}} (Pu)/\|Pu\|,$$

i.e.,

$$(Pe_{m+1})^{\mathrm{T}}(Pe_{m+1})/\|Pe_{m+1}\| \geq (Pe_{m+1})^{\mathrm{T}}(Pu)/\|Pu\|,$$

combining which, $P^2 = P,\ P^{\mathrm{T}} = P$, (23.12) and $\Delta y = Pe_{m+1}$ leads to (23.11). □

The preceding Proposition implies that $\Delta y \neq 0$ forms the most acute angle with objective gradient e_{m+1} of the subprogram, in the null of B^{T}. It is thereby termed *the steepest uphill.* In this sense, taking it as a search direction in y-space is the "best" choice.

To compute the search direction $(\Delta y, \Delta z_N)$, defined by (23.7) and (23.8), it is needed to solve $k \times k$ system

$$B^{\mathrm{T}}Bx_B = e_k. \tag{23.13}$$

Assume that the QR factorization of B is

$$B = QR_1 = Q\begin{pmatrix} L^{\mathrm{T}} \\ 0 \end{pmatrix}, \tag{23.14}$$

where $Q \in \mathcal{R}^{(m+1)\times(m+1)}$ is orthogonal, $L \in\in \mathcal{R}^{k\times k}$ is nonsingular lower triangular. Then it is verified that

$$B^{\mathrm{T}}B = LL^{\mathrm{T}}.$$

Thus, the L is just the Cholesky factor of $B^{\mathrm{T}}B$. Thereby, system (23.13) becomes

$$LL^{\mathrm{T}}x_B = e_k,$$

whose solution can be obtained by solving the following two triangular systems:

$$Lu = e_k, \quad L^{\mathrm{T}}x_B = u.$$

As the solution to the first system is $v = (1/\mu)e_k$, where μ is the kth diagonal of L, consequently only the following lower triangular system needs to be solved:

$$L^{\mathrm{T}}x_B = (1/\mu)e_k. \tag{23.15}$$

23.1.2 Updating Dual Solution

Assume that $(\bar{y}, \bar{z}_N) \in D_N$ and $\Delta y \neq 0$. The new dual solution is determined by the following line search scheme:

$$\hat{y} = \bar{y} + \beta \Delta y, \quad \hat{z}_N = \bar{z}_N - \beta N^{\mathrm{T}} \Delta y, \tag{23.16}$$

where the largest possible stepsize, maintaining $\bar{z}_N \geq 0$, is

$$\beta = \bar{z}_q/(a_q^{\mathrm{T}} \Delta y) = \min\{\bar{z}_j/(a_j^{\mathrm{T}} \Delta y) \mid a_j^{\mathrm{T}} \Delta y > 0, \ j \in N\} \geq 0. \tag{23.17}$$

If some components of $\bar{z}_N$ vanish, then dual face point $(\bar{y}, \bar{z}_N)$ is said to be *dual degenerate*, as is a case where stepsize β vanishes, yielding a point the same as the old.

Lemma 23.1.1. *Assume that $(\bar{y}, \bar{z}_N) \in D_N$ with $\Delta y \neq 0$.*

(i) If $N^{\mathrm{T}} \Delta y \not\leq 0$, then $\hat{z}_N$ is a boundary point, belonging to D_N, The objective value does not decrease, and strictly increases if $(\bar{y}, \bar{z}_N)$ is nondegenerate.
(ii) If $N^{\mathrm{T}} \Delta y \leq 0$, problem (23.1) is upper unbounded over D_N.

Proof. (i) In this case, the new point $(\hat{y}, \hat{z}_N)$ is well-defined. By $(\bar{y}, \bar{z}_N) \in D_N$, (23.17) and Proposition 23.1.1, it is known that $(\hat{y}, \hat{z}_N) \in D_N$. In addition, it is clear that $\hat{z}_q = 0$, hence the new iterate is on the boundary, associated with the objective value

$$\hat{y}_{m+1} = \bar{y}_{m+1} + \beta (\Delta y)_{m+1}. \tag{23.18}$$

Further, from $\beta \geq 0$ and Proposition 23.1.2, it is known that $(\Delta y)_{m+1} > 0$, hence $\hat{y}_{m+1} \geq \bar{y}_{m+1}$. If $(\bar{y}, \bar{z}_N)$ is nondegenerate, then $\beta > 0$, hence $\hat{y}_{m+1} > \bar{y}_{m+1}$, as indicates that the objective value increases strictly.

(ii) In this case, it is clear that $(\hat{y}, \hat{z}_N) \in D_N \ \forall \beta > 0$. It is then known by (23.18) and Proposition 23.1.2 that the objective value tends to ∞, as β tends to ∞. □

23.1.3 *Dual Face Contraction*

Assume that $N^{\mathrm{T}} \Delta y \not\leq 0$, in which case $\hat{z}_p = 0$. Update (B, N) by bring q from N to B. Denoting the new face matrix by $\breve{B} \in \mathcal{R}^{(m+1)\times(k+1)}$, we have the following result.

Proposition 23.1.4. *If* rank $B = k$, *then* rank $\breve{B} = k + 1$.

Proof. It is clear that $k \leq \text{rank } \breve{B} \leq k + 1$. Assume rank $\breve{B} \neq k + 1$. Then, there is a vector $u \neq 0$ such that $a_q = Bu$. Thus, it is known by Proposition 23.1.1 that $B^{\mathrm{T}} \Delta y = 0$, hence

$$a_q^{\mathrm{T}} \Delta y = u^{\mathrm{T}} B^{\mathrm{T}} \Delta y = 0.$$

Further, it follows from (23.17) that $a_q^{\mathrm{T}} \Delta y > 0$, as is a contradiction. Therefore rank $\breve{B} = k + 1$. □

According to the preceding Proposition, there exists the Cholesky factorization

$$\breve{B}^{\mathrm{T}}\breve{B} = \breve{L}\breve{L}^{\mathrm{T}}.$$

For sake of simplicity, it is advisable to insert a_q as the second last column of $\breve{B} \in \mathcal{R}^{(m+1)\times(k+1)}$ (followed by the last column $(-b^{\mathrm{T}}, 1)^{\mathrm{T}}$). Then, the new Cholesky factor can be computed by the following algorithm.

Algorithm 23.1.1. Assume that the QR factorization of B is given by (23.14). This algorithm finds the Cholesky factor $\breve{L}$ of $\breve{B}^{\mathrm{T}}\breve{B}$.

1. Compute $v = Q^{\mathrm{T}}a_q$.
2. Insert v before the last column of R_1, yielding matrix H.
3. Determine Givens rotation $J_{m+1}, \cdots, J_{k+1}$, where J_i is in (k, i)-plane, such that

$$J_{k+1}\cdots J_{m+1}H = \begin{pmatrix} \breve{L}^{\mathrm{T}} \\ 0 \end{pmatrix},$$

where $\breve{L} \in \mathcal{R}^{(k+1)\times(k+1)}$ is lower triangular.

Validity of Algorithm 23.1.1 can be verified as follows:
It is clear that

$$\begin{aligned} J_{k+1}\cdots J_{m+1}Q^{\mathrm{T}}\breve{B} &= J_{k+1}\cdots J_{m+1}Q^{\mathrm{T}}[a_1, \cdots, a_{k-1}, a_q, a_{n+1}] \\ &= J_{k+1}\cdots J_{m+1}H = \begin{pmatrix} \breve{L}^{\mathrm{T}} \\ 0 \end{pmatrix}. \end{aligned}$$

By Proposition 23.1.4, it is known that $\breve{L} \in \mathcal{R}^{(k+1)\times(k+1)}$ is nonsingular. Thus, from the preceding expression, it follows that

$$\breve{B}^{\mathrm{T}}\breve{B} = \breve{B}^{\mathrm{T}}QJ_{m+1}^{\mathrm{T}}\cdots J_{k+1}^{\mathrm{T}}J_{k+1}\cdots J_{m+1}Q^{\mathrm{T}}\breve{B} = (\breve{L}, 0)\begin{pmatrix} \breve{L}^{\mathrm{T}} \\ 0 \end{pmatrix} = \breve{L}\breve{L}^{\mathrm{T}}.$$

It is seen from step 1 of Algorithm 23.1.1 that it is now required to store the Givens rotations in product form, in contrast to the updating of the Cholesky factor in the face algorithm. For large-scale problems, therefore, the dual face algorithm should be restarted from scratch periodically.

23.1.4 Optimality Test

Assume that Δy vanishes, and can not be used as a search direction, e.g., when $k = m + 1$.

Lemma 23.1.2. *Assume that Δy is defined by (23.7). If $\Delta y = 0$, then D_N is a dual level face; and vice versa if there is $(\bar{y}, \bar{z}_N) \in D_N$ such that $\bar{z}_N > 0$.*

Proof. Assume that $\bar{x}_B$ is the solution to $B^{\mathrm{T}}Bx_B = e_k$. From $\Delta y = 0$ and the first expression of (23.7), it follows that

$$e_{m+1} = B\bar{x}_B.$$

Transposing the two sides of the preceding equality and postmultiplying the result by y (for any $(y, z_N) \in D_N$) gives

$$e_{m+1}^{\mathrm{T}}y = \bar{x}_B^{\mathrm{T}}B^{\mathrm{T}}y,$$

combining which and $B^{\mathrm{T}}y = c_B$ leads to

$$y_{m+1} = \bar{x}_B^{\mathrm{T}}c_B.$$

So the objective value is constant over D_N. Therefore, D_N is a level face.

Now assume that D_N is a level face but $\Delta y \neq 0$. If $N^{\mathrm{T}}\Delta y \leq 0$, it is known by Lemma 23.1.1(ii) that the objective value is upper unbounded over D_N, as contradicts that it is a level face; if, otherwise, $N^{\mathrm{T}}\Delta y \not\leq 0$, then since $(\bar{y}, \bar{z}_N)$ is nondegenerate, it is known from Lemma 23.1.1(i) that $(\hat{y}, \hat{z}_N)$, defined by (23.16) and (23.17), belongs to D_N, with a strictly increased objective value, as contradicts that D_N is a level face. Therefore $\Delta y = 0$. □

Now consider the primal problem of (23.1), i.e.,

$$\begin{array}{ll} \min & c^{\mathrm{T}}x, \\ \text{s.t.} & Ax = e_{m+1}, \quad x \geq 0. \end{array} \tag{23.19}$$

Theorem 23.1.3. *Assume that $(\bar{y}, \bar{z}_n) \in D_N$ and that $\Delta y = 0$ and $\bar{x}_B$ are determined by (23.7). If $\bar{x}_B \geq 0$, then $\bar{x}$ and $(\bar{y}, \bar{z})$ are a pair of primal and dual optimal solutions, and D_N is a dual optimal face to (23.1), with $\bar{x}_N = 0,\ \bar{z}_B = 0$.*

Proof. From (23.7) and $\Delta y = 0$, it follows that $e_{m+1} = B\bar{x}_B$. Also, it is easy to verify that $\bar{x}$ ($\bar{x}_N = 0$) is a feasible solution to (23.19). On the other hand, it is clear that $(\bar{y}, \bar{z})$ ($\bar{z}_B = 0$) is a feasible solution to (23.1). The two solutions exhibit slackness complementarity. Therefore, they are a pair of primal and dual optimal solutions. By Lemma 23.1.2, D_N is a (dual) level face. In fact, it a (dual) optimal face since including an optimal solution. □

23.1.5 Dual Face Expansion

If $\Delta y = 0$ but $\bar{x}_B \not\geq 0$, then it can not be declared that the level dual face D_N is optimal. In this case, determine index

$$p \in \arg\min_{j \in B} \bar{x}_j, \tag{23.20}$$

and update (B, N) by

$$\hat{B} = B \backslash \{p\}, \quad \hat{N} = N \cup \{p\}. \tag{23.21}$$

Proposition 23.1.5. *The index p determined by (23.20) is different from n + 1.*

Proof. It is know by (23.15) that $\bar{x}_{n+1} = (1/\mu^2) > 0$. As $\min_{j \in B} \bar{x}_j < 0$, it is not possible to select $n+1$ to be as p. □

Assume that $\hat{B} \in \mathcal{R}^{(m+1)\times(k-1)}$ is the matrix, resulting from dropping the pth column of B. Clearly, it is still of full column rank. The associated Cholesky factor can be obtained by the following steps.

Algorithm 23.1.2. Initial: Cholesky factorization $B^{\mathrm{T}}B = LL^{\mathrm{T}}$. This algorithm generates the Cholesky factor $\hat{L}$ of $\hat{B}\hat{B}^{\mathrm{T}}$.

1. Drop the pth column from R_1, yielding Hessenberg matrix H.
2. Determine Givens rotations $J_p, \cdots, J_{k-1}$, where J_i is in $(i, i+1)$-plane, such that

$$J_{k-1} \cdots J_p H = \begin{pmatrix} \hat{L}^{\mathrm{T}} \\ 0 \end{pmatrix},$$

where $\hat{L} \in \mathcal{R}^{(k-1)\times(k-1)}$ is lower triangular.

Validity of Algorithm 23.1.2 can be verified simply: since

$$\begin{aligned} J_{k-1} \cdots J_p Q^{\mathrm{T}} \hat{B} &= J_{k-1} \cdots J_p Q^{\mathrm{T}} [a_1, \cdots, a_{p-1}, a_{p+1}, \cdots, a_{n+1}] \\ &= J_{k-1} \cdots J_p H = \begin{pmatrix} \hat{L}^{\mathrm{T}} \\ 0 \end{pmatrix}, \end{aligned}$$

it holds that

$$\hat{B}^{\mathrm{T}} \hat{B} = \hat{B}^{\mathrm{T}} Q J_p^{\mathrm{T}} \cdots J_{k-1}^{\mathrm{T}} J_{k-1} \cdots J_p Q^{\mathrm{T}} \hat{B} = (\hat{L}, 0) \begin{pmatrix} \hat{L}^{\mathrm{T}} \\ 0 \end{pmatrix} = \hat{L}\hat{L}^{\mathrm{T}}.$$

A seemingly attractive way to expand the face is to replace (23.21) by

$$\hat{B} = B \backslash \{j \in B \mid \bar{x}_j < 0\}, \quad \hat{N} = A \backslash \hat{B}.$$

Unfortunately, updating of the associated Cholesky factor becomes cumbersome.

23.1.6 Dual Face Algorithm

To be an initial dual face matrix, a submatrix $B \in \mathcal{R}^{(m+1)\times k}$, $1 \le k \le m+1$ with small k seem to be preferable, though any one of full column rank is eligible.

The overall steps can be summarized into the following algorithm.

Algorithm 23.1.3 (Dual face algorithm). Initial: $B,\ N,\ 1 \leq k \leq m+1$; dual feasible solution $(\bar{y}, \bar{z}_N)$ and Cholesky factor L of $B^{\mathrm{T}}B$. This algorithm solves the pair of reduced problems (23.1) and (23.19).

1. Solve $L^{\mathrm{T}}\bar{x}_B = -(1/\mu)e_k$ for $\bar{x}_B$, where μ is the $(m+1)$th diagonal of L.
2. Compute $\Delta y = e_{m+1} - B\bar{x}_B$.
3. Go to step 10 if $\Delta y = 0$.
4. Stop if $N^{\mathrm{T}}\Delta y \leq 0$.
5. Determine index q and stepsize β such that

$$\beta = \bar{z}_q/(a_q^{\mathrm{T}}\Delta y) = \min\{\bar{z}_j/(a_j^{\mathrm{T}}\Delta y) \mid a_j^{\mathrm{T}}\Delta y > 0,\ j \in N\}.$$

6. If $\beta \neq 0$, update: $\bar{y} = \bar{y} + \beta\Delta y,\ \bar{z}_N = \bar{z}_N - \beta N^{\mathrm{T}}\Delta y$.
7. Call Algorithm 23.1.1 to update L.
8. Bring q from N and to B as its second last index.
9. Set $k = k + 1$, and go to step 1.
10. Stop if $\bar{x}_B \geq 0$.
11. Determine $p \in \arg\min_{j\in B} \bar{x}_j$.
12. Call Algorithm 23.1.2 to update L.
13. Update: $B = B\backslash\{p\},\ N = N \cup \{p\}$.
14. Set $k = k - 1$.
15. Go to step 1.

Note Initial B includes $n + 1$, which is the only index in case of $k = 1$.

Theorem 23.1.4. *Assume dual nondegeneracy through out the solution process, Algorithm 23.1.3 terminates either at*

(i) step 4, detecting upper unboundedness of (23.1); or at
(ii) step 10, generating a (dual) optimal face together with a pair of primal and dual optimal solutions.

Proof. Under the nondegeneracy assumption, the proof of finiteness of the Algorithm is the same as that for the simplex method. The meanings of its exits are derived from Lemma 23.1.1 and Theorem 23.1.3. □

Preliminary computational experiments were carried out in the same software and hardware environment, as described in Subsect. 22.1.6. The following two dense codes in FORTRAN 77 are compared:

1. RSA: The simplex algorithm (see Notation).
2. DFA: Dual face Algorithm 23.1.3, supported by Algorithm 23.2.1 being as Phase-I.

   The test set are the same 26 smallest Netlib problems (Appendix B: Table B.4, AFIRO-DEGEN2). The ratios of RSA to DFA are listed below: (excluding DEGEN2, which code RSA failed to solve):

Total iteration ratio: 1.19, Total time ratio: 10.04.

It is interesting to compare the performance of the dual face algorithm with the face algorithm presented in Sect. 22.1.6. It is seen that iterations ratios of the two codes are 1.51 and 1.19, respectively, and time ratios are as high as 4.91 and 10.04. So both methods outperformed the simplex algorithm with large margins. As for a comparison between the two face methods, the dual face method outperformed its primal counterpart with time ration 2.04 (see Appendix D for details). It is certainly advantageous that the dual face algorithm involves systems of order k, generally much less than the order, $m + 1$, of the systems in the face algorithm.

23.2 Dual Face Phase-I

To provide a dual feasible solution to get Algorithm 23.1.3 started, any dual Phase-I methods presented in Chap. 14 applies. Especially, the single-artificial-variable method described in Sect. 14.2 fit. In this section, an auxiliary program, slightly simpler than (14.9), is solved by the dual face algorithm.

Given any $(\bar{y}, \bar{z})$ with $\bar{z} \geq 0$. If $A^{\mathrm{T}}\bar{y} + \bar{z} = c$, then $(\bar{y}, \bar{z})$ is already a dual feasible solution. Now assume it is not the case.

Define

$$h = (\bar{z} - c + A^{\mathrm{T}}\bar{y})/\|\bar{z} - c + A^{\mathrm{T}}\bar{y}\|,$$

and introduce variable y_{m+1}. Construct auxiliary program

$$\begin{array}{ll} \max & y_{m+1}, \\ \text{s.t.} & A^{\mathrm{T}}y + hy_{m+1} + z = c, \quad z \geq 0, \quad y_{m+1} \leq 0. \end{array} \tag{23.22}$$

It is clear that this program has a feasible solution $(\bar{y}, \bar{y}_{m+1}, \bar{z})$, where

$$\bar{y}_{m+1} = -\|\bar{z} - c + A^{\mathrm{T}}\bar{y}\|,$$

and is upper bounded, hence there is an optimal solution. If its optimal value is equal to 0, then the part (y, z) of the optimal solution is a feasible solution to the original dual problem; else the dual problem is infeasible.

In order to decrease the possibility of zero stepsize, it is inadvisable to have many zero components of $\bar{z}$. A possible way is to set $\bar{y} = 0,\ \bar{z} = e$, thus

$$h = (e - c)/\|e - c\|, \tag{23.23}$$

and there is the feasible solution

$$\bar{y} = 0, \quad \bar{y}_{m+1} = -\|e - c\|, \quad \bar{z} = e, \tag{23.24}$$

to the auxiliary program. Then by setting $A^{\mathrm{T}} := (A^{\mathrm{T}} \vdots h)$, the auxiliary program is converted to a form, similar to (23.1). Consequently, the following Phase-I algorithm yields by modifying Algorithm 23.1.3 slightly.

Algorithm 23.2.1 (Dual face Phase-I algorithm). Initial: $B = \emptyset$, $N = A$, $k = 0$, $\bar{y} = \|e - c\|e_{m+1}$, $\bar{z}_N = e$. This algorithm finds a feasible solution to (23.1).

1. Set $\Delta y = e_{m+1}$, and go to step 4.
2. Solves $LL^{\mathrm{T}}\bar{x}_B = B^{\mathrm{T}}e_{m+1}$, where μ is the $(m+1)$th diagonal of L.
3. Compute $\Delta y = e_{m+1} - B\bar{x}_B$.
4. Go to step 10 if $\Delta y = 0$.
5. If $N^{\mathrm{T}}\Delta y \not\leq 0$, then:

   (1) determine q and β_1 such that

   $$\beta_1 = \bar{z}_q/(a_q^{\mathrm{T}}\Delta y) = \min\{\bar{z}_j/(a_j^{\mathrm{T}}\Delta y) \mid a_j^{\mathrm{T}}\Delta y > 0, \ j \in N\};$$

   (2) compute $\beta = \min\{-\bar{y}m + 1/\Delta y_{m+1}, \ \beta_1\}$;
   else
   (3) compute $\beta = -\bar{y}_{m+1}/\Delta y_{m+1}$.
6. Update: $\bar{y} = \bar{y} + \beta\Delta y$, $\bar{z}_N = \bar{z}_N - \beta N^{\mathrm{T}}\Delta y$.
7. Stop if $\bar{y}_{m+1} = 0$ (dual feasibility achieved).
8. Update L by Algorithm 23.1.1.
9. Update (B, N) by bringing q from N to B.
10. Set $k = k + 1$, and go to step 2.
11. Stop if $\bar{x}_B \geq 0$ (dual infeasible problem).
12. Determine $p \in \arg\min_{j \in B} \bar{x}_j$.
13. Update L by Algorithm 23.1.2.
14. Update (B, N) by bringing p from B to N.
15. Set $k = k - 1$.
16. Go to step 2.

Example 23.2.1. Solve the following problem by the two-phase dual face algorithm:

$$\begin{array}{lrcl}
\min & 4x_1 - x_2 + 2x_3, & & \\
\text{s.t.} & 2x_1 + x_2 + x_3 + x_4 & = & 2, \\
& x_1 - x_2 - x_3 - x_5 & = & -1, \\
& 2x_1 - x_3 + x_5 & = & 0, \\
& x_j \geq 0, \quad j = 1, \cdots, 5. & &
\end{array}$$

Answer The dual problem is

$$\begin{aligned}
\max\ & 2y_1 - y_2, \\
\text{s.t.}\ & 2y_1 + y_2 + 2y_3 + z_1 = 4, \\
& y_1 - y_2 + z_2 = -1, \\
& y_1 - y_2 - y_3 + z_3 = 2, \\
& y_1 + z_4 = 0, \\
& - y_2 + y_3 + z_5 = 0, \\
& z_j \geq 0, \quad j = 1, \cdots, 5.
\end{aligned}$$

Construct the following auxiliary program ($\|c - e\| = 4$):

$$\begin{aligned}
\max\ & y_4, \\
\text{s.t.}\ & 2y_1 + y_2 + 2y_3 - (3/4)y_4 + z_1 = 4, \\
& y_1 - y_2 + (1/2)y_4 + z_2 = -1, \\
& y_1 - y_2 - y_3 - (1/4)y_4 + z_3 = 2, \\
& y_1 + (1/4)y_4 + z_4 = 0, \\
& - y_2 + y_3 + (1/4)y_4 + z_5 = 0, \\
& z_j \geq 0, \quad j = 1, \cdots, 5.
\end{aligned}$$

Phase-I: Call Algorithm 23.2.1.

Initial: $k = 0$, $B = \emptyset$, $N = \{1, 2, 3, 4, 5\}$, $\bar{y} = (0, 0, 0, -4)^{\mathrm{T}}$, $\bar{z} = (1, 1, 1, 1, 1)^{\mathrm{T}}$.

Iteration 1:

1. $\Delta y = (0, 0, 0, 1)^{\mathrm{T}}$.
4. $\Delta y \neq 0$.
5. $N^{\mathrm{T}} \Delta y = (-3/4, 1/2, -1/4, 1/4, 1/4)^{\mathrm{T}} \not\leq 0$.

   (1) $\beta_1 = \min\{1/(1/2), 1/(1/4), 1/(1/4)\} = 2, q = 2$.
   (2) $\beta \ = \min\{4/1, 2\} = 2$.

6. $\bar{y} \ = (0, 0, 0, -4)^{\mathrm{T}} + 2(0, 0, 0, 1)^{\mathrm{T}} = (0, 0, 0, -2)^{\mathrm{T}}$.
   $\bar{z}_N = (1, 1, 1, 1, 1)^{\mathrm{T}} - 2(-3/4, 1/2, -1/4, 1/4, 1/4)^{\mathrm{T}}$
   $= (5/2, 0, 3/2, 1/2, 1/2)^{\mathrm{T}}$.
7. $\bar{y}_4 \ = -2 \neq 0$.
8. $L \ = [3/2]$.
9. $B \ = \{2\}$, $N = \{1, 3, 4, 5\}$.
10. $k \ = 1$.

Iteration 2:

2. $\bar{x}_B \ = (2/9)$.
3. $\Delta y \ = (0, 0, 0, 1)^{\mathrm{T}} - (1, -1, 0, 1/2)^{\mathrm{T}}(2/9) = (-2/9, 2/9, 0, 8/9)^{\mathrm{T}}$.

4. $\Delta y \neq 0$.
5. $N^{\mathrm{T}}\Delta y = (-8/9, -2/3, 0, 0)^{\mathrm{T}} \leq 0$.

   (3) $\beta = -(-2)/(8/9) = 9/4$.

6. $\bar{y} = (0, 0, 0, -2)^{\mathrm{T}} + (9/4)(-2/9, 2/9, 0, 8/9)^{\mathrm{T}} = (-1/2, 1/2, 0, 0)^{\mathrm{T}}$.
   $\bar{z}_N = (5/2, 3/2, 1/2, 1/2)^{\mathrm{T}} - (9/4)(-8/9, -2/3, 0, 0)^{\mathrm{T}}$
   $= (9/2, 3, 1/2, 1/2)^{\mathrm{T}}$.
7. $\bar{y}_4 = 0$, achieves dual feasible.

Convert the dual problem to a form of (23.1):

$$\begin{array}{llrrrrrrl} \max & y_4, & & & & & & & \\ \text{s.t.} & 2y_1 & + y_2 & + 2y_3 & & + z_1 & = & 4, \\ & y_1 & - y_2 & & & + z_2 & = & -1, \\ & y_1 & - y_2 & - y_3 & & + z_3 & = & 2, \\ & y_1 & & & & + z_4 & = & 0, \\ & & - y_2 & + y_3 & & + z_5 & = & 0, \\ & -2y_1 & + y_2 & & + y_4 & + z_6 & = & 0, \\ & & z_j \geq 0, & j = 1, \cdots, 6. & & & & \end{array} \tag{23.25}$$

Phase-II: Call Algorithm 23.1.3.
Initial: $k = 2$, $B = \{2, 6\}$, $N = \{1, 3, 4, 5\}$ $\bar{y} = (-1/2, 1/2, 0, -3/2)^{\mathrm{T}}$,
$\bar{z}_N = (9/2, 3, 1/2, 1/2)^{\mathrm{T}}$, $L = \begin{pmatrix} \sqrt{2} & 0 \\ -(3\sqrt{2})/2 & \sqrt{3/2} \end{pmatrix}$.

Iteration 3:

1. $\bar{x}_B = (1, 2/3)^{\mathrm{T}}$.
2. $\Delta y = \begin{pmatrix} 0 \\ 0 \\ 0 \\ 1 \end{pmatrix} - \begin{pmatrix} 1 & -2 \\ -1 & 1 \\ 0 & 0 \\ 0 & 1 \end{pmatrix} \begin{pmatrix} 1 \\ 2/3 \end{pmatrix} = \begin{pmatrix} 1/3 \\ 1/3 \\ 0 \\ 1/3 \end{pmatrix}$.
3. $\Delta y \neq 0$.
4. $N^{\mathrm{T}}\Delta y = \begin{pmatrix} 2 & 1 & 2 & 0 \\ 1 & -1 & -1 & 0 \\ 1 & 0 & 0 & 0 \\ 0 & -1 & 1 & 0 \end{pmatrix} \begin{pmatrix} 1/3 \\ 1/3 \\ 0 \\ 1/3 \end{pmatrix} = \begin{pmatrix} 1 \\ 0 \\ 1/3 \\ -1/3 \end{pmatrix}$.
5. $\beta = \min\{(9/2)/1, (1/2)/(1/3)\} = 3/2$, $q = 4$.
6. $\bar{y} = (-1/2, 1/2, 0, -3/2)^{\mathrm{T}} + (3/2)(1/3, 1/3, 0, 1/3)^{\mathrm{T}} = (0, 1, 0, -1)^{\mathrm{T}}$.
7. $L = \begin{pmatrix} \sqrt{2} & 0 & 0 \\ \sqrt{2}/2 & \sqrt{2}/2 & 0 \\ -3/\sqrt{2} & -\sqrt{2}/2 & 1 \end{pmatrix}$.
8. $B = \{2, 4, 6\}$, $N\{1, 3, 5\}$.
9. $k = 3$.

Iteration 4:

1. $\bar{x}_B = (1, 1, 1)^{\mathrm{T}}$.

2. $\Delta y = \begin{pmatrix} 0 \\ 0 \\ 0 \\ 1 \end{pmatrix} - \begin{pmatrix} 1 & 1 & -2 \\ -1 & 0 & 1 \\ 0 & 0 & 0 \\ 0 & 0 & 1 \end{pmatrix} \begin{pmatrix} 1 \\ 1 \\ 1 \end{pmatrix} = \begin{pmatrix} 0 \\ 0 \\ 0 \\ 0 \end{pmatrix}.$

3. $\Delta y = 0$.

10. $\bar{x}_B \geq 0$. The optimal solution and optimal value to (23.25):

$$\bar{y} = (0, 1, 0, -1)^{\mathrm{T}}, \quad \bar{z} = (3, 0, 3, 0, 1, 0)^{\mathrm{T}},$$

with the related optimal face

$$y_1 = 0, \quad y_2 = 1, \quad -3 \leq y_3 \leq 1, \quad y_4 = -1.$$

Finally, the optimal solution and optimal value to the original problem are

$$\bar{x} = (0, 1, 0, 1, 0)^{\mathrm{T}}, \quad \bar{f} = -1,$$

and dual optimal solution is

$$\bar{y} = (0, 1, 0)^{\mathrm{T}}, \quad \bar{z} = (3, 0, 3, 0, 1)^{\mathrm{T}}.$$

23.3 Dual Face Method Using Gauss-Jordan Elimination

In Algorithm 23.1.3, the QR factorization is used to produce the Cholesky factor L of $B^T B$ initially, and Givens notations used to update L in each iteration. Such manipulations by orthogonal matrices would not be amenable to sparse computations, however. In this section, dual face algorithms using the Gauss-Jordan elimination will be derived to remedy the situation.

We are concerned with the D-reduced problem (25.3), whose dual problem is

$$\begin{array}{lll} \max & g = y_r & \\ \text{s.t.} & A^T y + z = \bar{z}, & z \geq 0. \end{array} \tag{23.26}$$

Let a dual feasible solution $(\bar{y}, \bar{z})$ be available.

As the steepest uphill, the gradient e_r of the dual objective is taken as the search direction in y-space. Such doing leads to

$$\Delta y = e_r, \qquad \Delta z = -A^T e_r \equiv -\omega.$$

The new dual feasible solution is then determined by the following line search scheme:

$$\hat{y} = \bar{y} + \beta e_r, \quad \hat{z} = \bar{z} - \beta\omega, \tag{23.27}$$

where the largest possible stepsize maintaining $\bar{z}$ nonnegative is

$$\beta = \bar{z}_q/\omega_q = \min\{\bar{z}_j/\omega_j \mid \omega_j > 0,\ j \in A\} \geq 0.$$

The related computations can be arranged in a tableau form as follows.

Based on Proposition 4.7.1, we turn to the following problem instead of (25.3), equivalently.

$$\begin{array}{ll} \min & f = \bar{y}_r + \bar{z}^T x, \\ \text{s.t.} & Ax = e_r, \qquad x \geq 0. \end{array} \tag{23.28}$$

Denote by ω^T the so-called "datum" (rth) row of A, and by A_R the matrix consisting A's rows other than the datum row. The the initial tableau of (23.28) is then

x^{T}	f	RHS
A_R		
ω^T		1
$\bar{z}^{\mathrm{T}}$	-1	$-\bar{y}_r$

According to the second formula of (23.27), add $-\beta$ times of the datum row of the preceding tableau to its bottom line, giving the new $\bar{z}$ ($\bar{y}$ can be computed at the end if required). For the moment, assume that the q-indexed column, $a_q(R)$, of A_R is nonzero, and a_{pq} is the largest in module among all its components. Taken as the pivot, a_{pq}, is converted to 1 and the other nonzeros in the column is eliminated by elementary transformations. The q-indexed entry of the new datum row vanishes. The pivot entry 1 is moved to the position in the first row and first column by row and column exchanges, and the first iteration is then complete. It is noted that the first column of the resulting tableau is a unit vector with the first component 1.

In general, we initially set

$$B = \emptyset, \quad N = A, \qquad R = \emptyset, \quad R' = \{1, \cdots, m\}. \tag{23.29}$$

Assume that at some iteration, faced are

$$B = \{j_1, \cdots, j_k\}, \quad N = A \backslash B; \qquad R = \{i_1, \cdots, i_k\}, \quad R' = \{1, \cdots, m\} \backslash R,$$

and the following tableau:

$$\begin{array}{cc|c|c}
\hline
x_B^{\mathrm{T}} & x_N^{\mathrm{T}} & f & \mathrm{RHS} \\
\hline
I & \bar{N}_R & & \\
 & \bar{N}_{R'} & & \\
 & \bar{\omega}_N^{\mathrm{T}} & & 1 \\
\hline
 & \bar{z}_N^{\mathrm{T}} & -1 & -\bar{f} \\
\hline
\end{array} \tag{23.30}$$

where the north-west corner is the unit $k \times k$ matrix with $0 \leq k \leq m$, generated from elementary transformations, The tableau corresponds to the following dual program (see Sect. 4.7 for the relation between the dual problems associated with the initial and subsequent tableaus):

$$\begin{array}{ll}
\max & y_r \\
\text{s.t.} & \begin{pmatrix} I & 0 & 0 \\ \bar{N}_R^T & \bar{N}_{R'}^T & \bar{\omega}_N \end{pmatrix} \begin{pmatrix} y_R \\ y_{R'} \\ y_r \end{pmatrix} + \begin{pmatrix} z_B \\ z_N \end{pmatrix} = \begin{pmatrix} 0 \\ \bar{z}_N \end{pmatrix}, \qquad z_B, z_N \geq 0,
\end{array}$$

from which it is seen that $\Delta y = e_r$ is the steepest uphill in y-space, associated with the search direction in z-space, i.e.,

$$\Delta z = -\begin{pmatrix} 0 \\ \bar{\omega}_N \end{pmatrix}.$$

It is clear that the dual feasible solution $(\bar{y}, \bar{z})$, given by the bottom line of tableau (23.30), and search direction $(\Delta y, \Delta z)$ are both in the dual face

$$D_N = \{(y, z) \in \mathcal{R}^m \times \mathcal{R}^n \mid A^{\mathrm{T}} y + z = c,\ z_B = 0,\ z_N \geq 0\}.$$

Define

$$J = \{j \in N \mid \bar{\omega}_j > 0\}.$$

If $J = \emptyset$, the dual problem is unbounded, and hence the original problem is infeasible. Otherwise, the dual feasible solution $\bar{z}$ and associated objective value $\bar{g}$ are updated by the following line search scheme (the associated $\bar{y}$ is not needed if only a primal optimal solution is wanted; it can be computed at the end if required):

$$\bar{z}_B = 0, \qquad \hat{z}_N = \bar{z}_N - \beta \bar{\omega}_N, \qquad \hat{g} = \bar{g} + \beta. \tag{23.31}$$

where β is the largest possible stepsize such that

$$\beta = \bar{z}_q / \omega_q = \min_{j \in J} \bar{z}_j / \omega_j \geq 0.$$

But we will use the following instead:

$$q \in \begin{cases} \arg\max\{\bar{\omega}_j \mid j \in J_1\}, & \beta = 0, \qquad \text{if } J_1 \neq \emptyset, \\ \arg\min\{\bar{z}_j/\bar{\omega}_j \mid j \in J\}, & \beta = \bar{z}_q/\bar{\omega}_q, \qquad \text{otherwise,} \end{cases} \tag{23.32}$$

where $J_1 = \{j \mid \bar{z}_j = 0, \ j \in J\}$. The according new objective value is $\bar{f} + \beta$. These new quantities can be generated by adding $-\beta$ times of the ω row to the bottom in tableau (23.30).

Let $\bar{a}_q(R')$ be the q-indexed column of $\bar{N}_{R'}$ and let $\bar{a}_q(R)$ be that of $\bar{N}_R$. There will be the following two cases handled separately:

(i) $R' \neq \emptyset$ and $\bar{a}_q(R') \neq 0$.
Determine row index $p = \arg\max_{i\in R'} |\bar{a}_{i\,q}|$. Convert $\bar{a}_{p\,q}$ to 1 and eliminate the other nonzeros in the column by elementary transformations. Then bring column index q from N to B and row index p from R' to R. This is a face contraction iteration.

(ii) $R' = \emptyset$ or $\bar{a}_q(R') = 0$.
Determine row index $s \in \arg\max\{\bar{a}_{i_t,q} \mid t = 1,\ldots,k\}$. If $\bar{a}_{i_s\,q} \leq 0$ hence $\bar{a}_q(R) \leq 0$, achieved is then the basic optimal solution, with following basic components:

$$\bar{x}_q = 1/\bar{\omega}_q \geq 0, \qquad \bar{x}_B = -\bar{x}_q\bar{a}_q(R) \geq 0. \tag{23.33}$$

In this case, a so-called "dual level face" is reached. In fact, if $\bar{\omega}_q$ were deduced to 1 and the other nonzeros in the column were eliminated by elementary transformations, the tableau would become a normal simplex tableau, just giving the basic optimal solution (23.33). If row index sets I_1 and I_2 are defined to correspond to zero and nonzero basic components, respectively, then a dual face is determined over which the dual objective value is constant (see the last half of Sect. 25.2). It is an optimal dual face if $\bar{a}_{i_s\,q} \leq 0$. Otherwise, convert $\bar{a}_{i_s\,q}$ to 1 and eliminate the other nonzeros in the column by elementary transformations. Then exchange column indices j_s and q to complete an iteration.

The overall steps can be formulated as follows.

Algorithm 23.3.1 (Dual face algorithm using Gauss-Jordan elimination). Initial: tableau of form (23.30) with $B,\ N,\ R,\ R',\ k = 0$. This algorithm solves the D-reduced problem (25.3).

1. Stop if $J = \{j \in N \mid \bar{\omega}_j > 0\} = \emptyset$ (infeasible problem).
2. Determine column index q and β by (23.32).
3. If $\beta \neq 0$, add $-\beta$ times of the datum row to the bottom row.
4. If $k = 0$, go to step 10.
5. Determine $s \in \arg\max\{\bar{a}_{i_t,q} \mid t = 1,\ldots,k\}$.
6. Go to step 9 if $\bar{a}_{i_s,q} \leq 0$.

7. Convert $\bar{a}_{i_s,q}$ to 1, and eliminate the other nonzeros in the column by elementary transformations.
8. Update (B, N) by exchanging j_s and q, and go to step 1.
9. Go to step 13 if $R' = \emptyset$ or $\bar{a}_q(R') = 0$.
10. Determine row index $p \in \arg\max_{i \in R'} |\bar{a}_{i\,q}|$.
11. Convert $\bar{a}_{p\,q}$ to 1, and eliminate the other nonzeros in the column by elementary transformations.
12. Set $k = k + 1$, bring q from N to B and p from R' to R, and go to step 1.
13. Convert ω_q to 1, and eliminate the other nonzeros in the column by elementary transformations.
14. Stop (optimality achieved).

Example 23.3.1. Solve the following LP problem by Algorithm 23.3.1, supported by a Phase-I using auxiliary program (23.22):

$$\begin{array}{llllllllll} \min & f = 2x_1 + x_2 - 2x_3 + x_4 - 3x_5 + x_6 + 2x_7 - 2x_8 \\ \text{s.t.} & 2x_1 - 3x_2 + x_3 - 6x_4 + x_5 + 4x_6 + 6x_7 & = 0, \\ & 5x_1 + 6x_3 - 2x_4 + 10x_6 + 8x_7 - x_8 = 0, \\ & -x_1 + 8x_2 - 4x_5 + 2x_8 = 0, \\ & 4x_1 - 2x_3 + 5x_4 + 8x_5 - 4x_6 + 3x_7 = 1, \\ & x_j \ge 0, \quad j = 1, \ldots, 8. \end{array}$$

Answer Phase-I: Denote by c and A the costs and the coefficient matrix, respectively. Set

$$\begin{aligned} h &= (e - c)/\|e - c\| = (-1, 0, 3, 0, 4, 0, -1, 3)^T/6 \\ &= (-1/6, 0, 1/2, 0, 2/3, 0, -1/6, 1/2)^T, \end{aligned}$$

where e is the vector with all ones. Then the Phase-I auxiliary dual program is of form below:

$$\begin{array}{ll} \max & y_5, \\ \text{s.t.} & A^{\mathrm{T}} y + h y_5 + z = c, \quad z \ge 0, \quad y_5 \le 0, \end{array}$$

which has the feasible solution

$$\bar{y} = 0, \quad \bar{y}_5 = -\|e - c\| = -6, \bar{z} = e,$$

associated with auxiliary objective value -6. Based on Proposition 4.7.1, solve the according primal program

$$\begin{array}{lll} \min & e^T x - 6, \\ \text{s.t.} & Ax = 0, \\ & h^T x = 1, \quad x \ge 0, \end{array}$$

by Algorithm 23.3.1, where set J includes index 9 (associated with the auxiliary right-hand side RHS1), as the auxiliary objective value is less then and equal to zero.

Initial tableau (the right-hand side e_4 of the original problem is put as the last column to get the Phase-II start easier): $k = 0,\ B = \emptyset,\ N = A, R = \emptyset,\ R' = \{1, 2, 3, 4\}$.

x_1	x_2	x_3	x_4	x_5	x_6	x_7	x_8	RHS1	RHS
2	−3	1	−6	1	4	6			
5		6	−2		10	8	−1		
−1	8			−4			2		
4		−2	5	8*	−4	3			1
−1/6		1/2		2/3		−1/6	1/2	1	
1	1	1	1	1	1	1	1	6	

Iteration 1:

1. $J = \{3, 5, 8, 9\} \neq \emptyset$.
2. $J_1 = \emptyset, \beta = \min\{1/(1/2), 1/(2/3), 1/(1/2), 6/1\} = 3/2,\ q = 5$.
3. Add $-3/2$ times of row 5 to the bottom row.
4. $k = 0$.
10. $\max\{|1|, 0, |-4|, |8|\} = 8 > 0,\ p = 4$.
11. Multiply row 4 by 1/8, then add $-1, 4, -2/3$ times of row 4 to rows 1,3,5, respectively:
12. $k = 1, B = \{5\},\ N = A\backslash B, R = \{4\},\ R' = \{1, 2, 3\}$.

x_1	x_2	x_3	x_4	x_5	x_6	x_7	x_8	RHS1	RHS
3/2	−3	5/4	−53/8		9/2	45/8			−1/8
5		6*	−2		10	8	−1		
1	8	−1	5/2		−2	3/2	2		1/2
1/2		−1/4	5/8	1	−1/2	3/8			1/8
−1/2		2/3	−5/12		1/3	−5/12	1/2	1	−1/12
5/4	1	1/4	1		1	5/4	1/4	9/2	

Iteration 2:

1. $J = \{3, 6, 8, 9\} \neq \emptyset$.
2. $J_1 = \emptyset, \beta = \min\{(1/4)/(1/2), 1/(1/3), (1/4)/(1/2), (9/2)/1\} = 3/8$, $q = 3$.
3. Add $-3/8$ times of row 5 to the bottom row.
5. $\max\{-1/4\} \leq 0$.
9. $\bar{a}_q(R') \neq 0$.

10. $\max\{|5/4|, |6|, |-1|\} = 6 > 0,\ p = 2.$
11. Multiply row 2 by 1/6, then add $-5/4, 1, 1/4, -2/3$ times of row 2 to rows 1,3,4,5 respectively:
12. $k = 2,\ B = \{5, 3\},\ N = A\backslash B, R = \{4, 2\},\ R' = \{1, 3\}.$

x_1	x_2	x_3	x_4	x_5	x_6	x_7	x_8	RHS1	RHS
11/24	−3		−149/24		29/12	95/24	5/24		−1/8
5/6		1	−1/3		5/3	4/3	−1/6		
11/6	8		13/6		−1/3	17/6	11/6*		1/2
17/24			13/24	1	−1/12	17/24	−1/24		1/8
−19/18			−7/36		−7/9	−47/36	11/18	1	−1/12
23/16	1		37/32		7/8	45/32	1/16	33/8	1/32

Iteration 3:

1. $J = \{8, 9\} \neq \emptyset.$
2. $J_1 = \emptyset.\ \beta = \min\{(1/16)/(11/18), (33/8)/1\} = (1/16)/(11/18),\ q = 8.$
3. Add $-\beta$ times of row 5 to the bottom row.
5. $\max\{-1/6, -1/24\} \leq 0.$
9. $\bar{a}_q(R') \neq 0.$
10. $\max\{|5/24|, |11/6|\} = 11/6 > 0,\ p = 3.$
11. Multiply row 3 by 6/11, then add $-5/24, 1/6, 1/24, -11/18$ times of row 2 to rows 1,2,4,5 respectively:
12. $k = 3,\ B = \{5, 3, 8\},\ N = A\backslash B, R = \{4, 2, 3\},\ R' = \{1\}.$

x_1	x_2	x_3	x_4	x_5	x_6	x_7	x_8	RHS1	RHS
1/4	−43/11		−71/11		27/11	40/11			−2/11
1	8/11	1	−3/22		18/11	35/22			1/22
1	48/11		13/11		−2/11	17/11	1		3/11
3/4	2/11		13/22	1	−1/11	17/22			3/22
−5/3	−8/3		−11/12		−2/3	−9/4		1	−1/4
17/11	1		207/176		21/22	271/176		177/44	7/176

Iteration 4:

1. $J = \{9\}.$
2. $J_1 = \emptyset.\ \beta = (177/44)/1,\ q = 9.$
3. Add $-177/44$ times of row 5 to the bottom row. The auxiliary objective value vanishes.

   Phase-I terminates with the bottom row giving the dual feasible solution to the original problem. Obtain a dual feasible tableau to the original problem by dropping the second bottom row, and the RHS1 column:

x_1	x_2	x_3	x_4	x_5	x_6	x_7	x_8	RHS
1/4	−43/11		−71/11		27/11	40/11		−2/11
1	8/11	1	−3/22		18/11	35/22		1/22
1	48/11		13/11		−2/11	17/11	1	3/11
3/4	2/11		13/22	1	−1/11	17/22		3/22*
33/4	129/11		2,107/22		40/11	233/22		23/22

Phase-II (1):

Iteration 5:

To deduced the preceding tableau to a D-reduced one, multiply row 4 by $22/3$, and add $2/11, -1/22, -3/11$ times of row 4 to rows 1,2,3, respectively:

x_1	x_2	x_3	x_4	x_5	x_6	x_7	x_8	RHS
5/4	−11/3		−17/3	4/3*	7/3	14/3		
3/4	2/3	1	−1/3	−1/3	5/3	4/3		
−1/2	4			−2			1	
11/2	4/3		13/3	22/3	−2/3	17/3		1
33/4	129/11		107/22		40/11	233/22		23/22

$k = 2$, $B = \{3, 8\}$, $N = A \backslash B$, $R = \{2, 3\}$, $R' = \{1\}$.

Iteration 6:

1. $J = \{1, 2, 4, 5, 7\}$.
2. $J_1 = \{5\}$. $\beta = 0$, $q = 5$.
5. $\max\{-1/3, -2\} \leq 0$.
9. $\bar{a}_q(R') \neq 0$.
10. $\max\{|4/3|\}$, $p = 1$.
11. Multiply row 1 by $3/4$, then add $1/3, 1/6, 2, -22/31/24$ times of row 1 to rows 2,3,4, respectively:
12. $k = 3$, $B = \{3, 8, 5\}$, $N = A \backslash B$, $R = \{4, 2, 3, 1\}$, $R' = \emptyset$.

x_1	x_2	x_3	x_4	x_5	x_6	x_7	x_8	RHS
15/16	−11/4		−17/4	1	7/4	7/2		
17/16	−1/4	1	−7/4		9/4	5/2		
11/8	−3/2		−17/2		7/2	7	1	
−11/8	43/2		71/2*		−27/2	−20		1
33/4	129/11		107/22		40/11	233/22		23/22

Iteration 7:

1. $J = \{2, 4\}$.
2. $J_1 = \emptyset$. $\beta = \{(129/11)/(43/2), (107/22)/(71/2)\} = (107/22)/(71/2)$, $q = 4$.
3. Add $-\beta$ times of row 5 to the bottom row.
5. $\max\{-17/4, -7/4, -17/2\} \leq 0$.
9. $R' = \emptyset$.
13. Multiply row 4 by $2/71$, then add $17/4, 7/4, 17/2$ times of row 4 to rows 1,2,3, respectively:

x_1	x_2	x_3	x_4	x_5	x_6	x_7	x_8	RHS
439/568	−25/142			1	19/142	157/142		17/142
565/568	115/142	1			225/142	215/142		7/142
297/284	259/71				19/71	157/71	1	17/71
−11/284	43/71		1		−27/71	−40/71		2/71
4,793/568	1,247/142				779/142	1,893/142		129/142

14. The basic optimal solution:

$$\bar{x} = (0, 0, 7/142, 2/71, 17/142, 0, 0, 17/71)^T, \qquad \bar{f} = -129/142.$$

Phase-II (2):
Alternatively, Phase-II can start from the original data, except for the resulting bottom and datum rows (or the original datum row). The following tableau uses the original datum row: $B = \emptyset,\ N = A, R = \emptyset,\ R' = \{1, 2, 3\}$

x_1	x_2	x_3	x_4	x_5	x_6	x_7	x_8	RHS
2	−3	1	−6	1	4	6		
5		6	−2		10	8	−1	
−1	8			−4*	0		2	
4		−2	5	8	−4	3		1
33/4	129/11		107/22		40/11	233/22		23/22

Iteration 5:

1. $J = \{1, 4, 5, 7\}$.
2. $J_1 = \{5\}$. $\beta = 0$, $q = 5$.
5. $\max\{-17/4, -7/4, -17/2\} \leq 0$.
4. $k = 0$.
10. $\max\{|1|, 0, |-4|\} = 4$, $p = 3$.
11. Multiply row 3 by $-1/4$, then add $-1, -8$ times of row 3 to rows 1,4, respectively:
12. $k = 1$, $B = \{5\}$, $N = A \backslash B$, $R = \{3\}$, $R' = \{1, 2\}$.

x_1	x_2	x_3	x_4	x_5	x_6	x_7	x_8	RHS
7/4	−1	1	−6		4	6	1/2	
5		6	−2		10	8	−1*	
1/4	−2			1			−1/2	
2	16	−2	5		−4	3	4	1
33/4	129/11		107/22		40/11	233/22		23/22

Iteration 6:

1. $J = \{1, 2, 4, 7, 8\}$.
2. $J_1 = \{8\}$. $\beta = 0$, $q = 8$.
5. $\max\{-1/2\} \le 0$.
9. $\bar{a}_q(R') \neq 0$.
10. $\max\{|1/2|, |-1|\} = 1$, $p = 2$.
11. Multiply row 2 by −1, then add $-1/2, 1/2, -4$ times of row 2 to rows 1,3,4, respectively:
12. $k = 2$, $B = \{5, 8\}$, $N = A \backslash B$, $R = \{3, 2\}$, $R' = \{1\}$.

x_1	x_2	x_3	x_4	x_5	x_6	x_7	x_8	RHS
17/4	−1	4*	−7		9	10		
−5		−6	2		−10	−8	1	
−9/4	−2	−3	1	1	−5	−4		
22	16	22	−3		36	35		1
33/4	129/11		107/22		40/11	233/22		23/22

Iteration 7:

1. $J = \{1, 2, 3, 6, 7\}$.
2. $J_1 = \{3\}$. $\beta = 0$, $q = 3$.
5. $\max\{-6, -3\} \le 0$.
9. $\bar{a}_q(R') \neq 0$.
10. $\max\{|4|, |-1|\}$, $p = 1$.
11. Multiply row 1 by 1/4, then add $6, 3, -22$ times of row 1 to rows 2,3,4, respectively:
12. $k = 2$, $B = \{5, 8, 3\}$, $N = A \backslash B$, $R = \{3, 2, 1\}$, $R' = \emptyset$.

x_1	x_2	x_3	x_4	x_5	x_6	x_7	x_8	RHS
17/16	−1/4	1	−7/4		9/4	5/2		
11/8	−3/2		−17/2		7/2	7	1	
15/16	−11/4		−17/4	1	7/4	7/2		
−11/8	43/2		71/2*		−27/2	−20		1
33/4	129/11		107/22		40/11	233/22		23/22

Iteration 8:

1. $J = \{2, 4\}$.
2. $J_1 = \emptyset$. $\beta = \min\{(129/11)/(43/2), (107/22)/(71/2)\} = (107/22)/(71/2)$, $q = 4$.
5. $\max\{-7/4, -17/2, -17/4\} = -7/4 \le 0$.
9. $R' = \emptyset$.
13. Multiply row 4 by $2/71$, then add $7/4, 17/2, 17/4$ times of row 4 to rows 1,2,3, respectively:

x_1	x_2	x_3	x_4	x_5	x_6	x_7	x_8	RHS
565/568	115/142	1			225/142	215/142		7/142
297/284	259/71				19/71	157/71	1	17/71
439/568	−25/142			1	19/142	157/142		17/142
−11/284	43/71		1		−27/71	−40/71		2/71
4,793/568	1,247/142				779/142	1,893/142		129/142

14. Obtained is the same basic optimal solution as from Phase-II (1).

23.4 Variant of the Dual Face Algorithm

This section offers a variant of the dual face algorithm by updating $(B^T B)^{-1}$ rather than Cholesky factor. It is originally motivated by the idea of being amenable to large-scale sparse problems by getting rid of orthogonal transformations. Surprisingly, the variant turns out to be so simple that there is no need for solving any linear system though might be less stable than the dual face algorithm.

Firstly, we consider how to update the inverse in the contraction iteration.

Lemma 23.4.1. *Let $B \in \mathcal{R}^{(m+1)\times k}$, $1 \le k < m+1$ be a face matrix. If $0 \ne a_q \in \mathcal{R}^{m+1}$ satisfies $a_q \notin \text{range } B$, then μ is well-defined such that*

$$\mu = (a_q^T a_q - a_q^T B(B^T B)^{-1} B^T a_q)^{-1}. \tag{23.34}$$

Proof. Since B is of full column rank k, $(B^T B)^{-1} \in \mathcal{R}^{k\times k}$ is well-defined. Condition $a_q \notin \text{range } B$ implies that the orthogonal projection of a_q onto the orthogonal complement space of range B does not vanish, i.e.,

$$a_q - B(B^T B)^{-1} B^T)a_q = (I - B(B^T B)^{-1} B^T))a_q \ne 0,$$

hence it holds that

$$a_q^T a_q - a_q^T B(B^T B)^{-1} B^T a_q = a_q^T (I - B(B^T B)^{-1} B^T)a_q \ne 0.$$

Thus, (23.34) is well-defined. □

Theorem 23.4.1 *Let $B \in \mathcal{R}^{(m+1)\times k}, 1 \leq k < m+1$ be a face matrix and let $(B^T B)^{-1}$ be available. If $\breve{B} = (B \vdots a_q)$, then it holds that*

$$(\breve{B}^T \breve{B})^{-1} = \begin{pmatrix} (B^T B)^{-1} + \mu(B^T B)^{-1} B^T a_q a_q^T B(B^T B)^{-1} & -\mu(B^T B)^{-1} B^T a_q \\ -\mu a_q^T B(B^T B)^{-1} & \mu \end{pmatrix}, \tag{23.35}$$

where μ is defined by (23.34).

Proof. Proposition 23.1.4 implies that $a_q \notin \text{range } B$. According to Lemma 23.4.1, therefore, μ (23.34) is well-defined. Then, in view of

$$\breve{B}^T \breve{B} = \begin{pmatrix} B^T \\ a_q^T \end{pmatrix} \begin{pmatrix} B & a_q \end{pmatrix} = \begin{pmatrix} B^T B & B^T a_q \\ a_q^T B & a_q^T a_q \end{pmatrix},$$

the validity of (23.35) is easily verified. □

From the preceding Theorem, the formula for updating $(B^T B)^{-1}$ follows, i.e.,

$$(\breve{B}^T \breve{B})^{-1} = \begin{pmatrix} D & v \\ v^T & \mu \end{pmatrix}, \tag{23.36}$$

where

$$u = B^T a_q, \quad w = (B^T B)^{-1} u, \quad \mu = (a_q^T a_q - u^T w)^{-1}, \quad v = -\mu w,$$
$$D = (B^T B)^{-1} - v w^T. \tag{23.37}$$

Nevertheless, the column a_q should be inserted before the last column of B. So, the new face matrix $\hat{B}$ may be obtained from $\breve{B}$ by interchanging its last two columns. Denote the according column permutation matrix by P^T. Then, it follows that

$$(\hat{B}^T \hat{B})^{-1} = ((\breve{B} P^T)^T (\breve{B} P^T))^{-1} = P(\breve{B}^T \breve{B})^{-1} P^T.$$

Therefore, the wanted inverse can be obtained by interchanging the last two columns and rows of matrix (23.35).

Let us turn to the case of the expansion iteration. Let $B \in \mathcal{R}^{(m+1)\times k}, 1 < k \leq m+1$ be current face matrix and let $\hat{B}$ be the new face matrix, resulting from dropping column a_p from B. Then, there is a column permutation Q^T such that

$$BQ^T = (\hat{B} \vdots a_p) \triangleq \breve{B},$$

from which it follows that

$$(\breve{B}^T \breve{B})^{-1} = ((BQ^T)^T (BQ^T))^{-1} = Q(B^T B)^{-1} Q^T.$$

As $(B^T B)^{-1}$ is available, therefore, $(\breve{B}^T \breve{B})^{-1}$ is easily obtained.

Theorem 23.4.2 If $(\breve{B}^T \breve{B})^{-1}$ is of form (23.36), then it holds that

$$(\hat{B}^T \hat{B})^{-1} = D - \mu^{-1} v v^T. \tag{23.38}$$

Proof. From (23.36) and Theorem 23.4.1, it is known that

$$v = -\mu(\hat{B}^T \hat{B})^{-1} \hat{B}^T a_p, \qquad D = (\hat{B}^T \hat{B})^{-1} + \mu(\hat{B}^T \hat{B})^{-1} \hat{B}^T a_p a_p^T \hat{B}(\hat{B}^T \hat{B})^{-1},$$

from which, (23.38) follows. □

In practice, an integer vector is enough to record positions of columns to avoid moving them all around. The overall steps can be put into the following algorithm.

Algorithm 23.4.1. (Dual face algorithm: inverse updating) Initial : B, N, $1 \leq k \leq m+1$, $(B^T B)^{-1}$, dual feasible solution $(\bar{y}, \bar{z}_N)$. This algorithm solves the pair of reduced problems (23.1) and (23.19).

1. Compute $\bar{x}_B = (B^T B)^{-1} e_k$.
2. Compute $\Delta y = e_{m+1} - B\bar{x}_B$.
3. Go to step 11 if $\Delta y = 0$.
4. Stop if $N^{\mathrm{T}} \Delta y \leq 0$ (problem (23.1) is upper unbounded).
5. Determine index q and stepsize β such that

$$\beta = \bar{z}_q/(a_q^{\mathrm{T}} \Delta y) = \min\{\bar{z}_j/(a_j^{\mathrm{T}} \Delta y) \mid a_j^{\mathrm{T}} \Delta y > 0,\ j \in N\}.$$

6. If $\beta \neq 0$, update: $\bar{y} = \bar{y} + \beta \Delta y$, $\bar{z}_N = \bar{z}_N - \beta N^{\mathrm{T}} \Delta y$.
7. Update $(B^T B)^{-1}$ by (23.36) with adding permutation.
8. Bring q from N and insert before the last column of B.
9. Set $k = k + 1$, and go to step 1.
10. Stop if $\bar{x}_B \geq 0$ (optimality achieved).
11. Determine $p \in \arg\min_{j \in B} \bar{x}_j$.
12. Update $(B^T B)^{-1}$ by (23.38) with dropping permutation
13. Update: $B = B \backslash \{p\}$, $N = N \cup \{p\}$.
14. Set $k = k - 1$.
15. Go to step 1.

It is seen that no linear system has to be solved in the solution process if $k = 1$ and $B = (a_{n+1})$ are set initially.

Example 23.4.1. Solve the following problem by Algorithm 23.4.1:

$$
\begin{array}{llllllllll}
\min & f = 5x_1 + x_2 + 3x_4 + 6x_6 + 2x_7, \\
\text{s.t.} & 3x_1 & & - \;7x_3 & + \;5x_4 & - \;4x_5 & - \;6x_6 & - \;5x_7 & = & -18, \\
 & 1/3x_1 & + \;2x_2 & + \;1/3x_3 & + & - \;2/3x_5 & & + \;2x_7 & = & 0, \\
 & & - \;5x_2 & + \;11x_3 & - \;7x_4 & - \;3x_5 & + \;4x_6 & & = & 9, \\
 & & & & & x_j \geq 0, & j = 1, \cdots, 7.
\end{array}
$$

Answer

The problem is transformed to the following one in tableau form:

x_1	x_2	x_3	x_4	x_5	x_6	x_7	x_8	RHS
3		−7	5	−4	−6	−5	18	
1/3	2	1/3		−2/3		2		
	−5	11	−7	−3	4		−9	
							1	1
5	1		3		6	2		

$k = 1$, $B = a_8 = (18, 0, -9, 1)^T$, $N = A \backslash B$; $(B^T B)^{-1} = (1/406)$, $\bar{y} = (0, 0, 0, 0)^T$, $\bar{z} = (5, 1, 0, 3, 0, 6, 2, 0)^T$.

Iteration 1:

1. $\bar{x}_B = (1/406)$.
2. $\Delta y = (-9/203, 0, 9/406, 405/406)^T$.
4. $N^T \Delta y = (-27/203, -45/406, 225/406, -153/406, 45/406, 72/203, 45/203)^T \not\leq 0$.
5. $\beta = \min\{0/(225/406), 0/(45/406), 6/(72/203), 2/(45/203)\} = 0$,
   $q = 3$.
   $a_q = (-7, 1/3, 11, 0)^T$.
7. $(B^T B)^{-1} = \begin{pmatrix} 179/8130 & 91/7458 \\ 91/7458 & 517/56043 \end{pmatrix}$.
8. $B = \begin{pmatrix} -7 & 18 \\ 1/3 & \\ 11 & -9 \\ & 1 \end{pmatrix}$, $\quad N = \begin{pmatrix} 3 & & 5 & -4 & -6 & -5 \\ 1/3 & 2 & & -2/3 & & 2 \\ & -5 & -7 & -3 & 4 & \\ & & & & & \end{pmatrix}$.
9. $x_B = (91/7458, 517/56043)^T$.

Iteration 2:

2. $\Delta y = (-227/2815, -53/13031, -294/5743, 537/542)^T \not\leq 0$.
4. $N^T \Delta y = (-642/2639, 314/1267, -84/1873, 781/1631, 1541/5522, 4977/12598)^T$.

5. $\beta \;= \min\{1/(314/1267), 0/(781/1631), 6/(1541/5522), 2/(4977/12598)\} = 0, q = 5.$

$a_q \;= (-4, -2/3, -3, 0)^T.$

7. $$(B^T B)^{-1} = \begin{pmatrix} 1027/977 & 420/271 & 2223/2947 \\ 420/271 & 1447/620 & 827/740 \\ 2223/2947 & 827/740 & 1049/1927 \end{pmatrix}.$$

8. $$B = \begin{pmatrix} -7 & -4 & 18 \\ 1/3 & -2/3 & \\ 11 & -3 & -9 \\ & & 1 \end{pmatrix}, \quad N = \begin{pmatrix} 3 & & 5 & -6 & -5 \\ 1/3 & 2 & & & 2 \\ & -5 & -7 & 4 & \\ & & & & \end{pmatrix}.$$

9. $x_B \;= (2223/2947, 827/740, 1049/1927)^T.$

Iteration 3:

2. $\Delta y = (-125/2599, 733/1485, -134/2941, 878/1927)^T \not\leq 0.$
4. $N^T \Delta y = (325/16051, 1893/1558, 51/650, 37/348, 1224/997)^T.$
5. $\beta \;= \min\{5/(325/16051), 1/(1893/1558), 3/(51/650), 6/(37/348), 2/(1224/997)\} = 1/(1893/1558),\ q = 2. a_q = (0, 2, -5, 0)^T.$
6. $\bar{y} \;= (-217/5482, 13765/33883, -1165/31067, 33364/88971)^T,$

   $\bar{z}_N \;= (299/60, 0, 1591/542, 11689/1977, 1709/1727)^T.$

7. $$(B^T B)^{-1} = \begin{pmatrix} 2271/1000 & 1559/559 & 684/1115 & 2371/1581 \\ 1559/559 & 2407/670 & 1217/1953 & 778/415 \\ 684/1115 & 1217/1953 & 1106/3585 & 298/795 \\ 2371/1581 & 778/415 & 298/795 & 5281/5282 \end{pmatrix}.$$

8. $$B = \begin{pmatrix} -7 & -4 & & 18 \\ 1/3 & -2/3 & 2 & \\ 11 & -3 & -5 & -9 \\ & & & 1 \end{pmatrix}, \quad N = \begin{pmatrix} 3 & 5 & -6 & -5 \\ 1/3 & & & 2 \\ & -7 & 4 & \\ & & & \end{pmatrix}.$$

9. $\bar{x}_B \;= (2371/1581, 778/415, 298/795, 5281/5282)^T = (1.4997, 1.8747, 0.3748, 0.9998)^T.$

Iteration 4:

2. $\Delta y \approx 0.$
11. Optimal solution and value:

    $\bar{x} \;= (0, 0.3748, 0, 1.4997, 0, 1.8747, 0)^T \geq 0, \quad \bar{f} = c_B^T \bar{x}_B = 0.3748.$

    In a comparison, the optimal solution obtained by MALIB is

    $x^* \;= (0, 0.3750, 0, 1.5000, 0, 1.8750, 0), \qquad f^* = 0.3750.$

Chapter 24
Pivotal Interior-Point Method

The simplex method and the interior-point method are two diverging and competitive types of methods for solving LP problems. The former moves on the underlying polyhedron, from vertex to adjacent vertex, along edges until an optimal vertex is reached while the latter approaches an optimal point by moving across interior of the polyhedron.

Although the basic ideas, motivations and development tracks of the two methods appear quite different, attempts will be made in this chapter to combine the two methods to take advantages of both, in a natural manner.

In view of that the interior-point method has been seriously restricted in applications since it can not be "warmly started", and provides only an approximate optimal solution (see Sect. 9.5), three pivotal interior-point algorithms will be derived. These algorithms yield from standard interior-point algorithms by adding inner steps. Presented in the last section of this chapter, on the other hand, the so-called "feasible-point simplex algorithm" is established along another line, which might be the first simplex-like algorithm that acrosses the interior of the polyhedron.

24.1 Pivotal Affine Interior-Point Method

In this section, firstly an interior-point algorithms is designed by forming a search direction based on affine-scaling. This search direction is equivalent to that used in Dikin's affine algorithm in theory. However, the new framework allows to introduce inner pivotal steps to create a better interior-point algorithm.

Let $\bar{x}$ be the current interior-point. Consider the dual problem of (9.25), i.e.,

$$\begin{array}{ll} \max & g = b^{\mathrm{T}}y, \\ \text{s.t.} & (\bar{X}A^{\mathrm{T}} \vdots I)\begin{pmatrix} y \\ z \end{pmatrix} = \bar{X}c, \; z \geq 0. \end{array} \tag{24.1}$$

P.-Q. PAN, *Linear Programming Computation*, DOI 10.1007/978-3-642-40754-3_24,

Note that there is a 1-to-1 correspondence between columns of the unit matrix I and indices of z.

At first, we will realize the "dual elimination" by orthogonal transformations (Sect. 25.1.3). Since A is of full row rank, hence $\bar{X}A^{\mathrm{T}}$ is of full column rank, there exists the QR factorization

$$\bar{X}A^{\mathrm{T}} = (Q_1, Q_3)\begin{pmatrix} R_1 \\ 0 \end{pmatrix} = Q_1 R_1, \tag{24.2}$$

where (Q_1, Q_3) is orthogonal, partitioned as $Q_1 \in \mathcal{R}^{n\times m}$ and $Q_3 \in \mathcal{R}^{n\times(n-m)}$, and $R_1 \in \mathcal{R}^{m\times m}$ is nonsingular upper triangular.

The following result reveals that the search direction in x'-space, used in the Dikin's affine algorithm, can be obtained alternatively by using the matrix Q_3.

Proposition 24.1.1. *$\Delta x'$, defined by (9.27), is equal to $Q_3 Q_3^{\mathrm{T}} \bar{X} c$.*

Proof. Substituting (24.2) to (9.27) and noting $Q_1^{\mathrm{T}} Q_1 = I$ and $Q_1 Q_1^{\mathrm{T}} + Q_3 Q_3^{\mathrm{T}} = I$ gives

$$\Delta x' = (I - Q_1 R_1 (R_1^{\mathrm{T}} Q_1^{\mathrm{T}} Q_1 R_1)^{-1} R_1^{\mathrm{T}} Q_1^{\mathrm{T}})\bar{X}c = (I - Q_1 Q_1^{\mathrm{T}})\bar{X}c = Q_3(Q_3^{\mathrm{T}}\bar{X}c). \tag{24.3}$$

□

Thereby, we are led to the following variant of the affine algorithm.

Algorithm 24.1.1 (Variant of Algorithm 9.2.1). The same as Algorithm 9.2.1, except for its step 1 replaced by

1. Compute $\Delta x'$ by (24.3).

As they differ only in the way to compute the same search direction, Algorithm 24.1.1 and Dikin's algorithm are equivalent. The computational efforts involved in them depend on how to implement, the sparsity of A, and the number $n - m$, compared with m, and etc. We will not go into details here because, after all, what we are really interested in is not the algorithm itself but a variant, as derived as follows.

As the affine method with long step turned out to be superior to that with short step in practice, it is attractive to go further along this line by introducing inner pivotal steps to decrease the objective value as much as possible, with reasonable costs.

We begin with premultiplying the augmented matrix of the equality constraints of (24.1) by $Q^{\mathrm{T}} = [Q_1, Q_3]^{\mathrm{T}}$. Such doing leads to a so-called *triangular form*, i.e.,

$$Q^{\mathrm{T}}(\bar{X}A^{\mathrm{T}} \vdots I \mid \bar{X}c) = \begin{pmatrix} R_1 & Q_1^{\mathrm{T}} & \mid Q_1^{\mathrm{T}}\bar{X}c \\ 0 & Q_3^{\mathrm{T}} & \mid Q_3^{\mathrm{T}}\bar{X}c \end{pmatrix}, \tag{24.4}$$

which represents the linear system equivalent to the dual equality constraints. Based on Proposition 24.1.1, it is known that the south-east submatrix $(Q_3^{\mathrm{T}} \mid Q_3^{\mathrm{T}}\bar{X}c)$ gives the projection $\Delta x'$, defined by (24.3), i.e.,

$$\Delta x' = (I - Q_1R_1(R_1^{\mathrm{T}}Q_1^{\mathrm{T}}Q_1R_1)^{-1}R_1^{\mathrm{T}}Q_1^{\mathrm{T}})\bar{X}c = (I - Q_1Q_1^{\mathrm{T}})\bar{X}c = Q_3(Q_3^{\mathrm{T}}\bar{X}c). \tag{24.5}$$

Then, update $\bar{x}$ by (9.30), i.e.,

$$\hat{x} = \bar{x} - \lambda\bar{X}\Delta x'/\max(\Delta x'), \tag{24.6}$$

where $\lambda \in (0, 1)$ a *stepsize*. If it goes to the next iteration at this point, the resulting is just the same as Algorithm 24.1.1.

On the contrary, we will carry out a series of inner iterations in x'-space, starting from

$$N = \emptyset, \quad B = A$$

Assume that $\Delta x'_B = \Delta x' \not\leq 0$. Update the $\bar{x}'_B$ by the following formula:

$$\hat{x}'_B = e - \lambda\Delta x'_B/\max(\Delta x'_B),$$

and determine an index q such that

$$q = \arg\max\{\Delta x'_j \mid j = 1, \cdots, n\}.$$

It is not difficult to show that $(\bar{X}A^{\mathrm{T}} \vdots e_q)$ is of full column rank. If the QR factorization of it is available, then the orthogonal projection of the objective gradient $\bar{X}c$ onto the null space of

$$\begin{pmatrix} A\bar{X} \\ e_q^{\mathrm{T}} \end{pmatrix}$$

can be computed analogously as before, and it is thereby able to update the solution once more in x'-space. Note that the q-indexed component of the projection equals 0, hence the q-indexed component of the solution remains unchanged.

Assume that after $k < n - m$ inner iterations, there are index sets

$$N = \{1, \cdots, k\}, \quad B = \{k + 1, \cdots, n\}.$$

Let the QR factorization $(\bar{X}A^{\mathrm{T}} \vdots I_N) = QR$ be available. Premultiplying by Q^{T} the augmented matrix of equality constraints of (24.1) gives

$$\begin{aligned} &Q^{\mathrm{T}}(\bar{X}A^{\mathrm{T}} \vdots I_N \vdots I_B | \bar{X}c) \\ &= \begin{pmatrix} Q_1^{\mathrm{T}}\bar{X}A^{\mathrm{T}} & Q_1^{\mathrm{T}}I_N & Q_1^{\mathrm{T}}I_B & | \; Q_1^{\mathrm{T}}\bar{X}c \\ Q_2^{\mathrm{T}}\bar{X}A^{\mathrm{T}} & Q_2^{\mathrm{T}}I_N & Q_2^{\mathrm{T}}I_B & | \; Q_2^{\mathrm{T}}\bar{X}c \\ Q_3^{\mathrm{T}}\bar{X}A^{\mathrm{T}} & Q_3^{\mathrm{T}}I_N & Q_3^{\mathrm{T}}I_B & | \; Q_3^{\mathrm{T}}\bar{X}c \end{pmatrix} = \begin{pmatrix} R_{11} & R_{12} & Q_1^{\mathrm{T}}I_B & | \; Q_1^{\mathrm{T}}\bar{X}c \\ 0 & R_{22} & Q_2^{\mathrm{T}}I_B & | \; Q_2^{\mathrm{T}}\bar{X}c \\ 0 & 0 & Q_3^{\mathrm{T}}I_B & | \; Q_3^{\mathrm{T}}\bar{X}c \end{pmatrix}. \end{aligned} \tag{24.7}$$

where the orthogonal matrix $Q = (Q_1, Q_2, Q_3)$ is partitioned as $Q_1 \in \mathcal{R}^{n\times m}$, $Q_2 \in \mathcal{R}^{n\times k}$ and $Q_3 \in \mathcal{R}^{n\times(n-m-k)}$, and $R_{11} \in \mathcal{R}^{(m+k)\times(m+k)}$ is nonsingular upper triangular. The preceding is the kth triangular form, whose south-east submatrix gives projection

$$\Delta x'_B = (Q_3^{\mathrm{T}} I_B)^{\mathrm{T}}(Q_3^{\mathrm{T}} \bar{X} c). \tag{24.8}$$

Assume that $\Delta x'_B \not\leq 0$. Since $\Delta x'_N = (Q_3^{\mathrm{T}} I_N)^{\mathrm{T}}(Q_3^{\mathrm{T}} \bar{X} c) = 0$ and $\bar{x}'_N$ remains unchanged, what only needs to do is updating by

$$\hat{x}'_B = \bar{x}'_B - \lambda\alpha\Delta x'_B. \tag{24.9}$$

where

$$\alpha = \bar{x}'_q/\Delta x'_q = \min\{\bar{x}'_j/\Delta x'_j \mid \Delta x'_j > 0,\ j \in B\}, \tag{24.10}$$

which is the stepsize from the current solution to the nearest boundary. It is clear that the new solution is again an interior point. Then move q from B to N, and go to the $(k+1)$th inner iteration.

The forgoing process terminates when $k = n-m$ or $Q_3^{\mathrm{T}} \bar{X} c = 0$, hence the $\Delta x'_B$ defined by (24.8) vanishes. In this case, substituting $\bar{z}_B = 0$ to the dual equality constraints, represented by (24.7), leads to the upper triangular system

$$\begin{pmatrix} R_{11} & R_{12} \\ 0 & R_{22} \end{pmatrix}\begin{pmatrix} y \\ z_N \end{pmatrix} = \begin{pmatrix} Q_1^{\mathrm{T}} \bar{X} c \\ Q_2^{\mathrm{T}} \bar{X} c \end{pmatrix}. \tag{24.11}$$

Assume that $(\bar{y}, \bar{z}_N)$ is the solution to this system.

If $\bar{z}_N \geq 0$, it is not difficult to show that $\bar{x}$ is the optimal solution to the following problem:

$$\begin{array}{ll} \min & c^{\mathrm{T}} x, \\ \text{s.t.} & Ax = b, \\ & x_B \geq 0, \quad x_N \geq \bar{x}_N. \end{array}$$

If $\lambda \in (0,1)$ is sufficiently close to 1, then $\bar{x}_N$ can be arbitrarily close to 0, in principle. If λ is predetermined to be close to 1, therefore, $\bar{x}$ can be regarded as an approximate optimal solution to the original problem, and the solution process terminates.

In the other case when $\bar{z}_N \not\geq 0$, the inner iterations are finished. The related solution in the original space is computed by

$$\bar{x} = \bar{X}\bar{x}',$$

and a new $\bar{X}$ can be formed to be ready to go to the next outer iteration.

The trick of the algorithm lies in that the QR factors in each inner iteration can be obtained via recurrence, need not to compute from scratch. In fact, the last $n-m-k-1$ components of the q-indexed column of (24.7) can be eliminated by Givens rotations. Assume that $\tilde{Q} \in \mathcal{R}^{(n-m-k)\times(n-m-k)}$ is the product of Givens rotations such that

$$\tilde{Q}Q_3^{\mathrm{T}}e_q = \eta e_1,$$

then $\hat{Q}^{\mathrm{T}} = [I \ \vdots \ \tilde{Q}^{\mathrm{T}}]^{\mathrm{T}}Q^{\mathrm{T}}$ is just the wanted factor for the $(k+1)$th inner iteration. It is then seen that the $(k+1)$ and kth triangular forms are the same, except for the submatrix, associated with Q_3. On the other hand, nevertheless, at the beginning of each round of outer iterations, it is necessary to compute the QR factors from scratch since $\bar{X}$ changed.

The overall steps can be summarized to the following algorithm.

Algorithm 24.1.2 (Pivotal affine interior-point algorithm). Given $\lambda \in (0,1)$. Initial: interior point $\bar{x} > 0$. This algorithm solves the standard LP problem.

1. Set $k = 0$, and compute triangular form (24.4).
2. Compute $\Delta x' = Q_3(Q_3^{\mathrm{T}}\bar{X}c)$.
3. Stop if $\Delta x' \le 0$ (lower unbounded).
4. Determine α and q such that
   $\alpha = \bar{x}'_q/\Delta x'_q = \min\{\bar{x}'_j/\Delta x'_j \mid \Delta x'_j > 0, j \in B\}$.
5. If $\alpha \ne 0$, update: $\bar{x}'_B = \bar{x}'_B - \lambda\alpha\Delta x'_B$ (24.9).
6. Set $k = k+1$, and update (B, N) by bringing q from B to N.
7. Go to step 10 if $k = n-m$ or $Q_3^{\mathrm{T}}\bar{X}c = 0$.
8. Eliminate the $(m+k+1)$ to nth components of the q-indexed column of the triangular form by Givens rotations.
9. Go to step 2.
10. Solve the upper triangular system (24.11).
11. Stop if $\bar{z}_N \ge 0$ (approximate optimality achieved).
12. Set $\bar{x} = \bar{X}\bar{x}'$.
13. Go to step 1.

Note This Algorithm contains steps 2–9 as its inner steps.

The algorithm, developed by Pan (2013) is slightly different from the preceding algorithm, as the former uses update

$$\hat{x}'_B = \bar{x}'_B - \alpha\Delta x'_B$$

rather than (24.9). Thereby, the resulting iterate is not interior but boundary point. If the optimality condition is not satisfied after a round of inner iterations finished, it goes back to a nearby interior point to start the next round of outer iterations as follows.

Assume that $\bar{x}$ is the interior point at the beginning of the outer iteration, and $\hat{x}$ is the end boundary point of the inner iterations. The interior point used for the next outer iteration is determined by

$$\bar{x} = \bar{x} + \mu(\hat{x} - \bar{x}),$$

where $\mu \in (0,1)$ ($\mu = 0.95$ is taken by Pan).

Table 24.1 Iteration and time ratios

	AIP/VAIP		AIP/PAIP	
Problem	Iterations	Time	Iterations	Time
Set 1 (16)	0.98	0.32	6.32	1.47
Set 2 (10)	0.92	0.25	3.55	1.52
Average (26)	0.95	0.26	4.56	1.52

There is no available numerical results with Algorithm 24.1.2. As it is close to Pan's original algorithm, we cite his numerical results to give the reader a clue on its performance.

The associated computational experiments were carried out on a Pentium III 550E PC with Windows 98 operating system, 168 MB inner storage and about 16 decimal precision. Visual Fortran 5.0 compiler was used. There were following three dense codes involved:

1. AIP: affine Algorithm 9.2.1.
2. VAIP: Algorithm 24.1.1.
3. PAIP: Algorithm 24.1.2.

The preceding codes are tested on the 26 smallest (by $m + n$) Netlib standard problems. The first set involves 16 smaller problems, and the second set are the rest 10 problems (Appendix B: Table B.4, problems AFIRO-DEGEN2).

Table 24.1 lists iterations and time ratios:

From the bottom line of the preceding Table, it is seen that total iteration and time ratios of AIP to VAIP are 0.95 and 0.26, respectively. As expected, the latter performs worse than the former. However, PAIP outperforms AIP significantly: the total iteration and time ratios of AIP to PAIP are 4.56 and 1.52, respectively. Therefore, the pivotal inner iterations appear effective.

It is not surprising that the time ratio of AIP to PAIP is much less than their iteration ratio (1.52 vs. 4.56), since each iteration of the latter is more time consuming, due to the use of the orthogonal transformation. Fortunately, the so-called "dual elimination" allows to use the Gaussian elimination instead (see Lemma 25.1.1). In particular, such doing is advantageous for sparse computations. On the other hand, of course, the associated search direction will no longer be the desired *orthogonal* projection. It is not known how such an algorithm will perform.

24.2 Pivotal Affine Face Interior-Point Method

In this section, two interior-point variants of the affine face method (Sect. 22.4) will be derived. Firstly, an interior-point variant is designed by directly using the initial search direction of the affine face method. Then, it is modified further by introducing pivotal inner iterations.

Consider reduced problem (22.1). Introduce notation

$$\bar{X} = \mathrm{diag}(\bar{x}_1, \cdots, \bar{x}_n, 1). \tag{24.12}$$

The initial search direction is defined by (22.30) with $k = n + 1$ or $B = A$, or denoted by

$$\Delta = -e_{n+1} + \bar{X}^2 A^{\mathrm{T}} \bar{y}, \quad (A\bar{X}^2 A^{\mathrm{T}})\bar{y} = -e_{m+1}. \tag{24.13}$$

which is downhill with respect to the objective function, as well as points to the interior of the feasible region. It is suitable to be a search direction for designing an interior-point algorithm. What is needed is just starting from an interior point and taking λ times of the original stepsize, where λ is a positive number less than 1 (e.g., 95–99 %).

Using the preceding notation, the overall steps can be put into the following interior-point algorithm.

Algorithm 24.2.1 (Affine face interior-point algorithm). Given $\lambda \in (0, 1)$. Initial: interior point $\bar{x} > 0$. This algorithm approximately solves the reduced problem (22.1).

1. Compute $A\bar{X}^2 A^T = LL^T$.
2. Solve $L^T \bar{y} = -(1/\nu)e_{m+1}$ for $\bar{y}$, where ν is the $(m + 1)$th diagonal of L.
3. Compute $\Delta = -e_{n+1} + \bar{X}^2 A^T \bar{y}$.
4. Stop if $J = \{j \in A \mid \Delta_j < 0\} = \emptyset$ (lower unbounded).
5. Determine $\alpha = \lambda \min_{j \in J} -\bar{x}_j / \Delta_j$.
6. Update: $\bar{x} = \bar{x} + \alpha\Delta$.
7. Go to step 1.

Was astonished, the author found that the preceding algorithm is the same as Dikin's affine algorithm (hence Algorithm 24.1.1), essentially. The only difference lies in that the former solves the reduced problem while the latter solves the standard problem. In fact, if Dikin handled the reduced rather than standard problem, he would have derived Algorithm 24.2.1.

On the other hand, the two algorithms differ computationally. Algorithm 24.2.1 should be preferable, as it saves the computational effort by solving a triangular system in each iteration. It is more than that. In fact, Algorithm 24.2.1 can be improved by incorporate pivotal inner iterations.

The resulting algorithm can be obtained by modifying Algorithm 22.4.1 easily.

Algorithm 24.2.2 (Pivotal affine face interior-point algorithm). Given $\lambda \in (0, 1)$. Initial: interior point $\bar{x} > 0$. This algorithm solves the reduced problem (22.1).

The same as Algorithm 22.4.1, except for step 7 replaced by

7. Update $\bar{x}_B = \bar{x}_B + \lambda\alpha\Delta_B$.

It is not difficult to modify the preceding conformably by including some termination criterion on precision.

Note that the vertex optimal solution can be computed by setting $\bar{x}_N = 0$ at the end of the solution precess, if needed.

Example 24.2.1. Solve the following problem by Algorithm 24.2.2:

$$\begin{array}{lrcl} \min & f = x_6, & & \\ \text{s.t.} & -4x_1 + 3x_2 + x_3 - 2x_4 + 2x_5 & = & 5, \\ & 3x_1 - x_2 + 2x_3 - 3x_4 - 4x_5 & = & -8, \\ & x_1 + x_2 + 2x_3 + x_4 + 3x_5 & = & 12, \\ & -2x_1 + 3x_3 + 2x_4 + x_5 - x_6 & = & 0, \\ & x_j \geq 0, \quad j = 1, \cdots, 6. & & \end{array}$$

Set $\lambda = 99/100$. Initial interior point: $\bar{x} = (1, 2, 1, 1, 2, -1)^{\mathrm{T}}$.

Answer Iteration 1:

1. $k = 6$, $B = \{1, 2, 3, 4, 5, 6\}$, $N = \emptyset$. $\bar{X}_B = \mathrm{diag}(1, 2, 1, 1, 2, -1)^{\mathrm{T}}$, face matrix B and the Cholesky factor of $B\bar{X}_B^2 B^{\mathrm{T}}$ are

$$B = \begin{pmatrix} -4 & 3 & 1 & -2 & 2 & \\ 3 & -1 & 2 & -3 & -4 & \\ 1 & 1 & 2 & 1 & 3 & \\ -2 & & -3 & 2 & 1 & -1 \end{pmatrix},$$

$$L = \begin{pmatrix} 1{,}399/134 & & & \\ -2{,}362/411 & 7{,}176/919 & & \\ 1{,}985/471 & -1{,}327/373 & 916/207 & \\ 1{,}181/822 & -302/85 & -800/331 & 815/672 \end{pmatrix}.$$

2. $\bar{y} = (-100/5{,}011, -2{,}485/5{,}193, -215/579, -1{,}219/1{,}793)^{\mathrm{T}}$.
3. $\Delta_B = (-583/1,587, -2,127/2,168, 509/1,591, -356/1,393, 278/865, -1,385/3,339)^{\mathrm{T}} \neq 0$.
5. $J = \{1, 2, 4\} \neq \emptyset$.
6. $\alpha = (99/100)\min\{-1/(-583/1{,}587), -2/(-2{,}127/2{,}168), -1/(-356/1{,}393)\}$
$= (99/100)(1{,}745/856) = 2{,}333/1{,}156$, $p = 2$.
7. $\bar{x}_B = (1, 2, 1, 1, 2, -1)^{\mathrm{T}} + (2,333/1,156)(-583/1,587, -2,127/2{,}168, 509/1{,}591, -356/1,393, 278/865, -1,385/3,339)^{\mathrm{T}}$
$= (323/1,249, 750/37,499, 1,593/968, 261/539, 2,005/757, -2,019/1,099)^{\mathrm{T}}$.
8. $B = \{1, 3, 4, 5, 6\}$, $N = \{2\}$.
9. $$L = \begin{pmatrix} 882/145 & & & \\ -290/49 & 307/43 & & \\ 1{,}450/441 & -959/279 & 3{,}244/737 & \\ 145/98 & -7{,}775/2{,}199 & -1{,}184/473 & 2{,}500/2{,}447 \end{pmatrix}.$$
10. $k = 5$.

Iteration 2:

2. $\bar{y} = (-361/1{,}907, -1{,}405/1{,}907, -1{,}039/1{,}907, -845/882)^{\mathrm{T}}$.
3. $\Delta_B = (-156/1{,}907, 232/1{,}907, 244/1{,}907, -184/1{,}907, -277/6{,}603)^{\mathrm{T}} \neq 0$.
5. $J = \{1, 5\} \neq \emptyset$.
6. $\alpha = (99/100)\min\{-(323/1,249)/(-156/1,907), -(2,005/757)/(-184/1,907)\}$
   $= (99/100)(7{,}761/2{,}455) = 917/293, \ p = 1$.
7. $\bar{x}_B = (323/1,249, 1,593/968, 261/539, 2,005/757, -2,019/1{,}099)^{\mathrm{T}} + (917/293)(-156/1{,}907, 232/1{,}907, 244/1{,}907, -184/1{,}907, -277/6{,}603)^{\mathrm{T}}$
   $= (109/42{,}149, 1{,}688/833, 583/659, 1{,}117/476, -5{,}360/2{,}723)^{\mathrm{T}}$.
8. $B = \{3, 4, 5, 6\}, \ N = \{1, 2\}$.
9. $$L = \begin{pmatrix} 3{,}524/769 & & & \\ -1{,}435/274 & 1{,}211/172 & & \\ 1{,}435/274 & -3{,}433/1{,}235 & 1{,}187/491 & \\ 769/3{,}524 & -2{,}201/577 & -499/322 & 1 \end{pmatrix}.$$
10. $k = 4 = m + 1$.
11. $\bar{z}_N = (156/1{,}907, 717/1{,}907)^{\mathrm{T}} \geq 0$.
12. The approximate basic optimal solution and optimal value are

$$\bar{x} \approx (109/42{,}149, 750/37{,}499, 1{,}688/833, 583/659, 1{,}117/476)^{\mathrm{T}},$$
$$\bar{x}_6 \approx -5{,}360/2{,}723.$$

The outcome is close to the exact basic optimal solution and optimal value, i.e.,

$$x^* = (0, 0, 79/39, 34/39, 92/39)^{\mathrm{T}}, \ x_6^* = -77/99.$$

The error in components of the approximate optimal solution is about 0.01, while that in the approximate optimal value is about 0.05.

24.3 Pivotal D-Reduced Gradient Interior-Point Method

In this section, we derive a pivotal interior-point algorithm with the deficient-basis framework, without discussion on the related theoretical problems.

It is noted that the search direction, used in the D-reduced gradient method (Sect. 21.5), is uphill with respect to the dual objective function, as well as points to the interior of the feasible region. Thereby, the direction is suitable to be used to design an interior-point algorithm by the known trick: starting from an interior point, and cutting down the stepsize to λ times of the original, where λ is positive number less than 1 (e.g., 0.95–0.999).

The according steps are put in the following algorithm.

Algorithm 24.3.1 (D-reduced gradient interior-point algorithm). Given $\mu > 0$, $\lambda \in (0,1)$, $\epsilon > 0$. Initial: interior point $\bar{z} > 0$. This algorithm solves the D-reduced problem (17.1).

1. Compute ω by (21.15).
2. Stop if $J = \{j \in A \mid \omega_j > 0\} = \emptyset$ (infeasible problem).
3. Compute $\beta = \bar{z}_q/\bar{\omega}_q = \min_{j \in J} \bar{z}_j/\bar{\omega}_j$.
4. Stop if $\beta < \epsilon$ (approximate optimality achieved).
5. If $\beta \neq 0$, update $\bar{z} = \bar{z} - \lambda\beta\omega$.
6. Go to step 1.

Nevertheless, our preliminary test indicates that the preceding algorithm converges quite slow (if does).

As in the previous two sections, we incorporate pivotal inner iterations to it. The resulting algorithm can be obtained by modifying Algorithm 21.6.1 easily.

Algorithm 24.3.2 (Pivotal D-reduced gradient interior-point algorithm). Given $\mu > 0, \lambda \in (0,1)$. Initial: $\bar{z}_N > 0$. This algorithm solves D-reduced problem (17.1).

The same as Algorithm 21.6.1, except for steps 4 and 5 replaced respectively by

4. Determine β and q such that $\beta = \bar{z}_q/\bar{\omega}_q = \min_{j \in J} \bar{z}_j/\bar{\omega}_j$, and compute $\lambda\beta$.
5. If $\beta \neq 0$, update $\bar{z}_N = \bar{z}_N - \lambda\beta\bar{\omega}_N$.

It is not difficult to modify the preceding algorithm by introducing some precision tolerance.

Example 24.3.1. Solve the following problem by Algorithm 24.3.2:

$$
\begin{array}{ll}
\min & f = 2x_1 + x_2 + 3x_3 + x_4 + 2x_5 + 4x_6, \\
\text{s.t.} & \quad x_1 + 3x_2 - 2x_3 + 3x_4 - 4x_5 + 2x_6 = 0, \\
& \quad\quad - 2x_2 \quad\quad - x_4 - 2x_5 + x_6 = 0, \\
& -2x_1 + x_2 + 2x_3 - 4x_4 + 3x_5 - 3x_6 = 0, \\
& \quad\quad + 2x_2 + 3x_3 \quad\quad - 2x_5 - x_6 = 1, \\
& \quad\quad x_j \geq 0, \quad j = 1, \cdots, 6.
\end{array}
$$

Answer For convenience of comparison, the related tableau will be given for each iteration.

Initial tableau:

x_1	x_2	x_3	x_4	x_5	x_6	RHS
1	3	−2	3	−4	2	
	−2		−1	−2	1	
−2	1	2	−4	3	−3	
	2	3		−2	−1	1
2	1	3	1	2	4	

$\lambda = 95/100,\ \mu = 1.\ r = 4.$

$$N_{R'} = \begin{pmatrix} 1 & 3 & -2 & 3 & -4 & 2 \\ 0 & -2 & 0 & -1 & -2 & 1 \\ -2 & 1 & 2 & -4 & 3 & -3 \end{pmatrix}.$$

Iteration 1:

1. $k \ = 0;\ B, R{=}\emptyset,\ N{=}\{1,\cdots,6\},\ R' = \{1,2,3\}.\ \bar{z}_N = (2,1,3,1,2,4)^{\mathrm{T}}.$
2. $\omega_N = (0,2,3,0,-2,-1)^{\mathrm{T}} - N_{R'}^{\mathrm{T}} N_{R'}(1/2,1,1/3,1,1/2,1/4)^{\mathrm{T}}$
$= (-19/2,-191/12,101/6,-325/12,187/12,-41/3)^{\mathrm{T}},\ \bar{\omega}_N = \omega_N.$

x_1	x_2	x_3	x_4	x_5	x_6	RHS
1	3	−2	3	−4	2	
	−2		−1	−2	1	
−2	1	2	−4	3	−3	
−19/2	−191/12	101/6	−325/12	187/12	−41/3	1
2	1	3	1	2	4	

3. $J \ = \{3,5\}.$
4. $\beta \ = (95/100)\min\{3/(101/6), 2/(187/12)\} = (95/100)(24/187)$
$= 114/935,\ q = 5.$
5. $\bar{z}_N \ = (2,1,3,1,2,4)^{\mathrm{T}}$
$- (114/935)(-19/2,-191/12,101/6,-325/12,187/12,-41/3)^{\mathrm{T}}.$
6. $\bar{a}_q(R') = (-4,-2,3)^{\mathrm{T}}.$
7. $\bar{a}_q(R') \neq 0.$
8. $p \ = 1,\ \tau = -1/4,\ \hat{B}_{\hat{R}}^{-1} = (-1/4).$
9. $k \ = 1,\ B = \{5\},\ R = \{1\}.N = \{1,2,3,4,6\},\ R' = \{2,3\}.$
10. $\bar{\omega}_N = (-19/2,-191/12,101/6,-325/12,-41/3)^{\mathrm{T}}$
$- (1,3,-2,3,2)^{\mathrm{T}}(-1/4)(187/12)$
$= (-269/48,-203/48,217/24,-739/48,-47/8)^{\mathrm{T}}.$

x_1	x_2	x_3	x_4	x_5	x_6	RHS
−1/4	−3/4	1/2	−3/4	1	−1/2	
−1/2	−7/2	1	−5/2			
−5/4	13/4	1/2	−7/4		−3/2	
−269/48	−203/48	217/24	−739/48		−47/8	1
1,696/537	644/219	886/935	1,609/374	1/10	5,298/935	−114/935

Iteration 2:

3. $J = \{3\}.$
4. $\beta = (95/100)\min\{(886/935)/(217/24)\} = (95/100)(875/8{,}349)$
$= 205/2{,}059, q = 3.$

5. $\bar{z}_N = (1{,}696/537, 644/219, 886/935, 1{,}609/374, 5{,}298/935)^{\mathrm{T}}$
$\quad - (205/2{,}059)(-269/48, -203/48, 217/24, -739/48, -47/8)^{\mathrm{T}}$
$\quad = (1{,}349/363, 9{,}359/2{,}784, 47/992, 4{,}102/703, 2{,}513/402)^{\mathrm{T}}$.
6. $\bar{a}_q(R) = (-1/4)(-2) = 1/2,\ \bar{a}_q(R') = (0,2)^{\mathrm{T}} - (-2,3)^{\mathrm{T}}(-2) = (-4,8)^{\mathrm{T}}$.
7. $\bar{a}_q(R') \neq 0$.
8. $p = 3,\ \tau = (2 - 3(1/2))^{-1} = 2, v = -2(1/2) = -1, d^{\mathrm{T}} = -2(3)(-1/4) = 3/2$.

$$U = (-1/4) - (-1)3(-1/4) = -1, \quad \hat{B}_{\hat{R}}^{-1} = \begin{pmatrix} -1 & -1 \\ 3/2 & 2 \end{pmatrix}.$$

9. $k = 2,\ B = \{5,3\},\ R = \{1,3\}. N = \{1,2,4,6\},\ R' = \{2\}$.
10. $\bar{\omega}_N = (-19/2, -191/12, -325/12, -41/3)^{\mathrm{T}}$

$$- \begin{pmatrix} 1 & 3 & 3 & 2 \\ -2 & 1 & -4 & -3 \end{pmatrix}^{\mathrm{T}} \begin{pmatrix} -1 & -1 \\ 3/2 & 2 \end{pmatrix}^{\mathrm{T}} (187/12, 101/6)^{\mathrm{T}}$$

$= (17, -63, 65/4, 85/4)^{\mathrm{T}}$.

x_1	x_2	x_3	x_4	x_5	x_6	RHS
1	−4		1	1	1	
2	−10		1		3	
−5/2	13/2	1	−7/2		−3	
17	−63		65/4		85/4	1
1,349/363	9,359/2,784	47/992	4,102/703	1/10	2,513/402	−905/4,086

Iteration 3:

3. $J = \{1,4,6\}$.
4. $\lambda\beta = (95/100)\min\{(1{,}696/537)/17, (4{,}102/703)/(65/4), (2{,}513/402)/(85/4)\}$
$\quad = (95/100)(473/2{,}546) = 473/2{,}680,\ q = 1$.
5. $\bar{z}_N = (1{,}349/363, 9{,}359/2{,}784, 4{,}102/703, 2{,}513/402)^{\mathrm{T}}$
$\quad - (473/2{,}680)(17, -63, 65/4, 85/4)^{\mathrm{T}}$
$\quad = (577/806, 9{,}398/649, 1{,}528/515, 1{,}608/643)^{\mathrm{T}}$.
6. $$\bar{a}_q(R) = \begin{pmatrix} -1 & -1 \\ 3/2 & 2 \end{pmatrix} \begin{pmatrix} 1 \\ -2 \end{pmatrix} = \begin{pmatrix} 1 \\ -5/2 \end{pmatrix},$$
$\bar{a}_q(R') = 0 - (-2,0)(1,-2)^{\mathrm{T}} = 2$.
7. $\bar{a}_q(R') \neq 0$.
8. $p = 2.\ \tau = (0 - (-2,0)(1,-5/2)^{\mathrm{T}})^{-1} = 1/2$.
$v = -(1/2)(1,-5/2)^{\mathrm{T}} = (-1/2, 5/4)^{\mathrm{T}}$.
$d^{\mathrm{T}} = -(1/2)((-2,0)B_R^{-1}) = -(1/2)(2,2) = (-1,-1)$.

$$U = \begin{pmatrix} -1 & -1 \\ 3/2 & 2 \end{pmatrix} - (-1/2, 5/4)^{\mathrm{T}}(2,2) = \begin{pmatrix} 0 & 0 \\ -1 & -1/2 \end{pmatrix},$$

$$\hat{B}_{\hat{R}}^{-1} = \begin{pmatrix} 0 & 0 & -1/2 \\ -1 & -1/2 & 5/4 \\ -1 & -1 & 1/2 \end{pmatrix}.$$

9. $k = 2,\ B = \{5,3,1\},\ R = \{1,3,2\}.N = \{2,4,6\},\ R' = \emptyset.$
10. $\bar{\omega}_N = (-191/12, -325/12, -41/3)^{\mathrm{T}}$

$$- \begin{pmatrix} 3 & 3 & 2 \\ 1 & -4 & -3 \\ -2 & -1 & 1 \end{pmatrix}^{\mathrm{T}} \begin{pmatrix} 0 & 0 & -1/2 \\ -1 & -1/2 & 5/4 \\ -1 & -1 & 1/2 \end{pmatrix}^{\mathrm{T}} (187/12, 101/6, -19/2)^{\mathrm{T}}$$

$= (22, 31/4, -17/4)^{\mathrm{T}}.$

x_1	x_2	x_3	x_4	x_5	x_6	RHS
	1		1/2	1	−1/2	
1	−5		1/2		3/2	
	−6	1	−9/4		3/4	
	22		31/4		−17/4	1
577/806	9,398/649	47/992	1,528/515	1/10	1,608/643	−1,143/2,872

Iteration 4:

3. $J = \{2,4\}.$
4. $\lambda\beta = (95/100)\min\{(8{,}645/597)/22, (1{,}528/515)/(31/4)\}$
   $= (95/100)(1{,}557/4{,}067) = 559/1{,}537,\ q = 4.$
5. $\bar{z}_N = (9{,}398/649, 1{,}528/515, 1{,}608/643)^{\mathrm{T}} - (559/1{,}537)(22, 31/4, -17/4)^{\mathrm{T}}$
   $= (6,460/997, 301/2,029, 3,395/839)^{\mathrm{T}}.$

x_1	x_2	x_3	x_4	x_5	x_6	RHS
	1		1/2	1	−1/2	
1	−5		1/2		3/2	
	−6	1	−9/4		3/4	
	22		31/4		−17/4	1
577/806	6,460/997	47/992	301/2,029	1/10	3,395/839	−636/835

6. $$\bar{a}_q(R) = \begin{pmatrix} 0 & 0 & -1/2 \\ -1 & -1/2 & 5/4 \\ -1 & -1 & 1/2 \end{pmatrix} \begin{pmatrix} 3 \\ -4 \\ -1 \end{pmatrix} = \begin{pmatrix} 1/2 \\ -9/4 \\ 1/2 \end{pmatrix}.$$
7. $\bar{a}_q(R') = 0.$
11. $\max\{1/2, -9/4, 1/2\} = 1/2 > 0,\ s = 1,\ p = 1.$

Iteration 5:

1. $k \;= 0;\; B, R = \emptyset,\; N = \{1, \cdots, 6\},\; R' = \{1, 2, 3\}.$
2. $\omega_N = (0, 2, 3, 0, -2, -1)^{\mathrm{T}}$
 $\quad - N_{R'}^{\mathrm{T}} N_{R'}(806/577, 997/6{,}460, 992/47, 2{,}029/301{,}10, 839/3{,}395)^{\mathrm{T}}$
 $\quad = (9{,}606/67, 1{,}965/23, -22{,}001/110, 22{,}051/69, -23{,}926/57,$
 $\quad 6{,}767/25)^{\mathrm{T}},$
 $\bar{\omega}_N = \omega_N.$

x_1	x_2	x_3	x_4	x_5	x_6	RHS
1	3	−2	3	−4	2	
	−2		−1	−2	1	
−2	1	2	−4	3	−3	
9,606/67	1,965/23	−22,001/110	22,051/69	−23,926/57	6,767/25	1
577/806	6,460/997	47/992	301/2,029	1/10	3,395/839	−636/835

3. $J \;= \{1, 2, 4, 6\}.$
4. $\lambda\beta = (95/100)\min\{(577/806)/(9, 606/67), (6, 460/997)/(1, 965/23),$
 $\quad (301/2, 029)/(22, 051/69), (3, 395/839)/(6, 767/25)\}$
 $\quad = (95/100)(37/79, 707) = 8/18, 141,\; q = 4.$
5. $\bar{z}_N \;= (577/806, 6, 460/997, 47/992, 301/2, 029, 1/10, 3, 395/839)^{\mathrm{T}}$
 $\quad - (8/18, 141)(9, 606/67, 1, 965/23, -22, 001/110, 22, 051/69,$
 $\quad - 23, 926/57, 6, 767/25)^{\mathrm{T}}$
 $\quad = (295/452, 1, 604/249, 189/1, 394, 104/14, 021, 1, 074/3, 767, 1,$
 $\quad 347/343)^{\mathrm{T}}.$
6. $\bar{a}_q(R') = (3, -1, -4)^{\mathrm{T}}.$
7. $\bar{a}_q(R') \neq 0.$
8. $p \;= 3.\; \tau = -1/4. \hat{B}_{\hat{R}}^{-1} = (-1/4).$
9. $k \;= 1,\; B = \{4\},\; R = \{3\}. N = \{1, 2, 3, 5, 6\},\; R' = \{1, 2\}.$
10. $\bar{\omega}_N = (9, 606/67, 1, 965/23, -22, 001/110, -23, 926/57, 6, 767/25)^{\mathrm{T}}$
 $\quad - (-2, 1, 2, 3, -3)^{\mathrm{T}}(-1/4)(22, 051/69)$
 $\quad = (-25, 331/1, 543, 15, 045/91, -10, 457/260, -7, 743/43, 6,$
 $\quad 478/209)^{\mathrm{T}}.$

x_1	x_2	x_3	x_4	x_5	x_6	RHS
−1/2	15/4	−1/2		−7/4	−1/4	
1/2	−9/4	−1/2		−11/4	7/4	
1/2	−1/4	−1/2	1	−3/4	3/4	
−25,331/1,543	15,045/91	−10,457/260		−7,743/43	6,478/209	1
295/452	1,604/249	189/1,394	104/14,021	1,074/3,767	1,347/343	−1,022/1,341

Iteration 6:

3. $J = \{2, 6\}$.
4. $\beta = (95/100)\min\{(1{,}604/249)/(15{,}045/91), (1{,}347/343)/(6{,}478/209)\}$
$= (95/100)(239/6{,}134) = 311/8{,}402,\ q = 2.$
5. $\bar{z}_N = (295/452, 1{,}604/249, 189/1{,}394, 1{,}074/3{,}767, 1{,}347/343)^{\mathrm{T}}$
$- (311/8{,}402)(-25{,}331/1{,}543, 15{,}045/91, -10{,}457/260,$
$- 7{,}743/43, 6{,}478/209)^{\mathrm{T}}$
$= (1{,}893/1{,}502, 439/1{,}363, 2{,}019/1{,}243, 1{,}821/262, 1{,}957/704)^{\mathrm{T}}.$
6. $\bar{a}_q(R) = (-1/4)(1) = -1/4,\ \bar{a}_q(R') = (3, -2)^{\mathrm{T}} - (3, -1)^{\mathrm{T}}(-1/4) = (15/4,$
$- 9/4)^{\mathrm{T}}.$
7. $\bar{a}_q(R') \neq 0.$
8. $p = 1,\ \tau = (3 - 3(-1/4))^{-1} = 4/15, v = -(4/15)(-1/4) = 1/15,$
$d^{\mathrm{T}} = -(4/15)(3)(-1/4) = 1/5.$
$U = (-1/4) - (1/15)3(-1/4) = -1/5, \quad \hat{B}_{\hat{R}}^{-1} = \begin{pmatrix} -1/5 & 1/15 \\ 1/15 & 4/15 \end{pmatrix}.$
9. $k = 2,\ B = \{4, 2\},\ R = \{3, 1\}. N = \{1, 3, 5, 6\},\ R' = \{2\}.$
10. $\bar{\omega}_N = (9{,}606/67, -22{,}001/110, -23{,}926/57, 6{,}767/25)^{\mathrm{T}}$
$- \begin{pmatrix} -2 & 2 & 3 & -3 \\ 1 & -2 & -4 & 2 \end{pmatrix}^{\mathrm{T}} \begin{pmatrix} -1/5 & 1/15 \\ 1/15 & 4/15 \end{pmatrix}^{\mathrm{T}} (22{,}051/69, 1{,}965/23)^{\mathrm{T}}$
$= (1{,}570/279, -1{,}763/97, -9{,}777/95, 14{,}664/349)^{\mathrm{T}}.$

x_1	x_2	x_3	x_4	x_5	x_6	RHS
−2/15	1	−2/15		−7/15	−1/15	
1/5		−4/5		−19/5	8/5	
7/15		−8/15	1	−13/15	11/15	
1,570/279		−1,763/97		−9,777/95	14,664/349	1
1,893/1,502	439/1,363	2,019/1,243	104/14,021	1,821/262	1,957/704	−553/692

Iteration 7:

3. $J = \{1, 6\}$.
4. $\lambda\beta = (95/100)\min\{(1{,}893/1{,}502)/(1{,}570/279), (1{,}957/704)/$
$(14{,}664/349)\}$
$= (95/100)(426/6{,}439) = 436/6{,}937,\ q = 6.$
5. $\bar{z}_N = (1{,}893/1{,}502, 2{,}019/1{,}243, 1{,}821/262, 1{,}957/704)^{\mathrm{T}}$
$- (436/6{,}937)(1{,}570/279, -1{,}763/97, -9{,}777/95, 14{,}664/349)^{\mathrm{T}}$
$= (437/482, 3{,}118/1{,}127, 5{,}287/394, 113/813)^{\mathrm{T}}.$
6. $\bar{a}_q(R) = \begin{pmatrix} -1/5 & 1/15 \\ 1/15 & 4/15 \end{pmatrix} \begin{pmatrix} -3 \\ 2 \end{pmatrix} = \begin{pmatrix} 11/15 \\ -1/15 \end{pmatrix},$
$\bar{a}_q(R') = 1 - (-1, -2)(11/15, -1/15)^{\mathrm{T}} = 8/5.$
7. $\bar{a}_q(R') \neq 0.$

8. $p = 2,\ \tau = (1 - (-1,-2)(11/15,-1/15)^{\mathrm{T}})^{-1} = 5/8.$
$v = -(5/8)(11/15,-1/15)^{\mathrm{T}} = (-11/24, 1/24)^{\mathrm{T}}.$
$d^{\mathrm{T}} = -(5/8)((-1,-2)B_R^{-1}) = -(5/8)(-1/5,-3/5) = (1/8, 3/8).$

$$U = \begin{pmatrix} -1/5 & 1/15 \\ 1/15 & 4/15 \end{pmatrix} - (-11/24, 1/24)^{\mathrm{T}}(-1/5,-3/5) = \begin{pmatrix} -7/24 & -5/24 \\ 5/24 & 7/24 \end{pmatrix}.$$

$$\hat{B}_{\hat{R}}^{-1} = \begin{pmatrix} -7/24 & -5/24 & -11/24 \\ 5/24 & 7/24 & 1/24 \\ 1/8 & 3/8 & 5/8 \end{pmatrix}.$$

9. $k = 2,\ B = \{4,2,6\},\ R = \{3,1,2\}. N = \{1,3,5\},\ R' = \emptyset.$
10. $\bar{\omega}_N = (9{,}606/67, -22{,}001/110, -23{,}926/57)^{\mathrm{T}}$

$$- \begin{pmatrix} -2 & 2 & 3 \\ 1 & -2 & -4 \\ 0 & 0 & -2 \end{pmatrix}^{\mathrm{T}} \begin{pmatrix} -7/24 & -5/24 & -11/24 \\ 5/24 & 7/24 & 1/24 \\ 1/8 & 3/8 & 5/8 \end{pmatrix}^{\mathrm{T}} \cdot (22{,}051/69,$$
$$1{,}965/23, 6{,}767/25)^{\mathrm{T}}$$
$$= (515/1{,}373, 25{,}888/9{,}137, -9{,}428/3{,}017)^{\mathrm{T}}.$$

x_1	x_2	x_3	x_4	x_5	x_6	RHS
−1/8	1	−1/6		−5/8		
1/8		−1/2		−19/8	1	
3/8		−1/6	1	7/8		
515/1,373		25,888/9,137		−9,428/3,017		1
437/482	439/1,363	3,118/1,127	104/14,021	5,287/394	113/813	−1,755/2,036

Iteration 8:

3. $J = \{1,3\}.$
4. $\lambda\beta = (95/100)\min\{(437/482)/(515/1{,}373), (3{,}118/1{,}127)/(25{,}888/9{,}137)\}$
$= (95/100)(2{,}780/2{,}847) = 500/539,\ q = 3.$
5. $\bar{z}_N = (437/482, 3{,}118/1{,}127, 5{,}287/394)^{\mathrm{T}}$
$-(500/539)(515/1{,}373, 25{,}888/9{,}137, -9{,}428/3{,}017)^{\mathrm{T}}$
$= (733/1{,}312, 214/1{,}547, 9{,}350/573)^{\mathrm{T}}.$

x_1	x_2	x_3	x_4	x_5	x_6	RHS
−1/8	1	−1/6		−5/8		
1/8		−1/2		−19/8	1	
3/8		−1/6	1	7/8		
515/1,373		25,888/9,137		−9,428/3,017		1
733/1,312	439/1,363	214/1,547	104/14,021	9,350/573	113/813	−2,450/1,369

6. $\bar{a}_q(R) = \begin{pmatrix} -7/24 & -5/24 & -11/24 \\ 5/24 & 7/24 & 1/24 \\ 1/8 & 3/8 & 5/8 \end{pmatrix} \begin{pmatrix} 2 \\ -2 \\ 0 \end{pmatrix} = \begin{pmatrix} -1/6 \\ -1/6 \\ -1/2 \end{pmatrix}.$
7. $R' = \emptyset.$
11. $\max\{-1/6, -1/6, -1/2\} \leq 0.$
13. Setting $\bar{x}_3 = 9{,}137/25{,}888$ gives

$$\begin{aligned}\bar{x}_B &= (9{,}137/25{,}888)(1/6, 1/6, 1/2)^{\mathrm{T}} \\ &= (9{,}137/155{,}328, 9{,}137/155{,}328, 9{,}137/51{,}776)^{\mathrm{T}}.\end{aligned}$$

Thus, the approximate basic optimal solution and optimal value are

$$\begin{aligned}\bar{x} &\approx (0, 9{,}137/155{,}328, 9{,}137/25{,}888, 9{,}137/155{,}328, 0, 9{,}137/51{,}776)^{\mathrm{T}}, \\ \bar{f} &\approx (1, 3, 1, 4)(9{,}137/155{,}328, 9{,}137/25{,}888, 9{,}137/155{,}328, 9{,}137/ \\ &\quad 51{,}776)^{\mathrm{T}} = 9{,}137/4{,}854.\end{aligned}$$

On the other hand, the exact basic optimal solution and optimal value are

$$x^* = (0, 1/17, 6/17, 1/17, 0, 3/17)^{\mathrm{T}}, \quad f^* = 32/17.$$

The errors are less than 10^{-5}, which are accumulated from the computation of $\bar{\omega}$.

24.4 Feasible-Point Simplex Method

On the conventional simplex framework, the method presented in this section will utilize a new pivot rule to generates a sequence of feasible points, which are not necessarily vertices or interior points. If the initial is an interior point, however, it becomes an interior-point algorithm. It might be the first simplex-like method that may go across the interior of the feasible region.

We are concerned with the bounded-variable problem (7.13), i.e.,

$$\begin{array}{ll} \min & c^T x \\ s.t & Ax = b, \quad l \leq x \leq u, \end{array} \tag{24.14}$$

where $A \in R^{m \times n} (m < n)$; rank $A = m$. The values of components of l and u are assumed to be finite, hence the problems is bounded. For problems with infinite bounds, one may use numbers in large enough module instead.

Let B be the current (standard) basis and N the associated nonbasis, defined by

$$B = \{1, \cdots, m\}, \qquad N = A \backslash B.$$

24.4.1 Column Pivot Rule and Optimality Condition

Let $\bar{x}$ be current feasible solution, whose nonbasic components are not necessarily on either lower or upper bound. The reduced costs are

$$\bar{c}_B = 0, \qquad \bar{c}_N = c_N - N^T \bar{y}, \qquad B^T \bar{y} = c_B. \tag{24.15}$$

Introduce nonbasic index sets

$$N_1 = \{j \in N \mid \bar{c}_j < 0\}, \qquad N_2 = \{j \in N \mid \bar{c}_j > 0\}. \tag{24.16}$$

Recall that Algorithm 7.4.1 selects an entering index based on $\bar{c}$ only. In contrast, we not only consider $\bar{c}$, but also take into account the possible ranges the nonbasic components of the current feasible solution are allowed to change. To this end, introduce notation

$$\delta_j = \begin{cases} u_j - \bar{x}_j, & j \in N_1, \\ \bar{x}_j - l_j, & j \in N_2, \\ \min\{u_j - \bar{x}_j, \bar{x}_j - l_j\} & j \in N \setminus N_1 \cup N_2. \end{cases} \tag{24.17}$$

So, δ_j is the distance from $\bar{x}_j$ to one of the associated its bounds that will be violated if $\bar{x}_j$ changes to decrease the objective value.

Then the following rule is applicable.

Rule 24.4.1 (Column pivot rule) Select a nonbasic index q such that

$$|\bar{c}_q|\delta_q = \max \{ \, |\bar{c}_j|\delta_j \mid j \in N \, \}. \tag{24.18}$$

Under the preceding rule, the objective value will decrease the most (by amount $|\bar{c}_q|\delta_q$), ignoring presence of broking basic variables.

In this context, the following optimal condition is relevant.

Proposition 24.4.1. *$\bar{x}$ is an optimal solution if it holds that*

$$|\bar{c}_j|\delta_j = 0, \qquad \forall\, j \in N. \tag{24.19}$$

Proof. Note that quantities $|\bar{c}_j|\delta_j$ $(j \in N)$ are upper bounds of the amount by which the objective value can decrease as the value of x_j changes. Thus, condition (24.19) implies that the objective value can not decrease any further, and the proposition is valid. □

Furthermore, it is clear that $\bar{x}$ is a basic optimal solution if

$$\delta_j = 0, \qquad \forall\, j \in N. \tag{24.20}$$

The preceding optimality condition might be suitable for applications when a vertex solution is required.

24.4.2 Search Direction

If the optimality condition is not fulfilled, it is possible to decrease the objective value.

Assume that a nonbasic index q has been selected to enter the basis. For a search direction, consider vector Δx defined by

$$\Delta x_B = \text{sign}(\bar{c}_q)\bar{a}_q, \qquad B\bar{a}_q = a_q. \tag{24.21}$$

$$\Delta x_j = \begin{cases} 0, & j \in N;\ j \neq q \\ -\text{sign}(\bar{c}_q), & j = q. \end{cases} \tag{24.22}$$

We have the following Lemma, ensuring the eligibility of Δx to be a search direction.

Lemma 24.4.1. *Assume that q is determined by (24.18). Vector Δx is a downhill with respective to the objective in the null of A.*

Proof. Note that $\bar{c}_q \neq 0$, since, otherwise, $|\bar{c}_q|\delta_q = 0$ implies that the optimality condition (24.19) holds, as leading to a contradiction.

From (24.21) and (24.22), it follows that

$$A\Delta x = \text{sign}(\bar{c}_q)BB^{-1}a_q - \text{sign}(\bar{c}_q)a_q = 0 \tag{24.23}$$

Thus, Δx is in the null of A.

Further, it holds by (24.15) that

$$\bar{c}_q = c_q - c_B^T B^{-1} a_q,$$

which together with (24.21), (24.22), (24.18) and (24.16) gives

$$\begin{aligned} c^T\Delta x &= -\text{sign}(\bar{c}_q)c_q + \text{sign}(\bar{c}_q)c_B^T B^{-1}a_q \\ &= -\text{sign}(\bar{c}_q)(c_q - c_B^T B^{-1}a_q) \\ &= -\text{sign}(\bar{c}_q)\bar{c}_q \\ &< 0. \end{aligned} \tag{24.24}$$

Therefore, Δx is downhill with respect to the objective function. □

24.4.3 Stepsize and Row Pivot Rule

Using Δx as a search direction, we are led to the line search scheme below:

$$\hat{x} = \bar{x} + \alpha \Delta x,$$

or equivalently,

$$\hat{x}_B = \bar{x}_B + \alpha \Delta x_B, \tag{24.25}$$

$$\hat{x}_q = \bar{x}_q - \text{sign}(\bar{c}_q)\alpha, \tag{24.26}$$

$$\hat{x}_j = \bar{x}_j, \quad j \in N;\ j \neq q, \tag{24.27}$$

where stepsize α is to be determined.

Introduce notation

$$\gamma_i = \begin{cases} (u_i - \bar{x}_i)/\Delta x_i & \text{if } \Delta x_i > 0, \text{ and } i \in B, \\ (l_i - \bar{x}_i)/\Delta x_i & \text{if } \Delta x_i < 0, \text{ and } i \in B, \end{cases} \tag{24.28}$$

Then the largest possible value of α is derived subject to $l \leq \hat{x} \leq u$, i.e.,

$$\bar{\alpha} = \min\{\delta_q,\ \min\{\gamma_i \mid \Delta x_i \neq 0,\ i \in B\}\}. \tag{24.29}$$

However, the algorithm will not take α itself as a stepsize, but a smaller one instead, i.e.,

$$\alpha = \lambda\bar{\alpha}, \qquad 0 < \lambda < 1. \tag{24.30}$$

Therefore, a new solution $\hat{x}$ can be computed via (24.25)–(24.30).

There are two cases arising:

(i) $\alpha = \delta_q$. It is then clear that there is no need for any basis change.
(ii) $\alpha < \delta_q$. A basis change is performed. The following row rule is used to determine a leaving index to match the entering index q.

Rule 24.4.2 (Row pivot rule) Select a basic index p such that

$$\gamma_p = \alpha,\ p \in B,\ \Delta x_p \neq 0. \tag{24.31}$$

Lemma 24.4.2. *$\hat{x}$ defined by (24.25)–(24.27) with α given by (24.28)–(24.30) is a feasible solution.*

Proof. Firstly, it is easy to verify that $\hat{x}$ defined by (24.25)–(24.27) satisfies the equality constraints $A\hat{x} = b$ for any real number α. In addition, by the feasibility of $\bar{x}$, (24.25)–(24.27), and (24.30), $\hat{x}$ satisfies $l \leq \hat{x} \leq u$, and is hence feasible. □

However, if α vanishes, $\hat{x}$ is not really new but the same as $\bar{x}$. This happens only when the current solution is degenerate (see Definition 7.4.1).

Theorem 24.4.3. *If feasible solution $\bar{x}$ is nondegenerate, then the associated stepsize is positive, and the new iterate $\hat{x}$ is a nondegenerate feasible solution, associated with a strictly lower objective value than the old.*

Proof. From Lemma 24.4.2, it is clear that the new iterate $\hat{x}$ is a feasible solution.

Note that δ_q is positive, since, otherwise, $|\bar{c}_q|\delta_q = 0$ implies that the optimality condition (24.19) holds. Moreover, the nondegeneracy assumption implies that γ_i, defined by (24.28), are positive for all $i \in B$ with $\Delta x_i \neq 0$. These along with (24.29) and (24.30) leads to $0 < \alpha < \alpha$. For $\hat{x}$ defined by (24.25)–(24.27), therefore, the basic components are still not on their bounds, no matter whether the basis change is performed ($\alpha < \delta_q$) or not.

Furthermore, by $\alpha > 0$, (24.25)–(24.27) and Lemma 24.4.1 it holds that $c^T\hat{x} < c^T\bar{x}$, which completes the proof. □

24.4.4 Formulation of the Algorithm

Some computational considerations should be incorporated in the implementation of the algorithm. In practice, for instance, what required is often an *approximate* optimal solution only. Therefore, the algorithm will use the following condition in place of (24.19) instead.

Definition 24.4.1. $\bar{x}$ is an ϵ-optimal solution if

$$|\bar{c}_j| \leq \epsilon_1 \qquad \text{or} \qquad \delta_j \leq \epsilon_2, \qquad \forall\, j \in N, \tag{24.32}$$

where $\epsilon_1 > 0$ and $\epsilon_2 > 0$ are predetermined tolerances.

Accordingly, index sets N_1 and N_2, defined by (24.16), are now redefined by

$$N_1 = \{j \in N \mid \bar{c}_j < -\epsilon_1\}, \qquad N_2 = \{j \in N \mid \bar{c}_j > \epsilon_2\}. \tag{24.33}$$

The overall steps can be put into the following algorithm.

Algorithm 24.4.1 (Feasible-point simplex algorithm). Given $0 < \lambda < 1$, $\epsilon_1, \epsilon_2 > 0$ and $M \gg 1$. Initial: basis B, associated with feasible solution $\bar{x}$. This algorithm solves the bounded-variable problem.

1. Solve $B^T\bar{y} = c_B$ for $\bar{y}$.
2. Compute $\bar{c}_N = c_N - N^T\bar{y}$.
3. Compute $\delta_j, j \in N$ by (24.17).
4. Stop if (24.32) is satisfied.
5. Determine index q such that $|\bar{c}_q|\delta_q = \max\{\, |\bar{c}_j|\delta_j \mid j \in N\,\}$.

6. Solve $B\bar{a}_q = a_q$ for $\bar{a}_q$.
7. Compute Δx by (24.21) and (24.22).
8. Determine $\bar{\alpha} = \min\{\delta_q, \min\{\gamma_i \mid \Delta x_i \neq 0, i \in B\}\}$.
9. Compute $\alpha = \lambda\bar{\alpha}$.
10. Stop if $\alpha > M$.
11. Update $\bar{x}$ by (24.25)–(24.27).
12. Go to Step 1 if $\alpha = \delta_q$.
13. Determine index p such that $\gamma_p = \alpha,\ p \in B,\ \Delta x_p \neq 0$.
14. Update basis B by replacing its p-th column with q-th column.
15. Go to step 1.

Theorem 24.4.4. *Assume termination of Algorithm 24.4.1. It terminates at either*

(i) step 4, achieving an ϵ-optimal basic solution; or
(ii) step 10, declaring lower unboundedness.

The meanings of the exits of the preceding Algorithm is clear. At this stage, however, it has not been possible to rule out the possibility of infiniteness. As it solved a large number of test problems in our computational experiments (see below), we claim that Algorithm 24.4.1 should be regarded as finite practically, just like the standard simplex algorithm.

Even if it is still open whether Algorithm 24.4.1 is finite or not, on the other hand, the following result is valid.

Theorem 24.4.5. *Assume that the initial feasible solution is nondegenerate. Then all subsequent stepsizes are positive, and hence iterates are all nondegenerate feasible solutions.*

Proof. It is enough to consider a current iteration.

By Lemma 24.4.2, the new iterate $\hat{x}$ satisfies $A\hat{x} = b$. From the feasibility of $\bar{x}$, (24.25)–(24.27), and (24.30), it follows that $\bar{x}$ satisfies $l \leq \hat{x} \leq u$, and hence is a feasible solution.

By (3.3), moreover, δ_q is positive, since, otherwise, the iteration would have terminated at Step 3(3). In addition, the nondegeneracy assumption implies that γ_i, defined by (24.28), are positive for all $i \in B$ satisfying $\Delta x_i \neq 0$. This fact along with (24.29) and (24.30) gives $0 < \alpha < \alpha$. Consequently, from (24.25)–(24.27) it follows that $\hat{x}$ is again nondegenerate.

Furthermore, by $\alpha > 0$, (24.25)–(24.27) and Lemma 24.4.1, it holds that $c^T\hat{x} < c^T\bar{x}$. This completes the proof. □

If the initial feasible solution is nondegenerate, Theorem 24.4.5 implies that the objective value strictly decreases in the solution process. Moreover, the algorithm has the following feature.

Proposition 24.4.2. *Assume that the initial feasible solution is nondegenerate. An on-bound component of $\bar{x}$ could become an interior component; but any interior component never becomes an on-bound component.*

The preceding implies that Algorithm 24.4.1 is an interior-point algorithm if an initial interior-point is used. On the other hand, solutions generated by Algorithm 24.4.1 are not vertices in general, even if the initial one is. Consequently, the algorithm goes across the interior of the feasible region.

Finally, we introduce the concept of an *approximate* optimal *basic* solution in place of (24.20), alternatively:

Definition 24.4.2. $\bar{x}$ is an ϵ-optimal basic solution if

$$\delta_j < \epsilon, \qquad \forall\ j \in N, \tag{24.34}$$

where $\epsilon > 0$ is a predetermined tolerance.

It is noted that if condition (24.34) is used as termination criterion, $\bar{c}_q = 0$ could holds, and hence the associated Δx is not downhill. In this case, the algorithm continues using Δx to move to an optimal *basic* solution, even if optimality has already been achieved (see Proposition 24.4.1).

24.4.5 Phase-1 and Purification

Any Phase-I approach for the bounded-variable problem can be used to provide an initial feasible solution. If one wants the algorithm starting from an interior point, the approach described at the end of Sect. 7.4 applies.

As for obtaining an exact optimal basic solution, the following simple purification can be incorporated to Algorithm 24.4.1.

Assume that Algorithm 24.4.1 terminates at step 4 with an ϵ-optimal basic solution. The purification is done by moving nonbasic components of the solution onto their respective nearest bounds with the basic components unchanged. If the resulting solution, say x^0, satisfies $Ax^0 = b$ within some tolerance, it is clearly an optimal basic solution. In the other case, a standard two-phase simplex algorithm can be used to attain a basic optimal solution, hopefully within few iterations.

24.4.6 Computational Results

Computational experiments have been performed to gain an insight into the behavior of Algorithm 24.4.1. A summary of the associated numerical results are offered in this subsection.

Implemented, and compared are the following three codes:

1. MINOS: MINOS 5.51 with full pricing.
2. FPS: Two-Phase code based on Algorithm 24.4.1.
3. FPSP: Two-Phase code based on Algorithm 24.4.1 with the purification.

Table 24.2 Ratio

	MINOS/FPSP		MINOS/FPS		FPSP/FPS	
Problem	Itns	Time	Itns	Time	Itns	Time
Kennington(16)	6.6	3.2	9.0	3.5	1.4	1.1
BPMPD(15)	2.8	3.4	9.5	9.0	3.3	2.6
Average(31)	3.3	3.4	9.4	6.6	2.8	2.0

The first set of test problems included all 16 problems from Kennington and the second included all 17 problems from BPMPD that were more than 500KB in compressed form (Appendix B: Tables B.2–B.3).

In Table 24.2, a comparison between the three codes is made.

From the table, it is seen that FPSP and FPS outperformed MINOS remarkably, with average iteration ratios 6.6 and 9.0, and time ratios 3.2 and 3.5 for the 16 Kennington problems. They outperformed MINOS, by average iterations ratios 2.8 and 9.5, and time ratios 3.4 and 9.0 for BPMPD problems. For the entire set of the 31 test problems, FPSP and FPS defeated MINOS by average iterations ratios 3.3 and 9.4, and time ratios 3.4 and 6.6.

In summary, the feasible-point simplex algorithm is significantly superior to the standard simplex algorithm with the test set (see Appendix E for more details).

Chapter 25
Special Topics

In this final chapter, we highlight an assortment of several topics. Firstly, several special forms of the LP problem are introduced in the next section. Then, approaches to intercepting for primal and dual optimal sets are given in Sect. 25.2. Practical pricing schemes for computing reduced costs used by pivot rules are presented in Sect. 25.3. The relaxation principle and two algorithms based on it are described in Sect. 25.4. Local duality is introduced in Sect. 25.5. A decomposition principle and related two LP algorithms are highlighted in Sect. 25.6. Finally, an ILP method based on the reduced simplex framework is developed in Sect. 25.7.

All algorithms presented in this chapter are unpublished yet, except for the pricing schemes. The discussions made in the last three sections apply not only to LP problems but also more generale optimization problems, e.g., those with linear objective function and convex feasible region.

25.1 Special Forms of the LP Problem

In this section, we introduce several special forms of the LP problem, which can be more advantageous than the standard LP problem in some cases, although conventional algorithms are usually designed with respect to the latter.

Assume that vector c of the standard LP problem is not a linear combination of rows of A. Such an assumption is appropriate, as the problem is trivial otherwise: if the feasible region is nonempty, then there exists a vector y such that $c = A^{\mathrm{T}}y$, leading to

$$c^{\mathrm{T}}x = y^{\mathrm{T}}Ax = y^{\mathrm{T}}b,$$

which implies that the objective function is constant over the feasible region.

P.-Q. PAN, *Linear Programming Computation*, DOI 10.1007/978-3-642-40754-3_25,

25.1.1 Reduced Problem

This special form can be obtained by regarding f as a variable and taking $f = c^{\mathrm{T}}x$ as a constraint, i.e.,

$$\begin{aligned} &\min\ f, \\ &\text{s.t.}\ \begin{pmatrix} A & 0 \\ c^{\mathrm{T}} & -1 \end{pmatrix}\begin{pmatrix} x \\ f \end{pmatrix} = \begin{pmatrix} b \\ 0 \end{pmatrix}, \quad x \geq 0. \end{aligned} \tag{25.1}$$

It might be well to introduce

$$A := \begin{pmatrix} A \\ c^{\mathrm{T}} \end{pmatrix}, \quad b := \begin{pmatrix} b \\ 0 \end{pmatrix}.$$

Thereafter f is indexed by $n+1$, and deemed the same as x_{n+1}, So, $a_{n+1} = -e_{m+1}$, and the problems can be written as a so-called "reduced problem":

$$\begin{aligned} &\min\ f, \\ &\text{s.t.}\ \left(A \,\vdots\, a_{n+1}\right)\begin{pmatrix} x \\ f \end{pmatrix} = b, \quad x \geq 0, \end{aligned} \tag{25.2}$$

where $(A\ a_{n+1}) \in \mathcal{R}^{(m+1)\times(n+1)}$, $b \in \mathcal{R}^{m+1}$, rank $(A\ a_{m+1}) = m+1$, $m < n$. Note that the reduced problem is of a standard from, but its objective function involves only a single (free) objective variable f.

Computations can be somehow simplified with the reduced problem, compared with the standard problem. Consider the use of the simplex algorithm to solve (25.2). Taking f as the $(m + 1)$th basic variable, the reduced cost vector is

$$\bar{z}_N = N^{\mathrm{T}}\bar{y}, \quad B^{\mathrm{T}}\bar{y} = e_{m+1}.$$

If $B = LU$ is available, the second system of the preceding can be converted to two triangular systems below:

$$U^{\mathrm{T}}v = e_{m+1}, \quad L^{\mathrm{T}}\bar{y} = v.$$

The solution to the first system is readily available, i.e., $v = (1/\mu)e_{m+1}$, where μ is the $(m + 1)$th diagonal of U. Thereby, only a single triangular system

$$L^{\mathrm{T}}\bar{y} = (1/\mu)e_{m+1}.$$

needs to be solved to obtain the simplex multiplier $\bar{y}$. In other words, only three triangular systems are solved in each iteration, compared with the four solved in the conventional simplex algorithm.

In fact, handling the reduced problem is not really new. Some LP codes, such as MINOS, have already adopted this approach. More important applications of the reduced problem can be found in Chaps. 15 and 16.

25.1.2 D-Reduced Problem

This form yields from applying the idea behind the reduced form, presented in the forgoing section, to the dual side.

Assume that $b_r \neq 0$. Multiplying the rth equality by $1/b_r$, then adding appropriate multiple of it to the other equalities converts the right-hand side of the system to the unit vector e_r. If the coefficient matrix of the resulting system is still denoted by A, it becomes the so-called "D-reduced problem" below:

$$\begin{array}{ll} \min & c^{\mathrm{T}}x, \\ \text{s.t.} & Ax = e_r, \quad x \geq 0. \end{array} \tag{25.3}$$

The objective function of the associated dual problem involves a single variable.

There are multiple choices for row index r. Although any row index, associated with a nonzero component, can be taken as r, too small one is certainly not attractive. It seems natural to take a component equaling 1, if any. Considering numerical stability, however, it is reasonable to take the largest in magnitude. On the other hand, a row index that involves minimum nonzeros should be chosen to reduce fill-ins. Numerical stability and sparsity have to be balanced in large-scale sparse computations.

Finally, it is favorable to move the chosen row to as the mth row to simplify computations. If $B = LU$ is known, e.g., solving $Bx_B = e_m$ is converted to solving a single triangular system only. It is more than that. A more important use of the reduced problem of such type can be found in Chap. 17.

25.1.3 Dual Elimination

This form results from the standard dual problem via elimination. If the standard problem is of a coefficient matrix close to square, i.e., $n - m \ll m$, it would be favorable to turn to such a form, an $(n - m) \times m$ standard problem, with respect to dual variable z. It will be realized with Gauss elimination though the orthogonal triangularization also applies.

Consider the tableau of the dual problem (4.2), with $\max g = b^{\mathrm{T}}y$ replaced by $\min -g = -b^{\mathrm{T}}y$:

y^{T}	z^{T}	g	RHS		
A^{T}	I		c	n	(25.4)
$-b^{\mathrm{T}}$		1		1	

Assume that A is of full row rank. Through the Gauss elimination with row exchanges, the preceding can be converted to a tableau of the form

y^{T}	z^{T}	g	RHS	
U	G_1		d_1	m
	G_2		d_2	$n-m$
	$\bar{x}^{\mathrm{T}}$	1	$\bar{g}$	1

(25.5)

where $U \in \mathcal{R}^{m\times m}$ is nonsingular upper triangular, $G_1 \in \mathcal{R}^{m\times n}$, $G_2 \in \mathcal{R}^{(n-m)\times n}$. The preceding corresponds to an equivalent form of the dual problem, i.e.,

$$\begin{array}{ll} \min & -g = -\bar{g} + \bar{x}^{\mathrm{T}}z, \\ \text{s.t.} & G_2 z = d_2, \quad z \geq 0, \\ & Uy + G_1 z = d_1, \end{array} \tag{25.6}$$

where rank $G_2 = n - m$. Therefore, it is only needed to solve an $(n-m)\times n$ standard problem with respect to variable z, corresponding to the last $n-m+1$ rows of tableau (25.5), i.e.

z^{T}	g	RHS
G_2		d_2
$\bar{x}^{\mathrm{T}}$	1	$\bar{g}$

(25.7)

Starting from the preceding tableau, problem (25.6) (hence the original problem) can be solved by the two-phase primal or dual tableau simplex algorithm. In case when $n - m \ll m$, handling the tableau with $n - m$ rows is much cheaper than handling the original problem with m rows. When an dual optimal solution $\bar{z}$ is obtained, solving triangular system

$$Uy = d_1 - G_1\bar{z}$$

gives the related $\bar{y}$. When a non-tableau algorithm is used, the involved systems are of small order $(n-m)\times(n-m)$, compared with original order $m \times m$.

In addition, the associated primal basic optimal solution is readily available. To explain, let us give the following result first.

Lemma 25.1.1. *Let matrix $A \in \mathcal{R}^{m\times n}$, $m < n$, be of full row rank. If there is a nonsingular matrix $G \in \mathcal{R}^{n\times n}$ such that*

$$GA^{\mathrm{T}} = \begin{pmatrix} U \\ 0 \end{pmatrix}, \tag{25.8}$$

where $U \in \mathcal{R}^{m\times m}$ is nonsingular upper triangular, then the range space of G_2^{T} is just the null space of A, where G_2 consists of the last $n - m$ rows of G.

Proof. Assume that G_1 consists of the first m rows of G. It follows from (25.8) that

$$\left(AG_1^{\mathrm{T}} \vdots AG_2^{\mathrm{T}}\right) = A\left(G_1^{\mathrm{T}} \vdots G_2^{\mathrm{T}}\right) = AG^{\mathrm{T}} = \left(U^{\mathrm{T}} \vdots 0\right), \tag{25.9}$$

hence leading to $AG_2^{\mathrm{T}} = 0$. Since G is nonsingular, it holds that rank $G_2 = n - m$. Therefore, the range space of G_2^{T} is the null space of A. □

It is easy to generalize Lemma 25.1.1 to allow the case when A is not of full row rank. Also, we stress that the Lemma is valid for any $n \times n$ nonsingular matrix G. In particular, performing Gauss elimination with row exchanges on (25.4) amounts to premultiplying the first n rows by

$$G = \tilde{G}_m P_m \cdots \tilde{G}_1 P_1,$$

where $\tilde{G}_i \in \mathcal{R}^{n\times n}$, $i = 1, \cdots, m$ are Gauss transformations, and $P_i \in \mathcal{R}^{n\times n}$, $i = 1, \cdots, m$ are permutations.

Theorem 25.1.1. *If there is an optimal tableau to (25.7), then its bottom row gives an basic optimal solution to the primal problem.*

Proof. Let (25.7) be an optimal simplex tableau. It is known that it is equivalent to

z^{T}	g	RHS
G_2		$G_2 c$
$b^{\mathrm{T}}U^{-1}G_1$	1	$b^{\mathrm{T}}U^{-1}d_1$

(25.10)

from which it is obtained that

$$\bar{x} = G_1^{\mathrm{T}} U^{-\mathrm{T}} b.$$

Thus, it holds by (25.9) that

$$A\bar{x} = (AG_1^{\mathrm{T}})U^{-\mathrm{T}}b = U^{\mathrm{T}}U^{-\mathrm{T}}b = b.$$

Therefore, the bottom row of (25.5) gives a primal solution. By Lemma 25.1.1, in addition, the range space of G_2^{T} gives the null space of A. This is true for all G_2^{T}'s in the successive tableaus, generated in the process, and each bottom row comes from its predecessor plus a multiple of some row of a G_2^{T}. Therefore, bottom rows of all these simplex tableaus give primal solutions. Since the bottom row of the optimal simplex tableau is nonnegative, and exhibits complementarity with the dual basic optimal solution, it corresponds to a basic optimal solution to the original problem. □

Example 25.1.1. Solve the following LP problem by the preceding approach:

$$
\begin{array}{lrrrrrrrrrl}
\min & f = 3x_1 & - x_2 & + 2x_3 & - x_4 & + x_5 & & - x_7 & - 2x_8, & & \\
\text{s.t.} & x_1 & & + x_3 & & & - x_6 & & - 2x_8 & = & -2, \\
& & - x_2 & & + 4x_4 & & + 2x_6 & & - 3x_8 & = & 9, \\
& & & + x_3 & & - x_5 & + x_6 & - 2x_7 & & = & -5, \\
& 2x_1 & & & - 5x_4 & & & & - x_8 & = & -18, \\
& & + x_2 & & & - 3x_5 & & + x_7 & & = & 3, \\
& -2x_1 & & - x_3 & & + 4x_5 & + 2x_6 & & - 7x_8 & = & -13, \\
& & & & & & x_j \geq 0, & & j = 1, \cdots, 8. & &
\end{array}
$$

Answer Construct an initial tableau of form (25.4).

y_1	y_2	y_3	y_4	y_5	y_6	z_1	z_2	z_3	z_4	z_5	z_6	z_7	z_8	RHS
1			2		−2	1								3
	−2			1			1							−1
1		1			−1			1						2
	4		−5						1					−1
		−1		−3	4					1				1
−1	2	1			2						1			
		−2		1								1		−1
2	−3		−1		−7								1	−2
2	−9	5	18	−3	13									

Use the Gauss elimination to turn the first six columns to upper triangular. The resulting tableau, consisting of the first six rows, and the tableau of form (25.7) are respectively as follows:

y_1	y_2	y_3	y_4	y_5	y_6	z_1	z_2	z_3	z_4	z_5	z_6	z_7	z_8	RHS
1			2		−2	1								3
	−2			1			1							−1
		1	−2		1	−1		1						−1
			−5	2		0	2		1					−3
				−19/5	5	−1	−4/5	1	−2/5	1				6/5
					46/19	25/19	39/19	−6/19	10/19	13/19	1			27/19

z_1	z_2	z_3	z_4	z_5	z_6	z_7	z_8	RHS
−5/2	−5/2	2	−1	−1/2	−1/2	1		−3/2
761/92	867/92*	−36/23	143/46	289/92	433/92		1	947/92
−513/92	−723/92	−58/23	63/46	−333/92	−561/92			−1,739/92

There are only three rows in the preceding tableau. Here we use a two-phase dual simplex method.

Phase-I: Call Algorithm 14.3.2.

Iteration 1:

1. $\min\{-513/92, -723/92, -58/23, 63/46, -333/92, -561/92\} = -723/92,$
   $q = 2.$
4. $\max\{-5/2, 867/92\} = 867/92,\ p = 2.$
5. Multiply row 2 by 92/867, then add 5/2, 723/92 times of row 2 to rows 1,3, respectively:

z_1	z_2	z_3	z_4	z_5	z_6	z_7	z_8	RHS
−265/867		458/289*	−152/867	1/3	649/867	1	230/867	1,067/867
761/867	1	−48/289	286/867	1/3	433/867		92/867	947/867
382/289		−1,106/289	1,145/289	−1	−628/289		241/289	−2,982/289

Iteration 2:

1. $\min\{382/289, -1{,}106/289, 1{,}145/289, -1, -628/289, 241/289\} =$
   $-1{,}106/289,\ q = 3.$
4. $\max\{458/289, -48/289\} = 458/289,\ p = 1.$
5. Multiply row 1 by 289/458, then add 48/289, 1,106/289 times of row 1 to rows 2,3, respectively:

z_1	z_2	z_3	z_4	z_5	z_6	z_7	z_8	RHS
−265/1,374		1	−76/687	289/1,374	649/1,374	289/458	115/687	1,067/1,374
581/687	1		214/687	253/687	397/687*	24/229	92/687	839/687
401/687			2,431/687	−134/687	−251/687	553/229	1,013/687	−2,057/280

Iteration 3:

1. $\min\{401/687, 2{,}431/687, -134/687, -251/687, 553/229, 1{,}013/687\} =$
   $-251/687,\ q = 6.$
4. $\max\{649/1{,}374, 397/687\} = 397/687,\ p = 2.$
5. Multiply row 2 by 687/397, then add −649/1,374, 251/687 times of row 2 to rows 1,3, respectively:

z_1	z_2	z_3	z_4	z_5	z_6	z_7	z_8	RHS
−351/397	−649/794	1	−145/397	−36/397*		433/794	23/397	−88/397
581/397	687/397		214/397	253/397	1	72/397	92/397	839/397
444/397	251/397		1,483/397	15/397		985/397	619/397	−2,610/397

Phase-II: Call dual simplex Algorithm 4.4.1.

Iteration 4:

1. $\min\{-88/397, 839/397\} = -88/397,\ p = 1.$
4. $\min\{-(444/397)/(-351/397), -(251/397)/(-649/794), -(1{,}483/397)/(-145/397), -(15/397)/(-36/397)\} = 5/12.$
5. Multiply row 1 by $-397/36$, then add $-253/397, -15/397$ times of row 1 to rows 2,3, respectively:

z_1	z_2	z_3	z_4	z_5	z_6	z_7	z_8	RHS
39/4	649/72	−397/36	145/36	1		−433/72	−23/36	22/9
−19/4	−289/72	253/36	−73/36		1	289/72	23/36	5/9
3/4	7/24	5/12	43/12			65/24	19/12	−20/3

Optimal basic solution and related objective value:
$\bar{x} = (3/4, 7/24, 5/12, 43/12, 0, 0, 65/24, 19/12)^{\mathrm{T}},\ \bar{g} = -20/3.$
Associated dual basic optimal solution:
$\bar{z} = (0, 0, 0, 0, 22/9, 5/9, 0, 0)^{\mathrm{T}}.$

If the related $\bar{y}$ is needed, substituting $\bar{z}$ to the triangular system, associated with the second tableau of this example, leads to

$$\begin{pmatrix} 1 & & & 2 & & -2 \\ & -2 & & & 1 & \\ & & 1 & -2 & & 1 \\ & & & -5 & 2 & \\ & & & & -19/5 & 5 \\ & & & & & 46/19 \end{pmatrix} \begin{pmatrix} y_1 \\ y_2 \\ y_3 \\ y_4 \\ y_4 \\ y_6 \end{pmatrix} = \begin{pmatrix} 3 \\ -1 \\ -1 \\ -3 \\ 6/5 \\ 27/19 \end{pmatrix} - (22/9) \begin{pmatrix} 0 \\ 0 \\ 0 \\ 0 \\ 1 \\ 13/19 \end{pmatrix} \begin{pmatrix} 0 \\ 0 \\ 0 \\ 0 \\ 0 \\ 5/9 \end{pmatrix},$$

Then solving the preceding gives the wanted $\bar{y} = (11/9, 4/9, 4/9, 5/9, -1/9, -1/3)^{\mathrm{T}}$.

The dual tableau (25.4) is of $(n+1) \times (m+n+1)$ order, which is larger than $(m+1) \times (n+1)$, the order of the primal tableau. At a first glance, the approach described above seems to be impracticable.

It is different if one takes a closer look at it. In fact, the triangularization of the former amounts to performing LU-factorization of

$$\begin{pmatrix} A^{\mathrm{T}} \\ -b^{\mathrm{T}} \end{pmatrix}$$

(with row exchanges, except for the bottom row). Contents in Sect. 5.3 are still valid in this context. When $n - m \ll m$, the associated additional cost is not as much, but the return is great, as the scale of the linear systems, involved in subsequent iterations, is reduced considerably. As for obtaining (25.5), it is only needed to accumulate Gauss transformation and permutation factors on (25.4). For obtaining $\bar{y}$, it is only needed to accumulate rows, associated with tableau (25.7), hence only touching upon corresponding components of factors. Alternatively, these factors can be stored, and wanted vectors can be calculated from them and original data in subsequent simplex iterations. Note that G_2 includes the $(n-m)\times(n-m)$ unit matrix, corresponding to zeros in the bottom row. Therefore, there is no need for computing the associated $n-m$ columns. Taking the unit matrix as an initial basis, one can get a two-phase primal or dual simplex algorithm started directly.

25.1.4 Reduced Dual Elimination

This form results from combining the previous two approaches, that is, carrying out dual elimination with the reduced problem.

Consider the dual of the reduced program (25.2). Replacing objective row $\max g = b^{\mathrm{T}}y$ by $\min -g = -b^{\mathrm{T}}y$, its tableau form becomes

y^{T}	z^{T}	z_{n+1}	g	RHS	
A^{T}	I				n
$-e_{m+1}^{\mathrm{T}}$		1		1	1
$-b^{\mathrm{T}}$			1		1

Since rank $A = m+1$, performing Gauss elimination with row exchanges converts the submatrix, corresponding to y^{T}, to upper triangular (without touching the last two rows, hence the last three columns). Assume that the resulting tableau is

y^{T}	z^{T}	z_{n+1}	g	RHS	
U	G_1				$m+1$
	G_2				$n-m-1$
	d^{T}	1		1	1
	$\bar{x}^{\mathrm{T}}$		1		1

where $U \in \mathcal{R}^{(m+1)\times(m+1)}$ is nonsingular upper triangular, $G_1 \in \mathcal{R}^{(m+1)\times n}$, $G_2 \in \mathcal{R}^{(n-m-1)\times n}$. Then what next to do is to solve the following $(n-m)\times(n+1)$ standard problem with respect to z, associated with a tableau of form

z^{T}	z_{n+1}	g	RHS
G_2			
d^{T}	1		1
$\bar{x}^{\mathrm{T}}$		1	

(25.11)

If the simplex algorithm is applied, only a single triangular system needs to be solved to obtain $\bar{x}_B$. Moreover, since G_2 involves the $(n-m-1)\times(n-m-1)$ unit submatrix, and the right-hand side is the unit vector e_n, the tableau can be handled directly by D-reduced simplex method or deficient-basis D-reduced method.

25.2 Intercepting of Optimal Set and Bilevel LP

In some applications, there may be a need for not only a primal or dual optimal solution, but also the set of primal or dual optimal solutions (see, e.g., Megiddo 1989). It is shown in this section that it is not difficult to achieve this if an optimal simplex tableau (or basic optimal solution) is available. Firstly, an approach to intercepting for the primal optimal set is described, and then for the dual optimal set.

Let $B = \{j_1, \ldots, j_m\}$ and $N = A\backslash B$ be respectively optimal basis and nonbasis. Assume that the associated basic optimal solution is

$$\bar{x}_B = B^{-1}b \geq 0, \qquad \bar{x}_N = 0,$$

with optimal value $\bar{f} = c_B^T \bar{x}_B$, and that the reduced costs and the simplex multipliers are

$$\bar{z}_B = 0, \quad \bar{z}_N = c_N - N^T\bar{y} \geq 0, \qquad \bar{y} = B^{-T}c_B,$$

which are actually dual optimal solutions, as was shown in Sect. 4.1.

The original problem can be written

$$\begin{array}{ll} \min & \bar{z}_N^T x_N, \\ \text{s.t.} & x_B + B^{-1}Nx_N = \bar{x}_B, \qquad x_B, x_N \geq 0. \end{array} \tag{25.12}$$

Section the nonbasis N to (N_1, N_2), defined as

$$N_1 = \{j \in N \mid \bar{z}_j = 0\}, \qquad N_2 = \{j \in N \mid \bar{z}_j > 0\}. \tag{25.13}$$

Then the reduced objective function can be written

$$f = \bar{f} + \bar{z}_{N_2}^T x_{N_2}, \tag{25.14}$$

and the original constraint system is equivalent to

$$x_B = \bar{x}_B - B^{-1}N_1x_{N_1} - B^{-1}N_2x_{N_2}.$$

Using the preceding notations, we have the following result.

Theorem 25.2.1. *The optimal set is*

$$\begin{aligned} F &= \{x \in \mathcal{R}^n \mid x_B = B^{-1}b - B^{-1}N_1x_{N_1},\ x_B, x_{N_1} \geq 0,\ x_{N_2} = 0\} \\ &= \{x \in \mathcal{R}^n \mid Bx_B + N_1x_{N_1} = b,\ x_B, x_{N_1} \geq 0,\ x_{N_2} = 0\}. \end{aligned} \tag{25.15}$$

Proof. According to Lemma 2.3.2, F is the set of optimal solutions if and only if

$$F = P \cap \{x \in \mathcal{R}^n \mid c^T x = \bar{f}\}.$$

Since (25.14) is a reduced objective function, it holds that

$$c^T x = \bar{f} + \bar{z}_{N_2}^T x_{N_2}, \qquad x \in P,$$

which combined with (25.14) and the second expression of (25.13) leads to the first equality in (25.15), i.e.,

$$\begin{aligned} F &= \{x \in \mathcal{R}^n \mid Ax = b,\ x \geq 0;\ c^{\mathrm{T}}x = \bar{f}\} \\ &= \{x \in \mathcal{R}^n \mid x_B = B^{-1}b - B^{-1}N_1x_{N_1} - B^{-1}N_2x_{N_2},\ x \geq 0;\ \bar{z}_{N_2}^{\mathrm{T}}x_{N_2} = 0\} \\ &= \{x \in \mathcal{R}^n \mid x_B = B^{-1}b - B^{-1}N_1x_{N_1} - B^{-1}N_2x_{N_2},\ x \geq 0;\ x_{N_2} = 0\} \\ &= \{x \in \mathcal{R}^n \mid x_B = B^{-1}b - B^{-1}N_1x_{N_1}, x_B, x_{N_1} \geq 0;\ x_{N_2} = 0\}. \end{aligned}$$

The second equality is obvious. □

Set F is actually the largest optimal face of the problem, whose dimension is dependent on $|N_1|$ (or $|N_2|$) (Proposition 2.1.2). If any index $q \in N_1$ is selected to enter the basis, a row index p can be determined to leave the basis, just as in the simplex context. For the resulting basis B, system $x_B = B^{-1}b - B^{-1}N_1x_{N_1}$ renders a new basic optimal solution, which is truly new if its predecessor is nondegenerate. This method can be used to obtain multiple basic optimal solutions. It is clear that the number of all such solutions is no more than $C_{m+|N_1|}^m$.

Corollary 25.2.1. *If N_1 is empty, there exists an unique optimal solution or basis to the LP problem.*

The preceding implies that a primal optimal solution is unique if the associated dual optimal solution is nondegenerate, as coincides to Proposition 3.9.3.

As an application, consider the special *bilevel LP problem* below:

$$\begin{aligned} &\min\ f_1 = c_1^T x, \\ &\min\ f_2 = c_2^T x, \\ &\text{s.t.}\quad Ax = b, \qquad x \geq 0. \end{aligned} \tag{25.16}$$

The first level is to attain the optimal set to the problem with the first objective function f_1 subject to the constraints, and the second level minimizes the second objective f_2 over the optimal set.

Example 25.2.1. Solve the following bilevel LP problem:

$$
\begin{aligned}
\min\ & f_1 = -3x_5 + 5x_6, \\
\min\ & f_2 = -x_1 - 4x_2 + 2x_3 - 6x_4 + x_5 + 2x_6, \\
\text{s.t.}\ & -3x_1 - x_2 + x_3 + 6x_4 - 4x_6 + x_7 = 1, \\
& 5x_1 - 2x_3 + x_4 - x_5 + x_8 = 4, \\
& 2x_2 + 3x_5 + x_6 + x_9 = 3, \\
& x_j \geq 0, \quad j = 1, \ldots, 9.
\end{aligned}
$$

Answer The initial feasible simplex tableau is

x_1	x_2	x_3	x_4	x_5	x_6	x_7	x_8	x_9	RHS
−3	−1	1	6		−4	1			1
5		−2	1	−1			1		4
	2			3*	1			1	3
−1	−4	2	−6	1	2				
				−3	5				

where the bottom and second bottom rows correspond to the first and second objective functions, respectively.

Call Algorithm 3.2.1 to solve the first level problem.

Iteration 1:

1. $\min\{0, 0, 0, 0, -3, 5, 0, 0, 0\} = -3,\ q = 5$.
3. $I = \{3\} \neq \emptyset$.
4. $\min\{3/3\},\ p = 3$.
5. Multiply row 3 by 1/3, and then add $1, -1, 3$ times of row 3 to rows 2,4,5, respectively:

x_1	x_2	x_3	x_4	x_5	x_6	x_7	x_8	x_9	RHS
−3	−1	1	6*		−4	1			1
5	2/3	−2	1		1/3		1	1/3	5
	2/3			1	1/3			1/3	1
−1	−6	2	−6		1			−1	−1
	2				6			1	3

Iteration 2:

1. $\min\{0, 2, 0, 0, 6, 1\} \geq 0$. Therefore, the first level problem is minimized with optimal value $f_1 = -3$ ($N_1 = \{1, 3, 4\},\ N_2 = \{2, 6, 9\}$). The second objective value equals to $f_2 = 1$. Dropping columns corresponding to N_2 and the bottom row gives the feasible simplex tableau to the second level problem, i.e.,

x_1	x_3	x_4	x_5	x_7	x_8	RHS
−3	1	6*		1		1
5	−2	1			1	5
			1			1
−1	2	−6				−1

Call Algorithm 3.2.1.

Iteration 3:

1. $\min\{-1, 2, -6\} = -6,\ q = 4.$
3. $I = \{1, 2\} \neq \emptyset.$
4. $\min\{1/6, 5/1\} = 1/6,\ p = 1.$
5. Multiply row 1 by 1/6, and then add $-1, 6$ times of row 1 to rows 2,4, respectively:

x_1	x_3	x_4	x_5	x_7	x_8	RHS
−1/2	1/6	1		1/6		1/6
11/2*	−13/6			−1/6	1	29/6
			1			1
−4	3			1		

Iteration 4:

1. $\min\{-4, 3, 1\} = -4,\ q = 1.$
3. $I = \{2\} \neq \emptyset.$
4. $\min\{(29/6)/(11/2)\},\ p = 2.$
5. Multiply row 2 by 2/11, and then add $1/2, 4$ times of row 2 to rows 1,4, respectively:

x_1	x_3	x_4	x_5	x_7	x_8	RHS
	−1/33	1		5/33	1/11	20/33
1	−13/33			−1/33	2/11	29/33
			1			1
	47/33			29/33	8/11	116/33

Iteration 5:

1. $\min\{47/33, 29/33, 8/11\} \geq 0.$
2. The second level problem is solved with the basic optimal solution $\bar{x} = (29/33, 0, 0, 20/33, 1, 0, 0, 0, 0)^T$, corresponding to the first and second objective values $f_1 = -3$ and $f_2 = -116/33$, respectively.

On the other hand, the dual optimal set can be obtained in an analogous manner as follows.

Without loss of generality, assume that $B = \{1, \cdots, m\}$, $N = A \backslash B$ and that

$$\bar{x}_i = 0, \qquad i = 1, \cdots, t,$$
$$\bar{x}_i > 0, \qquad i = t+1, \cdots, m.$$

Define row index sets

$$I_1 = \{1, \cdots, t\}, \qquad I_2 = \{t+1, \cdots, m\}.$$

Bisection rows of $\bar{N} = B^{-1}N$ such that

$$\bar{N} = \begin{pmatrix} \bar{N}_{I_1}^T \\ \bar{N}_{I_2}^T \end{pmatrix},$$

Then, the dual problem of (25.12) can be written

$$\begin{aligned} \max \quad & \bar{x}_{I_2}^T y_{I_2}, \\ \text{s.t.} \quad & \begin{pmatrix} y_{I_1} \\ y_{I_2} \end{pmatrix} + \begin{pmatrix} z_{I_1} \\ z_{I_2} \end{pmatrix} = 0, \\ & (\bar{N}_{I_1}, \bar{N}_{I_2}) \begin{pmatrix} y_{I_1} \\ y_{I_2} \end{pmatrix} + z_N = \bar{z}_N, \\ & z_B, z_N \geq 0. \end{aligned}$$

Setting $y_{I_2}, z_{I_2} = 0$ in the constraints of the preceding program gives the set of dual optimal solutions.

Using these notations, we state the following result, the proof of which is omitted.

Theorem 25.2.2. *The dual optimal set is*

$$G = \{(y, z) \in \mathcal{R}^m \times \mathcal{R}^n \mid z_N = \bar{z}_N - \bar{N}_{I_1} y_{I_1} \geq 0;\ z_{I_1} = -y_{I_1} \geq 0;\ y_{I_2}, z_{I_2} = 0\}.$$

25.3 Pricing Scheme

All pivot rules are closely related to reduced costs. In fact, knowledge of all or part of reduced costs is a prerequisite for determination of a pivot in the simplex context. This section is devoted to approaches toward pricing.

Still assume that the current basis is $B = \{1, \cdots, m\}$, and nonbasis is $N = \{m+1, \cdots, n\}$. The standard formulas for computing reduced costs are

$$B^{\mathrm{T}} y = c_B, \quad z_N = c_N - N^{\mathrm{T}} y. \tag{25.17}$$

We will present several alternative schemes, some of which are being used in practice.

Assume that a pivot column index $q \in N$ and pivot row index $p \in B$ have been determined. In this section, the new basis and other associated quantities are denoted by a prime. Thus,

$$B' = (B \backslash p) \cup q, \quad N' = (N \backslash q) \cup p. \tag{25.18}$$

The relation between the current basis and the new basis can be expressed as

$$B' = B + (a_q - a_p)e_p^{\mathrm{T}}, \tag{25.19}$$

and y' satisfies

$$B'^{\mathrm{T}} y' = c_{B'}. \tag{25.20}$$

If h denotes the solution to the following system:

$$B^{\mathrm{T}} h = e_p, \tag{25.21}$$

then h' satisfies

$$B'^{\mathrm{T}} h' = e_p. \tag{25.22}$$

Thereby, the following recurrence formulas can be proved:

$$y' = y + z_q h', \tag{25.23}$$

$$h = \bar{a}_{p,q} h', \tag{25.24}$$

$$z'_{N'} = z_{N'} - z_q N'^{\mathrm{T}} h', \tag{25.25}$$

$$= z_{N'} - (z_q / \bar{a}_{p,q}) N'^{\mathrm{T}} h, \tag{25.26}$$

where $\bar{a}_{p,q}$ is the pivot, i.e., the pth component of the solution to $B\bar{a}_q = a_q$, as can be calculated independent of pricing. Note that $z_p = 0$ in the preceding formulas.

As an alternative scheme, reduced costs may be calculated based on (25.25) or (25.26). Zoutendijk (1960) uses formula (25.25) and (25.22). Bixby (1994) uses (25.26) and (25.21). An advantage of the latter lies in that the solution h or h' to (25.21) or (25.22) is usually much sparser than y, and especially suitable for the steepest-edge rule, which also needs to solve (25.21) or (25.22) (see (11.11)), so does the dual simplex algorithm (see step 3 of Algorithm 4.5.1).

However, recurrence formulas are not suitable for partial pricing. In contrast to the preceding schemes, Tomlin uses (25.23) and (25.22) to compute the simplex multipliers in a recurrence manner, as is clearly amenable to partial pricing.

As in the standard formulas, the preceding schemes need to solve two triangular systems. Assume that the LU factorization with row and column exchanges of the current basis matrix B is

$$PBQ = LU, \tag{25.27}$$

where L is unit lower triangular, U is nonsingular upper triangular, and P and Q are permutations. Recall that each system involved in simplex iterations can be converted to two triangular systems. For $B\bar{a}_q = a_q$, e.g., one may solve

$$Lv = Pa_q, \quad Uu = v, \tag{25.28}$$

and set $\bar{a}_q = Qu$. It is similar to solve (25.21) or (25.22).

Hu and Pan (2008b) modified Tomlin's scheme, so that only a single triangular system needs to be solved for pricing. It is found latter that Goldfarb (1977) proposed the same approach along another line earlier without offering any numerical results.

In the following, we will focus on this scheme, and report favorable results, obtained in extensive computational experiments. To this end, firstly we derive the Bartels-Golub update in a way, slightly different from that in Sect. 5.4 (Forrest-Tomlin's variant).

From (25.19) and (25.27) it follows that

$$\begin{aligned} PB'Q &= LU + P(a_q - a_p)e_p^{\mathrm{T}}Q, \\ &= L(U + (L^{-1}Pa_q - L^{-1}Pa_p)e_p^{\mathrm{T}}Q). \end{aligned}$$

By (25.27) and $Be_p = a_p$ it holds that

$$Pa_p = PBe_p = LUQ^{\mathrm{T}}e_p.$$

Besides, it is clear that there is an integer $1 \le r \le m$ such that

$$e_r = Q^{\mathrm{T}}e_p. \tag{25.29}$$

Denote by v the solution to the first triangular system of (25.28). Then, combining the preceding three expressions leads to

$$PB'Q = L(U + (L^{-1}Pa_q - UQ^{\mathrm{T}}e_p)e_p^{\mathrm{T}}Q) = LR, \tag{25.30}$$

where the matrix

$$R = U + (v - Ue_r)e_r^{\mathrm{T}}$$

is upper triangular, except for the rth column, as it results from U by replacing the $r$the column by v. Move the rth column backward to the end position and the $(r+1)$ to mth columns forward one column. Then put the rth row to the bottom row and the $(r+1)$ to mth rows up one row. If $\hat{Q}$ denotes the according permutation, then the resulting matrix $\hat{Q}^{\mathrm{T}}R\hat{Q}$ is upper triangular, except for entries in the r to $(m-1)$th columns in the mth row perhaps being nonzero ($m \ge 2$). These entries can be eliminated by a series of Gauss transformations; while proceeding, exchange the

mth row and the row at which the diagonal locates whenever a diagonal is too small in module, compared with the according nonzero entry. Thus, there are permutations P_i and lower triangular matrices $L_i, i = 1, \cdots, s(1 \le s \le m - r)$ such that

$$L_s^{-1} P_s \cdots L_1^{-1} P_1 \hat{Q}^{\mathrm{T}} R \hat{Q} \triangleq U' \tag{25.31}$$

is nonsingular upper triangular. On the other hand, it is clear that

$$L' = L \hat{Q} P_1^{\mathrm{T}} L_1 \cdots P_s^{\mathrm{T}} L_s \tag{25.32}$$

is lower triangular with row exchanges. From (25.31), (25.32) and (25.30), it is verified that the new basis matrix B' has the following LU factorization with row and column exchanges:

$$P' B' Q' = L' U', \tag{25.33}$$

where

$$P' = P, \quad Q' = Q \hat{Q}. \tag{25.34}$$

Fortunately, this process leads to simplification of computation of simplex multipliers.

Theorem 25.3.1. *Let y be the solution to $B^{\mathrm{T}} y = c_B$ and z_q be the reduced cost, related to the pivot column index q. If w' is the solution to*

$$L'^{\mathrm{T}} P w' = (1/u'_{mm}) e_m, \tag{25.35}$$

where u'_{mm} is the mth diagonal of U', then $y' = y + z_q w'$ is the solution to (25.20).

Proof. Based on relationship between quantities, related to the old and new bases, it is only required to prove that w' is the solution to (25.22).

Premultiplying (25.35) by U'^{T} gives

$$U'^{\mathrm{T}} L'^{\mathrm{T}} P w' = (1/u'_{mm}) U'^{\mathrm{T}} e_m,$$

combining which, (25.33),(25.34) and $U'^{\mathrm{T}} e_m = u'_{mm} e_m$ leads to

$$\hat{Q}^{\mathrm{T}} Q^{\mathrm{T}} B'^{\mathrm{T}} w' = e_m.$$

Then by (25.29) and the definition of permutation $\hat{Q}$, it is known that

$$e_p^{\mathrm{T}} Q \hat{Q} = e_r^{\mathrm{T}} \hat{Q} = e_m^{\mathrm{T}}.$$

Table 25.1 Iteration and time ratios of MINOS 5.51 to NEW

Problem	Small (38)	Medium (41)	Large (17)	Ken. (16)	BPMPD (17)	Total average (129)
Iterations	1.00	1.01	1.12	1.05	1.12	1.10
Time	1.03	1.17	1.28	1.28	1.23	1.24

Finally, it follows from the preceding two expressions that

$$B'^{\mathrm{T}} w' = Q\hat{Q}e_m = Q\hat{Q}\hat{Q}^{\mathrm{T}}Q^{\mathrm{T}}e_p^{\mathrm{T}} = e_p.$$

□

This theorem says that after new LU factors are obtained by Forrest-Tomlin update, one may solve triangular system (25.35) for w', calculate the new simplex multipliers y', and then compute new reduced costs by (25.25).

Our numerical experiments involve the following two codes (Hu and Pan 2008b):

1. MINOS 5.51.
2. NEW: MINOS 5.51 with the presented approach for pricing.

The preceding codes both adopted default sectional pricing option (partial price 10).

There were 129 Netlib, Kennington and BPMPD test problems involved (Appendix B: Tables B.1–B.3). All the 96 Netlib problems fall into three 3 sets: Small (38), Medium (41), Large (17) (in the order of increasing $m + n$). Table 25.1 lists total iteration and time ratios of MINOS 5.51 to NEW.

From the table, it is seen that the presented pricing scheme outperformed the standard scheme with each set of problems (except for small Netlib problems), in terms of either iterations or running time. The overall iteration and average time ratios attain 1.10 and 1.24, respectively.

25.4 Relaxation Principle

Problem obtained by dropping some constraints of a problem is called *relaxation problem* of the latter. Feasibility or optimality of the two problems is closely related.

Proposition 25.4.1. *Any feasible solution to the original problem is feasible to the relaxation problem, and an optimal solution to the former is optimal to the latter if dropped constraints are inactive at it.*

Proof. Since the feasible region of the original problem is a subset of that of the relaxation problem, the first half of the Proposition is clear. If dropped constraints are inactive at an optimal solution to the original, the solution is a local optimal solution to the relaxation problem. According to Theorem 4.3.4, it is a (global) optimal solution to the latter. □

It is favorable for solving the original problem if a relaxation problem of it was solved.

Proposition 25.4.2. *If the relaxation problem is infeasible, so is the original problem. If an optimal solution to the relaxation problem is feasible to the original problem, it is also optimal to the latter.*

Proof. The statement is easily drawn from the fact that the feasible region of the original problem is a subset of that of the relaxation problem. □

These results form a basis of the so-called "relaxation principle": a relaxation problem is solved first, which is usually easier to solve. If the obtained optimal solution satisfies the dropped constraints, then it is also optimal to the original; if not so, or unboundedness of the relaxation problem is detected, part or all of the dropped constraints are added to the relaxation problem, and the latter is then solved again. Since there are only finitely many constrains, the original problem can be solved in finitely many iterations, if each relaxation problem is solved finitely.

The relaxation principle may be applied in various ways, such as sensitivity analysis (Chap. 6), and the active set method (Sect. 2.4), including the simplex method itself, even ILP solvers, like branch-bound and cutting-plane methods.

25.4.1 Illustrative Application: Partial Pricing

It is possible to design various relaxation algorithms by dropping and adding constraints according to different criterions. In the following, we only bring up two algorithms, as they can be realized conveniently via partial pricing.

Assume that a feasible simplex tableau is available. For a given threshold $\sigma \geq 0$, section the nonbasis N as follows:

$$N_1 = \{j \in N \mid \bar{z}_j < -\sigma\}, \quad N_2 = N \backslash N_1. \tag{25.36}$$

Giving a priority to indices in N_1 in column selection leads to the following algorithm.

Algorithm 25.4.1 (Relaxation simplex algorithm: partial pricing). Given $\sigma \geq 0$. Initial: feasible simplex tableau of form (3.18), (N_1, N_2) defined by (25.36). This algorithm solves the standard LP problem.

1. Select pivot column index $q \in \arg\min_{j \in N_1} \bar{z}_j$.
2. Go to step 5 if $\bar{z}_q < 0$.
3. Select pivot column index $q \in \arg\min_{j \in N_2} \bar{z}_j$.
4. Stop if $\bar{z}_q \geq 0$ (optimality achieved).
5. Stop if $I = \{i = 1, \cdots, m \mid \bar{a}_{i\,q} > 0\} = \emptyset$ (unbounded problem).
6. Determine pivot row index $p \in \arg\min_{i \in I} \bar{b}_i / \bar{a}_{i\,q}$.

7. Convert $\bar{a}_{pq}$ to 1, and eliminate the other nonzeros in the column by elementary transformations.
8. If $q \in N_1$, set $N_1 = N_1\backslash\{q\} \cup \{j_p\}$; else, if $q \in N_2$, set

$$N_1 = \{j \in N_2 \mid \bar{z}_j < -\sigma\} \setminus \{q\} \cup \{j_p\}, \qquad N_2 = N \setminus \{q\} \setminus N_1.$$

9. Go to Step 1.

In fact, the preceding algorithm solves the dual relaxation program featured by N_1 firstly, ignoring dual nonnegative constrains associated with N_2. Once achieving optimality, it selects indices, associated with violated constraints, from N_2 to form a new N_1, and then does the same thing, again and again. Despite an initial (N_1, N_2) can be an arbitrary section of N, it seems to be preferable to form N_1, associated with dual nonnegative constraints, violated by current dual solution; if $\sigma = 0$, e.g., N_1 corresponds to all violated ones.

Example 25.4.1. Solve the following problem by Algorithm 25.4.1:

$$\begin{array}{llll}
\min & f = -5x_1 - 2x_3 - 3x_7 - x_8 + 4x_9, & & \\
\text{s.t.} & -2x_1 \quad + 4x_3 & + x_6 - 8x_7 - 5x_8 + 6x_9 = 1, & \\
& -4x_1 + x_2 + 5x_3 & - 3x_7 - \ x_8 - \ x_9 = 3, & \\
& 2x_1 \quad - 3x_3 \quad + x_5 & + 6x_7 + 3x_8 - 3x_9 = 4, & \\
& -3x_1 \quad + 8x_3 + x_4 & - 6x_7 - 4x_8 + 5x_9 = 5, & \\
& & x_j \geq 0, \quad j = 1, \cdots, 9. &
\end{array}$$

Answer Set $\sigma = 0$. Initial tableau

x_1	x_2	x_3	x_4	x_5	x_6	x_7	x_8	x_9	RHS
−2		4			1	−8	−5	6	1
−4	1	5				−3	−1	−1	3
2*		−3		1		6	3	−3	4
−3		8	1			−6	−4	5	5
−5		−2				−3	−1	4	

Iteration 1: $N_1 = \{1, 3, 7, 8\}$, $N_2 = \{9\}$.

1. $\min\{-5, -2, -3, -1\} = -5 < 0$, $q = 1$.
5. $I \ = \{3\} \neq \emptyset$.
6. $\min\{4/2\} = 2$, $p = 3$.
7. Multiply row 3 by 1/2, then add 2, 4, 3, 5 times of row 3 to rows 1,2,4,5, respectively:
8. $N_1 = \{3, 7, 8, 5\}$, $N_2 = \{9\}$.

x_1	x_2	x_3	x_4	x_5	x_6	x_7	x_8	x_9	RHS
		1		1	1	−2	3	5	5
	1	−1		2		9	−7	11	11
1		−3/2		1/2		3	−3/2	2	2
		7/2*	1	3/2		3	1/2	11	11
		−19/2		5/2		12	−7/2	10	10

Iteration 2:

1. $\min\{-19/2, 12, -7/2, 5/2\} = -19/2,\ q = 3.$
5. $I \ = \{1, 4\} \neq \emptyset.$
6. $\min\{5/1, 11/(7/2)\} = 22/7,\ p = 4.$
7. Multiply row 4 by $2/7$, then add $-1, 1, 3/2, 19/2$ times of row 4 to rows 1,2,3,5, respectively:
8. $N_1 = \{7, 8, 5, 4\},\ N_2 = \{9\}.$

x_1	x_2	x_3	x_4	x_5	x_6	x_7	x_8	x_9	RHS
			−2/7	4/7	1	−20/7	−15/7	20/7*	13/7
	1		2/7	17/7		69/7	36/7	−48/7	99/7
1			3/7	8/7		30/7	12/7	−9/7	47/7
		1	2/7	3/7		6/7	1/7	1/7	22/7
			19/7	46/7		141/7	55/7	−15/7	279/7

Iteration 3:

1. $\min\{141/7, 55/7, 46/7, 19/7\} \geq 0.$
3. $\min\{-15/7\} < 0,\ q = 9.$
5. $I \ = \{1, 4\} \neq \emptyset.$
6. $\min\{(13/7)/(20/7), (22/7)/(1/7)\} = (13/7)/(20/7),\ p = 1.$
7. Multiply row 1 by $7/201/3$, then add $48/7, 9/7, -1/7, 15/7$ times of row 1 to rows 2,3,4,5, respectively.
8. $N_1 = \{6\},\ N_2 = \{4, 5, 7, 8\}.$

x_1	x_2	x_3	x_4	x_5	x_6	x_7	x_8	x_9	RHS
			−1/10	1/5	7/20	−1	−3/4	1	13/20
	1		−2/5	19/5	12/5	3			93/5
1			3/10	7/5	9/20	3	3/4		151/20
		1	3/10	2/5	−1/20	1	1/4		61/20
			5/2	7	3/4	18	25/4		165/4

Iteration 4:

1. $\min\{3/4\} \geq 0.$
3. $\min\{5/2, 7, 18, 25/4\} \geq 0.$

4. Optimality achieved. The basic optimal solution and related objective value are

$$\bar{x} = (151/20, 93/5, 61/20, 0, 0, 0, 0, 0, 13/20)^{\mathrm{T}}, \quad \bar{f} = -165/4.$$

On the other hand, a relaxation dual simplex algorithm can be derived analogously. Given an initial dual feasible simplex tableau. For any threshold $\gamma \geq 0$, section row indices as follows:

$$I_1 = \{i = 1, \cdots, m \mid \bar{b}_i < -\gamma\}, \quad I_2 = \{1, \cdots, m\} \backslash I_1. \tag{25.37}$$

If the associated constraints are written

$$\bar{b} - \sum_{j \in N} N x_N \geq 0, \quad x_N \geq 0,$$

then selecting a row index firstly from I_1 means ignoring inequalities, associated with I_2 in the first formula. This leads to the following algorithm.

Algorithm 25.4.2 (Relaxation dual simplex algorithm: partial pricing). Given $\gamma \geq 0$. Initial: dual feasible simplex tableau of form (3.18), (I_1, I_2) are defined by (25.37). This algorithm solves the standard LP problem.

1. Select pivot row index $p \in \arg\min_{i \in I_1} \bar{b}_i$.
2. Go to step 5 if $\bar{b}_p < 0$.
3. Select pivot row index $p \in \arg\min_{i \in I_2} \bar{b}_i$.
4. Stop if $\bar{b}_p \geq 0$ (optimality achieved).
5. Stop if $N' = \{j \in N \mid \bar{a}_{pj} < 0\} = \emptyset$ (dual unbounded or primal infeasible).
6. Determine pivot column index $q \in \arg\min_{j \in N'} -\bar{z}_j/\bar{a}_{pj}$.
7. Convert $\bar{a}_{pq}$ to 1, and eliminate the other nonzeros in the column by elementary transformations.
8. If $p \in I_2$, set $I_2 = \{i \in I_2 \mid \bar{b}_i < -\gamma\}$ and $I_1 = \{1, \cdots, m\} \backslash I_2$.
9. Go to step 1.

As there is no numerical results available at this stage, however, it is not yet clear how preceding two algorithms perform. However, some relaxation simplex variants, resulting from embedding the most-obtuse-angle heuristics to criterion dropping and adding constraints, performed favorably in preliminary computational tests (Yang and Pan 2006; Zhou et al. 2009).

25.5 Local Duality

This section is devoted to localization of the optimality condition and a related LP solution strategy.

A relaxation problem, resulting from dropping all inactive constraints at a current iterate, is termed *local problem*. Thus, constraints of a local problem are the set of active constraints, which are binding or violated at the iterate (Sect. 2.4).

Proposition 25.5.1. *A current iterate is an optimal solution to the original problem if and only if it is optimal to the associated local problem.*

Proof. If it is an optimal solution to the associated local problem, the current iterate is optimal to the original problem since it is feasible to the latter, according to Proposition 25.4.2. Conversely, if the iterate is an optimal solution to the original problem, it is optimal to the local problem since the dropped constraints are inactive at it, according to Proposition 25.4.1. □

As a relaxation problem, a local problem is simple and easy to handle, compared to the original. As constraints of a local problem is a subset of those of the original problem, the associated optimality condition, termed *local*, involves less formulas than that for the original problem. If the iterate fulfil the local optimality condition, then we are done. Otherwise, a feasible downhill is determined as a search direction with respect to the local problem. Then a new iterate is created by taking a step from the iterate along the direction, subject to constraints of the original problem. These steps are repeated until achieving optimality. So, such a local duality strategy and the active set approach converge (Sect. 2.4).

In practice, it should be favorable to use ϵ-active constraints to form the local problem (Powell 1989).

Example 25.5.1. Assume that the following bounded-variable problem

$$\begin{aligned} \min \quad & c^{\mathrm{T}}x, \\ \text{s.t.} \quad & Ax = b, && (25.38) \\ & 0 \le x \le u, && (25.39) \end{aligned}$$

has a basic solution $\bar{x}$ such that $\bar{x}_N = 0$, but $\bar{x}_B$ may not fulfil (25.39). Consider the local problem below:

$$\begin{aligned} \min \quad & c^{\mathrm{T}}x, \\ \text{s.t.} \quad & Ax = b, \quad x \ge 0, \end{aligned}$$

which is a standard problem.

Let x^* and (y^*, z^*) be primal and dual optimal solutions to the preceding program, respectively. If it satisfies (25.39), then x^* is an optimal solution to the original problem. In this case, optimality is actually achieved, since x^* and (y^*, z^*) fulfil optimal condition (4.16) with $w^* = 0$ and $l = 0$.

Based on local duality, the generalized dual simplex algorithm was derived for solving the bounded-variable problem, in Sect. 7.6.

25.6 Decomposition Principle

As well-known, any LP problem can be handled via solving only a system of equations if an optimal basis is available. Unfortunately, it seems to be impossible to get an optimal basis directly. Proposition 2.5.1 is only heuristics, and the associated crash procedure is very unbelievable to provide an optimal basis (Sect. 5.5). Therefore, LP problems are only solved by iterative methods.

Have long been a challenging task, solving large-scale LP problems may require huge amount of iterations and storage. To solve such problems, as a result, it is natural to turn to decomposition methods, such as Dantzig-Wolfe decomposition and Benders decomposition. Unfortunately, applications of these methods are restricted to a certain type of structured problems. In fact, very large problems can not be handled at present, despite advance of computer's hardware.

On the reality side, interestingly, some decomposition arrangement for big events has been put into effect successfully for many years. For instance, World Cup actually follows a "parallel decomposition" competition system: firstly preliminary selection contest is undertaken in each of the six continents; then, winners take part in the final, and the champion stands out. In contrast, an arena contest actually follows a "serial decomposition" competition system: the champion is produced through a series of matches between interim winner and challenger in succession. Without a doubt, it is impossible to organize any big event without decomposition.

A close analogy with very large-scale optimization can be made. Belonging to the "arena contest" category, the so-called "decomposition principle" presented in this section finds an optimal solution (or basis) to a large-scale LP problem by solving a series of small LP problems, as would hopefully be applicable to more general optimization problems.

25.6.1 Subprograms and Algorithms

Firstly, we show that the standard problem can be solved by handling two smaller subprograms in succession.

Let B_0 be an initial feasible basis, associated with basic feasible solution $\bar{x}$. Partition nonbasis N to two sections $(N1, N2)$. Denote the according sections of c by $(c_{B_0}, c_{N_1}, c_{N_2})$. The first subprogram is of the following form:

$$\begin{array}{ll} \min & c_{B_0}^T x_{B_0} + c_{N_1}^T x_{N_1}, \\ \text{s.t.} & B_0 x_{B_0} + N_1 x_{N_1} = b, \quad x_j \geq 0,\ j \in B_0 \cup N_1, \end{array} \tag{25.40}$$

whose optimal basis (basic solution or value) is termed *suboptimal* with respect to the original problem. Based on the resulting suboptimal basis, say B_1, we construct a second subprogram as follows:

$$\begin{array}{ll} \min & c_{B_1}^T x_{B_1} + c_{N_2}^T x_{N_2}, \\ \text{s.t.} & B_1 x_{B_1} + N_2 x_{N_2} = b, \quad x_j \geq 0,\ j \in B_1 \cup N_2. \end{array} \tag{25.41}$$

It is clear that if either of the subprograms is lower unbounded, so is the original problem. Assume that it is not the case. Then each suboptimal solution corresponds to a basic feasible solution to the original problem; e.g., after the first subprogram is solved, the components, associated with $B_0 \cup N_1$, of the feasible solution $\bar{x}$, change to new values of the suboptimal solution, while the other components remain unchanged. Note that all nonbasic components of the suboptimal solutions are of default value zero, so are those of the corresponding feasible solutions to the original problem.

It will be shown that optimality of the original problem is achieved if the suboptimal values of the two subprograms are equal. Otherwise, take the suboptimal basis, obtained from solving the second subprogram, as a new B_0 to repeat the previous steps.

The preceding can be generalized by partitioning the nonbasis to multiple sections. Let B_0 be the initial feasible basis, and let $(N_1, \cdots, N_s),\ s \geq 2$ be a partition of the nonbasis. The tth subprogram ($t = 1, \ldots, s$) to be solved is of form

$$\begin{array}{ll} \min & c_{B_{t-1}}^T x_{B_{t-1}} + c_{N_t}^T x_{N_t}, \\ \text{s.t.} & B_{t-1} x_{B_{t-1}} + N_t x_{N_t} = b, \quad x_j \geq 0,\ j \in B_{t-1} \cup N_t, \end{array} \tag{25.42}$$

where B_{t-1} is a suboptimal basis, yielding from solving the $(t-1)$th subprogram $(t = 2, \cdots, s)$.

This scheme can be organized into the following model.

Algorithm 25.6.1 (Decomposition algorithm 1). Given integer $s \geq 2$. Initial: feasible basis B_0. This algorithm solves the standard LP problem.

1. Partition $A \backslash B_0$ to $(N_1, \cdots, N_s)$.
2. Set $t = 0$.
3. Set $t = t + 1$.
4. Solve subprogram (25.42) by the simplex Algorithm 3.5.1:
   (1) stop if lower unbounded;
   (2) if $t < s$, go to step 3 with the suboptimal basis B_t and objective value $\bar{f}_t$;
5. Stop if $\bar{f}_s = \bar{f}_1$;
6. Set $B_0 = B_s$
7. Go to step 1.

Iterations associated with $t = 1, \ldots, s$ form a "circle" (solving s subprograms). The solution process involves a series of circles, in general.

Lemma 25.6.1. *Algorithm 25.6.1 generates a sequence of feasible bases, with objective values decreasing (not necessarily strictly) monotonically.*

Proof. It might be well to assume that the algorithm terminates at the end of the second circle. Take the first circle. As the initial basis B_0 is feasible to the original problem, the suboptimal bases B_t are well-defined ($t = 1, \cdots, s$). It is clear that each basis B_t and associated objective value $\bar{f}_t$ are not only optimal to the tth subprogram but also feasible to the original problem. Since each $\bar{f}_t$ is optimal to the tth subprogram and feasible to the $(t+1)$th subprogram, the objective values decrease (not necessarily strictly) monotonically. Likewise, it can be shown that the objective values decrease monotonically within the second circle; therefore, so do all the objective values since B_0 for the second circle is just the end suboptimal basis B_s of the first circle. □

Theorem 25.6.1. *Under the nondegeneracy assumption, Algorithm 25.6.1 terminates either at*

(i) step 4(1), detecting lower unboundedness of the original problem (1.8); or at
(ii) step 5, generating an optimal basis to the original problem.

Proof. Assume at the moment that the algorithm terminates. By Lemma 25.6.1, the condition $\bar{f}_s = \bar{f}_1$ implies that all the feasible values are equal.

It is clear that termination at step 4(1) implies lower unboundedness of the original problem. Assume that the process terminates at step 5. We show (ii) by induction on the number s of sections.

Consider for $s = 1$. Since B_1 is an optimal basis to the first subprogram, the according optimal dual solution, say $(\tilde{y}, \tilde{z})$, satisfies

$$\tilde{z}_{B_0 \cup N_1} = c_{B_0 \cup N_1} - (B_0 \cup N_1)^T \tilde{y} \geq 0, \qquad \tilde{y} = B_1^{-T} c_{B_1}. \tag{25.43}$$

As $A = \{B_0, N_1, N_2\}$, it is therefore only needed to show that

$$\tilde{z}_{N_2} = c_{N_2} - N_2^T \tilde{y} \geq 0. \tag{25.44}$$

Since $\bar{f}_2 = \bar{f}_1$, B_1 is also optimal to the second subprogram. If (25.44) does not hold, then it follows from $\bar{x}_{B_1} = B_1^{-1} b > 0$ (nondegeneracy assumption) that the B_1 is not optimal to the second subprogram (Lemma 3.9.2), as is a contradiction. Therefore, (25.44) holds, which and (25.43) together imply that B_1 is an optimal basis to the original problem.

Assume that (ii) is valid for $1 \leq s \leq t$, $t \geq 1$, we will show its validity for $s = t + 1$. Since $\bar{f}_{t+1} = \bar{f}_1$, it holds that $\bar{f}_t = \bar{f}_1$ (Lemma 25.6.1). According to the assumption of induction, therefore, B_1 is an optimal basis to problem

$$\begin{array}{ll} \min & c_{B_0}^T x_{B_0} + c_{N'}^T x_{N'}, \\ \text{s.t.} & B_0 x_{B_0} + N' x_{N'} = b, \quad x_j \geq 0, \ j \in B_0 \cup N', \end{array} \tag{25.45}$$

where $N' = N_1 \cup N_2 \cup \cdots \cup N_t$. If we take a look at partition (B_0, N', N_{t+1}) of the original problem, and note that the statement is valid for $s = 2$, it is asserted that B_1 is an optimal basis to the original problem.

Let us turn to termination of Algorithm 25.6.1. All associated subprograms are feasible, hence the simplex algorithm utilized to solve them terminates under the nondegeneracy assumption. Thus, it does not terminate only if there are infinite many circles, and $\bar{f}_s < \bar{f}_1$ holds for each. This implies that there are infinitely many bases, as leads to a contradiction. Therefore, Algorithm 25.6.1 terminates. □

A drawback of subprogram (25.40) (or (25.42)) is that the entering set N_1 (or N_t) is predetermined without considering the associated reduced costs at all. This may be remedied as follows.

Let ϵ be the optimality tolerance. Given a real number $\sigma > \epsilon > 0$ and an integer, say $m' \leq m$. Denote by B_0 the current feasible basis and by N the candidate indices from which to choose the entering set N_1. Initially, just take the nonbasis as N, i.e., $N = A \backslash B_0$. Assume that $\bar{z}_N$ is the nobasic reduced costs. Then, N_1 is determined by

$$N_1 = \{j \in N \mid \bar{z}_j \leq -\sigma\}, \tag{25.46}$$

subject to $|N_1| \leq m'$. So the number of indices in N_1 is no more than m. If $|N_1| > 0$, subprogram (25.40) is solved. Then we set $N = A \backslash (B_0 \cup N_1)$ for the next iteration. In the solution process, σ is decreased by a percentage τ to maintain $|N_1|$ reasonably large whenever $|N_1| < m'$. Assuming $|N_1| = 0$, optimality of the original problem is achieved if $\sigma = \epsilon$; or N_1 is determined by (25.46) again with a reduced σ if $\sigma > \epsilon$, and so on.

The preceding discussions lead to a variant of Algorithm 25.6.1, as formulated as follows.

Algorithm 25.6.2 (Decomposition algorithm 2). Given optimality tolerance $0 < \epsilon \ll 1$, $0 < \tau < 1$ and integer $m' \leq m$. Initial: $\sigma > \epsilon$; feasible basis B_0 and nonbasis N. This algorithm solves the standard LP problem.

1. Compute $\bar{z}_N = c_N - N^T \bar{y}$, $B_0^T \bar{y} = c_{B_0}$.
2. Determine set $N_1 = \{j \in N \mid \bar{z}_j \leq -\sigma\}$.
3. If $|N_1| < m'$, set $\sigma = \max\{\epsilon, \tau\sigma\}$.
4. Go to step 7 if $|N_1| > 0$.
5. Stop if $\sigma = \epsilon$ (optimality achieved).
6. Go to step 2.
7. Solve subprogram (25.40) by the simplex Algorithm 3.5.1:

    (1) stop if lower unbounded;
    (2) set $N = A \backslash (B_0 \cup N_1)$, and then set the yielding suboptimal basis to B_0.

8. Go to step 1.

25.6.2 *The Basic Idea: "arena contest"*

The preceding decomposition algorithms work as if some "arena contest" is organized to produce the "team champion" via a series of "matches".

The number of players (columns) in the whole area (coefficient matrix) is too huge to match in the limited sports facilities (available hardware) to produce a team champion (optimal basis). Therefore, a system is formulated and executed to govern the competition (optimization), so that a series of relatively small matches can be proceeded in the limited facilities. Specific ideas behind the two decomposition algorithms may be explained as follows, respectively:

(i) Algorithm 25.6.1:
At the beginning, an initial team (basis B_0) is selected from players (columns) of the area (coefficient matrix A) under some criterion. The other players (nonbasic columns N) are divided to s teams $((N_1, \cdots, N_s))$. The first match takes place in the facilities among the initial team and the first team $((B_0, N_1))$, producing a first winner team (suboptimal basis B_1). Then, the second match takes place among the first winner and the second team $((B_1, N_2))$, and so on. Once the sth match is finished, the last winner team (suboptimal basis B_s) is declared as the ultimate champion ("optimal" basis) if it has the same record (suboptimal value) as the first winner; otherwise, the next round of matches are carried out by taking the last winner as a new initial team (setting $B_0 = B_s$).

(ii) Algorithm 25.6.2:
After an interim winner team (B_0) is selected, a challenging team (N_1) is determined in a tryout among all the other players (N) under a criterion, which will be relaxed whenever the member of the challenging team is less than a threshold. Then a match takes place among the interim winner and the challenging team $((B_0, N_1))$, producing a new interim winner team (setting the last suboptimal basis to B_0), and repeat. If no player wins out in the tryout under the criterion of the lowest standard ($\bar{z}_N \geq -\epsilon$), the last winner team (the end suboptimal basis) is declared as the ultimate team champion ("optimal" basis).

The decomposition principle and the relaxation principle (Sect. 25.4) converge in certain sense, as both may be embedded into partial pricing. By carefully considering steps of the tableau simplex method, it is seen that Algorithms 25.6.1 and 25.6.2 are, in theory, respectively equivalent to the following two variants of Algorithm 3.2.1, with relevant partial pricing, though they are entirely different from the decomposition Algorithms both conceptually and computationally.

Algorithm 25.6.3 (Simplex algorithm: partial-pricing 1). Initial: a feasible simplex tableau of form 3.1.1, associate with basis B. Given positive integer $s \geq 2$. This algorithm solves the standard LP problem.

1. Partition nonbasis N to $(N_1, \cdots, N_s)$.
2. Set $t = 0$ and set $\bar{f}_1 = \bar{f}$.
3. Set $t = t + 1$.
4. Determine column index $q \in \arg\min_{j \in N_t} \bar{z}_j$.
5. Go to step 9 if $\bar{z}_q < 0$.
6. Go to step 3 if $t < s$.
7. Stop if $\bar{f} = \bar{f}_1$ (optimality achieved).

8. Go to step 1.
9. Stop if $I = \{i = 1, \ldots, m \mid \bar{a}_{i\,q} > 0\} = \emptyset$ (lower unbounded).
10. Determine row index $p \in \arg\min_{i \in I} \bar{b}_i / \bar{a}_{iq}$.
11. Convert $\bar{a}_{p\,q}$ to 1, and eliminate the other nonzeros in the column by elementary transformations.
12. Go to step 4.

Note N_t at step 4 denotes the *current* nonbasis of the tth subprogram (25.42).

It turned out that the preceding algorithm with $s = 2$ is the same as Algorithm 25.4.2 essentially.

Algorithm 25.6.4 (Simplex algorithm: partial-pricing 2). Given optimality tolerance $0 < \epsilon \ll 1$, $0 < \tau < 1$ and integer $m' \leq m$. Initial: $\sigma > \epsilon$; feasible simplex tableau of form 3.1.1, associate with basis B and nonbasis N. This algorithm solves the standard LP problem.

1. Determine $N_1 = \{j \in N \mid \bar{z}_j \leq -\sigma\}$,
2. If $|N_1| < m'$, set $\sigma = \max\{\epsilon, \tau\sigma\}$.
3. Go to step 6 if $|N_1| > 0$.
4. Stop if $\sigma = \epsilon$ (optimality achieved).
5. Go to step 1.
6. Determine a column index $q \in \arg\min_{j \in N_1} \bar{z}_j$.
7. Go to step 10 if $\bar{z}_q < 0$.
8. Set $N = A \backslash (B \cup N_1)$.
9. Go to step 1.
10. Stop if $I = \{i = 1, \ldots, m \mid \bar{a}_{i\,q} > 0\} = \emptyset$ (lower unbounded).
11. Determine row index $p \in \arg\min_{i \in I} \bar{b}_i / \bar{a}_{iq}$.
12. Convert $\bar{a}_{p\,q}$ to 1, and eliminate the other nonzeros in the column by elementary transformations.
13. Go to step 6.

Note N_1 at steps 6 and 8 denotes a *current* nonbasis of the subprogram (25.40).

25.6.3 Practical Remarks

Many large-scale LP problems involve a large number of variables, compared with the number of constraint equations ($m \ll n$). In this case, it should be favorable to form subprograms involving columns fewer than m or less. Moreover, in case when $n - m \ll m$, the problem can be converted by the "dual elimination" (Sect. 25.1.3) to an $(n - m) \times n$ standard LP problem, so that a large-scale problem can be handled by solving small programs with the decomposition algorithm.

The partition $(N_1, \cdots, N_s)$ involved in Algorithm 25.6.1 is not unique. Of course, we hope an optimal basis be contained in the first few sections, in particular, in N_1 ideally; so is in Algorithm 25.6.2. How to achieve this is worth further investigating.

It might be favorable to give the nonbasic indices an order by modifying the pivoting-index (Sect. 2.5) by taking sparsity into account, so that sparser columns are in front of the line to have priority to enter N_1.

It is noted that both Algorithms 25.6.1 and 25.6.2 require an initial feasible basis to get started. So a two-phase procedure is needed for solving general LP problems. Any available Phase-I approach can be used to provide an initial feasible basis, though it is preferable to handle some Phase-I (e.g., the conventional artificial-variable) auxiliary program by the decomposition principle.

It should be pointed out that feasibility of the initial basis is not necessary. In fact, it suffices to ensure that the initial subprogram is itself feasible.

The potential use of the decomposition principle would be of great importance. In fact, Algorithms 25.6.1 and 25.6.2 could be generalized along the following two lines:

(i) Instead of the simplex algorithm, other basis-based pivotal algorithms, such as the dual simplex algorithm, the reduced simplex algorithm and etc, might be used as a subalgorithm, perhaps with slight modifications. Even the face algorithm, producing an optimal face, applies. But primal algorithms seem to be amenable, compared with dual algorithms.
(ii) Some hard optimization problems could be handled if they have linear constraints and separable variables; e.g., large integer or mixed ILP problems would be approximately solved by the Algorithms, supported by the branch-bound or controlled-branch algorithm amenable to small such problems. Further investigation is expected in this aspect.

Algorithms 25.6.1 and 25.6.2 are designed for solving standard LP problems, where all nonbasic components of solutions are of default value 0, and hence may be omitted in solution process. For other problems, such as the bounded-variable problem, a basic feasible solution should be available initially. In each iteration, the components, associated with the current subprogram, of the solution is updated while the other components remain unchanged. For integer or mixed LP problems (see Example 7.5.1), an initial feasible integer solution is certainly preferable, but not necessary. A feasible solution to the associated relaxed LP problem can be taken as an initial solution to get some variant of Algorithms 25.6.1 or 25.6.2 started, though there is a need for existence of an optimal solution to the initial integer subprogram. Of course, it is favorable to start from an optimal solution to the associated relaxed LP problem.

Finally, it should be pointed out that a parallel decomposition can be realized by some sort of imitation of parallel competition system.

25.6.4 *Illustration on the Standard LP Problem*

In this subsection, an instance is offered for illustration of the decomposition principle on a standard problem.

Example 25.6.1. Solve the following LP problem by Algorithm 25.6.1 with $s = 2$:

$$\begin{array}{llllllll} \min & f = -2x_1 + x_2 + 3x_3 - 5x_4 - 3x_6 - 7x_7 - 4x_8 - 2x_9, \\ \text{s.t.} & \quad - 3x_2 - 2x_3 + x_4 \quad + 4x_7 + \quad - 2x_9 = 1, \\ & 3x_1 \quad + 4x_3 \quad + x_5 \quad + 2x_7 + x_8 + 5x_9 = 2, \\ & -x_1 + x_2 + x_3 \quad + x_6 + 8x_7 - 6x_8 + 3x_9 = 8, \\ & x_j \geq 0, \ j = 1, \ldots, 9. \end{array}$$

Answer Initial feasible basis and nonbasis: $B_0 = \{4, 5, 6\}$, $N = \{1, 2, 3, 7, 8, 9\}$.
Outer iteration 1:
Subiteration 1:

1. $N_1 = \{1, 2, 3\}$, $N_2 = \{7, 8, 9\}$.
2. $t = 0$.
3. $t = 0 + 1 = 1$.
4. Call Algorithm 3.5.1 to solve the first subprogram ($B_0 = \{4, 5, 6\}$, $N_1 = \{1, 2, 3\}$)

$$\begin{array}{ll} \min & -2x_1 + x_2 + 3x_3 - 5x_4 - 3x_6, \\ \text{s.t.} & \quad - 3x_2 - 2x_3 + x_4 \quad = 1, \\ & 3x_1 \quad + 4x_3 \quad + x_5 \quad = 2, \\ & -x_1 + x_2 + x_3 \quad + x_6 = 8, \\ & x_j \geq 0, \ j = 1, \ldots, 6. \end{array}$$

6. The resulting optimal basis: $B_0 = \{1, 2, 4\}$, associated with $\bar{x}_{B_0} = (2/3, 26/3, 27)^T$ (ignoring nonbasic components. the same below) with objective value $\bar{f}_1 = (-2, 1, -5)(2/3, 26/3, 27)^T = -383/3$.
$t = 1 < s$.

Subiteration 2:

3. $t = 1 + 1 = 2$.
4. Call Algorithm 3.5.1 to solve the second subprogram ($B_0 = \{1, 2, 4\}$, $N_2 = \{7, 8, 9\}$)

$$\begin{array}{ll} \min & -2x_1 + x_2 - 5x_4 - 7x_7 - 4x_8 - 2x_9, \\ \text{s.t.} & \quad - 3x_2 + x_4 + 4x_7 \quad - 2x_9 = 1, \\ & 3x_1 \quad + 2x_7 + x_8 + 5x_9 = 2, \\ & -x_1 + x_2 \quad + 8x_7 - 6x_8 + 3x_9 = 8, \\ & x_j \geq 0, \ j = 1, 2, 4, 7, 8, 9. \end{array}$$

6. The resulting optima basis: $B_1 = \{2, 4, 8\}$, associated with $\bar{x}_{B_1} = (20, 61, 2)^T$ with $\bar{f}_2 = (1, -5, -4)(20, 61, 2)^T = -293$.
$t = s$

7. $\bar{f}_2 \neq \bar{f}_1$.
8. $B_0 = \{2, 4, 8\}$, $N = \{1, 3, 5, 6, 7, 9\}$.

Outer iteration 2: Set $B_0 = B_1 = \{2, 4, 8\}$.
Subiteration 1:

1. $N_1 = \{1, 3, 5\}$, $N_2 = \{6, 7, 9\}$.
2. $t \; = 0$.
3. $t \; = 0 + 1 = 1$.
4. Call Algorithm 3.5.1 to solve the first subprogram

$$\begin{array}{ll} \min & -2x_1 + x_2 + 3x_3 - 5x_4 - 4x_8, \\ \text{s.t.} & \quad - 3x_2 - 2x_3 + x_4 = 1, \\ & 3x_1 + 4x_3 + x_5 + x_8 = 2, \\ & -x_1 + x_2 + x_3 - 6x_8 = 8, \\ & x_j \geq 0, \; j = 1, 2, 3, 4, 5, 8. \end{array}$$

6. The resulting suboptimal basis is $B_0 = \{2, 4, 8\}$, associated with $\bar{x}_{B_0} = (20, 61, 2)^T$ with $\bar{f}_1 = -293$.
   $t < s$.

Subiteration 2:

3. $t = 1 + 1 = 2$.
4. Call Algorithm 3.5.1 to solve the second subprogram ($B_0 = \{2, 4, 8\}$, $N_2 = \{6, 7, 9\}$)

$$\begin{array}{ll} \min & x_2 - 5x_4 - 3x_6 - 7x_7 - 4x_8 - 2x_9, \\ \text{s.t.} & -3x_2 + x_4 + 4x_7 - 2x_9 = 1, \\ & 2x_7 + x_8 + 5x_9 = 2, \\ & x_2 + x_6 + 8x_7 - 6x_8 + 3x_9 = 8, \\ & x_j \geq 0, \; j = 2, 4, 6, 7, 8, 9. \end{array}$$

6. The resulting suboptimal basis is $B_1 = \{2, 4, 8\}$, associated with $\bar{x}_{B_1} = (20, 61, 2)^T$ with $\bar{f}_2 = (1, -5, -4)(20, 61, 2)^T = -293$.
   $t = s$
7. $\bar{f}_2 = \bar{f}_1$. The optimal solution and optimal value:

$$\bar{x} = (0, 20, 0, 61, 0, 0, 0, 2, 0)^T, \qquad \bar{f} = -293,$$

which can be verified by solving the original problem as a whole by the simplex algorithm.

25.6.5 Illustration on the Bounded-Variable LP Problem

Algorithm 25.6.1 can be modified to solve general LP problems, e.g., the bounded-variable problem, if the simplex algorithm called from step 4 is replace by a bounded-variable problem solver.

Example 25.6.2. Solve the following bounded-variable problem by a modified Algorithm 25.6.1 with $s = 2$, using Algorithm 7.4.1 as the subalgorithm:

$$\begin{array}{ll} \min & f = 2x_1 - x_2 + 3x_3 - 6x_4, \\ \text{s.t.} & -2x_1 + 3x_2 - 4x_3 + 2x_4 + x_5 \qquad\qquad\qquad = \ 14, \\ & -3x_1 + 4x_2 - 5x_3 + 6x_4 \qquad + x_6 \qquad\qquad = \ 16, \\ & \quad x_1 - 2x_2 + 2x_3 - 7x_4 \qquad\qquad + x_7 \qquad = -15, \\ & -15 \le x_1 \le 30, \quad -12 \le x_2 \le 20, \quad -17 \le x_3 \le 10, \\ & -8 \le x_4 \le 15, \quad -200 \le x_5 \le 26, \quad -200 \le x_6 \le 34, \\ & \ 0 \le x_7 \le 200. \end{array}$$

Answer Initial feasible basis and nonbasis: $B_0 = \{5, 6, 7\}$, $N = \{1, 2, 3, 4\}$.
Outer iteration 1:
Subiteration 1:

1. $N_1 = \{1, 2\}$, $N_2 = \{3, 4\}$.
2. $t \ = 0$.
3. $t \ = 0 + 1 = 1$.
4. Construct the first subalgorithm with $B_0 = \{5, 6, 7\}$, $N_1 = \{1, 2\}$ by setting $\bar{x}_{N_2} = (-17, 15)^T$, where the right-hand side is changed to

$$b - N_2\bar{x}_{N_2} = \begin{pmatrix} 14 \\ 16 \\ -15 \end{pmatrix} - \begin{pmatrix} -4 & 2 \\ -5 & 6 \\ 2 & -7 \end{pmatrix} \begin{pmatrix} -17 \\ 15 \end{pmatrix} = \begin{pmatrix} -84 \\ -159 \\ 124 \end{pmatrix},$$

that is

$$\begin{array}{ll} \min & 2x_1 - x_2, \\ \text{s.t.} & -2x_1 + 3x_2 + x_5 \qquad\qquad\qquad = \ -84, \\ & -3x_1 + 4x_2 \qquad + x_6 \qquad\qquad = -159, \\ & \quad x_1 - 2x_2 \qquad\qquad + x_7 \qquad = \ 124, \\ & -15 \le x_1 \le 30, \quad -12 \le x_2 \le 20, \quad -200 \le x_5 \le 26, \\ & -200 \le x_6 \le 34, \quad 0 \le x_7 \le 200. \end{array}$$

Call Algorithm 7.4.1 to solve the preceding program.

6. The resulting suboptimal basis is $B_0 = \{2, 5, 7\}$, associated with the basic solution and objective value

$$\bar{x} = (-15, -1, -17, 15 - 111, -200, 137)^T,$$

$$\bar{f}_1 = (2, -1, 3, -6, 0, 0, 0)\bar{x} = -170$$

$t = 1 < s$.

Subiteration 2:

3. $t = 1 + 1 = 2$.
4. Construct the second subalgorithm with $B_0 = \{2, 5, 7\}$, $N_2 = \{3, 4\}$ by setting $(\bar{x}_1, \bar{x}_6) = (-15, -200)$, where the right-hand side is changed to

$$b - N_2 \bar{x}_{N_2} = \begin{pmatrix} 14 \\ 16 \\ -15 \end{pmatrix} - \begin{pmatrix} -2 & 0 \\ -3 & 1 \\ 1 & 0 \end{pmatrix} \begin{pmatrix} -15 \\ -200 \end{pmatrix} = \begin{pmatrix} -16 \\ 171 \\ 0 \end{pmatrix},$$

yielding

$$\begin{array}{lllll} \min & -x_2 + 3x_3 - 6x_4, & & & \\ \text{s.t.} & \quad 3x_2 \; - \; 4x_3 + 2x_4 + x_5 & & = & -16, \\ & \quad 4x_2 \; - \; 5x_3 + 6x_4 & & = & 171, \\ & -2x_2 + 2x_3 \; - \; 7x_4 \qquad + x_7 & & = & 0, \\ & -12 \le x_2 \le 20, \quad -17 \le x_3 \le 10, & & & \\ & -8 \le x_4 \le 15, \quad -200 \le x_5 \le 26, \quad 0 \le x_7 \le 200. & & & \end{array}$$

Call Algorithm 7.4.1 to solve the preceding program.

6. The resulting suboptimal basis $B_1 = \{2, 5, 7\}$ is equal to B_0.
   $t = s$.
7. The optimal solution and optimal value are therefore:

$$\bar{x} = (-15, -1, -17, 15 - 111, -200, 137)^T,$$

$$\bar{f}_2 = (2, -1, 3, -6, 0, 0, 0)\bar{x} = -170,$$

which can be verified by solving the original problem as a whole by Algorithm 7.4.1.

25.6.6 *Illustration on the ILP Problem*

It is possible to handle large-scale ILP problems by some variant of Algorithm 25.6.1 if there is an optimal solution to the initial ILP subprogram. In general, however, this approach may not produce an optimal solution to the ILP problem, but an approximate one only. Starting from such a point, however, methods presented in Chap. 10, in particular the realization presented in the next section, is applicable for reaching an exact optimal solution. Besides, it is straightforward to extend the approach to solving large-scale mixed LP problems.

To start-up, one would solve the associated relaxed LP problem first, and then to construct an initial subprogram, based on the resulting optimal basis, if any. But we will do otherwise with the following instance.

Example 25.6.3. Solve the following ILP problem by a modified Algorithm 25.6.1 with $s = 2$, supported by some ILP solver:

$$\begin{aligned} \min \quad & f = x_1 - 3x_2 + 2x_3 - 5x_4 + x_5, \\ \text{s.t.} \quad & 2x_1 + x_2 - x_3 + 3x_4 - 2x_5 + x_6 = 7, \\ & 4x_1 - 3x_2 - 5x_4 - 4x_5 + x_7 = 2, \\ & -3x_1 + 2x_2 + x_3 + 2x_4 + x_5 + x_8 = 8, \\ & \text{integer } x_j \geq 0, \ j = 1, \cdots, 8. \end{aligned}$$

Answer Initial feasible basis and nonbasis: $B_0 = \{6, 7, 8\}$, $N = \{1, 2, 3, 4, 5\}$.
Outer iteration 1:
Subiteration 1:

1. $N_1 = \{1, 2\}$, $N_2 = \{3, 4, 5\}$.
2. $t = 0$.
3. $t = 0 + 1 = 1$.
4. Solve the first subalgorithm with $B_0 = \{6, 7, 8\}$, $N_1 = \{1, 2\}$, formed by setting $\bar{x}_{N_2} = (0, 0, 0)$, i.e.,

$$\begin{aligned} \min \quad & x_1 - 3x_2, \\ \text{s.t.} \quad & 2x_1 + x_2 + x_6 = 7, \\ & 4x_1 - 3x_2 + x_7 = 2, \\ & -3x_1 + 2x_2 + x_8 = 8, \\ & \text{integer } x_j \geq 0, \ j = 1, 2, 6, 7, 8. \end{aligned}$$

6. The resulting suboptimal basis is $B_0 = \{1, 2, 7\}$, associated with the solution and objective value

$$\bar{x}^{(1)} = (1, 5, 0, 0, 0, 0, 13, 1)^T, \qquad \bar{f}_1 = (1, -3, 0, 0, 0, 0, 0, 0)\bar{x}^{(1)} = -14$$

$t = 1 < s$.

Subiteration 2:

3. $t = 1 + 1 = 2$.
4. Construct the second subalgorithm with $B_0 = \{1, 2, 7\}$, $N_2 = \{3, 4, 5\}$ by setting $\bar{x}_{N_1} = (0, 1)^T$, where the right-hand side is changed to

$$b - N_1 \bar{x}_{N_1} = \begin{pmatrix} 7 \\ 2 \\ 8 \end{pmatrix} - \begin{pmatrix} 1 & 0 \\ 0 & 0 \\ 0 & 1 \end{pmatrix} \begin{pmatrix} 0 \\ 1 \end{pmatrix} = \begin{pmatrix} 7 \\ 2 \\ 7 \end{pmatrix},$$

yielding

$$\begin{aligned} \min \quad & x_1 - 3x_2 + 2x_3 - 5x_4 + x_5, \\ \text{s.t.} \quad & 2x_1 + x_2 - x_3 + 3x_4 - 2x_5 = 7, \\ & 4x_1 - 3x_2 - 5x_4 - 4x_5 + x_7 = 2, \\ & -3x_1 + 2x_2 + x_3 + 2x_4 + x_5 = 7, \\ & \text{integer } x_j \geq 0, \ j = 1, 2, 3, 4, 5, 7. \end{aligned}$$

Solve the preceding ILP subprogram.

6. The resulting suboptimal basis is $B_1 = \{4, 5, 7\}$.
7. The optimal solution and optimal value of the subprogram are

$$\bar{x}^{(2)} = (0, 0, 0, 3, 1, 0, 21, 1)^T, \qquad \bar{f}_2 = (1, -3, 2, -5, 1, 0, 0, 0)\bar{x}^{(2)} = -14$$

$t = s$.

7. $\bar{f}_2 = \bar{f}_1 = -14$. $\bar{x}^{(1)}$ or $\bar{x}^{(2)}$ would be an acceptable solution to the original ILP problem.

The exact optimal solution to the original problem are

$$x^* = (3, 0, 0, 5, 7, 0, 43, 0)^T \qquad f^* = -15.$$

25.7 ILP: Based on the Reduced Simplex Framework

As was shown, the ILP methods may be implemented via the dual simplex framework (Chap. 10). At first, each method solves the LP relaxation of the ILP problem by the simplex method. If the solution is noninteger, it handles a sequence of LP relaxations by the dual simplex method. Since each LP relaxation yields from its predecessor by adding a cut, the number of constraints increases in solution process, as is unfavorable for solving large-scale ILP problems.

The drawback can be purged if the branch-and-bound method is realized via the generalized simplex framework. But the latter is not very consistent with the controlled-branch method (Sect. 10.4) which allows the actualization of adding an integer cut by simply changing the associated bound. Fortunately, the generalized *reduced* simplex framework turns out to be desirable for this purpose.

Consider the variable-bounded ILP problem

$$\begin{array}{ll} \min & x_{n+1}, \\ \text{s.t.} & A\begin{pmatrix} x \\ x_{n+1} \end{pmatrix} = b, \quad l \leq \text{integer } x \leq u, \end{array} \tag{25.47}$$

which yields from the variable-bounded LP problem (19.1) by adding the integrality requirement.

To grow branches of its enumerate tree, the controlled-branch method carries out a depth-oriented step if a valid point (see Sect. 10.4) is found noninteger, whereas takes on a breadth-oriented step if absence of such a solution is, with current branch. To do so, we pursue primal feasibility with fixed $\bar{x}_{n+1} = f^+$ (while maintaining dual feasibility) via the generalized dual reduced simplex methodology (Sect. 19.4).

Let $\bar{x}$ be the current solution with $\bar{x}_{n+1} = f^+$, where f^+ is a suspected-optimal value. Denote the associated dual feasible reduced tableau by (16.6). If the "bound violation" defined by (7.30), i.e.,

$$\rho_i = \begin{cases} l_{j_i} - \bar{x}_{j_i}, & \text{if } \bar{x}_{j_i} < l_{j_i}, \\ u_{j_i} - \bar{x}_{j_i}, & \text{if } \bar{x}_{j_i} > u_{j_i}, \\ 0, & \text{if } l_{j_i} \leq \bar{x}_{j_i} \leq u_{j_i}, \end{cases} \qquad i = 1, \cdots, m+1, \tag{25.48}$$

are all zero, then $\bar{x}$ is a valid point to the current branch. Note however that a branch is usually generated from the solved predecessor by adding a valid cut, and therefore only a single constraint is violated at first.

To handle a branch, we describe the following subalgorithm in tableau form, which is yielded by a generalized modification of Algorithm 16.5.1.

Algorithm 25.7.1 (ILP subalgorithm: tableau form). Initial: Suspected-optimal value f^+ and $\bar{x}_{n+1} = f^+$. An improved reduced tableau of form (16.6), with basic solution $(\bar{x}, \bar{x}_{n+1})$. This procedure pursues a valid point of the branch.

1. Select row index

$$p \in \arg\max\{|\rho_i| \mid i = 1, \cdots, m+1\},$$

   where ρ_i is defined by (25.48).
2. Return if $\rho_p = 0$.
3. Go to step 8 if

$$J = \{j \in \Gamma \mid \text{sign}(\rho_p)\bar{a}_{p\,j} < 0\} \cup \{j \in \Pi \mid \text{sign}(\rho_p)\bar{a}_{p\,j} > 0\} = \emptyset.$$

   where Γ and Π are defined by (7.19)
4. Determine column index $q \in \arg\max_{j \in J} |\bar{a}_{p\,j}|$.
5. Convert $\bar{a}_{p\,q}$ to 1, and eliminate the other nonzeros in the column by elementary transformations (without touching the $\bar{x}_B$ column except for its pth component replaced by the value of $\bar{x}_q$).
6. Update $\bar{x}_B = \bar{x}_B - \rho_p \bar{a}_{j_p}, \bar{x}_{j_p} = \bar{x}_{j_p} + \rho_p$.
7. Go to step 1.
8. Compute $\alpha = -\rho_p / \bar{a}_{p,n+1}$.
9. Return if $\alpha < 0$.
10. If $\alpha = 0$, set $\alpha = 1$; otherwise, set $\alpha = \ulcorner \alpha \urcorner$.
11. Add $-\alpha$ times of x_{n+1} column to $\bar{x}_B$ column, and set $\bar{x}_{n+1} = \bar{x}_{n+1} + \alpha$.
12. Return.

The preceding Subalgorithm has the following three exits:

Step 2: The end $\bar{x}$ is a valid point. If it is integer, an optimal solution to the ILP problem is reached (corresponding to the so-called "integer basis"). If it is noninteger, execute the Subalgorithm on a deeper branch, yielded from adding an integer cut.
Step 9: The branch is infeasible, and so fathomed.
Step 12: There is no valid point to the branch, and stalled. The end $\bar{x}$ is dual feasible with integer objective value $\bar{x}_{n+1} > f^+$.

If the Subalgorithm returns from step 9 or 12, handle the nearest pending branch.

Assume that an integer valid point is not attained in the process eventually. If all the branches are infeasible, so is the original ILP problem. In the other case, determine the solution associated with the smallest value, say $\bar{x}_{n+1}$, among end objective values of the stalled branches. If it is integer, the solution is optimal to the original ILP problem; otherwise, add the objective cut with $f^+ = \bar{x}_{n+1}$ to the branch, and execute the Subalgorithm, and so on, until an optimal solution to the original ILP problem is found, or infeasibility is detected.

The following is an illustrative instance:

Example 25.7.1. Solve the following ILP problem:

$$\begin{aligned} \min \quad & x_6 = -5x_1 + 8x_2 - x_3, \\ \text{s.t.} \quad & 4x_1 - 5x_2 + x_3 + x_4 = 5, \\ & -3x_1 + 7x_2 - 2x_3 + x_5 = 2, \\ & 0 \le \text{integer } x_j \le 6, \quad j = 1, \cdots, 5. \end{aligned} \tag{25.49}$$

Answer

(1) Initial tableau:

	x_1	x_2	x_3	x_4	x_5	x_6	$\bar{x}_B$
	4	−5	1	1			5
	−3	7	−2		1		2
	−5*	8	−1			−1	0
u	6	6	6	6	6	$+\infty$	
$\bar{x}$	0	0	0	5	2	0	
l	0	0	0	0	0	$-\infty$	

Iteration 1:
Convert the preceding to reduced simplex tableau by pivoting on the entry in row 3 and column 1, and call generalized reduced simplex Algorithm 19.1.1, yielding an optimal reduced simplex tableau to the LP relaxation:

	x_1	x_2	x_3	x_4	x_5	x_6	$\bar{x}_B$
		7/5	1/5	1		−4/5	0
		11/5	−7/5		1	3/5	23/4
	1	−8/5	1/5			1/5	5/4
u	6	6	6	6	6	$+\infty$	
$\bar{x}$	5/4	0	0	0	23/4	−25/4	
l	0	0	0	0	0	$-\infty$	

with $p = 1$ and $\bar{x}_{n+1} = -6\frac{1}{4}$.

(2) Change the lower bound on the objective variable x_6 to the suspected-optimal value −6 in the tableau. Increase the objective value to −6 by giving

increment $1/4$, and accordingly update the solution by $\bar{x}_B = (023/45/4)^T - (1/4)(-4/53/51/5)^T = (1/5, 28/5, 6/5)^T$.

	x_1	x_2	x_3	x_4	x_5	x_6	$\bar{x}_B$
		7/5	1/5	1		−4/5	1/5
		11/5	−7/5		1	3/5	28/5
	1	−8/5*	1/5			1/5	6/5
u	6	6	6	6	6	$+\infty$	
$\bar{x}$	6/5	0	0	1/5	28/5	−6	
l	0	0	0	0	0	[−6]	

Call Subalgorithm 25.7.1. It returns from step 2 with a feasible tableau with the objective value $\bar{x}_6 = -6$.

(3) As $6/5c_1 > 1/5c_4 = 28/5c_5 = 0$, select x_1 as the cutting variable ($1 < \bar{x}_1 = 6/5 < 2$). Add integer cut $x_1 \geq 2$ by changing the lower bound on x_1 to 2 (while let the subprogram associated with $x_1 \leq 1$ pend).
Call Subalgorithm 25.7.1.

Iteration 2:

1. $p \;= 3, \rho_3 = 2 - 6/5 = 4/5$.
3. $J \;= \{2\}$.
4. $q \;= 2$.
5. Multiply row 3 by $-5/8$, and then add $-7/5, -11/5$ times of row 3 to rows 1 and 2, respectively (without touching the $\bar{x}_B$ column but its third component is replaced by value, 0, of $\bar{x}_2$).
6. $\bar{x}_B = (1/5, 28/5, 0)^T - (4/5)(7/8, 11/8, -5/8)^T = (-1/2, 9/2, 1/2)^T$.

	x_1	x_2	x_3	x_4	x_5	x_6	$\bar{x}_B$
	7/8		3/8	1		−5/8	−1/2
	11/8		−6/8		1	7/8	9/2
	−5/8	1	−1/8			−1/8	1/2
u	6	6	6	6	6	$+\infty$	
$\bar{x}$	2	1/2	0	−1/2	6/2	−6	
l	[2]	0	0	0	0	[−6]	

Iteration 3:

1. $p \quad = 1, \rho_1 = 1/2, j_p = 4$.
3. $J \quad = \emptyset$.
8. $\alpha \quad = -(1/2)/(-5/8) = 4/5$.
10. $\alpha \quad = 1$.
11. $\bar{x}_B \quad = (-1/2, 9/2, 1/2)^T - (-5/8, 7/8, -1/8)^T = (1/8, 29/8, 5/8)^T$;
$\bar{x}_{n+1} = -6 + 1 = -5$.
Returned from step 12. The ILP subprogram is stalled.

(4) Handle the pending ILP subprogram with adding cut $x_1 \leq 1$ to the end tableau of (2). Change the upper bound on x_1 to 1.
Call Subalgorithm 25.7.1.

Iteration 4:

1. $p = 3, \rho_3 = 1 - 6/5 = -1/5$.
3. $J = \{3\}$.
4. $q = 3$.

	x_1	x_2	x_3	x_4	x_5	f	$\bar{x}_B$
		7/5	1/5	1		−4/5	1/5
		11/5	−7/5		1	3/5	28/5
	1	−8/5	1/5*			1/5	6/5
u	[1]	6	6	6	6	$+\infty$	
$\bar{x}$	6/5	0	0	1/5	28/5	−6	
l	0	0	0	0	0	[−6]	

5. Multiply row 3 by 5, and then add $-1/5, 7/5$ times of row 3 to rows 1 and 2, respectively (without touching the $\bar{x}_B$ column but is third component replaced by the value, 0, of $\bar{x}_3$. (6) $\bar{x}_B = (1/528/50)^T + (1/5)(-1, 7, 5)^T = (0, 7, 1)^T$.

	x_1	x_2	x_3	x_4	x_5	x_6	$\bar{x}_B$
	−1	3		1		−1	0
	7	−9			1	2	7
	5	−8	1			1	1
u	[1]	6	6	6	6	$+\infty$	
$\bar{x}$	1	0	1	0	7	−6	
l	0	0	0	0	0	[−6]	

Iteration 5:

1. $p = 2, \rho_2 = 6 - 7 = -1, j_p = 5$.
3. $J = \emptyset$.
8. $\alpha = 1/2$.
10. $\alpha = 1$.
11. $\bar{x}_B = (0, 7, 1)^T - (-1, 2, 1)^T = (1, 5, 0)^T, \quad \bar{x}_{n+1} = -6 + 1 = -5$.
Returned from step 12. The ILP subprogram is stalled.

	x_1	x_2	x_3	x_4	x_5	x_6	$\bar{x}_B$
	-1	3		1		-1	1
	7	-9			1	2	5
	5	-8	1			1	0
u	[1]	6	6	6	6	$+\infty$	
$\bar{x}$	1	0	0	1	5	-5	
l	0	0	0	0	0	[-5]	

It is seen that (3) and (4) give the same lower objective bound -5. We finally grasp the end solution of (4) since it is feasible integer, and hence optimal to the original ILP problem, i.e.,

$$x^* = (1, 0, 0, 1, 5)^T, \quad f^* = -5.$$

On the other hand, more iterations would be required if a bad suspected-optimal value is used. See that following instance with binary (0 or 1) variables.

Example 25.7.2. Solve the following 0-1 LP problem:

$$\begin{array}{lrl} \min & f = 4x_1 + 2x_2 + 3x_3 + 3x_4 + 2x_5, & \\ \text{s.t.} & 3x_1 - 4x_2 - 7x_3 - 3x_4 & \leq -2, \\ & x_1 + 2x_2 - x_3 - x_4 - x_5 & \leq 1, \\ & -3x_2 + 11x_3 - 3x_4 - 6x_5 & \leq -1, \\ & x_j \in \{0, 1\}, \quad j = 1, \cdots, 5. & \end{array}$$

Answer Replacing $x_j \in \{0, 1\}$ by $0 \leq$ integer $x_j \leq 1$, the 0-1 problem is converted to an ILP problem, to which the controlled-branch method applies.

Solve its LP relaxation. Introduce slack variables x_j, $j = 6, 7, 8$.

Initial tableau: $\bar{x} = (0, 1, 0, 0, 1, 2, 0, 8)^T$, $\bar{x}_9 = 4$

	x_1	x_2	x_3	x_4	x_5	x_6	x_7	x_8	$x_9(f)$	$\bar{x}_B$
	3	-4	-7	-3		1				2
	1	2	-1	-1	-1		1			0
		-3	11	-3	-6			1		8
	4	2*	3	3	2				-1	
u	1	1	1	1	1	$+\infty$	$+\infty$	$+\infty$	$+\infty$	
$\bar{x}$	0	1	0	0	1	2	0	8	4	
l	0	0	0	0	0	0	0	0	$-\infty$	

Transfer the tableau to reduced one by multiplying row 4 by $1/2$, and adding $4, -2, 3$ times of row 4 to rows 1,2,3, respectively (without touching the last column–only its fourth component changed accordingly).

	x_1	x_2	x_3	x_4	x_5	x_6	x_7	x_8	$x_9(f)$	$\bar{x}_B$
	11		−1	3	4*	1			−2	2
	−3		−4	−4	−3		1		1	0
	6		31/2	3/2	−3			1	−3/2	8
	2	1	3/2	3/2	1				−1/2	1
u	1	1	1	1	1	$+\infty$	$+\infty$	$+\infty$	$+\infty$	
$\bar{x}$	0	1	0	0	1	2	0	8	4	
l	0	0	0	0	0	0	0	0	$-\infty$	

(1) Call generalized reduced simplex Algorithm 19.1.1.

Iteration 1:

1. $\alpha \; = 1, \; p = 1.$
2. $\bar{x}_B = (2, 0, 8, 1)^T + 1 \times (-2, 1, -3/2, -1/2) = (0, 1, 13/2, 1/2)^T,$
   $\bar{x}_9 \; = 4 - 1 = 3.$
4. $q \; = 5$
5. Multiplying row 1 by 1/4, and adding $3, 3, -1$ times of row 1 to rows 2,3,4, respectively (without touching the $\bar{x}_B$ column but its first component replaced by the value, 1, of $\bar{x}_5$).

	x_1	x_2	x_3	x_4	x_5	x_6	x_7	x_8	$x_9(f)$	$\bar{x}_B$
	11/4		−1/4*	3/4	1	1/4			−1/2	1
	21/4		−19/4	−7/4		3/4	1		−1/2	1
	57/4		59/4	15/4		3/4		1	−3	13/2
	−3/4	1	7/4	3/4		−1/4			0	1/2
u	1	1	1	1	1	$+\infty$	$+\infty$	$+\infty$	$+\infty$	
$\bar{x}$	0	1/2	0	0	1	0	1	13/2	3	
l	0	0	0	0	0	0	0	0	$-\infty$	

Iteration 2:

1. $\alpha \; = 2, \; p = 1.$
2. $\bar{x}_B = (1, 1, 13/2, 1/2)^T + 2 * (-1/2, -1/2, -3, 0) = (0, 0, 1/2, 1/2)^T,$
   $\bar{x}_9 \; = 3 - 2 = 1.$
4. $q \; = 3.$
5. Multiplying row 1 by -4, and adding $19/4, -59/4, -7/4$ times of row 1 to rows 2,3,4, respectively (without touching the $\bar{x}_B$ column but its first component replaced by the same value, 0, of $\bar{x}_3$).

It is clear that the optimal value of the LP relaxation is strictly greater than 0. Therefore, the current solution is valid with suspected-optimal value, $\bar{x}_9 = 1$. So, add objective cut $x_9 \geq f^+ = 1$. Change the lower bound on the objective variable x_9 to the suspected-optimal value 1 in the tableau.
(2) Take x_2 as the cutting variable. Add cut $x_2 \geq 1$ and let $x_2 \leq 0$ pend, as the gradient of the former forms the obtuse angle with the negative objective gradient. Change the lower bound on x_2 to 1 in the tableau.

	x_1	x_2	x_3	x_4	x_5	x_6	x_7	x_8	$x_9(f)$	$\bar{x}_B$
	-11		1	-3	-4	-1			2	0
	-47			-16	-19	-4	1		9	0
	$353/2$			48	59	$31/2$		1	$-65/2$	$1/2$
	$37/2$	1		6	7	$3/2$			$-7/2$	$1/2$
u	1	1	1	1	1	$+\infty$	$+\infty$	$+\infty$	$+\infty$	
$\bar{x}$	0	$1/2$	0	0	0	0	0	$1/2$	1	
l	0	[1]	0	0	0	0	0	0	[1]	

Call Subalgorithm 25.7.1 ($f^+ = 1,\ x_2 \geq 1$).

Iteration 3:

1. $p\ = 4, \rho_4 = 1 - 1/2 = 1/2,\ j_p = 2.$
3. $J\ = \emptyset.$
8. $\alpha\ = -(1/2)/(-7/2) = 1/7 > 0.$
10. $\alpha\ = \ulcorner 1/7 \urcorner = 1.$
11. $\bar{x}_B = (0, 0, 1/2, 1/2)^T - 1 \times (2, 9, -65/2, -7/2)^T = (-2, -9, 33, 4)^T$
    $\bar{x}_9\ = 1 + 1 = 2 > f^+ = 1.$

This branch is stalled.

(3) Handle the pending branch, created by adding $x_2 \leq 0$. Change the upper bound on x_2 to 0 in the tableau.

	x_1	x_2	x_3	x_4	x_5	x_6	x_7	x_8	$x_9(f)$	$\bar{x}_B$
	-11		1	-3	-4	-1			2	0
	-47			-16	-19	-4	1		9	0
	$353/2$			48	59	$31/2$		1	$-65/2$	$1/2$
	$37/2$*	1		6	7	$3/2$			$-7/2$	$1/2$
u	1	[0]	1	1	1	$+\infty$	$+\infty$	$+\infty$	$+\infty$	
$\bar{x}$	0	$1/2$	0	0	0	0	0	$1/2$	1	
l	0	0	0	0	0	0	0	0	[1]	

Call Subalgorithm 25.7.1 ($f^+ = 1,\ x_2 \leq 0$).

Iteration 4:

1. $p\ = 4, \rho_4 = 0 - 1/2 = -1/2,\ j_p = 2.$
3. $J\ = \{1, 4, 5, 6\}.$
4. $q\ = 1, \max\{|37/2|, |6|, |7|, |3/2|\} = 37/2.$
5. Multiply row 4 by $2/37$, and then add $11, 47, -353/2$ times of row 4 to rows $1, 2, 3$, respectively (without touching the $\bar{x}_B$ column but its fourth component is replaced by value, 0, of $\bar{x}_1$).
6. $\bar{x}_B = (0, 0, 1/2, 0)^T + (1/2)(22/37, 94/37, -353/37, 2/37)^T$
    $= (11/37, 47/37, -158/37, 1/37)^T$
    $\bar{x}_2\ = 1/2 + (-1/2) = 0.$

	x_1	x_2	x_3	x_4	x_5	x_6	x_7	x_8	$x_9(f)$	$\bar{x}_B$
		22/37	1	21/37	6/37	−4/37			−3/37	11/37
		94/37		−28/37	−45/37	−7/37	1		4/37	47/37
		−353/37		−342/37*	−288/37	44/37		1	33/37	−158/37
	1	2/37		12/37	14/37	3/37			−7/37	1/37
u	1	[0]	1	1	1	$+\infty$	$+\infty$	$+\infty$	$+\infty$	
$\bar{x}$	1/37	0	11/37	0	0	0	47/37	−158/37	1	
l	0	0	0	0	0	0	0	0	[1]	

Iteration 5:

1. $p \;= 3, \rho_3 = 0 - (-158/37) = 158/37,\ j_p = 8.$
3. $J \;= \{4, 5\}.$
4. $q \;= 4, \max\{|-342/37|, |6|, |-288/37|\} = 342/37.$
5. Multiply row 3 by $-37/342$, and then add $-21/37, 28/37, -12/37$ times of row 3 to rows $1, 2, 4$, respectively (without touching the $\bar{x}_B$ column but its second component is replaced by value, 0, of $\bar{x}_4$).
6. $\bar{x}_B = (11/37, 47/37, 0, 1/37)^T - (158/37)(7/114, -14/171, -37/342, 2/57)^T = (2/57, 277/171, 79/171, -7/57)^T,$
   $\bar{x}_8 = -158/37 + 158/37 = 0.$

	x_1	x_2	x_3	x_4	x_5	x_6	x_7	x_8	$x_9(f)$	$\bar{x}_B$
		1/114	1		−6/19	−2/57		7/114	−1/38	2/57
		568/171			−11/19	−49/171	1	−14/171	2/57	277/171
		353/342		1	16/19	−22/171		−37/342	−11/114	79/171
	1	−16/57			2/19	7/57		2/57	−3/19	−7/57
u	1	[0]	1	1	1	$+\infty$	$+\infty$	$+\infty$	$+\infty$	
$\bar{x}$	−7/57	0	2/57	79/171	0	0	277/171	0	1	
l	0	0	0	0	0	0	0	0	[1]	

Iteration 6:

1. $p \;= 4, \rho_4 = 0 - (-7/57) = 7/57,\ j_p = 1.$
3. $J \;= \emptyset.$
8. $\alpha \;= -(7/57)/(-3/19) = 7/9 > 0.$
10. $\alpha \;= \lceil 7/9 \rceil = 1.$
11. $\bar{x}_B = (2/57, 277/171, 79/171, -7/57)^T - 1 \times (-1/38, 2/57, -11/114, -3/19)^T = (7/114, 271/171, 191/342, 2/57)^T,$
    $\bar{x}_9 = 1 + 1 = 2 > f^+ = 1.$

The branch is stalled.

Comparing the end objective values reached in branches (2) and (3) gives a new suspected-optimal value $f^+ = 2$.

(4) We continue to handle (2) by adding objective cut $x_9 \geq f^+ = 2$ first. Change the lower bound on x_9 to 2 and the lower bound on x_2 to 1.

	x_1	x_2	x_3	x_4	x_5	x_6	x_7	x_8	$x_9(f)$	$\bar{x}_B$
	-11		1	-3	-4	-1			2	-2
	-47			-16	-19	-4	1		9	-9
	$353/2$			48	59	$31/2$		1	$-65/2$	33
	$37/2^*$	1		6	7	$3/2$			$-7/2$	4
u	1	1	1	1	1	$+\infty$	$+\infty$	$+\infty$	$+\infty$	
$\bar{x}$	0	4	-2	0	0	0	-9	33	2	
l	0	[1]	0	0	0	0	0	0	[2]	

Call Subalgorithm 25.7.1 ($f^+ = 2,\ x_2 \geq 1$).

Iteration 7:

1. $p = 4,\ \rho_3 = 1 - 4 = -3,\ j_p = 2.$
3. $J = \{1, 4, 5, 6\}$. (4) $q = 1, \max\{|37/2|, |6|, |7|, |3/2|\} = 37/2.$
5. Multiply row 4 by $2/37$, and then add $11, 47, -353/2$ times of row 4 to rows 1, 2, 3 respectively (without touching the $\bar{x}_B$ column but its second component is replaced by value, 0, of $\bar{x}_1$).
6. $\bar{x}_B = (-2, -9, 33, 0)^T + 3(22/37, 94/37, -353/37, 2/37)^T$
   $= (-8/37, -51/37, 162/37, 6/37)^T, \bar{x}_2 = 4 + (-3) = 1.$

	x_1	x_2	x_3	x_4	x_5	x_6	x_7	x_8	$x_9(f)$	$\bar{x}_B$
		$22/37$	1	$21/37$	$6/37$	$-4/37$			$-3/37$	$-8/37$
		$94/37$		$-28/37$	$-45/37^*$	$-7/37$	1		$4/37$	$-51/37$
		$-353/37$		$-342/37$	$-288/37$	$44/37$		1	$33/37$	$162/37$
	1	$2/37$		$12/37$	$14/37$	$3/37$			$-7/37$	$6/37$
u	1	1	1	1	1	$+\infty$	$+\infty$	$+\infty$	$+\infty$	
$\bar{x}$	$6/37$	1	$-8/37$	0	0	0	$-51/37$	$162/37$	2	
l	0	[1]	0	0	0	0	0	0	[2]	

Iteration 8:

1. $p = 2,\ \max\{|0 - (-8/37)|, |0 - (-51/37)|\} = 51/37,\ \rho_2 = 51/37,$
   $j_p = 7.$
3. $J = \{4, 5, 6\}.$
4. $q = 5, \max\{|-28/37|, |-45/37|, |-7/37|\} = 45/37.$
5. Multiply row 2 by $-37/45$, and then add $-6/37, 288/37, -14/37$ times of row 2 to rows 1, 3, 4, respectively (without touching the $\bar{x}_B$ column but its second component is replaced by value, 0, of $\bar{x}_5$).
6. $\bar{x}_B = (-8/37, 0, 162/37, 6/37)^T - (51/37)(2/15, -37/45, -32/5, 14/45)^T$
   $= (-2/5, 17/15, 66/5, -4/15)^T,\ \bar{x}_7 = 0.$

	x_1	x_2	x_3	x_4	x_5	x_6	x_7	x_8	$x_9(f)$	$\bar{x}_B$
		14/15	1	7/15		−2/15*	2/15		−1/15	−2/5
		−94/45		28/45	1	7/45	−37/45		−4/45	17/15
		−129/5		−22/5		12/5	−32/5	1	1/5	66/5
	1	38/45		4/45		1/45	14/45		−7/45	−4/15
u	1	1	1	1	1	$+\infty$	$+\infty$	$+\infty$	$+\infty$	
$\bar{x}$	−4/15	1	−2/5	0	17/15	0	0	66/5	2	
l	0	[1]	0	0	0	0	0	0	[2]	

Iteration 9:

1. $p = 1$, $\max\{|0-(-2/5)|, |1-(17/15)|, |0-(-4/15)|\} = 2/5$, $\rho_1 = 2/5$, $j_p = 3$.
3. $J = \{6\}$. (4) $q = 6$.
5. Multiply row 1 by $-15/2$, and then add $-7/45, -12/5, -1/45$ times of row 1 to rows $2, 3, 4$, respectively (without touching the $\bar{x}_B$ column but its first component is replaced by value, 0, of $\bar{x}_6$).
6. $\bar{x}_B = (0, 17/15, 66/5, -4/15)^T - (2/5)(-15/2, 7/6, 18, 1/6)^T$
   $= (3, 2/3, 6, -1/3)^T, \bar{x}_3 = 0.$

	x_1	x_2	x_3	x_4	x_5	x_6	x_7	x_8	$x_9(f)$	$\bar{x}_B$
		−7	−15/2	−7/2		1	−1		1/2	3
		−1	7/6	7/6	1		−2/3		−1/6	2/3
		−9	18	4			−4	1	−1	6
	1	1	1/6	1/6			1/3		−1/6	−1/3
u	1	1	1	1	1	$+\infty$	$+\infty$	$+\infty$	$+\infty$	
$\bar{x}$	−1/3	1	0	0	2/3	3	0	6	2	
l	0	[1]	0	0	0	0	0	0	[2]	

Iteration 10:

1. $p = 4$, $\rho_4 = 1/3$, $j_p = 1$.
3. $J = \emptyset$.
8. $\alpha = -(1/3)/(-1/6) = 2 > 0$.
10. $\alpha = 2$.
11. $\bar{x}_B = (3, 2/3, 6, -1/3)^T - 2(1/2, -1/6, -1, -1/6)^T = (2, 1, 8, 0)^T$,
    $\bar{x}_9 = 2 + 2 = 4 > f^+ = 2$.

This branch is stalled.

(5) We continue to handle (3) by adding objective cut $x_9 \geq f^+ = 2$. Change the lower bound on x_9 to 2 and the upper bound on x_2 to 0 in the tableau of (3).

	x_1	x_2	x_3	x_4	x_5	x_6	x_7	x_8	$x_9(f)$	$\bar{x}_B$
		1/114	1		−6/19	−2/57		7/114	−1/38	7/114
		568/171			−11/19	−49/171	1	−14/171	2/57	271/171
		353/342		1	16/19	−22/171		−37/342	−11/114	191/342
	1	−16/57			2/19	7/57		2/57	−3/19	2/57
u	1	[0]	1	1	1	$+\infty$	$+\infty$	$+\infty$	$+\infty$	
$\bar{x}$	2/57	0	7/114	191/342	0	0	271/171	0	2	
l	0	0	0	0	0	0	0	0	[2]	

Call Subalgorithm 25.7.1 ($f^+ = 2,\ x_2 \leq 0$).

Iteration 11:

1. $\rho_i = 0, i = 1, 2, 3, 4$. Returning from step (2) gives a valid point: $j = \arg\max\{4(2/57), 3(7/114), 3(191/342)\} = 3$.
   Add cut $x_4 \geq 1$ and let the program with $x_4 \leq 0$ pend. Change the lower bound on x_4 to 1 in the tableau.

Call Subalgorithm 25.7.1 ($f^+ = 2,\ x_2 \leq 0,\ x_4 \geq 1$).

Iteration 12:

1. $p = 3,\ \rho_3 = 151/342,\ j_p = 4$.
3. $J = \{6, 8\}$. (4) $q = 6$, $\max\{|-22/171|, |-37/342|\} = 22/171$.
5. Multiply row 3 by $-171/22$, and then add $2/57, 49/171, -7/57$ times of row 3 to rows 1, 2, 4, respectively (without touching the $\bar{x}_B$ column but its third component is replaced by value, 0, of $\bar{x}_6$).
6. $\bar{x}_B = (7/114, 271/171, 0, 2/57)^T - (151/342)(-3/11, -49/22, -171/22, 21/22)^T = (2/11, 113/44, 151/44, -17/44)^T$, $\bar{x}_4 = 1$.

	x_1	x_2	x_3	x_4	x_5	x_6	x_7	x_8	$x_9(f)$	$\bar{x}_B$
		−3/11	1	−3/11	−6/11			1/11	0	2/11
		45/44		−49/22	−27/11		1	7/44	1/4	113/44
		−353/44		−171/22	−72/11	1		37/44	3/4	151/44
	1	31/44		21/22	10/11			−3/44*	−1/4	−17/44
u	1	[0]	1	1	1	$+\infty$	$+\infty$	$+\infty$	$+\infty$	
$\bar{x}$	−17/44	0	2/11	1	0	151/44	113/44	0	2	
l	0	0	0	[1]	0	0	0	0	[2]	

Iteration 13:

1. $p = 4,\ \rho_4 = 17/44,\ j_p = 1$.
3. $J = \{8\}$.

4. $q = 8$.
5. Multiply row 4 by $-44/3$, and then add $-1/11, -7/44, -37/44$ times of row 4 to rows $(1, 2, 3$, respectively without touching the $\bar{x}_B$ column but its fourth component is replaced by value, 0, of $\bar{x}_8$).
6. $\bar{x}_B = (2/11, 113/44, 151/44, 0)^T - (17/44)(4/3, 7/3, 37/3, -44/3)^T$
   $= (-1/3, 5/3, -4/3, 17/3)^T$, $\bar{x}_1 = 0$.

	x_1	x_2	x_3	x_4	x_5	x_6	x_7	x_8	$x_9(f)$	$\bar{x}_B$
	4/3	2/3	1	1	2/3				−1/3	−1/3
	7/3	8/3		0	−1/3		1		−1/3	5/3
	37/3	2/3		4	14/3	1			−7/3	−4/3
	−44/3	−31/3		−14	−40/3			1	11/3	17/3
u	1	[0]	1	1	1	$+\infty$	$+\infty$	$+\infty$	$+\infty$	
$\bar{x}$	0	0	−1/3	1	0	−4/3	5/3	17/3	2	
l	0	0	0	[1]	0	0	0	0	[2]	

Iteration 14:

1. $p = 3$, $\rho_3 = 4/3$, $j_p = 6$.
3. $J = \emptyset$.
8. $\alpha = -(4/3)/(-7/3) = 4/7 > 0$.
10. $\alpha = 1$.
11. $\bar{x}_B = (-1/3, 5/3, -4/3, 17/3)^T - 1 \times (-1/3, -1/3, -7/3, 11/3)^T$
    $= (0, 2, 1, 2)^T$, $\bar{x}_9 = 2 + 1 = 3 > f^+ = 2$.

This branch is stalled.
(6) The end objective value 3 reached by branch (5) is less than value 4 by branch (4). So $f^+ = 3$ is a new suspected-optimal value. Continue to handle (5) by adding objective cut $x_9 \geq f^+ = 3$. Change the lower bound on x_9 to 3.
Call Subalgorithm 25.7.1.

1. $\rho_j = 0$, $j = 1, 2, 3, 4$. Since it is integer, the valid point is optimal to the original 0-1 problem:

$$x^* = (0, 0, 0, 1, 0)^T, \qquad f^* = 3.$$

In contrast, it is simpler to solve the 0-1 problem by the following "implicit enumeration" approach. First, $\bar{x} = (0, 0, 0, 0, 0)^T$ is ruled out to be optimal because it is infeasible. Then, compare objective values associated with solutions having only a single 1-valued component, i.e., the coefficients, $4, 2, 3, 3, 2$, of the objective function. Corresponding to value $\min\{4, 2, 3, 3, 2\} = 2$, both solutions $(0, 1, 0, 0, 0)^T$ and $(0, 0, 0, 0, 1)^T$ are ignored since they are infeasible. So, turn to $(0, 0, 1, 0, 0)^T$ and $(0, 0, 0, 1, 0)^T$, corresponding to the second minimum 3. It is then clear that the later is the only optimal solution to the 0-1 problem.

A good initial suspected-optimal value f^+ is important to the controlled-branch method; it is even desirable if the true optimal value of the ILP problem is available. If the initial f^+ had been set to 2 or 3 in the preceding Example, e.g., there would have been a short cut to the solution. This is of significance, as for certain type of ILP problems, like 0-1 problem, it would be possible to determine a tight integral lower bound on the optimal value in priori.

Appendix A
MPS File

MPS (Mathematical Programming System) is a kind of data file format, used to represent LP problems and mixed ILP problems. It is widely applied in academic as well as commercial optimization codes.

There exists several expansions of MPS format. This appendix presents the MPS format based on that used in MINOS (Murtagh and Saunders 1998). Nevertheless, we are only concerned here with the part of liner programming only, though MINOS is also used to handle nonlinear programming problems.

In the MPS format, data of the problem to be solved have to appear in specific columns. A MPS file is sectioned under several items as follows:

NAME
ROWS
.
COLUMNS
.
RHS
.
RANGES (optional)
.
BOUNDS (optional)
.
ENDDATA

Each item must start from the first column. Symbols "." under an item are of the following form:

Column	2–3	5–12	15–22	25–36	40–47	50–61
Content	Key	Name0	Name1	Value1	Name2	Value2

where note rows can be inserted, with symbol "*" at the first column, and any characters at the 2–22th columns.

P.-Q. PAN, *Linear Programming Computation*, DOI 10.1007/978-3-642-40754-3,

1. NAME of the problem

Example A.1. NAME row

```
1......4     15.......22
NAME         AFIRO
```

the name row is usually the first row of a MPS file.

Characters **NAME** are in the first to fourth columns. The name of the problem is filled in the 15–22th columns, which can be 1–8 any characters or blank spaces. The name may also be used to indicate outputs, at the first row of each output basis.

2. ROWS Section

Example A.2. ROWS

```
 2  5.....12
ROWS
 E  ROW4
 G  ROW1
 L  ROW7
 N  COST
```

This section lists given names of all (linear) constraints (e.g., ROW4,ROW1, ROW7, etc.), one row for each constraint, where a single character under **Key** gives the types of constraints, placed in columns 2–3. **Name0** gives a name of 8 characters, placed in columns 5–12. The types are

Key	Row-type
E	$=$
G	$\geq$
L	$\leq$
N	Objective
N	Free

Symbols E, G, and L respectively specify constraint types of "equal to", "greater than or equal to" and "less than or equal to"; N indicates "free" or without restriction.

3. COLUMNS Section

Example A.3. COLUMNS section

```
1   5.....12  15....22  25.......36  40....47  50........61
COLUMNS
    X1        ROW2      2.3          ROW3      -2.4
    X1        ROW1      -6.7         ROW6      5.22222
    X1        ROW7      15.88888
    X2        ROW2      1.0          ROW4      -4.1
    X2        ROW5      2.6666666
```

For each variable x_j, COLUMNS section gives names and lists nonzeros a_{ij} of the according columns of the coefficient matrix. Nonzero components of the first column must be listed before those of the seconde column, and so on. In case when there are multiple nonzeros in a column, these components are allowed to appear in any order.

Usually, **Key** is blank (except for comment line), **Name0** is the name of a column and **Name1**, **Value1** give the name of a row and the nonzero value of some component of the column. The name of another row of the same column and value can be listed in **Name2**, **Value2** of the same line, or put in the next row. **Name1** or **Name2** can be blank.

In the preceding example, the variable named by **X1** has 5 nonzero components located respectively at rows **ROW2, ROW3, ROW1, ROW6, ROW7**.

4. RHS Section

Example A.4. RHS section

```
1   5.....12   15....22   25.......36   40....47   50........61
RHS
    RHS1       ROW1       −3.5          ROW3       1.1111
    RHS1       ROW4       11.33333      ROW7       −2.4
    RHS1       ROW2       2.6
    RHS2       ROW4       17.3          ROW2       −5.6
    RHS2       ROW1       1.9
```

This section lists nonzeros of the right-hand side of the constraint system, with the same format as the COLUMN section. If components of the right-hand side are all zero, then only line **RHS** appears in the section. This section may include multiple right-hand sides. But only the first will be used unless indicated otherwise in the SPECS file.

5. RANGES Section (optional)

Example A.5. RANGES section

```
1   5.....12   15....22   25.......36   40....47   50........61
ROWS
 E  ROW2
 G  ROW4
 L  ROW5
 E  ROW1
 .
COLUMNS
 .
RHS
    RHS1       ROW5       2.0           ROW3       2.0
 .
RANGES
    RANGE1     ROW1       3.0           ROW3       4.5
    RANGE2     ROW2       1.4           ROW5       6.7
```

RANGES section expresses constraints of form

$$l \leq a^T x \leq u,$$

where l and u both are finite values. The range of the constraint is $r = u - l$. This section only gives r, not l or u. The according l or u will be determined by the row type of the constraint and the sign of r:

Type of row	Sign of r	Lower bound l	upper bound u
E	$+$	b	$b + \|r\|$
E	$-$	$b - \|r\|$	b
G	$+$ or $-$	b	$b + \|r\|$
L	$+$ or $-$	$b - \|r\|$	b

The format is the same as the COLUMN section, where **Name0** gives the name of a range.

6. BOUNDS Section (optional)

Example A.6. BOUNDS section

```
1   5.....12   15....22   25.......36
BOUNDS
 UP BOUND1     X1         10.5
 UP BOUND1     X2         6.0
 LO BOUND1     X3         -5.0
 UP BOUND1     X3         4.5
```

Default bounds of all variables x_j (except for slack variables) are $0 \leq x_j \leq \infty$. If necessary, the default values 0 and ∞ can be modified to $l \leq x_j \leq u$ via **LOWER** and **UPPER** in the SPECS file.

In this section, **Key** gives the type of bounds (UP and LO respectively indicate upper and lower bounds) **Name0** gives the name of the bound set, and **Name1** and **Value1** list the name of column and the value of bound, respectively.

Example A.7. The model of human arrangement problem in Example 1.3.4 is

$$\begin{array}{llllllll}
\min & f = & 0.1x_2 + 0.2x_3 + 0.3x_4 + 0.8x_5 + 0.9x_6 \\
\text{s.t.} & x_1 & & & & + x_6 & \geq 7 \\
 & x_1 + x_2 & & & & & \geq 15 \\
 & & x_2 + x_3 & & & & \geq 25 \\
 & & & x_3 + x_4 & & & \geq 20 \\
 & & & & x_4 + x_5 & & \geq 30 \\
 & & & & & x_5 + x_6 & \geq 7 \\
 & & & x_j \geq 0, \ j = 1, \ldots, 6. & & &
\end{array}$$

The MPS file is as follows:

```
1  5....12    15....22    25....36    40....47    50....61
NAME          ASSIGN.
ROWS
 G TEAM1
 G TEAM2
 G TEAM3
 G TEAM4
 G TEAM5
 G TEAM6
 N HANDS
COLUMNS
    X1        TEAM1          1.0      TEAM2          1.0
    X2        TEAM2          1.0      TEAM3          1.0
    X2        HANDS          0.1
    X3        TEAM3          1.0      TEAM4          1.0
    X3        HANDS          0.2
    X4        TEAM4          1.0      TEAM5          1.0
    X4        HANDS          0.3
    X5        TEAM5          1.0      TEAM6          1.0
    X5        HANDS          0.8
    X6        TEAM1          1.0      TEAM6          1.0
    X6        HANDS          0.9
RHS
    RHS1      TEAM1          7.0      TEAM2         15.0
    RHS1      TEAM3         25.0      TEAM4         20.0
    RHS1      TEAM5         30.0      TEAM6          7.0
ENDDATA
```

Appendix B
Test LP Problems

In this Appendix, test problems (Gay 1985) are listed according to increasing $m + n$ order.
Source:

NETLIB (http://www.netlib.org/lp/data)
Kennington (http://www-fp.mcs.anl.gov/otc/Guide/TestProblems/LPtest/)
BPMPD (http://www.sztaki.hu/meszaros/bpmpd/).

Notations in the tables:

m: the number of rows of the coefficient matrix.
n: the number of columns of the coefficient matrix.
Nonzeros: the number of nonzeros of the coefficient matrix.
BR: indicates whether there are BOUNDS and RANGES sections.
Optimal value: optimal values given by the file.

Table B.1 96 Netlib problems

No.	Name	$m+n$	m	n	Nonzeros	BR	Optimal value
1	AFIRO	60	28	32	88		−4.6475314286E + 02
2	KB2	85	44	41	291	B	−1.7499001299E + 03
3	SC50B	99	51	48	119		−7.0000000000E + 01
4	SC50A	99	51	48	131		−6.4575077059E + 01
5	ADLITTLE	154	57	97	465		2.2549496316E + 05
6	BLEND	158	75	83	521		−3.0812149846E + 01
7	SHARE2B	176	97	79	730		−4.1573224074E + 02
8	SC105	209	106	103	281		−5.2202061212E + 01
9	STOCFOR1	229	118	111	474		−4.1131976219E + 04
10	SCAGR7	270	130	140	553		−2.3313897524E + 06
11	RECIPE	272	92	180	752	B	−2.6661600000E + 02
12	BOEING2	310	167	143	1, 339	BR	−3.1501872802E + 02
13	ISRAEL	317	175	142	2, 358		−8.9664482186E + 05

(continued)

P.-Q. PAN, *Linear Programming Computation*, DOI 10.1007/978-3-642-40754-3,

Table B.1 (continued)

No.	Name	$m+n$	m	n	Nonzeros	BR	Optimal value
14	SHARE1B	343	118	225	1,182		−7.6589318579E+04
15	VTP.BASE	402	199	203	914	B	1.2983146246E+05
16	SC205	409	206	203	552		−5.2202061212E+01
17	BEACONFD	436	174	262	3,476		3.3592485807E+04
18	GROW7	442	141	301	2,633	B	−4.7787811815E+07
19	LOTFI	462	154	308	1,086		−2.5264706062E+01
20	BRANDY	470	221	249	2,150		1.5185098965E+03
21	E226	506	224	282	2,767		−1.1638929066E+01
22	BORE3D	549	234	315	1,525	B	1.3730803942E+03
23	FORPLAN	583	162	421	4,916	BR	−6.6421896127E+02
24	CAPRI	625	272	353	1,786	B	2.6900129138E+03
25	AGG	652	489	163	2,541		−3.5991767287E+07
26	BOEING1	736	352	384	3,865	BR	−3.3521356751E+02
27	SCORPION	747	389	358	1,708		1.8781248227E+03
28	BANDM	778	306	472	2,659		−1.5862801845E+02
29	SCTAP1	781	301	480	2,052		1.4122500000E+03
30	SCFXM1	788	331	457	2,612		1.8416759028E+04
31	AGG3	819	517	302	4,531		1.0312115935E+07
32	AGG2	819	517	302	4,515		−2.0239252356E+07
33	STAIR	824	357	467	3,857	B	−2.5126695119E+02
34	SCSD1	838	78	760	3,148		8.6666666743E+00
35	TUFF	921	334	587	4,523	B	2.9214775747E−01
36	GROW15	946	301	645	5,665	B	−1.0687094129E+08
37	SCAGR25	972	472	500	2,029		−1.4753433061E+07
38	DEGEN2	979	445	534	4,449		−1.4351780000E+03
39	FIT1D	1,051	25	1,026	14,430	B	−9.1463780924E+03
40	ETAMACRO	1,089	401	688	2,489	B	−7.5571521687E+02
41	FINNIS	1,112	498	614	2,714	B	1.7279096547E+05
42	FFFFF800	1,379	525	854	6,235		5.5567967533E+05
43	GROW22	1,387	441	946	8,318	B	−1.6083433648E+08
44	PILOT4	1,411	411	1,000	5,145	B	−2.5809984373E+03
45	STANDATA	1,435	360	1,075	3,038	B	1.2576995000E+03
46	SCSD6	1,498	148	1,350	5,666		5.0500000078E+01
47	STANDMPS	1,543	468	1,075	3,686	B	1.4060175000E+03
48	SEBA	1,544	516	1,028	4,874	BR	1.5711600000E+04
49	STANDGUB	1,546	362	1,184	3,147	B	1.2576995000E+03
50	SCFXM2	1,575	661	914	5,229		3.6660261565E+04
51	SCRS8	1,660	491	1,169	4,029		9.0430601463E+02
52	GFRD-PNC	1,709	617	1,092	3,467	B	6.9022359995E+06
53	BNL1	1,819	644	1,175	6,129		1.9776292440E+03
54	SHIP04S	1,861	403	1,458	5,810		1.7987147004E+06
55	PEROLD	2,002	626	1,376	6,026	B	−9.3807477973E+03
56	MAROS	2,290	847	1,443	10,006	B	−5.8063743701E+04
57	FIT1P	2,305	628	1,677	10,894	B	9.1463780924E+03

(continued)

Table B.1 (continued)

No.	Name	$m+n$	m	n	Nonzeros	BR	Optimal value
58	MODSZK1	2,308	688	1,620	4,158	B	3.2061972906E+02
59	SHELL	2,312	537	1,775	4,900	B	1.2088253460E+09
60	SCFXM3	2,362	991	1,371	7,846		5.4901254550E+04
61	25FV47	2,393	822	1,571	11,127		5.5018458883E+03
62	SHIP04L	2,521	403	2,118	8,450		1.7933245380E+06
63	QAP8	2,545	913	1,632	8,304		2.0350000002E+02
64	WOOD1P	2,839	245	2,594	70,216		1.4429024116E+00
65	PILOT.JA	2,929	941	1,988	43,220	B	−6.1130535369E+03
66	SCTAP2	2,971	1,091	1,880	8,124		1.7248071429E+03
67	GANGES	2,991	1,310	1,681	7,021	B	−1.0958577038E+05
68	SCSD8	3,148	398	2,750	11,334		9.0499999993E+02
69	PILOTNOV	3,148	976	2,172	13,129	B	−4.4972761882E+03
70	SHIP08S	3,166	779	2,387	9,501		1.9200982105E+06
71	SIERRA	3,264	1,228	2,036	9,252	B	1.5394392927E+07
72	DEGEN3	3,322	1,504	1,818	26,230		−9.8729400000E+02
73	PILOT.WE	3,512	723	2,789	9,218	B	−2.7201037711E+06
74	NESM	3,586	663	2,923	13,988	BR	1.4076065462E+07
75	SHIP12S	3,915	1,152	2,763	10,941		1.4892361344E+06
76	SCTAP3	3,961	1,481	2,480	10,734		1.4240000000E+03
77	STOCFOR2	4,189	2,158	2,031	9,492		−3.9024408538E+04
78	CZPROB	4,453	930	3,523	14,173	B	2.1851966989E+06
79	CYCLE	4,761	1,904	2,857	21,322	B	−5.2263930249E+00
80	SHIP08L	5,062	779	4,283	17,085		1.9090552114E+06
81	PILOT	5,094	1,442	3,652	14,706	B	−5.5728790853E+02
82	BNL2	5,814	2,325	3,489	16,124		1.8112404450E+03
83	SHIP12L	6,579	1,152	5,427	21,597		1.4701879193E+06
84	D6CUBE2	6,600	416	6,184	43,888		3.1473177974E+02
85	D6CUBE	6,600	416	6,184	43,888	B	3.1549166667E+02
86	PILOT87	6,914	2,031	4,883	73,804	B	3.0171261980E+02
87	D2Q06C	7,339	2,172	5,167	35,674		1.2278421653E+05
88	GREENBEA	7,798	2,393	5,405	31,499	B	−7.2461393630E+07
89	WOODW	9,504	1,099	8,405	37,478		1.3044763331E+00
90	FIT2D	10,526	26	10,500	138,018	B	−6.8464293007E+04
91	QAP12	12,049	3,193	8,856	44,244		5.2289435056E+02
92	80BAU3B	12,062	2,263	9,799	29,063	B	9.8724313086E+05
93	MAROS-R7	12,545	3,137	9,408	151,120		1.4971851665E+06
94	FIT2P	16,526	3,001	13,525	60,784	B	6.8464293232E+04
95	DFL001	18,302	6,072	12,230	41,873	B	1.1266503030E+07
96	STOCFOR3	32,371	16,676	15,695	74,004		−3.9976256537E+04

Table B.2 16 Kennington problems

No.	Name	$m+n$	m	n	Nonzeros	BR	Optimal value
1	KEN-07	6,029	2,427	3,602	11,981	B	−6.7952044338E+08
2	CRE-C	6,747	3,069	3,678	16,922		2.5275116141E+07
3	CRE-A	7,584	3,517	4,067	19,054		2.3595410589E+07
4	PDS-02	10,489	2,954	7,535	21,252	B	2.8857862010E+10
5	OSA-07	25,068	1,119	23,949	167,643		5.3572251730E+05
6	KEN-11	36,044	14,695	21,349	70,354	B	−6.9723822625E+09
7	PDS-06	38,537	9,882	28,655	82,269	B	2.7761037600E+10
8	OSA-14	54,798	2,338	52,460	367,220		1.1064628448E+06
9	PDS-10	65,322	16,559	48,763	140,063	B	2.6727094976E+10
10	KEN-13	71,292	28,633	42,659	139,834	B	−1.0257394789E+10
11	CRE-D	78,907	8,927	69,980	312,626		2.4454969898E+07
12	CRE-B	82,096	9,649	72,447	328,542		2.3129640065E+07
13	OSA-30	104,375	4,351	100,024	700,160		2.1421398737E+06
14	PDS-20	139,603	33,875	105,728	304,153	B	2.3821658640E+10
15	OSA-60	243,247	10,281	232,966	1,630,758		4.0440725060E+06
16	KEN-18	259,827	105,128	154,699	512,719	B	−5.2217025287E+10

Table B.3 17 BPMPD problems

No.	Name	$m+n$	m	n	Nonzeros	BR	Optimal value
1	RAT7A	12,545	3,137	9,408	275,180		2.0743714157E+06
2	NSCT1	37,883	22,902	14,981	667,499		−3.8922436000E+07
3	NSCT2	37,985	23,004	14,981	686,396		−3.7175082000E+07
4	ROUTING	44,818	20,895	23,923	210,025	B	5.9416502767E+03
5	DBIR1	46,160	18,805	27,355	1,067,815		−8.1067070000E+06
6	DBIR2	46,262	18,907	27,355	1,148,847		−6.1169165000E+06
7	T0331-4L	47,580	665	46,915	477,897	B	2.9730033352E+04
8	NEMSEMM2	49,077	6,944	42,133	212,793	B	6.2161463095E+05
9	SOUTHERN	54,160	18,739	35,421	148,318	B	1.8189401971E+09
10	RADIO.PRIM	66,919	58,867	8,052	265,975	B	1.0000000000E+00
11	WORLD.MOD2	67,393	35,665	31,728	220,116	B	4.3648258595E+07
12	WORLD	68,245	35,511	32,734	220,748	B	6.9133457165E+07
13	RADIO.DUAL	74,971	8,053	66,918	328,891	B	−1.0000000000E+00
14	NEMSEMM1	75,359	3,946	71,413	1,120,871	B	5.1442135978E+05
15	NW14	123,483	74	123,409	1,028,319	B	6.1844000000E+04
16	LPL1	164,952	39,952	125,000	462,127	B	6.7197548415E+10
17	DBIC1	226,436	43,201	183,235	1,217,046	B	−9.7689730000E+06

Table B.4 53 Netlib problems without bounds and ranges

No.	Name	$m+n$	m	n	Nonzeros	Optimal value
1	AFIRO	60	28	32	88	−4.6475314286E + 02
2	SC50B	99	51	48	119	−7.0000000000E + 01
3	SC50A	99	51	48	131	−6.4575077059E + 01
4	ADLITTLE	154	57	97	465	2.2549496316E + 05
5	BLEND	158	75	83	521	−3.0812149846E + 01
6	SHARE2B	176	97	79	730	−4.1573224074e + 02
7	SC105	209	106	103	281	−5.2202061212E + 01
8	STOCFOR1	229	118	111	474	−4.1131976219E + 04
9	SCAGR7	270	130	140	553	−2.3313897524E + 06
10	ISRAEL	317	175	142	2, 358	−8.9664482186E + 05
11	SHARE1B	343	118	225	1, 182	−7.6589318579E + 04
12	SC205	409	206	203	552	−5.2202061212E + 01
13	BEACONFD	436	174	262	3, 476	3.3592485807E + 04
14	LOTFI	462	154	308	1, 086	−2.5264706062E + 01
15	BRANDY	470	221	249	2, 150	1.5185098965E + 03
16	E226	506	224	282	2, 767	−1.1638929066E + 01
17	AGG	652	489	163	2, 541	−3.5991767287E + 07
18	SCORPION	747	389	358	1, 708	1.8781248227E + 03
19	BANDM	778	306	472	2, 659	−1.5862801845E + 02
20	SCTAP1	781	301	480	2, 052	1.4122500000E + 03
21	SCFXM1	788	331	457	2, 612	1.8416759028E + 04
22	AGG3	819	517	302	4, 531	1.0312115935E + 07
23	AGG2	819	517	302	4, 515	−2.0239252356E + 07
24	SCSD1	838	78	760	3, 148	8.6666666743E + 00
25	SCAGR25	972	472	500	2, 029	−1.4753433061E + 07
26	DEGEN2	979	445	534	4, 449	−1.4351780000E + 03
27	FFFFF800	1,379	525	854	6, 235	5.5567967533E + 05
28	SCSD6	1,498	148	1,350	5, 666	5.0500000078E + 01
29	SCFXM2	1,575	661	914	5, 229	3.6660261565E + 04
30	SCRS8	1,660	491	1,169	4, 029	9.0430601463E + 02
31	BNL1	1,819	644	1,175	6, 129	1.9776292440E + 03
32	SHIP04S	1,861	403	1,458	5, 810	1.7987147004E + 06
33	SCFXM3	2,362	991	1,371	7, 846	5.4901254550E + 04
34	25FV47	2,393	822	1,571	11, 127	5.5018458883E + 03
35	SHIP04L	2,521	403	2,118	8, 450	1.7933245380E + 06
36	QAP8	2,545	913	1,632	8, 304	2.0350000002E + 02
37	WOOD1P	2,839	245	2,594	70, 216	1.4429024116E + 00
38	SCTAP2	2,971	1,091	1,880	8, 124	1.7248071429E + 03
39	SCSD8	3,148	398	2,750	11, 334	9.0499999993E + 02
40	SHIP08S	3,166	779	2,387	9, 501	1.9200982105E + 06
41	DEGEN3	3,322	1,504	1,818	26, 230	−9.8729400000E + 02
42	SHIP12S	3,915	1,152	2,763	10, 941	1.4892361344E + 06
43	SCTAP3	3,961	1,481	2,480	10, 734	1.4240000000E + 03
44	STOCFOR2	4,189	2,158	2,031	9, 492	−3.9024408538E + 04
45	SHIP08L	5,062	779	4,283	17, 085	1.9090552114E + 06

(continued)

Table B.4 (continued)

No.	Name	$m+n$	m	n	Nonzeros	Optimal value
46	BNL2	5,814	2,325	3,489	16,124	1.8112404450E+03
47	SHIP12L	6,579	1,152	5,427	21,597	1.4701879193E+06
48	D6CUBE2	6,600	416	6,184	43,888	3.1473177974E+02
49	D2Q06C	7,339	2,172	5,167	35,674	1.2278421653E+05
50	WOODW	9,504	1,099	8,405	37,478	1.3044763331E+00
51	QAP12	12,049	3,193	8,856	44,244	5.2289435056E+02
52	MAROS-R7	12,545	3,137	9,408	151,120	1.4971851665E+06
53	STOCFOR3	32,371	16,676	15,695	74,004	−3.9976256537E+04

Table B.5 Kennington problems without bounds and ranges

No.	Name	$m+n$	m	n	Nonzeros	Optimal value
1	CRE-C	6,747	3,069	3,678	16,922	2.5275116141E+07
2	CRE-A	7,584	3,517	4,067	19,054	2.3595410589E+07
3	OSA-07	25,068	1,119	23,949	167,643	5.3572251730E+05
4	OSA-14	54,798	2,338	52,460	367,220	1.1064628448E+06
5	CRE-D	78,907	8,927	69,980	312,626	2.4454969898E+07
6	CRE-B	82,096	9,649	72,447	328,542	2.3129640065E+07
7	OSA-30	104,375	4,351	100,024	700,160	2.1421398737E+06
8	OSA-60	243,247	10,281	232,966	1,630,758	4.0440725060E+06

Table B.6 BPMPD problems without bounds and ranges

No.	Name	$m+n$	m	n	Nonzeros	Optimal value
1	RAT7A	12,545	3,137	9,408	275,180	2.0743714157E+06
2	NSCT1	37,883	22,902	14,981	667,499	−3.8922436000E+07
3	NSCT2	37,985	23,004	14,981	686,396	−3.7175082000E+07
4	DBIR1	46,160	18,805	27,355	1,067,815	−8.1067070000E+06
5	DBIR2	46,262	18,907	27,355	1,148,847	−6.1169165000E+06

Appendix C
Empirical Evaluation of Nested Pricing

This appendix offers detailed numerical results, obtained from our extensive computational experiments on nested pricing.

Compiled using Visual Fortran 5.0, all codes were run under a Windows XP system Home Edition Version 2002 on an IBM PC with an Intel(R) Pentium(R) processor 1.00 GB of 1.86 GHz memory, and about 16 digits of precision.

The test sets included three types of problems from Netlib, Kennington and BPMPD test sets (see Appendix B). These problems are ordered by their sizes, in terms of $m + n$, where m and n are the numbers of rows and columns of the constraint matrix, excluding slack variables.

In results listed below, all reported CPU times were measured in seconds with utility routine CPU_TIME, excluding the time spent on preprocessing and scaling. Final objective values reached are not listed. All runs were terminated within the primal and dual feasibility tolerance 10^{-6}.

MINOS was used as the testing benchmark and platform. To compare easy and fairly, each rule is implemented within MINOS 5.51, the latest version of MINOS at that time. Actually, all resulting codes are the same as MINOS 5.51, except for its pivot rule replaced by relevant ones. All codes used the default options, except for Rows 200,000; Columns 300,000; Elements 5,000,000; Iterations 20,000,000; Scale yes; Solution no; Log frequency 0; Print level 0.

(1) We first report results associated with the following five codes and make comparisons between one another:

1. Dantzig: MINOS 5.51 with the full pricing option.
2. Devex: Devex rule.
3. P-Dantzig: MINOS 5.51 with the default *partial* pricing option.
4. N-Devex: Nested-Devex Rule.
5. N-Dantzig: Rule 11.5.1.

P.-Q. PAN, *Linear Programming Computation*, DOI 10.1007/978-3-642-40754-3,

(A) Results for 48 Netlib problems

The first set of test problems included the 48 largest Netlib problems. Associated results are listed in Table C.1, where total iterations and time required for solving each problem are listed in columns labeled Iters and Time for the five codes. The totals listed in the second bottom line are for the 47 problems, excluding QAP15, as it is too time consuming to solve QAP15 with Dantzig and P-Dantzig, and results with the other three codes would heavily dominate the entire computations. Results associated with QAP15 are listed separately in the bottom row of the table.

To serve as a comparison between the codes, Table C.2 lists iterations and run time ratios of the first four codes to N-Dantzig. From the second bottom line, it is seen that N-Dantzig and N-Devex outperformed all the standard codes, in terms of both iterations and run time. N-Devex performed slightly better than N-Dantzig with time ratio 0.95. On the other hand, code Dantzig performed worst on this set with largest iterations ratio 5.16 and time ratio 6.00. It is also noted that Devex performed better than P-Dantzig.

(B) Results for 16 Kennington problems

The second set of problems includes all 16 problems in Kennington test set. These problems are larger than problems in sets 1, overall. Associated numerical results are listed in Table C.3 and compared in Table C.4. From the bottom line of the latter, it is seen that N-Dantzig and N-Devex again defeated all the standard codes in terms of both iterations and run time. N-Devex performed a little better again than N-Dantzig with time ratio 0.91. Code Dantzig performed worst on this set with largest iterations ratio 5.63 and time ratio 5.65. In contrast to with the first set, however, Devex performed worse than P-Dantzig with this set.

(C) Results for 17 BPMPD problems

The third set consists of the 17 largest BPMPD problems, in terms of more than 500 KB in compressed form. Associated numerical results are listed in Table C.5 and compared in Table C.6. From the bottom line of Table C.6, it is observed that the codes performed similarly to with the first two sets. Even with larger margins, N-Dantzig and N-Devex defeated the standard codes in terms of both iterations and run time. In fact, N-Dantzig performed best with this set while code Dantzig did worst with iterations ratio 8.29 and time ratio 12.86!

(D) A summary

Table C.7 offers ratios of the first four codes to N-Dantzig for each test set and for all (excluding QAP15). It is seen that the codes performed consistently with the three sets, overall. Further, N-Dantzig and N-Devex defeated the standard codes in terms of both total iterations and run time. N-Dantzig even outperformed Devex rule with total iterations ratio 3.48 and average time ratio 5.73. It can be asserted that N-Dantzig performed best overall while code Dantzig did worst with iterations ratio 6.78 and time ratio 9.75.

Table C.1 Statistics for 48 Netlib problems

		Dantzig		Devex		P-Dantzig		N-Devex		N-Dantzig	
Problem	m + n	Iters	Time	Iters	Time	Iters	Time	Iters	Time	Iters	Time
SCRS8	1,660	742	0.2	386	0.1	737	0.2	526	0.1	579	0.2
GFRD-PNC	1,709	668	0.2	608	0.2	668	0.2	755	0.2	713	0.2
BNL1	1,819	1,333	0.5	934	0.4	1, 236	0.4	1, 659	0.4	1, 509	0.4
SHIP04S	1,861	154	0.1	150	0.1	169	0.1	233	0.1	228	0.1
PEROLD	2,002	3,032	1.4	2, 200	1.1	3, 816	1.4	4, 653	1.8	4, 751	1.8
MAROS	2,290	1,642	0.8	1, 442	0.7	2, 377	0.9	2, 455	0.9	3, 194	1.3
FIT1P	2,305	890	0.5	554	0.3	839	0.4	919	0.4	945	0.5
MODSZK1	2,308	768	0.4	644	0.2	1, 025	0.3	951	0.3	980	0.3
SHELL	2,312	256	0.1	252	0.1	340	0.1	275	0.1	275	0.1
SCFXM3	2,362	936	0.5	927	0.5	1, 100	0.5	1, 351	0.5	1, 408	0.6
25FV47	2,393	7,146	4.3	3, 728	2.3	7, 129	3.4	5, 319	2.4	6, 018	2.7
SHIP04L	2,521	226	0.1	229	0.1	262	0.1	347	0.1	344	0.1
QAP8	2,545	11,274	11.6	5, 921	6.1	11, 055	10.4	9, 496	8.7	11,161	10.7
WOOD1P	2,839	681	0.9	610	0.9	816	0.6	818	0.6	1, 050	0.7
PILOT.JA	2,929	6,802	4.4	3, 379	2.3	6, 562	3.3	6, 944	3.7	6, 567	3.3
SCTAP2	2,971	663	0.3	731	0.4	767	0.3	760	0.3	733	0.3
GANGES	2,991	574	0.3	590	0.3	719	0.3	743	0.3	699	0.3
PILOTNOV	3,148	1,706	1.1	1, 310	0.9	2, 047	1.0	2, 795	1.5	2, 762	1.4
SCSD8	3,148	1,695	0.7	1, 900	0.8	2, 779	0.7	2, 277	0.5	2, 330	0.6
SHIP08S	3,166	249	0.2	235	0.2	256	0.1	262	0.1	258	0.1
SIERRA	3,264	1,275	0.7	1, 160	0.6	1, 107	0.5	1, 231	0.5	1, 221	0.5
DEGEN3	3,322	9,035	9.3	3, 599	3.5	7, 301	6.3	3, 541	2.9	4, 064	3.3
PILOT.WE	3,512	5,635	2.9	2, 090	1.2	4, 215	1.6	6, 358	2.4	5, 953	2.2
NESM	3,586	3,110	1.5	3, 035	1.5	3, 237	0.9	8, 556	2.6	6, 917	2.1
SHIP12S	3,915	379	0.3	376	0.3	414	0.2	422	0.2	418	0.2

(continued)

Table C.1 (continued)

Problem	$m+n$	Dantzig		Devex		P-Dantzig		N-Devex		N-Dantzig	
		Iters	Time	Iters	Time	Iters	Time	Iters	*Time*	Iters	Time
SCTAP3	3, 961	750	0.5	835	0.6	915	0.5	899	0.4	907	0.5
STOCFOR2	4, 189	1,490	1.2	1, 400	1.1	1,922	1.4	3, 303	2.2	3, 651	2.5
CZPROB	4, 453	1,492	0.8	1, 043	0.6	1,604	0.6	1, 504	0.6	1, 420	0.5
CYCLE	4, 761	2,462	2.4	2, 194	2.1	2,711	2.1	3, 098	2.3	3, 288	2.4
SHIP08L	5, 062	449	0.3	446	0.3	438	0.2	595	0.3	594	0.3
PILOT	5, 094	13,374	29.7	8, 884	21.2	16,865	31.2	21, 692	33.9	22, 138	38.3
BNL2	5, 814	5,454	6.2	3, 788	4.3	4,682	4.1	5, 105	4.2	5, 356	4.5
SHIP12L	6, 579	792	0.7	818	0.7	831	0.4	906	0.5	875	0.5
D6CUBE	6, 600	77,001	78.7	21, 235	23.6	106,878	44.4	17, 705	7.2	13, 792	5.8
D6CUBE2	6, 600	115,853	120.4	20, 824	23.2	136,114	57.0	22, 273	9.2	25, 157	11.1
PILOT87	6, 914	19,641	112.1	10, 028	55.9	20,055	98.9	26, 289	86.7	25, 615	97.2
D2Q06C	7, 339	49,302	77.5	11, 456	17.8	46,638	53.3	22, 956	23.9	18, 784	20.5
GREENBEA	7, 798	27,878	38.6	13, 201	18.3	24,005	24.5	13, 992	13.8	13, 032	13.2
WOODW	9, 504	3,968	4.4	3, 098	3.7	3,710	2.4	5, 548	3.1	5, 331	2.9
TRUSS	9, 807	33,065	42.1	9, 552	12.6	13,253	8.5	14, 607	8.3	14, 259	8.5
FIT2D	10, 526	46,627	83.3	10, 296	20.5	32,150	8.1	22, 278	9.5	23, 077	8.5
QAP12	12, 049	417,689	3,685.5	77, 356	696.7	254,428	1,966.8	169, 080	1, 336.3	170, 669	1, 454.1
80BAU3B	12, 062	14,514	17.0	8, 069	9.6	10,488	7.5	13, 472	8.6	13, 263	9.0
MAROS-R7	12, 545	2,515	16.0	2, 720	18.0	2,520	13.3	9, 026	39.3	7, 615	36.2
FIT2P	16, 526	13,653	44.6	9, 171	25.6	16,064	43.5	14, 395	33.6	14, 081	37.0
DFL001	18, 302	2,177,575	8,999.0	453, 186	1,682.6	2,019,465	6,569.3	156, 242	429.7	136, 487	398.6
STOCFOR3	32, 371	11,982	80.1	12, 220	78.3	14,189	77.4	15, 634	76.4	15, 603	78.8
Total(47)	275, 734	3,098,397	13,484.4	718, 810	2,742.4	2,790,898	9,049.6	624, 245	2, 161.6	600, 071	2, 264.9
QAP15	28, 606	–	–	596, 500	20,272.6	–	–	613, 641	21, 269.9	625, 953	23, 279.8

Table C.2 Ratio to N-Dantzig for 48 Netlib problems

	Dantzig		Devex		P-Dantzig		N-Devex	
Problem	Iters	Time	Iters	Time	Iters	Time	Iters	Time
SCRS8	1.28	1.00	0.67	0.50	1.27	1.00	0.91	0.50
GFRD-PNC	0.94	1.00	0.85	1.00	0.94	1.00	1.06	1.00
BNL1	0.88	1.25	0.62	1.00	0.82	1.00	1.10	1.00
SHIP04S	0.68	1.00	0.66	1.00	0.74	1.00	1.02	1.00
PEROLD	0.64	0.78	0.46	0.61	0.80	0.78	0.98	1.00
MAROS	0.51	0.62	0.45	0.54	0.74	0.69	0.77	0.69
FIT1P	0.94	1.00	0.59	0.60	0.89	0.80	0.97	0.80
MODSZK1	0.78	1.33	0.66	0.67	1.05	1.00	0.97	1.00
SHELL	0.93	1.00	0.92	1.00	1.24	1.00	1.00	1.00
SCFXM3	0.66	0.83	0.66	0.83	0.78	0.83	0.96	0.83
25FV47	1.19	1.59	0.62	0.85	1.18	1.26	0.88	0.89
SHIP04L	0.66	1.00	0.67	1.00	0.76	1.00	1.01	1.00
QAP8	1.01	1.08	0.53	0.57	0.99	0.97	0.85	0.81
WOOD1P	0.65	1.29	0.58	1.29	0.78	0.86	0.78	0.86
PILOT.JA	1.04	1.33	0.51	0.70	1.00	1.00	1.06	1.12
SCTAP2	0.90	1.00	1.00	1.33	1.05	1.00	1.04	1.00
GANGES	0.82	1.00	0.84	1.00	1.03	1.00	1.06	1.00
PILOTNOV	0.62	0.79	0.47	0.64	0.74	0.71	1.01	1.07
SCSD8	0.73	1.17	0.82	1.33	1.19	1.17	0.98	0.83
SHIP08S	0.97	2.00	0.91	2.00	0.99	1.00	1.02	1.00
SIERRA	1.04	1.40	0.95	1.20	0.91	1.00	1.01	1.00
DEGEN3	2.22	2.82	0.89	1.06	1.80	1.91	0.87	0.88
PILOT.WE	0.95	1.32	0.35	0.55	0.71	0.73	1.07	1.09
NESM	0.45	0.71	0.44	0.71	0.47	0.43	1.24	1.24
SHIP12S	0.91	1.50	0.90	1.50	0.99	1.00	1.01	1.00
SCTAP3	0.83	1.00	0.92	1.20	1.01	1.00	0.99	0.80
STOCFOR2	0.41	0.48	0.38	0.44	0.53	0.56	0.90	0.88
CZPROB	1.05	1.60	0.73	1.20	1.13	1.20	1.06	1.20
CYCLE	0.75	1.00	0.67	0.87	0.82	0.87	0.94	0.96
SHIP08L	0.76	1.00	0.75	1.00	0.74	0.67	1.00	1.00
PILOT	0.60	0.78	0.40	0.55	0.76	0.81	0.98	0.89
BNL2	1.02	1.38	0.71	0.96	0.87	0.91	0.95	0.93
SHIP12L	0.91	1.40	0.93	1.40	0.95	0.80	1.04	1.00
D6CUBE	5.58	13.57	1.54	4.07	7.75	7.66	1.28	1.24
D6CUBE2	4.61	10.85	0.83	2.09	5.41	5.14	0.89	0.83
PILOT87	0.77	1.15	0.39	0.58	0.78	1.02	1.03	0.89
D2Q06C	2.62	3.78	0.61	0.87	2.48	2.60	1.22	1.17
GREENBEA	2.14	2.92	1.01	1.39	1.84	1.86	1.07	1.05
WOODW	0.74	1.52	0.58	1.28	0.70	0.83	1.04	1.07
TRUSS	2.32	4.95	0.67	1.48	0.93	1.00	1.02	0.98

(continued)

Table C.2 (continued)

	Dantzig		Devex		P-Dantzig		N-Devex	
Problem	Iters	Time	Iters	Time	Iters	Time	Iters	Time
FIT2D	2.02	9.80	0.45	2.41	1.39	0.95	0.97	1.12
QAP12	2.45	2.53	0.45	0.48	1.49	1.35	0.99	0.92
80BAU3B	1.09	1.89	0.61	1.07	0.79	0.83	1.02	0.96
MAROS-R7	0.33	0.44	0.36	0.50	0.33	0.37	1.19	1.09
FIT2P	0.97	1.21	0.65	0.69	1.14	1.18	1.02	0.91
DFL001	15.95	22.58	3.32	4.22	14.80	16.48	1.14	1.08
STOCFOR3	0.77	1.02	0.78	0.99	0.91	0.98	1.00	0.97
Average	5.16	6.00	1.20	1.21	4.65	4.00	1.04	0.95
QAP15	–	–	0.95	0.87	–	–	0.98	0.91

The ratio of the number of iterations required by a code to the number of rows and columns should be an indicator for problem's difficulty with respect to the code: the smaller the ratio is, the easier the problem solved by the code. Following Forrest and Goldfarb (1992), we therefore list in Tables C.8 and C.9 normalized iterations required by the five codes for the Kennington and BPMPD test sets, respectively. From the bottom line of Table C.8 for Kennington problems, it is seen that the ratio of average normalized iterations required by Dantzig to N-Dantzig is $2.24/0.40 = 5.60$, whereas the ratio by Devex to N-Devex is $2.21/0.40 = 5.53$. From the bottom line of Table C.9 for BPMPD problems, it is seen that the ratio of average normalized iterations by Dantzig to N-Dantzig is $6.74/0.81 = 8.32$, whereas the ratio by Devex to N-Devex is $3.11/0.96 = 3.24$.

Overall, normalized iterations require by N-Dantzig and N-Devex are quite stable for the problems, while those required by their conventional counterparts fluctuate. This can be seen intuitively from associated plots in Figs. C.1 and C.2.

There is no doubt that nested pricing rules are amenable to the test sets, as is a case may be expected in general.

(2) Further testing

The forgoing performance comparison involves neither the steepest-edge and largest decrease rules nor their nested variants. We only offer results associated with the steepest-edge and its nested variant because, according to our experience, the largest decrease rule is out of the line of competitors.

We make a comparison between the following three codes:

1. Steepest-Edge(Stp): the Steepest-edge rule.
2. N-Steepest(N-Stp): Nested-steepest-edge rule.
3. N-Dantzig(N-D): Rule 11.5.1.

Results associated with Netlib, Kennington and BPMPD problems are contained in Tables C.10–C.15. The first two codes failed to solve QAP15 in Netlib, KEN-18 in Kennington, and LPL1 and DBIC1 in BPMPD within 48 h. So, we report results, excluding the four problems.

Table C.3 Statistics for 16 Kennington problems

		Dantzig		Devex		P-Dantzig		N-Devex		N-Dantzig	
Problem	m + n	Iters	Time	Iters	Time	Iters	Time	Iters	Time	Iters	Time
KEN-07	6,029	1,724	1.7	1,657	1.5	1,822	1.4	1,723	1.3	1,718	1.3
CRE-C	6,747	4,068	5.2	3,113	4.1	4,252	4.3	4,485	4.4	4,345	4.4
CRE-A	7,584	4,066	5.9	2,986	4.4	3,642	4.2	4,222	4.8	4,228	4.9
PDS-02	10,489	5,110	6.6	3,550	4.6	3,118	2.8	2,022	1.7	2,099	1.8
OSA-07	25,068	1,767	5.5	1,757	6.0	1,919	1.9	1,494	1.6	1,504	1.5
KEN-11	36,044	13,327	82.8	12,814	77.4	15,328	72.5	13,449	62.2	13,361	61.7
PDS-06	38,537	59,521	306.6	32,424	170.1	26,127	81.6	13,107	40.7	13,100	40.6
OSA-14	54,798	3,821	24.9	3,730	27.1	4,115	7.4	3,372	7.0	3,212	6.0
PDS-10	65,322	177,976	1,615.0	105,469	988.1	78,125	440.3	30,149	176.7	24,179	134.1
KEN-13	71,292	35,709	475.7	29,196	355.6	47,041	500.6	33,056	314.3	33,074	339.5
CRE-D	78,907	377,243	3,428.1	415,765	4,141.5	295,538	1,051.9	70,807	341.8	73,528	312.1
CRE-B	82,096	328,160	3,159.1	363,018	3,833.4	188,706	726.4	63,250	293.0	63,703	277.1
OSA-30	104,375	7,536	91.6	7,048	96.3	8,141	26.2	6,128	25.7	5,963	20.5
PDS-20	139,603	1,557,953	31,598.4	1,589,107	33,356.3	823,110	10,913.9	89,139	1,129.2	93,045	1,199.4
OSA-60	243,247	16,278	458.2	16,188	505.9	17,098	125.5	13,992	126.5	13,751	112.1
KEN-18	259,827	157,224	10,037.9	127,756	6,818.1	198,019	10,017.3	137,508	5,712.3	137,975	6,565.8
Total(16)	1,229,965	2,751,483	51,303.2	2,715,578	50390.4	1,716,101	23,978.2	487,903	8,243.2	488,785	9,082.8

Table C.4 Ratio to N-Dantzig for 16 Kennington problems

	Dantzig		Devex		P-Dantzig		N-Devex	
Problem	Iters	Time	Iters	Time	Iters	Time	Iters	Time
KEN-07	1.00	1.31	0.96	1.15	1.06	1.08	1.00	1.00
CRE-C	0.94	1.18	0.72	0.93	0.98	0.98	1.03	1.00
CRE-A	0.96	1.20	0.71	0.90	0.86	0.86	1.00	0.98
PDS-02	2.43	3.67	1.69	2.56	1.49	1.56	0.96	0.94
OSA-07	1.17	3.67	1.17	4.00	1.28	1.27	0.99	1.07
KEN-11	1.00	1.34	0.96	1.25	1.15	1.18	1.01	1.01
PDS-06	4.54	7.55	2.48	4.19	1.99	2.01	1.00	1.00
OSA-14	1.19	4.15	1.16	4.52	1.28	1.23	1.05	1.17
PDS-10	7.36	12.04	4.36	7.37	3.23	3.28	1.25	1.32
KEN-13	1.08	1.40	0.88	1.05	1.42	1.47	1.00	0.93
CRE-D	5.13	10.98	5.65	13.27	4.02	3.37	0.96	1.10
CRE-B	5.15	11.40	5.70	13.83	2.96	2.62	0.99	1.06
OSA-30	1.26	4.47	1.18	4.70	1.37	1.28	1.03	1.25
PDS-20	16.74	26.35	17.08	27.81	8.85	9.10	0.96	0.94
OSA-60	1.18	4.09	1.18	4.51	1.24	1.12	1.02	1.13
KEN-18	1.14	1.53	0.93	1.04	1.44	1.53	1.00	0.87
Average	5.63	5.65	5.56	5.55	3.51	2.64	1.00	0.91

Table C.16 offers ratios for each test set and for all the 77 problems as a whole. It is seen that N-Dantizig outperformed Steepest-Edge with average time ratio as high as 25.22 despite it required more total iterations (with iterations ratio 0.34). It is also noted that Nested-Steepest-Edge was inferior to Steepest-Edge.

In summary, nested pricing rules are unambiguously superior to standard ones, and the nested Dantzig rule is the best among nested rules tested.

Table C.5 Statistics for 17 BPMPD problems

		Dantzig		Devex		P-Dantzig		N-Devex		N-Dantzig	
Problem	m + n	Iters	Time	Iters	Time	Iters	Time	Iters	Time	Iters	Time
RAT7A	12,545	4,073	558.6	5,226	93.6	3,801	135.8	18,814	186.9	20,074	296.8
NSCT1	37,883	2,190	31.8	1,607	27.6	2,151	19.1	5,868	45.5	9,090	65.1
NSCT2	37,985	11,555	169.2	6,266	88.1	12,298	103.1	6,258	44.2	5,908	42.8
ROUTING	44,818	190,627	1,697.1	68,504	606.5	74,550	424.8	18,742	105.5	22,336	124.5
DBIR1	46,160	2,289	42.9	1,848	45.8	1,710	16.4	11,858	88.0	12,209	82.9
DBIR2	46,262	42,575	819.8	13,031	248.8	53,839	469.1	7,704	54.7	8,012	60.4
T0331-4L	47,580	62,333	638.0	24,435	249.7	74,755	277.4	45,302	276.4	49,932	282.1
NEMSEMM2	49,077	23,581	154.4	17,117	118.1	23,423	73.4	13,899	43.6	10,392	30.0
SOUTHERN	54,160	29,693	432.6	16,774	234.6	23,031	316.9	23,937	315.8	23,771	318.1
RADIO.PR	66,919	2	1.4	2	1.4	5	1.5	2	1.4	2	1.4
WORLD.MD	67,393	1,405,151	23,856.3	427,776	7,403.1	1,241,489	18,068.8	217,502	2,979.3	220,329	3,077.7
WORLD	68,245	2,627,208	46,257.3	501,004	8,747.2	2,182,226	31,763.3	261,210	3,590.0	278,284	3,899.0
RADIO.DL	74,971	3	0.9	3	1.0	7	1.0	3	0.9	3	0.9
NEMSEMM1	75,359	13,650	207.9	10,574	168.8	13,065	60.2	12,787	41.7	9,514	30.9
NW14	123,483	407	9.1	394	9.5	953	5.7	778	5.9	736	5.2
LPL1	164,952	3,343,112	89,309.0	2,337,341	60,073.4	1,047,339	17,532.1	445,616	7,320.2	262,076	4,474.3
DBIC1	226,436	624,907	25,384.9	437,658	18,252.6	342,960	7,404.3	102,935	2,822.4	78,334	1,949.6
Total(17)	1,244,228	8,383,356	189,571.2	3,869,560	96,369.8	5,097,602	76,672.9	1,193,215	17,922.4	1,011,002	14,741.7
Total(80)	14,233,236	254,358.8	7,303,948	149,502.6	9,604,601	109,700.7	2,305,363	28,327.2	2,099,858	26,089.4	

Table C.6 Ratio to N-Dantzig for 17 BPMPD problems

	Dantzig		Devex		P-Dantzig		N-Devex	
Problem	Iters	Time	Iters	Time	Iters	Time	Iters	Time
RAT7A	0.20	1.88	0.26	0.32	0.19	0.46	0.94	0.63
NSCT1	0.24	0.49	0.18	0.42	0.24	0.29	0.65	0.70
NSCT2	1.96	3.95	1.06	2.06	2.08	2.41	1.06	1.03
ROUTING	8.53	13.63	3.07	4.87	3.34	3.41	0.84	0.85
DBIR1	0.19	0.52	0.15	0.55	0.14	0.20	0.97	1.06
DBIR2	5.31	13.57	1.63	4.12	6.72	7.77	0.96	0.91
T0331-4L	1.25	2.26	0.49	0.89	1.50	0.98	0.91	0.98
NEMSEMM2	2.27	5.15	1.65	3.94	2.25	2.45	1.34	1.45
SOUTHERN	1.25	1.36	0.71	0.74	0.97	1.00	1.01	0.99
RADIO.PR	1.00	1.00	1.00	1.00	2.50	1.07	1.00	1.00
WORLD.MD	6.38	7.75	1.94	2.41	5.63	5.87	0.99	0.97
WORLD	9.44	11.86	1.80	2.24	7.84	8.15	0.94	0.92
RADIO.DL	1.00	1.00	1.00	1.11	2.33	1.11	1.00	1.00
NEMSEMM1	1.43	6.73	1.11	5.46	1.37	1.95	1.34	1.35
NW14	0.55	1.75	0.54	1.83	1.29	1.10	1.06	1.13
LPL1	12.76	19.96	8.92	13.43	4.00	3.92	1.70	1.64
DBIC1	7.98	13.02	5.59	9.36	4.38	3.80	1.31	1.45
Average	8.29	12.86	3.83	6.54	5.04	5.20	1.18	1.22

Table C.7 A ratio summary

	Dantzig		Devex		P-Dantzig		N-Devex	
Problem	Iters	Time	Iters	Time	Iters	Time	Iters	Time
Netlib(47)	5.16	5.95	1.20	1.21	4.65	4.00	1.04	0.95
Kennington(16)	5.63	5.65	5.56	5.55	3.51	2.64	1.00	0.91
BPMPD(17)	8.29	12.86	3.83	6.54	5.04	5.20	1.18	1.22
Average(80)	6.78	9.75	3.48	5.73	4.57	4.20	1.10	1.09

Table C.8 Normalized iterations for 16 kennington Problems

No	Problem	Dantzig	Devex	P-Dantzig	N-Devex	N-Dantzig
1	KEN-07	0.29	0.27	0.30	0.29	0.28
2	CRE-C	0.60	0.46	0.63	0.66	0.64
3	CRE-A	0.54	0.39	0.48	0.56	0.56
4	PDS-02	0.49	0.34	0.30	0.19	0.20
5	OSA-07	0.07	0.07	0.08	0.06	0.06
6	KEN-11	0.37	0.36	0.43	0.37	0.37
7	PDS-06	1.54	0.84	0.68	0.34	0.34
8	OSA-14	0.07	0.07	0.08	0.06	0.06
9	PDS-10	2.72	1.61	1.20	0.46	0.37
10	KEN-13	0.50	0.41	0.66	0.46	0.46
11	CRE-D	4.78	5.27	3.75	0.90	0.93
12	CRE-B	4.00	4.42	2.30	0.77	0.78
13	OSA-30	0.07	0.07	0.08	0.06	0.06
14	PDS-20	11.16	11.38	5.90	0.64	0.67
15	OSA-60	0.07	0.07	0.07	0.06	0.06
16	KEN-18	0.61	0.49	0.76	0.53	0.53
	Average	2.24	2.21	1.40	0.40	0.40

Table C.9 Normalized iterations for 17 BPMPD Problems

No	Problem	Dantzig	Devex	P-Dantzig	N-Devex	N-Dantzig
1	RAT7A	0.32	0.42	0.30	1.50	1.60
2	NSCT1	0.06	0.04	0.06	0.15	0.24
3	NSCT2	0.30	0.16	0.32	0.16	0.16
4	ROUTING	4.25	1.53	1.66	0.42	0.50
5	DBIR1	0.05	0.04	0.04	0.26	0.26
6	DBIR2	0.92	0.28	1.16	0.17	0.17
7	T0331-4L	1.31	0.51	1.57	0.95	1.05
8	NEMSEMM2	0.48	0.35	0.48	0.28	0.21
9	SOUTHERN	0.55	0.31	0.43	0.44	0.44
10	RADIO.PR	0.00	0.00	0.00	0.00	0.00
11	WORLD.MD	20.85	6.35	18.42	3.23	3.27
12	WORLD	38.50	7.34	31.98	3.83	4.08
13	RADIO.DL	0.00	0.00	0.00	0.00	0.00
14	NEMSEMM1	0.18	0.14	0.17	0.17	0.13
15	NW14	0.00	0.00	0.01	0.01	0.01
16	LPL1	20.27	14.17	6.35	2.70	1.59
17	DBIC1	2.76	1.93	1.51	0.45	0.35
	Average	6.74	3.11	4.10	0.96	0.81

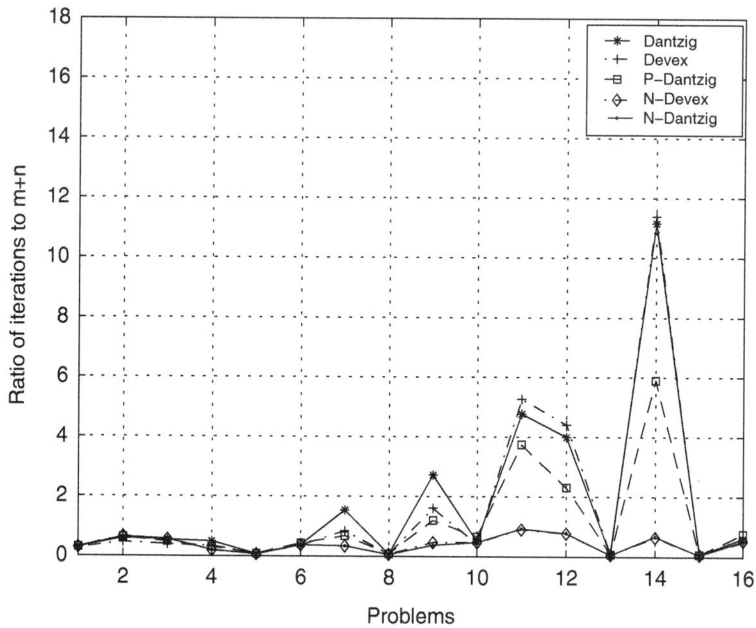

Fig. C.1 Comparison of normalized iterations for 16 Kennington problems

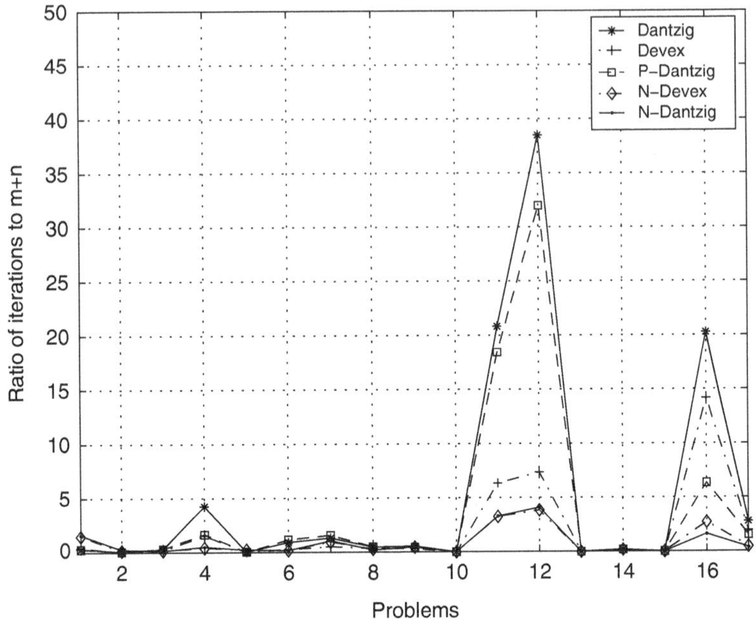

Fig. C.2 Comparison of normalized iterations for 17 BPMPD problems

Table C.10 Statistics for 47 Netlib problems

	Stp		N-Stp		N-D	
Problem	Iters	Time	Iters	Time	Iters	Time
SCRS8	6,372	567.4	486	0.4	579	0.2
GFRD-PNC	626	0.3	722	0.4	713	0.2
BNL1	624	0.8	958	1.0	1,509	0.4
SHIP04S	145	0.2	146	0.2	228	0.1
PEROLD	1,638	2.0	2,772	3.3	4,751	1.8
MAROS	1,115	1.8	1,725	2.7	3,194	1.3
FIT1P	396	0.6	804	1.1	945	0.5
MODSZK1	596	1.1	748	1.0	980	0.3
SHELL	232	0.2	244	0.2	275	0.1
SCFXM3	715	1.0	947	1.2	1,408	0.6
25FV47	1,577	3.0	3,716	6.6	6,018	2.7
SHIP04L	223	0.3	223	0.3	344	0.1
QAP8	3,315	9.9	4,351	13.9	11,161	10.7
WOOD1P	763	3.2	775	2.9	1,050	0.7
PILOT.J	1,862	3.8	4,114	8.0	6,567	3.3
SCTAP2	612	1.2	845	1.5	733	0.3
GANGES	571	0.6	727	0.8	699	0.3
PILOTN	1,108	2.6	2,256	5.0	2,762	1.4

(continued)

Table C.10 (continued)

Problem	Stp Iters	Stp Time	N-Stp Iters	N-Stp Time	N-D Iters	N-D Time
SCSD8	751	1.1	1,634	2.4	2,330	0.6
SHIP08S	237	0.6	267	0.6	258	0.1
SIERRA	1,005	2.0	1,018	2.0	1,221	0.5
DEGEN3	1,854	6.7	2,887	10.1	4,064	3.3
PILOT.W	1,339	3.3	3,736	8.2	5,953	2.2
NESM	2,200	4.0	5,558	10.5	6,917	2.1
SHIP12S	393	1.1	446	1.2	418	0.2
SCTAP3	771	2.2	960	2.7	907	0.5
STOCF.2	1,429	3.9	2,119	5.3	3,651	2.5
CZPROB	900	2.5	1,189	2.8	1,420	0.5
CYCLE	1,125	4.7	3,004	12.3	3,288	2.4
SHIP08L	408	1.8	538	1.8	594	0.3
PILOT	19,344	251.7	15,765	319.5	22,138	38.3
BNL2	2,032	11.3	3,118	17.0	5,356	4.5
SHIP12L	813	4.4	912	4.6	875	0.5
D6CUBE	7,793	40.9	9,387	45.0	13,792	5.8
D6CUBE2	9,585	48.8	12,911	62.4	25,157	11.1
PILOT87	9,224	304.4	19,541	759.1	25,615	97.2
D2Q06C	5,986	62.9	10,970	117.3	18,784	20.5
GREENBE	7,210	63.6	11,242	98.9	13,032	13.2
WOODW	2,524	21.5	3,907	33.1	5,331	2.9
TRUSS	8,877	83.0	12,743	118.7	14,259	8.5
FIT2D	5,970	33.0	11,558	57.3	23,077	8.5
QAP12	29,494	3,055.0	70,841	7,583.5	170,669	1,454.1
80BAU3B	4,565	48.1	8,006	95.1	13,263	9.0
MAROSR7	6,029	232.6	7,598	409.5	7,615	36.2
FIT2P	6,645	197.5	12,138	418.3	14,081	37.0
DFL001	24,704	1,020.6	52,010	2,153.4	136,487	398.6
STOCF.3	10,628	1,082.6	15,703	1,612.0	15,603	78.8
Total(47)	196,325	7,195.9	328,265	14,015.0	600,071	2,264.7

Table C.11 Ratio for 47 Netlib problems

Problem	m	n	Stp/N-Stp Iters	Stp/N-Stp Time	Stp/N-D Iters	Stp/N-D Time	N-Stp/N-D Iters	N-Stp/N-D Time
SCRS8	491	1,169	13.11	1,383.93	11.01	3,546.31	0.84	2.56
GFRDP	617	1,092	0.87	0.89	0.88	2.00	1.01	2.24
BNL1	644	1,175	0.65	0.74	0.41	1.70	0.63	2.32
SH04S	403	1,458	0.99	1.00	0.64	2.13	0.64	2.13
PEROL	626	1,376	0.59	0.61	0.34	1.14	0.58	1.86
MAROS	847	1,443	0.65	0.66	0.35	1.39	0.54	2.10
FIT1P	628	1,677	0.49	0.55	0.42	1.26	0.85	2.28
MODS1	688	1,620	0.80	1.09	0.61	4.04	0.76	3.70

(continued)

Table C.11 (continued)

			Stp/N-Stp		Stp/N-D		N-Stp/N-D	
Problem	m	n	Iters	Time	Iters	Time	Iters	Time
SHELL	537	1,775	0.95	1.00	0.84	2.75	0.89	2.75
SCFX3	991	1,371	0.76	0.79	0.51	1.76	0.67	2.24
25F47	822	1,571	0.42	0.46	0.26	1.10	0.62	2.41
SH04L	403	2,118	1.00	0.94	0.65	2.82	0.65	3.00
QAP8	913	1,632	0.76	0.71	0.30	0.93	0.39	1.30
WOOD1	245	2,594	0.98	1.10	0.73	4.40	0.74	4.00
PIL.J	941	1,988	0.45	0.47	0.28	1.14	0.63	2.42
SCTA2	1,091	1,880	0.72	0.78	0.83	4.30	1.15	5.48
GANGE	1,310	1,681	0.79	0.81	0.82	2.03	1.04	2.52
PLTNO	976	2,172	0.49	0.51	0.40	1.83	0.82	3.56
SCSD8	398	2,750	0.46	0.44	0.32	1.89	0.70	4.32
SH08S	779	2,387	0.89	1.04	0.92	4.46	1.03	4.31
SIERR	1,228	2,036	0.99	1.01	0.82	3.94	0.83	3.88
DEGE3	1,504	1,818	0.64	0.66	0.46	2.01	0.71	3.03
PIL.W	723	2,789	0.36	0.41	0.22	1.53	0.63	3.77
NESM	663	2,923	0.40	0.38	0.32	1.85	0.80	4.90
SH12S	1,152	2,763	0.88	0.88	0.94	4.91	1.07	5.59
SCTA3	1,481	2,480	0.80	0.82	0.85	4.74	1.06	5.79
STCF2	2,158	2,031	0.67	0.73	0.39	1.54	0.58	2.12
CZPRO	930	3,523	0.76	0.90	0.63	4.77	0.84	5.28
CYCLE	1,904	2,857	0.37	0.39	0.34	1.95	0.91	5.06
SH08L	779	4,283	0.76	1.00	0.69	7.24	0.91	7.24
PILOT	1,442	3,652	1.23	0.79	0.87	6.58	0.71	8.35
BNL2	2,325	3,489	0.65	0.67	0.38	2.51	0.58	3.77
SH12L	1,152	5,427	0.89	0.96	0.93	8.82	1.04	9.18
D6CUB	416	6,184	0.83	0.91	0.57	7.06	0.68	7.76
D6CU2	416	6,184	0.74	0.78	0.38	4.42	0.51	5.64
PIL87	2,031	4,883	0.47	0.40	0.36	3.13	0.76	7.81
D2Q06	2,172	5,167	0.55	0.54	0.32	3.07	0.58	5.73
GREEN	2,393	5,405	0.64	0.64	0.55	4.81	0.86	7.48
WOODW	1,099	8,405	0.65	0.65	0.47	7.38	0.73	11.32
TRUSS	1,001	8,806	0.70	0.70	0.62	9.76	0.89	13.96
FIT2D	26	10,500	0.52	0.58	0.26	3.88	0.50	6.74
QAP12	3,193	8,856	0.42	0.40	0.17	2.10	0.42	5.22
80BAU	2,263	9,799	0.57	0.51	0.34	5.36	0.60	10.59
MARR7	3,137	9,408	0.79	0.57	0.79	6.43	1.00	11.31
FIT2P	3,001	13,525	0.55	0.47	0.47	5.33	0.86	11.29
DFL00	6,072	12,230	0.47	0.47	0.18	2.56	0.38	5.40
STOO3	16,676	15,695	0.68	0.67	0.68	13.73	1.01	20.45
Ave.	–	–	0.60	0.51	0.33	3.18	0.55	6.19

Table C.12 Statistics for 15 Kennington problems

	Stp		N-Stp		N-D	
Problem	Iters	Time	Iters	Time	Iters	Time
KEN-07	1,699	6.0	1,723	6.1	1,718	1.3
CRE-C	1,391	13.7	2,249	22.5	4,345	4.4
CRE-A	1,497	18.4	2,325	29.2	4,228	4.9
PDS-02	911	10.6	1,133	12.8	2,099	1.8
OSA-07	1,507	39.3	1,570	38.2	1,504	1.5
KEN-11	11,931	1,068.0	12,247	1,090.8	13,361	61.7
PDS-06	3,394	467.2	4,032	583.7	13,100	40.6
OSA-14	3,180	334.2	3,227	336.8	3,212	6.0
PDS-10	6,562	2,605.2	8,377	3,440.4	24,179	134.1
KEN-13	23,769	9,181.5	25,464	9,861.8	33,074	339.5
CRE-D	17,782	8,227.1	22,075	10,265.2	73,528	312.1
CRE-B	16,411	8,534.1	24,482	12,764.0	63,703	277.1
OSA-30	6,254	2,264.3	7,727	2,720.3	5,963	20.5
PDS-20	24,854	52,549.0	28,443	59,942.4	93,045	1,199.4
OSA-60	12,809	24,627.3	16,731	30,928.0	13,751	112.1
Total(15)	133,951	109,945.9	161,805	132,042.2	350,810	2,517.0

Table C.13 Ratio for 15 Kennington problems

			Stp/N-Stp		Stp/N-D		N-Stp/N-D	
Problem	m	n	Iters	Time	Iters	Time	Iters	Time
CRE-C	3,069	3,678	0.62	0.61	0.32	3.10	0.52	5.10
CRE-A	3,517	4,067	0.64	0.63	0.35	3.73	0.55	5.91
PDS-02	2,954	7,535	0.80	0.83	0.43	5.80	0.54	7.02
OSA-07	1,119	23,949	0.96	1.03	1.00	27.11	1.04	26.34
KEN-11	14,695	21,349	0.97	0.98	0.89	17.32	0.92	17.69
PDS-06	9,882	28,655	0.84	0.80	0.26	11.50	0.31	14.37
OSA-14	2,338	52,460	0.99	0.99	0.99	55.69	1.00	56.13
PDS-10	16,559	48,763	0.78	0.76	0.27	19.42	0.35	25.65
KEN-13	28,633	42,659	0.93	0.93	0.72	27.04	0.77	29.05
CRE-D	8,927	69,980	0.81	0.80	0.24	26.36	0.30	32.89
CRE-B	9,649	72,447	0.67	0.67	0.26	30.80	0.38	46.07
OSA-30	4,351	100,024	0.81	0.83	1.05	110.72	1.30	133.02
PDS-20	33,875	105,728	0.87	0.88	0.27	43.81	0.31	49.98
OSA-60	10,281	232,966	0.77	0.80	0.93	219.63	1.22	275.82
Average	–	–	0.83	0.83	0.38	43.68	0.46	52.46

Table C.14 Statistics for 15 BPMPD problems

	Stp		N-Stp		N-D	
Problem	Iters	Time	Iters	Time	Iters	Time
RAT7A	6,372	567.4	11,779	1,498.3	20,074	296.8
NSCT1	1,956	541.1	1,981	573.0	9,090	65.1
NSCT2	3,867	1,023.8	4,211	1,126.5	5,908	42.8
ROUTING	7,956	2,350.6	8,350	2,459.9	22,336	124.5
DBIR1	2,544	1,134.8	2,443	1,090.4	12,209	82.9
DBIR2	1,980	801.3	2,085	839.6	8,012	60.4
T0331-4L	6,304	1,445.1	22,066	6,031.8	49,932	282.1
NEMSEMM2	11,680	2,485.7	9,282	1,913.5	10,392	30.0
SOUTHERN	16,472	4,008.5	21,608	5,491.0	23,771	318.1
RADIO.PR	2	29.7	2	29.7	2	1.4
WORLD.MD	74,418	92,591.3	111,230	132,327.3	220,329	3,077.7
WORLD	79,915	103,793.9	129,969	163,859.8	278,284	3,899.0
RADIO.DL	4	36.8	4	36.8	3	0.9
NEMSEMM1	7,459	2,374.7	7,215	2,275.8	9,514	30.9
NW14	215	12.8	313	14.2	736	5.2
Total(15)	221,144	213,197.3	332,538	319,567.4	670,592	8,317.8
Total(77)	551,420	330,339.1	822,608	465,624.6	1,621,473	13,099.5

Table C.15 Ratio for 15 BPMPD problems

			Stp/N-Stp		Stp/N-D		N-Stp/N-D	
Problem	m	n	Iters	Time	Iters	Time	Iters	Time
RAT7A	3,137	9,408	0.54	0.38	0.32	1.91	0.59	5.05
NSCT1	22,902	14,981	0.99	0.94	0.22	8.31	0.22	8.80
NSCT2	23,004	14,981	0.92	0.91	0.65	23.94	0.71	26.34
ROUTING	20,895	23,923	0.95	0.96	0.36	18.88	0.37	19.76
DBIR1	18,805	27,355	1.04	1.04	0.21	13.69	0.20	13.15
DBIR2	18,907	27,355	0.95	0.95	0.25	13.27	0.26	13.90
T0331-4L	665	46,915	0.29	0.24	0.13	5.12	0.44	21.38
NEMSEMM2	6,944	42,133	1.26	1.30	1.12	82.99	0.89	63.89
SOUTHERN	18,739	35,421	0.76	0.73	0.69	12.60	0.91	17.26
RADIO.PR	58,867	8,052	1.00	1.00	1.00	21.81	1.00	21.82
WORLD.MD	35,665	31,728	0.67	0.70	0.34	30.08	0.50	43.00
WORLD	35,511	32,734	0.61	0.63	0.29	26.62	0.47	42.03
RADIO.DL	8,053	66,918	1.00	1.00	1.33	38.68	1.33	38.75
NEMSEMM1	3,946	71,413	1.03	1.04	0.78	76.80	0.76	73.60
NW14	74	123,409	0.69	0.90	0.29	2.44	0.43	2.71
Average	–	–	0.67	0.67	0.33	25.63	0.50	38.42

Table C.16 A second ratio summary

	Stp/N-Stp		Stp/N-D		N-Stp/N-D	
Problem	Iters	Time	Iters	Time	Iters	Time
Netlib(47)	0.60	0.51	0.33	3.18	0.55	6.19
Kennington(15)	0.83	0.83	0.38	43.68	0.46	52.46
BPMPD(15)	0.67	0.67	0.33	25.63	0.50	38.42
Average(77)	0.67	0.71	0.34	25.22	0.51	35.55

Appendix D
Empirical Evaluation of Face Methods

To give an insight into the behavior of face methods, the following three codes in FORTRAN 77 were tested, and compared:

1. RSA: The simplex algorithm (see Notation).
2. FALP: Face Algorithm 22.1.3, supported by Algorithm 22.2.1, with components of the initial $\hat{x}$ being all ones.
3. DFA: Dual face Algorithm 23.1.3, supported by Algorithm 23.2.1.

All the preceding were dense codes in the sense that they did not exploit sparsity structure. Harris' two-pass practical tactic was used. Number 10^{-6} was taken as primal and dual feasibility tolerance, and 10^{-4} as equality relative tolerance. In FALP, $\|\Delta_B\|_\infty < 10^{-8}$ was used in place of $\Delta_B = 0$, while in DFA, $\|\Delta y\|_\infty < 10^{-8}$ used in place of $\Delta y = 0$.

Compiled using the Visual FORTRAN 5.0, all codes were run under a Windows XP system Home Edition Version 2002 on an IBM PC with an Intel(R) Pentium(R) processor 1.00 GB of 1.86 GHz memory, and about 16 digits of precision. All reported CPU times were measured in seconds with utility routine CPU_TIME.

Tested was a set of 26 standard problems from NETLIB that do not have BOUNDS and RANGES sections in their MPS files. Specifically, these were the first 26 such problems in the order of increasing sum of numbers of rows and columns of the constraint matrix, before adding slack variables (see Table B.4 in Appendix B).

Test statistics obtained with RSA, FALP and DFA are listed in Tables D.1–D.3, respectively. Total iterations and time, required for solution of each problem, are displayed in columns labeled Itn and Time respectively, and those associated with Phase-1 alone in columns labeled Itn1 and Time1. Percentages of total degenerate iterations are listed in columns labeled % Deg; those associated with Phase-1 are in the column labeled % Deg1. In the last column labeled k-m of Table D.2, listed are the dimensions of the optimal face finally reached, whereas the column labeled m-k in Table D.3 are dimensions of dual optimal faces. Objective values attained are not listed—all runs were terminated within the feasibility tolerances except for problem

P.-Q. PAN, *Linear Programming Computation*, DOI 10.1007/978-3-642-40754-3,

Table D.1 Statistics for Code RSA

	Total			Phase 1		
Problem	Itn	Time	%Deg	Itn1	Time1	%Deg1
AFIRO	29	0.03	69.0	28	0.03	71.4
SC50B	57	0.06	80.7	50	0.06	90.0
SC50A	55	0.09	74.5	51	0.09	78.4
ADLITTLE	134	0.13	12.7	69	0.13	24.6
BLEND	115	0.17	40.0	90	0.17	51.1
SHARE2B	196	0.28	57.1	149	0.27	67.8
SC105	123	0.36	73.2	110	0.36	77.3
STOCFOR1	174	0.50	68.4	150	0.48	79.3
SCAGR7	181	0.67	34.3	146	0.64	42.5
ISRAEL	507	1.41	0.8	188	0.97	1.1
SHARE1B	309	1.67	7.1	190	1.58	11.6
SC205	255	2.23	66.7	211	2.16	80.1
BEACONFD	213	2.61	51.2	159	2.53	57.9
LOTFI	354	3.13	16.4	190	2.91	22.6
BRANDY	356	3.86	33.7	238	3.63	49.2
E226	615	5.50	23.7	394	4.94	32.7
AGG	639	13.64	6.4	571	12.89	7.0
SCORPION	405	16.64	61.0	380	16.47	64.5
BANDM	666	20.42	21.8	467	19.39	30.0
SCTAP1	481	23.11	36.6	349	22.44	44.4
SCFXM1	587	26.98	35.1	429	26.03	45.5
AGG2	824	39.47	5.5	629	36.70	5.6
AGG3	837	52.14	5.5	627	49.17	6.1
SCSD1	186	52.31	81.7	80	52.23	97.5
SCAGR25	922	64.52	34.1	596	60.42	44.5
Total(25)	9,220	331.93	39.9	6,541	316.69	47.3
DEGEN2	10,000	177.83	48.7	1,990	87.84	45.7

DEGEN2, which RSA failed to solve within 10,000 iterations, and the associated results were listed separately, in the bottom lines of the tables.

(1) A comparison between RSA and FALP.
Table D.4 gives iteration and time ratios of RSA to FALP. From the second bottom line, it is seen that the total iteration ratio with the 25 test problems are 1.51, while the total time ratio is 4.91! As the time ratio with SCSD1 is exceptionally high (83.03), the bottom line lists total ratios with the 24 problems, excluding SCSD1. It is seen that the associated iterations and time ratio are 1.69 and 4.17, respectively, as indicates that results with SCSD1 alone do not dominate too much.

(2) A comparison between RSA and DFA.
Table D.5 gives iterations and time ratios of RSA to DFA. If is seen from the second bottom line that the total iteration ratio with the 25 test problems are 1.19, while the total time ratio is as high as 10.04! As the time ratio with SCSD1

Table D.2 Statistics for FALP

	Total			Phase 1			
Problem	Itn	Time	%Deg	Itn1	Time1	%Deg1	k-m
AFIRO	26	0.02	0.0	7	0.00	0.0	9
SC50B	30	0.02	0.0	1	0.00	0.0	20
SC50A	32	0.02	0.0	1	0.00	0.0	20
ADLITTLE	175	0.13	0.0	40	0.03	0.0	15
BLEND	49	0.08	0.0	11	0.03	0.0	44
SHARE2B	85	0.06	0.0	55	0.03	0.0	14
SC105	65	0.06	0.0	1	0.02	0.0	45
STOCFOR1	66	0.09	0.0	23	0.03	0.0	63
SCAGR7	134	0.23	0.0	91	0.13	0.0	84
ISRAEL	294	0.67	0.3	79	0.16	0.0	0
SHARE1B	243	0.34	0.0	140	0.16	0.0	89
SC205	139	0.53	0.0	1	0.08	0.0	91
BEACONFD	135	0.44	3.0	60	0.16	5.0	144
LOTFI	353	0.73	0.6	198	0.30	0.0	95
BRANDY	246	0.91	1.2	76	0.23	0.0	142
E226	486	1.81	0.2	192	0.53	0.0	37
AGG	160	4.45	0.0	94	1.81	0.0	49
SCORPION	119	2.36	3.4	23	0.61	17.4	249
BANDM	410	5.02	0.0	167	1.55	0.0	305
SCTAP1	449	3.55	2.9	112	1.02	0.0	120
SCFXM1	528	6.23	0.2	383	3.78	0.0	191
AGG2	282	7.97	0.0	67	2.14	0.0	76
AGG3	290	8.25	0.0	86	2.50	0.0	76
SCSD1	736	0.63	9.8	1	0.03	0.0	77
SCAGR25	556	23.06	0.0	361	13.47	0.0	300
Total(25)	6,088	67.66	0.9	2,270	28.80	0.9	94
DEGEN2	452	11.13	0.0	190	3.94	0.0	219

is as exceptionally high (1046.20), the bottom line lists total ratios with the other 24 problems: the associated iteration and time ratios are 1.17 and 8.47, respectively.

Considering that the time ratio is the only indication for efficiency, we conclude that the primal and dual face codes outperformed RSA unambiguously.

Such outcome might be astonishing, but not surprising:

On one hand, FALP solves a single triangular $(m + 1) \times (m + 1)$-system in each iteration, compared with RSA solving four triangular $m \times m$-systems.

On the other hand, it is seen from the last column of Table D.2 that all optimal faces, finally reached by FALP, are not vertices, except for ISREAL, and that the average dimensions of these faces is about 94. The ratio of average dimensions to average rows is as low as $94/213 = 0.44$. In general, a high-dimensional optimal face is easier to achieve than an optimal vertex.

Table D.3 Statistics for Code DFA

	Total			Phase 1			
Problem	Itn	Time	%Deg	Itn1	Time1	%Deg1	m-k
AFIRO	30	0.00	20.0	22	0.00	0.0	3
SC50B	50	0.00	0.0	48	0.00	0.0	2
SC50A	53	0.00	0.0	47	0.00	0.0	1
ADLITTLE	164	0.03	15.2	29	0.02	0.0	0
BLEND	104	0.03	2.9	75	0.02	0.0	0
SHARE2B	142	0.03	11.3	69	0.02	1.4	0
SC105	110	0.03	0.0	102	0.02	0.0	1
STOCFOR1	152	0.08	0.0	127	0.05	0.0	0
SCAGR7	254	0.23	11.8	110	0.06	10.9	0
ISRAEL	321	0.36	12.8	140	0.11	0.0	0
SHARE1B	245	0.13	4.5	119	0.06	0.0	0
SC205	209	0.27	0.0	202	0.20	0.0	2
BEACONFD	136	0.06	27.9	1	0.00	0.0	50
LOTFI	264	0.27	8.3	137	0.13	0.0	0
BRANDY	556	0.89	36.7	1	0.00	0.0	16
E226	381	0.78	7.3	156	0.23	0.0	9
AGG	542	3.59	9.0	122	0.63	0.8	11
SCORPION	362	1.44	22.1	1	0.00	0.0	21
BANDM	558	2.75	0.7	294	1.22	0.3	0
SCTAP1	366	0.75	79.0	1	0.02	0.0	33
SCFXM1	713	3.30	5.6	216	0.70	0.5	1
AGG2	571	4.97	5.3	161	1.25	0.0	0
AGG3	588	5.14	5.1	151	1.14	0.0	0
SCSD1	78	0.05	35.9	1	0.00	0.0	0
SCAGR25	822	7.89	16.8	360	2.75	13.3	8
Total(25)	7,771	33.07	13.5	2,692	8.63	1.1	6
DEGEN2	1,627	18.09	6.0	434	3.53	0.2	4

As for a comparison between the two face codes, DFA is even superior to FALP, partially due to that the former solves a smaller triangular $k \times k$ system in each iteration, where k varies dynamically but never exceeds m. From the last column of Table D.3, in fact, it is seen that k did not reach m for 14 out of the 26 test problems.

Table D.4 Ratio of RSA to FALP

				Total		Phase 1	
Problem	m	n	m + n	Itn	Time	Itn	Time
AFIRO	28	32	60	1.12	1.50	4.00	300.00
SC50B	51	48	99	1.90	3.00	50.00	600.00
SC50A	51	48	99	1.72	4.50	51.00	900.00
ADLITTLE	57	97	154	0.77	1.00	1.73	4.33
BLEND	75	83	158	2.35	2.13	8.18	5.67
SHARE2B	97	79	176	2.31	4.67	2.71	9.00
SC105	106	103	209	1.89	6.00	110.00	18.00
STOCFOR1	118	111	229	2.64	5.56	6.52	16.00
SCAGR7	130	140	270	1.35	2.91	1.60	4.92
ISRAEL	175	142	317	1.72	2.10	2.38	6.06
SHARE1B	118	225	343	1.27	4.91	1.36	9.88
SC205	206	203	409	1.83	4.21	211.00	27.00
BEACONFD	174	262	436	1.58	5.93	2.65	15.81
LOTFI	154	308	462	1.00	4.29	0.96	9.70
BRANDY	194	249	443	1.45	4.24	3.13	15.78
E226	224	282	506	1.27	3.04	2.05	9.32
AGG	489	163	652	3.99	3.07	6.07	7.12
SCORPION	358	358	716	3.40	7.05	16.52	27.00
BANDM	306	472	778	1.62	4.07	2.80	12.51
SCTAP1	301	480	781	1.07	6.51	3.12	22.00
SCFXM1	331	457	788	1.11	4.33	1.12	6.89
AGG2	517	302	819	2.92	4.95	9.39	17.15
AGG3	517	302	819	2.89	6.32	7.29	19.67
SCSD1	78	760	838	0.25	83.03	80.00	1,741.00
SCAGR25	472	500	972	1.66	2.80	1.65	4.49
Average(25)	213	248	461	1.51	4.91	2.88	11.00
Average(24)	218	226	445	1.69	4.17	2.85	9.19

Table D.5 Ratio of RSA to DFA

				Total		Phase 1	
Problem	m	n	m + n	Itn	Time	Itn	Time
AFIRO	28	32	60	0.97	–	1.27	–
SC50B	51	48	99	1.14	–	1.04	–
SC50A	51	48	99	1.04	–	1.09	–
ADLITTLE	57	97	154	0.82	4.33	2.38	6.50
BLEND	75	83	158	1.11	5.67	1.20	8.50
SHARE2B	97	79	176	1.38	9.33	2.16	13.50
SC105	106	103	209	1.12	12.00	1.08	18.00
STOCFOR1	118	111	229	1.14	6.25	1.18	9.60
SCAGR7	130	140	270	0.71	2.91	1.33	10.67
ISRAEL	175	142	317	1.58	3.92	1.34	8.82
SHARE1B	118	225	343	1.26	12.85	1.60	26.33
SC205	206	203	409	1.22	8.26	1.04	10.80
BEACONFD	174	262	436	1.57	43.50	159.00	–
LOTFI	154	308	462	1.34	11.59	1.39	22.38
BRANDY	194	249	443	0.64	4.34	238.00	–
E226	224	282	506	1.61	7.05	2.53	21.48
AGG	489	163	652	1.18	3.80	4.68	20.46
SCORPION	358	358	716	1.12	11.56	380.00	–
BANDM	306	472	778	1.19	7.43	1.59	15.89
SCTAP1	301	480	781	1.31	30.81	349.00	1,122.00
SCFXM1	331	457	788	0.82	8.18	1.99	37.19
AGG2	517	302	819	1.44	7.94	3.91	29.36
AGG3	517	302	819	1.42	10.14	4.15	43.13
SCSD1	78	760	838	2.38	1,046.20	80.00	–
SCAGR25	472	500	972	1.12	8.18	1.66	21.97
Average(25)	213	248	461	1.19	10.04	2.43	36.70
Average(24)	218	226	445	1.17	8.47	2.40	30.64

Appendix E
Empirical Evaluation of the Feasible-Point Simplex Method

Computational experiments have been performed, involving the following three codes:

1. MINOS: MINOS 5.51 with full pricing.
2. FPS: Two-Phase code based on Algorithm 24.4.1.
3. FPSP: Two-Phase code based on Algorithm 24.4.1 with the purification.

In both FPS and FPSP, $\epsilon_1 = 10^{-8}$ and $\epsilon_2 = 10^{-8}$ were used for Phase-1, and $\epsilon_1 = 10^{-3}$ and $\epsilon_2 = 10^{-6}$ for Phase-2.

Codes FPS and FPSP were developed using MINOS 5.51 as a platform. Therefore, the three codes shared such features as preprocessing, scaling, LUSOL (Saunders, 87), etc. Only the Mi50lp module was replaced by programs written by the author himself, with minor changes in few other modules. All codes used the default options, except for those listed in Appendix C.

Compiled using Visual Fortran 5.0, all codes were run under a Microsoft Windows XP Professional version 2002 on an ACER PC with an Intel(R) Pentium(R) 4 CPU 3.06 GHz, 1.00 GB memory, and about 16 digits of precision.

The first set of test problems included all 16 problems from Kennington and the second included all 17 problems from BPMPD that were more than 500 KB in compressed form (Appendix B: Tables B.2 and B.3).

In the tables, test problems are ordered by their sizes, in terms of m+n as before, where m and n are the numbers of rows and columns of the constraint matrix, excluding slack variables. All reported CPU times were measured in seconds with utility routine CPU_TIME, excluding the time spent on preprocessing and scaling. FPS and FPSP used 10^{-6} as the equality tolerance (for holding of $Ax = b$), compared with MINOS, which used 10^{-4}, as usual.

P.-Q. PAN, *Linear Programming Computation*, DOI 10.1007/978-3-642-40754-3,

Table E.1 Statistics for 16 Kennington problems

	MINOS		FPSP		FPS	
Problem	Itns	Time	Itns	Time	Itns	Time
KEN-07	1,724	1.97	1,754	2.02	1,750	2.02
CRE-C	4,068	6.81	7,487	11.89	2,751	4.66
CRE-A	4,066	7.67	8,972	16.77	4,069	8.09
PDS-02	5,110	9.06	1,171	1.84	1,025	1.61
OSA-07	1,767	7.16	1,949	8.00	1,204	5.02
KEN-11	13,327	111.39	13,549	113.67	13,542	113.59
PDS-06	59,521	433.67	8,860	57.41	6,317	40.36
OSA-14	3,821	33.48	3,581	32.16	1,747	16.19
PDS-10	177,976	2,253.64	11,191	122.13	7,110	75.47
KEN-13	35,709	637.92	35,301	629.66	35,162	627.05
CRE-D	377,243	4,796.42	28,042	358.75	1,144	13.23
CRE-B	328,160	4,406.19	32,228	430.81	1,422	17.91
OSA-30	7,536	126.30	5,886	101.59	2,778	49.33
PDS-20	1,557,953	44,803.02	88,068	2,258.83	64,777	1,643.14
OSA-60	16,278	638.50	10,976	449.45	4,714	199.36
KEN-18	157,224	19,395.88	155,355	19,315.88	154,801	19,253.55
Total	2,751,483	77,669.08	414,370	23,910.86	304,313	22,070.58

Numerical results obtained with Kennington and BPMPD problems are displayed, respectively, in Tables E.1 and E.2, where the total iterations and time required for solving each problem are listed in the columns labeled Itns and Time under MINOS, FPSP and FPS, separately. As MINOS solved problems RADIO.PR and RADIO.DU in few iterations, results for the two problems are listed separately in the bottom lines, and excluded from comparison with BPMPD set.

In Table E.3, a performance comparison between the three codes with Kennington set is made by giving iteration and time ratios of MINOS to FPSP, MINOS to FPS and FPSP to FPS for each problem. It is seen from the bottom line that FPSP and FPS outperformed MINOS remarkably, with average iterations ratios 6.6 and 9.0, and time ratios 3.2 and 3.5! It is noted that the difference between FPSP and FPS is small (with iterations ratio 1.4 and time ratio 1.1).

In Table E.4, a comparison between the three codes with BPMPD set is made. It is seen from the second bottom line that FPSP and FPS outperformed MINOS, by average iterations ratios 2.8 and 9.5, and time ratios 3.4 and 9.0! The margins between FPSP and FPS is smaller (with iterations ratio 3.3 and time ratio 2.6).

Table E.2 Statistics for 17 BPMPD problems

	MINOS		FPSP		FPS	
Problem	Itns	Time	Itns	Time	Itns	Time
RAT7A	4,073	573.59	4,102	52.06	4,100	52.03
NSCT1	2,190	35.83	13,198	197.36	95	1.44
NSCT2	11,555	196.63	11,971	182.23	92	1.45
RIUTING	190,627	2,143.70	75,759	813.08	72,774	780.00
DBIR1	2,289	47.44	12,781	242.64	34	0.61
DBIR2	42,575	966.77	12,750	235.83	12,745	235.72
T0331-4L	62,333	789.83	108,780	1,330.63	108,575	1,328.14
NEMSEMM2	23,581	204.59	18,109	136.98	7,327	58.00
SOUTHERN	29,693	462.67	27,550	314.17	27,550	314.16
WORLD.MO	1,405,151	30,953.42	822,287	19,193.38	94,542	2,929.00
WORLD	2,627,208	61,340.78	1,260,355	27,315.31	165,045	4,132.83
NEMSEMM1	13,650	263.20	14,575	255.17	14,491	253.64
NW14	407	8.45	899	19.20	690	14.42
LPL1	3,343,112	127,379.39	203,967	7,214.02	2,414	104.41
DBIC1	624,907	33,335.58	377,655	18,746.17	375,734	18,679.69
Total	8,383,351	258,701.88	2,964,738	76,248.23	886,208	28,885.54
RADIO.PR	2	0.50	639,702	14,371.16	623,810	14,009.16
RADIO.DU	3	0.09	5,055	56.11	5,018	55.66

Table E.3 Ratio for 16 Kennington problems

			MINOS/FPSP		MINOS/FPS		FPSP/FPS	
Problem	M	N	Itns	Time	Itns	Time	Itns	Time
KEN-07	2,427	3,602	1.0	1.0	1.0	1.0	1.0	1.0
CRE-C	3,069	3,678	0.5	0.6	1.5	1.5	2.7	2.6
CRE-A	3,517	4,067	0.5	0.5	1.0	0.9	2.2	2.1
PDS-02	2,954	7,535	4.4	4.9	5.0	5.6	1.1	1.1
OSA-07	1,119	23,949	0.9	0.9	1.5	1.4	1.6	1.6
KEN-11	14,695	21,349	1.0	1.0	1.0	1.0	1.0	1.0
PDS-06	9,882	28,655	6.7	7.6	9.4	10.7	1.4	1.4
OSA-14	2,338	52,460	1.1	1.0	2.2	2.1	2.0	2.0
PDS-10	16,559	48,763	15.9	18.5	25.0	29.9	1.6	1.6
KEN-13	28,633	42,659	1.0	1.0	1.0	1.0	1.0	1.0
CRE-D	8,927	69,980	13.5	13.4	329.8	362.5	24.5	27.1
CRE-B	9,649	72,447	10.2	10.2	230.8	246.0	22.7	24.1
OSA-30	4,351	100,024	1.3	1.2	2.7	2.6	2.1	2.1
PDS-20	33,875	105,728	17.7	19.8	24.1	27.3	1.4	1.4
OSA-60	10,281	232,966	1.5	1.4	3.5	3.2	2.3	2.3
KEN-18	105,128	154,699	1.0	1.0	1.0	1.0	1.0	1.0
Ave.(16)	–	–	6.6	3.2	9.0	3.5	1.4	1.1

Table E.4 Ratio for 15 BPMPD problems

			MINOS/FPSP		MINOS/FPS		FPSP/FPS	
Problem	M	N	Itns	Time	Itns	Time	Itns	Time
RAT7A	3,137	9,408	1.0	11.0	1.0	11.0	1.0	1.0
NSCT1	22,902	14,981	0.2	0.2	23.1	24.9	138.9	137.1
NSCT2	23,004	14,981	1.0	1.1	125.6	135.6	130.1	125.7
RIUTI	20,895	23,923	2.5	2.6	2.6	2.7	1.0	1.0
DBIR1	18,805	27,355	0.2	0.2	67.3	77.8	375.9	397.8
DBIR2	18,907	27,355	3.3	4.1	3.3	4.1	1.0	1.0
T0331	665	46,915	0.6	0.6	0.6	0.6	1.0	1.0
NEM.2	6,944	42,133	1.3	1.5	3.2	3.5	2.5	2.4
SOUTH	18,739	35,421	1.1	1.5	1.1	1.5	1.0	1.0
WOR.M	35,665	31,728	1.7	1.6	14.9	10.6	8.7	6.6
WORLD	35,511	32,734	2.1	2.2	15.9	14.8	7.6	6.6
NEM.1	3,946	71,413	0.9	1.0	0.9	1.0	1.0	1.0
NW14	74	123,409	0.5	0.4	0.6	0.6	1.3	1.3
LPL1	39,952	125,000	16.4	17.7	1,384.9	1,220.0	84.5	69.1
DBIC1	43,201	183,235	1.7	1.8	1.7	1.8	1.0	1.0
Ave.(15)	–	–	2.8	3.4	9.5	9.0	3.3	2.6
Ave.(31)	–	–	3.3	3.4	9.4	6.6	2.8	2.0

In the bottom line of Table E.4, listed are average ratios for the entire set of the 31 test problems. FPSP and FPS again defeated MINOS: the average iterations ratios are 3.3 and 9.4, while the time ratios are 3.4 and 6.6.

In summary, the feasible-point simplex algorithm is significantly superior to the standard simplex algorithm with the test sets.

References

Abadie J, Corpentier J (1969) Generalization of the Wolfe reduced gradient method to the case of mon-linear constrained optimization. In: Fletcher R (ed) Optimization. Academic, London, pp 37–48

Abel P (1987) On the choice of the pivot columns of the simplex method: gradient criteria. Computing 38:13–21

Adler I, Megiddo N (1985) A simplex algorithm whose average number of steps is bounded between two quadratic functions of the smaller dimension. J ACM 32:871–895

Adler I, Resende MGC, Veige G, Karmarkar N (1989) An implementation of Karmarkar's algorithm for linear programming. Math Program 44:297–335

Andersen E, Andersen K (1995) Presolving in linear programming. Math Program 71:221–245

Andersen ED, Gondzio J, Mészáros C, Xu X (1996) Implementation of interior-point methods for large scale linear programming. In: Terlaky T (ed) Interior point methods of mathematical programming. Kluwer, Dordrecht

Andrel N, Barbulescu M (1993) Balance constraints reduction of large-scale linear programming problems. Ann Oper Res 43:149–170

Anstreicher KM, Watteyne P (1993) A family of search directions for Karmarkar's algorithm. Oper Res 41:759–767

Arrow KJ, Hurwicz L (1956) Reduction of constrained maxima to saddle-point problems. In: Neyman J (ed) Proceedings of the third Berkeley symposium on mathematical statistics and probability, vol 5. University of California Press, Berkeley, pp 1–26

Avis D, Chvatal V (1978) Notes on Bland's pivoting rule. Math Program 8:24–34

Balas E (1965) An additive algorithm for solving linear programs with zero-one variables. Oper Res 13:517–546

Balinski ML, Gomory RE (1963) A mutual primal-dual simplex method. In: Graves RL, Wolfe P (eds) Recent advances in mathematical programming. McGraw-Hill, New York

Balinski ML, Tucker AW (1969) Duality theory of linear problems: a constructive approach with applications. SIAM Rev 11:347–377

Barnes ER (1986) A variation on Karmarkars algorithm for solving linear programming problems. Math Program 36:174–182

Bartels RH (1971) A stabilization of the simplex method. Numer Math 16:414–434

Bartels RH, Golub GH (1969) The simplex method of linear programming using LU decomposition. Commun ACM 12:266–268

Bartels RH, Stoer J, Zenger Ch (1971) A realization of the simplex method based on triangular decompositions. In: Wilkinson JH, Reinsch C (eds) Contributions I/II in handbook for automatic computation, volume II: linear algebra. Springer, Berlin/London

Bazaraa MS, Jarvis JJ, Sherali HD (1977) Linear programming and network flows, 2nd edn. Wiley, New York

P.-Q. PAN, *Linear Programming Computation*, DOI 10.1007/978-3-642-40754-3,

Beale EML (1954) An alternative method for linear programming. Proc Camb Philos Soc 50:513–523
Beale EML (1955) Cycling in the dual simplex algorithm. Nav Res Logist Q 2:269–275
Beale E (1968) Mathematical programming in practice. Topics in operations research. Pitman & Sons, London
Benders JF (1962) Partitioning procedures for solving mixed-variables programming problems. Numer Math 4:238–252
Benichou MJ, Cautier J, Hentges G, Ribiere G (1977) The efficient solution of large scale linear programming problems. Math Program 13:280–322
Bixby RE (1992) Implementing the simplex method: the initial basis. ORSA J Comput 4:287–294
Bixby RE (1994) Progress in linear programming. ORSA J Comput 6:15–22
Bixby RE (2002) Solving real-world linear problems: a decade and more of progress. Oper Res 150:3–15
Bixby RE, Saltzman MJ (1992) Recovering an optimal LP basis from the interior point solution. Technical report 607, Dapartment of Mathematical Sciences, Clemson University, Clemson
Bixby RE, Wagner DK (1987) A note on detecting simple redundancies in linear systems. Oper Res Lett 6:15–17
Bixby RE, Gregory JW, Lustig IJ, Marsten RE, Shanno DF (1992) Very large-scale linear programming: a case study in combining interior point and simplex methods. Oper Res 40:885–897
Björck A, Plemmons RJ, Schneider H (1981) Large-scale matrix problems. North-Holland, Amsterdanm
Bland RG (1977) New finite pivoting rules for the simplex method. Math Oper Res 2:103–107
Borgwardt K-H (1982a) Some distribution-dependent results about the asymptotic order of the arerage number of pivot steps of the simplex method. Math Oper Res 7:441–462.
Borgwardt K-H (1982b) The average number of pivot steps required by the simplex method is polynomial. Z Oper Res 26:157–177
Botsaris CA (1974) Differential gradient methods. J Math Anal Appl 63:177–198
Breadley A, Mitra G, Williams HB (1975) Analysis of mathematica problems prior to applying the simplex algorithm. Math Program 8:54–83
Brown AA, Bartholomew-Biggs MC (1987) ODE vs SQP methods for constrained optimization. Technical report, 179, The Numerical Center, Hatfield Polytechnic
Carolan MJ, Hill JE, Kennington JL, Niemi S, Wichmann SJ (1990) An empirical evaluation of the KORBX algorithms for military airlift applications. Oper Res 38:240–248
Cavalier TM, Soyster AL (1985) Some computational experience and a modification of the Karmarkar algorithm. ISME working paper, The Pennsylvania State University, pp 85–105
Chan TF (1985) On the existence and computation of LU factorizations with small pivots. Math Comput 42:535–548
Chang YY (1979) Least index resolution of degeneracy in linear complementarity problems. Technical Report 79–14, Department of OR, Stanford University
Charnes A (1952) Optimality and degeneracy in linear programming. Econometrica 20:160–170
Cheng MC (1987) General criteria for redundant and nonredundant linear inequalities. J Optim Theory Appl 53:37–42
Chvatal V (1983) Linear programming. W.H. Freeman, New York
Cipra BA (2000) The best of the 20th century: editors name top 10 algorithms. SIAM News 33: 1–2
Coleman TF, Pothen A (1987) The null space problem II. Algorithms. SIAM J Algebra Discret Methods 8:544–562
Cook AS (1971) The complexity of theorem-proving procedure. In: Proceedings of third annual ACM symposium on theory of computing, Shaker Heights, 1971. ACM, New York, pp 151–158
Cottle RW, Johnson E, Wets R (2007) George B. Dantzig (1914–2005). Not AMS 54:344–362
CPLEX ILOG (2007) 11.0 User's manual. ILOG SA, Gentilly, France
Curtis A, Reid J (1972) On the automatic scaling of mtrices for Gaussian elimination. J Inst Math Appl 10:118–124

Dantzig GB (1948) Programming in a linear structure, Comptroller, USAF, Washington, DC
Dantzig GB (1951a) Programming of interdependent activities, mathematical model. In: Koopmas TC (ed) Activity analysis of production and allocation. Wiley, New York, pp 19–32; Econometrica 17(3/4):200–211 (1949)
Dantzig GB (1951b) Maximization of a linear function of variables subject to linear inequalities. In: Koopmans TC (ed) Activity analysis of production and allocation. Wiley, New York, pp 339–347
Dantzig GB (1951c) A proof of the equivalence of the programming problem and the game problem. In: Koopmans T (ed) Activity analysis of production and allocation. Wiley, New York, pp 330–335
Dantzig GB (1963) Linear programming and extensions. Princeton University Press, Princeton
Dantzig GB (1991) linear programming. In: Lenstra JK, Rinnooy Kan AHG, Schrijver A (eds) History of mathematical programming. CWI, Amsterdam, pp 19–31
Dantzig GB, Ford LR, Fulkerson DR (1956) A primal-dual algorithm for linear programs. In: Kuhn HK, Tucker AW (eds) Linear inequalities and related systems. Princeton University Press, Princeton, pp 171–181
Dantzig GB, Orchard-Hays W (1953) Alternate algorithm for the revised simplex method using product form for the inverse. Notes on linear programming: part V, RM-1268. The RAND Corporation, Santa Monica
Dantzig GB, Orchard-Hayes W (1954) The product form for the inverse in the simplex method. Math Tables Other Aids Comput 8:64–67
Dantzig GB, Thapa MN (1997) Linear programming 1: introduction. Springer, New York
Dantzig GB, Thapa MN (2003) Linear programming 2: theory and extensions. Springer, New York
Dantzig GB, Wolfe P (1960) Decomposition principle for linear programs. Oper Res 8:101–111
Dantzig GB, Orden A, Wolfe P (1955) The generalized simplex method for minimizing a linear form under linear inequality constraints. Pac J Math 5:183–195
de Ghellinck G, Vial J-Ph, Polynomial (1986) Newton method for linear programming. Algorithnica 1:425–453. (Special issue)
Dikin I (1967) Iterative solution of problems of linear and quadratic programming. Sov Math Dokl 8:674–675
Dikin I (1974) On the speed of an iterative process. Upravlyaemye Sistemi 12:54–60
Dorfman R, Samuelson PA, Solow RM (1958) Linear programming and economic analysis. McGraw-Hill, New York
Duff IS, Erisman AM, Reid JK (1986) Direct methods for sparse matrices. Oxford University Press, Oxford
Evtushenko YG (1974) Two numerical methods for solving non-linear programming problems. Sov Math Dokl 15:420–423
Fang S-C (1993) Linear optimization and extensions: theory and algorithms. AT & T, Prentice-Hall, Englewood Cliffs
Farkas J (1902) Uber die Theorie der Einfachen Ungleichungen. Journal für die Reine und Angewandte Mathematik 124:1–27
Fiacco AV, Mccormick GP (1968) Nonlinear programming: sequential unconstrained minimization techniques. Wiley, New York
Fletcher R (1981) Practical methods of optimization. Volume 2: constrained optimization. Wiley, Chichester
Ford Jr LR, Fulkerson DR (1956) Maximal flow through a network. Can J Math 8:399–407
Forrest JJH, Goldfarb D (1992) Steepest edge simplex algorithm for linear programming. Math Program 57:341–374
Forrest J, Tomlin J (1972) Updating triangular factors of the basis to maintain sparsity in the product form simplex method. Math Program 2:263–278
Forsythe GE, Malcom MA, Moler CB (1977) Computer methods for mathematical computations. Pretice-Hall, EnglewoodCliffs
Fourer R (1979) Sparse Gaussian elimination of staircase linear systems. Tehchnical report ADA081856, Calif Systems Optimization LAB, Stanford University

Fourier JBJ (1823) Analyse des travaux de l'Academie Royale des Science, pendant l'Iannee, Partie mathematique, Histoire de l'Acanemie Royale des Sciences de l'nstitut de France 6 [1823] (1826), xxix–xli (partially reprinted as: Premier extrait, in Oeuvres de Fourier, Tome 11 (Darboux G, ed.), Gauthier-Villars, Paris, 1890, (reprinted: G. Olms, Hildesheim, 1970), pp 321–324

Frisch KR (1955) The logarithmic potentical method of convex programming. Memorandum, University Institute of Economics, Oslo

Fulkerson D, Wolfe P (1962) An algorithm for scaling matrices. SIAM Rev 4:142–146

Gale D, Kuhn HW, Tucker AW (1951) Linear programming and the theory of games. In: Koopmans T (ed) Activity analysis of production and allocation. Wiley, New York, pp 317–329

Gass SI (1985) Linear programming: methods and applications. McGraw-Hill, New York

Gass SI, Saaty T (1955) The computational algorithm for the parametric objective function. Nav Res Logist Q 2:39–45

Gay DM (1978) On combining the schemes of Reid and Saunders for sparse LP bases. In: Duff IS, Stewart GW (eds) Sparse matrix proceedings. SIAM, Philadelphia, pp 313–334

Gay D (1985) Electronic mail distribution of linear programming test problems. COAL Newsl 13:10–12

Gay DM (1987) A variant of Karmarkar's linear programming algorithm for problems in standard form. Math Program 37:81–90

Geoffrion AM (1972) Generalized Benders decomposition. JOTA 10:137–154

George A, Liu W-H (1981) Computing solution of large sparse positive definite systems. Prentice-Hall, Engleewood Cliffs

Gill PE, Murray W (1973) A numerically stable form of the simplex algorithm. Linear Algebra Appl 7:99–138

Gill PE, Murray W, Saunders MA, Tomlin JA, Wright MH (1985) On projected Newton barrier methods for linear programming and an equivalence to Karmarkar's projected method. Technical report SOL 85–11, Department of Operations Research, Stanford University

Gill PE, Murray W, Saunders MA, Tomlin JA, Wright MH (1986) On projected Newton methods for linear programming and an equivalence to Karmarkar's projected method. Math Program 36:183–209

Gill PE, Murray W, Saunders MA, Wright MH (1987) Maintaining LU factors of a general sparse matrix. Linear Algebra Appl 88/89:239–270

Gill PE, Murray W, Saunders MA, Wright MH (1989) A practical anti-cycling procedure for linearly constrainted optimization. Math Program 45:437–474

Goldfarb D (1977) On the Bartels-Golub decomposition for linear programming bases. Math Program 13:272–279

Goldfarb D, Reid JK (1977) A practicable steepest-edge simplex algorithm. Math Program 12:361–371

Goldman AJ, Tucker AW (1956a) Polyhedral convex cones. In: Kuhn HW, Tucker AW (eds) Linear inequalities and related systems. Annals of mathmatical studies, vol 38. Princeton University Press, Princeton, pp 19–39

Goldman AJ, Tucker AW (1956b) Theory of linear programming. In: Kuhn HW, Tucker AW (eds) Linear inequalities and related systems. Annals of mathmatical studies, vol 38. Princeton University Press, Princeton, pp 53–97

Golub GH (1965) Numerical methods for solving linear least squares problems. Numer Math 7:206–216

olub GH, Van Loan CF (1989) Matrix computations, 2edn. The Johns Hopkins University Press, Baltimore

mory AW (1958) Outline of an algorithm for integer solutions to linear programs. Bull Am Math Soc 64:275–278

zaga CC (1987) An algorithm for solving linear programming problems in $O(n^3L)$ operations. chnical Report UCB/ERL M87/10, Electronics Research Laboratory, University of Califor-, Berkeley

Gonzaga CC (1990) Convergence of the large step primal affine-scaling algorithm for primal non-degenerate linear problems. Technical report, Department of Systems Engineering and Computer Sciences, COPPE-Federal University of Riode Janeiro
Gould N, Reid J (1989) New crash procedure for large systems of linear constraints. Math Program 45:475–501
Greenberg HJ (1978) Pivot selection tactics. In: Greenberg HJ (ed) Design and implementation of optimization software. Sijthoff and Noordhoff, Alphen aan den Rijn, pp 109–143
Greenberg HJ, Kalan J (1975) An exact update for Harris' tread. Math Program Study 4:26–29
Guerrero-Garcia P, Santos-Palomo A (2005) Phase I cycling under the most-obtuse-angle pivot rule. Eur J Oper Res 167:20–27
Guerrero-Garcia P, Santos-Palomo A (2009) A deficient-basis dual counterpart of Pararrizos, Samaras ans Stephanides' primal-dual simplex-type algorithm. Optim Methods Softw 24:187–204
Güler O, Ye Y (1993) Convergence behavior of interior-point algorithms. Math Program 60:215–228
Hadley G (1972) Linear programming. Addison-Wesley, Reading
Hager WW (2002) The dual active set algorithm and its application to linear programming. Comput Optim Appl 21:263–275
Haimovich H (1996) The simplex method is very good!—on the expected number of pivot steps and related properties of random linear programs, Hebrew University of Jerusalem, No. 99
Hall LA, Vanderbei RJ (1993) Two-third is sharp for affine scaling. Oper Res Lett 13:197–201
Hamming RW (1971) Introduction to applied numerical analysis. McGraw-Hill, New York
Harris PMJ (1973) Pivot selection methods of the Devex LP code. Math Program 5:1–28
Hattersley B, Wilson J (1988) A dual approach to primal degeneracy. Math Program 42:135–145
He X-C, Sun W-Y (1991) An intorduction to generalized inverse matrix (in Chinese). Jiangsu Science and Technology Press, Nanjing
Hellerman E, Rarick DC (1971) Reinversion with the preassigned pivot procedure. Math Program 1:195–216
Hellerman E, Rarick DC (1972) The partitioned preassigned pivot procedure. In: Rose DJ, Willouhby RA (eds) Sparse matrices and their applications. Plenum, New York, pp 68–76
Hertog DD, Roos C (1991) A survey of search directions in interior point methods for linear programming. Math Program 52:481–509
Hoffman AJ (1953) Cycling in the simplex algorithm. Technical report 2974, National Bureau of Standards
Hu J-F (2007) A note on "an improved initial basis for the simplex algorithm". Comput Oper Res 34:3397–3401
Hu J-F, Pan P-Q (2006) A second note on 'A method to solve the feasible basis of LP' (in Chinese). Oper Res Manag Sci 15:13–15
Hu J-F, Pan P-Q (2008a) Fresh views on some recent developments in the simplex algorithm. J Southeast Univ 24:124–126
Hu J-F, Pan P-Q (2008b) An efficient approach to updating simplex multipliers in the simplex algorithm. Math Program Ser A 114:235–248
Jansen B, Terlakey T, Roos C (1994) The theory of linear programming: skew symmetric self-dual problems and the central path. Optimization 29:225–233
Jansen B, Roos C, Terlaky T (1996) Target-following methods for linear programming. In: Terlaky T (ed) Interior point methods of mathematical programming. Kluwer, Dordrecht
Jeroslow R (1973) The simplex algorithm with the pivot rule of maximizing criterion improvement. Discret Appl Math 4:367–377
Kalantari B (1990) Karmarkar's algorithm with improved steps. Math Program 46:73–78
Kallio M, Porteus EL (1978) A class of methods for linear programming. Math Program 14:161–169
Kantorovich LV (1960) Mathematical methods in the organization and planning of production. Manag Sci 6:550–559. Original Russian version appeared in 1939

Karmarkar N (1984) A new polynomial time algorithm for linear programming. Combinatorica 4:373–395

Karmarkar N, Ramakrishnan K (1985) Further developments in the new polynomial-time algorithm for linear progrmming. In: Talk given at ORSA/TIMES national meeting, Boston, Apr 1985

Khachiyan L (1979) A polynomial algorithm in linear programming. Doklady Academiia Nauk SSSR 244:1093–1096

Kirillova FM, Gabasov R, Kostyukova OI (1979) A method of solving general linear programming problems. Doklady AN BSSR (in Russian) 23:197–200

Klee V (1965) A class of linear problemminng problems requiring a larger number of iterations. Numer Math 7:313–321

Klee V, Minty GJ (1972) How good is the simplex algorithm? In: Shisha O (ed) Inequalities-III. Academic, New York, pp 159–175

Koberstein A (2008) Progress in the dual simplex algorithm for solving large scale LP problems: techniques for a fast and stable implementation. Comput Optim Appl 41:185–204

Koberstein A, Suhl UH (2007) Progress in the dual simplex method for large scale LP problems: practical dual phase 1 algorithms. Comput Optim Appl 37:49–65

Kojima M, Mizuno S, Yoshise A (1989) A primal-dual interior point algorithm for linear programming. In: Megiddo N (ed) Progress in mathematical programming. Springer, New York, pp 29–47

Kojima M, Megiddo N, Mizuno S (1993) A primal-dual infeasible-interior-point algorithm for linear programming. Math Program 61:263–280

Kortanek KO, Shi M (1987) Convergence results and numerical experiments on a linear programming hybrid algorithm. Eur J Oper Res 32:47–61

Kostina E (2002) The long step rule in the bounded-variable dual simplex method: numerical experiments. Math Methods Oper Res 55:413–429

Kotiah TCT, Steinberg DI (1978) On the possibility of cycling with the simplex method. Oper Res 26:374–376

Kuhn HW, Quandt RE (1953) An experimental study of the simplex method. In: Metropolis NC et al (eds) Eperimental arithmetic, high-speed computing and mathematics. Proceedings of symposia in applied mathematics XV. American Mathematical Society, Providence, pp 107–124

Land AH, Doig AG (1960) An automatic method of solving discrete programming problems. Econometrica 28:497–520

Leichner SA, Dantzig GB, Davis JW (1993) A strictly, improving linear programming phase I algorithm. Ann Oper Res 47:409–430

Lemke CE (1954) The dual method of solving the linear programming problem. Nav Res Logist Q 1:36–47

Li W (2004) A note on two direct methods in linear programming. Eur J Oper Res 158:262–265

Li C, Pan P-Q, Li W (2002) A revised simplex algorithm based on partial pricing pivotrule (in Chinese). J Wenzhou Univ 15:53–55

Li W, Guerrero-Garcia P, Santos-Palomo A (2006a) A basis-deficiency-allowing primal phase-1 algorithm using the most-obtuse-angle column rule. Comput Math Appl 51:903–914

Li W, Pan P-Q, Chen G (2006b) A combined projected gradient algorithm for linear programming. Optim Methods Softw 21:541–550

Llewellyn RW (1964) Linear programming. Holt, Rinehart and Winston, New York

Luo Z-Q, Wu S (1994) A modified predictor-corrector method for linear programming. Comput Optim Appl 3:83–91

Lustig IJ (1990) Feasibility issues in a prinal-dual interior-point method for linear programming. Math Program 49:145–162

Lustig IJ, Marsten RE, Shanno DF (1991) Computational exeperience with a primal-dual interior point method for linear programming. Linear Algebra Appl 152:191–222

Lustig IJ, Marsten RE, Shanno DF (1992) On implementing Mehrotras's predictor-corrector interior-point for linear programming. SIAM J Optim 2:435–449

Lustig IJ, Marsten R, Shanno D (1994) Interior point methods for linear programming: computational state of the art. ORSA J Comput 6:1–14
Markowitz HM (1957) The elimination form of the inverse and its application to linear programming. Manag Sci 3:255–269
Maros I (1986) A general phase-1 method in linear programming. Eur J Oper Res 23:64–77
Maros I (2003a) A generalied dual phase-2 simplex algorithm. Eur J Oper Res 149:1–16
Maros I (2003b) Computational techniques of the simplex method. International series in operations research and management, vol 61. Kluwer, Boston
Maros I, Khaliq M (2002) Advances in design and implementation of optimization software. Eur J Oper Res 140:322–337
Marshall KT, Suurballe JW (1969) A note on cycling in the simplex method. Nav Res Logist Q 16:121–137
Martin RK (1999) Large scale linear and integer optimization: a unified approach. Kluwer, Boston
Mascarenhas WF (1997) The affine scaling algorithm fails for $\lambda = 0.999$. SIAM J Optim 7:34–46
McShane KA, Monma CL, Shanno DF (1989) An implementation of a primal-dual method for linear programming. ORSA J Comput 1:70–83
Megiddo N (1986a) Introduction: new approaches to linear programming. Algorithmca 1:387–394. (Special issue)
Megiddo N (1986b) A note on degenerach in linear programming. Math Program 35:365–367
Megiddo N (1989) Pathways to the optimal set in linear programming. In: Megiddo N (ed) Progress in mathematical programming. Springer, New York, pp 131–158
Megiddo N, Shub M (1989) Boundary behavior of interior point algorithm in linear programming. Math Oper Res 14:97–146
Mehrotra S (1991) On finding a vertex solution using interior point methods. Linear Algebra Appl 152:233–253
Mehrotra S (1992) On the implementation of a primal-dual interior point method. SIAM J Optim 2:575–601
Mizuno S, Todd MJ, Ye Y (1993) On adaptive-step primal-dual interior-point algorithms for linear programming. Math Oper Res 18:964–981
Monteiro RDC, Adler I (1989) Interior path following primal-dual algorithms: Part I: linear programming. Math Program 44:27–41
Murtagh BA (1981) Advances in linear programming: computation and practice. McGraw-Hill, New York/London
Murtagh BA, Saunders MA (1998) MINOS 5.5 user's guid. Technical report SOL 83-20R, Department of Engineering Economics Systems & Operations Research, Stanford University, Stanford
Murty KG (1983) Linear programming. Wiley, New York
Nazareth JL (1987) Computer solutions of linear programs. Oxford University Press, Oxford
Nazareth JL (1996) The implementation of linear programming algorithms based on homotopies. Algorithmica 15:332–350
Nemhauser GL (1994) The age of optimizaiton: solving large-scale real-world problems. Oper Res 42:5–13
Nemhauser GL, Wolsey LA (1999) Integer and combinatorial optimization. Wiley, New York
Nocedal J, Wright SJ (1999) Numerical optimization. Springer, Berlin
Ogryczak W (1988) The simplex method is not always well behaved. Linear Algebra Appl 109:41–57
Orchard-Hays W (1954) Background development and extensions of the revised simplex method. Report RM 1433, The Rand Corporation, Santa Monica
Orchard-Hays W (1956) Evolution of computer codes for linear programming. Paper P-810, The RAND Corporation, p 2224
Orchard-Hays W (1971) Advanced linear programming computing techniques. McGraw-Hill, New York
Padberg MW (1995) Linear optimization and extensions. Springer, Berlin

Pan P-Q (1982) Differential equaltion methods for unconstarained optimization (in Chinese). Numer Math J Chin Univ 4:338–349
Pan P-Q (1990) Practical finite pivoting rules for the simplex method. OR Spektrum 12:219–225
Pan P-Q (1991) Simplex-like method with bisection for linear programming. Optimization 22:717–743
Pan P-Q (1992a) New ODE methods for equality constrained optimization (I) – equations. J Comput Math 10:77–92
Pan P-Q (1992b) New ODE methods for equality constrained optimization (II) – algorithms. J Comput Math 10:129–146
Pan P-Q (1992c) Modification of Bland's pivoting rule (in Chinese). Numer Math 14:379–381
Pan P-Q (1994a) A variant of the dual pivot rule in linear programming. J Inf Optim Sci 15:405–413
Pan P-Q (1994b) Composite phase-1 methods without measuring infeasibility. In: Yue M-Y (ed) Theory of optimization and its applications. Xidian University Press, Xian, pp 359–364.
Pan P-Q (1994c) Ratio-test-free pivoting rules for the bisection simplex method. In: Proceedings of national conference on decision making science, Shangrao, pp 24–29
Pan P-Q (1994d) Ratio-test-free pivoting rules for a dual phase-1 method. In: Xiao S-T, Wu F (eds) Proceeding of the third conference of Chinese SIAM. Tsinghua University press, Beijing, pp 245–249
Pan P-Q (1995) New non-monotone procedures for achieving dual feasibility. J Nanjing Univ Math Biquarterly 12:155–162
Pan P-Q (1996a) A modified bisection simplex method for linear programming. J Comput Math 14:249–255
Pan P-Q (1996b) New pivot rules for achieving dual feasibility. In: Wei Z (ed) Theory and applications of OR. Proceedings of the fifth conference of Chinese OR society, Xian, 10–14 Oct 1996. Xidian University Press, Xian, pp 109–113
Pan P-Q (1996c) Solving linear programming problems via appending an elastic constraint. J Southeast Univ (English edn) 12:253–265
Pan P-Q (1997) The most-obtuse-angle row pivot rule for achieving dual feasibility in linear programming: a computational study. Eur J Oper Res 101:164–176
Pan P-Q (1998a) A dual projective simplex method for linear programming. Comput Math Appl 35:119–135
Pan P-Q (1998b) A basis-deficiency-allowing variation of the simplex method. Comput Math Appl 36:33–53
Pan P-Q (1999a) A new perturbation simplex algorithm for linear programming. J Comput Math 17:233–242
Pan P-Q (1999b) A projective simplex method for linear programming. Linear Algebra Appl 292:99–125
Pan P-Q (2000a) A projective simplex algorithm using LU decomposition. Comput Math Appl 39:187–208
Pan P-Q (2000b) Primal perturbation simplex algorithms for linear programming. J Comput Math 18:587–596
Pan P-Q (2000c) On developments of pivot algorithms for linear programming. In: Proceedings of the sixth national conference of operations research society of China, Changsha, 10–15 Oct 2000. Global-Link Publishing, Hong Kong, pp 120–129
Pan P-Q (2004) A dual projective pivot algorithm for linear programming. Comput Optim Appl 29:333–344
Pan P-Q (2005) A revised dual projective pivot algorithm for linear programming. SIAM J Optim 16:49–68
Pan P-Q (2008a) A largest-distance pivot rule for the simplex algorithm. Eur J Oper Res 187:393–402
Pan P-Q (2008b) A primal deficient-basis algorithm for linear programming. Appl Math Comput 198:898–912
Pan P-Q (2008c) Efficient nested pricing in the simplex algorithm. Oper Res Lett 38:309–313

Pan P-Q (2010) A fast simplex algorithm for linear programming. J Comput Math 28(6):837–847
Pan P-Q (2013) An affine-scaling pivot algorithm for linear programming. Optimization 62: 431–445
Pan P-Q, Li W (2003) A non-monotone phase-1 method in linear programming. J Southeast Univ (English edn) 19:293–296
Pan P-Q, Ouiang Z-X (1993) Two variants of the simplex algorithm (in Chinese). J Math Res Expo 13(2):274–275
Pan P-Q, Ouyang Z-X (1994) Moore-Penrose inverse simplex algorithms based on successive linear subprogramming approach. Numer Math 3:180–190
Pan P-Q, Pan Y-P (2001) A phase-1 approach to the generalized simplex algorithm. Comput Math Appl 42:1455–1464
Pan P-Q, Li W, Wang Y (2004) A phase-1 algorithm using the most-obtuse-angle rule for the basis-deficiency-allowing dual simplex method. OR Trans 8:88–96
Pan P-Q, Li W, Cao J (2006a) Partial pricing rule simplex method with deficient basis. Numer Math J Chin Univ (English series) 15:23–30
Pan P-Q, Hu J-F, Li C (2006b) Feasible region contraction interior point algorithm. Appl Math Comput 182:1361–1368
Papadimitriou CH, Steiglitz K (1982) Combinatorial optimization: algorithms and complexity. Prentice-Hall, New Jersey
Perold AF (1980) A degeneracy exploiting LU factorization for the simplex method. Math Program 19:239–254
Powell MJD (1989) A tolerant algorithm for linearly constrained optimization calculations. Math Program 45:547–566
Reid JK (1982) A sparsity-exploiting variant of the Bartels-Golub decomposition for linear programming bases. Math Program 24:55–69
Rockafellar RT (1997) Convext analysis. Princeton University Press, Princeton
Roos C (1990) An exponential example for Terlaky's pivoting rule for the criss-cross simplex method. Math Program 46:79–84
Roos C, Vial J-Ph (1992) A polynomial method of approximate centers for linear programming. Math Program 54:295–305
Roos C, Terlaky T, Vial J-P (1997) Theory and algorithms for linear programming. Wiley, Chichester
Rothenberg RI (1979) Linear programming. North-Holland, New York
Ryan D, Osborne M (1988) On the solution of highly degenerate linear problems. Math Program 41:385–392
Saigal R (1995) Linear programming. Kluwer, Boston
Santos-Palomo A (2004) The sagitta method for solving linear problems. Eur J Oper Res 157:527–539
Saunders MA (1972) Large scale linear programming using the Cholesky factorization. Technical report STAN-CS-72-152, Stanford University
Saunders MA (1973) The complexity of LU updating in the simplex method. In: Andersen R, Brent R (eds) The complexity of computational problem solving. University Press, St. Lucia, pp 214–230
Saunders MA (1976) A fast and stable implementation of the simplex method using Bartels-Golub updating. In: Bunch J, Rose D (eds) Sparse matrix computation. Academic, New York, pp 213–226
Schrijver A (1986) Theory of linear and integer programming. Wiley, Chichester
Shanno DF, Bagchi A (1990) A unified view of interior point methods for linear programming. Ann Oper Res 22:55–70
Shen Y, Pan P-Q (2006) Dual besiction simplex algorithm (in Chinese). In: Preceedings of the national conference of operatios research society of China, Shenzhen. Globa-Link Informatics, Hong Kong, pp 168–174
Shi Y, Pan P-Q (2011) Higher order iteration schemes for unconstrained optimization. Am J Oper Res 1:73–83

Smale S (1983a) On the average number of steps of the simplex method of linear programming. Math Program 27:241–262

Smale S (1983b) The problem of the average speed of the simplex method. In: Bachem A, Grotschel M, Korte B, (eds) Mathematical programming, the state of the art. Springer, Berlin, pp 530–539

Srinath LS (1982) Linear programming: principles and applications. Affiliated East-West Press, New Delhi

Suhl UH (1994) Mathematical optimization system. Eur J Oper Res 72:312–322

Suhl LM, Suhl UH (1990) Computing sparse LU factorization for large-scale linear programming bases. ORSA J Comput 2:325–335

Suhl LM, Suhl UH (1993) A fast LU-update for linear programming. Ann Oper Res 43:33–47

Sun W, Yuan Y-X (2006) Optimization theory an methods: nonlinear programming. Springer, New York

Swietanowaki A (1998) A new steepest edge approximation for the simplex method for linear programming. Comput Optim Appl 10:271–281

Taha H (1975) Integer programming: theory, applications, and computations. Academic, Orlando

Talacko J V, Rockefeller RT (1960) A Compact Simplex Algorithm and a Symmetric Algorithm for General Linear Programs. Unpublished paper, Marquette University

Tanabe K (1977) A geometric method in non-linear programming. Technical Report 23343-AMD780, Brookhaven National Laboratory, New York

Tanabe K (1990) Centered Newton method for linear prgramming: interior and 'exterior' point method, (in Japannese). In: Tone K (ed) New methods for linear programming 3. The Institute of Statistical Mathematics, Tokeo, pp 98–100

Tapia RA, Zhang Y (1991) An optimal-basis identification technique for interior-point linear programming algorithms. Linear Algebra Appl 152:343–363

Terlaky T (1985) A covergent criss-cross method. Math Oper Stat Ser Optim 16:683–690

Terlaky T (1993) Pivot rules for linear programming: a survey on recent theoretical developments. Ann Oper Res 46:203–233

Terlaky T (ed) (1996) Interior point methods of mathematical programming. Kluwer, Dordrecht

Todd MJ (1982) An implementation of the simplex method for linear programming problems with variable upper bounds. Math Program 23:23–49

Todd MJ (1983) Large scale linear programming: geometry, working bases and factorizations. Math Program 26:1–23

Tomlin JA (1972) Modifying trangular factors of the basis in the simplex method. In: Rose DJ, Willoughby RA (eds) Sparse matrices and applications. Plenum, New York

Tomlin JA (1974) On pricing and backward transformation in linear programming. Math Program 6:42–47

Tomlin JA (1975) On scaling linear programming problems. Math Program Study 4:146–166

Tomlin JA (1987) An experimental approach to Karmarkar's projective method, for linear programming. Math Program 31:175–191

Tsuchiya T (1992) Global convergence property of the affine scaling method for primal degenerate linear programming problems. Math Oper Res 17:527–557

Tsuchiya T, Muramatsu M (1995) Global convergence of a long-step affine-scaling algorithm for degenrate linear programming problems. SIAM J Optim 5:525–551

Tucker AW (1956) Dual systems of homegeneous linear relations. In: Kuhn HW, Tucker AW, Dantzig GB (eds) Linear inequalities and related systems. Princeton University Press, Princeton, pp 3–18

Turner K (1991) Computing projections for the Karmarkar algorithtm. Linear Algebra Appl 152:141–154

Vanderbei RJ, Lagarias JC (1990) I.I. Dikin's convergence result for the affine-scaling algorithm. Contemp Math 114:109–119

Vanderbei RJ, Meketon M, Freedman B (1986) A modification of Karmarkars linear programming algorithm. Algorithmica 1:395–407

Vemuganti RR (2004) On gradient simplex methods for linear programs. J Appl Math Decis Sci 8:107–129
Wang Z (1987) A conformal elimination-free algorithm for oriented matroid programming. Chin Ann Math 8(B1):16–25
Wilkinson JH (1971) Moden error analysis. SIAM Rev 13:548–568
Wolfe P (1963) A technique for resolving degeneracy in linear programming. J Oper Res Soc 11:205–211
Wolfe P (1965) The composite simplex algorithm. SIAM Rev 7:42–54
Wolsey L (1998) Integer programming. Wiley, New York
Wright SJ (1997) Primal-dual interior-point methods. SIAM, Philadelphia
Xu X, Ye Y (1995) A generalized homogeneous and self-dual algorithm for linear prgramming. Oper Res Lett 17:181–190
Xu X, Hung P-F, Ye Y (1996) A simplified homogeneous and self-dual linear programming algorithm and its implementation. Ann Oper Res 62:151–171
Yan W-L, Pan P-Q (2001) Improvement of the subproblem in the bisection simplex algorithm (in Chinese). J Southeast Univ 31:324–241
Yan A, Pan P-Q (2005) Variation of the conventional pivot rule and the application in the deficient basis algorithm (in Chinese). Oper Res Manag Sci 14:28–33
Yan H-Y, Pan P-Q (2009) Most-obtuse-angle criss-cross algorithm for linear programming (in Chinese). Numer Math J Chin Univ 31:209–215
Yang X-Y, Pan P-Q (2006) Most-obtuse-angle dual relaxation algorithm (in Chinese). In: Preceedings of the national conference of operatios research society of China, Shenzhen. Globa-Link Informatics, Hong Kong, pp 150–155
Ye Y (1987) Eliminating columns in the simplex method for linear programming. Technical report SOL 87–14, Department of Operations Research, Stanford Univesity, Stanford
Ye Y (1990) A 'build-down' scheme for linear probleming. Math Program 46:61–72
Ye Y (1997) Interior point algorithms: theory and analysis. Wiley, New York
Ye Y, Todd MJ, Mizuno S (1994) An $o(\sqrt{n}l)$-iteration homogeneous and selfdual linear programming algorithm. Math Oper Res 19:53–67
Zhang H, Pan P-Q (2008) An interior-point algorithm for linear programming (in Chinese). In: Preceedings of the national conference of operatios research society of China, Nanjing. Globa-Link Informatics, Hong Kong, pp 183–187
Zhang J-Z, Xu S-J (1997) Linear programming (in Chinese). Science Press, Bejing
Zhang L-H, Yang W-H, Liao L-Z (2013) On an efficient implementation of the face algorithm for linear programming. J Comp Math 31:335–354
Zhou Z-J, Pan P-Q, Chen S-F (2009) Most-obtuse-angle relaxation algorithm (in Chinese). Oper Res Manag Sci 18:7–10
Zionts S (1969) The criss-cross method for solving linear programming problems. Manag Sci 15:420–445
Zlatev Z (1980) On some pivotal strategies in Gaussian elimination by sparse technique. SIAM J Numer Anal 17:12–30
Zörnig P (2006) Systematic construction of examples for cycling in the simplex method. Comput Oper Res 33:2247–2262
Zoutendijk G (1960) Methods of feasible directions. Elsevier, Amsterdam

Kolibri 360 GmbH
Ferdinand-Jühlke-Straße 7
99095 Erfurt, Deutschland
produktsicherheit@kolibri360.de